T0140068

Lecture Notes in Computer Science   13671

More information about this series at https://link.springer.com/bookseries/558

Shai Avidan · Gabriel Brostow ·
Moustapha Cissé · Giovanni Maria Farinella ·
Tal Hassner (Eds.)

Computer Vision – ECCV 2022

17th European Conference
Tel Aviv, Israel, October 23–27, 2022
Proceedings, Part XI

Springer

Editors
Shai Avidan
Tel Aviv University
Tel Aviv, Israel

Gabriel Brostow ⓘ
University College London
London, UK

Moustapha Cissé
Google AI
Accra, Ghana

Giovanni Maria Farinella ⓘ
University of Catania
Catania, Italy

Tal Hassner ⓘ
Facebook (United States)
Menlo Park, CA, USA

ISSN 0302-9743   ISSN 1611-3349 (electronic)
Lecture Notes in Computer Science
ISBN 978-3-031-20082-3   ISBN 978-3-031-20083-0 (eBook)
https://doi.org/10.1007/978-3-031-20083-0

This Springer imprint is published by the registered company Springer Nature Switzerland AG
The registered company address is: Gewerbestrasse 11, 6330 Cham, Switzerland

Foreword

Organizing the European Conference on Computer Vision (ECCV 2022) in Tel-Aviv during a global pandemic was no easy feat. The uncertainty level was extremely high, and decisions had to be postponed to the last minute. Still, we managed to plan things just in time for ECCV 2022 to be held in person. Participation in physical events is crucial to stimulating collaborations and nurturing the culture of the Computer Vision community.

There were many people who worked hard to ensure attendees enjoyed the best science at the 16th edition of ECCV. We are grateful to the Program Chairs Gabriel Brostow and Tal Hassner, who went above and beyond to ensure the ECCV reviewing process ran smoothly. The scientific program includes dozens of workshops and tutorials in addition to the main conference and we would like to thank Leonid Karlinsky and Tomer Michaeli for their hard work. Finally, special thanks to the web chairs Lorenzo Baraldi and Kosta Derpanis, who put in extra hours to transfer information fast and efficiently to the ECCV community.

We would like to express gratitude to our generous sponsors and the Industry Chairs, Dimosthenis Karatzas and Chen Sagiv, who oversaw industry relations and proposed new ways for academia-industry collaboration and technology transfer. It's great to see so much industrial interest in what we're doing!

Authors' draft versions of the papers appeared online with open access on both the Computer Vision Foundation (CVF) and the European Computer Vision Association (ECVA) websites as with previous ECCVs. Springer, the publisher of the proceedings, has arranged for archival publication. The final version of the papers is hosted by SpringerLink, with active references and supplementary materials. It benefits all potential readers that we offer both a free and citeable version for all researchers, as well as an authoritative, citeable version for SpringerLink readers. Our thanks go to Ronan Nugent from Springer, who helped us negotiate this agreement. Last but not least, we wish to thank Eric Mortensen, our publication chair, whose expertise made the process smooth.

October 2022

Rita Cucchiara
Jiří Matas
Amnon Shashua
Lihi Zelnik-Manor

Preface

Welcome to the proceedings of the European Conference on Computer Vision (ECCV 2022). This was a hybrid edition of ECCV as we made our way out of the COVID-19 pandemic. The conference received 5804 valid paper submissions, compared to 5150 submissions to ECCV 2020 (a 12.7% increase) and 2439 in ECCV 2018. 1645 submissions were accepted for publication (28%) and, of those, 157 (2.7% overall) as orals.

846 of the submissions were desk-rejected for various reasons. Many of them because they revealed author identity, thus violating the double-blind policy. This violation came in many forms: some had author names with the title, others added acknowledgments to specific grants, yet others had links to their github account where their name was visible. Tampering with the LaTeX template was another reason for automatic desk rejection.

ECCV 2022 used the traditional CMT system to manage the entire double-blind reviewing process. Authors did not know the names of the reviewers and vice versa. Each paper received at least 3 reviews (except 6 papers that received only 2 reviews), totalling more than 15,000 reviews.

Handling the review process at this scale was a significant challenge. To ensure that each submission received as fair and high-quality reviews as possible, we recruited more than 4719 reviewers (in the end, 4719 reviewers did at least one review). Similarly we recruited more than 276 area chairs (eventually, only 276 area chairs handled a batch of papers). The area chairs were selected based on their technical expertise and reputation, largely among people who served as area chairs in previous top computer vision and machine learning conferences (ECCV, ICCV, CVPR, NeurIPS, etc.).

Reviewers were similarly invited from previous conferences, and also from the pool of authors. We also encouraged experienced area chairs to suggest additional chairs and reviewers in the initial phase of recruiting. The median reviewer load was five papers per reviewer, while the average load was about four papers, because of the emergency reviewers. The area chair load was 35 papers, on average.

Conflicts of interest between authors, area chairs, and reviewers were handled largely automatically by the CMT platform, with some manual help from the Program Chairs. Reviewers were allowed to describe themselves as senior reviewer (load of 8 papers to review) or junior reviewers (load of 4 papers). Papers were matched to area chairs based on a subject-area affinity score computed in CMT and an affinity score computed by the Toronto Paper Matching System (TPMS). TPMS is based on the paper's full text. An area chair handling each submission would bid for preferred expert reviewers, and we balanced load and prevented conflicts.

The assignment of submissions to area chairs was relatively smooth, as was the assignment of submissions to reviewers. A small percentage of reviewers were not happy with their assignments in terms of subjects and self-reported expertise. This is an area for improvement, although it's interesting that many of these cases were reviewers hand-picked by AC's. We made a later round of reviewer recruiting, targeted at the list of authors of papers submitted to the conference, and had an excellent response which helped provide enough emergency reviewers. In the end, all but six papers received at least 3 reviews.

The challenges of the reviewing process are in line with past experiences at ECCV 2020. As the community grows, and the number of submissions increases, it becomes ever more challenging to recruit enough reviewers and ensure a high enough quality of reviews. Enlisting authors by default as reviewers might be one step to address this challenge.

Authors were given a week to rebut the initial reviews, and address reviewers' concerns. Each rebuttal was limited to a single pdf page with a fixed template.

The Area Chairs then led discussions with the reviewers on the merits of each submission. The goal was to reach consensus, but, ultimately, it was up to the Area Chair to make a decision. The decision was then discussed with a buddy Area Chair to make sure decisions were fair and informative. The entire process was conducted virtually with no in-person meetings taking place.

The Program Chairs were informed in cases where the Area Chairs overturned a decisive consensus reached by the reviewers, and pushed for the meta-reviews to contain details that explained the reasoning for such decisions. Obviously these were the most contentious cases, where reviewer inexperience was the most common reported factor.

Once the list of accepted papers was finalized and released, we went through the laborious process of plagiarism (including self-plagiarism) detection. A total of 4 accepted papers were rejected because of that.

Finally, we would like to thank our Technical Program Chair, Pavel Lifshits, who did tremendous work behind the scenes, and we thank the tireless CMT team.

October 2022

Gabriel Brostow
Giovanni Maria Farinella
Moustapha Cissé
Shai Avidan
Tal Hassner

Organization

General Chairs

Rita Cucchiara
Jiří Matas

Amnon Shashua
Lihi Zelnik-Manor

University of Modena and Reggio Emilia, Italy
Czech Technical University in Prague, Czech
  Republic
Hebrew University of Jerusalem, Israel
Technion – Israel Institute of Technology, Israel

Program Chairs

Shai Avidan
Gabriel Brostow
Moustapha Cissé
Giovanni Maria Farinella
Tal Hassner

Tel-Aviv University, Israel
University College London, UK
Google AI, Ghana
University of Catania, Italy
Facebook AI, USA

Program Technical Chair

Pavel Lifshits

Technion – Israel Institute of Technology, Israel

Workshops Chairs

Leonid Karlinsky
Tomer Michaeli
Ko Nishino

IBM Research, Israel
Technion – Israel Institute of Technology, Israel
Kyoto University, Japan

Tutorial Chairs

Thomas Pock
Natalia Neverova

Graz University of Technology, Austria
Facebook AI Research, UK

Demo Chair

Bohyung Han

Seoul National University, Korea

Social and Student Activities Chairs

Tatiana Tommasi   Italian Institute of Technology, Italy
Sagie Benaim   University of Copenhagen, Denmark

Diversity and Inclusion Chairs

Xi Yin   Facebook AI Research, USA
Bryan Russell   Adobe, USA

Communications Chairs

Lorenzo Baraldi   University of Modena and Reggio Emilia, Italy
Kosta Derpanis   York University & Samsung AI Centre Toronto,
                             Canada

Industrial Liaison Chairs

Dimosthenis Karatzas   Universitat Autònoma de Barcelona, Spain
Chen Sagiv   SagivTech, Israel

Finance Chair

Gerard Medioni   University of Southern California & Amazon,
                             USA

Publication Chair

Eric Mortensen   MiCROTEC, USA

Area Chairs

Lourdes Agapito   University College London, UK
Zeynep Akata   University of Tübingen, Germany
Naveed Akhtar   University of Western Australia, Australia
Karteek Alahari   Inria Grenoble Rhône-Alpes, France
Alexandre Alahi   École polytechnique fédérale de Lausanne,
                             Switzerland
Pablo Arbelaez   Universidad de Los Andes, Columbia
Antonis A. Argyros   University of Crete & Foundation for Research
                             and Technology-Hellas, Crete
Yuki M. Asano   University of Amsterdam, The Netherlands
Kalle Åström   Lund University, Sweden
Hadar Averbuch-Elor   Cornell University, USA

Hossein Azizpour — KTH Royal Institute of Technology, Sweden
Vineeth N. Balasubramanian — Indian Institute of Technology, Hyderabad, India
Lamberto Ballan — University of Padova, Italy
Adrien Bartoli — Université Clermont Auvergne, France
Horst Bischof — Graz University of Technology, Austria
Matthew B. Blaschko — KU Leuven, Belgium
Federica Bogo — Meta Reality Labs Research, Switzerland
Katherine Bouman — California Institute of Technology, USA
Edmond Boyer — Inria Grenoble Rhône-Alpes, France
Michael S. Brown — York University, Canada
Vittorio Caggiano — Meta AI Research, USA
Neill Campbell — University of Bath, UK
Octavia Camps — Northeastern University, USA
Duygu Ceylan — Adobe Research, USA
Ayan Chakrabarti — Google Research, USA
Tat-Jen Cham — Nanyang Technological University, Singapore
Antoni Chan — City University of Hong Kong, Hong Kong, China
Manmohan Chandraker — NEC Labs America, USA
Xinlei Chen — Facebook AI Research, USA
Xilin Chen — Institute of Computing Technology, Chinese Academy of Sciences, China
Dongdong Chen — Microsoft Cloud AI, USA
Chen Chen — University of Central Florida, USA
Ondrej Chum — Vision Recognition Group, Czech Technical University in Prague, Czech Republic
John Collomosse — Adobe Research & University of Surrey, UK
Camille Couprie — Facebook, France
David Crandall — Indiana University, USA
Daniel Cremers — Technical University of Munich, Germany
Marco Cristani — University of Verona, Italy
Canton Cristian — Facebook AI Research, USA
Dengxin Dai — ETH Zurich, Switzerland
Dima Damen — University of Bristol, UK
Kostas Daniilidis — University of Pennsylvania, USA
Trevor Darrell — University of California, Berkeley, USA
Andrew Davison — Imperial College London, UK
Tali Dekel — Weizmann Institute of Science, Israel
Alessio Del Bue — Istituto Italiano di Tecnologia, Italy
Weihong Deng — Beijing University of Posts and Telecommunications, China
Konstantinos Derpanis — Ryerson University, Canada
Carl Doersch — DeepMind, UK

Bohyung Han	Seoul National University, Korea
Tian Han	Stevens Institute of Technology, USA
Emily Hand	University of Nevada, Reno, USA
Bharath Hariharan	Cornell University, USA
Ran He	Institute of Automation, Chinese Academy of Sciences, China
Otmar Hilliges	ETH Zurich, Switzerland
Adrian Hilton	University of Surrey, UK
Minh Hoai	Stony Brook University, USA
Yedid Hoshen	Hebrew University of Jerusalem, Israel
Timothy Hospedales	University of Edinburgh, UK
Gang Hua	Wormpex AI Research, USA
Di Huang	Beihang University, China
Jing Huang	Facebook, USA
Jia-Bin Huang	Facebook, USA
Nathan Jacobs	Washington University in St. Louis, USA
C.V. Jawahar	International Institute of Information Technology, Hyderabad, India
Herve Jegou	Facebook AI Research, France
Neel Joshi	Microsoft Research, USA
Armand Joulin	Facebook AI Research, France
Frederic Jurie	University of Caen Normandie, France
Fredrik Kahl	Chalmers University of Technology, Sweden
Yannis Kalantidis	NAVER LABS Europe, France
Evangelos Kalogerakis	University of Massachusetts, Amherst, USA
Sing Bing Kang	Zillow Group, USA
Yosi Keller	Bar Ilan University, Israel
Margret Keuper	University of Mannheim, Germany
Tae-Kyun Kim	Imperial College London, UK
Benjamin Kimia	Brown University, USA
Alexander Kirillov	Facebook AI Research, USA
Kris Kitani	Carnegie Mellon University, USA
Iasonas Kokkinos	Snap Inc. & University College London, UK
Vladlen Koltun	Apple, USA
Nikos Komodakis	University of Crete, Crete
Piotr Koniusz	Australian National University, Australia
Philipp Kraehenbuehl	University of Texas at Austin, USA
Dilip Krishnan	Google, USA
Ajay Kumar	Hong Kong Polytechnic University, Hong Kong, China
Junseok Kwon	Chung-Ang University, Korea
Jean-Francois Lalonde	Université Laval, Canada

Ivan Laptev	Inria Paris, France
Laura Leal-Taixé	Technical University of Munich, Germany
Erik Learned-Miller	University of Massachusetts, Amherst, USA
Gim Hee Lee	National University of Singapore, Singapore
Seungyong Lee	Pohang University of Science and Technology, Korea
Zhen Lei	Institute of Automation, Chinese Academy of Sciences, China
Bastian Leibe	RWTH Aachen University, Germany
Hongdong Li	Australian National University, Australia
Fuxin Li	Oregon State University, USA
Bo Li	University of Illinois at Urbana-Champaign, USA
Yin Li	University of Wisconsin-Madison, USA
Ser-Nam Lim	Meta AI Research, USA
Joseph Lim	University of Southern California, USA
Stephen Lin	Microsoft Research Asia, China
Dahua Lin	The Chinese University of Hong Kong, Hong Kong, China
Si Liu	Beihang University, China
Xiaoming Liu	Michigan State University, USA
Ce Liu	Microsoft, USA
Zicheng Liu	Microsoft, USA
Yanxi Liu	Pennsylvania State University, USA
Feng Liu	Portland State University, USA
Yebin Liu	Tsinghua University, China
Chen Change Loy	Nanyang Technological University, Singapore
Huchuan Lu	Dalian University of Technology, China
Cewu Lu	Shanghai Jiao Tong University, China
Oisin Mac Aodha	University of Edinburgh, UK
Dhruv Mahajan	Facebook, USA
Subhransu Maji	University of Massachusetts, Amherst, USA
Atsuto Maki	KTH Royal Institute of Technology, Sweden
Arun Mallya	NVIDIA, USA
R. Manmatha	Amazon, USA
Iacopo Masi	Sapienza University of Rome, Italy
Dimitris N. Metaxas	Rutgers University, USA
Ajmal Mian	University of Western Australia, Australia
Christian Micheloni	University of Udine, Italy
Krystian Mikolajczyk	Imperial College London, UK
Anurag Mittal	Indian Institute of Technology, Madras, India
Philippos Mordohai	Stevens Institute of Technology, USA
Greg Mori	Simon Fraser University & Borealis AI, Canada

Vittorio Murino	Istituto Italiano di Tecnologia, Italy
P. J. Narayanan	International Institute of Information Technology, Hyderabad, India
Ram Nevatia	University of Southern California, USA
Natalia Neverova	Facebook AI Research, UK
Richard Newcombe	Facebook, USA
Cuong V. Nguyen	Florida International University, USA
Bingbing Ni	Shanghai Jiao Tong University, China
Juan Carlos Niebles	Salesforce & Stanford University, USA
Ko Nishino	Kyoto University, Japan
Jean-Marc Odobez	Idiap Research Institute, École polytechnique fédérale de Lausanne, Switzerland
Francesca Odone	University of Genova, Italy
Takayuki Okatani	Tohoku University & RIKEN Center for Advanced Intelligence Project, Japan
Manohar Paluri	Facebook, USA
Guan Pang	Facebook, USA
Maja Pantic	Imperial College London, UK
Sylvain Paris	Adobe Research, USA
Jaesik Park	Pohang University of Science and Technology, Korea
Hyun Soo Park	The University of Minnesota, USA
Omkar M. Parkhi	Facebook, USA
Deepak Pathak	Carnegie Mellon University, USA
Georgios Pavlakos	University of California, Berkeley, USA
Marcello Pelillo	University of Venice, Italy
Marc Pollefeys	ETH Zurich & Microsoft, Switzerland
Jean Ponce	Inria, France
Gerard Pons-Moll	University of Tübingen, Germany
Fatih Porikli	Qualcomm, USA
Victor Adrian Prisacariu	University of Oxford, UK
Petia Radeva	University of Barcelona, Spain
Ravi Ramamoorthi	University of California, San Diego, USA
Deva Ramanan	Carnegie Mellon University, USA
Vignesh Ramanathan	Facebook, USA
Nalini Ratha	State University of New York at Buffalo, USA
Tammy Riklin Raviv	Ben-Gurion University, Israel
Tobias Ritschel	University College London, UK
Emanuele Rodola	Sapienza University of Rome, Italy
Amit K. Roy-Chowdhury	University of California, Riverside, USA
Michael Rubinstein	Google, USA
Olga Russakovsky	Princeton University, USA

Mathieu Salzmann	École polytechnique fédérale de Lausanne, Switzerland
Dimitris Samaras	Stony Brook University, USA
Aswin Sankaranarayanan	Carnegie Mellon University, USA
Imari Sato	National Institute of Informatics, Japan
Yoichi Sato	University of Tokyo, Japan
Shin'ichi Satoh	National Institute of Informatics, Japan
Walter Scheirer	University of Notre Dame, USA
Bernt Schiele	Max Planck Institute for Informatics, Germany
Konrad Schindler	ETH Zurich, Switzerland
Cordelia Schmid	Inria & Google, France
Alexander Schwing	University of Illinois at Urbana-Champaign, USA
Nicu Sebe	University of Trento, Italy
Greg Shakhnarovich	Toyota Technological Institute at Chicago, USA
Eli Shechtman	Adobe Research, USA
Humphrey Shi	University of Oregon & University of Illinois at Urbana-Champaign & Picsart AI Research, USA
Jianbo Shi	University of Pennsylvania, USA
Roy Shilkrot	Massachusetts Institute of Technology, USA
Mike Zheng Shou	National University of Singapore, Singapore
Kaleem Siddiqi	McGill University, Canada
Richa Singh	Indian Institute of Technology Jodhpur, India
Greg Slabaugh	Queen Mary University of London, UK
Cees Snoek	University of Amsterdam, The Netherlands
Yale Song	Facebook AI Research, USA
Yi-Zhe Song	University of Surrey, UK
Bjorn Stenger	Rakuten Institute of Technology
Abby Stylianou	Saint Louis University, USA
Akihiro Sugimoto	National Institute of Informatics, Japan
Chen Sun	Brown University, USA
Deqing Sun	Google, USA
Kalyan Sunkavalli	Adobe Research, USA
Ying Tai	Tencent YouTu Lab, China
Ayellet Tal	Technion – Israel Institute of Technology, Israel
Ping Tan	Simon Fraser University, Canada
Siyu Tang	ETH Zurich, Switzerland
Chi-Keung Tang	Hong Kong University of Science and Technology, Hong Kong, China
Radu Timofte	University of Würzburg, Germany & ETH Zurich, Switzerland
Federico Tombari	Google, Switzerland & Technical University of Munich, Germany

James Tompkin	Brown University, USA
Lorenzo Torresani	Dartmouth College, USA
Alexander Toshev	Apple, USA
Du Tran	Facebook AI Research, USA
Anh T. Tran	VinAI, Vietnam
Zhuowen Tu	University of California, San Diego, USA
Georgios Tzimiropoulos	Queen Mary University of London, UK
Jasper Uijlings	Google Research, Switzerland
Jan C. van Gemert	Delft University of Technology, The Netherlands
Gul Varol	Ecole des Ponts ParisTech, France
Nuno Vasconcelos	University of California, San Diego, USA
Mayank Vatsa	Indian Institute of Technology Jodhpur, India
Ashok Veeraraghavan	Rice University, USA
Jakob Verbeek	Facebook AI Research, France
Carl Vondrick	Columbia University, USA
Ruiping Wang	Institute of Computing Technology, Chinese Academy of Sciences, China
Xinchao Wang	National University of Singapore, Singapore
Liwei Wang	The Chinese University of Hong Kong, Hong Kong, China
Chaohui Wang	Université Paris-Est, France
Xiaolong Wang	University of California, San Diego, USA
Christian Wolf	NAVER LABS Europe, France
Tao Xiang	University of Surrey, UK
Saining Xie	Facebook AI Research, USA
Cihang Xie	University of California, Santa Cruz, USA
Zeki Yalniz	Facebook, USA
Ming-Hsuan Yang	University of California, Merced, USA
Angela Yao	National University of Singapore, Singapore
Shaodi You	University of Amsterdam, The Netherlands
Stella X. Yu	University of California, Berkeley, USA
Junsong Yuan	State University of New York at Buffalo, USA
Stefanos Zafeiriou	Imperial College London, UK
Amir Zamir	École polytechnique fédérale de Lausanne, Switzerland
Lei Zhang	Alibaba & Hong Kong Polytechnic University, Hong Kong, China
Lei Zhang	International Digital Economy Academy (IDEA), China
Pengchuan Zhang	Meta AI, USA
Bolei Zhou	University of California, Los Angeles, USA
Yuke Zhu	University of Texas at Austin, USA

Todd Zickler   Harvard University, USA
Wangmeng Zuo   Harbin Institute of Technology, China

Technical Program Committee

Davide Abati
Soroush Abbasi
  Koohpayegani
Amos L. Abbott
Rameen Abdal
Rabab Abdelfattah
Sahar Abdelnabi
Hassan Abu Alhaija
Abulikemu Abuduweili
Ron Abutbul
Hanno Ackermann
Aikaterini Adam
Kamil Adamczewski
Ehsan Adeli
Vida Adeli
Donald Adjeroh
Arman Afrasiyabi
Akshay Agarwal
Sameer Agarwal
Abhinav Agarwalla
Vaibhav Aggarwal
Sara Aghajanzadeh
Susmit Agrawal
Antonio Agudo
Touqeer Ahmad
Sk Miraj Ahmed
Chaitanya Ahuja
Nilesh A. Ahuja
Abhishek Aich
Shubhra Aich
Noam Aigerman
Arash Akbarinia
Peri Akiva
Derya Akkaynak
Emre Aksan
Arjun R. Akula
Yuval Alaluf
Stephan Alaniz
Paul Albert
Cenek Albl

Filippo Aleotti
Konstantinos P.
  Alexandridis
Motasem Alfarra
Mohsen Ali
Thiemo Alldieck
Hadi Alzayer
Liang An
Shan An
Yi An
Zhulin An
Dongsheng An
Jie An
Xiang An
Saket Anand
Cosmin Ancuti
Juan Andrade-Cetto
Alexander Andreopoulos
Bjoern Andres
Jerone T. A. Andrews
Shivangi Aneja
Anelia Angelova
Dragomir Anguelov
Rushil Anirudh
Oron Anschel
Rao Muhammad Anwer
Djamila Aouada
Evlampios Apostolidis
Srikar Appalaraju
Nikita Araslanov
Andre Araujo
Eric Arazo
Dawit Mureja Argaw
Anurag Arnab
Aditya Arora
Chetan Arora
Sunpreet S. Arora
Alexey Artemov
Muhammad Asad
Kumar Ashutosh

Sinem Aslan
Vishal Asnani
Mahmoud Assran
Amir Atapour-Abarghouei
Nikos Athanasiou
Ali Athar
ShahRukh Athar
Sara Atito
Souhaib Attaiki
Matan Atzmon
Mathieu Aubry
Nicolas Audebert
Tristan T.
  Aumentado-Armstrong
Melinos Averkiou
Yannis Avrithis
Stephane Ayache
Mehmet Aygün
Seyed Mehdi
  Ayyoubzadeh
Hossein Azizpour
George Azzopardi
Mallikarjun B. R.
Yunhao Ba
Abhishek Badki
Seung-Hwan Bae
Seung-Hwan Baek
Seungryul Baek
Piyush Nitin Bagad
Shai Bagon
Gaetan Bahl
Shikhar Bahl
Sherwin Bahmani
Haoran Bai
Lei Bai
Jiawang Bai
Haoyue Bai
Jinbin Bai
Xiang Bai
Xuyang Bai

Yang Bai
Yuanchao Bai
Ziqian Bai
Sungyong Baik
Kevin Bailly
Max Bain
Federico Baldassarre
Wele Gedara Chaminda
  Bandara
Biplab Banerjee
Pratyay Banerjee
Sandipan Banerjee
Jihwan Bang
Antyanta Bangunharcana
Aayush Bansal
Ankan Bansal
Siddhant Bansal
Wentao Bao
Zhipeng Bao
Amir Bar
Manel Baradad Jurjo
Lorenzo Baraldi
Danny Barash
Daniel Barath
Connelly Barnes
Ioan Andrei Bârsan
Steven Basart
Dina Bashkirova
Chaim Baskin
Peyman Bateni
Anil Batra
Sebastiano Battiato
Ardhendu Behera
Harkirat Behl
Jens Behley
Vasileios Belagiannis
Boulbaba Ben Amor
Emanuel Ben Baruch
Abdessamad Ben Hamza
Gil Ben-Artzi
Assia Benbihi
Fabian Benitez-Quiroz
Guy Ben-Yosef
Philipp Benz
Alexander W. Bergman

Urs Bergmann
Jesus Bermudez-Cameo
Stefano Berretti
Gedas Bertasius
Zachary Bessinger
Petra Bevandić
Matthew Beveridge
Lucas Beyer
Yash Bhalgat
Suvaansh Bhambri
Samarth Bharadwaj
Gaurav Bharaj
Aparna Bharati
Bharat Lal Bhatnagar
Uttaran Bhattacharya
Apratim Bhattacharyya
Brojeshwar Bhowmick
Ankan Kumar Bhunia
Ayan Kumar Bhunia
Qi Bi
Sai Bi
Michael Bi Mi
Gui-Bin Bian
Jia-Wang Bian
Shaojun Bian
Pia Bideau
Mario Bijelic
Hakan Bilen
Guillaume-Alexandre
  Bilodeau
Alexander Binder
Tolga Birdal
Vighnesh N. Birodkar
Sandika Biswas
Andreas Blattmann
Janusz Bobulski
Giuseppe Boccignone
Vishnu Boddeti
Navaneeth Bodla
Moritz Böhle
Aleksei Bokhovkin
Sam Bond-Taylor
Vivek Boominathan
Shubhankar Borse
Mark Boss

Andrea Bottino
Adnane Boukhayma
Fadi Boutros
Nicolas C. Boutry
Richard S. Bowen
Ivaylo Boyadzhiev
Aidan Boyd
Yuri Boykov
Aljaz Bozic
Behzad Bozorgtabar
Eric Brachmann
Samarth Brahmbhatt
Gustav Bredell
Francois Bremond
Joel Brogan
Andrew Brown
Thomas Brox
Marcus A. Brubaker
Robert-Jan Bruintjes
Yuqi Bu
Anders G. Buch
Himanshu Buckchash
Mateusz Buda
Ignas Budvytis
José M. Buenaposada
Marcel C. Bühler
Tu Bui
Adrian Bulat
Hannah Bull
Evgeny Burnaev
Andrei Bursuc
Benjamin Busam
Sergey N. Buzykanov
Wonmin Byeon
Fabian Caba
Martin Cadik
Guanyu Cai
Minjie Cai
Qing Cai
Zhongang Cai
Qi Cai
Yancheng Cai
Shen Cai
Han Cai
Jiarui Cai

Bowen Cai
Mu Cai
Qin Cai
Ruojin Cai
Weidong Cai
Weiwei Cai
Yi Cai
Yujun Cai
Zhiping Cai
Akin Caliskan
Lilian Calvet
Baris Can Cam
Necati Cihan Camgoz
Tommaso Campari
Dylan Campbell
Ziang Cao
Ang Cao
Xu Cao
Zhiwen Cao
Shengcao Cao
Song Cao
Weipeng Cao
Xiangyong Cao
Xiaochun Cao
Yue Cao
Yunhao Cao
Zhangjie Cao
Jiale Cao
Yang Cao
Jiajiong Cao
Jie Cao
Jinkun Cao
Lele Cao
Yulong Cao
Zhiguo Cao
Chen Cao
Razvan Caramalau
Marlène Careil
Gustavo Carneiro
Joao Carreira
Dan Casas
Paola Cascante-Bonilla
Angela Castillo
Francisco M. Castro
Pedro Castro

Luca Cavalli
George J. Cazenavette
Oya Celiktutan
Hakan Cevikalp
Sri Harsha C. H.
Sungmin Cha
Geonho Cha
Menglei Chai
Lucy Chai
Yuning Chai
Zenghao Chai
Anirban Chakraborty
Deep Chakraborty
Rudrasis Chakraborty
Souradeep Chakraborty
Kelvin C. K. Chan
Chee Seng Chan
Paramanand Chandramouli
Arjun Chandrasekaran
Kenneth Chaney
Dongliang Chang
Huiwen Chang
Peng Chang
Xiaojun Chang
Jia-Ren Chang
Hyung Jin Chang
Hyun Sung Chang
Ju Yong Chang
Li-Jen Chang
Qi Chang
Wei-Yi Chang
Yi Chang
Nadine Chang
Hanqing Chao
Pradyumna Chari
Dibyadip Chatterjee
Chiranjoy Chattopadhyay
Siddhartha Chaudhuri
Zhengping Che
Gal Chechik
Lianggangxu Chen
Qi Alfred Chen
Brian Chen
Bor-Chun Chen
Bo-Hao Chen

Bohong Chen
Bin Chen
Ziliang Chen
Cheng Chen
Chen Chen
Chaofeng Chen
Xi Chen
Haoyu Chen
Xuanhong Chen
Wei Chen
Qiang Chen
Shi Chen
Xianyu Chen
Chang Chen
Changhuai Chen
Hao Chen
Jie Chen
Jianbo Chen
Jingjing Chen
Jun Chen
Kejiang Chen
Mingcai Chen
Nenglun Chen
Qifeng Chen
Ruoyu Chen
Shu-Yu Chen
Weidong Chen
Weijie Chen
Weikai Chen
Xiang Chen
Xiuyi Chen
Xingyu Chen
Yaofo Chen
Yueting Chen
Yu Chen
Yunjin Chen
Yuntao Chen
Yun Chen
Zhenfang Chen
Zhuangzhuang Chen
Chu-Song Chen
Xiangyu Chen
Zhuo Chen
Chaoqi Chen
Shizhe Chen

Xiaotong Chen
Xiaozhi Chen
Dian Chen
Defang Chen
Dingfan Chen
Ding-Jie Chen
Ee Heng Chen
Tao Chen
Yixin Chen
Wei-Ting Chen
Lin Chen
Guang Chen
Guangyi Chen
Guanying Chen
Guangyao Chen
Hwann-Tzong Chen
Junwen Chen
Jiacheng Chen
Jianxu Chen
Hui Chen
Kai Chen
Kan Chen
Kevin Chen
Kuan-Wen Chen
Weihua Chen
Zhang Chen
Liang-Chieh Chen
Lele Chen
Liang Chen
Fanglin Chen
Zehui Chen
Minghui Chen
Minghao Chen
Xiaokang Chen
Qian Chen
Jun-Cheng Chen
Qi Chen
Qingcai Chen
Richard J. Chen
Runnan Chen
Rui Chen
Shuo Chen
Sentao Chen
Shaoyu Chen
Shixing Chen

Shuai Chen
Shuya Chen
Sizhe Chen
Simin Chen
Shaoxiang Chen
Zitian Chen
Tianlong Chen
Tianshui Chen
Min-Hung Chen
Xiangning Chen
Xin Chen
Xinghao Chen
Xuejin Chen
Xu Chen
Xuxi Chen
Yunlu Chen
Yanbei Chen
Yuxiao Chen
Yun-Chun Chen
Yi-Ting Chen
Yi-Wen Chen
Yinbo Chen
Yiran Chen
Yuanhong Chen
Yubei Chen
Yuefeng Chen
Yuhua Chen
Yukang Chen
Zerui Chen
Zhaoyu Chen
Zhen Chen
Zhenyu Chen
Zhi Chen
Zhiwei Chen
Zhixiang Chen
Long Chen
Bowen Cheng
Jun Cheng
Yi Cheng
Jingchun Cheng
Lechao Cheng
Xi Cheng
Yuan Cheng
Ho Kei Cheng
Kevin Ho Man Cheng

Jiacheng Cheng
Kelvin B. Cheng
Li Cheng
Mengjun Cheng
Zhen Cheng
Qingrong Cheng
Tianheng Cheng
Harry Cheng
Yihua Cheng
Yu Cheng
Ziheng Cheng
Soon Yau Cheong
Anoop Cherian
Manuela Chessa
Zhixiang Chi
Naoki Chiba
Julian Chibane
Kashyap Chitta
Tai-Yin Chiu
Hsu-kuang Chiu
Wei-Chen Chiu
Sungmin Cho
Donghyeon Cho
Hyeon Cho
Yooshin Cho
Gyusang Cho
Jang Hyun Cho
Seungju Cho
Nam Ik Cho
Sunghyun Cho
Hanbyel Cho
Jaesung Choe
Jooyoung Choi
Chiho Choi
Changwoon Choi
Jongwon Choi
Myungsub Choi
Dooseop Choi
Jonghyun Choi
Jinwoo Choi
Jun Won Choi
Min-Kook Choi
Hongsuk Choi
Janghoon Choi
Yoon-Ho Choi

Yukyung Choi
Jaegul Choo
Ayush Chopra
Siddharth Choudhary
Subhabrata Choudhury
Vasileios Choutas
Ka-Ho Chow
Pinaki Nath Chowdhury
Sammy Christen
Anders Christensen
Grigorios Chrysos
Hang Chu
Wen-Hsuan Chu
Peng Chu
Qi Chu
Ruihang Chu
Wei-Ta Chu
Yung-Yu Chuang
Sanghyuk Chun
Se Young Chun
Antonio Cinà
Ramazan Gokberk Cinbis
Javier Civera
Albert Clapés
Ronald Clark
Brian S. Clipp
Felipe Codevilla
Daniel Coelho de Castro
Niv Cohen
Forrester Cole
Maxwell D. Collins
Robert T. Collins
Marc Comino Trinidad
Runmin Cong
Wenyan Cong
Maxime Cordy
Marcella Cornia
Enric Corona
Huseyin Coskun
Luca Cosmo
Dragos Costea
Davide Cozzolino
Arun C. S. Kumar
Aiyu Cui
Qiongjie Cui

Quan Cui
Shuhao Cui
Yiming Cui
Ying Cui
Zijun Cui
Jiali Cui
Jiequan Cui
Yawen Cui
Zhen Cui
Zhaopeng Cui
Jack Culpepper
Xiaodong Cun
Ross Cutler
Adam Czajka
Ali Dabouei
Konstantinos M. Dafnis
Manuel Dahnert
Tao Dai
Yuchao Dai
Bo Dai
Mengyu Dai
Hang Dai
Haixing Dai
Peng Dai
Pingyang Dai
Qi Dai
Qiyu Dai
Yutong Dai
Naser Damer
Zhiyuan Dang
Mohamed Daoudi
Ayan Das
Abir Das
Debasmit Das
Deepayan Das
Partha Das
Sagnik Das
Soumi Das
Srijan Das
Swagatam Das
Avijit Dasgupta
Jim Davis
Adrian K. Davison
Homa Davoudi
Laura Daza

Matthias De Lange
Shalini De Mello
Marco De Nadai
Christophe De
  Vleeschouwer
Alp Dener
Boyang Deng
Congyue Deng
Bailin Deng
Yong Deng
Ye Deng
Zhuo Deng
Zhijie Deng
Xiaoming Deng
Jiankang Deng
Jinhong Deng
Jingjing Deng
Liang-Jian Deng
Siqi Deng
Xiang Deng
Xueqing Deng
Zhongying Deng
Karan Desai
Jean-Emmanuel Deschaud
Aniket Anand Deshmukh
Neel Dey
Helisa Dhamo
Prithviraj Dhar
Amaya Dharmasiri
Yan Di
Xing Di
Ousmane A. Dia
Haiwen Diao
Xiaolei Diao
Gonçalo José Dias Pais
Abdallah Dib
Anastasios Dimou
Changxing Ding
Henghui Ding
Guodong Ding
Yaqing Ding
Shuangrui Ding
Yuhang Ding
Yikang Ding
Shouhong Ding

Haisong Ding
Hui Ding
Jiahao Ding
Jian Ding
Jian-Jiun Ding
Shuxiao Ding
Tianyu Ding
Wenhao Ding
Yuqi Ding
Yi Ding
Yuzhen Ding
Zhengming Ding
Tan Minh Dinh
Vu Dinh
Christos Diou
Mandar Dixit
Bao Gia Doan
Khoa D. Doan
Dzung Anh Doan
Debi Prosad Dogra
Nehal Doiphode
Chengdong Dong
Bowen Dong
Zhenxing Dong
Hang Dong
Xiaoyi Dong
Haoye Dong
Jiangxin Dong
Shichao Dong
Xuan Dong
Zhen Dong
Shuting Dong
Jing Dong
Li Dong
Ming Dong
Nanqing Dong
Qiulei Dong
Runpei Dong
Siyan Dong
Tian Dong
Wei Dong
Xiaomeng Dong
Xin Dong
Xingbo Dong
Yuan Dong

Samuel Dooley
Gianfranco Doretto
Michael Dorkenwald
Keval Doshi
Zhaopeng Dou
Xiaotian Dou
Hazel Doughty
Ahmad Droby
Iddo Drori
Jie Du
Yong Du
Dawei Du
Dong Du
Ruoyi Du
Yuntao Du
Xuefeng Du
Yilun Du
Yuming Du
Radhika Dua
Haodong Duan
Jiafei Duan
Kaiwen Duan
Peiqi Duan
Ye Duan
Haoran Duan
Jiali Duan
Amanda Duarte
Abhimanyu Dubey
Shiv Ram Dubey
Florian Dubost
Lukasz Dudziak
Shivam Duggal
Justin M. Dulay
Matteo Dunnhofer
Chi Nhan Duong
Thibaut Durand
Mihai Dusmanu
Ujjal Kr Dutta
Debidatta Dwibedi
Isht Dwivedi
Sai Kumar Dwivedi
Takeharu Eda
Mark Edmonds
Alexei A. Efros
Thibaud Ehret

Max Ehrlich
Mahsa Ehsanpour
Iván Eichhardt
Farshad Einabadi
Marvin Eisenberger
Hazim Kemal Ekenel
Mohamed El Banani
Ismail Elezi
Moshe Eliasof
Alaa El-Nouby
Ian Endres
Francis Engelmann
Deniz Engin
Chanho Eom
Dave Epstein
Maria C. Escobar
Victor A. Escorcia
Carlos Esteves
Sungmin Eum
Bernard J. E. Evans
Ivan Evtimov
Fevziye Irem Eyiokur
    Yaman
Matteo Fabbri
Sébastien Fabbro
Gabriele Facciolo
Masud Fahim
Bin Fan
Hehe Fan
Deng-Ping Fan
Aoxiang Fan
Chen-Chen Fan
Qi Fan
Zhaoxin Fan
Haoqi Fan
Heng Fan
Hongyi Fan
Linxi Fan
Baojie Fan
Jiayuan Fan
Lei Fan
Quanfu Fan
Yonghui Fan
Yingruo Fan
Zhiwen Fan

Zicong Fan
Sean Fanello
Jiansheng Fang
Chaowei Fang
Yuming Fang
Jianwu Fang
Jin Fang
Qi Fang
Shancheng Fang
Tian Fang
Xianyong Fang
Gongfan Fang
Zhen Fang
Hui Fang
Jiemin Fang
Le Fang
Pengfei Fang
Xiaolin Fang
Yuxin Fang
Zhaoyuan Fang
Ammarah Farooq
Azade Farshad
Zhengcong Fei
Michael Felsberg
Wei Feng
Chen Feng
Fan Feng
Andrew Feng
Xin Feng
Zheyun Feng
Ruicheng Feng
Mingtao Feng
Qianyu Feng
Shangbin Feng
Chun-Mei Feng
Zunlei Feng
Zhiyong Feng
Martin Fergie
Mustansar Fiaz
Marco Fiorucci
Michael Firman
Hamed Firooz
Volker Fischer
Corneliu O. Florea
Georgios Floros

Wolfgang Foerstner
Gianni Franchi
Jean-Sebastien Franco
Simone Frintrop
Anna Fruehstueck
Changhong Fu
Chaoyou Fu
Cheng-Yang Fu
Chi-Wing Fu
Deqing Fu
Huan Fu
Jun Fu
Kexue Fu
Ying Fu
Jianlong Fu
Jingjing Fu
Qichen Fu
Tsu-Jui Fu
Xueyang Fu
Yang Fu
Yanwei Fu
Yonggan Fu
Wolfgang Fuhl
Yasuhisa Fujii
Kent Fujiwara
Marco Fumero
Takuya Funatomi
Isabel Funke
Dario Fuoli
Antonino Furnari
Matheus A. Gadelha
Akshay Gadi Patil
Adrian Galdran
Guillermo Gallego
Silvano Galliani
Orazio Gallo
Leonardo Galteri
Matteo Gamba
Yiming Gan
Sujoy Ganguly
Harald Ganster
Boyan Gao
Changxin Gao
Daiheng Gao
Difei Gao

Chen Gao
Fei Gao
Lin Gao
Wei Gao
Yiming Gao
Junyu Gao
Guangyu Ryan Gao
Haichang Gao
Hongchang Gao
Jialin Gao
Jin Gao
Jun Gao
Katelyn Gao
Mingchen Gao
Mingfei Gao
Pan Gao
Shangqian Gao
Shanghua Gao
Xitong Gao
Yunhe Gao
Zhanning Gao
Elena Garces
Nuno Cruz Garcia
Noa Garcia
Guillermo
    Garcia-Hernando
Isha Garg
Rahul Garg
Sourav Garg
Quentin Garrido
Stefano Gasperini
Kent Gauen
Chandan Gautam
Shivam Gautam
Paul Gay
Chunjiang Ge
Shiming Ge
Wenhang Ge
Yanhao Ge
Zheng Ge
Songwei Ge
Weifeng Ge
Yixiao Ge
Yuying Ge
Shijie Geng

Zhengyang Geng
Kyle A. Genova
Georgios Georgakis
Markos Georgopoulos
Marcel Geppert
Shabnam Ghadar
Mina Ghadimi Atigh
Deepti Ghadiyaram
Maani Ghaffari Jadidi
Sedigh Ghamari
Zahra Gharaee
Michaël Gharbi
Golnaz Ghiasi
Reza Ghoddoosian
Soumya Suvra Ghosal
Adhiraj Ghosh
Arthita Ghosh
Pallabi Ghosh
Soumyadeep Ghosh
Andrew Gilbert
Igor Gilitschenski
Jhony H. Giraldo
Andreu Girbau Xalabarder
Rohit Girdhar
Sharath Girish
Xavier Giro-i-Nieto
Raja Giryes
Thomas Gittings
Nikolaos Gkanatsios
Ioannis Gkioulekas
Abhiram
    Gnanasambandam
Aurele T. Gnanha
Clement L. J. C. Godard
Arushi Goel
Vidit Goel
Shubham Goel
Zan Gojcic
Aaron K. Gokaslan
Tejas Gokhale
S. Alireza Golestaneh
Thiago L. Gomes
Nuno Goncalves
Boqing Gong
Chen Gong

Yuanhao Gong
Guoqiang Gong
Jingyu Gong
Rui Gong
Yu Gong
Mingming Gong
Neil Zhenqiang Gong
Xun Gong
Yunye Gong
Yihong Gong
Cristina I. González
Nithin Gopalakrishnan
    Nair
Gaurav Goswami
Jianping Gou
Shreyank N. Gowda
Ankit Goyal
Helmut Grabner
Patrick L. Grady
Ben Graham
Eric Granger
Douglas R. Gray
Matej Grcić
David Griffiths
Jinjin Gu
Yun Gu
Shuyang Gu
Jianyang Gu
Fuqiang Gu
Jiatao Gu
Jindong Gu
Jiaqi Gu
Jinwei Gu
Jiaxin Gu
Geonmo Gu
Xiao Gu
Xinqian Gu
Xiuye Gu
Yuming Gu
Zhangxuan Gu
Dayan Guan
Junfeng Guan
Qingji Guan
Tianrui Guan
Shanyan Guan

Denis A. Gudovskiy
Ricardo Guerrero
Pierre-Louis Guhur
Jie Gui
Liangyan Gui
Liangke Gui
Benoit Guillard
Erhan Gundogdu
Manuel Günther
Jingcai Guo
Yuanfang Guo
Junfeng Guo
Chenqi Guo
Dan Guo
Hongji Guo
Jia Guo
Jie Guo
Minghao Guo
Shi Guo
Yanhui Guo
Yangyang Guo
Yuan-Chen Guo
Yilu Guo
Yiluan Guo
Yong Guo
Guangyu Guo
Haiyun Guo
Jinyang Guo
Jianyuan Guo
Pengsheng Guo
Pengfei Guo
Shuxuan Guo
Song Guo
Tianyu Guo
Qing Guo
Qiushan Guo
Wen Guo
Xiefan Guo
Xiaohu Guo
Xiaoqing Guo
Yufei Guo
Yuhui Guo
Yuliang Guo
Yunhui Guo
Yanwen Guo

Akshita Gupta
Ankush Gupta
Kamal Gupta
Kartik Gupta
Ritwik Gupta
Rohit Gupta
Siddharth Gururani
Fredrik K. Gustafsson
Abner Guzman Rivera
Vladimir Guzov
Matthew A. Gwilliam
Jung-Woo Ha
Marc Habermann
Isma Hadji
Christian Haene
Martin Hahner
Levente Hajder
Alexandros Haliassos
Emanuela Haller
Bumsub Ham
Abdullah J. Hamdi
Shreyas Hampali
Dongyoon Han
Chunrui Han
Dong-Jun Han
Dong-Sig Han
Guangxing Han
Zhizhong Han
Ruize Han
Jiaming Han
Jin Han
Ligong Han
Xian-Hua Han
Xiaoguang Han
Yizeng Han
Zhi Han
Zhenjun Han
Zhongyi Han
Jungong Han
Junlin Han
Kai Han
Kun Han
Sungwon Han
Songfang Han
Wei Han

Xiao Han
Xintong Han
Xinzhe Han
Yahong Han
Yan Han
Zongbo Han
Nicolai Hani
Rana Hanocka
Niklas Hanselmann
Nicklas A. Hansen
Hong Hanyu
Fusheng Hao
Yanbin Hao
Shijie Hao
Udith Haputhanthri
Mehrtash Harandi
Josh Harguess
Adam Harley
David M. Hart
Atsushi Hashimoto
Ali Hassani
Mohammed Hassanin
Yana Hasson
Joakim Bruslund Haurum
Bo He
Kun He
Chen He
Xin He
Fazhi He
Gaoqi He
Hao He
Haoyu He
Jiangpeng He
Hongliang He
Qian He
Xiangteng He
Xuming He
Yannan He
Yuhang He
Yang He
Xiangyu He
Nanjun He
Pan He
Sen He
Shengfeng He

Songtao He
Tao He
Tong He
Wei He
Xuehai He
Xiaoxiao He
Ying He
Yisheng He
Ziwen He
Peter Hedman
Felix Heide
Yacov Hel-Or
Paul Henderson
Philipp Henzler
Byeongho Heo
Jae-Pil Heo
Miran Heo
Sachini A. Herath
Stephane Herbin
Pedro Hermosilla Casajus
Monica Hernandez
Charles Herrmann
Roei Herzig
Mauricio Hess-Flores
Carlos Hinojosa
Tobias Hinz
Tsubasa Hirakawa
Chih-Hui Ho
Lam Si Tung Ho
Jennifer Hobbs
Derek Hoiem
Yannick Hold-Geoffroy
Aleksander Holynski
Cheeun Hong
Fa-Ting Hong
Hanbin Hong
Guan Zhe Hong
Danfeng Hong
Lanqing Hong
Xiaopeng Hong
Xin Hong
Jie Hong
Seungbum Hong
Cheng-Yao Hong
Seunghoon Hong

Yi Hong
Yuan Hong
Yuchen Hong
Anthony Hoogs
Maxwell C. Horton
Kazuhiro Hotta
Qibin Hou
Tingbo Hou
Junhui Hou
Ji Hou
Qiqi Hou
Rui Hou
Ruibing Hou
Zhi Hou
Henry Howard-Jenkins
Lukas Hoyer
Wei-Lin Hsiao
Chiou-Ting Hsu
Anthony Hu
Brian Hu
Yusong Hu
Hexiang Hu
Haoji Hu
Di Hu
Hengtong Hu
Haigen Hu
Lianyu Hu
Hanzhe Hu
Jie Hu
Junlin Hu
Shizhe Hu
Jian Hu
Zhiming Hu
Juhua Hu
Peng Hu
Ping Hu
Ronghang Hu
MengShun Hu
Tao Hu
Vincent Tao Hu
Xiaoling Hu
Xinting Hu
Xiaolin Hu
Xuefeng Hu
Xiaowei Hu

Yang Hu
Yueyu Hu
Zeyu Hu
Zhongyun Hu
Binh-Son Hua
Guoliang Hua
Yi Hua
Linzhi Huang
Qiusheng Huang
Bo Huang
Chen Huang
Hsin-Ping Huang
Ye Huang
Shuangping Huang
Zeng Huang
Buzhen Huang
Cong Huang
Heng Huang
Hao Huang
Qidong Huang
Huaibo Huang
Chaoqin Huang
Feihu Huang
Jiahui Huang
Jingjia Huang
Kun Huang
Lei Huang
Sheng Huang
Shuaiyi Huang
Siyu Huang
Xiaoshui Huang
Xiaoyang Huang
Yan Huang
Yihao Huang
Ying Huang
Ziling Huang
Xiaoke Huang
Yifei Huang
Haiyang Huang
Zhewei Huang
Jin Huang
Haibin Huang
Jiaxing Huang
Junjie Huang
Keli Huang

Lang Huang
Lin Huang
Luojie Huang
Mingzhen Huang
Shijia Huang
Shengyu Huang
Siyuan Huang
He Huang
Xiuyu Huang
Lianghua Huang
Yue Huang
Yaping Huang
Yuge Huang
Zehao Huang
Zeyi Huang
Zhiqi Huang
Zhongzhan Huang
Zilong Huang
Ziyuan Huang
Tianrui Hui
Zhuo Hui
Le Hui
Jing Huo
Junhwa Hur
Shehzeen S. Hussain
Chuong Minh Huynh
Seunghyun Hwang
Jaehui Hwang
Jyh-Jing Hwang
Sukjun Hwang
Soonmin Hwang
Wonjun Hwang
Rakib Hyder
Sangeek Hyun
Sarah Ibrahimi
Tomoki Ichikawa
Yerlan Idelbayev
A. S. M. Iftekhar
Masaaki Iiyama
Satoshi Ikehata
Sunghoon Im
Atul N. Ingle
Eldar Insafutdinov
Yani A. Ioannou
Radu Tudor Ionescu

Umar Iqbal
Go Irie
Muhammad Zubair Irshad
Ahmet Iscen
Berivan Isik
Ashraful Islam
Md Amirul Islam
Syed Islam
Mariko Isogawa
Vamsi Krishna K. Ithapu
Boris Ivanovic
Darshan Iyer
Sarah Jabbour
Ayush Jain
Nishant Jain
Samyak Jain
Vidit Jain
Vineet Jain
Priyank Jaini
Tomas Jakab
Mohammad A. A. K.
    Jalwana
Muhammad Abdullah
    Jamal
Hadi Jamali-Rad
Stuart James
Varun Jampani
Young Kyun Jang
YeongJun Jang
Yunseok Jang
Ronnachai Jaroensri
Bhavan Jasani
Krishna Murthy
    Jatavallabhula
Mojan Javaheripi
Syed A. Javed
Guillaume Jeanneret
Pranav Jeevan
Herve Jegou
Rohit Jena
Tomas Jenicek
Porter Jenkins
Simon Jenni
Hae-Gon Jeon
Sangryul Jeon

Boseung Jeong
Yoonwoo Jeong
Seong-Gyun Jeong
Jisoo Jeong
Allan D. Jepson
Ankit Jha
Sumit K. Jha
I-Hong Jhuo
Ge-Peng Ji
Chaonan Ji
Deyi Ji
Jingwei Ji
Wei Ji
Zhong Ji
Jiayi Ji
Pengliang Ji
Hui Ji
Mingi Ji
Xiaopeng Ji
Yuzhu Ji
Baoxiong Jia
Songhao Jia
Dan Jia
Shan Jia
Xiaojun Jia
Xiuyi Jia
Xu Jia
Menglin Jia
Wenqi Jia
Boyuan Jiang
Wenhao Jiang
Huaizu Jiang
Hanwen Jiang
Haiyong Jiang
Hao Jiang
Huajie Jiang
Huiqin Jiang
Haojun Jiang
Haobo Jiang
Junjun Jiang
Xingyu Jiang
Yangbangyan Jiang
Yu Jiang
Jianmin Jiang
Jiaxi Jiang

Jing Jiang
Kui Jiang
Li Jiang
Liming Jiang
Chiyu Jiang
Meirui Jiang
Chen Jiang
Peng Jiang
Tai-Xiang Jiang
Wen Jiang
Xinyang Jiang
Yifan Jiang
Yuming Jiang
Yingying Jiang
Zeren Jiang
ZhengKai Jiang
Zhenyu Jiang
Shuming Jiao
Jianbo Jiao
Licheng Jiao
Dongkwon Jin
Yeying Jin
Cheng Jin
Linyi Jin
Qing Jin
Taisong Jin
Xiao Jin
Xin Jin
Sheng Jin
Kyong Hwan Jin
Ruibing Jin
SouYoung Jin
Yueming Jin
Chenchen Jing
Longlong Jing
Taotao Jing
Yongcheng Jing
Younghyun Jo
Joakim Johnander
Jeff Johnson
Michael J. Jones
R. Kenny Jones
Rico Jonschkowski
Ameya Joshi
Sunghun Joung

Felix Juefei-Xu
Claudio R. Jung
Steffen Jung
Hari Chandana K.
Rahul Vigneswaran K.
Prajwal K. R.
Abhishek Kadian
Jhony Kaesemodel Pontes
Kumara Kahatapitiya
Anmol Kalia
Sinan Kalkan
Tarun Kalluri
Jaewon Kam
Sandesh Kamath
Meina Kan
Menelaos Kanakis
Takuhiro Kaneko
Di Kang
Guoliang Kang
Hao Kang
Jaeyeon Kang
Kyoungkook Kang
Li-Wei Kang
MinGuk Kang
Suk-Ju Kang
Zhao Kang
Yash Mukund Kant
Yueying Kao
Aupendu Kar
Konstantinos Karantzalos
Sezer Karaoglu
Navid Kardan
Sanjay Kariyappa
Leonid Karlinsky
Animesh Karnewar
Shyamgopal Karthik
Hirak J. Kashyap
Marc A. Kastner
Hirokatsu Kataoka
Angelos Katharopoulos
Hiroharu Kato
Kai Katsumata
Manuel Kaufmann
Chaitanya Kaul
Prakhar Kaushik

Yuki Kawana
Lei Ke
Lipeng Ke
Tsung-Wei Ke
Wei Ke
Petr Kellnhofer
Aniruddha Kembhavi
John Kender
Corentin Kervadec
Leonid Keselman
Daniel Keysers
Nima Khademi Kalantari
Taras Khakhulin
Samir Khaki
Muhammad Haris Khan
Qadeer Khan
Salman Khan
Subash Khanal
Vaishnavi M. Khindkar
Rawal Khirodkar
Saeed Khorram
Pirazh Khorramshahi
Kourosh Khoshelham
Ansh Khurana
Benjamin Kiefer
Jae Myung Kim
Junho Kim
Boah Kim
Hyeonseong Kim
Dong-Jin Kim
Dongwan Kim
Donghyun Kim
Doyeon Kim
Yonghyun Kim
Hyung-Il Kim
Hyunwoo Kim
Hyeongwoo Kim
Hyo Jin Kim
Hyunwoo J. Kim
Taehoon Kim
Jaeha Kim
Jiwon Kim
Jung Uk Kim
Kangyeol Kim
Eunji Kim

Daeha Kim
Dongwon Kim
Kunhee Kim
Kyungmin Kim
Junsik Kim
Min H. Kim
Namil Kim
Kookhoi Kim
Sanghyun Kim
Seongyeop Kim
Seungryong Kim
Saehoon Kim
Euyoung Kim
Guisik Kim
Sungyeon Kim
Sunnie S. Y. Kim
Taehun Kim
Tae Oh Kim
Won Hwa Kim
Seungwook Kim
YoungBin Kim
Youngeun Kim
Akisato Kimura
Furkan Osman Kınlı
Zsolt Kira
Hedvig Kjellström
Florian Kleber
Jan P. Klopp
Florian Kluger
Laurent Kneip
Byungsoo Ko
Muhammed Kocabas
A. Sophia Koepke
Kevin Koeser
Nick Kolkin
Nikos Kolotouros
Wai-Kin Adams Kong
Deying Kong
Caihua Kong
Youyong Kong
Shuyu Kong
Shu Kong
Tao Kong
Yajing Kong
Yu Kong

Zishang Kong
Theodora Kontogianni
Anton S. Konushin
Julian F. P. Kooij
Bruno Korbar
Giorgos Kordopatis-Zilos
Jari Korhonen
Adam Kortylewski
Denis Korzhenkov
Divya Kothandaraman
Suraj Kothawade
Iuliia Kotseruba
Satwik Kottur
Shashank Kotyan
Alexandros Kouris
Petros Koutras
Anna Kreshuk
Ranjay Krishna
Dilip Krishnan
Andrey Kuehlkamp
Hilde Kuehne
Jason Kuen
David Kügler
Arjan Kuijper
Anna Kukleva
Sumith Kulal
Viveka Kulharia
Akshay R. Kulkarni
Nilesh Kulkarni
Dominik Kulon
Abhinav Kumar
Akash Kumar
Suryansh Kumar
B. V. K. Vijaya Kumar
Pulkit Kumar
Ratnesh Kumar
Sateesh Kumar
Satish Kumar
Vijay Kumar B. G.
Nupur Kumari
Sudhakar Kumawat
Jogendra Nath Kundu
Hsien-Kai Kuo
Meng-Yu Jennifer Kuo
Vinod Kumar Kurmi

Yusuke Kurose
Keerthy Kusumam
Alina Kuznetsova
Henry Kvinge
Ho Man Kwan
Hyeokjun Kweon
Heeseung Kwon
Gihyun Kwon
Myung-Joon Kwon
Taesung Kwon
YoungJoong Kwon
Christos Kyrkou
Jorma Laaksonen
Yann Labbe
Zorah Laehner
Florent Lafarge
Hamid Laga
Manuel Lagunas
Shenqi Lai
Jian-Huang Lai
Zihang Lai
Mohamed I. Lakhal
Mohit Lamba
Meng Lan
Loic Landrieu
Zhiqiang Lang
Natalie Lang
Dong Lao
Yizhen Lao
Yingjie Lao
Issam Hadj Laradji
Gustav Larsson
Viktor Larsson
Zakaria Laskar
Stéphane Lathuilière
Chun Pong Lau
Rynson W. H. Lau
Hei Law
Justin Lazarow
Verica Lazova
Eric-Tuan Le
Hieu Le
Trung-Nghia Le
Mathias Lechner
Byeong-Uk Lee

Chen-Yu Lee
Che-Rung Lee
Chul Lee
Hong Joo Lee
Dongsoo Lee
Jiyoung Lee
Eugene Eu Tzuan Lee
Daeun Lee
Saehyung Lee
Jewook Lee
Hyungtae Lee
Hyunmin Lee
Jungbeom Lee
Joon-Young Lee
Jong-Seok Lee
Joonseok Lee
Junha Lee
Kibok Lee
Byung-Kwan Lee
Jangwon Lee
Jinho Lee
Jongmin Lee
Seunghyun Lee
Sohyun Lee
Minsik Lee
Dogyoon Lee
Seungmin Lee
Min Jun Lee
Sangho Lee
Sangmin Lee
Seungeun Lee
Seon-Ho Lee
Sungmin Lee
Sungho Lee
Sangyoun Lee
Vincent C. S. S. Lee
Jaeseong Lee
Yong Jae Lee
Chenyang Lei
Chenyi Lei
Jiahui Lei
Xinyu Lei
Yinjie Lei
Jiaxu Leng
Luziwei Leng

Jan E. Lenssen
Vincent Lepetit
Thomas Leung
María Leyva-Vallina
Xin Li
Yikang Li
Baoxin Li
Bin Li
Bing Li
Bowen Li
Changlin Li
Chao Li
Chongyi Li
Guanyue Li
Shuai Li
Jin Li
Dingquan Li
Dongxu Li
Yiting Li
Gang Li
Dian Li
Guohao Li
Haoang Li
Haoliang Li
Haoran Li
Hengduo Li
Huafeng Li
Xiaoming Li
Hanao Li
Hongwei Li
Ziqiang Li
Jisheng Li
Jiacheng Li
Jia Li
Jiachen Li
Jiahao Li
Jianwei Li
Jiazhi Li
Jie Li
Jing Li
Jingjing Li
Jingtao Li
Jun Li
Junxuan Li
Kai Li

Kailin Li
Kenneth Li
Kun Li
Kunpeng Li
Aoxue Li
Chenglong Li
Chenglin Li
Changsheng Li
Zhichao Li
Qiang Li
Yanyu Li
Zuoyue Li
Xiang Li
Xuelong Li
Fangda Li
Ailin Li
Liang Li
Chun-Guang Li
Daiqing Li
Dong Li
Guanbin Li
Guorong Li
Haifeng Li
Jianan Li
Jianing Li
Jiaxin Li
Ke Li
Lei Li
Lincheng Li
Liulei Li
Lujun Li
Linjie Li
Lin Li
Pengyu Li
Ping Li
Qiufu Li
Qingyong Li
Rui Li
Siyuan Li
Wei Li
Wenbin Li
Xiangyang Li
Xinyu Li
Xiujun Li
Xiu Li

Xu Li
Ya-Li Li
Yao Li
Yongjie Li
Yijun Li
Yiming Li
Yuezun Li
Yu Li
Yunheng Li
Yuqi Li
Zhe Li
Zeming Li
Zhen Li
Zhengqin Li
Zhimin Li
Jiefeng Li
Jinpeng Li
Chengze Li
Jianwu Li
Lerenhan Li
Shan Li
Suichan Li
Xiangtai Li
Yanjie Li
Yandong Li
Zhuoling Li
Zhenqiang Li
Manyi Li
Maosen Li
Ji Li
Minjun Li
Mingrui Li
Mengtian Li
Junyi Li
Nianyi Li
Bo Li
Xiao Li
Peihua Li
Peike Li
Peizhao Li
Peiliang Li
Qi Li
Ren Li
Runze Li
Shile Li

Sheng Li
Shigang Li
Shiyu Li
Shuang Li
Shasha Li
Shichao Li
Tianye Li
Yuexiang Li
Wei-Hong Li
Wanhua Li
Weihao Li
Weiming Li
Weixin Li
Wenbo Li
Wenshuo Li
Weijian Li
Yunan Li
Xirong Li
Xianhang Li
Xiaoyu Li
Xueqian Li
Xuanlin Li
Xianzhi Li
Yunqiang Li
Yanjing Li
Yansheng Li
Yawei Li
Yi Li
Yong Li
Yong-Lu Li
Yuhang Li
Yu-Jhe Li
Yuxi Li
Yunsheng Li
Yanwei Li
Zechao Li
Zejian Li
Zeju Li
Zekun Li
Zhaowen Li
Zheng Li
Zhenyu Li
Zhiheng Li
Zhi Li
Zhong Li

Zhuowei Li
Zhuowan Li
Zhuohang Li
Zizhang Li
Chen Li
Yuan-Fang Li
Dongze Lian
Xiaochen Lian
Zhouhui Lian
Long Lian
Qing Lian
Jin Lianbao
Jinxiu S. Liang
Dingkang Liang
Jiahao Liang
Jianming Liang
Jingyun Liang
Kevin J. Liang
Kaizhao Liang
Chen Liang
Jie Liang
Senwei Liang
Ding Liang
Jiajun Liang
Jian Liang
Kongming Liang
Siyuan Liang
Yuanzhi Liang
Zhengfa Liang
Mingfu Liang
Xiaodan Liang
Xuefeng Liang
Yuxuan Liang
Kang Liao
Liang Liao
Hong-Yuan Mark Liao
Wentong Liao
Haofu Liao
Yue Liao
Minghui Liao
Shengcai Liao
Ting-Hsuan Liao
Xin Liao
Yinghong Liao
Teck Yian Lim

Che-Tsung Lin
Chung-Ching Lin
Chen-Hsuan Lin
Cheng Lin
Chuming Lin
Chunyu Lin
Dahua Lin
Wei Lin
Zheng Lin
Huaijia Lin
Jason Lin
Jierui Lin
Jiaying Lin
Jie Lin
Kai-En Lin
Kevin Lin
Guangfeng Lin
Jiehong Lin
Feng Lin
Hang Lin
Kwan-Yee Lin
Ke Lin
Luojun Lin
Qinghong Lin
Xiangbo Lin
Yi Lin
Zudi Lin
Shijie Lin
Yiqun Lin
Tzu-Heng Lin
Ming Lin
Shaohui Lin
SongNan Lin
Ji Lin
Tsung-Yu Lin
Xudong Lin
Yancong Lin
Yen-Chen Lin
Yiming Lin
Yuewei Lin
Zhiqiu Lin
Zinan Lin
Zhe Lin
David B. Lindell
Zhixin Ling

Zhan Ling
Alexander Liniger
Venice Erin B. Liong
Joey Litalien
Or Litany
Roee Litman
Ron Litman
Jim Little
Dor Litvak
Shaoteng Liu
Shuaicheng Liu
Andrew Liu
Xian Liu
Shaohui Liu
Bei Liu
Bo Liu
Yong Liu
Ming Liu
Yanbin Liu
Chenxi Liu
Daqi Liu
Di Liu
Difan Liu
Dong Liu
Dongfang Liu
Daizong Liu
Xiao Liu
Fangyi Liu
Fengbei Liu
Fenglin Liu
Bin Liu
Yuang Liu
Ao Liu
Hong Liu
Hongfu Liu
Huidong Liu
Ziyi Liu
Feng Liu
Hao Liu
Jie Liu
Jialun Liu
Jiang Liu
Jing Liu
Jingya Liu
Jiaming Liu

Jun Liu
Juncheng Liu
Jiawei Liu
Hongyu Liu
Chuanbin Liu
Haotian Liu
Lingqiao Liu
Chang Liu
Han Liu
Liu Liu
Min Liu
Yingqi Liu
Aishan Liu
Bingyu Liu
Benlin Liu
Boxiao Liu
Chenchen Liu
Chuanjian Liu
Daqing Liu
Huan Liu
Haozhe Liu
Jiaheng Liu
Wei Liu
Jingzhou Liu
Jiyuan Liu
Lingbo Liu
Nian Liu
Peiye Liu
Qiankun Liu
Shenglan Liu
Shilong Liu
Wen Liu
Wenyu Liu
Weifeng Liu
Wu Liu
Xiaolong Liu
Yang Liu
Yanwei Liu
Yingcheng Liu
Yongfei Liu
Yihao Liu
Yu Liu
Yunze Liu
Ze Liu
Zhenhua Liu

Zhenguang Liu
Lin Liu
Lihao Liu
Pengju Liu
Xinhai Liu
Yunfei Liu
Meng Liu
Minghua Liu
Mingyuan Liu
Miao Liu
Peirong Liu
Ping Liu
Qingjie Liu
Ruoshi Liu
Risheng Liu
Songtao Liu
Xing Liu
Shikun Liu
Shuming Liu
Sheng Liu
Songhua Liu
Tongliang Liu
Weibo Liu
Weide Liu
Weizhe Liu
Wenxi Liu
Weiyang Liu
Xin Liu
Xiaobin Liu
Xudong Liu
Xiaoyi Liu
Xihui Liu
Xinchen Liu
Xingtong Liu
Xinpeng Liu
Xinyu Liu
Xianpeng Liu
Xu Liu
Xingyu Liu
Yongtuo Liu
Yahui Liu
Yangxin Liu
Yaoyao Liu
Yaojie Liu
Yuliang Liu

Yongcheng Liu
Yuan Liu
Yufan Liu
Yu-Lun Liu
Yun Liu
Yunfan Liu
Yuanzhong Liu
Zhuoran Liu
Zhen Liu
Zheng Liu
Zhijian Liu
Zhisong Liu
Ziquan Liu
Ziyu Liu
Zhihua Liu
Zechun Liu
Zhaoyang Liu
Zhengzhe Liu
Stephan Liwicki
Shao-Yuan Lo
Sylvain Lobry
Suhas Lohit
Vishnu Suresh Lokhande
Vincenzo Lomonaco
Chengjiang Long
Guodong Long
Fuchen Long
Shangbang Long
Yang Long
Zijun Long
Vasco Lopes
Antonio M. Lopez
Roberto Javier
    Lopez-Sastre
Tobias Lorenz
Javier Lorenzo-Navarro
Yujing Lou
Qian Lou
Xiankai Lu
Changsheng Lu
Huimin Lu
Yongxi Lu
Hao Lu
Hong Lu
Jiasen Lu

Juwei Lu
Fan Lu
Guangming Lu
Jiwen Lu
Shun Lu
Tao Lu
Xiaonan Lu
Yang Lu
Yao Lu
Yongchun Lu
Zhiwu Lu
Cheng Lu
Liying Lu
Guo Lu
Xuequan Lu
Yanye Lu
Yantao Lu
Yuhang Lu
Fujun Luan
Jonathon Luiten
Jovita Lukasik
Alan Lukezic
Jonathan Samuel Lumentut
Mayank Lunayach
Ao Luo
Canjie Luo
Chong Luo
Xu Luo
Grace Luo
Jun Luo
Katie Z. Luo
Tao Luo
Cheng Luo
Fangzhou Luo
Gen Luo
Lei Luo
Sihui Luo
Weixin Luo
Yan Luo
Xiaoyan Luo
Yong Luo
Yadan Luo
Hao Luo
Ruotian Luo
Mi Luo

Tiange Luo
Wenjie Luo
Wenhan Luo
Xiao Luo
Zhiming Luo
Zhipeng Luo
Zhengyi Luo
Diogo C. Luvizon
Zhaoyang Lv
Gengyu Lyu
Lingjuan Lyu
Jun Lyu
Yuanyuan Lyu
Youwei Lyu
Yueming Lyu
Bingpeng Ma
Chao Ma
Chongyang Ma
Congbo Ma
Chih-Yao Ma
Fan Ma
Lin Ma
Haoyu Ma
Hengbo Ma
Jianqi Ma
Jiawei Ma
Jiayi Ma
Kede Ma
Kai Ma
Lingni Ma
Lei Ma
Xu Ma
Ning Ma
Benteng Ma
Cheng Ma
Andy J. Ma
Long Ma
Zhanyu Ma
Zhiheng Ma
Qianli Ma
Shiqiang Ma
Sizhuo Ma
Shiqing Ma
Xiaolong Ma
Xinzhu Ma

Gautam B. Machiraju
Spandan Madan
Mathew Magimai-Doss
Luca Magri
Behrooz Mahasseni
Upal Mahbub
Siddharth Mahendran
Paridhi Maheshwari
Rishabh Maheshwary
Mohammed Mahmoud
Shishira R. R. Maiya
Sylwia Majchrowska
Arjun Majumdar
Puspita Majumdar
Orchid Majumder
Sagnik Majumder
Ilya Makarov
Farkhod F.
    Makhmudkhujaev
Yasushi Makihara
Ankur Mali
Mateusz Malinowski
Utkarsh Mall
Srikanth Malla
Clement Mallet
Dimitrios Mallis
Yunze Man
Dipu Manandhar
Massimiliano Mancini
Murari Mandal
Raunak Manekar
Karttikeya Mangalam
Puneet Mangla
Fabian Manhardt
Sivabalan Manivasagam
Fahim Mannan
Chengzhi Mao
Hanzi Mao
Jiayuan Mao
Junhua Mao
Zhiyuan Mao
Jiageng Mao
Yunyao Mao
Zhendong Mao
Alberto Marchisio

Diego Marcos
Riccardo Marin
Aram Markosyan
Renaud Marlet
Ricardo Marques
Miquel Martí i Rabadán
Diego Martin Arroyo
Niki Martinel
Brais Martinez
Julieta Martinez
Marc Masana
Tomohiro Mashita
Timothée Masquelier
Minesh Mathew
Tetsu Matsukawa
Marwan Mattar
Bruce A. Maxwell
Christoph Mayer
Mantas Mazeika
Pratik Mazumder
Scott McCloskey
Steven McDonagh
Ishit Mehta
Jie Mei
Kangfu Mei
Jieru Mei
Xiaoguang Mei
Givi Meishvili
Luke Melas-Kyriazi
Iaroslav Melekhov
Andres Mendez-Vazquez
Heydi Mendez-Vazquez
Matias Mendieta
Ricardo A. Mendoza-León
Chenlin Meng
Depu Meng
Rang Meng
Zibo Meng
Qingjie Meng
Qier Meng
Yanda Meng
Zihang Meng
Thomas Mensink
Fabian Mentzer
Christopher Metzler

Gregory P. Meyer
Vasileios Mezaris
Liang Mi
Lu Mi
Bo Miao
Changtao Miao
Zichen Miao
Qiguang Miao
Xin Miao
Zhongqi Miao
Frank Michel
Simone Milani
Ben Mildenhall
Roy V. Miles
Juhong Min
Kyle Min
Hyun-Seok Min
Weiqing Min
Yuecong Min
Zhixiang Min
Qi Ming
David Minnen
Aymen Mir
Deepak Mishra
Anand Mishra
Shlok K. Mishra
Niluthpol Mithun
Gaurav Mittal
Trisha Mittal
Daisuke Miyazaki
Kaichun Mo
Hong Mo
Zhipeng Mo
Davide Modolo
Abduallah A. Mohamed
Mohamed Afham
    Mohamed Aflal
Ron Mokady
Pavlo Molchanov
Davide Moltisanti
Liliane Momeni
Gianluca Monaci
Pascal Monasse
Ajoy Mondal
Tom Monnier

Aron Monszpart
Gyeongsik Moon
Suhong Moon
Taesup Moon
Sean Moran
Daniel Moreira
Pietro Morerio
Alexandre Morgand
Lia Morra
Ali Mosleh
Inbar Mosseri
Sayed Mohammad
    Mostafavi Isfahani
Saman Motamed
Ramy A. Mounir
Fangzhou Mu
Jiteng Mu
Norman Mu
Yasuhiro Mukaigawa
Ryan Mukherjee
Tanmoy Mukherjee
Yusuke Mukuta
Ravi Teja Mullapudi
Lea Müller
Matthias Müller
Martin Mundt
Nils Murrugarra-Llerena
Damien Muselet
Armin Mustafa
Muhammad Ferjad Naeem
Sauradip Nag
Hajime Nagahara
Pravin Nagar
Rajendra Nagar
Naveen Shankar Nagaraja
Varun Nagaraja
Tushar Nagarajan
Seungjun Nah
Gaku Nakano
Yuta Nakashima
Giljoo Nam
Seonghyeon Nam
Liangliang Nan
Yuesong Nan
Yeshwanth Napolean

Dinesh Reddy
    Narapureddy
Medhini Narasimhan
Supreeth
    Narasimhaswamy
Sriram Narayanan
Erickson R. Nascimento
Varun Nasery
K. L. Navaneet
Pablo Navarrete Michelini
Shant Navasardyan
Shah Nawaz
Nihal Nayak
Farhood Negin
Lukáš Neumann
Alejandro Newell
Evonne Ng
Kam Woh Ng
Tony Ng
Anh Nguyen
Tuan Anh Nguyen
Cuong Cao Nguyen
Ngoc Cuong Nguyen
Thanh Nguyen
Khoi Nguyen
Phi Le Nguyen
Phong Ha Nguyen
Tam Nguyen
Truong Nguyen
Anh Tuan Nguyen
Rang Nguyen
Thao Thi Phuong Nguyen
Van Nguyen Nguyen
Zhen-Liang Ni
Yao Ni
Shijie Nie
Xuecheng Nie
Yongwei Nie
Weizhi Nie
Ying Nie
Yinyu Nie
Kshitij N. Nikhal
Simon Niklaus
Xuefei Ning
Jifeng Ning

Yotam Nitzan
Di Niu
Shuaicheng Niu
Li Niu
Wei Niu
Yulei Niu
Zhenxing Niu
Albert No
Shohei Nobuhara
Nicoletta Noceti
Junhyug Noh
Sotiris Nousias
Slawomir Nowaczyk
Ewa M. Nowara
Valsamis Ntouskos
Gilberto Ochoa-Ruiz
Ferda Ofli
Jihyong Oh
Sangyun Oh
Youngtaek Oh
Hiroki Ohashi
Takahiro Okabe
Kemal Oksuz
Fumio Okura
Daniel Olmeda Reino
Matthew Olson
Carl Olsson
Roy Or-El
Alessandro Ortis
Guillermo Ortiz-Jimenez
Magnus Oskarsson
Ahmed A. A. Osman
Martin R. Oswald
Mayu Otani
Naima Otberdout
Cheng Ouyang
Jiahong Ouyang
Wanli Ouyang
Andrew Owens
Poojan B. Oza
Mete Ozay
A. Cengiz Oztireli
Gautam Pai
Tomas Pajdla
Umapada Pal

Simone Palazzo
Luca Palmieri
Bowen Pan
Hao Pan
Lili Pan
Tai-Yu Pan
Liang Pan
Chengwei Pan
Yingwei Pan
Xuran Pan
Jinshan Pan
Xinyu Pan
Liyuan Pan
Xingang Pan
Xingjia Pan
Zhihong Pan
Zizheng Pan
Priyadarshini Panda
Rameswar Panda
Rohit Pandey
Kaiyue Pang
Bo Pang
Guansong Pang
Jiangmiao Pang
Meng Pang
Tianyu Pang
Ziqi Pang
Omiros Pantazis
Andreas Panteli
Maja Pantic
Marina Paolanti
Joao P. Papa
Samuele Papa
Mike Papadakis
Dim P. Papadopoulos
George Papandreou
Constantin Pape
Toufiq Parag
Chethan Parameshwara
Shaifali Parashar
Alejandro Pardo
Rishubh Parihar
Sarah Parisot
JaeYoo Park
Gyeong-Moon Park

Hyojin Park
Hyoungseob Park
Jongchan Park
Jae Sung Park
Kiru Park
Chunghyun Park
Kwanyong Park
Sunghyun Park
Sungrae Park
Seongsik Park
Sanghyun Park
Sungjune Park
Taesung Park
Gaurav Parmar
Paritosh Parmar
Alvaro Parra
Despoina Paschalidou
Or Patashnik
Shivansh Patel
Pushpak Pati
Prashant W. Patil
Vaishakh Patil
Suvam Patra
Jay Patravali
Badri Narayana Patro
Angshuman Paul
Sudipta Paul
Rémi Pautrat
Nick E. Pears
Adithya Pediredla
Wenjie Pei
Shmuel Peleg
Latha Pemula
Bo Peng
Houwen Peng
Yue Peng
Liangzu Peng
Baoyun Peng
Jun Peng
Pai Peng
Sida Peng
Xi Peng
Yuxin Peng
Songyou Peng
Wei Peng

Weiqi Peng
Wen-Hsiao Peng
Pramuditha Perera
Juan C. Perez
Eduardo Pérez Pellitero
Juan-Manuel Perez-Rua
Federico Pernici
Marco Pesavento
Stavros Petridis
Ilya A. Petrov
Vladan Petrovic
Mathis Petrovich
Suzanne Petryk
Hieu Pham
Quang Pham
Khoi Pham
Tung Pham
Huy Phan
Stephen Phillips
Cheng Perng Phoo
David Picard
Marco Piccirilli
Georg Pichler
A. J. Piergiovanni
Vipin Pillai
Silvia L. Pintea
Giovanni Pintore
Robinson Piramuthu
Fiora Pirri
Theodoros Pissas
Fabio Pizzati
Benjamin Planche
Bryan Plummer
Matteo Poggi
Ashwini Pokle
Georgy E. Ponimatkin
Adrian Popescu
Stefan Popov
Nikola Popović
Ronald Poppe
Angelo Porrello
Michael Potter
Charalambos Poullis
Hadi Pouransari
Omid Poursaeed

Shraman Pramanick
Mantini Pranav
Dilip K. Prasad
Meghshyam Prasad
B. H. Pawan Prasad
Shitala Prasad
Prateek Prasanna
Ekta Prashnani
Derek S. Prijatelj
Luke Y. Prince
Véronique Prinet
Victor Adrian Prisacariu
James Pritts
Thomas Probst
Sergey Prokudin
Rita Pucci
Chi-Man Pun
Matthew Purri
Haozhi Qi
Lu Qi
Lei Qi
Xianbiao Qi
Yonggang Qi
Yuankai Qi
Siyuan Qi
Guocheng Qian
Hangwei Qian
Qi Qian
Deheng Qian
Shengsheng Qian
Wen Qian
Rui Qian
Yiming Qian
Shengju Qian
Shengyi Qian
Xuelin Qian
Zhenxing Qian
Nan Qiao
Xiaotian Qiao
Jing Qin
Can Qin
Siyang Qin
Hongwei Qin
Jie Qin
Minghai Qin

Yipeng Qin
Yongqiang Qin
Wenda Qin
Xuebin Qin
Yuzhe Qin
Yao Qin
Zhenyue Qin
Zhiwu Qing
Heqian Qiu
Jiayan Qiu
Jielin Qiu
Yue Qiu
Jiaxiong Qiu
Zhongxi Qiu
Shi Qiu
Zhaofan Qiu
Zhongnan Qu
Yanyun Qu
Kha Gia Quach
Yuhui Quan
Ruijie Quan
Mike Rabbat
Rahul Shekhar Rade
Filip Radenovic
Gorjan Radevski
Bogdan Raducanu
Francesco Ragusa
Shafin Rahman
Md Mahfuzur Rahman
   Siddiquee
Hossein Rahmani
Kiran Raja
Sivaramakrishnan
   Rajaraman
Jathushan Rajasegaran
Adnan Siraj Rakin
Michaël Ramamonjisoa
Chirag A. Raman
Shanmuganathan Raman
Vignesh Ramanathan
Vasili Ramanishka
Vikram V. Ramaswamy
Merey Ramazanova
Jason Rambach
Sai Saketh Rambhatla

Clément Rambour
Ashwin Ramesh Babu
Adín Ramírez Rivera
Arianna Rampini
Haoxi Ran
Aakanksha Rana
Aayush Jung Bahadur
   Rana
Kanchana N. Ranasinghe
Aneesh Rangnekar
Samrudhdhi B. Rangrej
Harsh Rangwani
Viresh Ranjan
Anyi Rao
Yongming Rao
Carolina Raposo
Michalis Raptis
Amir Rasouli
Vivek Rathod
Adepu Ravi Sankar
Avinash Ravichandran
Bharadwaj Ravichandran
Dripta S. Raychaudhuri
Adria Recasens
Simon Reiß
Davis Rempe
Daxuan Ren
Jiawei Ren
Jimmy Ren
Sucheng Ren
Dayong Ren
Zhile Ren
Dongwei Ren
Qibing Ren
Pengfei Ren
Zhenwen Ren
Xuqian Ren
Yixuan Ren
Zhongzheng Ren
Ambareesh Revanur
Hamed Rezazadegan
   Tavakoli
Rafael S. Rezende
Wonjong Rhee
Alexander Richard

Christian Richardt
Stephan R. Richter
Benjamin Riggan
Dominik Rivoir
Mamshad Nayeem Rizve
Joshua D. Robinson
Joseph Robinson
Chris Rockwell
Ranga Rodrigo
Andres C. Rodriguez
Carlos Rodriguez-Pardo
Marcus Rohrbach
Gemma Roig
Yu Rong
David A. Ross
Mohammad Rostami
Edward Rosten
Karsten Roth
Anirban Roy
Debaditya Roy
Shuvendu Roy
Ahana Roy Choudhury
Aruni Roy Chowdhury
Denys Rozumnyi
Shulan Ruan
Wenjie Ruan
Patrick Ruhkamp
Danila Rukhovich
Anian Ruoss
Chris Russell
Dan Ruta
Dawid Damian Rymarczyk
DongHun Ryu
Hyeonggon Ryu
Kwonyoung Ryu
Balasubramanian S.
Alexandre Sablayrolles
Mohammad Sabokrou
Arka Sadhu
Aniruddha Saha
Oindrila Saha
Pritish Sahu
Aneeshan Sain
Nirat Saini
Saurabh Saini

Takeshi Saitoh
Christos Sakaridis
Fumihiko Sakaue
Dimitrios Sakkos
Ken Sakurada
Parikshit V. Sakurikar
Rohit Saluja
Nermin Samet
Leo Sampaio Ferraz
    Ribeiro
Jorge Sanchez
Enrique Sanchez
Shengtian Sang
Anush Sankaran
Soubhik Sanyal
Nikolaos Sarafianos
Vishwanath Saragadam
István Sárándi
Saquib Sarfraz
Mert Bulent Sariyildiz
Anindya Sarkar
Pritam Sarkar
Paul-Edouard Sarlin
Hiroshi Sasaki
Takami Sato
Torsten Sattler
Ravi Kumar Satzoda
Axel Sauer
Stefano Savian
Artem Savkin
Manolis Savva
Gerald Schaefer
Simone Schaub-Meyer
Yoni Schirris
Samuel Schulter
Katja Schwarz
Jesse Scott
Sinisa Segvic
Constantin Marc Seibold
Lorenzo Seidenari
Matan Sela
Fadime Sener
Paul Hongsuck Seo
Kwanggyoon Seo
Hongje Seong

Dario Serez
Francesco Setti
Bryan Seybold
Mohamad Shahbazi
Shima Shahfar
Xinxin Shan
Caifeng Shan
Dandan Shan
Shawn Shan
Wei Shang
Jinghuan Shang
Jiaxiang Shang
Lei Shang
Sukrit Shankar
Ken Shao
Rui Shao
Jie Shao
Mingwen Shao
Aashish Sharma
Gaurav Sharma
Vivek Sharma
Abhishek Sharma
Yoli Shavit
Shashank Shekhar
Sumit Shekhar
Zhijie Shen
Fengyi Shen
Furao Shen
Jialie Shen
Jingjing Shen
Ziyi Shen
Linlin Shen
Guangyu Shen
Biluo Shen
Falong Shen
Jiajun Shen
Qiu Shen
Qiuhong Shen
Shuai Shen
Wang Shen
Yiqing Shen
Yunhang Shen
Siqi Shen
Bin Shen
Tianwei Shen

Xi Shen
Yilin Shen
Yuming Shen
Yucong Shen
Zhiqiang Shen
Lu Sheng
Yichen Sheng
Shivanand Venkanna
    Sheshappanavar
Shelly Sheynin
Baifeng Shi
Ruoxi Shi
Botian Shi
Hailin Shi
Jia Shi
Jing Shi
Shaoshuai Shi
Baoguang Shi
Boxin Shi
Hengcan Shi
Tianyang Shi
Xiaodan Shi
Yongjie Shi
Zhensheng Shi
Yinghuan Shi
Weiqi Shi
Wu Shi
Xuepeng Shi
Xiaoshuang Shi
Yujiao Shi
Zenglin Shi
Zhenmei Shi
Takashi Shibata
Meng-Li Shih
Yichang Shih
Hyunjung Shim
Dongseok Shim
Soshi Shimada
Inkyu Shin
Jinwoo Shin
Seungjoo Shin
Seungjae Shin
Koichi Shinoda
Suprosanna Shit

Palaiahnakote
    Shivakumara
Eli Shlizerman
Gaurav Shrivastava
Xiao Shu
Xiangbo Shu
Xiujun Shu
Yang Shu
Tianmin Shu
Jun Shu
Zhixin Shu
Bing Shuai
Maria Shugrina
Ivan Shugurov
Satya Narayan Shukla
Pranjay Shyam
Jianlou Si
Yawar Siddiqui
Alberto Signoroni
Pedro Silva
Jae-Young Sim
Oriane Siméoni
Martin Simon
Andrea Simonelli
Abhishek Singh
Ashish Singh
Dinesh Singh
Gurkirt Singh
Krishna Kumar Singh
Mannat Singh
Pravendra Singh
Rajat Vikram Singh
Utkarsh Singhal
Dipika Singhania
Vasu Singla
Harsh Sinha
Sudipta Sinha
Josef Sivic
Elena Sizikova
Geri Skenderi
Ivan Skorokhodov
Dmitriy Smirnov
Cameron Y. Smith
James S. Smith
Patrick Snape

Mattia Soldan
Hyeongseok Son
Sanghyun Son
Chuanbiao Song
Chen Song
Chunfeng Song
Dan Song
Dongjin Song
Hwanjun Song
Guoxian Song
Jiaming Song
Jie Song
Liangchen Song
Ran Song
Luchuan Song
Xibin Song
Li Song
Fenglong Song
Guoli Song
Guanglu Song
Zhenbo Song
Lin Song
Xinhang Song
Yang Song
Yibing Song
Rajiv Soundararajan
Hossein Souri
Cristovao Sousa
Riccardo Spezialetti
Leonidas Spinoulas
Michael W. Spratling
Deepak Sridhar
Srinath Sridhar
Gaurang Sriramanan
Vinkle Kumar Srivastav
Themos Stafylakis
Serban Stan
Anastasis Stathopoulos
Markus Steinberger
Jan Steinbrener
Sinisa Stekovic
Alexandros Stergiou
Gleb Sterkin
Rainer Stiefelhagen
Pierre Stock

Ombretta Strafforello
Julian Straub
Yannick Strümpler
Joerg Stueckler
Hang Su
Weijie Su
Jong-Chyi Su
Bing Su
Haisheng Su
Jinming Su
Yiyang Su
Yukun Su
Yuxin Su
Zhuo Su
Zhaoqi Su
Xiu Su
Yu-Chuan Su
Zhixun Su
Arulkumar Subramaniam
Akshayvarun Subramanya
A. Subramanyam
Swathikiran Sudhakaran
Yusuke Sugano
Masanori Suganuma
Yumin Suh
Yang Sui
Baochen Sun
Cheng Sun
Long Sun
Guolei Sun
Haoliang Sun
Haomiao Sun
He Sun
Hanqing Sun
Hao Sun
Lichao Sun
Jiachen Sun
Jiaming Sun
Jian Sun
Jin Sun
Jennifer J. Sun
Tiancheng Sun
Libo Sun
Peize Sun
Qianru Sun

Shanlin Sun
Yu Sun
Zhun Sun
Che Sun
Lin Sun
Tao Sun
Yiyou Sun
Chunyi Sun
Chong Sun
Weiwei Sun
Weixuan Sun
Xiuyu Sun
Yanan Sun
Zeren Sun
Zhaodong Sun
Zhiqing Sun
Minhyuk Sung
Jinli Suo
Simon Suo
Abhijit Suprem
Anshuman Suri
Saksham Suri
Joshua M. Susskind
Roman Suvorov
Gurumurthy Swaminathan
Robin Swanson
Paul Swoboda
Tabish A. Syed
Richard Szeliski
Fariborz Taherkhani
Yu-Wing Tai
Keita Takahashi
Walter Talbott
Gary Tam
Masato Tamura
Feitong Tan
Fuwen Tan
Shuhan Tan
Andong Tan
Bin Tan
Cheng Tan
Jianchao Tan
Lei Tan
Mingxing Tan
Xin Tan

Zichang Tan
Zhentao Tan
Kenichiro Tanaka
Masayuki Tanaka
Yushun Tang
Hao Tang
Jingqun Tang
Jinhui Tang
Kaihua Tang
Luming Tang
Lv Tang
Sheyang Tang
Shitao Tang
Siliang Tang
Shixiang Tang
Yansong Tang
Keke Tang
Chang Tang
Chenwei Tang
Jie Tang
Junshu Tang
Ming Tang
Peng Tang
Xu Tang
Yao Tang
Chen Tang
Fan Tang
Haoran Tang
Shengeng Tang
Yehui Tang
Zhipeng Tang
Ugo Tanielian
Chaofan Tao
Jiale Tao
Junli Tao
Renshuai Tao
An Tao
Guanhong Tao
Zhiqiang Tao
Makarand Tapaswi
Jean-Philippe G. Tarel
Juan J. Tarrio
Enzo Tartaglione
Keisuke Tateno
Zachary Teed

Ajinkya B. Tejankar
Bugra Tekin
Purva Tendulkar
Damien Teney
Minggui Teng
Chris Tensmeyer
Andrew Beng Jin Teoh
Philipp Terhörst
Kartik Thakral
Nupur Thakur
Kevin Thandiackal
Spyridon Thermos
Diego Thomas
William Thong
Yuesong Tian
Guanzhong Tian
Lin Tian
Shiqi Tian
Kai Tian
Meng Tian
Tai-Peng Tian
Zhuotao Tian
Shangxuan Tian
Tian Tian
Yapeng Tian
Yu Tian
Yuxin Tian
Leslie Ching Ow Tiong
Praveen Tirupattur
Garvita Tiwari
George Toderici
Antoine Toisoul
Aysim Toker
Tatiana Tommasi
Zhan Tong
Alessio Tonioni
Alessandro Torcinovich
Fabio Tosi
Matteo Toso
Hugo Touvron
Quan Hung Tran
Son Tran
Hung Tran
Ngoc-Trung Tran
Vinh Tran

Phong Tran
Giovanni Trappolini
Edith Tretschk
Subarna Tripathi
Shubhendu Trivedi
Eduard Trulls
Prune Truong
Thanh-Dat Truong
Tomasz Trzcinski
Sam Tsai
Yi-Hsuan Tsai
Ethan Tseng
Yu-Chee Tseng
Shahar Tsiper
Stavros Tsogkas
Shikui Tu
Zhigang Tu
Zhengzhong Tu
Richard Tucker
Sergey Tulyakov
Cigdem Turan
Daniyar Turmukhambetov
Victor G. Turrisi da Costa
Bartlomiej Twardowski
Christopher D. Twigg
Radim Tylecek
Mostofa Rafid Uddin
Md. Zasim Uddin
Kohei Uehara
Nicolas Ugrinovic
Youngjung Uh
Norimichi Ukita
Anwaar Ulhaq
Devesh Upadhyay
Paul Upchurch
Yoshitaka Ushiku
Yuzuko Utsumi
Mikaela Angelina Uy
Mohit Vaishnav
Pratik Vaishnavi
Jeya Maria Jose Valanarasu
Matias A. Valdenegro Toro
Diego Valsesia
Wouter Van Gansbeke
Nanne van Noord

Simon Vandenhende
Farshid Varno
Cristina Vasconcelos
Francisco Vasconcelos
Alex Vasilescu
Subeesh Vasu
Arun Balajee Vasudevan
Kanav Vats
Vaibhav S. Vavilala
Sagar Vaze
Javier Vazquez-Corral
Andrea Vedaldi
Olga Veksler
Andreas Velten
Sai H. Vemprala
Raviteja Vemulapalli
Shashanka
    Venkataramanan
Dor Verbin
Luisa Verdoliva
Manisha Verma
Yashaswi Verma
Constantin Vertan
Eli Verwimp
Deepak Vijaykeerthy
Pablo Villanueva
Ruben Villegas
Markus Vincze
Vibhav Vineet
Minh P. Vo
Huy V. Vo
Duc Minh Vo
Tomas Vojir
Igor Vozniak
Nicholas Vretos
Vibashan VS
Tuan-Anh Vu
Thang Vu
Mårten Wadenbäck
Neal Wadhwa
Aaron T. Walsman
Steven Walton
Jin Wan
Alvin Wan
Jia Wan

Jun Wan
Xiaoyue Wan
Fang Wan
Guowei Wan
Renjie Wan
Zhiqiang Wan
Ziyu Wan
Bastian Wandt
Dongdong Wang
Limin Wang
Haiyang Wang
Xiaobing Wang
Angtian Wang
Angelina Wang
Bing Wang
Bo Wang
Boyu Wang
Binghui Wang
Chen Wang
Chien-Yi Wang
Congli Wang
Qi Wang
Chengrui Wang
Rui Wang
Yiqun Wang
Cong Wang
Wenjing Wang
Dongkai Wang
Di Wang
Xiaogang Wang
Kai Wang
Zhizhong Wang
Fangjinhua Wang
Feng Wang
Hang Wang
Gaoang Wang
Guoqing Wang
Guangcong Wang
Guangzhi Wang
Hanqing Wang
Hao Wang
Haohan Wang
Haoran Wang
Hong Wang
Haotao Wang

Hu Wang
Huan Wang
Hua Wang
Hui-Po Wang
Hengli Wang
Hanyu Wang
Hongxing Wang
Jingwen Wang
Jialiang Wang
Jian Wang
Jianyi Wang
Jiashun Wang
Jiahao Wang
Tsun-Hsuan Wang
Xiaoqian Wang
Jinqiao Wang
Jun Wang
Jianzong Wang
Kaihong Wang
Ke Wang
Lei Wang
Lingjing Wang
Linnan Wang
Lin Wang
Liansheng Wang
Mengjiao Wang
Manning Wang
Nannan Wang
Peihao Wang
Jiayun Wang
Pu Wang
Qiang Wang
Qiufeng Wang
Qilong Wang
Qiangchang Wang
Qin Wang
Qing Wang
Ruocheng Wang
Ruibin Wang
Ruisheng Wang
Ruizhe Wang
Runqi Wang
Runzhong Wang
Wenxuan Wang
Sen Wang

Shangfei Wang
Shaofei Wang
Shijie Wang
Shiqi Wang
Zhibo Wang
Song Wang
Xinjiang Wang
Tai Wang
Tao Wang
Teng Wang
Xiang Wang
Tianren Wang
Tiantian Wang
Tianyi Wang
Fengjiao Wang
Wei Wang
Miaohui Wang
Suchen Wang
Siyue Wang
Yaoming Wang
Xiao Wang
Ze Wang
Biao Wang
Chaofei Wang
Dong Wang
Gu Wang
Guangrun Wang
Guangming Wang
Guo-Hua Wang
Haoqing Wang
Hesheng Wang
Huafeng Wang
Jinghua Wang
Jingdong Wang
Jingjing Wang
Jingya Wang
Jingkang Wang
Jiakai Wang
Junke Wang
Kuo Wang
Lichen Wang
Lizhi Wang
Longguang Wang
Mang Wang
Mei Wang

Min Wang
Peng-Shuai Wang
Run Wang
Shaoru Wang
Shuhui Wang
Tan Wang
Tiancai Wang
Tianqi Wang
Wenhai Wang
Wenzhe Wang
Xiaobo Wang
Xiudong Wang
Xu Wang
Yajie Wang
Yan Wang
Yuan-Gen Wang
Yingqian Wang
Yizhi Wang
Yulin Wang
Yu Wang
Yujie Wang
Yunhe Wang
Yuxi Wang
Yaowei Wang
Yiwei Wang
Zezheng Wang
Hongzhi Wang
Zhiqiang Wang
Ziteng Wang
Ziwei Wang
Zheng Wang
Zhenyu Wang
Binglu Wang
Zhongdao Wang
Ce Wang
Weining Wang
Weiyao Wang
Wenbin Wang
Wenguan Wang
Guangting Wang
Haolin Wang
Haiyan Wang
Huiyu Wang
Naiyan Wang
Jingbo Wang

Jinpeng Wang
Jiaqi Wang
Liyuan Wang
Lizhen Wang
Ning Wang
Wenqian Wang
Sheng-Yu Wang
Weimin Wang
Xiaohan Wang
Yifan Wang
Yi Wang
Yongtao Wang
Yizhou Wang
Zhuo Wang
Zhe Wang
Xudong Wang
Xiaofang Wang
Xinggang Wang
Xiaosen Wang
Xiaosong Wang
Xiaoyang Wang
Lijun Wang
Xinlong Wang
Xuan Wang
Xue Wang
Yangang Wang
Yaohui Wang
Yu-Chiang Frank Wang
Yida Wang
Yilin Wang
Yi Ru Wang
Yali Wang
Yinglong Wang
Yufu Wang
Yujiang Wang
Yuwang Wang
Yuting Wang
Yang Wang
Yu-Xiong Wang
Yixu Wang
Ziqi Wang
Zhicheng Wang
Zeyu Wang
Zhaowen Wang
Zhenyi Wang

Zhenzhi Wang
Zhijie Wang
Zhiyong Wang
Zhongling Wang
Zhuowei Wang
Zian Wang
Zifu Wang
Zihao Wang
Zirui Wang
Ziyan Wang
Wenxiao Wang
Zhen Wang
Zhepeng Wang
Zi Wang
Zihao W. Wang
Steven L. Waslander
Olivia Watkins
Daniel Watson
Silvan Weder
Dongyoon Wee
Dongming Wei
Tianyi Wei
Jia Wei
Dong Wei
Fangyun Wei
Longhui Wei
Mingqiang Wei
Xinyue Wei
Chen Wei
Donglai Wei
Pengxu Wei
Xing Wei
Xiu-Shen Wei
Wenqi Wei
Guoqiang Wei
Wei Wei
XingKui Wei
Xian Wei
Xingxing Wei
Yake Wei
Yuxiang Wei
Yi Wei
Luca Weihs
Michael Weinmann
Martin Weinmann

Congcong Wen
Chuan Wen
Jie Wen
Sijia Wen
Song Wen
Chao Wen
Xiang Wen
Zeyi Wen
Xin Wen
Yilin Wen
Yijia Weng
Shuchen Weng
Junwu Weng
Wenming Weng
Renliang Weng
Zhenyu Weng
Xinshuo Weng
Nicholas J. Westlake
Gordon Wetzstein
Lena M. Widin Klasén
Rick Wildes
Bryan M. Williams
Williem Williem
Ole Winther
Scott Wisdom
Alex Wong
Chau-Wai Wong
Kwan-Yee K. Wong
Yongkang Wong
Scott Workman
Marcel Worring
Michael Wray
Safwan Wshah
Xiang Wu
Aming Wu
Chongruo Wu
Cho-Ying Wu
Chunpeng Wu
Chenyan Wu
Ziyi Wu
Fuxiang Wu
Gang Wu
Haiping Wu
Huisi Wu
Jane Wu

Jialian Wu
Jing Wu
Jinjian Wu
Jianlong Wu
Xian Wu
Lifang Wu
Lifan Wu
Minye Wu
Qianyi Wu
Rongliang Wu
Rui Wu
Shiqian Wu
Shuzhe Wu
Shangzhe Wu
Tsung-Han Wu
Tz-Ying Wu
Ting-Wei Wu
Jiannan Wu
Zhiliang Wu
Yu Wu
Chenyun Wu
Dayan Wu
Dongxian Wu
Fei Wu
Hefeng Wu
Jianxin Wu
Weibin Wu
Wenxuan Wu
Wenhao Wu
Xiao Wu
Yicheng Wu
Yuanwei Wu
Yu-Huan Wu
Zhenxin Wu
Zhenyu Wu
Wei Wu
Peng Wu
Xiaohe Wu
Xindi Wu
Xinxing Wu
Xinyi Wu
Xingjiao Wu
Xiongwei Wu
Yangzheng Wu
Yanzhao Wu

Yawen Wu
Yong Wu
Yi Wu
Ying Nian Wu
Zhenyao Wu
Zhonghua Wu
Zongze Wu
Zuxuan Wu
Stefanie Wuhrer
Teng Xi
Jianing Xi
Fei Xia
Haifeng Xia
Menghan Xia
Yuanqing Xia
Zhihua Xia
Xiaobo Xia
Weihao Xia
Shihong Xia
Yan Xia
Yong Xia
Zhaoyang Xia
Zhihao Xia
Chuhua Xian
Yongqin Xian
Wangmeng Xiang
Fanbo Xiang
Tiange Xiang
Tao Xiang
Liuyu Xiang
Xiaoyu Xiang
Zhiyu Xiang
Aoran Xiao
Chunxia Xiao
Fanyi Xiao
Jimin Xiao
Jun Xiao
Taihong Xiao
Anqi Xiao
Junfei Xiao
Jing Xiao
Liang Xiao
Yang Xiao
Yuting Xiao
Yijun Xiao

Yao Xiao
Zeyu Xiao
Zhisheng Xiao
Zihao Xiao
Binhui Xie
Christopher Xie
Haozhe Xie
Jin Xie
Guo-Sen Xie
Hongtao Xie
Ming-Kun Xie
Tingting Xie
Chaohao Xie
Weicheng Xie
Xudong Xie
Jiyang Xie
Xiaohua Xie
Yuan Xie
Zhenyu Xie
Ning Xie
Xianghui Xie
Xiufeng Xie
You Xie
Yutong Xie
Fuyong Xing
Yifan Xing
Zhen Xing
Yuanjun Xiong
Jinhui Xiong
Weihua Xiong
Hongkai Xiong
Zhitong Xiong
Yuanhao Xiong
Yunyang Xiong
Yuwen Xiong
Zhiwei Xiong
Yuliang Xiu
An Xu
Chang Xu
Chenliang Xu
Chengming Xu
Chenshu Xu
Xiang Xu
Huijuan Xu
Zhe Xu

Jie Xu
Jingyi Xu
Jiarui Xu
Yinghao Xu
Kele Xu
Ke Xu
Li Xu
Linchuan Xu
Linning Xu
Mengde Xu
Mengmeng Frost Xu
Min Xu
Mingye Xu
Jun Xu
Ning Xu
Peng Xu
Runsheng Xu
Sheng Xu
Wenqiang Xu
Xiaogang Xu
Renzhe Xu
Kaidi Xu
Yi Xu
Chi Xu
Qiuling Xu
Baobei Xu
Feng Xu
Haohang Xu
Haofei Xu
Lan Xu
Mingze Xu
Songcen Xu
Weipeng Xu
Wenjia Xu
Wenju Xu
Xiangyu Xu
Xin Xu
Yinshuang Xu
Yixing Xu
Yuting Xu
Yanyu Xu
Zhenbo Xu
Zhiliang Xu
Zhiyuan Xu
Xiaohao Xu

Yanwu Xu
Yan Xu
Yiran Xu
Yifan Xu
Yufei Xu
Yong Xu
Zichuan Xu
Zenglin Xu
Zexiang Xu
Zhan Xu
Zheng Xu
Zhiwei Xu
Ziyue Xu
Shiyu Xuan
Hanyu Xuan
Fei Xue
Jianru Xue
Mingfu Xue
Qinghan Xue
Tianfan Xue
Chao Xue
Chuhui Xue
Nan Xue
Zhou Xue
Xiangyang Xue
Yuan Xue
Abhay Yadav
Ravindra Yadav
Kota Yamaguchi
Toshihiko Yamasaki
Kohei Yamashita
Chaochao Yan
Feng Yan
Kun Yan
Qingsen Yan
Qixin Yan
Rui Yan
Siming Yan
Xinchen Yan
Yaping Yan
Bin Yan
Qingan Yan
Shen Yan
Shipeng Yan
Xu Yan

Yan Yan
Yichao Yan
Zhaoyi Yan
Zike Yan
Zhiqiang Yan
Hongliang Yan
Zizheng Yan
Jiewen Yang
Anqi Joyce Yang
Shan Yang
Anqi Yang
Antoine Yang
Bo Yang
Baoyao Yang
Chenhongyi Yang
Dingkang Yang
De-Nian Yang
Dong Yang
David Yang
Fan Yang
Fengyu Yang
Fengting Yang
Fei Yang
Gengshan Yang
Heng Yang
Han Yang
Huan Yang
Yibo Yang
Jiancheng Yang
Jihan Yang
Jiawei Yang
Jiayu Yang
Jie Yang
Jinfa Yang
Jingkang Yang
Jinyu Yang
Cheng-Fu Yang
Ji Yang
Jianyu Yang
Kailun Yang
Tian Yang
Luyu Yang
Liang Yang
Li Yang
Michael Ying Yang

Yang Yang
Muli Yang
Le Yang
Qiushi Yang
Ren Yang
Ruihan Yang
Shuang Yang
Siyuan Yang
Su Yang
Shiqi Yang
Taojiannan Yang
Tianyu Yang
Lei Yang
Wanzhao Yang
Shuai Yang
William Yang
Wei Yang
Xiaofeng Yang
Xiaoshan Yang
Xin Yang
Xuan Yang
Xu Yang
Xingyi Yang
Xitong Yang
Jing Yang
Yanchao Yang
Wenming Yang
Yujiu Yang
Herb Yang
Jianfei Yang
Jinhui Yang
Chuanguang Yang
Guanglei Yang
Haitao Yang
Kewei Yang
Linlin Yang
Lijin Yang
Longrong Yang
Meng Yang
MingKun Yang
Sibei Yang
Shicai Yang
Tong Yang
Wen Yang
Xi Yang

Xiaolong Yang
Xue Yang
Yubin Yang
Ze Yang
Ziyi Yang
Yi Yang
Linjie Yang
Yuzhe Yang
Yiding Yang
Zhenpei Yang
Zhaohui Yang
Zhengyuan Yang
Zhibo Yang
Zongxin Yang
Hantao Yao
Mingde Yao
Rui Yao
Taiping Yao
Ting Yao
Cong Yao
Qingsong Yao
Quanming Yao
Xu Yao
Yuan Yao
Yao Yao
Yazhou Yao
Jiawen Yao
Shunyu Yao
Pew-Thian Yap
Sudhir Yarram
Rajeev Yasarla
Peng Ye
Botao Ye
Mao Ye
Fei Ye
Hanrong Ye
Jingwen Ye
Jinwei Ye
Jiarong Ye
Mang Ye
Meng Ye
Qi Ye
Qian Ye
Qixiang Ye
Junjie Ye

Sheng Ye
Nanyang Ye
Yufei Ye
Xiaoqing Ye
Ruolin Ye
Yousef Yeganeh
Chun-Hsiao Yeh
Raymond A. Yeh
Yu-Ying Yeh
Kai Yi
Chang Yi
Renjiao Yi
Xinping Yi
Peng Yi
Alper Yilmaz
Junho Yim
Hui Yin
Bangjie Yin
Jia-Li Yin
Miao Yin
Wenzhe Yin
Xuwang Yin
Ming Yin
Yu Yin
Aoxiong Yin
Kangxue Yin
Tianwei Yin
Wei Yin
Xianghua Ying
Rio Yokota
Tatsuya Yokota
Naoto Yokoya
Ryo Yonetani
Ki Yoon Yoo
Jinsu Yoo
Sunjae Yoon
Jae Shin Yoon
Jihun Yoon
Sung-Hoon Yoon
Ryota Yoshihashi
Yusuke Yoshiyasu
Chenyu You
Haoran You
Haoxuan You
Yang You

Quanzeng You
Tackgeun You
Kaichao You
Shan You
Xinge You
Yurong You
Baosheng Yu
Bei Yu
Haichao Yu
Hao Yu
Chaohui Yu
Fisher Yu
Jin-Gang Yu
Jiyang Yu
Jason J. Yu
Jiashuo Yu
Hong-Xing Yu
Lei Yu
Mulin Yu
Ning Yu
Peilin Yu
Qi Yu
Qian Yu
Rui Yu
Shuzhi Yu
Gang Yu
Tan Yu
Weijiang Yu
Xin Yu
Bingyao Yu
Ye Yu
Hanchao Yu
Yingchen Yu
Tao Yu
Xiaotian Yu
Qing Yu
Houjian Yu
Changqian Yu
Jing Yu
Jun Yu
Shujian Yu
Xiang Yu
Zhaofei Yu
Zhenbo Yu
Yinfeng Yu

Zhuoran Yu
Zitong Yu
Bo Yuan
Jiangbo Yuan
Liangzhe Yuan
Weihao Yuan
Jianbo Yuan
Xiaoyun Yuan
Ye Yuan
Li Yuan
Geng Yuan
Jialin Yuan
Maoxun Yuan
Peng Yuan
Xin Yuan
Yuan Yuan
Yuhui Yuan
Yixuan Yuan
Zheng Yuan
Mehmet Kerim Yücel
Kaiyu Yue
Haixiao Yue
Heeseung Yun
Sangdoo Yun
Tian Yun
Mahmut Yurt
Ekim Yurtsever
Ahmet Yüzügüler
Edouard Yvinec
Eloi Zablocki
Christopher Zach
Muhammad Zaigham
    Zaheer
Pierluigi Zama Ramirez
Yuhang Zang
Pietro Zanuttigh
Alexey Zaytsev
Bernhard Zeisl
Haitian Zeng
Pengpeng Zeng
Jiabei Zeng
Runhao Zeng
Wei Zeng
Yawen Zeng
Yi Zeng

Yiming Zeng
Tieyong Zeng
Huanqiang Zeng
Dan Zeng
Yu Zeng
Wei Zhai
Yuanhao Zhai
Fangneng Zhan
Kun Zhan
Xiong Zhang
Jingdong Zhang
Jiangning Zhang
Zhilu Zhang
Gengwei Zhang
Dongsu Zhang
Hui Zhang
Binjie Zhang
Bo Zhang
Tianhao Zhang
Cecilia Zhang
Jing Zhang
Chaoning Zhang
Chenxu Zhang
Chi Zhang
Chris Zhang
Yabin Zhang
Zhao Zhang
Rufeng Zhang
Chaoyi Zhang
Zheng Zhang
Da Zhang
Yi Zhang
Edward Zhang
Xin Zhang
Feifei Zhang
Feilong Zhang
Yuqi Zhang
GuiXuan Zhang
Hanlin Zhang
Hanwang Zhang
Hanzhen Zhang
Haotian Zhang
He Zhang
Haokui Zhang
Hongyuan Zhang

Hengrui Zhang
Hongming Zhang
Mingfang Zhang
Jianpeng Zhang
Jiaming Zhang
Jichao Zhang
Jie Zhang
Jingfeng Zhang
Jingyi Zhang
Jinnian Zhang
David Junhao Zhang
Junjie Zhang
Junzhe Zhang
Jiawan Zhang
Jingyang Zhang
Kai Zhang
Lei Zhang
Lihua Zhang
Lu Zhang
Miao Zhang
Minjia Zhang
Mingjin Zhang
Qi Zhang
Qian Zhang
Qilong Zhang
Qiming Zhang
Qiang Zhang
Richard Zhang
Ruimao Zhang
Ruisi Zhang
Ruixin Zhang
Runze Zhang
Qilin Zhang
Shan Zhang
Shanshan Zhang
Xi Sheryl Zhang
Song-Hai Zhang
Chongyang Zhang
Kaihao Zhang
Songyang Zhang
Shu Zhang
Siwei Zhang
Shujian Zhang
Tianyun Zhang
Tong Zhang

Tao Zhang
Wenwei Zhang
Wenqiang Zhang
Wen Zhang
Xiaolin Zhang
Xingchen Zhang
Xingxuan Zhang
Xiuming Zhang
Xiaoshuai Zhang
Xuanmeng Zhang
Xuanyang Zhang
Xucong Zhang
Xingxing Zhang
Xikun Zhang
Xiaohan Zhang
Yahui Zhang
Yunhua Zhang
Yan Zhang
Yanghao Zhang
Yifei Zhang
Yifan Zhang
Yi-Fan Zhang
Yihao Zhang
Yingliang Zhang
Youshan Zhang
Yulun Zhang
Yushu Zhang
Yixiao Zhang
Yide Zhang
Zhongwen Zhang
Bowen Zhang
Chen-Lin Zhang
Zehua Zhang
Zekun Zhang
Zeyu Zhang
Xiaowei Zhang
Yifeng Zhang
Cheng Zhang
Hongguang Zhang
Yuexi Zhang
Fa Zhang
Guofeng Zhang
Hao Zhang
Haofeng Zhang
Hongwen Zhang

Hua Zhang
Jiaxin Zhang
Zhenyu Zhang
Jian Zhang
Jianfeng Zhang
Jiao Zhang
Jiakai Zhang
Lefei Zhang
Le Zhang
Mi Zhang
Min Zhang
Ning Zhang
Pan Zhang
Pu Zhang
Qing Zhang
Renrui Zhang
Shifeng Zhang
Shuo Zhang
Shaoxiong Zhang
Weizhong Zhang
Xi Zhang
Xiaomei Zhang
Xinyu Zhang
Yin Zhang
Zicheng Zhang
Zihao Zhang
Ziqi Zhang
Zhaoxiang Zhang
Zhen Zhang
Zhipeng Zhang
Zhixing Zhang
Zhizheng Zhang
Jiawei Zhang
Zhong Zhang
Pingping Zhang
Yixin Zhang
Kui Zhang
Lingzhi Zhang
Huaiwen Zhang
Quanshi Zhang
Zhoutong Zhang
Yuhang Zhang
Yuting Zhang
Zhang Zhang
Ziming Zhang

Zhizhong Zhang
Qilong Zhangli
Bingyin Zhao
Bin Zhao
Chenglong Zhao
Lei Zhao
Feng Zhao
Gangming Zhao
Haiyan Zhao
Hao Zhao
Handong Zhao
Hengshuang Zhao
Yinan Zhao
Jiaojiao Zhao
Jiaqi Zhao
Jing Zhao
Kaili Zhao
Haojie Zhao
Yucheng Zhao
Longjiao Zhao
Long Zhao
Qingsong Zhao
Qingyu Zhao
Rui Zhao
Rui-Wei Zhao
Sicheng Zhao
Shuang Zhao
Siyan Zhao
Zelin Zhao
Shiyu Zhao
Wang Zhao
Tiesong Zhao
Qian Zhao
Wangbo Zhao
Xi-Le Zhao
Xu Zhao
Yajie Zhao
Yang Zhao
Ying Zhao
Yin Zhao
Yizhou Zhao
Yunhan Zhao
Yuyang Zhao
Yue Zhao
Yuzhi Zhao

Bowen Zhao
Pu Zhao
Bingchen Zhao
Borui Zhao
Fuqiang Zhao
Hanbin Zhao
Jian Zhao
Mingyang Zhao
Na Zhao
Rongchang Zhao
Ruiqi Zhao
Shuai Zhao
Wenda Zhao
Wenliang Zhao
Xiangyun Zhao
Yifan Zhao
Yaping Zhao
Zhou Zhao
He Zhao
Jie Zhao
Xibin Zhao
Xiaoqi Zhao
Zhengyu Zhao
Jin Zhe
Chuanxia Zheng
Huan Zheng
Hao Zheng
Jia Zheng
Jian-Qing Zheng
Shuai Zheng
Meng Zheng
Mingkai Zheng
Qian Zheng
Qi Zheng
Wu Zheng
Yinqiang Zheng
Yufeng Zheng
Yutong Zheng
Yalin Zheng
Yu Zheng
Feng Zheng
Zhaoheng Zheng
Haitian Zheng
Kang Zheng
Bolun Zheng

Haiyong Zheng
Mingwu Zheng
Sipeng Zheng
Tu Zheng
Wenzhao Zheng
Xiawu Zheng
Yinglin Zheng
Zhuo Zheng
Zilong Zheng
Kecheng Zheng
Zerong Zheng
Shuaifeng Zhi
Tiancheng Zhi
Jia-Xing Zhong
Yiwu Zhong
Fangwei Zhong
Zhihang Zhong
Yaoyao Zhong
Yiran Zhong
Zhun Zhong
Zichun Zhong
Bo Zhou
Boyao Zhou
Brady Zhou
Mo Zhou
Chunluan Zhou
Dingfu Zhou
Fan Zhou
Jingkai Zhou
Honglu Zhou
Jiaming Zhou
Jiahuan Zhou
Jun Zhou
Kaiyang Zhou
Keyang Zhou
Kuangqi Zhou
Lei Zhou
Lihua Zhou
Man Zhou
Mingyi Zhou
Mingyuan Zhou
Ning Zhou
Peng Zhou
Penghao Zhou
Qianyi Zhou

Shuigeng Zhou
Shangchen Zhou
Huayi Zhou
Zhize Zhou
Sanping Zhou
Qin Zhou
Tao Zhou
Wenbo Zhou
Xiangdong Zhou
Xiao-Yun Zhou
Xiao Zhou
Yang Zhou
Yipin Zhou
Zhenyu Zhou
Hao Zhou
Chu Zhou
Daquan Zhou
Da-Wei Zhou
Hang Zhou
Kang Zhou
Qianyu Zhou
Sheng Zhou
Wenhui Zhou
Xingyi Zhou
Yan-Jie Zhou
Yiyi Zhou
Yu Zhou
Yuan Zhou
Yuqian Zhou
Yuxuan Zhou
Zixiang Zhou
Wengang Zhou
Shuchang Zhou
Tianfei Zhou
Yichao Zhou
Alex Zhu
Chenchen Zhu
Deyao Zhu
Xiatian Zhu
Guibo Zhu
Haidong Zhu
Hao Zhu
Hongzi Zhu
Rui Zhu
Jing Zhu

Jianke Zhu
Junchen Zhu
Lei Zhu
Lingyu Zhu
Luyang Zhu
Menglong Zhu
Peihao Zhu
Hui Zhu
Xiaofeng Zhu
Tyler (Lixuan) Zhu
Wentao Zhu
Xiangyu Zhu
Xinqi Zhu
Xinxin Zhu
Xinliang Zhu
Yangguang Zhu
Yichen Zhu
Yixin Zhu
Yanjun Zhu
Yousong Zhu
Yuhao Zhu
Ye Zhu
Feng Zhu
Zhen Zhu
Fangrui Zhu
Jinjing Zhu
Linchao Zhu
Pengfei Zhu
Sijie Zhu
Xiaobin Zhu
Xiaoguang Zhu
Zezhou Zhu
Zhenyao Zhu
Kai Zhu
Pengkai Zhu
Bingbing Zhuang
Chengyuan Zhuang
Liansheng Zhuang
Peiye Zhuang
Yixin Zhuang
Yihong Zhuang
Junbao Zhuo
Andrea Ziani
Bartosz Zieliński
Primo Zingaretti

Nikolaos Zioulis
Andrew Zisserman
Yael Ziv
Liu Ziyin
Xingxing Zou
Danping Zou
Qi Zou

Shihao Zou
Xueyan Zou
Yang Zou
Yuliang Zou
Zihang Zou
Chuhang Zou
Dongqing Zou

Xu Zou
Zhiming Zou
Maria A. Zuluaga
Xinxin Zuo
Zhiwen Zuo
Reyer Zwiggelaar

Contents – Part XI

A Simple Approach and Benchmark for 21,000-Category Object Detection

Yutong Lin[1,2], Chen Li[1,2], Yue Cao[2], Zheng Zhang[2], Jianfeng Wang[2],
Lijuan Wang[2], Zicheng Liu[2], and Han Hu[2(✉)]

[1] Xi'an Jiaotong University, Xi'an, China
{yutonglin,edward82}@stu.xjtu.edu.cn
[2] Microsoft, Redmond, USA
{yuecao,zhez,jianfw,hanhu}@microsoft.com

Abstract. Current object detection systems and benchmarks typically
handle a limited number of categories, up to about a thousand categories.
This paper scales the number of categories for object detection systems
and benchmarks up to 21,000, by leveraging existing object detection and
image classification data. Unlike previous efforts that usually transfer
knowledge from base detectors to image classification data, we propose
to rely more on a reverse information flow from a base image classifier to object detection data. In this framework, the large-vocabulary
classification capability is first learnt thoroughly using only the image
classification data. In this step, the image classification problem is reformulated as a special configuration of object detection that treats the
entire image as a special RoI. Then, a simple multi-task learning approach is used to join the image classification and object detection data,
with the backbone and the RoI classification branch shared between two
tasks. This two-stage approach, though very simple without a sophisticated process such as multi-instance learning (MIL) to generate pseudo
labels for object proposals on the image classification data, performs
rather strongly that it surpasses previous large-vocabulary object detection systems on a standard evaluation protocol of tailored LVIS.

Considering that the tailored LVIS evaluation only accounts for a few
hundred novel object categories, we present a new evaluation benchmark
that assesses the detection of all 21,841 object classes in the ImageNet-
21K dataset. The baseline approach and evaluation benchmark will be
publicly available at https://github.com/SwinTransformer/Simple-21K-
Detection. We hope these would ease future research on large-vocabulary
object detection.

Keywords: Large-vocabulary object detection · Benchmark ·
Multi-task learning

Equal Contribution. The work is done when Yutong Lin and Chen Li are interns at
MSRA.

S. Avidan et al. (Eds.): ECCV 2022, LNCS 13671, pp. 1–18, 2022.
https://doi.org/10.1007/978-3-031-20083-0_1

1   Introduction

Current object detection datasets typically have a limited number of categories, for example, 80 classes of COCO datasets [26], 200 classes of ImageNet-DET [5], 365 classes of Objects365 [32], 600 classes of OpenImage [22] , 1,203 classes of LVIS [14], 1,594 classes of Visual Genome [20], and so on. Limited by object detection datasets, existing object detection systems typically detect up to a thousand categories or use up to a thousand categories for evaluation.

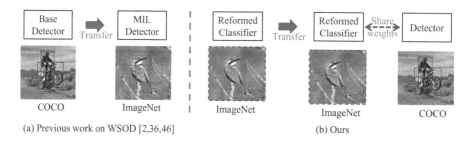

(a) Previous work on WSOD [2,36,46]   (b) Ours

Fig. 1. (a) The knowledge flow of previous works is from a base object detector to image classification data using a weakly supervised object detection (WSOD) framework; (b) The knowledge flow of our approach is from a thoroughly-trained reformulated visual classifier, to a simple multi-task learning framework that combines reformulated classification and object detection.

This paper aims to scale the number of categories in an object detection system up to more than 21,000. We note that existing image classification datasets involve much more object categories, such as the 21,841-category ImageNet-21K image classification dataset [5], which can serve as a complementary source to help achieve the goal of large-vocabulary object detection. Therefore, we are devoted to combining the object detection datasets that have limited object categories, with image classification datasets that have a large number of object categories, for large-vocabulary object detection. In previous works [6,40,46,50], such combination usually starts from a base detector that learns good foreground/background classification and localization capabilities, e.g., a good region proposal network (RPN), and then transfers these capabilities to the image classification data using a multi-instance learning (MIL) framework [2,40] (see Fig. 1(a)). We argue that this knowledge flow may be sub-optimal when the number of object categories is large, as the large-vocabulary classification capability is more difficult to be seized at scale. For example, training a good classifier for an ImageNet-1K dataset typically requires long training iterations with strong augmentation, such as traversing about 300 million image samples, however, iterating about 4 million images with weak augmentation is enough to train a good detector on the COCO object detection dataset. In fact, a general foreground/background separation capability that works well to some extent is usually easier to obtain, even when trained on a detection dataset with limited categories, e.g., COCO [13].

Table 1. Three experimental setups for the evaluation of large-vocabulary object detection methods based on joining of the object detection and image classification datasets.

Setups	DET dataset			CLS dataset			Evaluation metric
	Notation	#cat.	#im.	Notation	#cat.	#im.	
S1	LVIS-997-base	720	11K	IN-997	997	1.23M	LVIS-997-novel mAP
S2	COCO	80	11.5K	IN-1K	1,000	1.28M	IN-1K loc. acc.
S3	Object365v2	365	1.7M	IN-21K	21,841	14M	IN-21K loc. acc. LVIS finetune mAP

Based on this view, we propose to make a reverse information flow (see Fig. 1(b)), that we start from a good image classifier and then transfer the gained large-vocabulary classification capability into the object detection data. To enable smooth knowledge transferring, we reformulate the standard image classification approach, which uses a linear classification head on top of a backbone network, as a special configuration of object detection. In this new formulation, an image is represented by a special RoI corresponding to the entire image (red bounding box with dashed lines), and a heavier RoI classification head like that in an object detection framework is applied on this special RoI to realize image classification. By this reformulation, the two tasks of image classification and object detection are better aligned.

After having gained the large-vocabulary image classification capability by training on the image classification data alone, we employ a simple multi-task learning framework to join the two tasks of image classification and object detection. The image classification and object detection tasks will share the backbone, RPN network, as well as the RoI classification head. The weights of the shared networks are initialized from that of the first step, such that the initial network has owned a strong large-vocabulary capability.

As shown in Fig. 1(b), the above two-stage approach does not employ an explicit foreground/background separation mechanism on the image data like that in previous weakly supervised object detection (WSOD) works. This degeneration strategy simplifies training, yet performs surprisingly well in detecting object categories that do not appear in the object detection data. We hypothesize that the sharing of RPN and RoI classification head has been able to help the framework dig out the capabilities required by a large-vocabulary object detector.

We conducted three experimental setups, named S1, S2, and S3, described in Table 1. The first setup S1 aims to evaluate the developed framework with a standard protocol of tailored LVIS. It mainly follows [13] and a concurrent work of [53], which use the 997 intersected categories of the LVIS and ImageNet-21K datasets for experiments. The categories are divided into 720 *base* ("common" and "frequent" of LVIS) and 277 *novel* sets ("rare" of LVIS). The training uses an LVIS-*base* object detection set and an IN-997 image classification set. The evaluation is conducted on LVIS-*novel* set. In the second setup S2, the number of categories for evaluation is extended to 1,000. This setup follows the previous weakly supervised object localization field [3,10,43,49,52] to evaluate the localization accuracy on ImageNet-1K. In this setup, we allow the use of additional COCO object detection datasets to facilitate localization on ImageNet-1K. The third setup S3 allows training of object detectors with the number of categories as large as 21,000. This setup adopts the Object365v2 object detection dataset [32] and ImageNet-21K image classification dataset for training. To be able to evaluate a large number of object categories, we randomly select 5 validation images for each object category in the ImageNet-21K dataset and annotate all ground-truth bounding boxes on these images. This results in approximately 100,000 images being annotated for evaluation.

On Setup S1, the proposed approach has a much smaller performance gap between the base set and the novel set than recent efforts based on pre-trained visual language models (such as CLIP [29]) [13], or single-stage joint detection/classification [53]. On Setup S2, our approach achieves 75.1% top-1 accuracy with GT category labels known, and 68.3% top-1 accuracy with GT category labels unknown, which are absolutely 6.6%/13.5% higher than previous best weakly supervised object localization approaches [49]. This indicates the benefits of an additional base detector. On Setup S3, we demonstrate a 21,000-category object detector. All these experiments demonstrate the effectiveness of the proposed method. In addition, as a by-product, the proposed approach shows to learn representations that have better transferability to downstream tasks such as object detection on a standard LVIS, than methods that training on two datasets alone or successively.

We hope our simple approach, along with the new 21,000-category evaluation benchmark, will facilitate future research on large-vocabulary object detection.

2   Related Work

Image Classification is a visual task that assigns category labels to images. This problem has largely driven the development of visual backbone architectures, such as convolutional neural networks [16,21,33] and vision Transformers [8,27]. Image classification datasets can be made with large-vocabulary, for example, the ImageNet-21K dataset [5] contains 21,841 categories; Google's JFT dataset contains 18,291 categories and 3 billion images. These large vocabulary image classification datasets serve as a powerful visual pre-training and semantic concept learning basis for a variety of vision problems. A common practice [16,33] for image classification is to apply a simple linear head on top of the backbone architecture to obtain the classification. In training, it typically [8,21] employs strong augmentations to enhance the networks' invariance property, and a long learning scheduler to train the network thoroughly, such as to distinguish subtle differences between different object categories when the vocabulary size is large.

Object Detection is a vision task that simultaneously localizes objects and performs categorical classification for each object. This is a basic task that provides localized objects for the following additional recognition or analysis aim. Unlike classification datasets, object detection datasets are typically much smaller in terms of the number of classes and images. COCO [26] is the most widely used dataset for evaluating detection methods, with 80 object categories and 115K training images. Object365 [32]/OpenImages [22] scale the number of categories and images up to 365/600 categories and 1.7/1.9 million images. LVIS [14] is a re-annotation of COCO images with about 1,200 object categories that, along with Visual Genome [20], are the two largest publicly available object detection datasets on the number of categories. In general, object detection annotations are very expensive, which limit the scale in the number of images and categories.

Weakly Supervised Object Detection (WSOD) and Localization (WSOL) are two problems that learn to use image classification data for object detection and localization, respectively. There have been extensive studies on these topics [2, 3, 7, 10, 28, 34–37, 43, 47–49, 52]. The WSOD methods [2, 7, 35–37] usually first use unsupervised proposal methods such as EdgeBoxes [54] or Selective Search [41] to generate candidate bounding boxes, and then learns from the image label annotations by multi-instance learning (MIL). The WSOL methods [3, 10, 28, 34, 43, 47–49, 52] are mostly based on CAM [52] with the class activation maps as an indicator of the object area. Previous WSOD and WSOL methods are usually evaluated on relatively small datasets, such as COCO/VOC for WSOD and ImageNet-1K for WSOL. In addition, they try to solve difficult detection problems using only image labels without any box annotation, and therefore the accuracy of these systems is often too low for practical use.

This paper studies how image classification and object detection data can be combined to achieve large-vocabulary object detection for a more realistic scenario. There is a more relevant family of work as below.

WSOD with Base Detectors. This family of work transfers knowledge in base detectors to aid in the weakly supervised object detection of images with category label annotations. The knowledge transferred from a base detector is either an objectness predictor [6, 46], an object proposal [40, 50], or a universal bounding box regressor [23].

While most of these efforts are done on small-scale datasets such as Pascal VOC [9] and COCO [26], there are also some works that use these techniques to enable large-vocabulary or open-vocabulary object detection as below.

Large-Vocabulary Object Detection. YOLO9000 [30] combines detection and classification data to obtain a 9,000-category object detector. It jointly learns a standard detection loss on regular box annotations and a weakly supervised detection loss that assigns the classification labels to the anchor with the highest prediction score. [42] detects 11K object categories by exploiting semantic relationships between categories. A concurrent work to ours, Detic [53], learns about standard detection and weakly supervised detection similar to YOLO9000, but assigns classification labels to the largest object proposal. Our approach also attempts to transfer knowledge between tasks. However, unlike previous efforts that typically transfer knowledge from base detectors to image classification data, we emphasize the opposite knowledge flow and show that it is very beneficial to transfer the powerful large-vocabulary classifier learnt on the image classification datasets to object detection. In addition, unlike the weakly supervised detection methods, we show a fairly simple multi-task learning that combines classification and detection to already achieve very good performance.

Open-Vocabulary Object Detection. Another line is to perform open-vocabulary object detection. Early works expand the classifier of a base detector to be able to handle new categories by an already learnt word embedding [1]. A recent fashion is to use image text contrastive learning, such as CLIP [29], to help extend the classifier in a base detector to open-vocabulary scenario [13,24,44]. Our work is basically complementary to these works, and the text embeddings learnt in CLIP [29] also help to extend our approach to open-vocabulary scenario. We leave this as our future research.

3   Approach

3.1   Image Classification and Object Detection Practices

This paper aims to combine image classification data and object detection data towards large-vocabulary object detection. In this subsection, we review the common practice for image classification and object detection.

Image Classification Practice. In image classification models [8,16,21,27,33], the resolution of input images is usually small, such as 224×224. In training, the images go through a series of strong augmentations: random cropping [21], color jittering [21], mix up [45], random erasing [51], and so on, before they are fed into the encoder. The strong augmentations show to be very crucial for image classification training [27,39], probably because a good classifier needs to possess strong deformation invariance. After extracting features with the encoder, classification task usually uses the last layer output of the encoder with average pooling, as the input feature of the classification head. A cross-entropy loss is widely used to drive the training of classification tasks.

Object Detection Practice. In the object detection methods [12,31,38], the resolution of input images is usually set high to be able to detect tiny objects, e.g., 800×1333 is a common setting in a widely used baseline detector [15]. In training, it usually employs weak augmentation like random resizing. Similar to image classification, images are also fed into an encoder to extract image features. But in order to detect objects with various scales, feature maps are collected from more than one layer of the encoder, for example FPN [25]. In addition to that, an RoIAlign operator [15] is widely used to extract region features from the image feature maps to maintain the equivariance of region features, instead of the average pooling which usually sacrifice equivariance for invariance. In addition, as the object detector both needs to localize and recognize the objects, both a cross-entropy loss and a bounding box regression loss are adopted in the optimization process.

3.2   A Two-Stage Approach

Unlike most of previous works which typically start with a base detector and transfer the knowledge of this base detector to image classification data using a multi-instance learning framework, we argue for making a reverse information flow that transfers knowledge from an image classifier to detection.

The underline reason is that when the number of object categories is large, the large-vocabulary classification capability is very difficult to be seized. In fact, 300 million images need to be traversed to train a good classifier on ImageNet-1K using common vision Transformers [8,27]. To train a good classifier on the large-vocabulary ImageNet-21K dataset, even 4 times longer iterations are required [27]. The training also relies on strong data augmentation to perform well. On the other hand, 4 million images with weak data augmentations have been enough to train a good detector on COCO. The region proposal network (RPN) trained on COCO, which distinguishes foreground objects with the backgrounds, has been general enough for common objects to be also effect beyond the 80 categories annotated in COCO [13].

In this sense, we thus propose a reverse knowledge flow that transfers information from a good image classifier to object detection. There are two stages of training, as shown in Fig. 2. In the first stage, a large-vocabulary image classifier is thoroughly trained. There is a reformulation of previous standard image classification approach to be aligned well with the object detection framework. This reformulation facilitates a smooth transfer of the knowledge to the next stage of training. In the second stage, we join the capabilities of image classification and object detection through a simple multi-task learning framework.

In the following, we present the details of these two stages.

Stage I: Image Classification with Reformulation. In the first stage, we reformulate the traditional image classification task to make it as close as possible to object detection. In standard practice, the image classification is achieved using a simple linear classification head at the top of the backbone network, while object detection relies on heavier heads for object localization and classification.

To bridge these two tasks, we have two modifications, firstly, treating the entire image area as a proposal RoI to represent the image and performing RoIAlign instead of the previous average pooling operator on this RoI; and secondly, taking the same object classification head in object detection task to replace the traditional linear head. The reformulation is illustrated in Fig. 2(a). With this reformulation, we can still maintain the advantages of long/fully trained image classification and strong image augmentations, while have more shared layers and fewer gaps with the object detection task.

Stage II: A Simple Multi-task Learning Framework to Combine Object Detection and Reformulated Image Classification. In the second stage, we perform a joint training framework for image classification and object detection. The training pipeline of this stage is illustrated in Fig. 2(b).

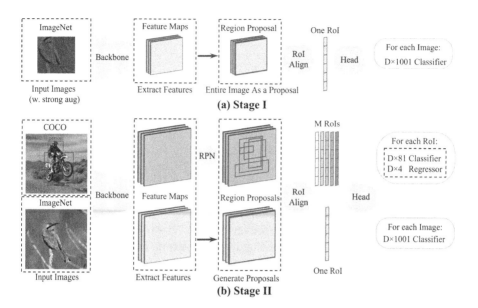

Fig. 2. Illustration on two stages of our approach: image classification with reformulation (stage I); joint object detection and reformulated image classification (stage II).

Firstly, the image processing pipeline is made unified between the object detection and the reformulated image classification tasks. Since the image classifier has been thoroughly trained at the first stage, the image processing pipeline is leaned to be friendly for the detection task: all images are resized to a large resolution such as $[800, 1333]$ as a common settings, and weak augmentations are employed.

Secondly, most modules in architecture of the reformulated image classification are shared to the object detection framework, including the backbone, FPN [25], and the RoI classification head except for the last layer to perform linear classification. The model weights of these components learnt during the first stage are used in initialization, with other new layers randomly initialized.

Since there are no background samples on the image classification branch, we decouple foreground/background classification from object category classification and regard all samples in image classification datasets as foreground. Specifically, we introduce an extra foreground/background classifier on top of the classification head to align it to the classification branch of the object detector,

$$S_i^{\text{fg}} = \text{sigmoid}(f(\mathbf{x}_i)), \qquad (1)$$

in additional to the original C-class classifiers,

$$S_i^c = \text{softmax}(g_c(\mathbf{x}_i)), \qquad (2)$$

$$\hat{S}_i^c = \text{softmax}(\hat{g}_c(\mathbf{x}_i)), \qquad (3)$$

where $f(\cdot) \in \mathbf{R}^{d \times 1}$, $g(\cdot) \in \mathbf{R}^{d \times C}$, and $\hat{g}(\cdot) \in \mathbf{R}^{d \times \hat{C}}$ are linear classifiers of the shared foreground/background classification, category classification on the detection dataset, and category classification on the classification dataset; i is an RoI index; c is a category index; $\mathbf{x}_i$ represents the region feature of RoI i.

The final category score for each foreground object corresponding to category c is:

$$S_i'^c = S_i^{\text{fg}} \cdot S_i^c, \hat{S}_i'^c = S_i^{\text{fg}} \cdot \hat{S}_i^c, \tag{4}$$

and the classification loss is performed on this final score.

In training, the foreground classifier is trained only with detection data, and the category classification classifiers are trained with both detection and classification data. Since image classification data is only used for training category classification, using even the very inaccurate entire image as RoI is sufficient for training the category classifier. In inference, proposals that overlap with objects are typically observed to have high scores on this object category, but only the proposal that is close to the object's ground-truth box will have higher foreground score, resulting in better overall scores.

Inference Method. In inference, we remove the image classification branch, except that the final linear classifier layer is remained and put on top of the RoI classification head of the object detection branch. For categories that overlap between object detection and image classification, we empirically use the weighted geometric average of scores from two branches:

$$S_i'^c = (S_i'^c)^{\frac{2}{3}} \cdot (\hat{S}_i'^c)^{\frac{1}{3}}. \tag{5}$$

For the categories appear in only one dataset, we use the score using the corresponding classifier to produce the final category score for RoI i.

4 Experiments

4.1 Experimental Settings

We conduct experiments using three setups as shown in Table 1.

Setup S1. The 997 categories intersected between LVIS [14] and ImageNet-21K [5] are considered, resulting in two tailored datasets noted as LVIS-997 and IN-997. We divide the categories into a *base* set and a *novel* set, respectively, according to its frequency on LVIS dataset. Specifically, the categories belong to "common" and "frequent" are set as *base* classes. The categories belong to "rare" are set as *novel* classes. Using this division, the *base* set involves 720 categories, while the *novel* set involves 277 categories.

In training, the LVIS-997-base and IN-997 training images are used. In evaluation, both the LVIS-997-*base* and LVIS-997-*novel* validation set are used. In addition to the absolute accuracy, the performance gap between LVIS-997-*novel* and LVIS-997-*base* can be a good indicator to evaluate the transferring performance of a method. In this setup, Mask R-CNN [15] is used as the base detector.

Setup S2. The COCO object detection dataset and the ImageNet-1K dataset are used. The COCO dataset contains 118K training images and 80 object categories. The ImageNet-1K dataset contains 1.28M training images and 1000 object categories. The goal of this setup is to evaluate the performance of weakly object localization on ImageNet-1K, which are widely used in previous WSOL literature.

In this setup, we do not map categories of the two dataset, but treat them as independent sets. In evaluation, we only use the linear classification weights trained on the ImageNet-1K dataset. In this setup, Faster R-CNN [31] with FPN [25] is adopted as the base detector.

Setup S3. This setup is used to develop our 21,000-vocabulary detector, as well as building our evaluation benchmark for large-vocabulary object detection.

The Objects365 [32] object detection dataset and the ImageNet-21K image classification datasets are used for training. The Objects365 [32] object detection dataset contains around 1.7 million images, 365 object categories and over 29 million bounding boxes annotations in the training split. The ImageNet-21K image classification dataset [5] has over 14 million images, covering 21,841 categories.

For the ImgaeNet-21K dataset, we first divide the dataset into training split and validation split. Specifically, 5 images of each category are randomly selected for evaluation. For the classes with images less than 25, we sample 20% of the images from each class. As some of the categories are rare in ImageNet-21K (less than 5 images), we exclude them from the validation set. After filtering, we obtain a benchmarking dataset with 101,625 images and 21,077 categories. Then we annotate these images with ground-truth bounding boxes. Given an image and its ground-truth category, we annotate all objects belonging to this category. For each object, the smallest possible box that contains all visible parts of the object is annotated. This benchmarking dataset has been made publicly available to facilitate future research on large-vocabulary object detection.

In addition to evaluating large-vocabulary object detectors, we also verify whether the proposed framework can serve as a good representation pretraining method. To this end, we test its fine-tuing performance on the LVIS dataset [14]. We use the full 1,203 object categories for evaluation, which are distributed in long-tail, that there exists categories that *rare* with less than 10 training samples.

In this setup, Faster R-CNN [31] with FPN [25] is used as the base detector. In evaluation, both the metrics of localization accuracy and mAP are included.

Training and Implementation Details

– *First stage.* We follow common training recipes for image classification training. For ResNet architectures, we basically follow [16] to use SGD as the optimizer, with a base learning rate of 0.1, and a cosine learning rate scheduler. The weight decay is set as 1e-4, and the training length is set 100 epoch (on ImageNet-1K/997). For data augmentation, the random crop and color jittering is employed. For vision Transformers such as Swin Transformer, we follow [27,39] and employ all the regularization and augmentation strategies in [27], including RandAugment [4], Mixup [45], CutMix [43], random erasing [51] and stochastic depth [17]. The training length is 300 epochs for ImageNet-1K/997, and 100 epochs for ImageNet-21K.

– *Second stage.* In this stage, we employ a large jittering augmentation [11] with a resolution of 1024 × 1024 resolution and a scale range of [0.1, 2.0]. The random horizontal flipping is also employed. The ratio of detection images and classification images are set as 1:3 in each iteration. The 3× training scheduler is conducted for COCO and LVIS. The learning rate is searched in 3×10^{-4}, 1×10^{-3}, 3×10^{-3} and the weight decay is 0.05. The loss weight of the classification branch is searched in 0.01, 0.1 and 1.0.

4.2 Object Detection on LVIS-997-Novel in Setup S1

Table 2 shows the results of our approach compared to previous methods in Setup S1. We note the previous approaches are built by a different implementation and training settings than ours, and thus the absolute accuracy numbers are not directly comparable, for example, the training length of previous methods are much longer than ours.

Nevertheless, the performance gaps Δ between AP_{mask}^{novel} and AP_{mask} can act as a good indicator to evaluate the detectors' transferring ability from *base* categories to *novel* categories. The proposed approach has marginal AP degradations when transferred from *base* categories to *novel* ones, significantly better than previous methods. Compared to the variant without reformulation of classification method, or without Stage I, the performance is all degraded significantly.

These results indicate that our two-stage approach that employs an inverse information flow from a good classifier to object detection is crucial to seize the powerful classification capability.

These results also indicates surprising effectiveness of the simple multi-task learning approach, which has no explicit mechanism to attend to the detailed foreground and background classification on the image data.

Also note our approach is complementary to previous MIL approaches, by applying it afterwards the second stage. We will leave this as our future research.

4.3 Object Localization on ImageNet-1K in Setup S2

We evaluate the proposed method on the ImageNet-1K localization benchmark with the metrics of localization accuracy and GT-Known accuracy. For the localization accuracy, a predicted box is correct when it satisfies the following two conditions. First, the predicted class matches the ground truth category. Second, the predicted box has over 50% IoU with at least one of the ground truth boxes. GT-Known accuracy assumes that the ground truth category is given and only requires the predicted box to have over 50% IoU with ground truth boxes.

Table 2. Setup S1 experiments. The comparison of different approaches on LVIS-997-base and LVIS-997-novel. *denotes that a pre-training model using IN-21K dataset, a framework of CenterNet2, and a stronger training recipe is used.

Method	Backbone	AP_{mask}	AP_{mask}^{base}	AP_{mask}^{novel}	Δ (novel, all)
ViLD-ens. [13]	ResNet-50	25.5	–	16.6	−8.9
Detic [53]	ResNet-50	26.8	–	17.8	−9.0
Detic* [53]	ResNet-50	32.4	–	24.6	−7.8
Ours	ResNet-50	27.6	29.1	22.8	−4.8
Ours (-reform.)	ResNet-50	26.8	28.4	21.8	−5.0
Ours (-stage I)	ResNet-50	15.5	16.3	13.0	−2.5
Ours	Swin-T	30.4	30.9	28.8	−1.6
Ours	Swin-S	35.1	36.5	30.6	−4.5

As summarized in Table 3, our approach outperforms other WSOL methods by a large margin. Specifically, on the basis of ResNet-50, we achieve the top-1 localization accuracy of 64.7% and the GT-Known accuracy of 73.8%, also surpassing previous state-of-the-art methods by a gap of 9.9% and 5.3%, respectively. These results indicate that an additional base object detection datasets can significantly benefit the weakly supervised object localization.

Also note by this experiment, we do not aim for fair comparisons, but to encourage the use of this settings for object localization, rather than the previous settings that use only image classification data.

The ablation with the variant that does not use a reformulated image classification model, or the on that does not involve this first stage of training, indicates the importance of our design.

4.4   21K-Category Object Detection in Setup S3

We benchmark the performance of 21,000-vocabulary object detection with the ImageNet-21K localization dataset. In addition to the localization accuracy, we evaluate different methods with the challenging Average Precision (AP) metric.

To test with the AP metric, we allow predicting up to 100 detection boxes per image and adopt a confidence score threshold of 1e-4 for our approach. We compare our approach with Detic [53], which is concurrent with our approach, and is the only paper to develop a detector that can handle more than 20,000 object categories. For Detic, we find that a small confidence score threshold like 1e-6 would still hurt the performance and thus set it as 0.

As shown in Table 4, the proposed method achieves a Top-1 localization accuracy of 19.2% and AP_{50} of 6.9%, revealing the difficulty of discriminating over 21,000 categories. Note that our method outperforms Detic [53] by a large margin by +17.5% on Top-1 localization accuracy and +5.6% on AP_{50}. Comparing to our approach, Detic has a much worse classification performance on the 21,000-category dataset. This indicates it is crucial to have a reverse information flow

Table 3. Comparison of joint training model with other state-of-the-art methods on ImageNet-1K validation set.

Methods	Backbones	Loc. acc		GT-Known
		Top-1	Top-5	Top-1
CAM [52]	VGG16	42.8	54.9	–
CutMix [43]	VGG16	43.5	–	–
I^{2}C [49]	VGG16	48.4	58.5	63.9
SPA [28]	VGG16	49.6	61.3	65.1
CAM [52]	InceptionV3	46.3	58.2	62.7
SPG [48]	InceptionV3	48.6	60.0	64.7
I^{2}C [49]	InceptionV3	53.1	64.1	68.5
SPA [28]	InceptionV3	52.7	64.3	68.3
CAM [52]	ResNet-50	46.2	–	–
CutMix [43]	ResNet-50	47.3	–	–
I^{2}C [49]	ResNet-50	54.8	64.6	68.5
TS-CAM [10]	DeiT-S	53.4	64.3	67.6
Ours	ResNet-50	64.7	69.9	73.8
Ours (-reform.)	ResNet-50	60.8	62.1	69.9
Ours (-stage I)	ResNet-50	50.4	51.3	71.0
Ours	Swin-T	66.5	72.6	73.9
Ours	Swin-S	**68.3**	**78.5**	**75.1**

Table 4. Results of the joint training model on ImageNet-21K large-vocabulary object localization benchmark. Swin-B is adopted as the backbone for both methods. For Detic [53], we test with the publicly released model Detic_C2_SwinB_896_4x_IN-21K.

Methods	AP$_{50}$	Loc. acc	
		Top-1	Top-5
Detic [53]	1.3	2.7	5.2
Ours	6.9	19.2	19.7

for large-vocabulary object detection, that a large-vocabulary classifier is well trained first before transferring the knowledge to object detection. We hope that these results can serve as a baseline for future studies.

4.5 The Proposed Approach as Pre-training

The proposed framework can also serve as a pre-trained model for down-stream tasks. To evaluate this, we fine-tune different pre-trained models on LVIS dataset for 24 epochs using the Faster R-CNN [31] framework. We use the AdamW [18] optimizer, with the learning rate of 1e−4, the weight decay of 0.05 and the batch size of 16. Repeat Factor Sampling (RFS) [14] with a factor of 0.001 is utilized to handle the class imbalance problem in LVIS.

Table 5 summarizes the results of the three methods. Successive training on classification and detection outperforms classification pre-training by a gap of 3.0% on box AP. This may be due to that the classification pre-trained models lack the ability to localize, while successive training models somewhat maintain the classification and localization ability at the same time. However, successive training methods may suffer from the problem of *catastrophic forgetting* [19]. In successive learning, as it would be fine-tuned on the object detection dataset, the model may only focus on categories of the detection dataset and *forget* other concepts in the classification pre-training. This would limit the model to be transferred to other datasets with different classes (Fig. 3).

The proposed joint training method could address this problem because it performs supervised classification and object detection in a joint manner. As shown in Table 5, when compared with the strong baseline of successive training, the proposed method still has a large performance boost of 1.3% and 2.0% on AP and AP_r, respectively. The good performance on *rare* categories verifies that the joint training model could learn richer category semantics.

Fig. 3. Visualization of object detection results on LVIS dataset. The rich category semantics learnt by joint training model makes it easy to transfer to other datasets.

Table 5. Comparison of different pre-training methods on LVIS. IN-21K denotes the supervised classification model trained on ImageNet-21K. IN-21K → O365 denotes the successive training of ImageNet-21K classification and Objects365 object detection.

Pre-training	AP	AP_r	AP_c	AP_f
IN-21K	35.8	23.6	36.1	40.9
IN-21K → O365	38.8	26.8	38.9	44.0
Ours	**40.1**	**28.8**	**40.2**	**45.0**

5   Conclusion

In this paper, we propose a simple two-stage approach for large-vocabulary object detection. Unlike previous approaches which transfers knowledge from a base detector to image classification data, we start from image classification and transfer the large-vocabulary capability to object detection. For better transferring, the image classification problem is reformulated as a special configuration of object detection. The joining of two datasets is conducted using a simple multi-task learning framework in the second stage. Though without sophisticated process to explicitly attend to region proposals on the image data, the proposed multi-task learning approach performs rather strongly using three experimental setups.

We also obtained a 21,000-category object detector, built using a combination of the ImageNet-21K image classification dataset and the Object365v2 object detection dataset. We also built a new benchmarking dataset for the evaluation of object detection on large-vocabularies. We hope the simple baseline approach and evaluation benchmark can ease the future study in this area.

References

1. Bansal, A., Sikka, K., Sharma, G., Chellappa, R., Divakaran, A.: Zero-shot object detection. In: Proceedings of the European Conference on Computer Vision (ECCV), pp. 384–400 (2018)
2. Bilen, H., Vedaldi, A.: Weakly supervised deep detection networks. In: Proceedings of the IEEE Conference on Computer Vision and Pattern Recognition, pp. 2846–2854 (2016)
3. Choe, J., Shim, H.: Attention-based dropout layer for weakly supervised object localization. In: Proceedings of the IEEE/CVF Conference on Computer Vision and Pattern Recognition (CVPR) (2019)
4. Cubuk, E.D., Zoph, B., Shlens, J., Le, Q.V.: RandAugment: practical automated data augmentation with a reduced search space. In: Proceedings of the IEEE/CVF Conference on Computer Vision and Pattern Recognition Workshops, pp. 702–703 (2020)
5. Deng, J., Dong, W., Socher, R., Li, L.J., Li, K., Fei-Fei, L.: ImageNet: a large-scale hierarchical image database. In: 2009 IEEE Conference on Computer Vision and Pattern Recognition, pp. 248–255. IEEE (2009)
6. Deselaers, T., Alexe, B., Ferrari, V.: Weakly supervised localization and learning with generic knowledge. Int. J. Comput. Vis. **100**(3), 275–293 (2012). https://doi.org/10.1007/s11263-012-0538-3
7. Dong, B., Huang, Z., Guo, Y., Wang, Q., Niu, Z., Zuo, W.: Boosting weakly supervised object detection via learning bounding box adjusters. In: Proceedings of the IEEE/CVF International Conference on Computer Vision, pp. 2876–2885 (2021)
8. Dosovitskiy, A., et al.: An image is worth 16×16 words: transformers for image recognition at scale. In: International Conference on Learning Representations (2020)
9. Everingham, M., Gool, L., Williams, C.K., Winn, J., Zisserman, A.: The pascal visual object classes (VOC) challenge. IJCV **88**(2), 303–338 (2010). https://doi.org/10.1007/s11263-009-0275-4

10. Gao, W., et al.: TS-CAM: token semantic coupled attention map for weakly supervised object localization (2021)
11. Ghiasi, G., et al.: Simple copy-paste is a strong data augmentation method for instance segmentation. In: Proceedings of the IEEE/CVF Conference on Computer Vision and Pattern Recognition, pp. 2918–2928 (2021)
12. Girshick, R., Donahue, J., Darrell, T., Malik, J.: Rich feature hierarchies for accurate object detection and semantic segmentation. In: Proceedings of the IEEE Conference on Computer Vision and Pattern Recognition, pp. 580–587 (2014)
13. Gu, X., Lin, T.Y., Kuo, W., Cui, Y.: Open-vocabulary object detection via vision and language knowledge distillation. arXiv preprint arXiv:2104.13921 (2021)
14. Gupta, A., Dollar, P., Girshick, R.: LVIS: a dataset for large vocabulary instance segmentation. In: Proceedings of the IEEE/CVF Conference on Computer Vision and Pattern Recognition, pp. 5356–5364 (2019)
15. He, K., Gkioxari, G., Dollár, P., Girshick, R.: Mask R-CNN. In: Proceedings of the IEEE International Conference on Computer Vision, pp. 2961–2969 (2017)
16. He, K., Zhang, X., Ren, S., Sun, J.: Deep residual learning for image recognition. In: Proceedings of the IEEE Conference on Computer Vision and Pattern Recognition, pp. 770–778 (2016)
17. Huang, G., Sun, Yu., Liu, Z., Sedra, D., Weinberger, K.Q.: Deep networks with stochastic depth. In: Leibe, B., Matas, J., Sebe, N., Welling, M. (eds.) ECCV 2016. LNCS, vol. 9908, pp. 646–661. Springer, Cham (2016). https://doi.org/10.1007/978-3-319-46493-0_39
18. Kingma, D.P., Ba, J.: Adam: a method for stochastic optimization. arXiv preprint arXiv:1412.6980 (2014)
19. Kirkpatrick, J., et al.: Overcoming catastrophic forgetting in neural networks. Proc. Natl. Acad. Sci. **114**(13), 3521–3526 (2017)
20. Krishna, R., et al.: Visual genome: connecting language and vision using crowdsourced dense image annotations. Int. J. Comput. Vis. **123**(1), 32–73 (2017)
21. Krizhevsky, A., Sutskever, I., Hinton, G.E.: ImageNet classification with deep convolutional neural networks. Adv. Neural Inf. Process. Syst. **25**, 1097–1105 (2012)
22. Kuznetsova, A., et al.: The open images dataset V4. Int. J. Comput. Vis. **128**(7), 1956–1981 (2020). https://doi.org/10.1007/s11263-020-01316-z
23. Lee, S., Kwak, S., Cho, M.: Universal bounding box regression and its applications. In: Jawahar, C.V., Li, H., Mori, G., Schindler, K. (eds.) ACCV 2018. LNCS, vol. 11366, pp. 373–387. Springer, Cham (2019). https://doi.org/10.1007/978-3-030-20876-9_24
24. Li, L.H., et al.: Grounded language-image pre-training. arXiv preprint arXiv:2112.03857 (2021)
25. Lin, T.Y., Dollár, P., Girshick, R., He, K., Hariharan, B., Belongie, S.: Feature pyramid networks for object detection. In: Proceedings of the IEEE Conference on Computer Vision and Pattern Recognition, pp. 2117–2125 (2017)
26. Lin, T.Y., et al.: Microsoft COCO: common objects in context. In: Fleet, D., Pajdla, T., Schiele, B., Tuytelaars, T. (eds.) ECCV 2014. LNCS, vol. 8693, pp. 740–755. Springer, Cham (2014). https://doi.org/10.1007/978-3-319-10602-1_48
27. Liu, Z., et al.: Swin transformer: hierarchical vision transformer using shifted windows. CoRR abs/2103.14030 (2021). https://arxiv.org/abs/2103.14030
28. Pan, X., et al.: Unveiling the potential of structure preserving for weakly supervised object localization. In: Proceedings of the IEEE/CVF Conference on Computer Vision and Pattern Recognition (CVPR), pp. 11642–11651 (2021)
29. Radford, A., et al.: Learning transferable visual models from natural language supervision (2021)

30. Redmon, J., Farhadi, A.: YOLO9000: better, faster, stronger. In: Proceedings of the IEEE Conference on Computer Vision and Pattern Recognition, pp. 7263–7271 (2017)
31. Ren, S., He, K., Girshick, R., Sun, J.: Faster R-CNN: towards real-time object detection with region proposal networks. Adv. Neural Inf. Process. Syst. **28**, 91–99 (2015)
32. Shao, S., et al.: Objects365: a large-scale, high-quality dataset for object detection. In: Proceedings of the IEEE/CVF International Conference on Computer Vision (ICCV) (2019)
33. Simonyan, K., Zisserman, A.: Very deep convolutional networks for large-scale image recognition. In: International Conference on Learning Representations (2015)
34. Singh, K.K., Lee, Y.J.: Hide-and-seek: Forcing a network to be meticulous for weakly-supervised object and action localization. In: International Conference on Computer Vision (ICCV) (2017)
35. Song, H.O., Girshick, R., Jegelka, S., Mairal, J., Harchaoui, Z., Darrell, T.: On learning to localize objects with minimal supervision. In: International Conference on Machine Learning, pp. 1611–1619 (2014)
36. Tang, P., Wang, X., Bai, S., Shen, W., Bai, X., Liu, W., Yuille, A.L.: PCL: proposal cluster learning for weakly supervised object detection. IEEE Trans. Pattern Anal. Mach. Intell. **42**, 176–191 (2018)
37. Tang, P., Wang, X., Bai, X., Liu, W.: Multiple instance detection network with online instance classifier refinement. In: Proceedings of the IEEE Conference on Computer Vision and Pattern Recognition, pp. 2843–2851 (2017)
38. Tian, Z., Shen, C., Chen, H., He, T.: FCOS: fully convolutional one-stage object detection. In: Proceedings of the IEEE/CVF International Conference on Computer Vision, pp. 9627–9636 (2019)
39. Touvron, H., Cord, M., Douze, M., Massa, F., Sablayrolles, A., Jegou, H.: Training data-efficient image transformers distillation through attention. In: International Conference on Machine Learning, vol. 139, pp. 10347–10357 (2021)
40. Uijlings, J., Popov, S., Ferrari, V.: Revisiting knowledge transfer for training object class detectors. In: Proceedings of the IEEE Conference on Computer Vision and Pattern Recognition, pp. 1101–1110 (2018)
41. Uijlings, J.R., Van De Sande, K.E., Gevers, T., Smeulders, A.W.: Selective search for object recognition. Int. J. Comput. Vis. **104**(2), 154–171 (2013). https://doi.org/10.1007/s11263-013-0620-5
42. Yang, H., Wu, H., Chen, H.: Detecting 11k classes: large scale object detection without fine-grained bounding boxes. In: Proceedings of the IEEE/CVF International Conference on Computer Vision, pp. 9805–9813 (2019)
43. Yun, S., Han, D., Oh, S.J., Chun, S., Choe, J., Yoo, Y.: CutMix: regularization strategy to train strong classifiers with localizable features. In: International Conference on Computer Vision (ICCV) (2019)
44. Zareian, A., Rosa, K.D., Hu, D.H., Chang, S.F.: Open-vocabulary object detection using captions. In: Proceedings of the IEEE/CVF Conference on Computer Vision and Pattern Recognition, pp. 14393–14402 (2021)
45. Zhang, H., Cisse, M., Dauphin, Y.N., Lopez-Paz, D.: mixup: beyond empirical risk minimization. In: International Conference on Learning Representations (2018)
46. Huang, K., Zhang, J., Zhang, J., et al.: Mixed supervised object detection with robust objectness transfer. IEEE Trans. Pattern Anal. Mach. Intell. **41**(3), 639–653 (2018)

47. Zhang, X., Wei, Y., Feng, J., Yang, Y., Huang, T.: Adversarial complementary learning for weakly supervised object localization. In: IEEE CVPR (2018)
48. Zhang, X., Wei, Y., Kang, G., Yang, Y., Huang, T.: Self-produced guidance for weakly-supervised object localization. In: Ferrari, V., Hebert, M., Sminchisescu, C., Weiss, Y. (eds.) ECCV 2018. LNCS, vol. 11216, pp. 610–625. Springer, Cham (2018). https://doi.org/10.1007/978-3-030-01258-8_37
49. Zhang, X., Wei, Y., Yang, Y.: Inter-image communication for weakly supervised localization. In: Vedaldi, A., Bischof, H., Brox, T., Frahm, J.-M. (eds.) ECCV 2020. LNCS, vol. 12364, pp. 271–287. Springer, Cham (2020). https://doi.org/10.1007/978-3-030-58529-7_17
50. Zhong, Y., Wang, J., Peng, J., Zhang, L.: Boosting weakly supervised object detection with progressive knowledge transfer. In: Vedaldi, A., Bischof, H., Brox, T., Frahm, J.-M. (eds.) ECCV 2020. LNCS, vol. 12371, pp. 615–631. Springer, Cham (2020). https://doi.org/10.1007/978-3-030-58574-7_37
51. Zhong, Z., Zheng, L., Kang, G., Li, S., Yang, Y.: Random erasing data augmentation. In: Proceedings of the AAAI Conference on Artificial Intelligence, vol. 34, pp. 13001–13008 (2020)
52. Zhou, B., Khosla, A., Lapedriza, A., Oliva, A., Torralba, A.: Learning deep features for discriminative localization. In: Computer Vision and Pattern Recognition (2016)
53. Zhou, X., Girdhar, R., Joulin, A., Krähenbühl, P., Misra, I.: Detecting twenty-thousand classes using image-level supervision. arXiv preprint arXiv:2201.02605 (2021)
54. Zitnick, C.L., Dollár, P.: Edge boxes: locating object proposals from edges. In: Fleet, D., Pajdla, T., Schiele, B., Tuytelaars, T. (eds.) ECCV 2014. LNCS, vol. 8693, pp. 391–405. Springer, Cham (2014). https://doi.org/10.1007/978-3-319-10602-1_26

Knowledge Condensation Distillation

Chenxin Li[1], Mingbao Lin[2], Zhiyuan Ding[1], Nie Lin[4], Yihong Zhuang[1],
Yue Huang[1,3(✉)], Xinghao Ding[1,3], and Liujuan Cao[1]

[1] School of Informatics, Xiamen University, Xiamen, China
{chenxinli,dingzhiyuan}@stu.xmu.edu.cn,
{zhuangyihong,yhuang2010,dxh,caoliujuan}@xmu.edu.cn
[2] Tencent Youtu Lab, Shanghai, China
[3] Institute of Artificial Intelligence, Xiamen University, Xiamen, China
[4] Hunan University, Changsha, China
nielin@hnu.edu.cn

Abstract. Knowledge Distillation (KD) transfers the knowledge from a
high-capacity teacher network to strengthen a smaller student. Existing
methods focus on excavating the knowledge hints and transferring the
whole knowledge to the student. However, the knowledge redundancy
arises since the knowledge shows different values to the student at different learning stages. In this paper, we propose Knowledge Condensation
Distillation (KCD). Specifically, the knowledge value on each sample is
dynamically estimated, based on which an Expectation-Maximization
(EM) framework is forged to iteratively condense a compact knowledge
set from the teacher to guide the student learning. Our approach is easy
to build on top of the off-the-shelf KD methods, with no extra training parameters and negligible computation overhead. Thus, it presents
one new perspective for KD, in which the student that actively identifies teacher's knowledge in line with its aptitude can learn to learn
more effectively and efficiently. Experiments on standard benchmarks
manifest that the proposed KCD can well boost the performance of student model with even higher distillation efficiency. Code is available at
https://github.com/dzy3/KCD.

Keywords: Knowledge distillation · Active learning · Efficient training

1 Introduction

Though deep neural networks (DNNs) have achieved great success in computer
vision, most advanced models are too computationally expensive to be deployed
on the resource-constrained devices. To address this, the light-weight DNNs have
been explored in the past decades. Typical methods include network pruning [18],
parameter quantization [36] and neural architecture search [2], *etc.* Among all
these methods, knowledge distillation [10] is widely integrated into their learning frameworks, whereby the original cumbersome model (teacher) transfers its

Supplementary Information The online version contains supplementary material
available at https://doi.org/10.1007/978-3-031-20083-0_2.

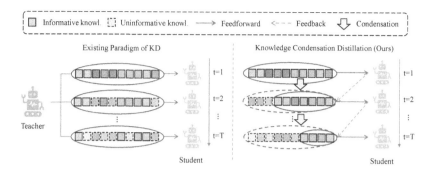

Fig. 1. Comparison between existing KD paradigm and our KCD. **Left:** Existing paradigm transfers the complete knowledge points from teacher model across the entire training process, regardless of the varying values to the student at different stages. **Right:** Knowledge points are first estimated based on the current capacity of the student, and then condensed to a compact yet informative sub-part for student model.

knowledge to enhance the recognition capacity of its compressed version, *a.k.a.* student model. Due to its flexibility, KD has received ever-increasing popularity in varieties of vision tasks.

In most existing studies on KD [4,10,13,16,17,22,25,30], the knowledge hints of the whole sample space, such as soft predictions [10], intermediate representations [25], attention maps [13], *etc*, are transferred to the student model across the entire training process as illustrated in Fig. 1(**Left**). However, these methods neglect the changing capacity of the student model at different learning stages. Specifically, all the knowledge points from teacher model are informative enough for the student model at its infant learning stage. However, as the learning proceeds, the value of different knowledge points starts to vary for the student. For example, the "well-memorized" knowledge points have a relatively limited impact to the student at the later training stages. Consequently, the concern regarding redundancy of knowledge transfer arises in existing studies, whereby the student model passively receives all the knowledge points from the teacher. This further poses two severe issues: (**1**) Training burden. The redundant knowledge requires not only additional memory storage, but also prolongs the training time. (**2**) Poor performance. The redundancy prevents the student model from concentrating enough on the more informative knowledge, which weakens the learning efficacy of the student model.

To overcome the above challenge, as shown in Fig. 1(**Right**), this paper presents a new perspective for KD, the cores of which are two main folds: (**1**) A feedback mechanism is introduced to excavate various values of the teacher's knowledge for the student at different training stages. (**2**) The student actively identifies the informative knowledge points and progressively condenses a core knowledge set for distillation. To this end, we propose a Knowledge Condensation Distillation (KCD) paradigm, whereby the label of knowledge value to student model is encoded as a latent variable, and an Expectation-Maximization (EM) framework is forged to iteratively condense the teacher's knowledge set and distill the student model. Furthermore, given the local-batch training fashion in student learning, we propose an Online Global Value Estimation (OGVE) module to dynamically estimate the knowledge value over the global knowledge space. To generate a compact yet effective encoding of teacher's knowledge, we develop a Value-Adaptive Knowledge Summary (VAKS) module to adaptively preserve high-value knowledge points, remove the valueless points as well as augment the intermediate ones. We conduct extensive experiments on two benchmarks, CIFAR100 [14] and ImagetNet [5], and many representative settings of teacher-student networks in KD. We show that our KCD can be well built upon majorities of existing KD pipelines as a plug-and-play solution, without bringing extra training parameters and computation overhead.

Our contributions are summarized as follows:

- We propose a novel KD paradigm of knowledge condensation, wherein the knowledge to transfer is actively determined by the student model, and a concise encoding condensed from the whole knowledge set is utilized in KD.
- We derive an Expectation-Maximization framework to accomplish our knowledge condensation distillation by iteratively performing knowledge condensation and model distillation.
- We propose an OGVE module to acquire an approximate global estimator of knowledge value while utilizing only local training statistics. We further present a VAKS module to harmonize the trade-off between compactness and informativeness of knowledge condensation encoding.

2   Related Work

Knowledge Distillation. The pioneering KD work dates back to [10], where the soft probability distribution from the teacher is distilled to facilitate the student's training. Since then, abundant developments have been committed to excavating richer knowledge hints, such as intermediate representation [9,25], attention maps [13], instance relation [23,32], self-supervised embedding [30,34] and so on. All these methods transfer the knowledge on all training instances to the student regardless of different training stages. Differently, we study the redundancy of the teacher's knowledge and emphasize the significance of making the student model actively condense an efficient knowledge set for learning.

One more recent study [35] considers the efficiency issue of KD by identifying the most informative samples in each training batch. Our method differs from the following aspects. First, the study [35] explores the difference of computation overheads in the forward passes of the teacher and student models, and fixes the knowledge set during the distillation process. As a comparison in our method, the knowledge set is dynamically condensed and explicitly encodes the patterns of the student model during training. Second, we estimate the knowledge value over the complete sample space rather than every single batch, which is more accurate and comprehensive.

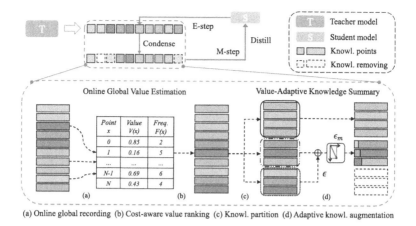

(a) Online global recording (b) Cost-aware value ranking (c) Knowl. partition (d) Adaptive knowl. augmentation

Fig. 2. Overview of the proposed KCD framework. The knowledge condensation and student distillation are optimized iteratively in an EM framework.

Coreset Construction. Another related literature is the problem of coreset construction [8,26]. The main idea behind them is that a learning agent can still perform well with fewer training samples by selecting data itself. Most existing works [12,20,31,33,40] construct this coreset by importance sampling. For example, In [12], sample importance is estimated via the magnitude of its loss gradient *w.r.t.* model parameters. CRAIG [20] selects a weighted coreset of training data that closely estimates the full gradient by maximizing a submodular function. Wang *et al.* [33] distilled the knowledge from the entire dataset to generate a synthetic smaller one. These ideas inspire us to seek a core component of the whole knowledge set from the teacher to realize an efficient KD.

3 Methodology

3.1 Preliminaries

In the task of knowledge distillation (KD), we are given a training dataset $\mathcal{X}$, a pre-trained teacher model $\mathcal{T}$ and a to-be-learned student model $\mathcal{S}$. Hinton *et al.* [10] proposed to minimize the cross-entropy loss between the output probability $p_{\mathcal{T}}(x)$ of the teacher and that $p_{\mathcal{S}}(x)$ of the student:

$$\mathcal{L}_{KD} = -\sum_{x \in \mathcal{X}} p_{\mathcal{T}}(x) \log \left(p_{\mathcal{S}}(x) \right). \tag{1}$$

Denoting each pair $\left(x, p_{\mathcal{T}}(x)\right)$ as a knowledge point, the teacher $\mathcal{T}$ in essence provides a knowledge set $K = \{\left(x, p_{\mathcal{T}}(x)\right) | x \in \mathcal{X}\}$, which is then transferred to the student $\mathcal{S}$. In conventional KD, the knowledge set K is fixed across the whole distillation process, despite different learning stages of the student model. As a core distinction, we propose to transfer simply a concise knowledge encoding $\hat{K}$

with $|\hat{K}| < |K|$, where the knowledge points are most valuable and adapt to the demand of the student model at different periods.

In what follows, we show that the efficient coding $\hat{K}$ can be deduced by the Expectation-Maximization (EM) framework that encodes the knowledge value for the student model as a latent variable Y, with which, we can identify the most valuable components in knowledge set K. Figure 2 shows the overview of the proposed method.

3.2 Knowledge Condensation Distillation

The goal of KD in Eq. (1) is to learn the parameter θ of student model in order to maximize the negative cross-entropy between the teacher $\mathcal{T}$ and student $\mathcal{S}$:

$$\hat{\theta} = \arg\max_{\theta} \sum_{x \in \mathcal{X}} \sum_{c \in \mathcal{C}} p_{\mathcal{T}}(x, c) \log p_{\mathcal{S}}(x, c; \theta), \tag{2}$$

where $\mathcal{C}$ denotes the class space. Instead of transferring the complete knowledge set $K = \{(x, p_{\mathcal{T}}(x)) | x \in \mathcal{X}\}$, we introduce a binary value variable $\mathcal{Y} \in \{0,1\}^{|K|}$, the i-th value of which indicates if the i-th knowledge point is valuable to the student. In this way, the traditional optimization of Eq. (2) in our setting becomes:

$$\hat{\theta} = \arg\max_{\theta} \sum_{x \in \mathcal{X}} \sum_{c \in \mathcal{C}} p_{\mathcal{T}}(x, c) \log \sum_{y \in \mathcal{Y}} p_{\mathcal{S}}(x, c, y; \theta). \tag{3}$$

To maximize this objective, we consider its low-bound surrogate:

$$\sum_{x \in \mathcal{X}} \sum_{c \in \mathcal{C}} p_{\mathcal{T}}(x, c) \log \sum_{y \in \mathcal{Y}} p_{\mathcal{S}}(x, c, y; \theta)$$

$$= \sum_{x \in \mathcal{X}} \sum_{x \in \mathcal{C}} p_{\mathcal{T}}(x, c) \log \sum_{y \in \mathcal{Y}} Q(y) \frac{p_{\mathcal{S}}(x, c, y; \theta)}{Q(y)} \tag{4}$$

$$\geq \sum_{x \in \mathcal{X}} \sum_{c \in \mathcal{C}} p_{\mathcal{T}}(x, c) \sum_{y \in \mathcal{Y}} Q(y) \log \frac{p_{\mathcal{S}}(x, c, y; \theta)}{Q(y)},$$

where $Q(y)$ denotes the distribution over the space of value labels $\mathcal{Y}$ so that $\sum_{y \in \mathcal{Y}} Q(y) = 1$. Note that we derive the last step based on Jensen's inequality where the equality holds if and only if $\frac{p_{\mathcal{S}}(x,c,y;\theta)}{Q(y)}$ is a constant [15]. Under this condition, the distribution $Q(y)$ should be:

$$Q(y) = \frac{p_{\mathcal{S}}(x, c, y; \theta)}{\sum_{y \in \mathcal{Y}} p_{\mathcal{S}}(x, c, y; \theta)} = \frac{p_{\mathcal{S}}(x, c, y; \theta)}{p_{\mathcal{S}}(x, c; \theta)} = p_{\mathcal{S}}(y; x, c, \theta). \tag{5}$$

Removing the constant term $-\sum_{y \in \mathcal{Y}} Q(y) \log Q(y)$ in Eq. (4) and combining Eq. (5) lead to our final optimization:

$$\sum_{x \in \mathcal{X}} \sum_{c \in \mathcal{C}} p_{\mathcal{T}}(x, c) \sum_{y \in \mathcal{Y}} p_{\mathcal{S}}(y; x, c, \theta) \log p_{\mathcal{S}}(x, c, y; \theta). \tag{6}$$

The maximization of the above problem can be realized by Expectation-Maximization (EM) algorithm, as elaborated below:

E-Step. In this step, we aim to evaluate value distribution $Q(y) = p_S(y; x, c, \theta)$. Before that, we first discuss how to measure the value of each knowledge point $(x, p_{T(x)})$, insight of which is two-fold: First, it has been verified that the average prediction entropy loss decreases drastically if a model is distilled by knowledge hints from a teacher model, instead of being trained solely [21]. This reflects the contribution of knowledge points to the training of the student model. Second, as discussed in [27], the knowledge which encodes informative semantic structure tends to require more training time for the student model to fit well.

These two insights indicate that the prediction entropy loss can be an option to measure the knowledge value. Besides, informative knowledge tends to cause a larger entropy loss. Therefore, given a knowledge point $(x, p_{T(x)})$, we utilize its prediction entropy to measure its value:

$$V(x) = -\sum_{c \in C} p_S(x, c) \log p_S(x, c). \tag{7}$$

With the prediction entropy, in order to estimate $p_S(y; x, c, \theta)$, we further conduct a ranking operation in an decreasing order *w.r.t.* $V(x)$ over $\mathcal{X}$. Then, based on the ranking position $\mathcal{R}_V(x) \in \{0, 1, \cdots, N\}$, we derive the relative likelihood probability about knowledge value:

$$p_{\mathcal{R}_V}(y; x, \theta) = 1 - \frac{\mathcal{R}_V(x)}{|\mathcal{X}|}. \tag{8}$$

Then the likelihood of value label $p_S(y; x, c, \theta)$ can be determined by a threshold τ: $p_S(y; x, c, \theta) = 1$ if $p_{\mathcal{R}_V}(y; x, \theta) \geq \tau$, and 0 otherwise.

M-Step. After E-step, the optimized object of Eq. (6) can be re-written as:

$$\sum_{x \in \mathcal{X}} \sum_{c \in C} p_T(x, c) \sum_{y \in \mathcal{Y}} p_S(y; x, c, \theta) \log p_S(x, c, y; \theta)$$

$$= \sum_{x \in \mathcal{X}} \sum_{c \in C} p_T(x, c) \sum_{y \in \mathcal{Y}} \mathbb{I}(p_{\mathcal{R}_V}(y; x, \theta) \geq \tau) \log p_S(x, c, y; \theta), \tag{9}$$

where $\mathbb{I}(\cdot)$ returns 1 if its input is true, and 0 otherwise. When the training samples are not provided, we assume a uniform priori over y (0 or 1):

$$p_S(x, c, y; \theta) = p_S(x, c; y, \theta) p_S(y; \theta) = \frac{1}{2} \cdot p_S(x, c; y, \theta), \tag{10}$$

where $p_S(y; \theta) = \frac{1}{2}$ due to the premise of uniform distribution. Then the distribution $p_S(x, c; y, \theta)$ is only conditioned on the estimated value label Y, *i.e.*, $\mathbb{I}(p_{\mathcal{R}_V}(y; x, \theta) \geq \tau)$. We conduct distillation only upon the knowledge points with label $y = 1$. Thus, we can re-write the maximum estimation in Eq. (2) as:

$$\hat{\theta} = \arg\min_{\theta} \sum_{x \in \mathcal{X}|Y(x)=1} \sum_{c \in C} -p_T(x, c) \log p_S(x, c; \theta), \tag{11}$$

where $x \in \mathcal{X}$ can be used for distillation only when the condition $y = 1$.

Consequently, our KCD iteratively performs E-step and M-step. The former aims to find the distribution of label Y and the concise knowledge encoding $\hat{K}$ comprises these knowledge points with $y = 1$, while the latter implements efficient distillation upon the concise set $\hat{K}$. However, current neural networks are trained in a batch fashion where a small portion of samples are fed forward each time. These local sample batches barricade a direct extraction of the concise knowledge set $\hat{K}$ from the whole training dataset $\mathcal{X}$. In what follows, we further propose an online global value estimation to solve this problem.

3.3 Online Global Value Estimation

In order to condense a valuable knowledge set $\hat{K}$ in a global fashion, we design an online global value estimation (OGVE) to derive the global statistics of the whole training dataset $\mathcal{X}$ which consists of an online value recording and cost-aware ranking below:

Online Value Recording. The estimation of $p_S(y; x, c, \theta)$ is conducted by $p_{\mathcal{R}_V}(y; x, \theta)$ over the whole (global) sample space $\mathcal{X}$ at the E-step. However, only a small-portion sub-set (local) of knowledge can be available at each training iteration. Besides, the same sample x might appear frequently at different training stages. To alleviate this issue, we propose to consider the historical statistics of x. Technically, when x is fed to the network at a particular training iteration, we first count the frequency of x ever involving in the training, denoted as $F(x)$. Also, we calculate its prediction entropy $V(x)$ at current training iteration using Eq. (7). Then, the global value of a knowledge point $(x, p_{T(x)})$ is updated in an online moving-average fashion as:

$$V^{F(x)}(x) = \frac{F(x) - 1}{F(x)} \times V^{F(x)-1} + \frac{1}{F(x)} V(x). \tag{12}$$

Cost-Aware Ranking. Based on the recorded global statistics of $V^{F(x)}(x)$, we can obtain a more accurate ranking of $R_V(x)$ without introducing any additional overhead. However, in the current design, the ranking order of two knowledge points with a similar value might be the same even though their training frequencies are very different, which is counter-intuitive, given the fact that the neural network tends to memorize and gives low prediction entropy to these samples that have ever seen more times [7]. Therefore, the knowledge points with similar $V(x)$ but a higher training cost $F(x)$ should be more critical for the student model and assigned with a top ranking. Considering this, in the ranking operation, we re-weight $V(x)$ using the training frequency $F(x)$ as:

$$\mathcal{R}_V(x) = \underset{x \in \mathcal{X}}{\arg \text{ sort}} \ V^{F(x)}(x) \times \left(F(x)\right)^\alpha, \tag{13}$$

where α controls the weighted effect of $F(x)$. Eq. (13) considers not only the status of $V(x)$, but also the cost $F(x)$ to achieve this status.

Combining $\mathcal{R}_V(x)$ in Eq. (13) and $p_{\mathcal{R}_V}(y; x, \theta)$ in Eq. (8), we can estimate the value label Y. Accordingly, the concise knowledge encoding $\hat{K}$ consists of the knowledge points with $y = 1$. As the training proceeds, many well learned knowledge points becomes less valuable to the student model. However, the relative likelihood probability $p_{\mathcal{R}_V}(y; x, \theta) \neq 0$ indicates a possibility to be selected again. Instead, we further propose a value-adaptive knowledge summary by solving this issue in a divide-and-conquer manner.

3.4   Value-Adaptive Knowledge Summary

Our value-adaptive knowledge summary (VAKS) performs concise knowledge encoding in a two-step fashion including a knowledge partition and an adaptive knowledge augmentation.

Knowledge Partition. According to our OGVE, we can obtain an explicit label set Y. Then, the original knowledge set can be divided into K_1 with $y = 1$ and K_0 with $y = 0$. For knowledge points in K_0, they are deemed to be valueless thus we choose to directly discard them. As for K_1, based on the relative likelihood probability $\mathcal{R}_V(x)$, we further partition it into a set K_{1H}, element of which has a relatively high $\mathcal{R}_V(x)$, and a set K_{1L}, element of which has relatively low $\mathcal{R}_V(x)$, as shown in Fig. 2. Besides, our partition also requires K_{1L} to be in the same size with K_0, i.e., $|K_{1L}| = |K_0|$, reason of which will be given in the following adaptive knowledge augmentation.

The knowledge points in K_{1H} are of high possibility to be valuable for the student, thus they can be safely transferred to the student as the conventional KD does. However, the knowledge in K_{1L} falls into a "boundary status". Though considered valuable, they are prone to being less valuable than knowledge in K_{1H} and easily absorbed by the student. This motivates us to enhance the knowledge points in K_{1L}. One straightforward approach is to introduce the gradient-based distillation [33,37,41] to generate new knowledge contents. However, the heavy time consumption barricades its application. In what follows, we introduce an adaptive knowledge augmentation to reach this goal in a training-free fashion.

Adaptive Knowledge Augmentation. Our insight of knowledge augmentation comes from the field of adversarial examples [6,29], where a subtle perturbation can greatly confuse the model recognition. Likewise, we also seek a perturbation on the knowledge points in K_{1L}. It is noteworthy that rather than to find the most disruptive disturbance in adversarial examples, our goal is to use some knowledge-wise perturbation to augment the knowledge points, making them more informative for the student model.

Concretely, denoting $S = \{|K_1|, |K_1| - 1, ..., |K_{1L}|\}$, as shown in Fig. 2, we propose to make full use of the removed valueless knowledge in K_0 to augment knowledge points in K_{1L} with a very small perturbation ratio ϵ. as:

$$K_{Aug} = \text{Ordered}(K_{1L}) \oplus \text{Ordered}(K_0) \otimes \epsilon(S), \tag{14}$$

where $\text{Ordered}(\cdot)$ reorders its input set in descending according to the value of knowledge point, and $\oplus$ denotes the element-wise adding operation. Recall that

---

Algorithm 1. Knowledge Condensation Distillation

Input: Training dataset $\mathcal{X}$; a student model $\mathcal{S}$ with learnable parameters θ; a full knowledge set K generated by a pre-trained teacher model $\mathcal{T}$.

Required: Number of epochs in a learning stage T; Desired final knowledge condensation ratio ρ.

Output: Distilled student model with parameter $\hat{\theta}$; condensed knowledge encoding $\hat{K}$ ($|\hat{K}| = |K| \cdot \rho$).

1: Init. $\hat{K} = K$;
2: **for** $i = 0, ..., I$ **epoch do**
3:     # M-step: Knowledge distillation
4:     Distill $\hat{\theta}$ of student $\mathcal{S}$ on the condensed knowledge $\hat{K}$ via Eq. (11);
5:     # E-step: Knowledge condensation
6:     ## Estimate knowledge value over K via proposed OGIE (Sect. 3.3)
7:     Cal. knowledge value $V(x)$ over compact (local) knowledge space $\hat{K}$ via Eq. (7); online update historical recording $V^{F(x)}(x)$ over complete (global) knowledge space K via Eq. (12);
8:     **if** $i \% T = 0$ **then**
9:         Cal. ranking position of knowledge value $\mathcal{R}_V(x)$ via Eq. (13); cal. ranking-based likelihood probability $p_{\mathcal{R}_V}(x)$ via Eq. (8);
10:         Binarize $p_{\mathcal{R}_V}(y; x)$ with the threshold τ^t at current stage $t = i/T$; determine value label Y ($y = 1$ or 0) over complete knowledge space K;
11:         ## Summarize knowledge encoding K via proposed VAKS (Sect. 3.4);
12:         Partition K into K_1 and K_0 via label Y; partition K_1 into K_{1H} and K_{1L}, s.t. $|K_{1L}| = |K_0|$;
13:         Augment K_{1L} via Eq. (14) and Eq. (15);
14:         Summarize compact knowledge encoding via $\hat{K} = K_{1H} \cup K_{Aug}$.
15:     **end if**
16: **end for**

$|K_{1L}| = |K_0|$ in our setting, thus, the $\oplus$ is applicable. $\epsilon(\cdot)$ is defined as:

$$\epsilon(x') = \frac{\epsilon_m}{|K_0|}(x' - |K_1|) + \epsilon_m. \tag{15}$$

Thus $\epsilon(S)$ is a set linearly increasing from 0 to a pre-given ϵ_m (see Fig. 2). The intuition of $\epsilon(x')$ is to make the knowledge points with lower-ranking positions $w.r.t.$ knowledge value to get more augmentation effect while the ones with higher positions to maintain more their original knowledge contents.

Finally, we obtain the knowledge condensation $\hat{K} = K_{1H} \cup K_{Aug}$.

3.5   Overall Procedure

The overall procedure of our proposed KCD is depicted in Algorithm 1. The proposed framework iteratively performs knowledge condensation in E-step and knowledge distillation in M-step, which can be practically formulated as a stage-based learning framework. The total I training epochs are equally divided into I/T learning stages, each with T epochs. Within each stage, the distillation is conducted for T epochs on the fixed knowledge set, followed by the recording of knowledge value in every training batch (Eq. (12)). At the end of each stage, we perform a ranking step across the whole knowledge set $w.r.t.$ knowledge value (Eq. (13)) and knowledge summary (Eq. (14)). to condense a smaller informative knowledge set. The condensed one is then used in the next stage.

It is noteworthy that the reduction of computation overhead mainly comes from using more compact knowledge encoding $\hat{K}$ during KD. To quantitatively portray this, absolute cost C_a is calculated by the number of knowledge points used, $e.g.$, $C_a = |K| \cdot I$ for conventional KD. We further calculate the relative cost C as the rate of C_a between our KCD and the conventional KD baseline:

$$C = \frac{|K| \cdot (\tau^0 + \tau^1 + \cdots + \tau^t + \cdots + \tau^{I/T}) \cdot T}{|K| \cdot I} \tag{16}$$

where τ^t denotes the threshold of the ranking-based probability $p_{\mathcal{R}_V}$ (Eq. (8)) for the value label Y at the t-th stage. It controls the condensation ratio, as $|\hat{K}| = |K| \cdot \tau^t$ at t-th stage and the final condensation rate $\rho = \tau^{I/T}$.

4   Experiments

Datasets. We conduct experiments on two benchmark datasets for KD, namely CIFAR100 [14] and ImageNet [5]. CIFAR100 contains 50K training images with 500 images per class and 10K test images with 100 images per class. The image size is 32×32. ImageNet is a large-scale classification dataset, containing 1.2 million images over 1K classes for training and 50K for validation. The image size is 224×224.

Implementation Details. Following the common practice in [30,34], we adopt the stochastic gradient descent (SGD) optimizer with a momentum of 0.9, weight decay of 5×10^{-4}. Batch size is set as 64 for CIFAR-100 and 256 for ImageNet. For CIFAR100 [14], the learning rate is initialized as 0.05, and decayed by 0.1 every 30 epochs after the first 150 epochs until the last 240 epochs. For ImageNet [5], the learning rate is initialized as 0.1, and decayed by 0.1 every 30 epochs. Without specification, the hyper-parameters in Algorithm 1 is set as follows: We set $I = 240$, $T = 40$ for CIFAR100 and $I = 90$, $T = 15$ for ImageNet. We set final condensation rate $\rho = 0.7$. The intermediate value of condensation threshold τ is set as exponential decay after every learning stage, with the initial value of $\tau^0 = \sqrt[T/I]{\rho} = 0.9423$. We set $\alpha = 0.03$ in Eq. (13), and the perturbation rate ϵ in Eq. (15) as linear growth from 0 to 0.3 ($\epsilon_m = 0.3$).

4.1   Comparisons with State-of-the Arts

Results on CIFAR100. We make comparison to various representative state-of-the-art KD methods, including vanilla KD [10], FitNet [25], AT [13], SP [32], VID [1], RKD [23], PKT [24], CRD [30], WCoRD [3], ReviewKD [4],

Table 1. Test Acc. (%) of the student networks on CIFAR100. **Bold** and underline denote the best and second best results. The comparison of whether to equip modern methods with our KCD is provided as $(+/-)$. Same-architecture and cross-architecture experiments are shown in two groups of columns.

Teacher	W40-2	W40-2	R56	R32x4	V13	V13	R50	R50	R32×4	R32×4	W40-2
Student	W16-2	W40-1	R20	R8×4	V8	MN2	MN2	V8	SN1	SN2	SN1
Teacher	75.61	75.61	72.34	79.42	74.64	74.64	79.34	79.34	79.42	79.42	75.61
Student	73.26	73.26	69.06	72.50	70.36	64.60	64.60	70.36	70.50	71.82	70.50
KD [10]	74.92	73.54	70.66	73.33	72.98	67.37	67.35	73.81	74.07	74.45	74.83
FitNet [25]	73.58	72.24	69.21	73.50	71.02	64.14	63.16	70.69	73.59	73.54	73.73
AT [13]	74.08	72.77	70.55	73.44	71.43	59.40	58.58	71.84	71.73	72.73	73.32
SP [32]	73.83	72.43	69.67	72.94	72.68	66.30	68.08	73.34	73.48	74.56	74.52
VID [1]	74.11	73.30	70.38	73.09	71.23	65.56	67.57	70.30	73.38	73.40	73.61
RKD [23]	73.35	72.22	69.61	71.90	71.48	64.52	64.43	71.50	72.28	73.21	72.21
PKT [24]	74.54	73.45	70.34	73.64	72.88	67.13	66.52	73.01	74.10	74.69	73.89
CRD [30]	75.64	74.38	71.63	75.46	74.29	69.94	69.54	74.58	75.12	76.05	76.27
WCoRD [3]	76.11	74.72	<u>71.92</u>	<u>76.15</u>	74.72	70.02	70.12	74.68	75.77	76.48	76.68
ReviewKD [4]	<u>76.12</u>	75.09	71.89	75.63	74.84	70.37	69.89	–	77.45	77.78	<u>77.14</u>
SSKD [34]	75.66	<u>75.27</u>	70.96	75.80	**75.12**	<u>70.92</u>	<u>71.14</u>	**75.72**	<u>77.91</u>	<u>78.37</u>	76.92
KC-KD	75.70	73.84	70.75	74.05	73.44	68.61	67.94	74.41	74.33	75.19	75.60
	(+0.78)	(+0.30)	(+0.09)	(+0.72)	(+0.46)	(+1.24)	(+0.59)	(+0.60)	(+0.26)	(+0.74)	(+0.77)
KC-PKT	75.01	74.12	72.08	74.45	72.82	67.99	67.92	73.32	74.60	75.79	75.78
	(+0.47)	(+0.67)	(+1.74)	(+0.81)	(−0.06)	(+0.86)	(+1.40)	(+0.31)	(+0.50)	(+1.10)	(+1.89)
KC-CRD	75.93	74.60	**72.11**	75.78	74.38	69.90	69.82	74.49	75.74	76.44	76.40
	(+0.29)	(+0.22)	(+0.48)	(+0.32)	(+0.09)	(-0.04)	(+0.28)	(-0.09)	(+0.62)	(+0.39)	(+0.13)
KC-SSKD	**76.24**	**75.35**	71.31	**76.48**	<u>74.93</u>	**71.32**	**71.29**	<u>75.65</u>	**78.28**	**78.59**	**77.61**
	(+0.58)	(+0.08)	(+0.35)	(+0.68)	(−0.21)	(+0.40)	(+0.15)	(−0.07)	(+0.37)	(+0.22)	(+0.69)

Table 2. Test Acc. (%) with computation cost C on CIFAR100, compared with the only existing method UNIX [35] that focuses on distillation efficiency.

Teacher	WRN-40-2	WRN-40-2	resnet56	VGG13	VGG13	ResNet50
Student	WRN-16-2	WRN-40-1	resnet20	VGG8	MobileNetV2	VGG8
KD	74.92 (100%)	73.54 (100%)	70.66 (100%)	72.98 (100%)	67.37 (100%)	73.81 (100%)
UNIX-KD	75.19 (75.3%)	73.51 (73.1%)	70.06 (76.0%)	73.18 (76.4%)	68.47 (77.5%)	73.62 (68.9%)
UNIX-KD†	75.25 (81.6%)	**74.18** (81.6%))	70.19 (81.6%)	73.27 (81.6%)	68.58 (81.6%)	74.24 (81.6%)
KC-KD	**75.70** (81.6%)	73.84 (81.6%)	**70.75** (81.6%)	**73.44** (81.6%)	**68.61** (81.6%)	**74.41** (81.6%)

SSKD [34]. We directly cite the quantitative results reported in their papers [3,4,22,30]. For the network of teacher and student models, we use Wide residual networks [38] (abbreviated as WRNd-w), MobileNetV2 [11] (MN2), ShuffleNetV1 [39]/ShuffleNetV2 [19] (SN1/SN2), and VGG13/VGG8 [28] (V13/V8). R110, R56 and R20 denote CIFAR-style residual networks, while R50 denotes an ImageNet-style ResNet50. **Teacher** and **Student** stand for the performance of teacher and student models when they are trained individually.

The experimental results on 11 teacher-student pairs are depicted in Table 1. We can see that constructing the proposed knowledge condensation (KC) on vanilla KD shows an impressive improvement. In addition, our KCD on top of various modern KD approaches all demonstrates an obvious accuracy gain. More

Table 3. Top-1/-5 error (%) on ImageNet, from ResNet34 to ResNet18. Equipping our method reduces computation cost C:100%→81.61%, C_a:114M→81M.

	Tea.	Stu.	AT [13]	SP [32]	OnlineKD [42]	SSKD [34]	KD [10]	KC-KD (Ours)	CRD [30]	KC-CRD (Ours)	ReKD [4]	KC-ReKD (Ours)
Top-1	26.69	30.25	29.30	29.38	29.45		28.38	29.34	28.61	28.83	28.46	28.39	**27.87**
Top-5	8.58	10.93	10.00	10.20	10.41		9.33	10.12	9.62	9.87	9.53	9.49	**9.08**

Table 4. Ablation study of the proposed KCD on four KD processes (%).

Teacher	WRN-40-2	VGG13	VGG13	resnet32×4
Student	WRN-16-2	VGG8	MobileNetV2	ShuffleNetV2
OGVE w/ random	74.54	73.01	67.76	74.92
OGVE w/o OVR	75.01	73.26	67.73	74.56
OGVE w/o CAR	75.27	73.04	67.79	74.88
OGVE -Full	75.48	73.08	68.23	75.16
OGVE + VAKS w/ KA ($\epsilon = \epsilon_m$)	75.57	73.20	68.34	75.14
OGVE + VAKS (Full)	**75.70**	**73.44**	**68.61**	**75.19**

importantly, the proposed KCD utilizes only the condensed knowledge, which enjoys the merits of both accuracy and efficiency.

We also compare KCD with the only existing work that focuses on the computation cost of KD, namely UNIX [35]. Table 2 displays the results on accuracy and computation cost C^1 (see Eq. (16)). It is noteworthy that the computation C of our KCD is unrelated to the network, thus keeps fixed across different teacher-student pairs. In contrast, C in UNIX [35] is dependent on the rate among the number of sample pass in teacher forward, student forward, and student backward, thus it presents diverse values across different pairs. KD denotes the vanilla baseline, with C set as 100%. UNIX denotes citing the accuracy results of the models reported in the original works which have the most similar C with our KCD. UNIX† denotes using their public code[2] to run and evaluate their methods with the same cost setting C with our method. It appears that the accuracy of the proposed KCD outperforms UNIX at the same level of computation cost.

Results on ImageNet. Following common practice [30,34], the experiments on ImageNet are conducted using ResNet34 (teacher) and ResNet18 (student). Table 3 displays the results of both Top-1 and Top-5 error. We can see that building the proposed knowledge condensation upon KD, CRD and ReviewKD (abbreviated as ReKD) all reduces the testing error significantly. Moreover, the proposed KC leads to the reduction of relative computation C to 81.61%, and the absolute computation C_a from 114M to 81M, which reveals an obvious gain of training efficiency in the large-scale benchmark.

---

[1] Calculation process of C of our method is detailed in supplementary materials.
[2] https://github.com/xuguodong03/UNIXKD.

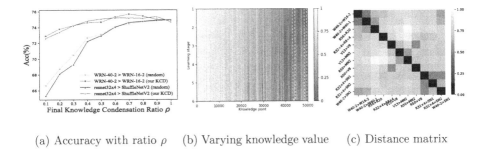

(a) Accuracy with ratio ρ (b) Varying knowledge value   (c) Distance matrix

Fig. 3. (a) Accuracy with the variation of condensation ratio ρ. (b) Pattern of varying knowledge value across the training process. (c) Hamming Distance matrix of the value label across KD processes.

4.2   Further Empirical Analysis

Ablation Study. We verify the effect of each component in the proposed framework by conducting ablation studies. The results are provided in Table 4. **(1)** OGVE w/ random denotes we randomly allocate ranking position $\mathcal{R}_V$ as well as value label Y instead of using OGVE. **(2)** OGVE w/o OVR, w/o CAR, -full denotes that we remove online value recording (*i.e.*, estimate the value during the mini-batch training), cost-aware ranking (*i.e.*, discard the weight $(F(x))^\alpha$ in Eq. (13), and keep the full setting of OGVE. Note that the above variants of OGVE are combined with a direct selection on label $y = 1$. **(3)** OGVE + VAKS w/ KA ($\epsilon = \epsilon_m$) denotes using non-adaptive knowledge augmentation with ϵ keeping its maximum $\epsilon_m = 0.3$ for all points. OGVE + VAKS denotes the full structure of our proposed KCD. When any component is removed, it appears that the performance drops accordingly, revealing the effectiveness of our design.

Influence of Knowledge Condensation Ratio ρ. Figure 3(a) displays the performance of different models (*i.e.*, random selection baseline and our KCD) on different KD processes (*i.e.*, W40-2 > W16-2 and R32×4 > SN2), with the variation of final knowledge condensation ratio ρ. We can see that our KCD outperforms the random baseline by a significant margin, especially as ρ decreasing. It is noteworthy that the proposed KCD achieves better results on ρ ranging between 0.6–0.8 than full-knowledge setting with $\rho = 1$, and nearly maintains the accuracy among a wide range ρ as 0.3–1.0, which implies that our KCD can identify and summarize the compact yet effective knowledge encoding that is robust to the size reduction of knowledge set.

Pattern of Knowledge Value. Figure 3(b) displays the varying ranking-based probability $w.r.t$ knowledge value during the entire training process. We can see that the value of knowledge points differs to the student at different learning stages. As marked in red, some knowledge points are valueless at the beginning stage while become more and more critical in the later stages. Figure 3(c) depicts

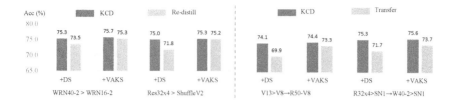

Fig. 4. The performance of reusing the condensed knowledge. **Left:** We utilize the condensed knowledge to directly re-train the student model in the original KD process from scratch. **Right:** We transfer the condensed knowledge encoding to facilitate another KD process.

the Hamming distance matrix about the estimated value label Y in the final stage across various KD processes, wherein the distance indicates the number of different elements in two masks. We can see that the value label represents a relatively strong correlation (small distance) when two KD processes have the same student architectures (*e.g.*, V13 > V8 and R50 > V8) or similar ones (*e.g.*, R32×4 > SN1 and R32×4 > SN2), revealing that the identified knowledge value really encodes some "patterns" of the student model.

Reuse of Condensed Knowledge. The observed similarity of knowledge value across KD processes inspires us to investigate on reusing the condensed knowledge for efficient training. As shown in Fig. 4(**Left**), we first use the ready-made condensed knowledge to re-distill the student from a scratch. "+DS" (direct selection) and "+VAKS" (value-adaptive knowledge summary) denote two variants of our KCD. It appears that compared with our standard KCD, the performance of the re-distilled student drops significantly when equipped with "+DS" while achieves comparable results with "+VAKS". As shown in Fig. 4(**Right**), we further evaluate the transferability of the knowledge condensation, where we transfer the knowledge encoding condensed in a source KD process to a target KD process to improve the efficiency. As can be seen, the performance of transferring the condensed knowledge degrades dramatically compared with the standard KCD. In comparison, when equipping the transfer process with our VASK module, the performance gap with standard KCD is reduced a lot. These observations demonstrate the potential of our KCD method for promoting the efficient training by reusing and transferring the condensed knowledge.

5   Conclusion

This paper proposes Knowledge Condensation Distillation (KCD) to address the knowledge redundancy during KD. Instead of relying on the whole knowledge set from teacher model, the key idea is to first identify the informative knowledge components and then summarize a compact knowledge encoding to perform KD efficiently. Specially, we forge an iterative optimization framework between condensing the compact knowledge encoding and compressing the student model based on the EM algorithm. We further present two collaborative modules to perform the proposed KCD, as online global value estimation (OGVE) and value-adaptive knowledge summary (VAKS). Extensive experiments demonstrate the effectiveness of the proposed KCD against the state-of-the-arts.

Acknowledgement. The study is supported partly by the National Natural Science Foundation of China under Grants 82172033, U19B2031, 61971369, 52105126, China, in part of Science and Technology Key Project of Fujian Province (No. 2019HZ020009).

References

1. Ahn, S., Hu, S.X., Damianou, A., Lawrence, N.D., Dai, Z.: Variational information distillation for knowledge transfer. In: Proceedings of the IEEE Conference on Computer Vision and Pattern Recognition (CVPR), pp. 9163–9171 (2019)
2. Cai, H., Zhu, L., Han, S.: ProxylessNAS: direct neural architecture search on target task and hardware. In: Proceedings of the International Conference of Learning Representation (ICLR) (2019)
3. Chen, L., Wang, D., Gan, Z., Liu, J., Henao, R., Carin, L.: Wasserstein contrastive representation distillation. In: Proceedings of the IEEE Conference on Computer Vision and Pattern Recognition (CVPR), pp. 16296–16305 (2021)
4. Chen, P., Liu, S., Zhao, H., Jia, J.: Distilling knowledge via knowledge review. In: Proceedings of the IEEE Conference on Computer Vision and Pattern Recognition (CVPR), pp. 5008–5017 (2021)
5. Deng, J., Dong, W., Socher, R., Li, L.J., Li, K., Fei-Fei, L.: ImageNet: a large-scale hierarchical image database. In: Proceedings of the IEEE Conference on Computer Vision and Pattern Recognition (CVPR), pp. 248–255 (2009)
6. Goodfellow, I.J., Shlens, J., Szegedy, C.: Explaining and harnessing adversarial examples. arXiv preprint arXiv:1412.6572 (2014)
7. Han, B., et al.: Co-teaching: robust training of deep neural networks with extremely noisy labels. In: Proceedings of the Advances in Neural Information Processing Systems (NeurIPS), vol. 31 (2018)
8. Har-Peled, S., Kushal, A.: Smaller coresets for k-median and k-means clustering. Discrete Comput. Geomet. **37**(1), 3–19 (2007)
9. Heo, B., Kim, J., Yun, S., Park, H., Kwak, N., Choi, J.Y.: A comprehensive overhaul of feature distillation. In: Proceedings of the IEEE Conference on Computer Vision and Pattern Recognition (CVPR), pp. 1921–1930 (2019)
10. Hinton, G., Vinyals, O., Dean, J.: Distilling the knowledge in a neural network. arXiv preprint arXiv:1503.02531 (2015)
11. Howard, A.G., et al.: MobileNets: efficient convolutional neural networks for mobile vision applications. arXiv preprint arXiv:1704.04861 (2017)
12. Katharopoulos, A., Fleuret, F.: Not all samples are created equal: deep learning with importance sampling. In: Proceedings of the International Conference on Machine Learning (ICML), pp. 2525–2534 (2018)
13. Komodakis, N., Zagoruyko, S.: Paying more attention to attention: improving the performance of convolutional neural networks via attention transfer. In: Proceedings of the International Conference of Learning Representation (ICLR) (2017)
14. Krizhevsky, A., Hinton, G., et al.: Learning multiple layers of features from tiny images (2009)

15. Li, J., Zhou, P., Xiong, C., Hoi, S.C.: Prototypical contrastive learning of unsupervised representations. In: Proceedings of the International Conference of Learning Representation (ICLR)
16. Li, S., Lin, M., Wang, Y., Fei, C., Shao, L., Ji, R.: Learning efficient gans for image translation via differentiable masks and co-attention distillation. IEEE Trans. Multimed. (TMM) (2022)
17. Li, S., Lin, M., Wang, Y., Wu, Y., Tian, Y., Shao, L., Ji, R.: Distilling a powerful student model via online knowledge distillation. IEEE Trans. Neural Netw. Learn. Syst. (TNNLS) (2022)
18. Lin, M., et al.: HRank: filter pruning using high-rank feature map. In: Proceedings of the IEEE Conference on Computer Vision and Pattern Recognition (CVPR), pp. 1529–1538 (2020)
19. Ma, N., Zhang, X., Zheng, H.T., Sun, J.: ShuffleNet V2: practical guidelines for efficient CNN architecture design. In: Proceedings of the European Conference on Computer Vision (ECCV), pp. 116–131 (2018)
20. Mirzasoleiman, B., Bilmes, J., Leskovec, J.: Coresets for data-efficient training of machine learning models. In: Proceedings of the International Conference on Machine Learning (ICML), pp. 6950–6960 (2020)
21. Müller, R., Kornblith, S., Hinton, G.E.: When does label smoothing help? In: Proceedings of the Advances in Neural Information Processing Systems (NeurIPS) (2019)
22. Olvera-López, J.A., Carrasco-Ochoa, J.A., Martínez-Trinidad, J.F., Kittler, J.: A review of instance selection methods. Artif. Intell. Rev. **34**(2), 133–143 (2010)
23. Park, W., Kim, D., Lu, Y., Cho, M.: Relational knowledge distillation. In: Proceedings of the IEEE Conference on Computer Vision and Pattern Recognition (CVPR), pp. 3967–3976 (2019)
24. Passalis, N., Tefas, A.: Learning deep representations with probabilistic knowledge transfer. In: Proceedings of the European Conference on Computer Vision (ECCV), pp. 268–284 (2018)
25. Romero, A., Ballas, N., Kahou, S.E., Chassang, A., Gatta, C., Bengio, Y.: FitNets: hints for thin deep nets. arXiv preprint arXiv:1412.6550 (2014)
26. Sener, O., Savarese, S.: Active learning for convolutional neural networks: a core-set approach. arXiv preprint arXiv:1708.00489 (2017)
27. Shen, Z., Liu, Z., Xu, D., Chen, Z., Cheng, K.T., Savvides, M.: Is label smoothing truly incompatible with knowledge distillation: an empirical study. In: Proceedings of the International Conference of Learning Representation (ICLR) (2020)
28. Simonyan, K., Zisserman, A.: Very deep convolutional networks for large-scale image recognition. arXiv preprint arXiv:1409.1556 (2014)
29. Szegedy, C., et al.: Intriguing properties of neural networks. arXiv preprint arXiv:1312.6199 (2013)
30. Tian, Y., Krishnan, D., Isola, P.: Contrastive representation distillation. In: Proceedings of the International Conference of Learning Representation (ICLR) (2019)
31. Toneva, M., Sordoni, A., des Combes, R.T., Trischler, A., Bengio, Y., Gordon, G.J.: An empirical study of example forgetting during deep neural network learning. In: Proceedings of the International Conference of Learning Representation (ICLR) (2018)
32. Tung, F., Mori, G.: Similarity-preserving knowledge distillation. In: Proceedings of the IEEE International Conference on Computer Vision (ICCV), pp. 1365–1374 (2019)
33. Wang, T., Zhu, J.Y., Torralba, A., Efros, A.A.: Dataset distillation. arXiv preprint arXiv:1811.10959 (2018)

34. Xu, G., Liu, Z., Li, X., Loy, C.C.: Knowledge distillation meets self-supervision. In: Vedaldi, A., Bischof, H., Brox, T., Frahm, J.-M. (eds.) ECCV 2020. LNCS, vol. 12354, pp. 588–604. Springer, Cham (2020). https://doi.org/10.1007/978-3-030-58545-7_34

35. Xu, G., Liu, Z., Loy, C.C.: Computation-efficient knowledge distillation via uncertainty-aware mixup. arXiv preprint arXiv:2012.09413 (2020)

36. Yamamoto, K.: Learnable companding quantization for accurate low-bit neural networks. In: Proceedings of the IEEE Conference on Computer Vision and Pattern Recognition (CVPR), pp. 5029–5038 (2021)

37. Yin, H., et al.: Dreaming to distill: data-free knowledge transfer via deepinversion. In: Proceedings of the IEEE Conference on Computer Vision and Pattern Recognition (CVPR) (2020)

38. Zagoruyko, S., Komodakis, N.: Wide residual networks. arXiv preprint arXiv:1605.07146 (2016)

39. Zhang, X., Zhou, X., Lin, M., Sun, J.: ShuffleNet: an extremely efficient convolutional neural network for mobile devices. In: Proceedings of the IEEE Conference on Computer Vision and Pattern Recognition (CVPR), pp. 6848–6856 (2018)

40. Zhang, Z., Chen, X., Chen, T., Wang, Z.: Efficient lottery ticket finding: less data is more. In: Proceedings of the International Conference on Machine Learning (ICML), pp. 12380–12390 (2021)

41. Zhao, B., Mopuri, K.R., Bilen, H.: Dataset condensation with gradient matching. In: Proceedings of the International Conference of Learning Representation (ICLR) (2020)

42. Zhu, X., Gong, S., et al.: Knowledge distillation by on-the-fly native ensemble. In: Proceedings of the Advances in Neural Information Processing Systems (NeurIPS) (2018)

Reducing Information Loss for Spiking Neural Networks

Yufei Guo[1], Yuanpei Chen[1], Liwen Zhang[1], YingLei Wang[1], Xiaode Liu[1],
Xinyi Tong[1], Yuanyuan Ou[2], Xuhui Huang[1], and Zhe Ma[1(✉)]

[1] Intelligent Science and Technology Academy of CASIC, Beijing 100854, China
yfguo@pku.edu.cn, mazhe_thu@163.com
[2] Chongqing University, Chongqing 400044, China

Abstract. The Spiking Neural Network (SNN) has attracted more and more attention recently. It adopts binary spike signals to transmit information. Benefitting from the information passing paradigm of SNNs, the multiplications of activations and weights can be replaced by additions, which are more energy-efficient. However, its "Hard Reset" mechanism for the firing activity would ignore the difference among membrane potentials when the membrane potential is above the firing threshold, causing information loss. Meanwhile, quantifying the membrane potential to 0/1 spikes at the firing instants will inevitably introduce the quantization error thus bringing about information loss too. To address these problems, we propose to use the "Soft Reset" mechanism for the supervised training-based SNNs, which will drive the membrane potential to a dynamic reset potential according to its magnitude, and Membrane Potential Rectifier (MPR) to reduce the quantization error via redistributing the membrane potential to a range close to the spikes. Results show that the SNNs with the "Soft Reset" mechanism and MPR outperform their vanilla counterparts on both static and dynamic datasets.

Keywords: Spiking neural network · Information loss · Soft reset · Quantization error · Membrane potential rectificater

1 Introduction

Deep Neural Networks (DNNs) have greatly improved many applications in computational vision, *e.g.*, object detection and recognition [16], object segmentation [40], object tracking [2], etc. In pursuit of models with better performance, more and more complex networks are proposed. However, the increasing complexity poses a new challenge to model deployment on power-constrained devices, thus becoming an impediment to the applications of these advanced complex models. There have been several approaches to address this problem, such as quantization [12,27,28], pruning [17], knowledge distillation [37], spiking neural networks (SNNs) [11,13,26,29,43], and so on. Among these approaches, the

---

Y. Guo and Y. Chen—Equal contribution.

S. Avidan et al. (Eds.): ECCV 2022, LNCS 13671, pp. 36–52, 2022.
https://doi.org/10.1007/978-3-031-20083-0_3

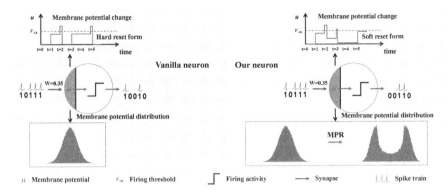

Fig. 1. The difference of our "Soft Reset"-based neuron and vanilla "Hard Reset"-based neuron. The membrane potential will be redistributed to reduce the quantization error in our neuron with MPR while not in the vanilla neuron.

biology-inspired method, SNNs provide a unique way to reduce energy consumption by mimicking the spiking nature of brain neurons. A spiking neuron integrates the inputs over time and fires a spike output whenever the membrane potential exceeds the firing threshold. And using 0/1 spike to transmit information makes SNNs enjoy the advantage of multiplication-free inference by converting multiplication to additions. Furthermore, SNNs are energy-efficient on neuromorphic hardwares, such as SpiNNaker [18], TrueNorth [1], Darwin [32], Tianjic [36], and Loihi [5].

Despite the attractive benefits, there is still a huge performance gap between existing SNN models and their DNN counterparts. We argue that the reason for the low accuracy is there exists information loss in SNNs. First, the information processing of neurons in supervised training-based SNNs are generally following the rules of the Integrate-and-Fire (IF) model or Leaky IF (LIF) model, where once a membrane potential exceeds the firing threshold, a "Hard Reset" operation will force the "residual" potential to be set to 0, *i.e.*, once fired, all the information will be taken away. Obviously, this mechanism of "residual" membrane potential-ignored reset mode would fail to preserve the diversity of various membrane potentials. Hence the information encoding capacity of the network is compromised, such that the risk of information loss increases accordingly. Second, although the 0/1 spike information processing paradigm enables SNNs to enjoy the advantage of high efficiency, quantifying the real-valued membrane potential to 0/1 spikes will inevitably introduce the quantization error, which also brings about information loss.

To address the information loss problem, we propose a "Soft Reset"-based IF (SRIF) neuron model that retains the "residual" membrane potential from subtracting its spike value at the firing instants. Hence the diversity of the membrane potentials that exceed the firing threshold will be preserved. Though "Soft Reset" is commonly used in converting methods from ANN to SNN (ANN2SNN) [14,15,20,26] methods, rarely applied in supervised SNNs [23], and has not been discussed in SNN enhancement from the perspective of information loss reducing. In addition, for alleviating quantization error, the Membrane Potential Rectifier (MPR) is proposed, which is performed before the firing activity to adjust the membrane potentials towards the spike values (*i.e.*, 0/1). With MPR, the membrane potential will be decoupled as an original one and a modulated one. The original one can keep the mechanism of a neuron and the modulated one enjoys less quantization error than the original one without suffering from any negative effects. The difference between our neuron and the vanilla neuron is illustrated in Fig. 1. Our main contributions are as follows:

- We propose using the SRIF model for supervised training-based SNNs. By retaining the "residual" membrane potential, SRIF enables the networks to distinguish the differences among those membrane potentials that exceed the firing threshold via subtracting their spike values thus enhancing the information encoding capacity of supervised training-based SNNs.
- We present MPR to mitigate the quantization error. By utilizing a non-linear function to modulate the membrane potential close to 0/1 before firing activity triggers, the gap between the potential and its corresponding 0/1 spike value is minified while maintaining the sparse spike activation mechanism of SNNs. To our best knowledge, few works have noticed the quantization error in SNNs, and a simple but effective method for addressing this problem is presented.
- Extensive experiments on both static and dynamic datasets were conducted to verify our method. Results show that the SNN trained with the proposed method is highly effective and efficient compared with other state-of-the-art SNN models, *e.g.*, 96.49% top-1 accuracy and 79.41% top-1 accuracy are achieved on the CIFAR-10 and CIFAR-100. These results of our models even outperform their DNN counterparts surprisingly, and it is very rare that SNNs may have a chance to surpass their DNN counterparts.

2   Related Work

2.1   Learning Methods of Spiking Neural Networks

The training methods of SNNs can be divided into two categories. The first one is ANN2SNN [14,15,20,26]. ANN2SNN yields the same input-output mapping for the ANN-SNN pair via approximating the continuous activation values of an ANN using ReLU by averaging the firing rate of an SNN under the rate-coding scheme. Since the ANN has achieved great success in many fields, ANN2SNN can maintain the smallest gap with ANNs in terms of performance and can be generalized to large-scale structures. However, being restricted to rate-coding, ANN2SNN usually requires dozens or even hundreds of timesteps to obtain well-performed networks. Lots of efforts have been done to reduce the long inference time, such as weight normalization [9], threshold rescaling [41], soft reset [15], threshold shift [26], and the quantization clip-floor-shift activation function [3], it is still hard to obtain high-performance SNNs with ultra-low latency.

The second one is supervised learning-based SNNs. SNNs quantize the real-valued membrane potentials into 0/1 spikes via the firing activity. Since the gradient of the firing activity function is zero almost everywhere, the gradient descent-based optimizer can not be directly used for the training of SNNs. To alleviate the optimization difficulty, the approximate gradient-based strategy is commonly used, and some related approaches had been proposed to achieve trainable SNNs with high performance. For example, by regarding the SNN as a special RNN, a training method of back-propagation through time with different kinds of surrogate gradient was proposed [33]. The spatio-temporal back-propagation (STBP) [42] method enables SNNs to be trained on the ANN programming platform, which also significantly promotes the direct training research of SNNs. Differentiable spike which can match the finite difference gradient of SNNs well was proposed in [29]. The temporal efficient training (TET) [7] method with a novel loss and a gradient descent regime that succeeds in obtaining more generalized SNNs, has also attracted much attention. In RecDis-SNN [13], a new perspective to understand the difficulty of training SNNs by analyzing undesired membrane potential shifts is presented and the MPD-Loss to penalize the undesired shifts is proposed. Numerous works verify that supervised learning can greatly reduce the number of timesteps and handle dynamic datasets. It has increasingly aroused researchers' interest in recent years. In this work, we focus on improving the performance of the supervised learning-based SNNs by repressing information loss, which is rarely mentioned in other works.

2.2 Threshold-Dependent Batch Normalization

Batch Normalization (BN) is one of the most widely used normalization technologies, which is initially designed for very deep Convolutional Neural Networks (CNNs). As it only focuses on normalizing the spatial feature maps, directly applying BN to SNNs would damage the temporal characteristic of SNNs, which stand with spatio-temporal feature maps, leading to low accuracy. To address this issue, some specially-designed normalization methods for SNNs were proposed recently. Typically, to simultaneously balance neural selectivity and normalize the neuron activity, NeuNorm [42] was proposed. Then, a more effective normalization technique that can take good care of the firing threshold, named threshold-dependent Batch Normalization (tdBN) was further proposed in [45]. It can normalize the feature maps of SNNs in both spatial and temporal domains [45]. Specifically, let $\mathbf{X}_t \in \mathbb{R}^{B \times C \times H \times W}$ represent the input maps at each timestep, where $t = 1, \ldots, T$ (B: batch size; C: channel; (H, W): spatial domain). Then for each channel c, the spatio-temporal sequence $\mathbf{X}^{(c)} = \{\mathbf{X}_1^{(c)}, \cdots, \mathbf{X}_T^{(c)}\}$ is normalized by tdBN as follows,

$$\tilde{\mathbf{X}}^{(c)} = \lambda \cdot \frac{\alpha V_{th}(\mathbf{X}^{(c)} - \bar{x}^{(c)})}{\sqrt{mean((\mathbf{X}^{(c)} - \bar{x}^{(c)})^2) + \epsilon}} + \beta, \tag{1}$$

where V_{th} is the firing threshold, α is a network-structure-dependent hyperparameter, ϵ is a tiny constant, λ and β are two learnable parameters, $\bar{x}^{(c)} =$

$mean(\mathbf{X}^{(c)})$ is the mean value of $\mathbf{X}^{(c)}$, $\tilde{\mathbf{X}}^{(c)}$ is the normalized maps. In this paper, tdBN is also adopted considering its spatio-temporal normalization mechanism.

3   Preliminary and Methodology

To avoid the information loss in supervised training-based SNNs, we propose the "Soft Reset" IF (SRIF) model and Membrance Potential Rectificater (MPR).

3.1   "Soft Reset" if Model

An SNN adopts a biology-inspired spiking neuron that accumulates inputs along the time dimension as its membrane potential and fires a spike when the potential exceeds the firing threshold. This mechanism makes it much different from its DNN counterpart. For better introducing the proposed SRIF neuron, a unified form defined by a recent work [11], is given to describe the dynamics of all kinds of spiking neurons as follows,

$$H[t] = f(U[t-1], X[t]), \tag{2}$$

$$O[t] = \Theta(H[t] - V_{th}), \tag{3}$$

$$U[t] = H[t](1 - O[t]) + V_{reset}O[t], \tag{4}$$

where $X[t]$, $H[t]$, $U[t]$, and $O[t]$ are the input, membrane potentials before and after the trigger of a spike, and output spike at the timestep t, respectively. V_{th} is the firing threshold, and is usually set to 0.5. $\Theta(\cdot)$ is the step function defined by $\Theta(x) = 1$ for $x \geq 0$ and $\Theta(x) = 0$ for $x < 0$. V_{reset} denotes the reset potential, which is set as 0. The function $f(\cdot)$ describes the neuronal dynamics of spiking neuron models, for the commonly used IF neuron and LIF neuron, $f(\cdot)$ can be respectively defined as follows,

$$H[t] = U[t-1] + X[t], \tag{5}$$

$$H[t] = \tau U[t-1] + X[t], \tag{6}$$

where τ denotes the membrane time constant.

Both LIF and IF neurons have some unique advantages, with decay characteristics introduced by the membrane time constant, LIF neuron behaves more biologically compared with IF neuron, while IF neuron is more efficient due to its addition-only processing manner. In terms of accuracy performance, neither of them show an overwhelming advantage, and more detailed experimental results of these two neurons are provided in Sect. 4. Considering the subtle gap in performance, we prefer to use LIF model due to its neurodynamic characteristic, from the perspective of brain science research. Conversely, from the perspective of computer science research, we recommend using IF model, since it is more friendly to hardwares.

However, both the IF model and LIF model might undertake a greater or lesser risk of information loss by the "Hard Reset" mechanism, i.e., when the input membrane potentials exceed the firing threshold, the neurons will force the membrane potentials to a fixed value. Such mechanism ignores the "residual" parts of those fired membrane potentials. These "residual" parts contain the diversity of the input potentials, and we argue that a neuron model which can preserve the diversity or differences of these membrane potentials that cause the firing is more suitable.

To this end, along with the consideration of efficiency, we propose using a "Soft Reset" mechanism-based IF neuron, SRIF, which can keep the diversity of the membrane potentials by subtracting their firing spike values from themselves at the time where the threshold is exceeded. Though this similar "Soft Reset" mechanism has been widely used in ANN2SNN [14,15,20,26], there are few works to use it in supervised learning-based SNNs [23]. We found its value in this field from a new perspective to reduce information loss. In SRIF neuron, Eq. (4) is updated as

$$U[t] = H[t](1 - O[t]) + (H[t] - O[t])O[t]. \tag{7}$$

It can be further simplified as

$$U[t] = H[t] - O[t]. \tag{8}$$

It can be seen that, similar to IF neuron, SRIF is also an addition-only model, thus enjoying computational efficiency when implementing on hardwares. Figure 2 compares the difference between IF neuron and SRIF neuron in an intuitive way. Suppose that both models receive weighted input sequence of $1.5V_{th}$, $1.2V_{th}$, $1.5V_{th}$, $0.9V_{th}$, and $1.4V_{th}$ across 5 consecutive timesteps. Our SRIF neuron will produce three spikes by retaining the residual potentials at the firing instants as depicted in Fig. 2. Whereas, the IF neuron will produce four spikes.

3.2   Membrane Potential Rectificater

To further mitigate the information loss, we present a non-linear function, called MPR by reducing the quantization error. MPR aims to redistribute the membrane potential before it is operated by the step function. It only modulates the membrane potential that is presented to the step function but does not modify the value of membrane potential, which receives and accumulates spikes from other neurons. Specifically, we further distinguish the membrane potentials as the original one, H as in Eq. (2) and the modulated one, $\hat{H}$, which is the membrane potential that will be presented to the step function. In all previous works, H and $\hat{H}$ are treated as the same. While in this paper, we would like to provide a new perspective that using a decoupling function to separate H and $\hat{H}$ can be helpful. Specifically, H manages the original tasks as in other work, $\hat{H}$ derives from H with a non-linear function, $\varphi(\cdot)$, and it will be fed into the step function with a modulated form that can shrink the quantization error. With this decoupling mechanism, a neuron model can not only keep the membrane potential updating rule but also enjoy less quantization error.

Before giving the full details of the MPR, we try to formulate the quantization error first. It is clear that the quantization errors corresponding to different

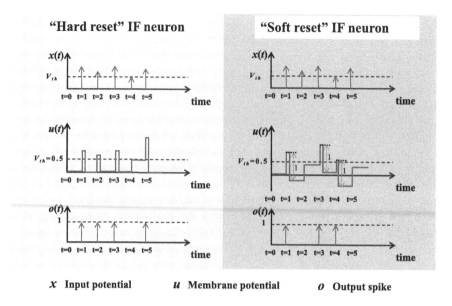

Fig. 2. The difference of "Hard Reset" IF neuron and "Soft Reset" IF (SRIF) neuron.

membrane potentials should be different. Hence, a value closer to its quantization spike, o, enjoys less quantization error. In specific, the firing threshold divides the membrane potentials into two parts, the part with smaller values is assigned to "0" spike, and the other with larger values is assigned to "1" spike. Then the quantization error depends on the margin between the membrane potential and its corresponding spike. Therefore, the quantization error can be defined as the square of the difference between the membrane potential and its corresponding quantization spike value as follows:

$$\mathcal{L}_q = (u - o)^2, \tag{9}$$

where u is the membrane potential and $o \in \{0, 1\}$. when u is below the firing threshold, o is 0, otherwise, 1.

Hence, the design of MPR should obey the following two principles:

- **Spike-approaching**: the modulated membrane potential, $\hat{H}$ should be closer to the 0/1 spikes than the original membrane potential, H. This principle ensures quantization error reduction.
- **Firing-invariance**: for the H less than V_{th}, the MPR should not produce the $\hat{H}$ greater than V_{th} and vice versa. This principle ensures the neuron output be consistent with or without using MPR.

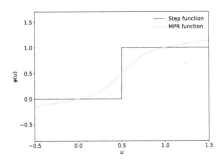

Fig. 3. The MPR function.

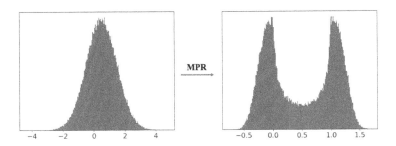

Fig. 4. The effect of the MPR. The original membrane potential distribution (left). The redistributed membrane potential distribution by MPR (right).

Based on the above two principles, we define the MPR as the following symmetrical function:

$$\varphi(u) = \begin{cases} -(1-u)^{1/3} + 1, & u<0, \\ \frac{1}{2tanh(3/2)}tanh(3(u-1/2)) + 1/2, & 0 \le u \le 1, \\ (u)^{1/3}, & u>1. \end{cases} \tag{10}$$

Figure 3 shows the response curve of the designed MPR function following the principles of spike-approaching and firing-invariance.

According to [45], the membrane potential follows a Gaussian distribution, $\mathcal{N}(\mu;\sigma)$. Hence, to visualize the effect of the MPR, we sample 1000,00 values from a Gaussian distribution with $\mathcal{N}(1/2;1)$, and present them to the MPR. Then the distribution of these 1000,00 MPR outputs is drawn in Fig. 4. It can be seen that the unimodal distribution, $\mathcal{N}(1/2;1)$ is adjusted to a bimodal distribution which is with less quantization error since it can naturally gather the membrane potentials near "0" and "1".

Moreover, it is worth noting that, the redistributed membrane potential, $\hat{H}$ by MPR is only used for narrowing the gap between the true membrane potential, H and its quantization spike. It will not replace the original H in our

---

Algorithm 1. Feed-Forward procedures for the "soft reset" IF neuron with MPR.

Input: the input current, X.
Output: the output spike train, O.
Feed-Forward:

1: **for** for all $t = 1, 2, \ldots, T$-th timesteps **do**
2:     Update the membrane potential, $H(t)$ by Eq. (11), which represents the membrane potential accumulating the input current.
3:     Redistribute the membrane potential, $H(t)$ by Eq. (12) and denote the redistributed membrane potential as $\hat{H}[t]$.
4:     Calculate the output spike, $O(t)$ by Eq. (13) using the new membrane potential, $\hat{H}[t]$.
5:     Update the membrane potential, $U(t)$ by Eq. (14), which represents the membrane potential after the trigger of a spike.
6: **end for**

SRIF neuron model. Then the complete new dynamics of the SRIF model can be described as follows,

$$H[t] = U[t-1] + X[t], \tag{11}$$

$$\hat{H}[t] = \varphi(H[t]), \tag{12}$$

$$O[t] = \Theta(\hat{H}[t] - V_{th}), \tag{13}$$

$$U[t] = H[t] - O[t]. \tag{14}$$

The detailed Feed-Forward procedure for the SRIF neuron with MPR is given in Algorithm 1.

4   Experiment

The proposed methods were evaluated on various static datasets (CIFAR-10 [21], CIFAR-100 [21], ImageNet [6]) and one neuromorphic dataset (CIFAR10-DVS [25]) with widely-used spiking archetectures including ResNet20 [38,41], VGG16 [38], ResNet18 [10], ResNet19 [45], and ResNet34 [10].

4.1   Datasets and Settings

Datasets. The CIFAR-10(100) dataset consists of 60,000 images in 10(100) classes with 32×32 pixels. The number of the training images is 50,000, and that of the test images is 10,000. The CIFAR10-DVS dataset is the neuromorphic version of the CIFAR-10 dataset. It is composed of 10,000 images in 10 classes, with 1000 images per class. ImageNet dataset has more than 1,250,000 training images and 50,000 test images.

Preprocessing. Data normalization is applied on all static datasets to ensure that input images have 0 mean and 1 variance. Besides, the random horizontal flipping and cropping on these datasets were conducted to avoid overfitting.

For CIFAR-10, the AutoAugment [4] and Cutout [8] were used for data augmentation. For the neuromorphic dataset, since the CIFAR10-DVS dataset does not separate data into training and testing sets, we split the dataset into 9000 training images and 1000 test images similar to [43]. For data preprocessing and augmentation, we resized the training image frames to 48×48 as in [45] and adopted random horizontal flip and random roll within 5 pixels. And the test images are just resized to 48×48 without any additional processing.

Training Setup. For all the datasets, the firing threshold V_{th} was set as 0.5 and V_{reset} as 0. For static image datasets, the images were encoded to binary spike using the first layer of the SNN, as in recent works [10,11,38]. This is similar to rate-coding. For the neuromorphic image dataset, we used the 0/1 spike format directly. The neuron models in the output layer accumulated the incoming inputs without generating any spike as the output like in [38]. For CIFAR-10(100) and CIFAR10-DVS datasets, the SGD optimizer with the momentum of 0.9 and learning rate of 0.01 with cosine decayed [30] to 0. All models were trained within 400 epochs with the same batch size of 128. For the ImageNet dataset, the SGD optimizer with a momentum set as 0.9 and a learning rate of 0.1 with cosine decayed [30] to 0. All models are trained within 320 epochs as in [10]. The batch size is set to 64.

4.2 Ablation Study for Different Neuron Models

We first conducted a set of ablation experiments to verify the effectiveness of the proposed SRIF model on CIFAR-10(100) using ResNet20 as the backbone under various timesteps without MPR. The results are shown in Table 1.

It can be seen that whether on CIFAR-10 or CIFAR-100, the SRIF neuron always obtains the best result ranging from 2 timesteps to 8 timesteps. This indicates the superiority of the SRIF neuron. On the other hand, the LIF neuron performs better than the "Hard Reset" IF neuron on CIFAR-10, while the IF neuron performs better on CIFAR-100, even though the LIF neuron is more like a biological neuron. This comparison also shows that, although SNNs are proposed to imitate the biological neural networks, for the implementation of large-scale networks, they still need to rely on computer hardwares. Hence, the characteristics of computational science should also be considered. In this respect, the SRIF neuron is more suitable for its advantage of low power consumption and capacity of reducing information loss.

4.3 Addition of MPR

Then, a set of ablation experiments for the MPR were conducted on CIFAR-10(100) using ResNet20 and ResNet19 as backbones within 4 timesteps. Results in Table 2 show that the MPR can greatly improve performance. Especially on CIFAR-100, where ResNet20 with MPR increases the accuracy by 2.73%. These results verify the effectiveness of MPR in terms of performance improvement.

Table 1. Ablation study for different neuron models without MPR.

Dataset	Neuron model	Timestep	Accuracy
CIFAR-10	"Hard Reset" LIF	2	90.36%
	"Hard Reset" IF	2	90.07%
	"Soft Reset" IF (SRIF)	2	**90.38%**
	"Hard Reset" LIF	4	92.22%
	"Hard Reset" IF	4	92.04%
	"Soft Reset" IF (SRIF)	4	**92.46%**
	"Hard Reset" LIF	6	92.66%
	"Hard Reset" IF	6	92.26%
	"Soft Reset" IF (SRIF)	6	**93.40%**
	"Hard Reset" LIF	8	92.90%
	"Hard Reset" IF	8	92.86%
	"Soft Reset" IF (SRIF)	8	**94.09%**
CIFAR-100	"Hard Reset" LIF	2	62.67%
	"Hard Reset" IF	2	63.43%
	"Soft Reset" IF (SRIF)	2	**63.85%**
	"Hard Reset" LIF	4	66.00%
	"Hard Reset" IF	4	66.95%
	"Soft Reset" IF (SRIF)	4	**67.90%**
	"Hard Reset" LIF	6	67.44%
	"Hard Reset" IF	6	68.31%
	"Soft Reset" IF (SRIF)	6	**69.59%**
	"Hard Reset" LIF	8	67.85%
	"Hard Reset" IF	8	69.14%
	"Soft Reset" IF (SRIF)	8	**69.90%**

We also computed the average quantization error of the first layer of the second block in the ResNet20/19 before and after MPR on the test set of CIFAR-10(100), respectively. Results in Table 3 show that the quantization error is obviously reduced by the MPR. The overall original membrane potential distribution and modulated membrane potential distribution by MPR of the first layer of the second block in ResNet20 on CIFAR-10 and CIFAR-100 test sets are shown in Fig. 5. It shows that the MPR adjusts the membrane potential distribution near "0" and "1", which is closer to its quantization spike. Put together, these results quantitatively support the effectiveness of MPR in reducing quantization error.

4.4 Comparisons with Other Methods

Our method was further compared with other state-of-the-art SNNs on static and neuromorphic datasets. Results are shown in Table 4, where for each run,

Table 2. Ablation study for MPR.

Dataset	Architecture	Method	Timestep	Accuracy
CIFAR-10	ResNet20	SRIF w/o MPR	4	92.46%
		SRIF w/ MPR	4	**92.94%**
	ResNet19	SRIF w/o MPR	4	95.44%
		SRIF w/ MPR	4	**96.27%**
CIFAR-100	ResNet20	SRIF w/o MPR	4	67.90%
		SRIF w/ MPR	4	**70.63%**
	ResNet19	SRIF w/o MPR	4	77.85%
		SRIF w/ MPR	4	**78.42%**

Table 3. Quantization error.

Dataset	Architecture	Method	Timestep	Avg. error
CIFAR-10	ResNet20	Before MPR	4	0.28
		After MPR	4	**0.04**
	ResNet19	Before MPR	4	0.20
		After MPR	4	**0.03**
CIFAR-100	ResNet20	Before MPR	4	0.38
		After MPR	4	**0.05**
	ResNet19	Before MPR	4	0.32
		After MPR	4	**0.04**

the mean accuracy and standard deviation of 3 trials are listed. For simplification, **InfLoR** (*i.e.*, short for **Inf**ormation **Lo**ss **R**educing) is used to denote the combination of SRIF and MPR.

CIFAR-10(100). For CIFAR-10, our method improves network performance across all commonly used backbones in SNNs. ResNet19-based InfLoR-SNN achieved 96.49% top-1 accuracy with 6 timesteps, which outperforms its STBP-tdBN counterpart with 3.33% higher accuracy and its ANN counterpart 0.20% higher accuracy even. The ResNet20-based InfLoR-SNN can reach to 93.65%, while only 92.54% in [38]. And our VGG16-based network also shows higher accuracy than other methods with fewer timesteps. On CIFAR-100, InfLoR-SNN also performs better and achieves a 1.89% increment on VGG16. Noteworthy, InfLoR-SNN significantly surpasses Diet-SNN [38] with 7.12% higher accuracy, which is not easy to achieve in the SNN field. Again, our ResNet19 also outperforms its ANN counterpart. To our best knowledge, it is the first time that the SNN can outperform its ANN counterpart.

ImageNet. For the ImageNet dataset, ResNet18 and ResNet34 were used as the backbones. Results show that our ResNet18 achieves a 1.60% increment on SEW ResNet18 and a 2.46% increment on Spiking ResNet18. The accuracy of

Table 4. Comparison with SoTA methods.* denotes self-implementation results.

Dataset	Method	Type	Architecture	Timestep	Accuracy
CIFAR-10	SpikeNorm [41]	ANN2SNN	VGG16	2500	91.55%
	Hybrid-Train [39]	Hybrid	VGG16	200	92.02%
	Spike-basedBP [24]	SNN training	ResNet11	100	90.95%
	STBP [43]	SNN training	CIFARNet	12	90.53%
	TSSL-BP [44]	SNN training	CIFARNet	5	91.41%
	PLIF [11]	SNN training	PLIFNet	8	93.50%
	Diet-SNN [38]	SNN training	VGG16	5	92.70%
				10	93.44%
			ResNet20	5	91.78%
				10	92.54%
	STBP-tdBN [45]	SNN training	ResNet19	2	92.34%
				4	92.92%
				6	93.16%
	ANN*	ANN	ResNet19	1	96.29%
	InfLoR-SNN	SNN training	ResNet19	2	**94.44%** ± 0.08
				4	**96.27%** ± 0.07
				6	**96.49%** ± 0.08
			ResNet20	5	**93.01%** ± 0.06
				10	**93.65%** ± 0.04
			VGG16	5	**94.06%** ± 0.08
				10	**94.67%** ± 0.07
CIFAR-100	BinarySNN [31]	ANN2SNN	VGG15	62	63.20%
	Hybrid-Train [39]	Hybrid	VGG11	125	67.90%
	T2FSNN [35]	ANN2SNN	VGG16	680	68.80%
	Burst-coding [34]	ANN2SNN	VGG16	3100	68.77%
	Phase-coding [19]	ANN2SNN	VGG16	8950	68.60%
	Diet-SNN [38]	SNN training	ResNet20	5	64.07%
			VGG16	5	69.67%
	ANN*	ANN	ResNet19	1	78.61%
	InfLoR-SNN	SNN training	ResNet20	5	**71.19%** ± 0.09
			VGG16	5	**71.56%** ± 0.10
				10	**73.17%** ± 0.08
			ResNet19	2	**75.56%** ± 0.11
				4	**78.42%** ± 0.09
				6	**79.51%** ± 0.06
ImageNet	Hybrid-Train [39]	Hybrid	ResNet34	250	61.48%
	SpikeNorm [41]	ANN2SNN	ResNet34	2500	69.96%
	STBP-tdBN [45]	SNN training	ResNet34	6	63.72%
	SEW ResNet [10]	SNN training	ResNet18	4	63.18%
			ResNet34	4	67.04%
	Spiking ResNet [10]	SNN training	ResNet18	4	62.32%
			ResNet34	4	61.86%
	InfLoR-SNN	SNN training	ResNet18	4	**64.78%** ± 0.07
			ResNet34	4	65.54% ± 0.08

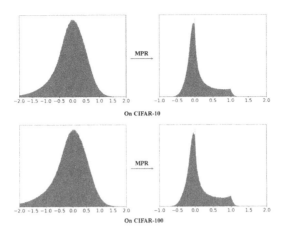

Fig. 5. The effect of MPR. The overall original membrane potential distribution (left) and the redistributed membrane potential distribution by MPR (right) of the first layer of the second block in ResNet20 on CIFAR-10 and CIFAR-100 test sets.

Table 5. Training Spiking Neural Networks on CIFAR10-DVS.

Dataset	Method	Type	Architecture	Timestep	Accuracy
CIFAR10-DVS	Rollout [22]	Rollout	DenseNet	10	66.80%
	STBP-tdBN [45]	SNN training	ResNet19	10	67.80%
	InfLoR	SNN training	ResNet19	10	**75.50% ± 0.12**
			ResNet20	10	**75.10% ± 0.09**

our ResNet34 does not exceed SEW ResNet34. However, SEW ResNet34 [10] transmits information with integers, which is not a typical SNN. For a fair comparison, we also report the result of Spiking ResNet34 in [10] which is worse than our method. Moreover, our InfLoR-based ResNet34 with 4 timesteps still obviously outperforms STBP-tdBN-based RersNet34 with 6 timesteps (Table 5).

CIFAR10-DVS. For the neuromorphic dataset, CIFAR10-DVS, InfLoR-SNN achieves the best performance with 75.50% and 75.10% top-1 accuracy in 10 timesteps with ResNet19 and ResNet18 as backbones, and obtains 7.80% improvement compared with STBP-tdBN for ResNet19. It's worth noting that, as a more complex model, ResNet19 only performs a little better than ResNet20 on CIFAR10-DVS. It might be that this neuromorphic dataset suffers much more noise than static ones, thus a more complex model is easier to overfit.

5   Conclusions

This work aims at addressing the information loss problem caused by the "Hard Reset" mechanism of neurons and the 0/1 spike quantification. Then, the SRIF model, which will drive the membrane potential to a dynamic reset potential, and the MPR that can adjust the membrane potential to a new value closer to quantification spikes than itself are proposed. A detailed analysis of why the SRIF and MPR can reduce the information loss is provided. Furthermore, abundant ablation studies of the proposed methods are given. Combining these two methods, our SNNs outperform other state-of-the-art methods.

References

1. Akopyan, F., et al.: Truenorth: design and tool flow of a 65 mw 1 million neuron programmable neurosynaptic chip. IEEE Trans. Comput. Aided Des. Integr. Circ. Syst. **34**(10), 1537–1557 (2015)
2. Bewley, A., Ge, Z., Ott, L., Ramos, F., Upcroft, B.: Simple online and realtime tracking. In: 2016 IEEE International Conference on Image Processing (ICIP), pp. 3464–3468. IEEE (2016)
3. Bu, T., Fang, W., Ding, J., Dai, P., Yu, Z., Huang, T.: Optimal ANN-SNN conversion for high-accuracy and ultra-low-latency spiking neural networks. In: International Conference on Learning Representations (2021)
4. Cubuk, E.D., Zoph, B., Mane, D., Vasudevan, V., Le, Q.V.: Autoaugment: learning augmentation policies from data. arXiv preprint arXiv:1805.09501 (2018)
5. Davies, M., et al.: Loihi: a neuromorphic manycore processor with on-chip learning. IEEE Micro **38**(1), 82–99 (2018)
6. Deng, J., Dong, W., Socher, R., Li, L.J., Li, K., Fei-Fei, L.: ImageNet: a large-scale hierarchical image database. In: 2009 IEEE Conference on Computer Vision and Pattern Recognition, pp. 248–255. IEEE (2009)
7. Deng, S., Li, Y., Zhang, S., Gu, S.: Temporal efficient training of spiking neural network via gradient re-weighting. arXiv preprint arXiv:2202.11946 (2022)
8. DeVries, T., Taylor, G.W.: Improved regularization of convolutional neural networks with cutout. arXiv preprint arXiv:1708.04552 (2017)
9. Diehl, P.U., Neil, D., Binas, J., Cook, M., Liu, S.C., Pfeiffer, M.: Fast-classifying, high-accuracy spiking deep networks through weight and threshold balancing. In: 2015 International Joint Conference on Neural Networks (IJCNN), pp. 1–8. IEEE (2015)
10. Fang, W., Yu, Z., Chen, Y., Huang, T., Masquelier, T., Tian, Y.: Deep residual learning in spiking neural networks. Adv. Neural. Inf. Process. Syst. **34**, 21056–21069 (2021)
11. Fang, W., Yu, Z., Chen, Y., Masquelier, T., Huang, T., Tian, Y.: Incorporating learnable membrane time constant to enhance learning of spiking neural networks. In: Proceedings of the IEEE/CVF International Conference on Computer Vision, pp. 2661–2671 (2021)
12. Gong, R., et al.: Differentiable soft quantization: bridging full-precision and low-bit neural networks. In: Proceedings of the IEEE/CVF International Conference on Computer Vision, pp. 4852–4861 (2019)
13. Guo, Y., et al.: Recdis-SNN: rectifying membrane potential distribution for directly training spiking neural networks. In: Proceedings of the IEEE/CVF Conference on Computer Vision and Pattern Recognition (CVPR), pp. 326–335 (2022)
14. Han, B., Roy, K.: Deep spiking neural network: energy efficiency through time based coding. In: Vedaldi, A., Bischof, H., Brox, T., Frahm, J.-M. (eds.) ECCV 2020. LNCS, vol. 12355, pp. 388–404. Springer, Cham (2020). https://doi.org/10.1007/978-3-030-58607-2_23

15. Han, B., Srinivasan, G., Roy, K.: RMP-SNN: residual membrane potential neuron for enabling deeper high-accuracy and low-latency spiking neural network. In: Proceedings of the IEEE/CVF Conference on Computer Vision and Pattern Recognition, pp. 13558–13567 (2020)
16. He, K., Zhang, X., Ren, S., Sun, J.: Deep residual learning for image recognition. In: Proceedings of the IEEE Conference on Computer Vision and Pattern Recognition, pp. 770–778 (2016)
17. He, Y., Zhang, X., Sun, J.: Channel pruning for accelerating very deep neural networks. In: Proceedings of the IEEE International Conference on Computer Vision, pp. 1389–1397 (2017)
18. Khan, M.M., et al.: Spinnaker: mapping neural networks onto a massively-parallel chip multiprocessor. In: 2008 IEEE International Joint Conference on Neural Networks (IEEE World Congress on Computational Intelligence), pp. 2849–2856. IEEE (2008)
19. Kim, J., Kim, H., Huh, S., Lee, J., Choi, K.: Deep neural networks with weighted spikes. Neurocomputing **311**, 373–386 (2018)
20. Kim, S., Park, S., Na, B., Yoon, S.: Spiking-yolo: spiking neural network for energy-efficient object detection. In: Proceedings of the AAAI Conference on Artificial Intelligence, vol. 34, pp. 11270–11277 (2020)
21. Krizhevsky, A., Nair, V., Hinton, G.: Cifar-10 (Canadian institute for advanced research) **5**(4), 1 (2010). http://www.cs.toronto.edu/kriz/cifar.html
22. Kugele, A., Pfeil, T., Pfeiffer, M., Chicca, E.: Efficient processing of spatio-temporal data streams with spiking neural networks. Front. Neurosci. **14**, 439 (2020)
23. Ledinauskas, E., Ruseckas, J., Juršėnas, A., Burachas, G.: Training deep spiking neural networks (2020)
24. Lee, C., Sarwar, S.S., Panda, P., Srinivasan, G., Roy, K.: Enabling spike-based backpropagation for training deep neural network architectures. Front. Neurosci. **119** (2020)
25. Li, H., Liu, H., Ji, X., Li, G., Shi, L.: Cifar10-DVS: an event-stream dataset for object classification. Front. Neurosci. **11**, 309 (2017)
26. Li, Y., Deng, S., Dong, X., Gong, R., Gu, S.: A free lunch from ANN: towards efficient, accurate spiking neural networks calibration. In: International Conference on Machine Learning, pp. 6316–6325. PMLR (2021)
27. Li, Y., Dong, X., Wang, W.: Additive powers-of-two quantization: an efficient non-uniform discretization for neural networks. arXiv preprint arXiv:1909.13144 (2019)
28. Li, Y., et al.: Brecq: pushing the limit of post-training quantization by block reconstruction. arXiv preprint arXiv:2102.05426 (2021)
29. Li, Y., Guo, Y., Zhang, S., Deng, S., Hai, Y., Gu, S.: Differentiable spike: rethinking gradient-descent for training spiking neural networks. Adv. Neural. Inf. Process. Syst. **34**, 23426–23439 (2021)
30. Loshchilov, I., Hutter, F.: SGDR: stochastic gradient descent with warm restarts. arXiv preprint arXiv:1608.03983 (2016)
31. Lu, S., Sengupta, A.: Exploring the connection between binary and spiking neural networks. Front. Neurosci. **14**, 535 (2020)
32. Ma, D., et al.: Darwin: a neuromorphic hardware co-processor based on spiking neural networks. J. Syst. Architect. **77**, 43–51 (2017)
33. Neftci, E.O., Mostafa, H., Zenke, F.: Surrogate gradient learning in spiking neural networks: bringing the power of gradient-based optimization to spiking neural networks. IEEE Signal Process. Mag. **36**(6), 51–63 (2019)

34. Park, S., Kim, S., Choe, H., Yoon, S.: Fast and efficient information transmission with burst spikes in deep spiking neural networks. In: 2019 56th ACM/IEEE Design Automation Conference (DAC), pp. 1–6. IEEE (2019)
35. Park, S., Kim, S., Na, B., Yoon, S.: T2FSNN: deep spiking neural networks with time-to-first-spike coding. In: 2020 57th ACM/IEEE Design Automation Conference (DAC), pp. 1–6. IEEE (2020)
36. Pei, J., et al.: Towards artificial general intelligence with hybrid Tianjic chip architecture. Nature **572**(7767), 106–111 (2019)
37. Polino, A., Pascanu, R., Alistarh, D.: Model compression via distillation and quantization. arXiv preprint arXiv:1802.05668 (2018)
38. Rathi, N., Roy, K.: Diet-SNN: direct input encoding with leakage and threshold optimization in deep spiking neural networks. arXiv preprint arXiv:2008.03658 (2020)
39. Rathi, N., Srinivasan, G., Panda, P., Roy, K.: Enabling deep spiking neural networks with hybrid conversion and spike timing dependent backpropagation. arXiv preprint arXiv:2005.01807 (2020)
40. Ronneberger, O., Fischer, P., Brox, T.: U-Net: convolutional networks for biomedical image segmentation. In: Navab, N., Hornegger, J., Wells, W.M., Frangi, A.F. (eds.) MICCAI 2015. LNCS, vol. 9351, pp. 234–241. Springer, Cham (2015). https://doi.org/10.1007/978-3-319-24574-4_28
41. Sengupta, A., Ye, Y., Wang, R., Liu, C., Roy, K.: Going deeper in spiking neural networks: VGG and residual architectures. Front. Neurosci. **13**, 95 (2019)
42. Wu, Y., Deng, L., Li, G., Zhu, J., Shi, L.: Spatio-temporal backpropagation for training high-performance spiking neural networks. Front. Neurosci. **12**, 331 (2018)
43. Wu, Y., Deng, L., Li, G., Zhu, J., Xie, Y., Shi, L.: Direct training for spiking neural networks: faster, larger, better. In: Proceedings of the AAAI Conference on Artificial Intelligence, vol. 33, pp. 1311–1318 (2019)
44. Zhang, W., Li, P.: Temporal spike sequence learning via backpropagation for deep spiking neural networks. Adv. Neural. Inf. Process. Syst. **33**, 12022–12033 (2020)
45. Zheng, H., Wu, Y., Deng, L., Hu, Y., Li, G.: Going deeper with directly-trained larger spiking neural networks. In: Proceedings of the AAAI Conference on Artificial Intelligence, vol. 35, pp. 11062–11070 (2021)

Masked Generative Distillation

Zhendong Yang[1,2], Zhe Li[2], Mingqi Shao[1], Dachuan Shi[1], Zehuan Yuan[2], and Chun Yuan[1(✉)]

[1] Tsinghua Shenzhen International Graduate School, Shenzhen, China
{yangzd21,smq21,sdc21}@mails.tsinghua.edu.cn, yuanzehuan@bytedance.com
[2] ByteDance Inc., Beijing, China
yuanc@sz.tsinghua.edu.cn

Abstract. Knowledge distillation has been applied to various tasks successfully. The current distillation algorithm usually improves students' performance by imitating the output of the teacher. This paper shows that teachers can also improve students' representation power by guiding students' feature recovery. From this point of view, we propose Masked Generative Distillation (MGD), which is simple: we mask random pixels of the student's feature and force it to generate the teacher's full feature through a simple block. MGD is a truly general feature-based distillation method, which can be utilized on various tasks, including image classification, object detection, semantic segmentation and instance segmentation. We experiment on different models with extensive datasets and the results show that all the students achieve excellent improvements. Notably, we boost ResNet-18 from 69.90% to 71.69% ImageNet top-1 accuracy, RetinaNet with ResNet-50 backbone from 37.4 to 41.0 Boundingbox mAP, SOLO based on ResNet-50 from 33.1 to 36.2 Mask mAP and DeepLabV3 based on ResNet-18 from 73.20 to 76.02 mIoU. Our codes are available at https://github.com/yzd-v/MGD.

Keywords: Knowledge distillation · Image classification · Object detection · Semantic segmentation · Instance segmentation

1 Introduction

Deep Convolutional Neural Networks (CNNs) have been widely applied to various computer vision tasks. Generally, a larger model has a better performance but a lower inference speed, making it hard to be deployed with a limited source. To get over this, knowledge distillation has been proposed [18]. It can be divided into two types according to the location of distillation. The first is specially designed for the different tasks, such as logit-based distillation [18,40] for classification and head-based distillation [10,39] for detection. The second is feature-based distillation [4,17,28]. Since only the head or projector after the feature is different among various networks, theoretically, the feature-based distillation

---

Z. Yang—This work was performed while Zhendong worked as an intern at ByteDance.
Z. Yang and Z. Li—Equal Contribution.

S. Avidan et al. (Eds.): ECCV 2022, LNCS 13671, pp. 53–69, 2022.
https://doi.org/10.1007/978-3-031-20083-0_4

method can be used in various tasks. However, distillation methods designed for a specific task are often unavailable for other tasks. For example, OFD [17] and KR [4] bring limited improvement for detectors. FKD [37] and FGD [35], specifically designed for detectors, cannot be utilized in other tasks due to the lack of neck.

The previous feature-based distillation methods usually make students imitate the teacher's output as closely as possible because the teacher's feature has a stronger representation power. However, we believe that it is unnecessary to mimic the teacher directly to improve the representation power of students' features. The features used for distillation are generally high-order semantic information through deep networks. The feature pixels already contain the information of adjacent pixels to a certain extent. Therefore, if we can use partial pixels to restore the teacher's full feature through a simple block, the representational power of these used pixels can also be improved. From this point of view, we present Masked Generative Distillation (MGD), which is a simple and efficient feature-based distillation method. As shown in Fig. 2, we first mask random pixels of the student's feature and then generate the teacher's full feature with the masked feature through a simple block. Since random pixels are used in each iteration, all pixels will be used throughout the training process, which means the feature will be more robust and its representation power of it will be improved. In our method, the teacher only serves as a guide for students to restore features and does not require the student to imitate it directly.

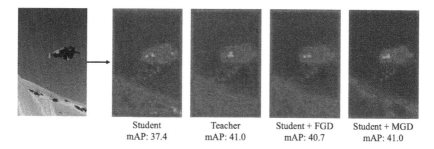

Student	Teacher	Student + FGD	Student + MGD
mAP: 37.4	mAP: 41.0	mAP: 40.7	mAP: 41.0

Fig. 1. Visualization of the feature from the first layer of FPN outputs. **Teacher**: RetinaNet-ResNeXt101. **Student**: RetinaNet-ResNet50. FGD [35] is a distillation method for detectors which forces the student to mimic the teacher's feature.

In order to confirm our hypothesis that without directly imitating the teacher, masked feature generation can improve students' feature representation power, we do the visualization of the feature attention from student's and teacher's neck. As Fig. 1 shows, the features of student and teacher are quite different. Compared with the teacher, the background of the student's feature has higher responses. The teacher's mAP is also significantly higher than the student's, 41.0 vs. 37.4. After distillation with a state-of-the-art distillation method FGD [35], which forces the student to mimic the teacher's feature with attention, the student's feature becomes more similar to the teacher's, and the mAP is greatly improved to 40.7. While after training with MGD, there is still a significant difference between the feature of the student and teacher, but the response to the background of the student is greatly reduced. We are also surprised that the student's performance exceeds FGD and even reaches the same mAP as the teacher. This also shows that training with MGD can improve the representation power of students' features. Besides, we also do abundant experiments on image classification and dense prediction tasks. The results show that MGD can bring considerable improvement to various tasks, including image classification, object detection, semantic segmentation and instance segmentation. MGD can also be combined with other logit-based or head-based distillation methods for even greater performance gains. To sum up, the contributions of this paper are:

1. We introduce a new way for feature-based knowledge distillation, which makes the student generate the teacher's feature with its masked feature instead of mimicking it directly.
2. We propose a novel feature-based distillation method, Masked Generative Distillation, which is simple and easy to use with only two hyper-parameters.
3. We verify the effectiveness of our method on various models via extensive experiments on different datasets. For both image classification and dense prediction tasks, the students achieve significant improvements with MGD.

2   Related Work

2.1   Knowledge Distillation for Classification

Knowledge distillation was first proposed by Hinton et al. [18], where the student is supervised by the labels and the soft labels from the teacher's last linear layer. However, more distillation methods are based on the feature map besides logit. FitNet [28] distills the semantic information from the intermediate layer. AT [36] summaries the values across the channel dimension and transfers the attention knowledge to the student. OFD [17] proposes margin ReLU and designs a new function to measure the distance for distillation. CRD [30] utilizes contrastive learning to transfer the knowledge to students. More recently, KR [4] builds a review mechanism and utilizes multi-level information for distillation. SRRL [33] decouples representation learning and classification, utilizing the teacher's classifier to train the student's penultimate layer feature. WSLD [40] proposes the weighted soft labels for distillation from a perspective of bias-variance trade-off.

2.2   Knowledge Distillation for Dense Prediction

There is a big difference between classification and dense prediction. Many distillation works for classification have failed on dense prediction. Theoretically, the feature-based distillation method should be helpful for both classification and dense prediction tasks, which is also the goal of our method.

Knowledge Distillation for Object Detection. Chen et al. [1] first calculate the distillation loss on the detector's neck and head. The key to distillation for object detection is where to distill due to the extreme imbalance between foreground and background. To avoid introducing noise from the background, FGFI [31] utilizes the fine-grained mask to distill the regions near objects. However, Defeat [13] points out that information from foreground and background are both important. GID [10] chooses the areas where the student and teacher perform differently for distillation. FKD [37] uses the sum of the teacher's and student's attention to make the student focus on changeable areas. FGD [35] proposes focal distillation which forces the student to learn the teacher's crucial parts and global distillation which compensates for missing global information.

Knowledge Distillation for Semantic Segmentation. Liu et al. [23] propose pair-wise and holistic distillation, enforcing pair-wise and high-order consistency between the outputs of the student and teacher. He et al. [16] reinterpret the output from the teacher network to a re-represented latent domain and capture long-term dependencies from the teacher network. CWD [29] minimizes the Kullback-Leibler (KL) divergence between the probability map which is calculated by normalizing the activation map of each channel.

3   Method

The architectures of models vary greatly for different tasks. Moreover, most distillation methods are designed for specific tasks. However, the feature-based distillation can be applied to both classification and dense prediction. The basic method for distillation on features can be formulated as:

$$L_{fea} = \sum_{k=1}^{C} \sum_{i=1}^{H} \sum_{j=1}^{W} \left(F_{k,i,j}^{T} - f_{align}(F_{k,i,j}^{S}) \right)^2 \tag{1}$$

where F^T and F^S denote the teacher's and student's feature, respectively, and f_{align} is the adaptation layer to align student's feature F^S with teacher's feature F^T. C, H, W denotes the shape of the feature map.

This method helps the student to mimic the teacher's features directly. However, we propose masked generative distillation (MGD), which aims at forcing the student to generate the teacher's feature instead of mimicking it, bringing the student significant improvements in both classification and dense prediction. The architecture of MGD is shown in Fig. 2 and we will introduce it specifically in this section.

3.1   Generation with Masked Feature

For CNN-based models, features of deeper layers have a larger receptive field and better representation of the original input image. In other words, the feature

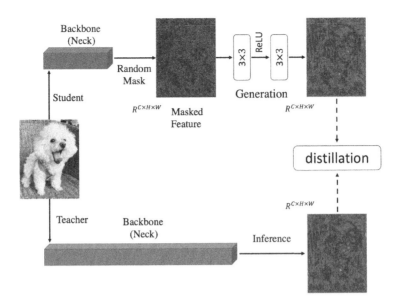

Fig. 2. An illustration of MGD, Masked Generative Distillation. We first randomly mask the student's feature. And then we use the projector layer to force the student to generate the teacher's feature with the masked feature.

map pixels already contain the information of adjacent pixels to a certain extent. Therefore, we can use partial pixels to recover the complete feature map. Our method aims at generating the teacher's feature by student's masked feature, which can help the student achieve a better representation.

We denote by the $T^l \in R^{C \times H \times W}$ and $S^l \in R^{C \times H \times W} (l = 1, .., L)$ the l-th feature map of the teacher and student, respectively. Firstly we set l-th random mask to cover the student's l-th feature, which can be formulated as:

$$M_{i,j}^l = \begin{cases} 0, & \text{if } R_{i,j}^l < \lambda \\ 1, & \text{Otherwise} \end{cases} \qquad (2)$$

where $R_{i,j}^l$ is a random number in $(0, 1)$ and i, j are the the horizontal and vertical coordinates of the feature map, respectively. λ is a hyper-parameter that denotes the masked ratio. The $l - th$ feature map is covered by the l-th random mask.

Then we use the corresponding mask to cover the student's feature map and try to generate teacher's feature maps with the left pixels, which can be formulated as follows:

$$\mathcal{G}\big(f_{align}(S^l) \cdot M^l\big) \longrightarrow T^l \qquad (3)$$

$$\mathcal{G}(F) = W_{l2}(ReLU(W_{l1}(F))) \qquad (4)$$

$\mathcal{G}$ denotes the projector layer which includes two convolutional layers: W_{l1} and W_{l2}, one activation layer $ReLU$. In this paper, we adopt 1×1 convolutional layers for the adaptation layer f_{align}, 3×3 convolutional layers for projector layer W_{l1} and W_{l2}.

According to this method, we design the distillation loss L_{dis} for MGD:

$$L_{dis}(S,T) = \sum_{l=1}^{L}\sum_{k=1}^{C}\sum_{i=1}^{H}\sum_{j=1}^{W}\left(T_{k,i,j}^{l} - \mathcal{G}\left(f_{align}(S_{k,i,j}^{l}) \cdot M_{i,j}^{l}\right)\right)^{2} \tag{5}$$

where L is the sum of layers for distillation and C, H, W denote the shape of the feature map. S and T denote the feature of the student and teacher, respectively.

3.2   Overall Loss

With the proposed distillation loss L_{dis} for MGD, we train all the models with the total loss as follows:

$$L_{all} = L_{original} + \alpha \cdot L_{dis} \tag{6}$$

where $L_{original}$ is the original loss for the models among all the tasks and α is a hyper-parameter to balance the loss.

MGD is a simple and effective method for distillation and can be applied to various tasks easily. The process of our method is summarized in Algorithm 1.

Algorithm 1. Masked Generative Distillation

Input: Teacher: T, Student: S, Input: x, label: y, hyper-parameter: α, λ

1: Using S to get the feature fea^{S} and output $\hat{y}$ of Input x
2: Using T to get the feature fea^{T} of Input x
3: Calculating the original loss of the model: $L_{original}(\hat{y}, y)$
4: Calculating the distillation loss in Equation 5: $L_{dis}(fea^{S}, fea^{T})$
5: Using $L_{all} = L_{original} + \alpha \cdot L_{dis}$ to update S

Output: S

---

4   Main Experiments

MGD is a feature-based distillation that can easily be applied to different models for various tasks. In this paper, we conduct experiments on various tasks, including classification, object detection, semantic segmentation and instance segmentation. We experiment with different models and datasets for different tasks, and all the models achieve excellent improvements with MGD.

4.1   Classification

Datasets. For classification task, we evaluate our knowledge distillation method on ImageNet [11], which contains 1000 object categories. We use the 1.2 million images for training and 50k images for testing for all the classification experiments. We use accuracy to evaluate the models.

Implementation Details. For the classification task, we calculate the distillation loss on the last feature map from the backbone. The ablation study about this is shown in Sect. 5.5. MGD uses a hyper-parameter α to balance the distillation loss in Eq. 6. The other hyper-parameter λ is used to adjust the masked ratio in Eq. 2. We adopt the hyper-parameters $\{\alpha = 7 \times 10^{-5}, \lambda = 0.5\}$ for all the classification experiments. We train all the models for 100 epochs with SGD optimizer, where the momentum is 0.9 and the weight decay is 0.0001. We initialize the learning rate to 0.1 and decay it for every 30 epochs. This setting is based on 8 GPUs. The experiments are conducted with MMClassification [6] and MMRazor [7] based on Pytorch [26].

Table 1. Results of different distillation methods on ImageNet dataset. **T** and **S** mean the teacher and student, respectively.

Type	Method	Top-1	Top-5	Method	Top-1	Top-5
	ResNet-34(T)	73.62	91.59	ResNet-50(T)	76.55	93.06
	ResNet-18(S)	69.90	89.43	MobileNet(S)	69.21	89.02
Logit	KD [18]	70.68	90.16	KD [18]	70.68	90.30
	WSLD [40]	71.54	90.25	WSLD [40]	72.02	90.70
Feature	AT [36]	70.59	89.73	AT [36]	70.72	90.03
	OFD [17]	71.08	90.07	OFD [17]	71.25	90.34
	RKD [25]	71.34	90.37	RKD [25]	71.32	90.62
	CRD [30]	71.17	90.13	CRD [30]	71.40	90.42
	KR [4]	71.61	90.51	KR [4]	72.56	**91.00**
	Ours	71.58	90.35	Ours	72.35	90.71
Feature + Logit	SRRL [33]	71.73	**90.60**	SRRL [33]	72.49	90.92
	Ours+WSLD	**71.80**	90.40	Ours+WSLD	**72.59**	90.94

Classification Results. We conduct experiments with two popular distillation settings for classification, including homogeneous and heterogeneous distillation. The first distillation setting is from ResNet-34 [15] to ResNet-18, the other setting is from ResNet-50 to MobileNet [19]. As shown in Table 1, we compare with various knowledge distillation methods [4,17,18,25,30,33,36,40], including feature-based methods, logit-based methods and the combination. The student ResNet-18 and MobileNet gain 1.68 and 3.14 Top-1 accuracy improvement with our method, respectively. Besides, as described above, MGD just needs to calculate the distillation loss on the feature maps and can be combined with other logit-based methods for image classification. So we try to add the logit-based distillation loss in WSLD [40]. In this way, the two students achieve 71.80 and 72.59 Top-1 accuracy, getting another 0.22 and 0.24 improvement, respectively.

4.2   Object Detection and Instance Segmentation

Datasets. We conduct experiments on COCO2017 dataset [22], which contains 80 object categories. We use the 120k train images for training and 5k val images for testing. The performances of models are evaluated in Average Precision.

Table 2. Results of different distillation methods for object detection on COCO.

Teacher	Student	mAP	AP_S	AP_M	AP_L
RetinaNet ResNeXt101 (41.0)	RetinaNet-Res50	37.4	20.6	40.7	49.7
	FKD [37]	39.6	22.7	43.3	52.5
	CWD [29]	40.8	22.7	44.5	55.3
	FGD [35]	40.7	22.9	45.0	54.7
	Ours	**41.0**	**23.4**	**45.3**	**55.7**
Cascade Mask RCNN ResNeXt101 (47.3)	Faster RCNN-Res50	38.4	21.5	42.1	50.3
	FKD [37]	41.5	23.5	45.0	55.3
	CWD [29]	41.7	23.3	45.5	55.5
	FGD [35]	42.0	**23.8**	46.4	55.5
	Ours	**42.1**	23.7	**46.4**	**56.1**
RepPoints ResNeXt101 (44.2)	RepPoints-Res50	38.6	22.5	42.2	50.4
	FKD [37]	40.6	23.4	44.6	53.0
	CWD [29]	42.0	24.1	46.1	55.0
	FGD [35]	42.0	24.0	45.7	55.6
	Ours	**42.3**	**24.4**	**46.2**	**55.9**

Implementation Details. We calculate the distillation loss on all the feature maps from the neck. We adopt the hyper-parameters $\{\alpha = 2 \times 10^{-5}, \lambda = 0.65\}$ for all the one-stage models, $\{\alpha = 5 \times 10^{-7}, \lambda = 0.45\}$ for all the two-stage models. We train all the models with SGD optimizer, where the momentum is 0.9 and the weight decay is 0.0001. Unless specified, we train the models for 24 epochs. We use inheriting strategy [20,35] which initializes the student with the teacher's neck and head parameters to train the student when they have the same head structure. The experiments are conducted with MMDetection [2].

Object Detection and Instance Segmentation Results. For object detection, we conduct experiments on three different types of detectors, including a two-stage detector (Faster RCNN [27]), an anchor-based one-stage detector (RetinaNet [21]) and an anchor-free one-stage detector (RepPoints [34]). We compare MGD with three recent state-of-the-art distillation methods for detectors [29,35,37]. For instance segmentation, we conduct experiments on two models, SOLO [32] and Mask RCNN [14]. As shown in Table 2 and Table 3, our method surpasses the other state-of-the-art methods for both object detection and instance segmentation. The students gain significant AP improvements with MGD, *e.g.* the ResNet-50 based RetinaNet and SOLO gets 3.6 Boundingbox mAP and 3.1 Mask mAP improvement on COCO dataset, respectively.

Table 3. Results of different distillation methods for instance segmentation on COCO. **MS** means multi-scale training. Here the AP means Mask AP.

Teacher	Students	mAP	AP_S	AP_M	AP_L
SOLO-Res101 3x,MS(37.1)	SOLO-Res50(1×)	33.1	12.2	36.1	50.8
	FGD [35]	36.0	14.5	39.5	54.5
	Ours	**36.2**	**14.2**	**39.7**	**55.3**
Cascade Mask RCNN ResNeXt101(41.1)	Mask RCNN-Res50	35.4	16.6	38.2	52.5
	FGD [35]	37.8	17.1	40.7	56.0
	Ours	**38.1**	**17.1**	**41.1**	**56.3**

4.3 Semantic Segmentation

Datasets. For the semantic segmentation task, we evaluate our method on CityScapes dataset [9], which contains 5000 high-quality images (2975, 500, and 1525 images for the training, validation, and testing). We evaluate all the models with mean Intersection-over-Union (mIoU).

Implementation Details. For all the models, we calculate the distillation loss on the last feature map from the backbone. We adopt the hyper-parameters $\{\alpha = 2 \times 10^{-5}, \lambda = 0.75\}$ for all the experiments. We train all the models with SGD optimizer, where the momentum is 0.9 and the weight decay is 0.0005. We run all the models on 8 GPUs. The experiments are conducted with MMSegmentation [8].

Semantic Segmentation Results. For the semantic segmentation task, we conduct experiments on two settings. In both settings, we use PspNet-Res101 [38] as the teacher and train it for 80k iterations with 512×1024 input size. We use PspNet-Res18 and DeepLabV3-Res18 [3] as students and train them for 40k iterations with 512×1024 input size. As shown in Table 4, our method surpasses the state-of-the-art distillation method for semantic segmentation. Both the homogeneous and heterogeneous distillation bring the students significant improvements, *e.g.* the ResNet-18 based PspNet gets 3.78 mIoU improvement. Besides, MGD is a feature-based distillation method and can be combined with other logit-based distillation methods. As the results show, the student PspNet and DeepLabV3 get another 0.47 and 0.29 mIoU improvement by adding the logit distillation loss of the head in CWD [29].

Table 4. Results of the semantic segmentation task on CityScapes dataset. **T** and **S** mean teacher and student, respectively. The results are the average value of three runs. * means adding the distillation loss of the head in CWD [29]

Method	Input size	mIoU
PspNet-Res101(T)	512×1024	78.34
PspNet-Res18(S)	512×512	69.85
SKDS [23]	512×512	72.70
CWD [29]	512×512	73.53
Ours	512×512	73.63
Ours*	512×512	74.10
PspNet-Res101(T)	512×1024	78.34
DeepLabV3-Res18(S)	512×512	73.20
SKDS [23]	512×512	73.87
CWD [29]	512×512	75.93
Ours	512×512	76.02
Ours*	512×512	76.31

5   Analysis

5.1   Better Representation with MGD

MGD forces the student to generate the teacher's complete feature map with its masked feature instead of mimicking it directly. It helps the students get a better representation of the input image. In this subsection, we study this by using the student to teach itself. We first train ResNet-18 directly as a teacher and the baseline. Then we use the trained ResNet-18 to distill itself with MGD. For comparison, we also distill the student by forcing the student to mimic the teacher directly. The distillation loss for mimicking is the square of L2 distance between the student's feature map and the teacher's feature map.

As shown in Table 5, the student also gains 1.01 accuracy improvement with MGD even when the teacher is itself. In contrast, the improvement is very limited when forcing the student to mimic the teacher's feature map directly. The comparison indicates that the student's feature map achieves better representation than the teacher's after distillation.

Furthermore, we visualize the training loss curves for distillation with MGD and mimicking the teacher, which is shown in Fig. 3. The **difference** in the figure means the square of L2 distance between the last feature map of student and teacher, which is also the distillation loss for mimicking the teacher. As the figure shows, the **difference** keeps decreasing during mimicking the teacher directly and finally the student gets a similar feature to the teacher. However, the improvement with this method is minimal. In contrast, the **difference** becomes larger after training with MGD. Although the student gets a different feature from the teacher, it gets higher accuracy, also indicating the student's feature obtains stronger representation power.

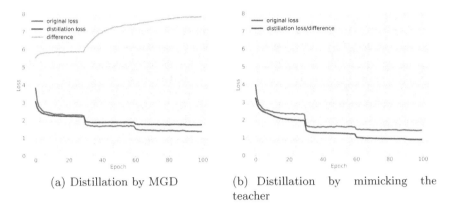

(a) Distillation by MGD   (b) Distillation by mimicking the teacher

Fig. 3. The training loss curves of ResNet-18 distilling ResNet-18. **Difference** means the square of L2 distance between the last feature map of student and teacher. It is also the distillation loss for mimicking the teacher.

Table 5. The results of distillation for Rse18-18 on ImageNet. We train ResNet-18 directly as the teacher and student baseline. **T** and **S** mean teacher and student.

	Top-1	Top-5
ResNet-18(T,S)	69.90	89.43
+mimicking feature	70.05	89.41
+MGD	70.91	89.82

5.2 Distillation by Masking Random Channels

For image classification, the models usually utilize a pooling layer to reduce the spatial dimension of the feature map. This layer makes the model more sensitive to the channels than spatial pixels. So in this subsection, we try to apply MGD by masking random channels instead of spatial pixels for classification. We adopt the masked ratio $\beta = 0.15$ and hyper-parameter $\alpha = 7 \times 10^{-5}$ for the experiments. As shown in Table 6, the student can get better performance by masking random channels instead of spatial for image classification. The student Res-18 and MobileNet achieve 0.13 and 0.14 Top-1 accuracy gains, respectively.

Table 6. Results of masking random channels on ImageNet dataset.

	Acc	MGD (Spatial)	MGD (Channel)
Res34-18	Top-1	71.58	**71.69**
	Top-5	90.35	**90.42**
Res50-mv1	Top-1	72.35	**72.49**
	Top-5	90.71	**90.94**

5.3   Distillation with Different Teachers

Cho et al. [5] show the teacher with higher accuracy may not be the better teacher for knowledge distillation on image classification. The conclusion is based on the logit-based distillation method. However, our method just needs to calculate the distillation loss on the feature maps. In this subsection, we study this conclusion by using different kinds of teachers to distill the same student ResNet-18, which is shown in Fig. 4.

As shown in Fig. 4, the better teacher benefits more to the student when the teacher and student have a similar architecture, *e.g.* ResNet-18 achieves 70.91 and 71.8 accuracy with ResNet-18 and ResNetV1D-152 as the teacher, respectively. However, when the teacher and student have a different architecture, it is hard for the student to generate the teacher's feature map and the improvement by distillation is limited. Moreover, the distillation performs worse with a larger difference between the architectures. For example, although Res2Net-101 [12] and ConvNeXt-T [24] has 79.19 and 82.05 accuracy, they just bring 1.53 and 0.88 accuracy improvement to the student, which is even lower than the ResNet-34 based teacher(73.62 accuracy).

The results in Fig. 4 indicate that the stronger teacher is better for feature-based distillation when they have similar architecture. Besides, the homogeneous teacher is much better for feature-based distillation than the teacher with high accuracy but a heterogeneous architecture.

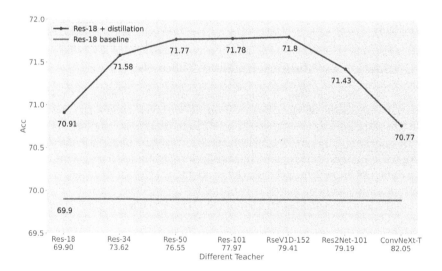

Fig. 4. The results of training ResNet-18 with our method by different teachers on ImageNet dataset.

5.4   The Generative Block

MGD uses a simple block to restore the feature, called the generative block. In Eq. 4, we use two 3×3 convolutional layers and one activation layer *ReLU* to accomplish this. In this subsection, we explore the effects of the generative block with different compositions, shown in Table 7.

As the results show, the student gets the slightest improvement when there is only one convolutional layer. However, when there are three convolutional layers, the student gets a worse Top-1 but better Top-5 accuracy. As for the kernel size, 5×5 convolutional kernel needs more compute resources, while it gets worse performance. Based on the results, we choose the architecture in Eq. 4 for MGD, which includes two convolutional layers and one activation layer.

Table 7. The results of distillation about generative part. The **Conv Layers** mean the sum of convolutional layers and the **kernel size** belongs to the convolutional layer. We add one activation layer *ReLU* between every two convolutional layers.

Conv. layers	Kernel size	Top-1	Top-5
1	3×3	71.28	90.30
2	3×3	**71.58**	90.35
3	3×3	71.49	**90.44**
2	5×5	71.32	90.28

5.5   Distillation on Different Stages

Our method can also be applied at other stages of the model. In this subsection, we explore distillation at different stages on ImageNet. We calculate the distillation loss on the corresponding layers of teacher and student. As shown in Table 8, distilling the shallower layers is also helpful to the student but very limited. While distilling the deeper stage which contains more semantic information benefits the student more. Furthermore, the features from the early stages are not directly used for classification. Therefore, Distilling such features with the last stage feature together may hurt the student's accuracy.

5.6   Sensitivity Study of Hyper-parameters

In this paper, we use α in Eq. 6 and λ in Eq. 2 to balance the distillation loss and adjust the mask ratio, respectively. In this subsection, we do the sensitivity study of the hyper-parameters by using ResNet-34 to distill ResNet-18 on ImageNet dataset. The results are shown in Fig. 5.

As shown in Fig. 5, MGD is not sensitive to the hyper-parameter α which is just used for balancing the loss. As for the mask ratio λ, the accuracy is 71.41

Table 8. The results of distillation about different stages for Rse34-18 on ImageNet.

	Top-1	Top-5
Stage 1	70.09	89.40
Stage 2	70.21	89.38
Stage 3	70.37	89.42
Stage 4	**71.58**	**90.35**
Stage 2+3+4	71.47	90.31

when it is 0, which means there are no masked parts for the generation. The student gets higher performance with larger ratio when $\lambda < 0.5$. However, when λ is too large, *e.g.* 0.8, the left semantic information is too poor to generate teacher's complete feature map and the performance improvement is also affected.

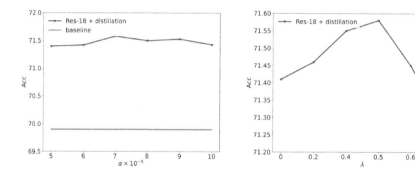

Fig. 5. Sensitivity study of hyper-parameters α and λ with ResNet34-ResNet18

6 Conclusions

In this paper, we propose a new way for knowledge distillation, which forces the student to generate the teacher's feature with its masked feature instead of mimicking it directly. Based on this way, we propose a new knowledge distillation method, Masked Generative Distillation (MGD). The students can obtain stronger representation power with MGD. Furthermore, our method is just based on the feature map so that MGD can easily be applied to various tasks such as image classification, object detection, semantic segmentation, and instance segmentation. Extensive experiments on various models with different datasets prove that our method is simple and efficient.

Acknowledgement. This work was supported by the SZSTC project Grant No.JCYJ20190809172201639 and Grant No. WDZC20200820200655001, Shenzhen Key Laboratory ZDSYS20210623092001004.

References

1. Chen, G., Choi, W., Yu, X., Han, T., Chandraker, M.: Learning efficient object detection models with knowledge distillation. Adv. Neural Inf. Process. Syst. **30** (2017)
2. Chen, K., et al.: MMDetection: open mmlab detection toolbox and benchmark. arXiv preprint arXiv:1906.07155 (2019)
3. Chen, L.C., Papandreou, G., Schroff, F., Adam, H.: Rethinking atrous convolution for semantic image segmentation. arXiv preprint arXiv:1706.05587 (2017)
4. Chen, P., Liu, S., Zhao, H., Jia, J.: Distilling knowledge via knowledge review. In: Proceedings of the IEEE/CVF Conference on Computer Vision and Pattern Recognition, pp. 5008–5017 (2021)
5. Cho, J.H., Hariharan, B.: On the efficacy of knowledge distillation. In: Proceedings of the IEEE/CVF International Conference on Computer Vision, pp. 4794–4802 (2019)
6. Contributors, M.: Openmmlab's image classification toolbox and benchmark (2020). https://github.com/open-mmlab/mmclassification
7. Contributors, M.: Openmmlab model compression toolbox and benchmark (2021). https://github.com/open-mmlab/mmrazor
8. Contributors, M.: MMSegmentation: Openmmlab semantic segmentation toolbox and benchmark (2020). https://github.com/open-mmlab/mmsegmentation
9. Cordts, M., et al.: The cityscapes dataset for semantic urban scene understanding. In: Proceedings of the IEEE Conference on Computer Vision and Pattern Recognition, pp. 3213–3223 (2016)
10. Dai, X., Jiang, Z., Wu, Z., Bao, Y., Wang, Z., Liu, S., Zhou, E.: General instance distillation for object detection. In: Proceedings of the IEEE/CVF Conference on Computer Vision and Pattern Recognition, pp. 7842–7851 (2021)
11. Deng, J., Dong, W., Socher, R., Li, L.J., Li, K., Fei-Fei, L.: Imagenet: a large-scale hierarchical image database. In: 2009 IEEE Conference on Computer Vision and Pattern Recognition, pp. 248–255. IEEE (2009)
12. Gao, S.H., Cheng, M.M., Zhao, K., Zhang, X.Y., Yang, M.H., Torr, P.: Res2net: a new multi-scale backbone architecture. IEEE TPAMI (2021). https://doi.org/10.1109/TPAMI.2019.2938758
13. Guo, J., et al.: Distilling object detectors via decoupled features. In: Proceedings of the IEEE/CVF Conference on Computer Vision and Pattern Recognition, pp. 2154–2164 (2021)
14. He, K., Gkioxari, G., Dollár, P., Girshick, R.: Mask r-cnn. In: Proceedings of the IEEE International Conference on Computer Vision, pp. 2961–2969 (2017)
15. He, K., Zhang, X., Ren, S., Sun, J.: Deep residual learning for image recognition. In: Proceedings of the IEEE Conference on Computer Vision and Pattern Recognition, pp. 770–778 (2016)
16. He, T., Shen, C., Tian, Z., Gong, D., Sun, C., Yan, Y.: Knowledge adaptation for efficient semantic segmentation. In: Proceedings of the IEEE/CVF Conference on Computer Vision and Pattern Recognition, pp. 578–587 (2019)
17. Heo, B., Kim, J., Yun, S., Park, H., Kwak, N., Choi, J.Y.: A comprehensive overhaul of feature distillation. In: Proceedings of the IEEE/CVF International Conference on Computer Vision, pp. 1921–1930 (2019)
18. Hinton, G., Vinyals, O., Dean, J., et al.: Distilling the knowledge in a neural network. arXiv preprint arXiv:1503.02531 2(7) (2015)

19. Howard, A.G., et al.: Mobilenets: efficient convolutional neural networks for mobile vision applications. arXiv preprint arXiv:1704.04861 (2017)
20. Kang, Z., Zhang, P., Zhang, X., Sun, J., Zheng, N.: Instance-conditional knowledge distillation for object detection. In: In Proceedings of the Thirty-Fifth Conference on Neural Information Processing Systems (NeurIPS) (2021)
21. Lin, T.Y., Goyal, P., Girshick, R., He, K., Dollár, P.: Focal loss for dense object detection. In: Proceedings of the IEEE International Conference on Computer Vision, pp. 2980–2988 (2017)
22. Lin, T.-Y., et al.: Microsoft COCO: common objects in context. In: Fleet, D., Pajdla, T., Schiele, B., Tuytelaars, T. (eds.) ECCV 2014. LNCS, vol. 8693, pp. 740–755. Springer, Cham (2014). https://doi.org/10.1007/978-3-319-10602-1_48
23. Liu, Y., Chen, K., Liu, C., Qin, Z., Luo, Z., Wang, J.: Structured knowledge distillation for semantic segmentation. In: Proceedings of the IEEE/CVF Conference on Computer Vision and Pattern Recognition, pp. 2604–2613 (2019)
24. Liu, Z., Mao, H., Wu, C.Y., Feichtenhofer, C., Darrell, T., Xie, S.: A convnet for the 2020s. arXiv preprint arXiv:2201.03545 (2022)
25. Park, W., Kim, D., Lu, Y., Cho, M.: Relational knowledge distillation. In: Proceedings of the IEEE/CVF Conference on Computer Vision and Pattern Recognition, pp. 3967–3976 (2019)
26. Paszke, A., et al.: Pytorch: an imperative style, high-performance deep learning library. Adv. Neural Inf. Process. Syst. **32** (2019)
27. Ren, S., He, K., Girshick, R., Sun, J.: Faster r-cnn: towards real-time object detection with region proposal networks. Adv. Neural Inf. Process. Syst. **28** (2015)
28. Romero, A., Ballas, N., Kahou, S.E., Chassang, A., Gatta, C., Bengio, Y.: Fitnets: hints for thin deep nets. arXiv preprint arXiv:1412.6550 (2014)
29. Shu, C., Liu, Y., Gao, J., Yan, Z., Shen, C.: Channel-wise knowledge distillation for dense prediction. In: Proceedings of the IEEE/CVF International Conference on Computer Vision, pp. 5311–5320 (2021)
30. Tian, Y., Krishnan, D., Isola, P.: Contrastive representation distillation. In: International Conference on Learning Representations (2019)
31. Wang, T., Yuan, L., Zhang, X., Feng, J.: Distilling object detectors with fine-grained feature imitation. In: Proceedings of the IEEE/CVF Conference on Computer Vision and Pattern Recognition, pp. 4933–4942 (2019)
32. Wang, X., Kong, T., Shen, C., Jiang, Y., Li, L.: SOLO: segmenting objects by locations. In: Vedaldi, A., Bischof, H., Brox, T., Frahm, J.-M. (eds.) ECCV 2020. LNCS, vol. 12363, pp. 649–665. Springer, Cham (2020). https://doi.org/10.1007/978-3-030-58523-5_38
33. Yang, J., Martinez, B., Bulat, A., Tzimiropoulos, G.: Knowledge distillation via softmax regression representation learning. In: International Conference on Learning Representations (2020)
34. Yang, Z., Liu, S., Hu, H., Wang, L., Lin, S.: Reppoints: point set representation for object detection. In: Proceedings of the IEEE/CVF International Conference on Computer Vision, pp. 9657–9666 (2019)
35. Yang, Z., et al.: Focal and global knowledge distillation for detectors. arXiv preprint arXiv:2111.11837 (2021)
36. Zagoruyko, S., Komodakis, N.: Paying more attention to attention: Improving the performance of convolutional neural networks via attention transfer. arXiv preprint arXiv:1612.03928 (2016)
37. Zhang, L., Ma, K.: Improve object detection with feature-based knowledge distillation: towards accurate and efficient detectors. In: International Conference on Learning Representations (2020)

38. Zhao, H., Shi, J., Qi, X., Wang, X., Jia, J.: Pyramid scene parsing network. In: Proceedings of the IEEE Conference on Computer Vision and Pattern Recognition, pp. 2881–2890 (2017)
39. Zhixing, D., Zhang, R., Chang, M., Liu, S., Chen, T., Chen, Y., et al.: Distilling object detectors with feature richness. Adv. Neural Inf. Process. Syst. **34** (2021)
40. Zhou, H., et al.: Rethinking soft labels for knowledge distillation: a bias-variance tradeoff perspective. In: International Conference on Learning Representations (2020)

Fine-grained Data Distribution Alignment for Post-Training Quantization

Yunshan Zhong[1,2], Mingbao Lin[3], Mengzhao Chen[2], Ke Li[3], Yunhang Shen[3], Fei Chao[2], Yongjian Wu[3], and Rongrong Ji[1,2(✉)]

[1] Institute of Artificial Intelligence, Xiamen University, Xiamen, China
zhongyunshan@stu.xmu.edu.cn, rrji@xmu.edu.cn
[2] Media Analytics and Computing Lab, Department of Artificial Intelligence,
School of Informatics, Xiamen University, Xiamen, China
cmzxmu@stu.xmu.edu.cn, fchao@xmu.edu.cn
[3] Tencent Youtu Lab., Shanghai, China
littlekenwu@tencent.com

Abstract. While post-training quantization receives popularity mostly due to its evasion in accessing the original complete training dataset, its poor performance also stems from scarce images. To alleviate this limitation, in this paper, we leverage the synthetic data introduced by zero-shot quantization with calibration dataset and propose a fine-grained data distribution alignment (FDDA) method to boost the performance of post-training quantization. The method is based on two important properties of batch normalization statistics (BNS) we observed in deep layers of the trained network, *i.e.*, inter-class separation and intra-class incohesion. To preserve this fine-grained distribution information: 1) We calculate the per-class BNS of the calibration dataset as the BNS centers of each class and propose a BNS-centralized loss to force the synthetic data distributions of different classes to be close to their own centers. 2) We add Gaussian noise into the centers to imitate the incohesion and propose a BNS-distorted loss to force the synthetic data distribution of the same class to be close to the distorted centers. By utilizing these two fine-grained losses, our method manifests the state-of-the-art performance on ImageNet, especially when both the first and last layers are quantized to the low-bit. Code is at https://github.com/zysxmu/FDDA.

Keywords: Batch normalization statistics · Post-training quantization · Synthetic data

1 Introduction

Recent years have witnessed the rising of deep neural networks (DNNs) in computer vision. Nevertheless, the increasing model size barricades the deployment of DNNs on resource-limited platforms such as mobile phones, embedding devices, *etc*. To overcome this dilemma, varieties of methods [9,17] are

Supplementary Information The online version contains supplementary material available at https://doi.org/10.1007/978-3-031-20083-0_5.

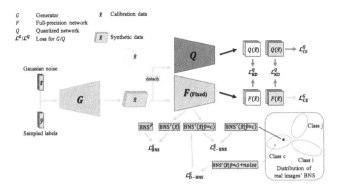

Fig. 1. Framework of our FDDA. We deploy a generator G to produce synthetic data supervised by the coarse BNS alignment loss ($\mathcal{L}^G_{\text{BNS}}$), and two proposed fine-grained distribution losses ($\mathcal{L}^G_{\text{C-BNS}}$ and $\mathcal{L}^G_{\text{D-BNS}}$).

explored to reduce the complexity of DNNs. Network quantization, which represents full-precision DNNs in a low-precision format, emerges as a promising direction [1,17,21,26,27,41].

By tuning the quantized DNNs using a small calibration dataset, post-training quantization, a sub-topic of low-precision quantization, has received increasing popularity from both academia and industries. Recent studies manifest that a post-training quantized model in high precision, such as 8-bit, can reach performance on par with its full-precision counterpart [1,17]. However, performance drops severely if being quantized to lower precision such as 4-bit [19]. For example, as reported in LAPQ [29], quantizing ResNet-18 [11] to 8-bit can well retain the accuracy of the full-precision network (around 71.5%), but only 60.3% top-1 accuracy can be observed when quantized to 4-bit. To alleviate this problem, many studies are explored to enhance the low-bit performance. The mainstream can be outlined into two folds. The first group designs sophisticated quantization methods, such as linear combination of multiple low-bit vectors [22], weight region separation [6], mixed-precision quantization [22], partial quantization [18], *etc.* The second group reformulates the rounding function or loss constraint from an analytical perspective. For example, Nagel *et al.* [28] derived an adaptive rounding by modeling rounding problem as quadratic constrained binary optimization. By the second-order analysis on rebuilding intermediate outputs, Li *et al.* [19] showed that the best output reconstruction lies in a block unit.

Though great efforts have been made, improvements of these studies are still limited. Besides, the performance gains are usually built on the premise that the first and last layers are quantized to 8-bit [14,36], or even retained in full-precision states [28,29]. However, severe performance degradation occurs when all layers are quantized to very low-bit integers (see Table 3). We consider the root cause of significant accuracy degradation in post-training quantization. We attribute it to a lack of training data. Specifically, the low-bit network bears poor representation ability, and a very small calibration dataset cannot support

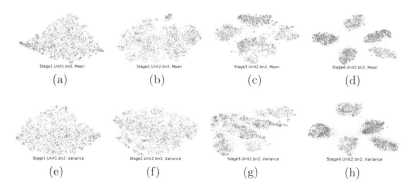

Fig. 2. t-SNE visualization (five classes) of BNS in different layers of pre-trained ResNet-18 on ImageNet. BNS in shallow layers are overlapping. For deep layers, different classes have varying BNS and there exists small distortion within data from the same class. Similar observations can be found in other networks as well (See the supplementary materials). Best viewed in color.

the quantized model to fit well the real data distribution. Many researches on zero-shot quantization are indicated to synthesizing fake images using data optimizer [2,10,38,39] or data generator [5,23,37]. The synthetic data is then used to train the quantized model. Though this manner partly alleviates data lacking, performance drops greatly if simply using synthetic data. For example, Li *et al.* [19] observed only 21.71% top-1 accuracy when quantizing ResNet-18 to 4-bit in a single zero-shot manner [2]. So far, combining real calibration dataset with synthetic data remains unexplored in post-training quantization, and we believe, might be a promise of boosting the low-bit performance.

Motivated by the above analyses, in this paper, we propose fine-grained data distribution alignment (termed FDDA) for post-training quantization, as illustrated in Fig. 1. FDDA is inspired by the fact that the representation of synthetic data has been demonstrated to be feasible in many other tasks such as image super-resolution [40], light-weight student network [3] and zero-shot quantization [37]. Thus, for the first time, we apply it to post-training quantization to tackle the insufficient data problem, providing a new perspective for post-training quantization. Following GDFQ [37], we take full use of the pre-trained model to guide the generator to synthesize fake data. Except for the distillation of output logits between the pre-trained model and its quantized version to improve the quantization performance, GDFQ also retains the distribution information of training data, modeled by the batch normalization statistics (BNS) in the pre-trained model. To this end, the mean and variance of the synthetic data distribution are constrained to be the same as those of the real data distribution. However, we realize that, based on our two insightful observations on BNS, this information retaining manner is very coarse for deep layers.

Specifically, we calculate the mean and variance *w.r.t.* each image and the results are visualized in Fig. 2. As can be seen, the BNS of different classes are overlapping in shallow layers. However, two properties of deep BNS including inter-class separation and intra-class incohesion are observed. The former indicates different classes possess varying BNS while the latter indicates a small distortion of BNS among data from the same class. The BNS captured in the pre-trained model only reflects the distribution of the whole dataset, which are applicable to shallow layers with mixed class-wise BNS. While these BNS in the pre-trained model are coarse for deep layers with separable class-wise BNS. Thus, the synthetic data needs a fine-grained BNS alignment for these deep layers.

To this end, we further introduce a BNS-centralized loss and a BNS-distorted loss, respectively to align the fine-grained BNS properties of inter-class separation and intra-class incohesion. In contrast to zero-shot quantization [2], an additional calibration dataset, usually comprising one image per class, can be available in post-training quantization. To fully utilize this bonus, we derive means and variances of each image in deep layers of the pre-trained model, and then define the computed means and variances as the BN centers of each class. To preserve inter-class separation, our BNS-centralized loss forces the synthetic data distributions of different classes closely to their own centers. To preserve intra-class incohesion, we add Gaussian noise into the centers to mimic the distortion and the BNS-distorted loss forces the synthetic data distribution of the same class closely to the distorted centers. Through a fine-grained BNS alignment, our proposed FDDA significantly improves the quantization performance over existing methods on ImageNet [14,19,36], particularly when the first and last layers are also quantized to low-bit. Our contributions are three-fold:

- To our best knowledge, we are the first to explore combining calibration dataset with synthetic data in post-training quantization, which might provide a new perspective for post-training quantization.
- We observe properties of inter-class separation and intra-class incohesion in deep BNS. Besides, we devise a BNS-centralized loss and a BNS-distorted loss to preserve these two properties in synthetic data.
- Extensive experiments demonstrate that our FDDA can well improve the performance on ImageNet. For example, our FDDA outperforms the current SOTA, BRECQ [19], by 6.64% in the top-1 accuracy when all layers of MobileNet-V1 are quantized to 4-bit.

2   Related Work

In this section, we briefly discuss the most related work to ours including post-training quantization and zero-shot quantization. A more comprehensive survey is referred to [7].

Post-training Quantization . Most existing post-training quantization methods attempt to alleviate the accuracy deterioration problem from two perspectives: designing more sophisticated quantization methods and introducing a new rounding function or loss function. From the first perspective, Liu *et al.* [22]

proposed to close the gap between the full-precision weight vector and its low-bit version by using the linear combination of multiple low-bit vectors. Wang *et al.* [36] finished quantization in a two-stage manner of bit-split and bit-stitching. In the bit-split stage, the K-bit constraint of integer is split into $(K-1)$ ternary learning problems, and each bit is then separately solved in an iterative optimization procedure. In the bit-stitching stage, the K-bit integer is recovered by the linear combination with a base of 2^{k-1} for the k-th bit. In addition, they also use filter-wise quantizer for activations and integrate the scaling factor into corresponding 2D kernels to avoid extra storage. Piece-wise linear quantization [6] splits the whole weights into two non-overlapping areas, including one dense region comprising low-magnitude weights, and one sparse region comprising high-magnitude weights. On top of the splitting, both areas are respectively quantized into the same low-bit. To model the quantization parameters, a linear regressor is constructed to predict the α-quantile of activations [18], which eliminates the involvement of complex sorting algorithm. From the second perspective, AdaRound [28] analyzes that it is not advisable to simply round full precision weight to its nearest fixed-point value. Alternatively, the rounding problem is formulated as a per-layer quadratic unconstrained binary optimization problem, based on which, a continuous relaxation is introduced to find an adaptive rounding. BRECQ [19], one of the state-of-the-art methods, builds a block-wise reconstruction between the outputs of the full precision network and quantized network to achieve a balance between cross-layer dependency and generalization error. Besides, trainable clipping [4] for activations is also considered by BRECQ. Similar motivation can also be found in earlier works [1,15,36].

Zero-Shot Quantization . Zero-shot quantization is not permitted to access the training dataset. Thus, synthetic samples become an alternative to calibrate and fine-tune the quantized models. According to the methodology of data synthesis, we categorize existing studies into two groups: data optimizer [2,10,20,38,39] and data generator [5,23,37]. Data optimizer based methods produce synthetic images from Gaussian noise [38]. ZeroQ [2] forwards propagated the Gaussian inputs to collect BNS. Then the optimization that minimizes the difference between the collected BNS and the BNS in the pre-trained model is constructed to update the Gaussian inputs to ensure that the synthetic data does not deviate from the real data distribution. Except for aligning the BNS, a Domain Prior loss and an Inception loss are introduced in [10]. The former encourages nearby pixels between the input image and its Gaussian-smoothed variant to be similar. The latter prevents the model from producing inputs that lead to exploding outputs. To break the data homogenization, DSG [39] slacks the alignment of BNS and introduces a layer-wise enhancement to enhance diverse data samples. As for data generator, this group is featured with a generator in Generative Adversarial Networks (GAN) [8] to synthesize images. GDFQ [37] exploits the classification boundary knowledge and distribution information in the pre-trained model, and then devises a knowledge matching generator to produce synthetic data for model quantization. To diversify generated data, ZAQ [23] trains the quantized model and generator in an adversarial fashion by adopting an elaborated two-level discrepancy. To capture the distribution of the original data lies on the decision boundaries, Qimera [5] introduces superposed latent embeddings to produce boundary supporting samples.

3   Methodology

3.1   Preliminaries

Quantizer. Following [2,37], we adopt asymmetric uniform quantization in this paper. Given the data $\boldsymbol{x}$ (weights or activations), bit-width b, lower bound l and upper bound u, the quantizer is defined as:

$$\mathbf{q} = round(\frac{clip(\boldsymbol{x}, l, u)}{s}), \tag{1}$$

where $clip(\boldsymbol{x}, l, u) = min(max(\boldsymbol{x}, l), u)$, $round(\cdot)$ rounds its input to the nearest integer, $s = \frac{u-l}{2^b-1}$ is the scaling factor that projects a floating-point number to a fixed-point integer, and $\mathbf{q}$ is the quantized fixed-point number. The corresponding de-quantized item $\bar{\boldsymbol{x}}$ can be obtained as:

$$\bar{\boldsymbol{x}} = \mathbf{q} \cdot s. \tag{2}$$

We use layer-wise quantizer and filter-wise quantizer for activations and weights, respectively. The lower bound l and upper bound u are set to the minimum and maximum of per-layer activations (per-channel weights).

Data Synthesis. Ideally, post-training quantization completes network compression with a small calibration dataset $D = \{(\hat{\boldsymbol{x}}, \hat{y})\}$, typically consisting of one image per class[1]. However, this small calibration dataset fails to retain performance when quantizing the network to very low precision, such as 4-bit. Inspired by zero-shot quantization, we resort to data synthesis. As shown in Fig. 1, we deploy a generator G to synthesize an image $\tilde{\boldsymbol{x}}$ from a random Gaussian noise $\tilde{\boldsymbol{z}}$ conditioned on the target label $\tilde{y}$, i.e., $\tilde{\boldsymbol{x}} = G(\tilde{\boldsymbol{z}}|\tilde{y})$. We expect that the synthetic data to be similar to the real data. Despite the inaccessibility of the whole training data, we can turn to the data distribution information captured by the batch normalization statistics (BNS) in the pre-trained model F. Following [37], the BNS loss can be adopted to preserve the distribution:

$$\mathcal{L}_{\text{BNS}}^G = \sum_{l=1}^{L} \|\boldsymbol{\mu}_l'(\tilde{\boldsymbol{x}}) - \boldsymbol{\mu}_l^F\|^2 + \|\boldsymbol{\sigma}_l'(\tilde{\boldsymbol{x}}) - \boldsymbol{\sigma}_l^F\|_2^2, \tag{3}$$

where $\boldsymbol{\mu}_l^F$ and $\boldsymbol{\sigma}_l^F$ are the running mean and variance in the l-th layer of pre-trained F. $\boldsymbol{\mu}_l'(\cdot)$ and $\boldsymbol{\sigma}_l'(\cdot)$ return the mean and variance of input data in the l-th layer of F. This loss can be regarded as matching the first and second moment of real data and synthetic data.

---

[1] Occasionally, the label $\hat{y}$ is not available. In this case, it can be predicted by the pre-trained full-precision model.

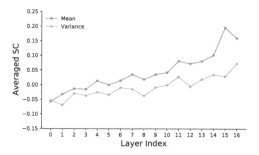

Fig. 3. The average of silhouette coefficient values *w.r.t.* the BNS in different layers.

Classification. We also use cross-entropy loss to ensure synthetic data can be correctly classified by the pre-trained model F:

$$\mathcal{L}_{\mathrm{CE}}^{G} = \mathbb{E}_{(x,y)\sim\{(\tilde{x},\tilde{y})\}}\big[\mathrm{CE}\big(F(x),y\big)\big]. \tag{4}$$

Note that, we fix F during the whole training process, and the generator G is updated instead.

3.2 Our Insights

The BNS in pre-trained F are calculated by a series of averages of different batches of the full training set, *a.k.a.*, moving average. Consequently, they capture the distribution of the whole dataset. However, it is unclear whether they can be a representative of per-class BNS or even per-image BNS. To verify this, we feed the whole ImageNet to ResNet-18 [11] and calculate per-image mean vector μ_l' and variance vector σ_l' in each layer. Figure 2 visualizes several examples via t-SNE [25].

As can be seen, the BNS over different classes vary a lot across different layers. Specifically, the BNS of different classes are overlapping in shallow layers while there is regularity in deep layers, which, we outline into two properties: **inter-class separation** and **intra-class incohesion**. The former indicates that the BNS within the same class are formed into one cluster and different classes are separable. The latter refers to a small distortion of BNS among data from the same class. To quantitatively measure these two properties, we introduce Silhouette Coefficient (SC) [33], a value of which reflects how similar an object is to its own cluster (cohesion) in comparison with other clusters (separation). SC value of one sample $\mathbf{v}$ is defined as:

$$\mathrm{SC}(\mathbf{v}) = \frac{b(\mathbf{v}) - a(\mathbf{v})}{max\big(a(\mathbf{v}), b(\mathbf{v})\big)}, \tag{5}$$

where $a(\mathbf{v})$ denotes the average of intra-cluster distance for sample $\mathbf{v}$, and $b(\mathbf{v})$ is the average of nearest-cluster distance for sample $\mathbf{v}$. Note that $b(\mathbf{v})$ is the distance between $\mathbf{v}$ and the nearest cluster that $\mathbf{v}$ is not a part of. The value of SC($\mathbf{v}$) ranging from -1 (separation) to $+1$ (cohesion). Values near 0 indicate overlapping clusters. Negative values generally indicate that a sample has been assigned to the wrong cluster, as a different cluster is more similar. With Eq. (5), we can obtain SC values for each mean vector $\boldsymbol{\mu}'_l$ and variance vector $\boldsymbol{\sigma}'_l$. Figure 3 displays the average of all input samples in each layer. As can be seen, the SC values are very small, even negative, in shallow layers, which indicates overlapping clusters. On the contrary, SC values increase in deep layers, which indicates inter-class separation. However, the increase of SC is very limited (no more than 0.25), which indicates a relatively large intra-cluster distance, *i.e.*, intra-class incohesion. These analyses are consistent with the observations in Fig. 2.

In-Depth Analysis. The observed inter-class separation and intra-class incohesion can be explained by the fact that networks extract class-unrelated universal low-level features in shallow layers such as edges and curves. While in deep layers, networks are learned to extract class-related semantic features distinguishable from other classes in deep layers, leading to inter-class separation. The intra-class incohesion results from the varying image contents though these images are from the same class. For shallow layers, the overlapping clusters hardly model a subtle per-class distribution, thus the BNS in the pre-trained model can be an alternative. However, the BNS in the pre-trained model are very coarse and the constraint of Eq. (3) cannot model the properties of per-class separation and intra-class incohesion. Thus, in addition to the coarse-grained alignment, a fine-grained BNS alignment is also necessary for deep layers.

3.3 Fine-grained BNS Alignment

To preserve the properties of inter-class separation and intra-class incohesion, in this subsection, we respectively introduce a BNS-centralized loss and a BNS-distorted loss. Details are presented below.

BNS-Centralized Loss. Since the BNS of each class are formed into one cluster, we can place a centroid as an explicit supervisory signal for each class and force the per-class distribution of synthetic images to be close to the assigned centroid. Recall that a real image $(\hat{\boldsymbol{x}}, \hat{y})$ per class can be available from the calibration dataset D in post-training quantization. It is reasonable to use the BNS of image $\hat{\boldsymbol{x}}$ as the corresponding centroid of class $\hat{y}$ since its BNS already fall into the target cluster and are separable from BNS of other classes. To this end, given $\hat{\boldsymbol{x}}$ with its label $\hat{y} = c$, we define the following BNS-centralized loss such that synthetic images can be further aligned to their corresponding centroids:

$$\mathcal{L}^G_{\text{C-BNS}} = \sum_{l=K}^{L} \|\boldsymbol{\mu}'_l(\tilde{\boldsymbol{x}}|\tilde{y} = c) - \boldsymbol{\mu}'_l(\hat{\boldsymbol{x}}|\hat{y} = c)\|^2 \\ + \|\boldsymbol{\sigma}'_l(\tilde{\boldsymbol{x}}|\tilde{y} = c) - \boldsymbol{\sigma}'_l(\hat{\boldsymbol{x}}|\hat{y} = c)\|^2, \tag{6}$$

where K is a pre-given hyper-parameter denoting the start of deep layers. In all experiments, we set $K = \text{ceil}(\frac{L}{2}) - 2$ where $\text{ceil}(\cdot)$ is the rounding up function.

BNS-Distorted Loss. Our BNS-centralized loss ensures the fine-grained separableness across different classes. However, how to retain the incohesion within the same class remains an issue. To solve this, we further propose to distort the centroid of per-class BNS by introducing Gaussian noise and define the following BNS-distroted loss:

$$\mathcal{L}_{\text{D-BNS}}^{G} = \sum_{l=K}^{L} \left\| \boldsymbol{\mu}_l'(\tilde{\boldsymbol{x}}|\tilde{y} = c) - \mathcal{N}\big(\boldsymbol{\mu}_l'(\hat{\boldsymbol{x}}|\hat{y} = c), \boldsymbol{v}_{\mu}\big) \right\|^2 \tag{7}$$
$$+ \left\| \boldsymbol{\sigma}_l'(\tilde{\boldsymbol{x}}|\tilde{y} = c) - \mathcal{N}\big(\boldsymbol{\sigma}_l'(\hat{\boldsymbol{x}}|\hat{y} = c), \boldsymbol{v}_{\sigma}\big) \right\|^2,$$

where $\boldsymbol{v}_{\mu} = 0.5$ and $\boldsymbol{v}_{\sigma} = 1.0$ are used to control the distortion degrees of mean and variance. For each synthetic data $\tilde{\boldsymbol{x}}$, its target is sampled from a Gaussian distribution centered on the class centroid of $\tilde{\boldsymbol{x}}$. Thus, our BNS-distorted loss provides diverse distorted centroids which prevent the BNS of per synthetic data from overfitting its centroid. This further retain the intra-class incohesion.

Our experimental results in Sect. 4.3 show that the BNS-distorted loss can retain the inter-class separation to some extent since the distorted centroid for synthetic data is centered on the corresponding class centroid. However, it is hard to manually model the Gaussian noise exactly such that the inter-class separableness and intra-class incohesion can be well preserved in the synthetic images simultaneously. Thus, both BNS-centralized loss and BNS-distorted loss are necessary as verified in the experiment.

3.4   Model Quantization

Classification. To take full use of the available data, both calibration images and synthetic images are used to fine-tune the quantized model Q, which can be realized through the cross-entropy loss:

$$\mathcal{L}_{\text{CE}}^{Q} = \mathbb{E}_{(\boldsymbol{x},y)\sim D\cup\{(\tilde{\boldsymbol{x}},\tilde{y})\}} \big[\text{CE}\big(Q(\boldsymbol{x}), y\big)\big]. \tag{8}$$

Distillation. It is possible that the synthetic image does not include corresponding class-specific features. As a result, $\tilde{y}$ may be unreliable. Thus, we apply knowledge distillation (KD) [12] to transfer the outputs of full-precision model F to quantized model Q so that even though the synthetic image may have an inaccurate label, Q can still be correctly optimized by learning the soft target provided by F. Moreover, KD is also beneficial to the learning of calibration data. The KD loss is defined by the Kullback-Leibler distance $\text{KL}(\cdot,\cdot)$ as:

$$\mathcal{L}_{\text{KD}}^{Q} = \mathbb{E}_{(\boldsymbol{x},y)\sim D\cup\{(\tilde{\boldsymbol{x}},\tilde{y})\}} \big[\text{KL}\big(Q(\boldsymbol{x}), F(\boldsymbol{x})\big)\big]. \tag{9}$$

3.5   Training Process

The training of our method consists of updating the generator G and the quantized model Q, where Q is obtained by quantizing the pre-trained full-precision model F. G produces a set of synthetic images while Q is trained with the aid of synthetic images and calibration images. We also emphasize that the F is fixed without any updating during the whole training process.

Updating Generator G. With a random Gaussian noise $\tilde{z}$ conditional on label $\tilde{y}$ as its input, the generator G synthesize an image $\tilde{x} = G(\tilde{z}|\tilde{y})$, which is then used for classification of Eq. (4) and preserving distribution of training set including the coarse alignment of Eq. (3), inter-class separation of Eq. (6) and intra-class incohesion of Eq. (7). Thus, the overall loss for the generator G is derived as:

$$\mathcal{L}^G = \alpha_1 \cdot \mathcal{L}_{CE}^G + \alpha_2 \cdot \mathcal{L}_{BNS}^G + \alpha_3 \cdot \mathcal{L}_{D\text{-}BNS}^G + \alpha_4 \cdot \mathcal{L}_{C\text{-}BNS}^G, \tag{10}$$

where the $\alpha_1, \alpha_2, \alpha_3$ and α_4 are the trade-off parameters.

Updating Quantized Model Q. The quantized model Q takes the synthetic data and calibration data as its inputs, and then the classification loss of Eq. (8) and distillation loss of Eq. (9) are constructed to retain the performance. Thus, the overall loss for the quantized model Q is derived as:

$$\mathcal{L}^Q = \mathcal{L}_{CE}^Q + \alpha_5 \cdot \mathcal{L}_{KD}^Q, \tag{11}$$

where α_5 is a trade-off parameter.

4 Experimentation

4.1 Implementation Details

We choose to quantize ResNet-18 [11], MobileNetV1 [13], MobileNetV2 [35] and RegNet-600MF [32]. All experiments are conducted on ImageNet with 1.2 million training images and 50,000 validation images from 1,000 classes [34]. The calibration dataset consists of 1,000 images including one image per class. We report the top-1 accuracy and the code is implemented using Pytorch [31]. Following the GDFQ [37], all full-precision models are downloaded from *pytorchcv*.

For ease of implementation, we directly import the generator from GDFQ [37] to produce synthetic images. The initial learning rates for the generator and quantized network are set to 10^{-3} and 10^{-6} respectively. For the generator, the optimizer is Adam [16] with 0.9 as the momentum and the learning rate are multiplied by 0.1 every 100 epochs. For the quantized network, the optimizer is SGD with Nesterov [30] with 10^{-4} as the weight decay and we adjust the learning rate using the cosine annealing [24]. Before formal training, we set up a warm-up updating of the generator G for 50 epochs. Then, a total of 350 epochs are used to update the generator G and quantized model Q.

4.2 Experimental Results

Comparison with Zero-Shot Methods. We first compare with the zero-shot GDFQ [37], since the data synthesis of our FDDA is built upon the framework of GDFQ. Table 1 displays our experimental results when all layers of ResNet-18 [11] are quantized to 4-bit. The accuracy of the full-precision model

Table 1. Comparison with GDFQ [37] by quantizing all layers of ResNet-18 to 4-bit. TAQ denotes training-aware quantization. "C" indicates the calibration dataset. "F" represents fine-grained data distribution alignment. Note that GDFQ + C + F = FDDA.

Method	Acc. (%)
Full precision	71.47
TAQ	68.24
GDFQ	60.60
GDFQ + C	65.64
GDFQ + C + F (**FDDA**)	**68.88**
FDDA + w/o label	**68.68**

Table 2. Comparison with zero-shot methods when all layers of ResNet-18 are quantized to 4-bit. "C" indicates the calibration dataset.

Method	Acc. (%)
Full precision	71.47
DI [38] + C	65.68
ADI [38] + C	66.30
ZeroQ [2] + C	66.89
ZAQ [23] + C	29.50
GDFQ [37] + C	65.64
Qimera [5] + C	66.19
DSG [39] + C	66.83
FDDA	**68.88**

decreases from 71.47% to around 68.24% when the training-aware quantization, which requires all training data, is applied. Regarding GDFQ which only considers the synthetic data for fine-tuning the quantized model, performance severely degenerates to 60.60%. Such poor performance disables the application of GDFQ. Given the calibration dataset, GDFQ can increase to 65.48%, well demonstrating the correctness of our motive in combining real calibration dataset with synthetic data. Nevertheless, the coarse BNS alignment in GDFQ fails to model the fine-grained properties of inter-class separableness and intra-class incohesion. In contrast, our FDDA increases the performance to 68.88%, better than the training-aware quantization. This result shows the efficacy of our BNS-centralized loss and BNS-distorted loss in synthesizing better images.

In our settings, we assume to have access to the image labels. However, these labels are sometimes not available in real applications. Luckily, the pre-trained full-precision model can be used to predict these labels. In Table 1, we also report the performance of our FDDA, *i.e.*, 68.68%, using predicted labels. The slight drops are attributed to some of the misclassified labels. Nevertheless, our FDDA without real labels still maintains better performance than GDFQ with real images, well demonstrating the importance of preserving the fine-grained inter-class separableness and intra-class incohesion in learning to synthesize images.

Table 2 further shows the comparison between our FDDA with advanced zero-shot studies including data optimizer based methods [2,38,39] and data generator based methods [5,23,37]. For the former group, we use the calibration dataset as well as 10,000 synthetic images to train the quantized model, and the training process is the same with ours. For the latter group, we insert the calibration dataset into their training process. In Table 2, when the calibration dataset is applied to all methods, our FDDA still outperforms the advance of DSG by a large margin of 2.05%, which again demonstrates the efficacy of our BNS-centralized loss and BNS-distorted loss.

Table 3. Comparisons with existing post-training quantization methods. WBAB indicates the weights and activations are quantized to B-bit while FBLB indicates the first layers and last layers are quantized to B-bit.

	Methods	ResNet-18	MobileNetV1	MobileNetV2	RegNet-600MF
Settings	Full precision	71.47	73.39	72.49	73.71
W5A5, F8L8	ACIQ-Mix [1]	68.34	52.34	61.74	69.53
	AdaQuant [14]	68.56	–	65.19	–
	Bit-Split [36]	69.10	–	–	–
	BRECQ [19]	70.60	70.16	70.83	73.38
	FDDA(Ours)	**70.86**	**71.16**	**71.99**	**73.99**
W4A4, F8L8	ACIQ-Mix [1]	67.0	5.06	39.49	54.22
	AdaQuant [14]	67.50	–	34.95	–
	Bit-Split [36]	67.56	–	–	–
	BRECQ [19]	69.60	63.66	66.57	68.33
	FDDA(Ours)	**69.76**	**65.76**	**69.32**	**70.33**
W5A5, F5L5	ACIQ-Mix [1]	66.80	51.65	60.42	69.13
	AdaQuant [14]	68.19	–	63.61	–
	Bit-Split [36]	68.88	–	–	–
	BRECQ [19]	70.27	66.51	70.26	72.78
	FDDA(Ours)	**70.56**	**70.26**	**71.63**	**73.62**
W4A4, F4L4	ACIQ-Mix [1]	57.47	4.68	34.84	51.74
	AdaQuant [14]	63.45	–	34.64	–
	Bit-Split [36]	67.49	–	–	–
	BRECQ [19]	67.94	57.11	63.64	66.17
	FDDA(Ours)	**68.88**	**63.75**	**68.38**	**68.96**

Comparison with Competitors. We compare with the recent studies on post-training quantization [1,14,19,36]. The quantized networks include ResNet-18 [11], MobileNetV1 [13], MobileNetV2 [35] and RegNet-600MF [32]. All networks are quantized to the low precision 5-bit and 4-bit. Besides, to show the advantage of our FDDA, we quantize the first and last layers to 8-bit and lower precision (5-bit or 4-bit). Table 3 shows the experimental results.

If the first and last layers of full-precision models are quantized to 8-bit (F8L8), both our FDDA and recent SOTA BRECQ [19] can retain a high performance of the full-precision models regardless of 5-bit (W5A5) or 4-bit (W4A4) weights and activations in other layers. Comparing to BRECQ, FDDA obtains performance gains by 0.26%, 1.00%, 1.16% and 0.61% when quantizing ResNet-18, MobileNetV1, MobileNetV2 and RegNet-600MF to W5A5, while they are 0.16%, 2.10%, 2.75% and 2.00% when quantized to W4A4. Our FDDA retains better performance than BRECQ when quantizing light-weight models such as MobileNets, particularly when lower precision, such as 4-bit, is performed.

When the first and last layers of full-precision models are quantized to lower precision (F5L5 or F4L4) as well, we notice that FDDA outperforms BRECQ by

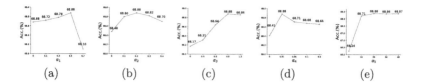

(a)   (b)   (c)   (d)   (e)

Fig. 4. Influence of the trade-off parameters.

margins. Specifically, our FDDA increases the performance of BRECQ in W5A5 by 0.29%, 3.75%, 1.37% and 0.84% *w.r.t.* ResNet-18, MobileNetV2, MobileNetV2 and RegNet-600MF, and the performance gains are 0.94%, 6.64%, 4.74%, 2.79% in the case of W4A4. These results verify our statement in the introduction that the performance improvements of existing studies are on the premise that the first and last layers are quantized to higher precision, and also demonstrates the efficacy of our combining real calibration dataset with synthetic data.

4.3   Ablation Studies

We further study the influence of the trade-off parameters in Eq. (10) and Eq. (11), and the number of available images in the calibration dataset. All experiments are conducted by quantizing all layers of ResNet-18 to 4-bit.

Trade-off Parameters. We first display the influence of different trade-off parameters in Fig. 4. The α_1, α_2, α_3, and α_4 from Eq. (10) balance different losses in updating the generator while α_5 from Eq. (11) balances the losses in updating the quantized model. Each α_i is first empirically initialized. Then, for α_i, we search its optimal value using the grid search with others fixed. From Fig. 4, we can see that the optimal configurations of these three parameters are $\alpha_1 = 0.5$, $\alpha_2 = 0.2$, $\alpha_3 = 0.9$, $\alpha_4 = 0.05$ and $\alpha_5 = 20$, which are also our settings for all the aforementioned experiments. Though they might not be the optimal for all networks, we find these configurations already bring better performance than the recent state-of-the-arts. Also, we observe that $\alpha_4 \ll \alpha_3$, in which α_3 and α_4 respectively balance the importance of the proposed BNS-distorted loss and BNS-centralized loss. This is due to the distorted centroid for synthetic data is centered on the corresponding class centroid, thus the BNS-distorted loss can retain the inter-class separation to some extent, which partly relieves the involvement of BNS-centralized loss and leads to a small α_4.

Effect of Available Classes. The calibration dataset consists of 1,000 images including one image per class by default. However, some images might be missing in real-world applications. Consequently, the class information of corresponding image is not available for our fine-grained BNS alignment. In Fig. 5, we further excavate the influence of available classes on our final performance. For unavailable classes, we omit their C-BNS and D-BNS loss when computing the Eq. (10).

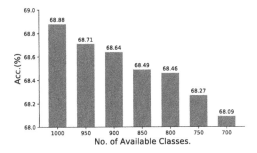

Fig. 5. Effect of available classes.

Note that in this case, the size of the calibration dataset is equal to the number of available classes. It can be seen from Fig. 5 that performance drops as the available classes decrease. Nevertheless, comparing to the recent advance, BRECQ, which obtains only 67.94% top-1 accuracy (see Table 3), our FDDA still maintains a higher performance of 68.09% even when only 700 classes are available. The good performance can be attributed to two reasons. On one hand, synthetic data benefits post-training quantization even though some classes are missing. On the other hand, the fine-grained data alignment helps to synthesize better images for fine-tuning the quantized model.

5   Conclusion

In this paper, we proposed a fine-grained data distribution alignment (FDDA) to solve the insufficient data problem in post-training quantization. We observed two important BNS properties of the inter-class separation and intra-class incohesion in the deep layers of neural network. To retain these two fine-grained distribution information, we respectively proposed the BNS-centralized loss and BNS-distorted loss. Using a real image from the calibration dataset as the centroid of each class, the BNS-centralized loss constrains the BNS of synthetic data to be close to the BNS of its class centroid, while the BNS-distorted loss introduces Gaussian noise to distort the class centroid for the purpose of incohesion. By retaining these two properties in the synthetic data, our FDDA shows its superiority over the state-of-the-art competitors on ImageNet, particularly in the hardware-friendly case where the first and last layers of networks are also quantized to low precision.

Acknowledgements. . This work was supported by the National Science Fund for Distinguished Young Scholars (No. 62025603), the National Natural Science Foundation of China (No. U21B2037, No. 62176222, No. 62176223, No. 62176226, No. 62072386, No. 62072387, No. 62072389, and No. 62002305), Guangdong Basic and Applied Basic Research Foundation (No. 2019B1515120049), and the Natural Science Foundation of Fujian Province of China (No. 2021J01002).

References

1. Banner, R., Nahshan, Y., Soudry, D., et al.: Post training 4-bit quantization of convolutional networks for rapid-deployment. In: Proceedings of the Advances in Neural Information Processing Systems (NeurIPS), pp. 7950–7958 (2019)
2. Cai, Y., Yao, Z., Dong, Z., Gholami, A., Mahoney, M.W., Keutzer, K.: Zeroq: a novel zero shot quantization framework. In: Proceedings of the IEEE/CVF Conference on Computer Vision and Pattern Recognition (CVPR), pp. 13169–13178 (2020)
3. Chen, H., et al.: Data-free learning of student networks. In: Proceedings of the IEEE/CVF International Conference on Computer Vision (ICCV), pp. 3514–3522 (2019)
4. Choi, J., Wang, Z., Venkataramani, S., Chuang, P.I.J., Srinivasan, V., Gopalakrishnan, K.: Pact: parameterized clipping activation for quantized neural networks. arXiv preprint arXiv:1805.06085 (2018)
5. Choi, K., Hong, D., Park, N., Kim, Y., Lee, J.: Qimera: data-free quantization with synthetic boundary supporting samples. In: Proceedings of the Advances in Neural Information Processing Systems (NeurIPS) (2021)
6. Fang, J., Shafiee, A., Abdel-Aziz, H., Thorsley, D., Georgiadis, G., Hassoun, J.H.: Post-training piecewise linear quantization for deep neural networks. In: Proceedings of the European Conference on Computer Vision (ECCV), pp. 69–86 (2020)
7. Gholami, A., Kim, S., Dong, Z., Yao, Z., Mahoney, M.W., Keutzer, K.: A survey of quantization methods for efficient neural network inference. arXiv preprint arXiv:2103.13630 (2021)
8. Goodfellow, I.J., et al.: Generative adversarial nets. In: Proceedings of the Advances in Neural Information Processing Systems (NeurIPS), pp. 2672–2680 (2014)
9. Han, S., Pool, J., Tran, J., Dally, W.J., et al.: Learning both weights and connections for efficient neural network. In: Proceedings of the Advances in Neural Information Processing Systems (NeurIPS), pp. 1135–1143 (2015)
10. Haroush, M., Hubara, I., Hoffer, E., Soudry, D.: The knowledge within: methods for data-free model compression. In: Proceedings of the IEEE/CVF Conference on Computer Vision and Pattern Recognition (CVPR), pp. 8494–8502 (2020)
11. He, K., Zhang, X., Ren, S., Sun, J.: Deep residual learning for image recognition. In: Proceedings of the IEEE/CVF Conference on Computer Vision and Pattern Recognition (CVPR), pp. 770–778 (2016)
12. Hinton, G., Vinyals, O., Dean, J.: Distilling the knowledge in a neural network. arXiv preprint arXiv:1503.02531 (2015)
13. Howard, A.G., et al.: Mobilenets: efficient convolutional neural networks for mobile vision applications. arXiv preprint arXiv:1704.04861 (2017)
14. Hubara, I., Nahshan, Y., Hanani, Y., Banner, R., Soudry, D.: Improving post training neural quantization: layer-wise calibration and integer programming. arXiv preprint arXiv:2006.10518 (2020)
15. Hubara, I., Nahshan, Y., Hanani, Y., Banner, R., Soudry, D.: Accurate post training quantization with small calibration sets. In: Proceedings of the International Conference on Machine Learning (ICML), pp. 4466–4475 (2021)
16. Kingma, D.P., Ba, J.: Adam: a method for stochastic optimization. In: Proceedings of the International Conference on Learning Representations (ICLR) (2014)
17. Krishnamoorthi, R.: Quantizing deep convolutional networks for efficient inference: a whitepaper. arXiv preprint arXiv:1806.08342 (2018)

18. Kryzhanovskiy, V., Balitskiy, G., Kozyrskiy, N., Zuruev, A.: Qpp: real-time quantization parameter prediction for deep neural networks. In: Proceedings of the IEEE/CVF Conference on Computer Vision and Pattern Recognition (CVPR), pp. 10684–10692 (2021)
19. Li, Y., et al.: Brecq: pushing the limit of post-training quantization by block reconstruction. In: Proceedings of the International Conference on Learning Representations (ICLR) (2021)
20. Li, Y., et al.: Mixmix: all you need for data-free compression are feature and data mixing. In: Proceedings of the IEEE/CVF International Conference on Computer Vision (ICCV), pp. 4410–4419 (2021)
21. Lin, M., et al.: Rotated binary neural network. In: Proceedings of the Advances in Neural Information Processing Systems (NeurIPS), pp. 7474–7485 (2020)
22. Liu, X., Ye, M., Zhou, D., Liu, Q.: Post-training quantization with multiple points: mixed precision without mixed precision. In: Proceedings of the AAAI Conference on Artificial Intelligence (AAAI), pp. 8697–8705 (2021)
23. Liu, Y., Zhang, W., Wang, J.: Zero-shot adversarial quantization. In: Proceedings of the IEEE/CVF Conference on Computer Vision and Pattern Recognition (CVPR), pp. 1512–1521 (2021)
24. Loshchilov, I., Hutter, F.: SGDR: stochastic gradient descent with warm restarts. In: Proceedings of the International Conference on Learning Representations (ICLR) (2016)
25. van der Maaten, L., Hinton, G.: Visualizing data using t-sne. J. Mach. Learn. Res. (JMLR) **9**, 2579–2605 (2008)
26. Martinez, J., Shewakramani, J., Liu, T.W., Bârsan, I.A., Zeng, W., Urtasun, R.: Permute, quantize, and fine-tune: efficient compression of neural networks. In: Proceedings of the IEEE/CVF Conference on Computer Vision and Pattern Recognition (CVPR), pp. 15699–15708 (2021)
27. Martinez, J., Zakhmi, S., Hoos, H.H., Little, J.J.: Lsq++: lower running time and higher recall in multi-codebook quantization. In: Proceedings of the European Conference on Computer Vision (ECCV), pp. 491–506 (2018)
28. Nagel, M., Amjad, R.A., Van Baalen, M., Louizos, C., Blankevoort, T.: Up or down? adaptive rounding for post-training quantization. In: Proceedings of the International Conference on Machine Learning (ICML), pp. 7197–7206 (2020)
29. Nahshan, Y., et al.: Loss aware post-training quantization. arXiv preprint arXiv:1911.07190 (2019)
30. Nesterov, Y.E.: A method of solving a convex programming problem with convergence rate o(k^2). In: Proceedings of the Russian Academy of Sciences (RAS), pp. 543–547 (1983)
31. Paszke, A., et al.: Pytorch: an imperative style, high-performance deep learning library. In: Proceedings of the Advances in Neural Information Processing Systems (NeurIPS), pp. 8026–8037 (2019)
32. Radosavovic, I., Kosaraju, R.P., Girshick, R., He, K., Dollár, P.: Designing network design spaces. In: Proceedings of the IEEE/CVF Conference on Computer Vision and Pattern Recognition (CVPR), pp. 10428–10436 (2020)
33. Rousseeuw, P.J.: Silhouettes: a graphical aid to the interpretation and validation of cluster analysis. J. Comput. Appl. Math. (JCAM) **20**, 53–65 (1987)
34. Russakovsky, O., et al.: Imagenet large scale visual recognition challenge. Int. J. Comput. Vision (IJCV) **115**, 211–252 (2015)
35. Sandler, M., Howard, A., Zhu, M., Zhmoginov, A., Chen, L.C.: Mobilenetv 2: inverted residuals and linear bottlenecks. In: Proceedings of the IEEE/CVF Conference on Computer Vision and Pattern Recognition (CVPR), pp. 4510–4520 (2018)

36. Wang, P., Chen, Q., He, X., Cheng, J.: Towards accurate post-training network quantization via bit-split and stitching. In: Proceedings of the International Conference on Machine Learning (ICML), pp. 9847–9856 (2020)
37. Xu, S., Li, H., Zhuang, B., Liu, J., Cao, J., Liang, C., Tan, M.: Generative low-bitwidth data free quantization. In: Proceedings of the European Conference on Computer Vision (ECCV), pp. 1–17 (2020)
38. Yin, H., et al.: Dreaming to distill: Data-free knowledge transfer via deepinversion. In: Proceedings of the IEEE/CVF Conference on Computer Vision and Pattern Recognition (CVPR), pp. 8715–8724 (2020)
39. Zhang, X., et al.: Diversifying sample generation for accurate data-free quantization. In: Proceedings of the IEEE/CVF Conference on Computer Vision and Pattern Recognition (CVPR), pp. 15658–15667 (2021)
40. Zhang, Y., Chen, H., Chen, X., Deng, Y., Xu, C., Wang, Y.: Data-free knowledge distillation for image super-resolution. In: Proceedings of the IEEE/CVF Conference on Computer Vision and Pattern Recognition (CVPR), pp. 7852–7861 (2021)
41. Zhong, Y., et al.: Intraq: learning synthetic images with intra-class heterogeneity for zero-shot network quantization. In: Proceedings of the IEEE/CVF Conference on Computer Vision and Pattern Recognition (CVPR), pp. 12339–12348 (2022)

Learning with Recoverable Forgetting

Jingwen Ye[1], Yifang Fu[1], Jie Song[2], Xingyi Yang[1], Songhua Liu[1], Xin Jin[3], Mingli Song[2], and Xinchao Wang[1(✉)]

[1] National University of Singapore, Queenstown, Singapore
{jingweny,xinchao}@nus.edu.sg, {e0724403,xyang,songhua.liu}@u.nus.edu
[2] Zhejiang University, Hangzhou, China
{sjie,brooksong}@zju.edu.cn
[3] Eastern Institute of Advanced Study, Ningbo, China
jinxin@eias.ac.cn

Abstract. Life-long learning aims at learning a sequence of tasks without forgetting the previously acquired knowledge. However, the involved training data may not be life-long legitimate due to privacy or copyright reasons. In practical scenarios, for instance, the model owner may wish to enable or disable the knowledge of specific tasks or specific samples from time to time. Such flexible control over knowledge transfer, unfortunately, has been largely overlooked in previous incremental or decremental learning methods, even at a problem-setup level. In this paper, we explore a novel learning scheme, termed as **L**earning w**I**th **R**ecoverable **F**orgetting (LIRF), that explicitly handles the task- or sample-specific knowledge removal and recovery. Specifically, LIRF brings in two innovative schemes, namely knowledge *deposit* and *withdrawal*, which allow for isolating user-designated knowledge from a pre-trained network and injecting it back when necessary. During the knowledge deposit process, the specified knowledge is extracted from the target network and stored in a deposit module, while the insensitive or general knowledge of the target network is preserved and further augmented. During knowledge withdrawal, the taken-off knowledge is added back to the target network. The deposit and withdraw processes only demand for a few epochs of finetuning on the removal data, ensuring both data and time efficiency. We conduct experiments on several datasets, and demonstrate that the proposed LIRF strategy yields encouraging results with gratifying generalization capability.

Keywords: Life-long learning · Incremental learning · Machine unlearning · Knowledge transfer

1 Introduction

Life-long learning finds its application across a wide spectrum of domains and has been a long-standing research task. Its main goal is to update a network to adapt

Supplementary Information The online version contains supplementary material available at https://doi.org/10.1007/978-3-031-20083-0_6.

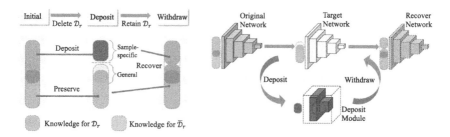

Fig. 1. Illustration of the proposed LIRF framework, comprising the knowledge deposit process and knowledge withdrawal process.

to new data, such as new instances or samples from a new class, without forgetting the learned knowledge on the past data. In some scenarios, on the contrary, we wish to deliberately forget or delete specified knowledge stored in the model, due to privacy or copyright issues. This task, known as, machine unlearning, has also attracted increasing attentions from the industry and research community due to its practical value.

Nevertheless, prior attempts in machine unlearning have been mostly focused on deleting the specified knowledge for good, meaning that once removed, it is not possible to revert the knowledge back. Despite the absolute IP protection, such knowledge deletion scheme indeed introduces much inconvenience in terms of the user control and largely hinders the flexibility of model interaction.

In this paper, we explore a novel learning scenario, which explicitly allows for the extracted knowledge from a pre-trained networked to be deposited and, whenever needed, injected back to the model. Such a flexible learning strategy grants users a maximum degree of freedom in terms of control over task- or sample-specific knowledge, and meanwhile ensures the network IP protection. Admittedly, this ambitious goal inevitably leads to a more challenging problem to tackle, since again we seek a portable modulation of knowledge on and off a pre-trained network.

To this end, we propose a dedicated scheme, termed as Learning with Recoverable Forgetting (LIRF). We illustrate the overall pipeline of LIRF in Fig. 1, When there is a request for deleting specified knowledge, denoted as $\mathcal{D}_r$ (with $\overline{\mathcal{D}}_r$ preserved), due to for example IP issues, LIRF isolates such knowledge from the pre-trained original network and stores it in a *deposit module*; the remaining network with $\mathcal{D}_r$ extracted is then denoted as the *target network*. When the IP issue is resolved and the model owner requests to revert the knowledge back or re-enables $\mathcal{D}_r$, LIRF withdraws the deposited knowledge and amalgamates it with the target network to produce the *recover network*. Specifically, during the knowledge deposit process, we partition the knowledge of the original networks, trained using full data, into sample-specific and general part. The former is deposited to a deposit module consisting of pruned blocks from the original network, while the latter is preserved in the target network.

Our contributions are therefore summarized as follows.

- We introduce a novel yet practical life-long learning setup, recoverable knowledge forgetting. In contrast to machine unlearning settings that delete specified knowledge for good, recoverable forgetting enables knowledge isolation and recovery from a pre-trained network, which brings in network IP protection alongside user flexibility and control.
- We develop the LIRF framework that explicitly allows for knowledge deposit and withdrawal, to achieve recoverable knowledge forgetting. LIRF is time- and data-efficient, as the deposit process requires only a few epochs to fine-tuning on the specified samples.
- Experimental results have verified the effectiveness of the proposed method, under various settings including class-incremental learning and machine unlearning.

2   Related Work

2.1   Life-Long Learning

Life-long/online/incremental learning, which is capable of learning, retaining and transferring knowledge over a lifetime, has been a long-standing research area in many fields [24,36,41]. As the pioneer work, Li *et al.* [23] propose Learning without Forgetting (LwF) by using only the new-coming examples for the new task's training, while preserving the responses on the existing tasks to prevent catastrophic forgetting. Peng *et al.* [29] present to train the hierarchical softmax function for deep language models for the new-coming tasks. FSLL [25] is proposed to perform on the few-shot setting by selecting very few parameters from the model. Apart from those works that still need part of the old data, many researchers are devoted to developing the methods without storing the old data by synthesizing old data [35,39] or even without referring to any old data [27,36,37]. In addition to the above works that tend to forbid the catastrophic forgetting of the old tasks, some researchers [13–15,56] pay more attention on the decremental cases where some features may vanish while feature evolving. Hou *et al.* [15] attempt to compress important information of vanished features into functions of survived features, and then expand to include the augmented features in the one-pass learning way. Zhang *et al.* [56] propose discrepancy measure for data with evolving feature space and data distribution.

Different from the current researches on life-long learning, we propose the more flexible learning scheme, which is capable of dealing with both the data adding and deleting cases.

2.2   Knowledge Transfer

Knowledge transfer aims at transferring knowledge from networks to networks. Here, we mainly discuss the related works in knowledge distillation [10,11,46], which trains a student model of a compact size by learning from a larger teacher model or a set of teachers handling the same task. It has been successfully conducted in deep model compression [53], incremental learning [32], continual learning [20,49] and other tasks other than classification [4,17,18,31,42,43,45,47,52].

In addition to the above methods that transfer knowledge from one network to another, it could happen in plenty forms. Such as for combining or amalgamating multi-source knowledge, Gao *et al.* [8] introduce a multi-teacher and single-student knowledge concentration approach. And in order to handle multi-task problems in one single network, knowledge amalgamation [48] is proposed to train the student network on multiple scene understanding tasks, leading to better performance than the teachers. To make it further, Ye *et al.* [50] apply a two-step filter strategy to customize the arbitrary task set on TargetNet. Besides, the multi-stage knowledge transfer is enabled by Yuan *et al.* [54] to design a multi-stage knowledge distillation paradigm to decompose the distillation process.

Knowledge distillation could also be a reliable method to transfer knowledge from old data to new data, and there are also some distillation-based works [6,7, 16,38] for solving the coming new data in life-long learning setting. Cheraghian *et al.* [6] address the problem of few-shot class incremental learning by utilizing the semantic information. Hu *et al.* [16] derive a distillation method to retain the old effect overwhelmed by the new data effect, and thus alleviate the forgetting of the old class in testing.

These knowledge transfer methods transfer knowledge from networks to networks, we make the first work to filter and deposit the knowledge.

2.3   Machine Unlearning

The concept of unlearning is firstly introduced by Bourtoule *et al.* [2] to eliminate the effect of data point(s) on the already trained model. Along this line, Neel *et al.* [26] give the first data deletion algorithms. To minimize the retraining time, data removal-enabled forests [3] are introduced as a variant of random forests that enables the removal of training data. Sekhari *et al.* [33] initiate a rigorous study of generalization in machine unlearning, where the goal is to perform well on previously unseen datapoints and the focus is on both computational and storage complexity. Gupta *et al.* [9] give a general reduction from deletion guarantees against adaptive sequences to deletion guarantees against non-adaptive sequences, using differential privacy and its connection to max information. Nguyen *et al.* [28] study the problem of approximately unlearning a Bayesian model from a small subset of the training data to be erased. As machine unlearning is studied for data privacy purpose, Chen *et al.* [5] firstly study on investigating the unintended information leakage caused by machine unlearning.

The previous works only consider the data deletion with the optimization objective of getting the same model as re-training without the deletion data. The proposed LIRF framework only deleted sample-specific knowledge, which can be stored for future use.

3   Knowledge Deposit and Withdrawal

The proposed LIRF framework focuses on the class-level life-long problem, in which the samples from multiple classes may be deposited or withdrawn.

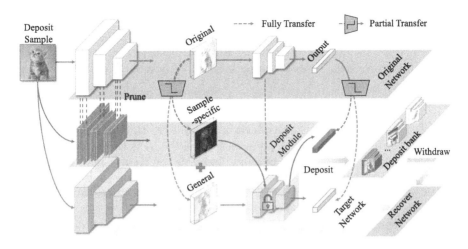

Fig. 2. The proposed LIRF framework. The knowledge is transferred fully and partially from the original network to the deposit module and the target network. The recover network is withdrawn from the target net and the deposit module.

We define our new problem illustrated in Fig. 1. Let $\mathcal{D}$ be the full original dataset, and the original network directly trained on $\mathcal{D}$ is donated as $\mathcal{T}_0$. In this problem, each of the learned samples is assigned to either deposit set or preservation set. Formally,

- **Deposit set** $\mathcal{D}_r$: A set of samples that should be forgotten at the target network $\mathcal{T}$, and remembered at the deposit module $\mathcal{T}_r$;
- **Preservation set** $\overline{\mathcal{D}}_r$: A set of samples that should be memorized at the target network (the complement of $\mathcal{D}_r$).

For clarity, we discuss on the case that one deposit set is required for deposit and withdrawal, which could be definitely generalized to multiple deposit sets.

Definition 1 (Deposit Problem). The Learning with knowledge deposit problem is defined as follows: Learn two models, one is $\mathcal{T} : \mathcal{X} \to \mathcal{Y}$ that should map an input x to its correct class label y if $x \subset \overline{\mathcal{D}}_r$, while map x to a wrong class label if $x \subset \mathcal{D}_r$; the other one is $\mathcal{T}_r : \mathcal{X} \to \mathcal{F}$ that stores the knowledge of set $\mathcal{D}_r$. *Constraints*: Only the original network $\mathcal{T}_0$ and deposit set $\mathcal{D}_r$ are available.

Definition 2 (Withdraw Problem). The Learning with knowledge withdraw problem is defined as follows: Recover a model $\widetilde{\mathcal{T}} : \mathcal{X} \to \mathcal{Y}$ that should map an input x to its correct class label y for both $x \subset \mathcal{D}_r$, and $x \subset \overline{\mathcal{D}}_r$. *Constraints*: Only the target network $\mathcal{T}$ and deposit module $\mathcal{T}_r$ are available.

4   Learning with Recoverable Forgetting

The essence of this work is to deposit and withdraw the sample-specific knowledge for the deleted data in the learning with recoverable forgetting way, which, we call LIRF framework. LIRF consists of two processes, one is knowledge deposit that transfers knowledge from original network to target network and deposit module, the other is knowledge withdrawal that recovers the knowledge back to the recover net. These two processes can be described as:

$$\mathcal{T}_0 \xrightarrow[\mathcal{D}_r]{\text{Deposit}} \{\mathcal{T}, \mathcal{T}_r\} \xrightarrow{\text{Withdraw}} \widetilde{\mathcal{T}}, \tag{1}$$

where $\mathcal{T}_0$ is the original network trained on the full set $\mathcal{D}$, $\mathcal{T}$ is the target network specified for the task of the preservation set $\overline{\mathcal{D}}_r$, $\mathcal{T}_r$ is the deposit module that only works as a knowledge container and $\widetilde{\mathcal{T}}$ is the recover network that is expected to recover all the prediction capacity of the full data set $\mathcal{D}$.

Now, given the original network $\mathcal{T}_0$ and the deposit set $\mathcal{D}_r$, the goal of LIRF is to learn $\mathcal{T}$, $\mathcal{T}_r$ and $\widetilde{\mathcal{T}}$, which includes three steps. Firstly, LIRF filters knowledge out of the original network to get the target net, at the meanwhile, it deposits the filtered sample-specific knowledge to the deposit module, and finally for recover request, LIRF withdraws the knowledge from the deposit module to recover net. Figure 2 provides an overall sketch of LIRF framework.

4.1   Filter Knowledge Out of Target Net

In the process of knowledge deposit, the objective of target net is to remove the sample-specific knowledge of $\mathcal{D}_r$ while maintaining the performance on $\mathcal{D}_r$.

To begin with, we divide the original network $\mathcal{T}_0$ into two modules at the n-th block, which are denoted as $\mathcal{T}_0^{(-n)}$ and $\mathcal{T}_0^{(n-)}$, respectively. And the target network is divided in the same way as $\mathcal{T} = \mathcal{T}^{(-n)} \circ \mathcal{T}^{(n-)}$. As has been discussed in the previous work [21] that upper layers are preferable for transfer in life-long learning setting, $\mathcal{T}_0^{(n-)}$ is fully transferred to the target network. That is, we fix the last few blocks ($\mathcal{T}^{(n-)} = \mathcal{T}_0^{(n-)}$) and expect this transfer configuration to benefit tasks that share high-level concepts but have low-level feature differences. Thus, we fully transfer $\mathcal{T}_0^{(n-)}$ to $\mathcal{T}^{(n-)}$, and partially transfer $\mathcal{T}_0^{(-n)}$ to $\mathcal{T}^{(-n)}$, as the lower layers of the network contain more sample-specific knowledge.

Sample-specific Knowledge Removal. This removal is conducted in two aspects. One is the logit level that the target network is incapable of making reliable prediction on the deposit set $\mathcal{D}_r$, the other is the feature level that the knowledge of $\mathcal{D}_r$ can't be distilled from $\mathcal{T}$. Thus, for each input $x \subset \mathcal{D}_r$, we assign a ***random*** label y_r, and force $\mathcal{T}$ to randomly predict on $\mathcal{D}_r$. And the loss to maximize attention transfer on the intermediate features is applied to the output of $\mathcal{T}^{(-n)}$, which makes $\mathcal{T}$ undistillable for $\mathcal{D}_r$. Thus, the loss $\mathcal{L}_{kr}$ for knowledge removal is calculated as:

$$\mathcal{L}_{kr} = \mathcal{L}_{ce}\big(\mathcal{T}(x), y_r\big) - \lambda_{at}\mathcal{L}_{at}\big(\mathcal{T}^{(-n)}(x), \mathcal{T}_0^{(-n)}(x)\big), \tag{2}$$

where λ_{at} is the weight, $\mathcal{L}_{ce}(\cdot, \cdot)$ is the cross-entropy loss, and $\mathcal{L}_{at}$ is the filtered attention distillation loss item [55] that calculates the activated feature-wise similarity of the intermediate features:

$$\mathcal{L}_{at}(\mathcal{F}_1, \mathcal{F}_2) = \left\| f\left(\frac{A(\mathcal{F}_1)}{\|A(\mathcal{F}_1)\|_2}\right) - f\left(\frac{A(\mathcal{F}_2)}{\|A(\mathcal{F}_2)\|_2}\right) \right\|^2,$$

$$A(\mathcal{F}) = \sum_{i=1}^{C} \|\mathcal{F}_i\|^2, \quad f(a(i)) = \begin{cases} 0 & a(i) < \epsilon \\ a(i) & \text{otherwise} \end{cases}, \tag{3}$$

where $\mathcal{F}_i \in \mathbb{R}^{H \times W}$ represents the feature $\mathcal{F} \in \mathbb{R}^{H \times W \times C}$ with the size of $H \times W \times C$ at depth i, with which, the l_2-normalized attention maps are obtained. And before calculating the attention similarity with $\mathcal{L}_{at}$, a filter function f is applied to set 0 to the deactivate regions, which enables the intermediate knowledge undisillable only for $x \subset \mathcal{D}_r$.

The knowledge removal loss $\mathcal{L}_{kr}$ is calculated on the deposit set $\mathcal{D}_r$ to fine-tune $\mathcal{T}^{(-n)}$, which is initialized with $\mathcal{T}_0^{(-n)}$. The former loss item of $\mathcal{L}_{kr}$ enables the knowledge forgetting in the logit-level, the latter item of $\mathcal{L}_{kr}$ enables the forgetting in the feature-level, which unlearns $\mathcal{D}_r$ from $\mathcal{T}$ and removes the privacy information of $\mathcal{D}_r$.

General Knowledge Preservation. As is stated in Fig. 1, there are two kinds of knowledge that need to be preserved by the target network. One is the knowledge coming from the preservation $\overline{\mathcal{D}}_r$, the other is the general knowledge from the $\mathcal{D}_r$. Since the target network $\mathcal{T}$ is initialized with the original network and the last few blocks of $\mathcal{T}$ keep fixed while fine-tuning, part of the knowledge has already been preserved by fully transferred from $\mathcal{T}_0^{(n-)}$ to $\mathcal{T}^{(n-)}$. In addition to it, the partial knowledge transfer with filter g is proposed on the $\overline{\mathcal{D}}_r$-related knowledge so as to prevent catastrophic forgetting on $\overline{\mathcal{D}}_r$, which is:

$$\mathcal{L}_{kp} = \mathcal{L}_{kd}\left(g\left(\frac{z_{\mathcal{T}}(x)}{T}\right), g\left(\frac{z_{\mathcal{T}_0}(x)}{T}\right)\right), \tag{4}$$

where $\mathcal{L}_{kd}$ is the KL-divergence loss, T is the temperature, and $z_{\mathcal{T}}$ and $z_{\mathcal{T}_0}$ are the output logits of $\mathcal{T}$ and $\mathcal{T}_0$, respectively. The filter g selects the logits that correspond to the class of the preservation set, in which way the knowledge is partially transferred to target net by minimizing $\mathcal{L}_{kp}$. Note that only the deposit samples are accessible in the whole LIRF framework, the output probabilities on the preservation class are thought to be low and may not be enough to maintain the performance on the preservation set. Thus, we set a higher temperature weight to transfer more knowledge for the preserved tasks.

4.2 Deposit Knowledge to Deposit Module

The key difference between the proposed LIRF with the traditional unlearning problem is that we store the sample-specific knowledge to the deposit module, which is directly vanished in previous unlearning methods. The deposit module should have two vital characteristics: firstly, it should be withdrawn easily to the recover network with the withdrawal request; Secondly, it should be light-weight to be stored.

To get a better knowledge container, we initialized the deposit module with the pruned original network:

$$\mathcal{T}_r \xleftarrow{\text{initialize}} \mathcal{P}rune\big[\mathcal{T}_0^{(-n)}\big], \tag{5}$$

where we use the simple ranking method by calculating the sum of its absolute kernel weights for pruning [22]. The detail of pruning is given in the supplementary.

Here, the deposit module is designed as the pruned version mainly for the following purposes: one is for model efficiency that the light-weight deposit module is more space-saving for storage (20% parameters of the original network); the other is for knowledge filtering that pruning would be better described as 'selective knowledge damage' [12], where only the activated filters are kept such that we only deposit the sample-specific knowledge of $\mathcal{D}_r$ rather than the whole knowledge. Also, similar as $\mathcal{L}_{kp}$, the partial knowledge transfer loss $\mathcal{L}_{pt}$ with the filter $\bar{g}$ is applied to augment this sample-specific knowledge by:

$$\mathcal{L}_{pt} = \mathcal{L}_{kd}\big(\bar{g}(\frac{z_{\mathcal{T}_r \circ \mathcal{T}^{(n-)}}(x)}{T}), \ \bar{g}(\frac{z_{\mathcal{T}_0}(x)}{T})\big), \tag{6}$$

where $\mathcal{L}_{kd}$ and T are previously defined in Eq. (4) and the logits produced by the deposit module are processed by $\mathcal{T}_r$ and $\mathcal{T}^{(n-)}$. The filter $\bar{g}$ selects the logits that correspond to the class of the deposit set, which transfers the $\mathcal{D}_r$-related knowledge from the original network to the deposit module.

By minimizing the loss $\mathcal{L}_{pt}$, the sample-specific knowledge is transferred to the deposit module, at the meanwhile we also finetune $\mathcal{T}_r$ to the easy-to-withdraw module, which means that the knowledge is recoverable for the recover network $\widetilde{\mathcal{T}}$. Hence the recovered performance on $\mathcal{D}_r$ is considered ahead in the deposit process, which is to minimize the classification loss of the recover net $\mathcal{L}_{re}$:

$$\mathcal{L}_{re} = \mathcal{L}_{ce}\big(\widetilde{\mathcal{T}}(x), y\big), \tag{7}$$

where y is the groundtruth label of input x. And the deposit module obtained here only works for storing the knowledge, which can't be used as a independent prediction model. Thus, $\mathcal{D}_r$ is much more safer form for storage than the original images.

4.3   Withdraw Knowledge to Recover Net

Once the knowledge has been successfully deposited, the proposed LIRF framework is completed, where the knowledge can be withdrawn directly without any fine-tuning, let alone no need for any data.

The recover net is re-organized without fine-tuning, which is in the form as:

$$\widetilde{\mathcal{T}}(x) = g\big(\mathcal{T}(x)\big) + \bar{g}\big(\mathcal{T}_r \circ \mathcal{T}^{(n-)}(x)\big), \tag{8}$$

where the filter functions g and $\bar{g}$ are doing the selection operation, which are the same in Eq. (4) and Eq. (6), respectively. Thus, the overall loss function to update the LIRF framework is:

$$\mathcal{L}_{all} = \mathcal{L}_{kr} + \lambda_{kp}\mathcal{L}_{kp} + \lambda_{re}\mathcal{L}_{re} + \lambda_{pt}\mathcal{L}_{pt}, \qquad (9)$$

where λ_{kp}, λ_{pt} and λ_{re} are the balancing weights. LIRF is trained by minimizing the overall $\mathcal{L}_{all}$ on the deposit set $\mathcal{D}_r$, where the preservation set $\overline{\mathcal{D}}_r$ doesn't participate in the whole process.

More Discussions. Note that the optimization objective $\mathcal{L}_{all}$ of knowledge deposit is different from machine unlearning, which aims at obtaining a target network that approximates the one trained from scratch with data $\overline{\mathcal{D}}_r$. In the proposed LIRF, the knowledge capacity of target net is larger than the network only trained with $\overline{\mathcal{D}}_r$, for it contains the general knowledge filtered from the delete set $\mathcal{D}_r$. And only the sample-specific knowledge that is privacy-related should be stored in the deposit module. In the process of withdrawal, the recover network $\widetilde{\mathcal{T}}$ built in Eq. (7) isn't forced to approach the original network: $\widetilde{\mathcal{T}} \neq \mathcal{T}_0$. Actually the recover network works better than the original network with the existence of full and partial knowledge transfer.

5 Experiments

5.1 Experimental Settings

Datasets. We use three widely used benchmark datasets for life-long learning, which are CIFAR-10, CIFAR-100 and CUB200-2011 datasets [40]. For CIFAR-10 and CIFAR-100 datasets, we are using input size of 32×32. For CUB200-2011 dataset, we are using input size of 256×256. In the normal knowledge deposit and withdrawal setting, the first 30% of classes are selected for the deposit set, while the rest classes belong to the preservation set.

Training Details. We used PyTorch framework for the implementation. We apply the experiments on the ResNet-18 backbone. For optimizing the target network and the deposit module, we use stochastic gradient descent with momentum of 0.9 and learning rate of 0.01 for 20 epochs. We employed a standard data augmentation strategy: random crop, horizontal flip, and rotation. For applying distillation, we set $T = 10$ for CIFAR10 dataset and $T = 20$ for CUB200-2011 dataset. For the weights balancing the loss items in $\mathcal{L}_{all}$, we set $\lambda_{kp} = \lambda_{pt} = \lambda_{re} = 10$. For the normal LIRF setting, the pruning rate is set as 50% and the original network $\mathcal{T}_0$ is divided into 4 blocks, where the last 2 blocks as well as the fc layers are formed as $\mathcal{T}_0^{(n-)}$.

Evaluation Metrics. For evaluation, we need to evaluate the performance of both target net and recover net. For recover net, we use the average accuracy for the preservation set (Pre Acc.), the average accuracy for the deposit set (Dep

Acc.), the average accuracy for the full set (Avg Acc.). And for target net, we use the average accuracy for the preservation set (Pre Acc.) and the the average accuracy for the deposit set (Dep Acc.) for the deposit set $\overline{\mathcal{D}}_r$. In addition, following the setting of LWSF [34], we use the harmonic mean (H Mean) of the two standard evaluation measures for life-long learning, which is computed by: $HMean = \frac{2 \cdot PreAcc \cdot F}{PreAcc + F}$, where the forgetting measure 'F' is computed for the deposit set by the accuracy drop (decrease) before and after knowledge deposit. For testing the withdrawal performance, all the metrics show better performance with higher values, which are similar for testing the deposit performance on target net, except that 'Dep Acc.' is better with lower values.

5.2　Experimental Results

Overall Performance. Table 1 shows overall performance of knowledge deposit (target network) and withdrawal (recover network) on CIFAR-10 and CUB200-2011 datasets. Besides, we compare the proposed LIRF with the 'Independent' networks independently trained on the two sub datasets (preservation set and deposit set) and the 'Original' network $\mathcal{T}_0$ trained on the full dataset. From Table 1, several observations are obtained:

- In the original network, the accuracy on preservation set ('Pre Acc.') is higher than the performance trained dependently ('93.77' vs '92.92' on CIFAR-10), which means that there exits the positive knowledge transfer from the deposit $\mathcal{D}_r$ to the preservation $\overline{\mathcal{D}}_r$. Thus, it is necessary to partial transfer the general knowledge to the preserved tasks.
- As the accuracy for randomly predicting on CIFAR-10 and CUB200-2011 is 10% and 0.5%, respectively, the 'Pre Acc.' while depositing decreases to 15% and 1.18%. This large accuracy drop demonstrates the logit-level forgetting of the deposit set in the target net $\mathcal{T}$.
- The recover network gains higher accuracy on both the preservation set and the deposit set than on the original network, which proves the knowledge has been augmented in LIRF with the partial and full knowledge transfer, which demonstrates the discussions in Sect. 4.3.

Sensitive Analysis of LIRF. Here we give a deeper analysis of the proposed LIRF by the ablation study of each loss items in $\mathcal{L}_{all}$. The comparative results are applied on, 'Scratch Train': train each net with the corresponding set from scratch; 'IL Train': train the target net from scratch and train recover net with KD loss in the incremental learning setting; $\mathcal{L}_{kr}, \mathcal{L}_{kp}, \mathcal{L}_{pt}, \mathcal{L}_{re}$ are the loss items defined in the LIRF framework; $\mathcal{L}_{kp[w/o\mathcal{L}_{at}]}$ is the loss without the attention distillation $\mathcal{L}_{at}$; 'Prune' denotes the pruning operation to initialize the deposit module. The experimental results are displayed in Table 2. As can be observed from the table: (1) The full setting with all the loss functions and the pruning strategy achieves almost the best on each metrics. (2) The attention loss item $\mathcal{L}_{at}$ would not affect the accuracy a lot ('row 4' and 'row 5'), but it is of vital

Table 1. Experimental results of the proposed LIRF on CIFAR-10 dataset and CUB200-2011 dataset. For each dataset, we randomly select 30% of classes for deposit (Dep Set), while the rest is kept in the preservation set (Pre Set).

Dataset	Metrics	Independent	Original	Deposit	Withdrawal
CIFAR-10	Pre Acc.↑	92.92	93.77	93.41	94.56
CIFAR-10	Dep Acc	96.61	94.60	15.00	97.92
CIFAR-10	F↑	–	0	79.60	–
CIFAR-10	H Mean ↑	–	0	85.95	–
CIFAR-10	Avg Acc. ↑	94.02	94.06	–	95.57
CUB200-2011	Pre Acc.↑	48.15	50.33	51.64	53.21
CUB200-2011	Dep Acc	52.73	48.60	1.18	55.89
CUB200-2011	F↑	–	0	47.42	–
CUB200-2011	H Mean ↑	–	0	49.44	–
CUB200-2011	Avg Acc. ↑	49.52	49.81	–	54.01

Table 2. Experimental results of the ablation study on the proposed LIRF framework.

Method	#Target Net		#Recover Net		
	Pre Acc. ↑	Dep Acc. ↓	Pre Acc. ↑	Dep Acc. ↑	Avg Acc. ↑
Scratch Train	92.92	96.61	93.77	94.60	94.06
IL Train	92.92	96.61	90.87	**98.37**	93.12
$\mathcal{L}_{kr}, \mathcal{L}_{kp[w/o\mathcal{L}_{at}]}$	93.38	15.55	–	–	–
$\mathcal{L}_{kr}, \mathcal{L}_{kp[w/o\mathcal{L}_{at}]}, \mathcal{L}_{pt}, \mathcal{L}_{re}$	93.25	14.81	94.26	97.03	95.09
$\mathcal{L}_{kr}, \mathcal{L}_{kp}, \mathcal{L}_{pt}, \mathcal{L}_{re}$	93.25	15.29	94.33	97.07	95.15
$\mathcal{L}_{kr}, \mathcal{L}_{kp}, \mathcal{L}_{pt}, \mathcal{L}_{re}$+Prune	**93.42**	**14.15**	**94.55**	97.67	**95.49**

importance to prevent the information leakage, which is discussed in the following experiment. (3) The pruning strategy on the deposit module is proved to be effective since the pruned deposit module can be withdrawn to recover net with the best Avg Acc.('95.49').

The visualization of the t-SNE plots is depicted in Fig. 3, where the features on the final layer of original net, target net and recover net are visualized. As is shown in the figure, both the original net and the recover net can produce discriminative features on all the 10 categories. And for the target net where the sample-specific knowledge of the deposit set is removed, the visualization proves that the target net produces highly discriminative features for the preservation set while vanishing the predicting capacity for the deposit set. And for visualization the t-SNE plots of the deposit module, we construct a network as $\mathcal{T}_r \circ \mathcal{T}_0^{(n-)}$. As can been seen in the right part of the figure, the pruned deposit module produces more 'narrow' features, which are thought as the sample-specific knowledge we want to deposit, proving the 'selective knowledge damage' scheme we mentioned in Sect. 4.2.

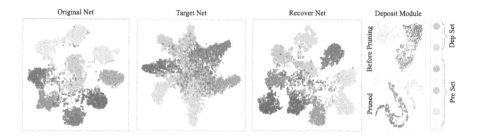

Fig. 3. t-SNE plots of the features obtained from the last layer of the network are shown. Each color in the t-SNE plot represents one category, where 3 categories are deposited and the rest 7 categories are in the preservation set (best viewed in color).

Table 3. Experimental results of the knowledge transferability to downstream networks. This experiment is conducted on the CIFAR-10 dataset.

Student	Distillation	#Original		#Target (w/0 $\mathcal{L}_{at}$)		#Target (w $\mathcal{L}_{at}$)	
		Pre Acc.	Dep Acc.	Pre Acc.↑	Dep Acc.↓	Pre Acc.↑	Dep Acc.↓
CNN	Logit-based	85.38	86.15	85.97(+0.59)	84.26(−1.89)	85.70(+0.32)	82.75(−3.40)
CNN	AT-based	85.27	85.94	86.01(+0.74)	85.72(−0.22)	85.54(+0.27)	81.83(−4.11)
ResNet18	Logit-based	94.26	95.73	94.55(+0.29)	92.70(−3.03)	94.64(+0.38)	91.49(−4.24)
ResNet18	AT-based	94.09	95.24	94.15(+0.06)	94.61(−0.63)	93.85(−0.24)	88.76(−6.48)

Knowledge Transferability to Downstream Networks. We use two evaluations to prove the success of sample-specific knowledge removal in the target net: One is the accuracy drop on the deposit set, which has been proved in the former experiments; The other is the knowledge transferability of the deposit set from the target network to downstream networks, which test the knowledge leakage risk by knowledge distillation. We have conducted the logit-based distillation (KL-divergence loss of the output logits) and the attention-based distillation (MSE loss of the attention maps of the intermediate features). The results are displayed in Table 3, where we choose the plain CNN and the ResNet-18 as the student. And we have also evaluated the necessity of the loss item $\mathcal{L}_{at}$ in $\mathcal{L}_{kr}$. Note that the groundtruth label of the training data is included for training with distillation, the accuracy wouldn't drop largely even when the knowledge is nontransferable. Thus, from the table, we observe that: (1) The knowledge transferability on the preservation set is guaranteed on both the original and the target networks, which is slightly influenced by target net with $\mathcal{L}_{at}$ distilled in the attention-based way; (2) When training the LIRF framework with loss item $\mathcal{L}_{at}$, the knowledge for the deposit set is hard to be distilled both in the attention-based and the logit-based way. It is much safer with $\mathcal{L}_{at}$, since without it, the knowledge of $\mathcal{D}_r$ is likely to be leaked through distillation in the attention-based manner. The privacy protection on the deposit set is further evaluated in the supplementary tested by the data-free distillation.

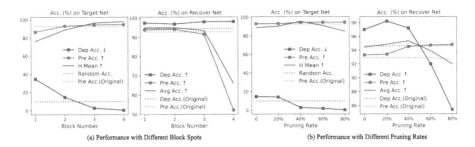

(a) Performance with Different Block Spots   (b) Performance with Different Pruning Rates

Fig. 4. The performance on knowledge deposit and withdrawal with different division block numbers and pruning rates.

The Influence of the Scale of the Deposit Module. There are two factors corresponding to the scale of the deposit module $\mathcal{T}_r$: one is the block number used to divide the original network; the other is the pruning rate. The influence on this two factors is depicted in Fig. 4. When the division block number n increases, the size of the deposit module becomes larger and the fully transferred part of the original network $(\mathcal{T}_0^{(n-)})$ becomes smaller. In this way, the deposit accuracy on the target network ('Dep Acc.' in the first sub figure) and the average accuracy on the recover network ('Avg Acc.' in the second sub figure) drop due to the less knowledge directly transferred from $\mathcal{T}_0^{(n-)}$ to $\mathcal{T}$, which is also completely transferred back while recovering. Considering the performance on both target net and recover net, dividing at $n = 2$ and $n = 3$ satisfies the demand, and we choose $n = 2$ for smaller knowledge deposit storage. When the pruning rate increases (the percent of filters to be pruned out), the size of the deposit module becomes smaller, which doesn't influence the deposit performance largely (the third sub figure). The average accuracy on the recover network('Avg Acc.' in the forth sub figure) increases at first due to selective knowledge damage on the deposit module, but drops at last due to the limit size for knowledge storage. So the pruning rate near 50% is a better choice.

Comparing with Incremental Learning and Unlearning. The proposed LIRF can be also conducted on incremental learning task and the machine unlearning task. Here we tested LIRF on these two tasks, following the setting of LWSF [34], whose task is to unlearn several classes of samples while dealing with the new classes. The individual experiments on incremental learning and machine unlearning are given in the supplementary. To begin with, we train the original network with the full dataset and then deposit each sub class into a deposit module set, then withdraw each in each incremental step. Table. 4 shows the comparative results of all the methods, which include: 'Baseline' (trained only with the classification loss), 'LwF' [23], 'EWC' [19], 'MAS' [1], LWSF [34] and the modified 'LwF*' and 'EWC*' that are modified to enable partial forgetting by the work [34]. The metrics of 'H Mean' (H), 'Pre Acc' (A) and 'F' are averaged until the last incremental step. The proposed LIRF works almost the best

Table 4. Comparative results of incremental learning and unlearning on CIFAR-100. Each column represents a different number of classes per incremental step.

	# Task:2, # Class:50		# Task:5, # Class:20		# Task:20, # Class:5	
	H ↑	(A↑, F↑)	H ↑	(A↑, F↑)	H ↑	(A↑, F↑)
Baseline	55.87	(55.21,56.55)	51.79	(39.66,74.62)	37.88	(25.41,74.41)
LwF	9.02	(74.69,4.80)	17.23	(79.05,9.67)	22.50	(80.74,13.07)
LwF*	54.64	(76.44,42.52)	68.24	(81.32,58.79)	63.62	(82.29,51.85)
EWC	58.58	(56.73,60.55)	48.57	(36.54,72.42)	34.91	(23.07,71.70)
EWC*	57.17	(56.25,58.13)	49.61	(36.58,77.08)	36.90	(23.68,83.52)
EWC*+LwF*	53.51	(77.11,40.98)	67.64	(81.20,57.96)	69.17	(74.11, 64.85)
MAS	55.44	(54.42,56.49)	47.46	(34.89,74.17)	35.26	(23.25,72.96)
MAS+LwF*	56.54	(76.85,44.72)	66.35	(81.83,55.79)	70.83	(74.63,67.41)
LWSF	70.08	(74.89,65.84)	**73.21**	(72.61,73.83)	71.63	(68.56,75.00)
LIRF	**77.69**	(79.24,76.19)	73.08	(78.24, 68.56)	**79.48**	(80.41,78.57)

among all the listed methods, especially on the incremental performance ('A') which is due to partial and fully knowledge transfer in the framework.

6   Conclusions

In this paper, we study a novel life-long learning task, recoverable knowledge forgetting. Unlike prior life-long learning tasks that either aim to prevent the forgetting of old knowledge or delete specified knowledge for good, the investigated setting enables flexible knowledge extraction and inserting, which in turn largely enriches the user control and meanwhile ensures network IP protection. To this end, we introduce a dedicated approach, termed as LIRF, where the innovative operation of knowledge deposit and knowledge withdrawal are proposed. During deposit, the sample-specific knowledge that may lead to privacy leakage is extracted from original network and maintained in the deposit module. Whenever needed, the deposited knowledge can be readily withdrawn to recover the original model. Experimental results demonstrate the effectiveness of the proposed LIRF under various settings, including incremental learning and machine unlearning.

Acknowledgements. This work is supported by NUS Advanced Research and Technology Innovation Centre (Project Reference: ECT-RP2), Centre for Advanced Robotics Technology Innovation (CARTIN) of Singapore, NUS Faculty Research Committee (WBS: A-0009440-00-00), National Natural Science Foundation of China (No. 62002318), Zhejiang Provincial Science and Technology Project for Public Welfare (LGF21F020020) and Ningbo Natural Science Foundation (202003N4318).

References

1. Aljundi, R., Babiloni, F., Elhoseiny, M., Rohrbach, M., Tuytelaars, T.: Memory aware synapses: learning what (not) to forget. In: Proceedings of the European Conference on Computer Vision (2018)
2. Bourtoule, L., et al.: Machine unlearning. In: IEEE Symposium on Security and Privacy, pp. 141–159 (2021)
3. Brophy, J., Lowd, D.: Machine unlearning for random forests. In: International Conference on Machine Learning (2021)
4. Chen, G., Choi, W., Yu, X., Han, T.X., Chandraker, M.: Learning efficient object detection models with knowledge distillation. In: Neural Information Processing Systems (2017)
5. Chen, M., Zhang, Z., Wang, T., Backes, M., Humbert, M., Zhang, Y.: When machine unlearning jeopardizes privacy. In: Proceedings of the 2021 ACM SIGSAC Conference on Computer and Communications Security, pp. 896–911 (2021)
6. Cheraghian, A., Rahman, S., Fang, P., Roy, S.K., Petersson, L., Harandi, M.T.: Semantic-aware knowledge distillation for few-shot class-incremental learning. In: Conference on Computer Vision and Pattern Recognition, pp. 2534–2543 (2021)
7. Dong, S., Hong, X., Tao, X., Chang, X., Wei, X., Gong, Y.: Few-shot class-incremental learning via relation knowledge distillation. In: AAAI Conference on Artificial Intelligence (2021)
8. Gao, J., Guo, Z., Li, Z., Nevatia, R.: Knowledge concentration: learning 100k object classifiers in a single CNN. arXiv (2017)
9. Gupta, V., Jung, C., Neel, S., Roth, A., Sharifi-Malvajerdi, S., Waites, C.: Adaptive machine unlearning. In: Advances Neural Information Processing System, vol. 34 (2021)
10. Han, X., Song, X., Yao, Y., Xu, X.S., Nie, L.: Neural compatibility modeling with probabilistic knowledge distillation. IEEE Trans. Image Process. **29**, 871–882 (2020)
11. Hinton, G.E., Vinyals, O., Dean, J.: Distilling the knowledge in a neural network. In: Advances Neural Information Processing System, vol. 2 (2015)
12. Hooker, S., Courville, A.C., Clark, G., Dauphin, Y., Frome, A.: What do compressed deep neural networks forget (2020)
13. Hou, B.J., Yan, Y.H., Zhao, P., Zhou, Z.H.: Storage fit learning with feature evolvable streams. In: AAAI Conference on Artificial Intelligence (2021)
14. Hou, B.J., Zhang, L., Zhou, Z.H.: Learning with feature evolvable streams. IEEE Trans. Knowl. Data Eng. **33**, 2602–2615 (2021)
15. Hou, C., Zhou, Z.H.: One-pass learning with incremental and decremental features. IEEE Trans. Pattern Anal. Mach. Intell. **40**, 2776–2792 (2018)
16. Hu, X., Tang, K., Miao, C., Hua, X., Zhang, H.: Distilling causal effect of data in class-incremental learning. In: Conference on Computer Vision and Pattern Recognition, pp. 3956–3965 (2021)
17. Huang, M., You, Y., Chen, Z., Qian, Y., Yu, K.: Knowledge distillation for sequence model. In: INTERSPEECH (2018)
18. Jing, Y., et al.: Dynamic instance normalization for arbitrary style transfer. In: AAAI (2020)
19. Kirkpatrick, J., et al.: Overcoming catastrophic forgetting in neural networks. Proc. Natl. Acad. Sci. **114**(13), 3521–3526 (2017)
20. Lange, M.D., et al.: A continual learning survey: defying forgetting in classification tasks. IEEE Trans. Pattern Anal. Mach. Intell. **44**, 3366–3385 (2021)

21. Lee, S., Behpour, S., Eaton, E.: Sharing less is more: lifelong learning in deep networks with selective layer transfer. In: International Conference on Machine Learning (2021)
22. Li, H., Kadav, A., Durdanovic, I., Samet, H., Graf, H.P.: Pruning filters for efficient convnets (2016)
23. Li, Z., Hoiem, D.: Learning without forgetting. In: European Conference on Computer Vision (2016)
24. Liu, H., Yang, Y., Wang, X.: Overcoming catastrophic forgetting in graph neural networks. In: AAAI Conference on Artificial Intelligence (2021)
25. Mazumder, P., Singh, P., Rai, P.: Few-shot lifelong learning. In: AAAI Conference on Artificial Intelligence (2021)
26. Neel, S., Roth, A., Sharifi-Malvajerdi, S.: Descent-to-delete: gradient-based methods for machine unlearning. In: International Conference on Algorithmic Learning Theory (2021)
27. Nekoei, H., Badrinaaraayanan, A., Courville, A.C., Chandar, S.: Continuous coordination as a realistic scenario for lifelong learning. In: International Conference on Machine Learning (2021)
28. Nguyen, Q.P., Low, B.K.H., Jaillet, P.: Variational bayesian unlearning. In: Advances Neural Information Processing System, vol. 33 (2020)
29. Peng, H., Li, J., Song, Y., Liu, Y.: Incrementally learning the hierarchical softmax function for neural language models. In: AAAI Conference on Artificial Intelligence (2017)
30. Rebuffi, S.A., Kolesnikov, A., Sperl, G., Lampert, C.H.: iCaRl: incremental classifier and representation learning. In: IEEE Conference on Computer Vision and Pattern Recognition, pp. 5533–5542 (2017)
31. Ren, S., Zhou, D., He, S., Feng, J., Wang, X.: Shunted self-attention via multi-scale token aggregation. In: Proceedings of the IEEE/CVF Conference on Computer Vision and Pattern Recognition (2022)
32. Rosenfeld, A., Tsotsos, J.K.: Incremental learning through deep adaptation. IEEE Trans. Pattern Anal. Mach. Intell. **42**, 651–663 (2020)
33. Sekhari, A., Acharya, J., Kamath, G., Suresh, A.T.: Remember what you want to forget: algorithms for machine unlearning. In: Advances Neural Information Processing System, vol. 34 (2021)
34. Shibata, T., Irie, G., Ikami, D., Mitsuzumi, Y.: Learning with selective forgetting. In: International Joint Conference on Artificial Intelligence (2021)
35. Shin, H., Lee, J.K., Kim, J., Kim, J.: Continual learning with deep generative replay. In: Conference on Neural Information Processing Systems (2017)
36. Shmelkov, K., Schmid, C., Alahari, K.: Incremental learning of object detectors without catastrophic forgetting. In: IEEE International Conference on Computer Vision, pp. 3420–3429 (2017)
37. Sun, G., Cong, Y., Xu, X.: Active lifelong learning with "watchdog". In: AAAI Conference on Artificial Intelligence (2018)
38. Tao, X., Hong, X., Chang, X., Dong, S., Wei, X., Gong, Y.: Few-shot class-incremental learning. In: Conference on Computer Vision and Pattern Recognition, pp. 12180–12189 (2020)
39. Venkatesan, R., Venkateswara, H., Panchanathan, S., Li, B.: A strategy for an uncompromising incremental learner. ArXiv abs/1705.00744 (2017)
40. Wah, C., Branson, S., Welinder, P., Perona, P., Belongie, S.J.: The caltech-UCSD birds-200-2011 dataset (2011)

41. Wu, C., Herranz, L., Liu, X., Wang, Y., van de Weijer, J., Raducanu, B.: Memory replay GANs: learning to generate new categories without forgetting. In: Conference and Workshop on Neural Information Processing Systems (2018)
42. Xu, D., Ouyang, W., Wang, X., Sebe, N.: PAD-Net: multi-tasks guided prediction-and-distillation network for simultaneous depth estimation and scene parsing. In: Computer Vision and Pattern Recognition, pp. 675–684 (2018)
43. Xue, Y., et al.: Point2Seq: detecting 3D objects as sequences. In: Proceedings of the IEEE/CVF Conference on Computer Vision and Pattern Recognition (2022)
44. Yang, X., Ye, J., Wang, X.: Factorizing knowledge in neural networks. In: European Conference on Computer Vision (2022)
45. Yang, Y., Feng, Z., Song, M., Wang, X.: Factorizable graph convolutional networks. In: Conference on Neural Information Processing Systems (2020)
46. Yang, Y., Qiu, J., Song, M., Tao, D., Wang, X.: Distilling knowledge from graph convolutional networks. In: Proceedings of the IEEE/CVF Conference on Computer Vision and Pattern Recognition (2020)
47. Ye, J., Ji, Y., Wang, X., Gao, X., Song, M.: Data-free knowledge amalgamation via group-stack dual-GAN. In: Proceedings of the IEEE/CVF Conference on Computer Vision and Pattern Recognition (2020)
48. Ye, J., Ji, Y., Wang, X., Ou, K., Tao, D., Song, M.: Student becoming the master: knowledge amalgamation for joint scene parsing, depth estimation, and more. In: Computer Vision and Pattern Recognition, pp. 2829–2838 (2019)
49. Ye, J., Mao, Y., Song, J., Wang, X., Jin, C., Song, M.: Safe distillation box. In: Proceedings of the AAAI Conference on Artificial Intelligence, vol. 36, pp. 3117–3124 (2022)
50. Ye, J., Wang, X., Ji, Y., Ou, K., Song, M.: Amalgamating filtered knowledge: learning task-customized student from multi-task teachers. In: International Joint Conference on Artificial Intelligence (2019)
51. Yin, H., et al.: Dreaming to distill: data-free knowledge transfer via deepinversion. In: Conference on Computer Vision and Pattern Recognition, pp. 8712–8721 (2020)
52. Yu, W., et al.: MetaFormer is actually what you need for vision. In: Proceedings of the IEEE/CVF Conference on Computer Vision and Pattern Recognition (2022)
53. Yu, X., Liu, T., Wang, X., Tao, D.: On compressing deep models by low rank and sparse decomposition. In: Computer Vision and Pattern Recognition, pp. 67–76 (2017)
54. Yuan, M., Peng, Y.: CKD: cross-task knowledge distillation for text-to-image synthesis. IEEE Trans. Multimed. **22**, 1955–1968 (2020)
55. Zagoruyko, S., Komodakis, N.: Paying more attention to attention: improving the performance of convolutional neural networks via attention transfer. arXiv preprint arXiv:1612.03928 (2016)
56. Zhang, Z.Y., Zhao, P., Jiang, Y., Zhou, Z.H.: Learning with feature and distribution evolvable streams. In: International Conference on Machine Learning (2020)

Efficient One Pass Self-distillation with Zipf's Label Smoothing

Jiajun Liang$^{(\boxtimes)}$ [ID], Linze Li [ID], Zhaodong Bing, Borui Zhao, Yao Tang, Bo Lin, and Haoqiang Fan

MEGVII Technology, Beijing, China
{liangjiajun,lilinze,bingzhaodong,zhaoborui,
tangyao02,linbo,fhq}@megvii.com

Abstract. Self-distillation exploits non-uniform soft supervision from itself during training and improves performance without any runtime cost. However, the overhead during training is often overlooked, and yet reducing time and memory overhead during training is increasingly important in the giant models' era. This paper proposes an efficient self-distillation method named Zipf's Label Smoothing (Zipf's LS), which uses the on-the-fly prediction of a network to generate soft supervision that conforms to Zipf distribution without using any contrastive samples or auxiliary parameters. Our idea comes from an empirical observation that when the network is duly trained the output values of a network's final softmax layer, after sorting by the magnitude and averaged across samples, should follow a distribution reminiscent to Zipf's Law in the word frequency statistics of natural languages. By enforcing this property on the sample level and throughout the whole training period, we find that the prediction accuracy can be greatly improved. Using ResNet50 on the INAT21 fine-grained classification dataset, our technique achieves +3.61% accuracy gain compared to the vanilla baseline, and 0.88% more gain against the previous label smoothing or self-distillation strategies. The implementation is publicly available at https://github.com/megvii-research/zipfls.

Keywords: Knowledge distillation · Self distillation · Label smoothing · Image classification · Zipf's Law

1 Introduction

A major trend in the study of multi-class classification models is to replace the one-hot encoding label with more informative supervision signal. This line of thought witnessed proliferation of great training techniques to improve the accuracy of a network without any runtime cost, the Knowledge Distillation

Supplementary Information The online version contains supplementary material available at https://doi.org/10.1007/978-3-031-20083-0_7.

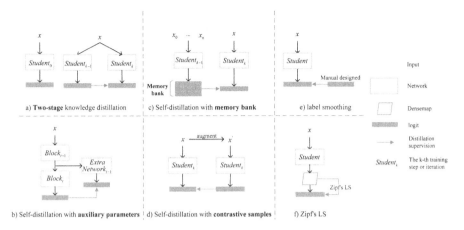

Fig. 1. Comparison between different soft label generation methods. a) **Two-stage** knowledge distillation method [7,12]. b) **Auxiliary parameters** [8,13,33]. c) Progressive distillation with **memory bank** storing past predictions of the entire dataset [2,14,31]. d) Contrastive samples [27,29] with **twice training iterations**. e) Label smoothing methods [19,28], with **uninformative** manual designed distributions. f) **Efficient one-pass** Zipf's Label Smoothing method, generating sample-level non-uniform soft labels almost without additional cost during training

being the most famous of them. Perhaps the most counter-intuitive discovery in this direction is the effectiveness of Self Distillation methods [7,27–29,32] where a network even benefits from predictions of its own.

Self distillation simplifies the two-stage knowledge distillation framework by distilling knowledge from itself instead of from the pretrained teacher, and still improves performance significantly without extra cost in inference time. However, the overhead during training in self-distillation is often overlooked, and yet reducing time and memory overhead in training is increasingly important in today's giant model era. Figure 1 shows several knowledge distillation paradigms, and self distillation methods rely on additional contrastive training instances, auxiliary parameters or intermediate dumped results for each training sample, which could double the training time and bring non-negligible memory overhead.

This paper aims to find efficient techniques which generate non-uniform supervision signals as informative as expensive self distillation approaches. The starting point of our construction is the observation of the network's soft-max output values. The class that corresponds to the final categorical output usually has the highest value, but the scores for other classes (which we call the non-target classes), containing important information of the network's understanding of the input image, also play an important role. Indeed, we surmise that **a good-performing network should obey certain laws in the non-zero prediction values they make on the non-target classes**.

We postulate that a significant part of the efficacy of the distillation techniques comes from enforcing the prediction scores into a shape that best balances

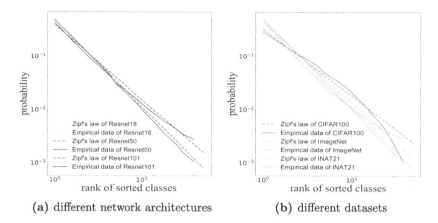

(a) different network architectures   **(b)** different datasets

Fig. 2. Distribution of sorted softmax scores from duly-trained networks of a) different architectures on INAT-21 and b) ResNet-50 on different datasets. The average distribution of sorted softmax values (the solid lines) well follows Zipf's law (the dashed lines of the same color). The probability-rank relations form straight lines in the log-log plot. This inspires us to design the soft supervision similar to this shape

between being "sharp" (so that the final prediction is unambiguous) and being "soft" (so that inter-class correlation are respected). We test this hypothesis by inventing a technique, which only uses the on-the-fly prediction of a network to generate a soft supervision label that conforms to our designated distribution law, and showing that this simple strategy already harvests or even surpasses the performance gains of many other more complex distillation techniques.

Specifically, we rank the output classes according to the feature output of the multi-class classification network, and assign to each class a target value according to Zipf's Law Distribution:

$$p \propto r^{-\alpha}$$

where p is the confidence distribution of different classes, r is our sorted rank index of classes (integer values starting from 1), and α is a hyper-parameter that controls the rate of decay. Divergence to this smoothed label is added as a loss term in complement to the usual cross-entropy with the one-hot encoded hard label.

The choice of Zipf's Law is not arbitrary. Indeed, we experimentally find out that when a network is trained to its convergence state, the average distribution of sorted softmax values well follows this law (see Fig. 2). By explicitly enforcing this shape of the distribution and giving supervision from the very beginning of training, the performance of the network boots by a significant amount.

Thus we propose an efficient and general plug-and-play technique for self distillation, named Zipf's Label Smoothing (Zipf's LS). Compared to other techniques (see Table 1), our method enjoys the advantage of incurring almost zero additional cost (during inference or training) as label smoothing and at the

Table 1. Comparison between different soft label generation methods. Our Zipf's Label Smoothing generates sample-level non-uniform soft labels with little cost during training. For simplicity, this table only shows top1-accuracy(Acc) from ResNet-18 on TinyImageNet(Tiny) and ResNet-50 on ImageNet(IMT). More comprehensive results on other models and datasets are reported in the remaining sections of this paper. Memory cost and training time test experiments are conducted on ImageNet using 4 2080Ti GPUs with batch size 16

Method	Non-uniform	w/o Pretrain teacher	w/o Contrastive samples	w/o Auxiliary parameters	CPU memory cost	GPU memory cost	Training time per epoch	Tiny Acc (gain)	IMT Acc (gain)
Baseline	✗				18 G	6.8 G	1.82 h	56.41	76.48
BAN [7]		✗		✗	18.9 G	7.4 G	2.67 h	(+2.24)	(+0.05)
BYOT [33]				✗	18.2 G	38.5 G	10.36 h	(+1.43)	(+0.51)
PS-KD [14]			✗		18 G	6.8 G	2.86 h	(+1.81)	(+0.18)
DDGSD [27]			✗		18.2 G	7.3 G	4.30 h	(+2.11)	(+0.46)
CS-KD [29]			✗		27.4 G	9.8 G	2.18 h	(+1.97)	(+0.30)
LS [24]	✗				18 G	6.8 G	1.83 h	(+0.48)	(+0.19)
TF-KD [28]	✗				18 G	6.8 G	1.82 h	(+0.26)	(+0.08)
Ours					18 G	6.8 G	1.83 h	**(+2.84)**	**(+0.77)**

same time strongly preserves the performance gains of the self-distillation. To summarize, our contributions are as follows:

- We find that distribution of non-target soft-max values of duly trained model fits well to Zipf's law, which could be used as a regularization criterion in the entire self-training process.
- We propose Zipf's Label Smoothing method, an efficient self distillation training technique without relying on additional contrastive training instances or auxiliary parameters.
- We verify our method on comprehensive combinations of models and datasets (including popular ResNet and DenseNet models, CIFAR, ImageNet and INAT classification tasks) and show strong results.

2   Related Work

2.1   Label Smoothing

The one-hot label is sub-optimal because objects from more than one class occur in the same image. Label Smoothing [24] (LS) is proposed to smooth the hard label to prevent over-confident prediction and improve classification performance. Müller et al. [19] found that intra-class distance in feature space is more compact when LS is used, which improves generalization. To obtain non-uniform soft labels, Zhang et al. [31] proposed the online label smoothing method (OLS) by maintaining the historical predictions to obtain the class-wise soft label. Yuan et al. [28] discussed the relationship between LS and knowledge distillation, and proposed a teacher-free knowledge distillation (Tf-KD) method to get better performance than LS. Label smoothing has become one of the best practices in the current deep learning community [11], but the paradigm of using uniform distribution for the non-target classes limits further improvement of performance.

2.2 Knowledge Distillation

Instead of imposing a fixed prior distribution, knowledge distillation was first proposed by Hinton in [12] to provide sample-level non-uniform soft labels. They demonstrated that the "dark knowledge" lies in the output distributions from a large capacity teacher network and benefits the student's representation learning. Recent works mainly explored to better transfer the "dark knowledge" and improve the efficiency from various aspects, such as reducing the difference between the teacher and student [3,5,18,34], designing student-friendly architecture [16,20], improving the distillation efficiency [7,14,27,29] and explaining the distillation's working mechanism [1,23].

In this work, we focus on how to transfer the "dark knowledge" in an almost free manner. Furlanello et al. [7] proposed to improve the performance of the student network by distilling a teacher network with the same architecture. However, it is still a two-stage approach, which first trains the teacher and then distills knowledge to the student. To reduce the training time, many self-distillation methods were proposed. They gain soft label supervision on the fly without the pretraining step.

2.3 Self Distillation

There are two categories of self-distillation techniques, namely the auxiliary parameter methods [2,8,13,30,33] and contrastive sample methods [14,27,29, 31]. Auxiliary model methods exploit additional branches to get extra predictions besides the main-branch prediction for soft label supervision at the cost of more parameters overhead. For example, Knowledge Distillation via Collaborative Learning (KDCL) [8] trained multiple parallel student networks at the same time and ensemble the output as extra soft label supervision for each parallel student network. On the other hand, contrastive sample methods get soft label supervision at the cost of additional data augmentation, enlarged batch size, or complex sampling strategy. The examples are Data-distortion Guided Self-Distillation (DDGSD) [27] which gains soft labels from different augmented views from the same instance and Regularizing Class-wise Predictions via Self-knowledge Distillation (CS-KD) [29] which gathers data from other samples of the same class.

As summarized above, Label Smoothing and Knowledge Distillation are two major techniques to acquire informative soft labels. However, the Label Smoothing methods are limited by the uniform hypothesis, while the Knowledge Distillation methods require much more memory or computation overhead. Our work aims to improve upon these issues.

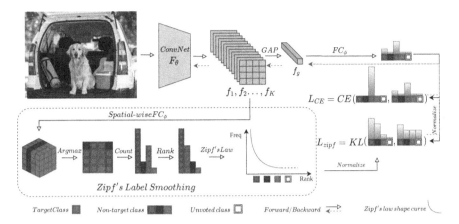

Fig. 3. The overall Zipf's soft label generation and training framework. The blue dashed box outlines the soft label generation process. We apply a shared-classifier on dense feature maps and count the number of argmax values from the dense prediction, which could provide ranking information to Zipf's law distribution generation. For the non-occurred classes with zero count, we give uniform constant energy to them. L_{zipf} is Kullback-Leibler divergence between the prediction of non-target classes and zipf's soft label, combined with L_{CE} of the hard label to provide gradient to representation learning

3   Method

Zipf's Label Smoothing aims to combine the best of the two worlds, the efficiency of teacher-free label smoothing and the informative soft label from self distillation. It generates non-uniform supervision signals from on-the-fly prediction of the network as shown in Fig. 3. Our method is inspired by the observation that the value and rank of the softmax output from the duly trained network follow a distribution reminiscent to the Zipf's Law on average as shown in Fig. 2, which could be applied as a shape prior to the softmax prediction during the whole training period. To apply Zipf's Law to soft-label generation, ranking information of output categories is needed. We propose Dense Classification Ranking which utilizes local classification results to rank the categories. Finally, KL-divergence between prediction and Zipf's soft label within non-target classes is measured to provide more informative gradients for representation learning.

3.1   Zipf's Law

Zipf's law is an empirical law which states that the normalized frequency of an element should be inversely proportional to the rank of the element, firstly discovered by G.Zipf on linguistic materials [21]. It can be described by an equation as:

$$f(r) = \frac{r^{-\alpha}}{\sum_{r=1}^{N} r^{-\alpha}}$$

$$log f(r) = -\alpha \log r - \log(\sum_{r=1}^{N} r^{-\alpha})$$

(1)

where r is the rank of the element, N is the total number of elements, f is the frequency and α is a constant larger than zero which controls the decay rate.

An interesting discovery is that the outputs of softmax networks follow Zipf's Law [21] when the network is trained to its convergence state, and this pattern consistently emerges in different datasets and models as shown in Fig. 2. We could exploit this shape prior to design a simple and efficient self-distillation algorithm. To generate a soft label using Zipf's distribution, ranking information of categories is a must which is almost impossible to get from annotation.

3.2   Dense Classification Ranking

To realize the Zipf distribution we need to find a way to properly rank the output categories. The naive thought is to directly sort the softmax prediction of a sample, which we call the logit-based ranking method. Though this method already generates performance gain (as shown in Sect. 4.4), we find a finer treatment of the relative ranking of the top classes to be beneficial.

Common image classification convolution networks extract a feature map $\boldsymbol{F}$ from an image sample. A global average pooling (GAP) is then applied to $\boldsymbol{F}$. The fully-connected layer FC and soft-max operation will output the logits $\boldsymbol{z}$ and the final prediction $\boldsymbol{p}$.

If we used FC directly on every pixel of the dense feature map $\boldsymbol{F}$, we cat get local classification results $\boldsymbol{p}^L$:

$$\boldsymbol{p}_k^L = Softmax(FC_\phi(\boldsymbol{F}_k)), k = 1, 2, ..., H \times W$$

(2)

where L is just a mark to show the difference between global prediction $\boldsymbol{p}$ and local predictions $\boldsymbol{p}^L$. These predictions give a more complete description of what the image contains, as the object of the target class usually only occupies part of the image. We take this information into account by identifying the individual top-1 class from each location as a vote and aggregate the votes into a histogram. The classes are finally ranked by their frequency of appearance in the histogram. The remaining classes will share the same lowest rank.

3.3   Zipf's Loss

Zipf's loss L_{Zipf} of image sample $\boldsymbol{x}$ and label y is defined as KL-divergence D_{KL} between normalized non-target prediction $\hat{\boldsymbol{p}}$ and the synthetic Zipf's soft label $\tilde{\boldsymbol{p}}$ generated from dense classification ranking:

$$\hat{p}_c = \frac{\exp(z_c)}{\sum_{m=1,m\neq y}^{C} \exp(z_m)}$$

$$L_{Zipf}(x,y) = D_{KL}(\tilde{p}||\hat{p}) = \sum_{c=1,c\neq y}^{C} \tilde{p}_c log\frac{\tilde{p}_c}{\hat{p}_c} \tag{3}$$

The synthetic Zipf's label $\tilde{p}$ of non-target class c should follow the Eq. 4 of Zipf's law with the corresponding rank r_c:

$$\tilde{p}_c = \frac{r_c^{-\alpha}}{\sum_{m=1,m\neq y}^{C} r_m^{-\alpha}} \tag{4}$$

where α is a hyper-parameter that controls the shape of the distribution.

The L_{CE} is the standard cross entropy loss with one-hot ground-truth label. The combined loss function is as follow:

$$Loss(x,y) = L_{CE}(x,y) + \lambda L_{Zipf}(x,y) \tag{5}$$

L_{CE} encourages the prediction to be sharp and confident in the target class, while L_{Zipf} regularized the prediction to be soft within non-target classes.λ is a hyper-parameter which control the regularization strength.

Comparison with Uniform Label Smoothing. The gradient with respect to non-target logits for L_{zipf} and L_{LS} are shown as:

$$\frac{\partial L_{Zipf}(x,y)}{\partial z_c} = \begin{cases} 0 & c = y \\ \hat{p}_c - \tilde{p}_c & c \neq y \end{cases}$$

$$\frac{\partial L_{LS}(x,y)}{\partial z_c} = \begin{cases} p_c - (1-\beta) & c = y \\ p_c - \frac{\beta}{C-1} & c \neq y \end{cases} \tag{6}$$

Label smoothing generates a soft label with a uniform value $\frac{\beta}{C-1}$ for all non-target classes and $1-\beta$ for the target class, where C is the total number of classes. However, label smoothing suppresses predictions of high-ranked classes or promotes predictions of low-ranked classes to the same level since β is constant and rank-irrelevant, which is conceptually sub-optimal.

Our Zipf's loss is rank-relevant compared with label smoothing. In non-target classes, it encourages the high-ranked classes to keep larger predictions than low-ranked ones. Zipf's law distribution shows empirical success in our experiments against other rank-relevant distributions such as linear decay, more details are shown in Sect. 4.4.

Table 2. Top-1 accuracy (%) on CIFAR100, TinyImageNet image classification tasks with various model architectures. We report the mean and standard deviation over five runs with different random seeds. Vanilla indicates baseline results from the standard cross-entropy, the best results are indicated in bold, and the second-best results are indicated by underline. The performances of state-of-the-art methods are reported for comparison

Method	CIFAR100	TinyImageNet	CIFAR100	TinyImageNet
	DenseNet121		ResNet18	
Vanilla	77.86 ± 0.26	60.31 ± 0.36	75.51 ± 0.28	56.41 ± 0.20
BAN [7]	78.39 ± 0.14	59.34 ± 0.60	76.96 ± 0.04	58.65 ± 0.83
BYOT [33]	78.93 ± 0.05	60.54 ± 0.02	77.15 ± 0.03	57.84 ± 0.15
PS-KD [14]	78.82 ± 0.10	61.64 ± 0.12	76.74 ± 0.06	58.22 ± 0.17
DDGSD [27]	78.18 ± 0.02	60.80 ± 0.30	76.48 ± 0.13	58.52 ± 0.12
CS-KD [29]	78.31 ± 0.49	62.04 ± 0.09	**78.01 ± 0.13**	58.38 ± 0.38
LS [19]	78.12 ± 0.45	61.25 ± 0.18	77.31 ± 0.28	56.89 ± 0.16
TF-KD [28]	77.68 ± 0.21	60.17 ± 0.57	77.29 ± 0.15	56.67 ± 0.05
Zipf's LS	**79.03 ± 0.32**	**62.64 ± 0.30**	77.38 ± 0.32	**59.25 ± 0.20**

4   Experiment

4.1   Experimental Detail

Datasets. We conduct experiments in various image classification tasks to demonstrate our method's effectiveness and universality. Specifically, we use CIFAR100 [15] and TinyImageNet[1] for small-scale classification tasks, and ImageNet [6] for large-scale classification task. We also verify fine-grained classification performance with INAT21 [25] using the "mini" training dataset.

Training Setups. We followed the setups in recent related works [4,8,28,29] and the popular open-source work[2]. All experiments use MSRA initialization [10], SGD optimizer with 0.9 momentum, 0.1 initial learning rate, 1e-4 weight decay, and standard augmentations including random cropping, and flipping. For small-scale CIFAR100 and TinyImageNet datasets, we use 32×32 resized input images, 128 batch size, and step learning rate policy which decreased to 1/10 of its previous value at 100th and 150th in the overall 200 epochs. All small-scale experiments are trained with single GPU. For large-scale ImageNet and INAT21 datasets, we use 224×224 resized input images, 256 batch size, and step learning rate policy which decreased to 1/10 of its previous value at 30th, 60th, and 90th in the overall 100 epochs. All large-scale experiments are trained with 4 GPUs.

Hyper Parameters. Our method has two hyperparameters in general, λ and α. λ controls regularization strength and α controls the decay shape of Zipf's

---

[1] https://www.kaggle.com/c/tiny-imagenet.
[2] https://github.com/facebookresearch/pycls.

Table 3. Top-1 accuracy (%) comparison with state-of-the-art works. Experiments conduct on ImageNet, INAT21 image classification tasks with ResNet50, and on CIFAR100, TinyImageNet with DenseNet121.

Method	CIFAR100	TinyImageNet	ImageNet	INAT21
Vanilla	77.86	60.31	76.48	62.43
CS-KD [29]	78.31	62.04	76.78	65.45
LS [19]	78.12	61.25	76.67	65.16
TF-KD [28]	77.68	60.17	76.56	62.61
Zipf's LS	**79.03**	**62.64**	**77.25**	**66.04**

distribution, which is set to 1.0 in all experiments. β is only recommended for small resolution datasets such as CIFAR100 and TinyImageNet, to exploit the higher resolution intermediate feature map and make a more reliable ranking. A detailed hyperparameters ablation study is shown in the supplement.

4.2   General Image Classification Tasks

First, we conduct experiments on CIFAR100 dataset and TinyImageNet dataset to compare with other related state-of-the-art methods, including self-knowledge distillation methods (BAN [7]), online knowledge distillation methods (DDGSD [27], CS-KD [29]), and label smoothing regularization method (label smoothing [19], TF-KD [28]). Table 2 shows the classification results of each method based on different network architectures. All experiments of the above methods keep the same setups for a fair comparison, details can be seen in 4.1 **Training setups**. For other hyper-parameters, we keep their original settings.

Comparison with Two-Stage Knowledge Distillation. Two-stage knowledge distillation methods improve model accuracy using their previous models' dark knowledge. These methods rely on a pretrained model, which means they take twice or more training time than our method. We make a big advantage beyond BAN [7] (one step) as Table 2 shows, specifically, Zipf's LS surpasses BAN [7] by 0.64% and 3.28% respectively on CIFAR100 and TinyImageNet based on DenseNet121.

Comparison with Self-distillation. As one type self-distillation utilizing contrastive samples, DDGSD and CS-KD realize respectively exploiting instance-wise and class-wise consistency regularization techniques. The data process of DDGSD or the pair sample strategy of CS-KD brings double iterations per training epoch. As shown in Table 2, Zipf's LS achieves 0.9% and 0.73% gains compared with DDGSD [27] respectively on CIFAR100 and TinyImageNet based on ResNet18 without more training iterations. As another type self-distillation with auxiliary parameters, BYOT [33] squeeze deeper knowledge into lower network. Zipf's LS surpasses BYOT obviously on TinyImageNet.

Comparison with Label Smoothing Regularization. Label smoothing [19] is a general effective regularizing method with manually designed soft targets. In Table 2, Zipf's LS beats label smoothing by 2.36% and 1.39% advantage on TinyImageNet respectively based on ResNet18 and DenseNet121.

Table 4. Top-1 accuracy (gain) (%) on ImageNet and INAT21 image classification tasks with various model architectures. Vanilla indicates baseline results from the cross-entropy, and the best results are indicated in bold

Architecture	Method	ImageNet	INAT21
ResNet18	Vanilla	70.47	54.31
	Label Smooth	70.53(+0.06)	55.17(+0.86)
	Zipf's LS	**70.73 (+0.26)**	**56.36 (+2.03)**
ResNet50	Vanilla	76.48	62.43
	Label Smooth	76.67(+0.19)	65.16(+2.73)
	Zipf's LS	**77.25 (+0.77)**	**66.04 (+3.61)**
ResNet101	Vanilla	77.83	65.60
	Label Smooth	78.12(+0.29)	67.14(+1.54)
	Zipf's LS	**78.58 (+0.75)**	**68.45 (+2.85)**
ResNeXt50_32 × 4d	Vanilla	77.54	66.36
	Label Smooth	77.72(+0.18)	67.51(+1.15)
	Zipf's LS	**78.07 (+0.53)**	**69.24 (+2.88)**
ResNeXt101_32 × 8d	Vanilla	79.51	70.35
	Label Smooth	79.69(+0.18)	71.52(+1.17)
	Zipf's LS	**80.03 (+0.52)**	**72.18 (+1.83)**
DenseNet121	Vanilla	75.56	63.75
	Label Smooth	75.59(+0.03)	64.60(+0.85)
	Zipf's LS	**75.98 (+0.42)**	**66.60 (+2.85)**
MobileNetV2	Vanilla	65.52	55.75
	Label Smooth	65.71(+0.19)	56.29(+0.54)
	Zipf's LS	**66.03 (+0.51)**	**56.45 (+0.70)**

4.3 Large-Scale and Fine-Grained Image Classification Tasks

Comparison with State-of-the-Art Methods. Our method is one-pass with almost zero extra computational or memory cost. Label smoothing and TF-KD are the two most related works with ours as shown in Fig. 1 and Table 1. And CS-KD is the most superior method on small-scale datasets besides ours as shown in Table 2. So we further compare our method with CS-KD, label smoothing and TF-KD on large-scale and fine-grained datasets. As shown in Table 3, our method shows much more superior performance while label smoothing and CS-KD methods have already improved baseline with significant margins. For instance, we surpass the second-best method by 0.47% and 0.59% respectively on ImageNet and INAT21 based on ResNet50.

Improvements on Various Architectures. We evaluate our method on various network architectures on ImageNet and INAT21. Not only the widely used families of ResNet [9] and ResNeXt [26] are considered, but also the lighter architectures (such as MobileNetV2 [22]) are evaluated. Table 4 shows our significant improvements compared with vanilla cross-entropy training on various network architectures based on ImageNet and INAT21 datasets. For instance, our method boosts baseline by 0.75% and 2.85% respectively on ImageNet and INAT21 with ResNet101.

Table 5. Dense classification ranking v.s. logits-based ranking on TinyImageNet with ResNet18. CE indicates the standard cross-entropy loss. LR indicates that using the logits ranks. Dense1 indicates that using the last dense feature map of the last stage. Dense2 indicates that using the last dense feature map of the penultimate stage

Method	CE	LR	Dense1	Dense2	Top-1 acc(%)(gain)
Vanilla					56.41
Logits-based					(+1.70)
Voting-based					(+2.40)
Voting-based					**(+2.84)**

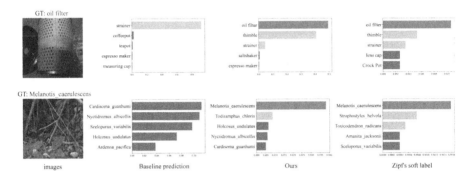

Fig. 4. Comparison of top-5 predictions visualization. The dark green, light green, and red colors denote ground-truth, similar and irrelevant categories respectively. More informative supervision and prediction are gained since categories that are more similar to ground truth arise top in Zipf's soft label (thimble and strainer vs oil filter). (Color figure online)

4.4   Ablation Study

Dense Classification Ranking v.s. Logits-Based Ranking. We introduced two ranking metrics in Sect. 3.2, logits-based ranking and dense classification ranking. While logits-based ranking indeed boosts performance compared to the baseline (1.7% as shown in the second row of Table 5), we still find our strategy, the dense classification ranking, is necessary for optimal performance. When we replace the rank by the dense votes from the last few feature maps, the accuracy improvement goes to as high as 2.84% (last row in Table 5).

Comparison of Different Distributions. We constraint the logits rank or dense vote rank to obey Zipf's law due to the discovery that the output of deep neural network trained for a classification task follows the law as well. To demonstrate Zipf's priority, we conduct constant style, random style and decay style distributions on various datasets. As shown in Table 6, Zipf makes the best performance among these distributions. It's worth noting that, although inferior to Zipf's distribution, constant distribution also achieves satisfying performance which benefits from regularizing only non-target classes different from normal label smoothing. We speculate that label smoothing coupled with the target class that might hurt performance.

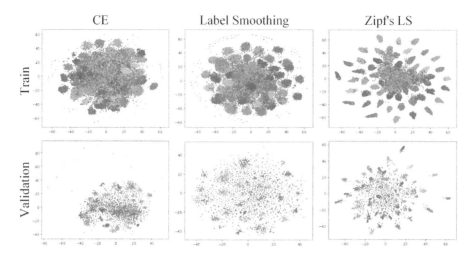

Fig. 5. T-SNE [17] visualization on 50 random sampled classes of TinyImageNet for CE, Label Smoothing and Zipf's label smoothing.

5   Discussion

Zipf's Soft Labels Results in More Reasonable Predictions. Figure 4 illustrates the top-5 predictions of our proposed method compared with the baseline method. The images are sampled from ImageNet and INAT21 respectively. Our method makes more reasonable predictions. Not only the top-1 prediction is correct, but also more similar concepts arise in the top-5 prediction. This results from more informative soft labels from Zipf's LS, which provides meaningful representations as knowledge distillation for the network to better grasp the concept of similar categories.

Table 6. Comparison of the top-1 accuracy (%) of different distributions and label smoothing. Experiments are conducted on TinyImageNet, ImageNet and INAT21. Zipf makes the best performance among these distributions

Distribution	TinyImageNet	ImageNet	INAT21
Vanilla	56.41	76.48	62.43
LS	56.89	76.67	65.16
Constant	58.76	77.09	65.86
Random_Uniform	58.24	76.89	65.61
Random_Pareto	58.52	76.61	65.9
Linear_Decay	58.39	76.87	65.86
Zipf	**59.25**	**77.25**	**66.04**

Zipf's Label Smoothing Achieves Better Representation Learning for Generalization. We compare our Zipf's LS techniques with cross-entropy training and uniform label smoothing training on the TinyImageNet dataset. As

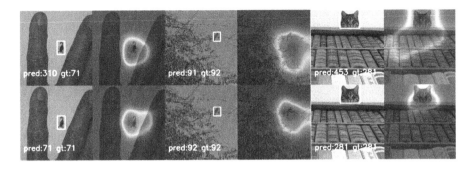

Fig. 6. Comparison of small objects results between CE(above) and Zipf's LS(below).

shown in Fig. 5, the intra-class distance in feature space learned from Zipf's label smoothing is more compact and the inter-class distance is more separate. Zipf's Label Smoothing achieves better representation learning for generalization. More discussion for generalization is shown in supplements.

Non-target Class Dense Classification Ranking Design. To better rank the classes in the soft label, we exploit dense classification ranking instead of logits-based ranking. Thus, larger objects are favored and small object classification performance might degrade. In our design, the target class is excluded in L_{Zipf} and included in L_{CE} only, ensuring the target class provides the correct gradient regardless of the object size. To verify the small object performance, we collect the five smallest objects from each class[3], Zipf's LS still beat CE(69.25% vs 67.72%). The cam visualization for the small object case is shown in Fig 6.

Limitation. Zipf's LS does not help in binary classification cases, since the L_{zipf} only considers non-target class and would always be zero. Further, as mentioned in the method section, we exploit dense classification ranking to get more reliable ranking information than logit-based ranking, which is only available in image data but not in modals such as speech and language. To make Zipf's LS work in multi-modal data is considered as future work.

6   Conclusion

In this work, we propose an efficient and effective one pass self-distillation method named Zipf's Label Smoothing, which not only generates soft-label supervision in a teacher-free manner as efficient as label smoothing but also generates non-uniform ones as informative as more expensive self distillation approaches. Zipf's Label Smoothing consistently performs better than the uniform label smoothing method and other parameters-free one pass self-distillation methods, it could be one of the plug-and-play self-distillation techniques in your deep learning toolbox.

---

[3] https://image-net.org/data/bboxes_annotations.tar.gz.

References

1. Allen-Zhu, Z., Li, Y.: Towards understanding ensemble, knowledge distillation and self-distillation in deep learning. arXiv preprint. arXiv:2012.09816 (2020)
2. Bagherinezhad, H., Horton, M., Rastegari, M., Farhadi, A.: Label refinery: improving imagenet classification through label progression. arXiv preprint. arXiv:1805.02641 (2018)
3. Beyer, L., Zhai, X., Royer, A., Markeeva, L., Anil, R., Kolesnikov, A.: Knowledge distillation: a good teacher is patient and consistent. arXiv preprint. arXiv:2106.05237 (2021)
4. Chen, D., Mei, J.P., Wang, C., Feng, Y., Chen, C.: Online knowledge distillation with diverse peers. In: Proceedings of the AAAI Conference on Artificial Intelligence, vol. 34, pp. 3430–3437 (2020)
5. Cho, J.H., Hariharan, B.: On the efficacy of knowledge distillation. In: Proceedings of the IEEE/CVF International Conference on Computer Vision, pp. 4794–4802 (2019)
6. Deng, J., Dong, W., Socher, R., Li, L.J., Li, K., Fei-Fei, L.: Imagenet: a large-scale hierarchical image database. In: 2009 IEEE Conference on Computer Vision and Pattern Recognition, pp. 248–255. IEEE (2009)
7. Furlanello, T., Lipton, Z., Tschannen, M., Itti, L., Anandkumar, A.: Born again neural networks. In: International Conference on Machine Learning, pp. 1607–1616. PMLR (2018)
8. Guo, Q., Wang, X., Wu, Y., Yu, Z., Liang, D., Hu, X., Luo, P.: Online knowledge distillation via collaborative learning. In: Proceedings of the IEEE/CVF Conference on Computer Vision and Pattern Recognition, pp. 11020–11029 (2020)
9. He, K., Zhang, X., Ren, S., Sun, J.: Deep residual learning for image recognition (2015)
10. He, K., Zhang, X., Ren, S., Sun, J.: Delving deep into rectifiers: surpassing human-level performance on imagenet classification (2015)
11. He, T., Zhang, Z., Zhang, H., Zhang, Z., Xie, J., Li, M.: Bag of tricks for image classification with convolutional neural networks. In: Proceedings of the IEEE/CVF Conference on Computer Vision and Pattern Recognition, pp. 558–567 (2019)
12. Hinton, G., Vinyals, O., Dean, J.: Distilling the knowledge in a neural network. arXiv preprint. arXiv:1503.02531 (2015)
13. Ji, M., Shin, S., Hwang, S., Park, G., Moon, I.C.: Refine myself by teaching myself: feature refinement via self-knowledge distillation. In: Proceedings of the IEEE/CVF Conference on Computer Vision and Pattern Recognition, pp. 10664–10673 (2021)
14. Kim, K., Ji, B., Yoon, D., Hwang, S.: Self-knowledge distillation with progressive refinement of targets. In: Proceedings of the IEEE/CVF International Conference on Computer Vision, pp. 6567–6576 (2021)
15. Krizhevsky, A., Hinton, G.: Learning multiple layers of features from tiny images. Technical Report, University of Toronto, Toronto, Ontario (2009)
16. Ma, H., Chen, T., Hu, T.K., You, C., Xie, X., Wang, Z.: Undistillable: making a nasty teacher that cannot teach students. ArXiv abs/2105.07381 (2021)
17. Van der Maaten, L., Hinton, G.: Visualizing data using t-sne. J. Mach. Learn. Res. 9(11), 2579–2605 (2008)
18. Mirzadeh, S.I., Farajtabar, M., Li, A., Levine, N., Matsukawa, A., Ghasemzadeh, H.: Improved knowledge distillation via teacher assistant. In: Proceedings of the AAAI Conference on Artificial Intelligence, vol. 34, pp. 5191–5198 (2020)

19. Müller, R., Kornblith, S., Hinton, G.: When does label smoothing help? arXiv preprint. arXiv:1906.02629 (2019)
20. Park, D.Y., Cha, M.H., Kim, D., Han, B., et al.: Learning student-friendly teacher networks for knowledge distillation. In: Advances in Neural Information Processing Systems, vol. 34 (2021)
21. Powers, D.M.: Applications and explanations of zipf's law. In: New Methods in Language Processing and Computational Natural Language Learning (1998)
22. Sandler, M., Howard, A., Zhu, M., Zhmoginov, A., Chen, L.C.: Mobilenetv 2: inverted residuals and linear bottlenecks (2019)
23. Stanton, S., Izmailov, P., Kirichenko, P., Alemi, A.A., Wilson, A.G.: Does knowledge distillation really work? In: Advances in Neural Information Processing Systems, vol. 34 (2021)
24. Szegedy, C., Vanhoucke, V., Ioffe, S., Shlens, J., Wojna, Z.: Rethinking the inception architecture for computer vision. In: Proceedings of the IEEE Conference on Computer Vision and Pattern Recognition, pp. 2818–2826 (2016)
25. Van Horn, G., Cole, E., Beery, S., Wilber, K., Belongie, S., Mac Aodha, O.: Benchmarking representation learning for natural world image collections. In: Proceedings of the IEEE/CVF Conference on Computer Vision and Pattern Recognition, pp. 12884–12893 (2021)
26. Xie, S., Girshick, R., Dollár, P., Tu, Z., He, K.: Aggregated residual transformations for deep neural networkcs (2017)
27. Xu, T.B., Liu, C.L.: Data-distortion guided self-distillation for deep neural networks. In: Proceedings of the AAAI Conference on Artificial Intelligence, vol. 33, pp. 5565–5572 (2019)
28. Yuan, L., Tay, F.E., Li, G., Wang, T., Feng, J.: Revisiting knowledge distillation via label smoothing regularization. In: Proceedings of the IEEE/CVF Conference on Computer Vision and Pattern Recognition, pp. 3903–3911 (2020)
29. Yun, S., Park, J., Lee, K., Shin, J.: Regularizing class-wise predictions via self-knowledge distillation. In: Proceedings of the IEEE/CVF Conference on Computer Vision and Pattern Recognition, pp. 13876–13885 (2020)
30. Zagoruyko, S., Komodakis, N.: Paying more attention to attention: improving the performance of convolutional neural networks via attention transfer. arXiv preprint. arXiv:1612.03928 (2016)
31. Zhang, C.B., Jiang, P.T., Hou, Q., Wei, Y., Han, Q., Li, Z., Cheng, M.M.: Delving deep into label smoothing. IEEE Trans. Image Process. **30**, 5984–5996 (2021)
32. Zhang, L., Bao, C., Ma, K.: Self-distillation: towards efficient and compact neural networks. IEEE Trans. Pattern Anal. Mach. Intell. **44**(8), 4388–4403 (2021)
33. Zhang, L., Song, J., Gao, A., Chen, J., Bao, C., Ma, K.: Be your own teacher: Improve the performance of convolutional neural networks via self distillation. In: Proceedings of the IEEE/CVF International Conference on Computer Vision, pp. 3713–3722 (2019)
34. Zhu, Y., Wang, Y.: Student customized knowledge distillation: Bridging the gap between student and teacher. In: Proceedings of the IEEE/CVF International Conference on Computer Vision, pp. 5057–5066 (2021)

Prune Your Model Before Distill It

Jinhyuk Park[ID] and Albert No[(✉)][ID]

Hongik University, Seoul 04066, Korea
c0292601@g.hongik.ac.kr, albertno@hongik.ac.kr

Abstract. Knowledge distillation transfers the knowledge from a cumbersome teacher to a small student. Recent results suggest that the student-friendly teacher is more appropriate to distill since it provides more transferrable knowledge. In this work, we propose the novel framework, "prune, then distill," that prunes the model first to make it more transferrable and then distill it to the student. We provide several exploratory examples where the pruned teacher teaches better than the original unpruned networks. We further show theoretically that the pruned teacher plays the role of regularizer in distillation, which reduces the generalization error. Based on this result, we propose a novel neural network compression scheme where the student network is formed based on the pruned teacher and then apply the "prune, then distill" strategy. The code is available at https://github.com/ososos888/prune-then-distill.

Keywords: Knowledge distillation · Label smoothing regularization (LSR) · Neural network compression · Pruning

1 Introduction

Recent progress in neural networks (NN) in various tasks highly depends on its over-parameterization, such as classification [21,55], language understanding [7,8], and self-supervised learning [3,15]. This leads to extensive computational cost and even causes environmental issues [38]. Therefore, neural network compression techniques have received increasing attention, such as knowledge distillation [20,40,51] and pruning [9,13,26,31].

Knowledge distillation (KD) [20] is a model compression tool that transfers the features from a cumbersome network to a smaller network. At first glance, a powerful teacher with higher accuracy may show better distillation results; however, Cho and Hariharan [4] showed that the less-trained teacher teaches better when the student network does not have enough capability. Lately, a line of works has proposed distillation schemes that focus on a "student-friendly" teacher, which provides more transferrable knowledge to the student network with limited capacity [36,37].

On the other hand, network pruning [26] is another network compression technique that effectively removes networks' weights or neurons while

Supplementary Information The online version contains supplementary material available at https://doi.org/10.1007/978-3-031-20083-0_8.

maintaining accuracy. Since pruning simplifies the neural network, we naturally conjecture that the pruned teacher provides student-friendly knowledge that is easier to transfer. This intuition leads us to our main question: *can pruning boost the performance of knowledge distillation?*

To answer this question, we propose a new framework, "prune, then distill," consisting of three steps: 1) train the (teacher) network, 2) prune the (teacher) network, and 3) distill the pruned network to the smaller (student) network. We examine several simple experiments to verify the proposed idea that compares the test accuracy of student networks with and without (unstructured) pruning on the teachers' side. More precisely, We conduct three experiments: 1) distill VGG19 [43] to VGG11, 2) distill VGG19 and ResNet18 [16] to itself (self distillation), and 3) distill ResNet18 to VGG16 and MobileNetV2 [42]. In all three cases, we observe that the student learned from the pruned teacher generally outperforms the student learned from an unpruned teacher.

We then provide theoretical support to answer why the pruned teacher is better in distillation. Knowledge distillation can be viewed as a label smoothing regularization (LSR) [54,58], which regularizes training by providing a smoother label. We find that a teacher trained with regularization provides a smoother label than the original teacher. This implies that the distillation with a regularized teacher is equivalent to LSR with smoother labels. Since pruning can be viewed as a regularized model with a sparsity-inducing regularizer [28], we conclude that the pruned teacher regularizes the distillation process.

Based on the observation that pruned teacher provides a better knowledge in distillation, we then suggest a novel network compression scheme. When a cumbersome network is provided, we want to compress the network by applying the "prune, then distill" strategy. However, since the distillation transfers knowledge to a *given* student network, the student network architecture design is required. The main idea of student network construction is matching the teacher and the student layerwise. We propose a student network with the same depth but fewer neurons so that the number of weights per layer matches the number of nonzero weights of the pruned network in the corresponding layer. We evaluate the proposed compression scheme with extensive experiments.

We summarize our contributions as:

- We propose a novel framework, "prune, then distill," that prunes teacher networks before distillation.
- We examine experiments that verify unstructured pruning on the teacher can boost the performance of knowledge distillation.
- We also provide a theoretical analysis that the distillation from a pruned teacher is effectively a label smoothing regularization with smoother labels.
- We propose a novel network compression that constructs the student network based on the pruned teacher, then apply the "prune, then distill" strategy.

2   Related Works

This section is devoted to prior works on neural network (NN) compression that are related to our work. In particular, we focus on knowledge distillation and network pruning. Note that there are other NN compression techniques such as quantization [2,29], coding [14,50], and matrix factorization [22,41].

2.1 Knowledge Distillation

Knowledge distillation (KD) [20] transfers the knowledge from the strong teacher network to a smaller student network. The student network is trained with soft targets provided by the teacher network and some intermediate features [40,52,56]. There are variations of KD such as KD using GAN [51], Jacobian matching KD [5,44], distillation of activation boundaries [19], contrastive distillation [48], and distillation from graph neural networks [23,52].

Recently, many works have reported that the large gap between student and teacher causes degradation in student network performance [36]. Cho and Hariharan showed that the less-trained network transfers better knowledge to a small network [4]. Park et al. [37] proposed a student-aware teacher learning to transfer the teacher's knowledge effectively. In this paper, we provide an extremely simple way to generate a student-friendly teacher using unstructured pruning.

2.2 Pruning

There are two main branches of pruning: 1) unstructured pruning, which prunes individual weights, and 2) structured pruning, which prunes neurons (in most cases, channels of convolutional neural networks). Although both approaches share a similar idea, these two strategies have been developed independently.

Unstructured Pruning: Unstructured pruning [26] removes NN components in weight-level while maintaining the number of neurons in the network. A general pruning pipeline consists of three steps: 1) train a large network, 2) prune weights (or neurons) based on its own rule, then 3) fine-tune the pruned model. The iterative magnitude pruning (IMP) technique, which iteratively applies magnitude-based pruning and fine-tuning, shows remarkable performance [13]. Lottery ticket rewinding (LTR), an iterative magnitude pruning method with weight rewinding, is highly successful [9,10]. Recently, IMP with learning rate (LR) rewinding, which repeats the learning rate schedule, shows better results in bigger networks [39]. However, the network architecture after unstructured pruning remains the same (i.e., number of channels per layer). It is hard to fully enjoy the benefit of a pruned network without dedicated hardware [12].

Structured Pruning: Structured pruning removes NN parameters at the level of neurons (mostly channels) [1,31,32,35,46]. It provides a smaller network with efficient network architecture, and we can save computational resources without designing dedicated hardware or libraries. Like magnitude-based unstructured pruning, the most naive method is to prune filters based on weights [17,31]. Another approach is adding an extra regularizer that induces sparsity while training [18,49,57]. Liu et al. [33] and Ye et al. [53] proposed the structured pruning scheme based on batch normalization (BN) scale factor of filters. Zhuang et al. [59] adds polarization regularizer to structured pruning with BN scale

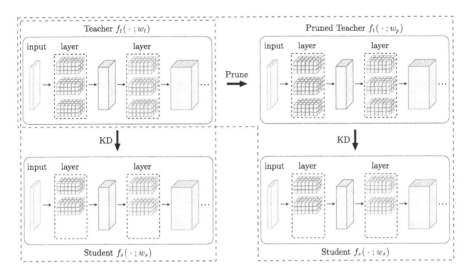

Fig. 1. Overview of the "prune, then distill" strategy. Instead of distilling directly from the teacher to the student (blue dotted box), we prune the teacher first, then distill from the pruned teacher to the student (red dotted box). (Color figure online)

factor. However, due to the structural constraint, the pruned network has more weights (parameters) than unstructured pruning [34].

3 Prune, Then Distill

3.1 Exploratory Experiments

We conduct experiments to verify the effectiveness of pruned teachers in KD. Instead of distilling the teacher network directly (dotted-blue block in Fig. 1), we first (unstructured) prune the teacher network and then distill to the student network (dotted-red block in Fig. 1).

Setups: We mainly considers VGG [43] and ResNet [16] for the teacher network, where VGG is trained on the CIFAR100 dataset [24] and ResNet is trained on the TinyImageNet dataset [25]. The TinyImageNet dataset is a subset of resized ($3 \times 64 \times 64$) ImageNet dataset [6]. We reserve 10% of the data as a validation set in all training. We apply unstructured pruning that removes more weights, more precisely LR rewinding [39], to prune the teacher model. In LR rewinding, we set the ratio of epochs by 0.65 for VGG-CIFAR100. In other words, we train the VGG19 for 200 epochs initially, then rewind the learning rates and retrains (fine-tuning) the network for 130 epochs (65%). Note that the different ratios from 0.6 to 0.9 do not make significant differences in pruning, and we use the ratio of 0.5 for ResNet-TinyImageNet. For a fair comparison, we train (and distill) networks with enough epochs and halt the training at their best performance

Table 1. Knowledge distillation from VGG19 to VGG11 on CIFAR100 with teacher pruning. VGG19DBL is the VGG19 with 2× more filters per layer. Teacher "None" indicates the student is trained without a teacher, while the pruning ratio "None" means the distillation from the unpruned teacher.

Teacher	Pruning Ratio	Teacher Accuracy	Student	Student Accuracy
None	–	–	VGG11	69.51 ± 0.24
VGG19	None	73.13	VGG11	72.02 ± 0.27
	36%	73.30	VGG11	72.76 ± 0.10
	59%	72.25	VGG11	72.59 ± 0.32
	79%	73.43	VGG11	72.67 ± 0.34
VGG19DBL	None	74.44	VGG11	71.81 ± 0.29
	36%	73.46	VGG11	72.01 ± 0.11
	59%	73.24	VGG11	72.40 ± 0.25
	79%	73.50	VGG11	72.48 ± 0.19

on validation dataset. All test accuracies are the average of three independent experiments, and we also provide the standard deviation.

For simplicity, we use the vanilla KD [20] with appropriate balancing parameter α and temperature τ. The balancing parameter α represents the ratio of two objectives (distill loss and hard-target loss). The temperature τ is a softening parameter, where higher τ produces a softer target. In the experiment, we fix the parameters by $\alpha = 0.95$ and $\tau = 10$. More detailed training parameters are provided in Appendix. Note that the purpose of experiments is not achieving the best test accuracy but to compare between *distilling from a pruned network* and *distilling from an unpruned network*. Thus, hyperparameters, as well as network architectures, are not optimized for test accuracies. Instead, we use as-is settings for a *fair* comparison between the pruned teacher and the unpruned teacher. For example, we follow default settings for MobileNetV2 [42] and ResNet18 [16] optimized for ImageNet dataset [6], while we use TinyImageNet dataset [25].

Distill VGG19 to VGG11: We set VGG19 [43] as a teacher network and VGG11 as a student network. The network architecture of VGG is unchanged except for the number of fully connected (FC) layers, where our VGG has a single FC layer (which is commonly used for CIFAR10 data). Then, we compare the KD results on the CIFAR100 dataset [24] between the regular VGG19 teacher and the pruned VGG19 teacher. We prune the teacher network with three sparsity levels: 36% sparsity (36% of weights are removed), 59% sparsity, and 79% sparsity.

Surprisingly, as shown in Table 1, VGG11 with pruned VGG19 consistently outperforms the one with the unpruned teacher. Table 1 also provides results when the teacher network is VGG19DBL, with 2× many channels in each layer. In both cases, the pruned teacher shows better performance.

Self Distillation: Motivated by [11,54], we conduct the self distillation experiment, where the teacher and the student share the same model. We con-

Table 2. Self distillation of VGG19 and ResNet18 with teacher pruning. DBL model has the same model structure with 2× more filters per layer. Teacher "None" indicates the student is trained without a teacher, while the pruning ratio "None" means the distillation from the unpruned teacher.

Teacher	Pruning ratio	Teacher accuracy	Student	Student accuracy
None	–	–	VGG19	72.76 ± 0.33
VGG19	None	73.13	VGG19	73.74 ± 0.20
	36%	73.30	VGG19	74.10 ± 0.26
	59%	72.25	VGG19	74.26 ± 0.37
	79%	73.43	VGG19	74.35 ± 0.10
None	–	–	VGG19DBL	74.62 ± 0.21
VGG19DBL	None	74.44	VGG19DBL	74.78 ± 0.37
	36%	73.46	VGG19DBL	75.16 ± 0.44
	59%	73.24	VGG19DBL	75.26 ± 0.77
	79%	73.50	VGG19DBL	75.05 ± 0.92
None	–	–	ResNet18	57.75 ± 0.24
ResNet18	None	57.75	ResNet18	57.97 ± 0.10
	36%	57.66	ResNet18	59.39 ± 0.21
	59%	57.58	ResNet18	58.99 ± 0.26
	79%	57.32	ResNet18	59.33 ± 0.18
None	–	–	ResNet18DBL	60.21 ± 0.24
ResNet18DBL	None	60.46	ResNet18DBL	61.35 ± 0.02
	36%	61.97	ResNet18DBL	63.03 ± 0.38
	59%	61.80	ResNet18DBL	63.19 ± 0.21
	79%	61.66	ResNet18DBL	63.16 ± 0.02

sider VGG19 and VGG19DBL with CIFAR100 dataset, where ResNet18 and ResNet18DBL are trained on the TinyImageNet dataset. Table 2 shows the test accuracies of 1) the model without KD, 2) the model learned from the unpruned teacher, and 3) the model learned from the pruned teacher. Similar to other experiments, we also observe the consistent result where the pruned model teaches better than the unpruned teacher. Note that learning from unpruned network also increases the test accuracy (compared to the one without a teacher); however, the gain with the pruned teacher is more significant.

Distill ResNet18 to VGG and MobileNet: We also investigate the KD from the pruned teacher when the student and the teacher have different network architectures. Specifically, we consider the TinyImageNet dataset, where the teacher is ResNet18 and students are VGG16 and MobileNetV2 [42]. Table 3 compares the test accuracies of 1) student without a teacher, 2) student learned from the unpruned teacher, and 3) student learned from the pruned teacher.

Table 3. Distillation from ResNet18 to MobileNetV2 and VGG16 with teacher pruning. Teacher "None" indicates the student is trained without a teacher, while the pruning ratio "None" means the distillation from the unpruned teacher.

Teacher	Pruning ratio	Teacher accuracy	Student	Student accuracy
None	–	–	VGG16	53.31 ± 0.45
ResNet18	None	57.75	VGG16	54.75 ± 0.29
	36%	57.66	VGG16	56.35 ± 0.35
	59%	57.58	VGG16	55.86 ± 0.04
	79%	57.32	VGG16	56.49 ± 0.15
None	–	–	MobileNetV2	50.79 ± 0.44
ResNet18	None	57.75	MobileNetV2	56.10 ± 0.23
	36%	57.66	MobileNetV2	56.73 ± 0.24
	59%	57.58	MobileNetV2	56.73 ± 0.43
	79%	57.32	MobileNetV2	57.20 ± 0.25

Consistently, we observe the better KD performance when the teacher is pruned. This implies that the better distillation is not limited to the case of the similar architecture between teacher and student networks.

Remark: One might suspect that better distillation result is due to higher accuracy of the teacher, where the pruned model often achieves better accuracy [9]. However, the higher accuracy of the teacher network does not guarantee better results in distillation [45]. Also, the pruned teacher works better even when test accuracy is lower than the unpruned teacher. For example, pruning decreases the test accuracy of the teacher network in ResNet18-TinyImageNet, where we observe that the pruned teacher transfers the knowledge better. This implies that the pruned teacher is better not because it has higher accuracy, but it provides better transferable knowledge.

We also investigate the agreement between the teacher and the student's prediction (details provided in Appendix). As shown by Stanton et al. [45], we observe that the agreement and the accuracy behave independently. For example, in VGG19 self distillation experiments, the pruned teacher provides a higher agreement, and the corresponding student has a higher accuracy; however, in ResNet18 self distillation, the pruned teacher shows lower agreement although the student's accuracy is higher. It implies that some students mimic the teacher better but perform worse. This result supports our theory that distillation indirectly helps the training student models with additional regularization.

3.2 Pruned Teacher as a Regularizer

In this section, we provide a theoretical analysis on the pruned teacher in KD. We first point out that the teacher trained with a regularizer provides an additional regularization during distillation.

Let $\{(x_i, y_i)\}_{i=1}^N$ be the dataset where the label y_i takes value from the set $\{1, 2, \ldots, K\}$. We are interested in a classification model which outputs a K-dimensional probability distributions. Let $f_{true}(x_i) \in \mathbb{R}^K$ be the one-hot encoded vector where $f_{true}(x_i)[y_i] = 1$ for the ground-truth label y_i and $f_{true}(x_i)[y'] = 0$ for all $y' \neq y_i$. We further let $f_t(x; w)$ be the output of the teacher network when the input is x and the weight is w. Then, we train the teacher $f_t(\cdot; w)$ and achieve w_t that minimizes the cross entropy loss

$$L_{CE}(w) = \frac{1}{N} \sum_{i=1}^N H(f_{true}(x_i), f_t(x_i; w)), \tag{1}$$

where the cross-entropy loss is defined by $H(p_1, p_2) = -\sum_{k=1}^K p_1[k] \log p_2[k]$.

Similarly, $f_s(x; \tilde{w})$ is the output of the student network when the input is x and the weight is $\tilde{w}$. For the temperature $\tau = 1$, the knowledge distillation loss is given by

$$L_{KD}(\tilde{w}) = \frac{1}{N} \sum_{i=1}^N (1 - \alpha) H(f_{true}(x_i), f_s(x_i; \tilde{w})) + \alpha H(f_t(x; w_t), f_s(x; \tilde{w})). \tag{2}$$

Yuan et al. [54] showed that the KD is equivalent to label smoothing regularization (LSR). More precisely, the author showed that

$$L_{KD}(\tilde{w}) = \frac{1}{N} \sum_{i=1}^N H(f_m^{(\alpha)}(x_i; w_t), f_s(x_i; \tilde{w})), \tag{3}$$

where $f_m^{(\alpha)}(x; w_t) = (1 - \alpha) f_{true}(x) + \alpha f_t(x; w_t)$, and therefore KD is equivalent to label smoothing regularization with smoothed label distribution $f_m^{(\alpha)}(x; w_t)$.

We then consider the case where the teacher is trained with a regularizer $R(w)$. The regularized teacher $f_t(\cdot; w_p)$ is obtained by minimizing

$$L_{REG}(w) = \frac{1}{N} \sum_{i=1}^N H(f_{true}(x_i), f_t(x_i; w)) + R(w), \tag{4}$$

i.e., $L_{REG}(w_p) = \min_w L_{REG}(w)$. Since $L_{CE}(w_t) = \min_w L_{CE}(w)$, we have

$$\frac{1}{N} \sum_{i=1}^N H(f_{true}(x_i), f_t(x_i; w_t)) \leq \frac{1}{N} \sum_{i=1}^N H(f_{true}(x_i), f_t(x_i; w_p)) \tag{5}$$

$$\frac{1}{N} \sum_{i=1}^N H(f_{true}(x_i), f_t(x_i; w_p)) + R(w_p) \leq \frac{1}{N} \sum_{i=1}^N H(f_{true}(x_i), f_t(x_i; w_t)) + R(w_t) \tag{6}$$

which implies

$$0 \leq \frac{1}{N} \sum_{i=1}^N \log \frac{f_t(x_i; w_t)[y_i]}{f_t(x_i; w_p)[y_i]} \leq R(w_t) - R(w_p) \tag{7}$$

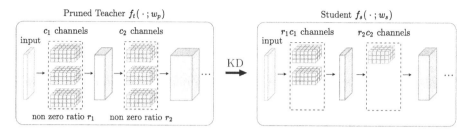

Fig. 2. Student network design. The number of channels of the student network is adjusted so that each layer's parameters match the number of nonzero parameters in each layer of the pruned teacher.

Thus, $f_t(x_i; w_t)[y_i]$ is larger than $f_t(x_i; w_p)[y_i]$ on average.

Recall that the distillation from $f_t(x_i; w_t)$ is equivalent to label smoothing regularization with smoothed label distribution $f_m^{(\alpha)}(x; w_t) = (1 - \alpha)f_{true}(x) + \alpha f_t(x, w_t)$. If we distill from $f_t(x_i; w_p)$ to the student, then it is essentially label smoothing regularization with a new smoothed label distribution $f_m^{(\alpha)}(x; w_p) = (1 - \alpha)f_{true}(x) + \alpha f_t(x, w_p)$. Since Eq. (7) implies that the new smoothed distribution $f_m^{(\alpha)}(x_i; w_p)$ has a smaller weight at the true label y_i on average, we can conclude that $f_m^{(\alpha)}(x_i; w_p)$ is *smoother*[1] than $f_m^{(\alpha)}(x_i; w_t)$. In other words, the regularization in teacher training also regularizes student distillation further. Note that Eq. (7) provides an upper bound of the ratio between the teacher's output and the regularized teacher's output at the true label. This effectively measures the smoothness of a smoothed label in label smoothing regularization.

The pruning can be viewed as a solution of the empirical risk minimization problem with sparsity-inducing regularization [28]. Thus, the distillation from the pruned teacher is a label smoothing regularization with smoother label distribution, which reduces a generalization error.

4   Transferring Knowledge of Sparsity

Based on the observation that the pruned teacher transfers the better knowledge, we propose a novel network compression framework that learns from the (unstructured) pruned network. The critical challenge is a student network architecture design to learn effectively from the pruned teacher.

More formally, let $f_t(\cdot; w_t)$ be a cumbersome network to compress, and the goal is to compress it to a smaller network $f_s(\cdot; w_s)$. In the previous section, we considered the distillation to a given student network. On the other hand, in this section, we provide a detailed architecture design for a student network f_s based on the pruned teacher $f_t(\cdot; w_p)$.

On top of the "prune, then distill" as described in Fig. 1, we add student network architecture design. The key idea of student network design is that the

---

[1] Instead of label's self-entropy, we measure the smoothness with true label's weight.

pruned teacher can also provide sparsity knowledge. We construct the narrower student where each layer matches the corresponding layer of the pruned teacher. More precisely, the student network has the same depth, but the number of channels per layer is reduced so that the number of weights is (approximately) equal to the number of remaining parameters in the pruned teacher (as described in Fig. 2). The intuition is to build a student network where each layer has enough capacity to learn from the pruned teacher. The rigorous construction of the student network is described in Appendix. Thus, the proposed compression algorithm has four steps:

1. Train the original network and obtain $f_t(\cdot; w_t)$.
2. Apply the unstructured pruning and obtain pruned network $f_t(\cdot; w_p)$.
3. Construct f_s based on each layer's sparsity of the pruned network $f_t(\cdot; w_p)$.
4. Distill the pruned network $f_t(\cdot; w_p)$ to the student $f_s(\cdot; w_s)$.

Note that the above framework does not depend on the specific choice of distillation or pruning method. In Sect. 5, we apply LR rewinding [39] to prune the model, and apply the vanilla KD [20] to distill the pruned teacher.

The proposed scheme transfers knowledge from the sparse network (from unstructured pruning) to a network with fewer channels to reduce the number of channels further. This is similar to residual distillation [30] which removes unwanted parts (residual connections) of residual networks. In our setting, we remove unwanted parts (more channels) of unstructured pruning by merging sparse filters into fewer filters via KD.

Note that our compression framework can be viewed as structured pruning since it effectively removes neurons (channels) of a given network. Since structured pruning is nearly an architecture search algorithm [34], the proposed framework suggests a novel network architecture search algorithm that learns from unstructured pruning. Recall that recent global unstructured pruning algorithms [27] (where the pruning scheme actively determines the pruning ratio for each layer) outperform precisely designed layerwise sparsity selection schemes.

5   Experiments

In this section, we present our experimental results verifying the proposed algorithm. Similar to Sect. 3, we compare test accuracies of three scenarios: 1) train student network without a teacher, 2) distill the pruned teacher to the student network, and 3) distill the original (unpruned) teacher to the student network. To maintain the consistency of experiments, we use the same training, pruning, and distillation procedure and the same network hyperparameters for all three scenarios (mostly from Sect. 3). All test accuracies are the average of three independent experiments, and we also provide the standard deviation.

5.1   Results

For the VGG-CIFAR100 experiment, we use VGG19 with batch normalization as a teacher. In the proposed framework, we apply LR rewinding to obtain

Table 4. Performance of the proposed compression algorithm on VGG19 with CIFAR100. VGG19-ST(X) is the constructed student network based on the proposed algorithm from X% pruned teacher. Teacher "None" indicates the student is trained without a teacher, while the pruning ratio "None" means the distillation from the unpruned teacher.

Teacher	Pruning ratio	Teacher accuracy	Student	Student accuracy
None	–	–	VGG19-ST36	72.32 ± 0.12
VGG19	None	73.13	VGG19-ST36	73.52 ± 0.20
	36%	73.30	VGG19-ST36	73.77 ± 0.16
None	–	–	VGG19-ST59	71.80 ± 0.18
VGG19	None	73.13	VGG19-ST59	73.18 ± 0.10
	59%	72.25	VGG19-ST59	73.81 ± 0.10
None	–	–	VGG19-ST79	70.89 ± 0.14
VGG19	None	73.13	VGG19-ST79	72.42 ± 0.16
	79%	73.43	VGG19-ST79	73.39 ± 0.11
None	–	–	VGG19DBL-ST36	74.39 ± 0.02
VGG19DBL	None	74.44	VGG19DBL-ST36	74.62 ± 0.34
	36%	73.46	VGG19DBL-ST36	75.40 ± 0.18
None	–	–	VGG19DBL-ST59	74.06 ± 0.22
VGG19DBL	None	74.44	VGG19DBL-ST59	74.67 ± 0.24
	59%	73.24	VGG19DBL-ST59	75.09 ± 0.23
None	–	–	VGG19DBL-ST79	73.81 ± 0.45
VGG19DBL	None	74.44	VGG19DBL-ST79	74.16 ± 0.04
	79%	73.50	VGG19DBL-ST79	75.19 ± 0.31

the pruned VGG19s with target sparsity 36%, 59%, and 79%. The test accuracy of the pruned teacher is similar to the baseline model (VGG19) or slightly higher. We construct the student network as described in the previous section. Let VGG19-ST36, VGG19-ST59, and VGG19-ST79 denote the student networks with fewer channels that correspond to pruned teachers with pruning ratios 36%, 59%, and 79%, respectively. We also run the same experiment with VGG19DBL (with 2× more channels per layer). Similar to VGG19, let VGG19DBL-ST36, VGG19DBL-ST59, and VGG19DBL-ST79 denote student networks that correspond to pruned teachers with pruning ratios 36%, 59%, and 79%, respectively.

For the ResNet-TinyImageNet experiment, we use ResNet18 as a teacher. The base ResNet18 is an unpruned teacher model where the test accuracy is 57.75%. The pruned ResNet18 is a teacher in the proposed framework where we apply LR rewinding with target sparsity 36%, 59%, and 79%. Notably, the pruned teacher's test accuracy is lower than the unpruned network, unlike the VGG-CIFAR100 setup. Similar to VGG-CIFAR100, let ResNet18-ST36, ResNet18-

Table 5. Performance of the proposed compression algorithm on ResNet18 with Tiny-ImageNet. ResNet18-ST(X) is the constructed student network based on the proposed algorithm from X% pruned teacher. Teacher "None" indicates the student is trained without a teacher, while the pruning ratio "None" means the distillation from the unpruned teacher.

Teacher	Pruning ratio	Teacher accuracy	Student	Student accuracy
None	–	–	ResNet18-ST36	56.44 ± 0.26
ResNet18	None	57.75	ResNet18-ST36	57.74 ± 0.22
	36%	57.66	ResNet18-ST36	58.75 ± 0.19
None	–	–	ResNet18-ST59	55.93 ± 0.32
ResNet18	None	57.75	ResNet18-ST59	56.70 ± 0.35
	59%	57.58	ResNet18-ST59	57.76 ± 0.31
None	–	–	ResNet18-ST79	54.48 ± 0.53
ResNet18	None	57.75	ResNet18-ST79	55.65 ± 0.24
	79%	57.32	ResNet18-ST79	56.23 ± 0.16
None	–	–	ResNet18DBL-ST36	59.88 ± 0.30
ResNet18DBL	None	60.46	ResNet18DBL-ST36	61.02 ± 0.15
	36%	61.97	ResNet18DBL-ST36	62.33 ± 0.21
None	–	–	ResNet18DBL-ST59	58.81 ± 0.28
ResNet18DBL	None	60.46	ResNet18DBL-ST59	60.99 ± 0.27
	59%	61.80	ResNet18DBL-ST59	62.41 ± 0.52
None	–	–	ResNet18DBL-ST79	57.79 ± 0.14
ResNet18DBL	None	60.46	ResNet18DBL-ST79	60.60 ± 0.26
	79%	61.66	ResNet18DBL-ST79	61.87 ± 0.27

ST59, and ResNet18-ST79 denote the student networks that correspond to the pruned teacher with pruning ratios 36%, 59%, and 79%, respectively.

Table 4 and Table 5 show the test accuracies of the student network. For comparison, we also provide test accuracies when the same student network is trained without a teacher. In all settings, the proposed scheme outperforms the student network trained from scratch by huge margin.

5.2   Ablation Study

Learning From the Unpruned Teacher: Table 4 and Table 5 also provide the KD result from the unpruned teacher with the same student networks. Similar to Sect. 3, it is consistent that the pruned teacher (with matching sparsity) provide better KD.

Alternative Student Network Design: For VGG19(DBL) teacher, we manually designed students VGG19-CL1 and VGG19-CL2. These networks have the

Table 6. Knowledge distillation to manually designed student networks. VGG19DBL is the VGG19 with 2× more filters per layer. Teacher "None" indicates the student is trained without a teacher, while the pruning ratio "None" means the distillation from the unpruned teacher.

Teacher	Pruning ratio	Teacher accuracy	Student	Student accuracy
None	–	–	VGG19-CL1	69.51 ± 0.24
VGG19	None	73.13	VGG19-CL1	70.47 ± 0.25
	36%	73.30	VGG19-CL1	71.52 ± 0.50
	59%	72.25	VGG19-CL1	71.43 ± 0.24
	79%	73.43	VGG19-CL1	71.82 ± 0.16
None	–	–	VGG19-CL1	69.51 ± 0.24
VGG19DBL	None	74.44	VGG19-CL1	70.38 ± 0.25
	36%	73.46	VGG19-CL1	70.84 ± 0.23
	59%	73.24	VGG19-CL1	70.52 ± 0.03
	79%	73.50	VGG19-CL1	71.00 ± 0.34
None	–	–	VGG19-CL2	71.36 ± 0.29
VGG19	None	73.13	VGG19-CL2	72.75 ± 0.60
	36%	73.30	VGG19-CL2	73.52 ± 0.22
	59%	72.25	VGG19-CL2	73.39 ± 0.21
	79%	73.43	VGG19-CL2	73.67 ± 0.09
None	–	–	VGG19-CL2	71.36 ± 0.29
VGG19DBL	None	74.44	VGG19-CL2	72.29 ± 0.12
	36%	73.46	VGG19-CL2	72.73 ± 0.41
	59%	73.24	VGG19-CL2	72.94 ± 0.37
	79%	73.50	VGG19-CL2	72.88 ± 0.20

same depth, but the number of channels is adjusted, where the number of network parameters is (approximately) half of the original network. VGG19-CL1 removes channels uniformly across the layer, and VGG19-CL2 removes channels unevenly. The detailed network architecture is provided in Appendix.

Table 6 compares the test accuracies of student networks with pruned and unpruned teachers. The number of parameters of VGG19-CL1 and VGG19-CL2 are 11.0 M and 9.9 M, respectively, which are comparable to VGG19-ST59 that has 8.2 M parameters (see Appendix for details). However, the test accuracy of VGG19-ST69 with the proposed framework is higher than accuracies of VGG19-CL1 and VGG19-CL2. The result justifies the proposed student network construction based on the pruned teacher.

Also, the student network with pruned teachers outperforms the student with the unpruned teacher. This implies that the surprising performance of pruned teachers does not rely on the architecture of the student. Note that VGG19DBL

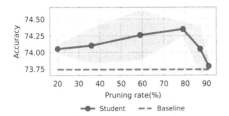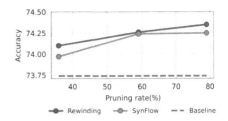

Fig. 3. Effect of pruning ratios and algorithms. The left plot shows the student's accuracies with various pruning ratios of pruned teachers. The right plot shows the student's accuracies when different pruning algorithms (LR rewinding [39] and SynFlow [47]) are applied to the teacher. In both cases, baseline is the student distilled from unpruned teacher.

has better test accuracy compared to VGG19, where the margin is about 1%. There is no significant difference in test accuracy when unpruned VGG19 and unpruned VGG19DBL are being used as teacher networks in KD. However, in KD, pruned VGG19 teaches better than pruned VGG19DBL with the same sparsity. It coincides with what we observed in the previous section, where the teacher with better accuracy does not guarantee better KD.

5.3 Discussions

Effect of Pruning Ratio and Pruning Algorithm: Fig. 3 shows the effect of pruning ratio and pruning algorithm. For VGG19 on CIFAR100, we apply the proposed scheme with additional pruning ratio 20%, 87%, and 91%. In the current setting, the 79% point is the optimal pruning ratio, and the student's performance is degraded if the pruning ratio is too high. We also applied another pruning algorithm, SynFlow [47]. Our result shows that the effectiveness of proposed compression scheme does not depend on the choice of pruning algorithm.

Large Scale Experiments: We also applied the proposed idea to the larger model (ResNet50) and the larger dataset (ImageNet). We consistently observe that the "prune, then distill" strategy is effective in large scale setups as well. We refer to the Appendix for a detailed setup and results of large-scale experiments.

6 Conclusion

Our experiments showed that the pruned teacher can be more effective than the original teacher in KD. We further showed theoretically that the pruned teacher provides an additional regularization in distillation. Based on this observation, we proposed a novel network compression scheme that distills a pruned teacher network to the student network whose architecture is based on an (unstructured) pruned network. The proposed network compression is effectively a structured pruning algorithm that utilizes the knowledge of sparsity from unstructured pruning, and therefore our work bridges two main pruning approaches.

Acknowledgments. JP and AN were supported by Basic Science Research Program through the National Research Foundation of Korea (NRF) funded by the Ministry of Education (2021R1F1A1059567). We thank Minhyeok Cho for giving valuable comments. We also thank anonymous reviewers for providing constructive feedback.

References

1. Anwar, S., Hwang, K., Sung, W.: Structured pruning of deep convolutional neural networks. ACM J. Emerg. Technol. Comput. Syst. (JETC) **13**(3), 1–18 (2017)
2. Banner, R., Hubara, I., Hoffer, E., Soudry, D.: Scalable methods for 8-bit training of neural networks. In: NeurIPS (2018)
3. Chen, T., Kornblith, S., Swersky, K., Norouzi, M., Hinton, G.E.: Big self-supervised models are strong semi-supervised learners. In: NeurIPS (2020)
4. Cho, J.H., Hariharan, B.: On the efficacy of knowledge distillation. In: ICCV (2019)
5. Czarnecki, W.M., Osindero, S., Jaderberg, M., Swirszcz, G., Pascanu, R.: Sobolev training for neural networks. In: NeurIPS (2017)
6. Deng, J., Dong, W., Socher, R., Li, L.J., Li, K., Fei-Fei, L.: Imagenet: a large-scale hierarchical image database. In: CVPR (2009)
7. Devlin, J., Chang, M.W., Lee, K., Toutanova, K.: Bert: pre-training of deep bidirectional transformers for language understanding. In: NAACL-HLT (2019)
8. Floridi, L., Chiriatti, M.: GPT-3: Its Nature, Scope, Limits, and Consequences. Mind. Mach. **30**(4), 681–694 (2020). https://doi.org/10.1007/s11023-020-09548-1
9. Frankle, J., Carbin, M.: The lottery ticket hypothesis: finding sparse, trainable neural networks. In: ICLR (2019)
10. Frankle, J., Dziugaite, G.K., Roy, D., Carbin, M.: Linear mode connectivity and the lottery ticket hypothesis. In: ICML (2020)
11. Grill, J.B., et al.: Bootstrap your own latent - a new approach to self-supervised learning. In: NeurIPS (2020)
12. Han, S., et al.: Eie: efficient inference engine on compressed deep neural network. ACM SIGARCH Comput. Archit. News **44**(3), 243–254 (2016)
13. Han, S., Mao, H., Dally, W.J.: Deep compression: compressing deep neural network with pruning, trained quantization and huffman coding. In: ICLR (2016)
14. Havasi, M., Peharz, R., Hernandez-Lobato, J.M.: Minimal random code learning: getting bits back from compressed model parameters. In: ICLR (2019)
15. He, K., Fan, H., Wu, Y., Xie, S., Girshick, R.: Momentum contrast for unsupervised visual representation learning. In: CVPR (2020)
16. He, K., Zhang, X., Ren, S., Sun, J.: Deep residual learning for image recognition. In: CVPR (2016)
17. He, Y., Kang, G., Dong, X., Fu, Y., Yang, Y.: Soft filter pruning for accelerating deep convolutional neural networks. In: IJCAI (2018)
18. He, Y., Zhang, X., Sun, J.: Channel pruning for accelerating very deep neural networks. In: ICCV (2017)
19. Heo, B., Lee, M., Yun, S., Choi, J.Y.: Knowledge transfer via distillation of activation boundaries formed by hidden neurons. In: AAAI (2019)
20. Hinton, G., Vinyals, O., Dean, J.: Distilling the knowledge in a neural network. In: NeurIPS Workshop (2015)
21. Huang, Y., et al.: Gpipe: efficient training of giant neural networks using pipeline parallelism. In: NeurIPS (2019)
22. Idelbayev, Y., Carreira-Perpinan, M.A.: Low-rank compression of neural nets: learning the rank of each layer. In: CVPR (2020)

23. Jing, Y., Yang, Y., Wang, X., Song, M., Tao, D.: Amalgamating knowledge from heterogeneous graph neural networks. In: CVPR (2021)
24. Krizhevsky, A., Hinton, G., et al.: Learning multiple layers of features from tiny images. (Technical report) (2009)
25. Le, Y., Yang, X.: Tiny imagenet visual recognition challenge. (Technical report) (2015)
26. LeCun, Y., Denker, J.S., Solla, S.A.: Optimal brain damage. In: NeurIPS (1990)
27. Lee, J., Park, S., Mo, S., Ahn, S., Shin, J.: Layer-adaptive sparsity for the magnitude-based pruning. In: ICLR (2021)
28. LeJeune, D., Javadi, H., Baraniuk, R.: The flip side of the reweighted coin: duality of adaptive dropout and regularization. In: NeurIPS (2021)
29. Li, F., Zhang, B., Liu, B.: Ternary weight networks. arXiv:1605.04711 (2016)
30. Li, G., et al.: Residual distillation: towards portable deep neural networks without shortcuts. In: NeurIPS (2020)
31. Li, H., Kadav, A., Durdanovic, I., Samet, H., Graf, H.P.: Pruning filters for efficient convnets. In: ICLR (2017)
32. Liu, Z., et al.: Metapruning: meta learning for automatic neural network channel pruning. In: ICCV (2019)
33. Liu, Z., Li, J., Shen, Z., Huang, G., Yan, S., Zhang, C.: Learning efficient convolutional networks through network slimming. In: ICCV (2017)
34. Liu, Z., Sun, M., Zhou, T., Huang, G., Darrell, T.: Rethinking the value of network pruning. In: ICLR (2018)
35. Luo, J.H., Wu, J., Lin, W.: Thinet: a filter level pruning method for deep neural network compression. In: ICCV (2017)
36. Mirzadeh, S.I., Farajtabar, M., Li, A., Levine, N., Matsukawa, A., Ghasemzadeh, H.: Improved knowledge distillation via teacher assistant. In: AAAI (2020)
37. Park, D.Y., Cha, M.H., Jeong, C., Kim, D., Han, B.: Learning student-friendly teacher networks for knowledge distillation. In: NeurIPS (2021)
38. Patterson, D., et al.: Carbon emissions and large neural network training. arXiv:2104.10350 (2021)
39. Renda, A., Frankle, J., Carbin, M.: Comparing rewinding and fine-tuning in neural network pruning. In: ICLR (2020)
40. Romero, A., Ballas, N., Kahou, S.E., Chassang, A., Gatta, C., Bengio, Y.: Fitnets: hints for thin deep nets. In: ICLR (2015)
41. Sainath, T.N., Kingsbury, B., Sindhwani, V., Arisoy, E., Ramabhadran, B.: Low-rank matrix factorization for deep neural network training with high-dimensional output targets. In: ICASSP (2013)
42. Sandler, M., Howard, A., Zhu, M., Zhmoginov, A., Chen, L.C.: Mobilenetv 2: inverted residuals and linear bottlenecks. In: CVPR (2018)
43. Simonyan, K., Zisserman, A.: Very deep convolutional networks for large-scale image recognition. In: ICLR (2015)
44. Srinivas, S., Fleuret, F.: Knowledge transfer with jacobian matching. In: ICML (2018)
45. Stanton, S., Izmailov, P., Kirichenko, P., Alemi, A.A., Wilson, A.G.: Does knowledge distillation really work? In: NeurIPS (2021)
46. Su, X., You, S., Wang, F., Qian, C., Zhang, C., Xu, C.: Bcnet: searching for network width with bilaterally coupled network. In: CVPR (2021)
47. Tanaka, H., Kunin, D., Yamins, D.L., Ganguli, S.: Pruning neural networks without any data by iteratively conserving synaptic flow. In: NeurIPS (2020)
48. Tian, Y., Krishnan, D., Isola, P.: Contrastive representation distillation. In: ICLR (2019)

49. Wen, W., Wu, C., Wang, Y., Chen, Y., Li, H.: Learning structured sparsity in deep neural networks. In: NeurIPS (2016)
50. Wiedemann, S., et al.: Deepcabac: a universal compression algorithm for deep neural networks. IEEE J. Sel. Top. Sig.l Process. **14**(4), 700–714 (2020)
51. Xu, Z., Hsu, Y.C., Huang, J.: Training shallow and thin networks for acceleration via knowledge distillation with conditional adversarial networks. In: ICLR Workshop (2017)
52. Yang, Y., Qiu, J., Song, M., Tao, D., Wang, X.: Distilling knowledge from graph convolutional networks. In: CVPR (2020)
53. Ye, J., Lu, X., Lin, Z., Wang, J.Z.: Rethinking the smaller-norm-less-informative assumption in channel pruning of convolution layers. In: ICLR (2018)
54. Yuan, L., Tay, F.E., Li, G., Wang, T., Feng, J.: Revisiting knowledge distillation via label smoothing regularization. In: CVPR (2020)
55. Zagoruyko, S., Komodakis, N.: Wide residual networks. In: BMVC (2016)
56. Zagoruyko, S., Komodakis, N.: Paying more attention to attention: improving the performance of convolutional neural networks via attention transfer. In: ICLR (2017)
57. Zhou, H., Alvarez, J.M., Porikli, F.: Less is more: towards compact cnns. In: Leibe, B., Matas, J., Sebe, N., Welling, M. (eds.) ECCV 2016. LNCS, vol. 9908, pp. 662–677. Springer, Cham (2016). https://doi.org/10.1007/978-3-319-46493-0_40
58. Zhou, H., et al.: Rethinking soft labels for knowledge distillation: a bias–variance tradeoff perspective. In: ICLR (2021)
59. Zhuang, T., Zhang, Z., Huang, Y., Zeng, X., Shuang, K., Li, X.: Neuron-level structured pruning using polarization regularizer. In: NeurIPS (2020)

Deep Partial Updating: Towards Communication Efficient Updating for On-Device Inference

Zhongnan Qu[1]([✉]), Cong Liu[2], and Lothar Thiele[1]

[1] ETH Zurich, Zurich, Switzerland
{quz,thiele}@ethz.ch
[2] UT Dallas, Richardson, TX, USA
cong@utdallas.edu

Abstract. Emerging edge intelligence applications require the server to retrain and update deep neural networks deployed on remote edge nodes to leverage newly collected data samples. Unfortunately, it may be impossible in practice to continuously send fully updated weights to these edge nodes due to the highly constrained communication resource. In this paper, we propose the weight-wise deep partial updating paradigm, which smartly selects a small subset of weights to update in each server-to-edge communication round, while achieving a similar performance compared to full updating. Our method is established through analytically upper-bounding the loss difference between partial updating and full updating, and only updates the weights which make the largest contributions to the upper bound. Extensive experimental results demonstrate the efficacy of our partial updating methodology which achieves a high inference accuracy while updating a rather small number of weights.

Keywords: Partial updating · Communication constraints · Parameter reuse · Server-to-edge · Deep neural networks

1 Introduction

On-device inference is a new disruptive technology that enables new intelligent applications, *e.g.*, mobile assistants, Internet of Things and augmented reality. Compared to traditional cloud inference, on-device inference is subject to severe limitations in terms of storage, energy, computing power and communication. On the other hand, it has many advantages, *e.g.*, it enables fast and stable inference even with low communication bandwidth or interrupted communication, and can save energy by avoiding the transfer of data to the cloud, which often costs significant amounts of energy than sensing and computation [8,17,35]. To deploy deep neural networks (DNNs) on resource-constrained edge devices, extensive research has been done to compress a well pre-trained model via pruning [7,9,29]

Supplementary Information The online version contains supplementary material available at https://doi.org/10.1007/978-3-031-20083-0_9.

and quantization [5,26,28]. During on-device inference, compressed DNNs may achieve a good balance between model performance and resource demand.

However, due to a possible lack of relevant training data at the time of initial deployment or due to an unknown sensing environment, pre-trained DNNs may either fail to perform satisfactorily or be significantly improved after the initial deployment. In other words, re-training the models by using newly collected data (from *edge devices* or *other sources*) is typically required to achieve the desired performance during the lifetime of devices.

Because of the resource-constrained nature of edge devices in terms of memory and computing power, on-device re-training (or federated learning) is typically restricted to tiny batch size, small inference (sub-)networks or limited optimization steps, all resulting in a performance degradation. Instead, retraining often occurs on a remote server with sufficient resources. One possible strategy to allow for a continuous improvement of the model performance is a two-stage iterative process: (*i*) at each round, edge devices collect new data samples and send them to the server, and (*ii*) the server retrains the model using all collected data, and sends updates to each edge device [4]. The first stage may be even not necessary if new data are collected in other ways and made directly available to the server.

Example Scenarios. Example application scenarios of relevance include vision robotic sensing in an unknown environment (*e.g.*, Mars) [20], local translators of low-resource languages on mobile phones [3,36], and sensor networks mounted in alpine areas [21], automatic wildlife monitoring [34]. We detail two specific scenarios. *Hazard alarming on mountains:* Researchers in [21] mounted tens of sensor nodes at different scarps in high alpine areas with cameras, geophones and high-precision GPS. The purpose is to achieve fast, stable, and energy-efficient hazard monitoring for early warning to protect people and infrastructure. To this end, a DNN is deployed on each node to on-device detect rockfalls and debris flows. The nodes regularly collect and send data to the server for labeling and retraining, and the server sends the updated model back through a low-power wireless network. Retraining during deployment is essential for a highly reliable hazard warning. *Endangered species monitoring:* To detect endangered species, researchers often deploy some audio or image sensor nodes in virgin rainforests [34]. Edge nodes are supposed to classify the potential signal from endangered species and send these relevant data to the server. Due to the limited prior information from environments and species, retraining the initially classifier with received data or data from other sources (*e.g.*, other areas) is necessary.

Challenges. An essential challenge herein is that the transmissions in the server-to-edge stage are highly constrained by the limited communication resource (*e.g.*, bandwidth, energy [2]) in comparison to the edge-to-server stage, if necessary at all. Typically, state-of-the-art DNNs often require tens or even hundreds of mega-Bytes (MB) to store parameters, whereas a single batch of data samples (a number of samples that lead to reasonable updates in batch training) needs a relatively smaller amount of data. For example, for CIFAR10 dataset [15], the weights of a popular VGGNet require 56.09 MB storage, while one batch of 128 samples only uses around 0.40 MB [28,33]. As an alternative approach, the server sends a full update of the inference model once or rarely. But in this case, every node will suffer from a low performance until such an update occurs. Besides, edge devices could decide on and send only critical samples by using active learning schemes [1]. The server may also receive data from other sources, *e.g.*, through data augmentation based on the data collected in previous rounds or new data collection campaigns. These considerations indicate that the updated weights that are sent to edge devices by the server become a major bottleneck.

Facing the above challenges, we ask the following question: *Is it possible to update only a small subset of weights while reaching a similar performance as updating all weights?* Doing such a *partial updating* can significantly reduce the server-to-edge communication overhead. Furthermore, fewer parameter updates also lead to less memory access on edge devices, which in turn results in smaller energy consumption than full updating [11].

Why Partial Updating Works. Since the model deployed on edge devices is trained with the data collected beforehand, some learned knowledge can be reused. In other words, we only need to distinguish and update the weights which are critical to the newly collected data.

How to Select Weights. Our key concept for partial updating is based on the hypothesis, that *a weight shall be updated only if it has a large contribution to the loss reduction* during the retraining given newly collected data samples. Specially, we define a binary mask m to describe which weights are subject to update and which weights are fixed (also reused). For any m, we establish the analytical upper bound on the difference between the loss value under partial updating and that under full updating. We determine an optimized mask m by combining two different view points: (i) measuring each weight's "global contribution" to the upper bound through computing the Euclidean distance, and (ii) measuring each weight's "local contribution" to the upper bound using gradient-related information. The weights to be updated according to m will be further sparsely fine-tuned while the remaining weights are rewound to their initial values.

Our contributions can be summarized as follows.

- We formalize the deep partial updating paradigm, *i.e.*, how to iteratively perform partial updating of inference models on remote edge devices, if newly collected training samples are available at the server. This reduces the computation and communication demand on edge devices substantially.
- We propose a novel approach that determines the optimized subset of weights that shall be selected for partial updating, through measuring each weight's contribution to the analytical upper bound on the loss reduction. This simple yet effective metric can be applied to any models that are trained with gradient-based optimizers.
- Experimental results on public vision datasets show that, under the similar accuracy level along the rounds, our approach can reduce the size of the

transmitted data by 95.3% on average (up to 99.3%), namely can update the model averagely 21 times more frequently than full updating.

2   Related Work

Partial Updating. Although partial updating has been adopted in some prior works, it is conducted in a fairly coarse-grained manner, *e.g.*, layer-wise or neuron-wise, and targets at completely different objectives. Especially, under continual learning settings, [13,37] propose to freeze all weights related to the neurons which are more critical in performing prior tasks than new ones, to preserve existing knowledge. Under adversarial attack settings, [32] updates the weights in the first several layers only, which yield a dominating impact on the extracted features, for better attack efficacy. Under meta learning settings, [27,31] reuse learned representations by only updating a subset of layers for efficiently learning new tasks. Unfortunately, such techniques do not focus on reducing the number of updated weights, and thus cannot be applied in our problem settings.

Federated Learning. Communication-efficient federated learning [14,18,19] studies how to compress multiple gradients calculated on different sets of non-*i.i.d.* local data, such that the aggregation of these compressed gradients could result in a similar convergence performance as centralized training on all data. Such compressed updates are fundamentally different from our setting, where (*i*) updates are not transmitted in each optimization step; (*ii*) training data are incrementally collected; (*iii*) centralized training is conducted. Our typical scenarios focus on outdoor areas, which generally do not involve data privacy issues, since these collected data are not personal data. In comparison to federated learning, our pipeline has the following advantages: (*i*) we do not conduct resource-intensive gradient backward propagation on edge devices; (*ii*) the collected data are not continuously accumulated and stored on memory-constrained edge nodes; (*iii*) we also avoid the difficult but necessary labeling process on each edge node in supervised learning tasks; (*iv*) if few events occur on some nodes, the centralized training may avoid degraded updates in local training, *e.g.*, batch normalization.

Compression. The communication cost could also be reduced through some compression techniques, *e.g.*, quantizing/encoding the updated weights and the transmission signal [9]. But note that these techniques are orthogonal to our approach and could be applied in addition, see Appendix H.

Unstructured Pruning. Deep partial updating is inspired by recent unstructured pruning methods, *e.g.*, [6,7,9,24,29,38]. Traditional pruning methods aim at reducing the number of operations and storage consumption by setting some weights to zero. Sending a pruned DNN with only non-zero weights may also reduce the communication cost, but to a much lesser extent as shown in the experimental results, see Sect. 5.2. Since our objective namely reducing the server-to-edge communication cost when updating the deployed DNN is fundamentally different from pruning, we can leverage some learned knowledge by retaining weights (partial updating) instead of zero-outing weights (pruning).

Domain Adaptation. Domain adaptation targets reducing domain shift to transfer knowledge into new learning tasks [39]. This paper mainly considers the scenario where the inference task is not explicitly changed along the rounds, *i.e.*, the overall data distribution maintains the same along the data collection rounds. Thus, selecting critical weights (features) by measuring their impact on domain distribution discrepancy is invalid herein. Applying deep partial updating on streaming tasks where the data distribution varies along the rounds would be also worth studying, and we leave it for future works.

3   Notation and Setting

In this section, we define the notations used throughout this paper, and provide a formalized problem setting. We consider a set of remote edge devices that implement on-device inference. They are connected to a host server that is able to perform DNN training and retraining. We consider the necessary amount of information that needs to be communicated to each edge device to update its inference model.

Assume there are R rounds of model updates. The model deployed in the r-th round is represented with its weight vector $\boldsymbol{w}^r$. The training data used to update the model for the r-th round is represented as $\mathcal{D}^r = \delta\mathcal{D}^r \cup \mathcal{D}^{r-1}$. Also, newly collected data samples $\delta\mathcal{D}^r$ are made available to the server in round $r-1$.

To reduce the amount of information that needs to be sent to edge devices, only partial weights of $\boldsymbol{w}^{r-1}$ shall be updated when determining $\boldsymbol{w}^r$. The overall optimization problem for weight-wise partial updating in round $r-1$ is thus,

$$\min_{\delta\boldsymbol{w}^r} \ \ell\left(\boldsymbol{w}^{r-1} + \delta\boldsymbol{w}^r; \mathcal{D}^r\right) \tag{1}$$

$$\text{s.t. } \|\delta\boldsymbol{w}^r\|_0 \leq k \cdot I \tag{2}$$

where ℓ denotes the loss function, $\|.\|_0$ denotes the L0-norm, k denotes the updating ratio that is determined by the communication constraints in practical scenarios, and $\delta\boldsymbol{w}^r$ denotes the increment of $\boldsymbol{w}^{r-1}$. Note that both $\boldsymbol{w}^{r-1}$ and $\delta\boldsymbol{w}^r$ are drawn from $\mathbb{R}^I$, where I is the total number of weights.

In this case, only a fraction of $k \cdot I$ weights and the corresponding index information need to be communicated to each edge device for updating the model, namely the partial updates $\delta\boldsymbol{w}^r$. It is worth noting that the index information is relatively small in size compared to the partially updated weights (see Sect. 5). On each edge device, the weight vector is updated as $\boldsymbol{w}^r = \boldsymbol{w}^{r-1} + \delta\boldsymbol{w}^r$. To simplify the notation, we will only consider a single update, *i.e.*, from weight vector $\boldsymbol{w}$ (corresponding to $\boldsymbol{w}^{r-1}$) to weight vector $\widetilde{\boldsymbol{w}}$ (corresponding to $\boldsymbol{w}^r$) with $\widetilde{\boldsymbol{w}} = \boldsymbol{w} + \widetilde{\delta\boldsymbol{w}}$.

4 Deep Partial Updating

We developed a two-step approach for resolving the partial updating optimization problem in Eq. (1)–Eq. (2). The overall approach is depicted in Fig. 1.

The First Step. The first step not only determines the subset of weights that are allowed to change their values, but also computes the initial values for the second step. In particular, we first optimize the loss function ℓ by updating all weights from the initialization $\boldsymbol{w}$ with a standard optimizer, *e.g.*, SGD or its variants. We thus obtain the minimized loss $\ell\left(\boldsymbol{w}^{\mathrm{f}}\right)$ with $\boldsymbol{w}^{\mathrm{f}} = \boldsymbol{w} + \delta\boldsymbol{w}^{\mathrm{f}}$, where the superscript f denotes "full updating". To consider the constraint of Eq. (2), the information gathered during this optimization is used to determine the subset of weights that will be changed, also that are communicated to the edge devices.

In the explanation of the method in Sect. 4.1, we use the binary mask $\boldsymbol{m}$ with $\boldsymbol{m} \in \{0,1\}^{I}$ to describe which weights are subject to change and which ones are not. The weights with $m_i = 1$ are trainable, whereas the weights with $m_i = 0$ will be rewound from the values in $\boldsymbol{w}^{\mathrm{f}}$ to their initial values in $\boldsymbol{w}$, *i.e.*, unchanged. Obviously, we find $\|\boldsymbol{m}\|_0 = \sum_i m_i = k \cdot I$.

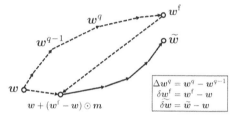

Fig. 1. The figure depicts the overall approach that consists of two steps. The first step is depicted with dotted arrows and starts from the deployed model $\boldsymbol{w}$. In Q optimization steps, all weights are trained to the optimum $\boldsymbol{w}^{\mathrm{f}}$. Based on the collected information, a binary mask $\boldsymbol{m}$ is determined that characterizes the set of weights that are rewound to the ones of $\boldsymbol{w}$. Therefore, the second step (solid arrows) starts from $\boldsymbol{w} + \delta\boldsymbol{w}^{\mathrm{f}} \odot \boldsymbol{m}$. According to the mask, this solution is sparsely fine-tuned to the final weights $\widetilde{\boldsymbol{w}}$, *i.e.*, $\widetilde{\delta\boldsymbol{w}}$ has only non-zero values where the mask value is 1

The Second Step. In the second step we start a sparse fine-tuning from a DNN with $k \cdot I$ weights from the optimized model $\boldsymbol{w}^{\mathrm{f}}$ and $(1-k) \cdot I$ weights from the previous, still deployed model $\boldsymbol{w}$. In other words, the initial weights for the second step are $\boldsymbol{w} + \delta\boldsymbol{w}^{\mathrm{f}} \odot \boldsymbol{m}$, where $\odot$ denotes an element-wise multiplication. To determine the final solution $\widetilde{\boldsymbol{w}} = \boldsymbol{w} + \widetilde{\delta\boldsymbol{w}}$, we conduct a sparse fine-tuning (still with a standard optimizer), *i.e.*, we keep all weights with $m_i = 0$ constant during the optimization. Therefore, $\widetilde{\delta\boldsymbol{w}}$ is zero wherever $m_i = 0$, and only weights where $m_i = 1$ are updated.

4.1 Metrics for Rewinding

We will now describe a new metric that determines the weights that should be kept constant, *i.e.*, with $m_i = 0$. Like most learning methods, we focus on minimizing a loss function. The two-step approach relies on the following assumption: the better the loss $\ell(\boldsymbol{w} + \delta\boldsymbol{w}^{\mathrm{f}} \odot \boldsymbol{m})$ of the initial solution for the second step, the better the final performance. Therefore, the first step should select a mask $\boldsymbol{m}$ such that the loss difference $\ell(\boldsymbol{w} + \delta\boldsymbol{w}^{\mathrm{f}} \odot \boldsymbol{m}) - \ell(\boldsymbol{w}^{\mathrm{f}})$ is as small as possible.

To determine an optimized mask $\boldsymbol{m}$, we propose to upper-bound the above loss difference in two view points, and measure each weight's contribution to the bounds. The "global contribution" uses the norm information of incremental weights $\delta \boldsymbol{w}^{\mathrm{f}} = \boldsymbol{w}^{\mathrm{f}} - \boldsymbol{w}$. The "local contribution" takes into account the gradient-based information that is gathered during the optimization in the first step, $i.e.$, in the path from $\boldsymbol{w}$ to $\boldsymbol{w}^{\mathrm{f}}$. Both contributions will be combined to determine the mask $\boldsymbol{m}$.

The two view points are based on the concept of smooth differentiable functions. A function $f(x)$ with $f : \mathbb{R}^d \to \mathbb{R}$ is called L-smooth if it has a Lipschitz continuous gradient $g(x)$: $\|g(x) - g(y)\|_2 \leq L\|x - y\|_2$ for all x, y. Note that Lipschitz continuity of gradients is essential to ensuring convergence of many gradient-based algorithms. Under such a condition, one can derive the following bounds, see also [22]:

$$|f(y) - f(x) - g(x)^{\mathrm{T}} \cdot (y - x)| \leq L/2 \cdot \|y - x\|_2^2 \quad \forall x, y \qquad (3)$$

Global Contribution. One would argue that a large absolute value in $\delta \boldsymbol{w}^{\mathrm{f}} = \boldsymbol{w}^{\mathrm{f}} - \boldsymbol{w}$ indicates that this weight has moved far from its initial value in $\boldsymbol{w}$, and thus should not be rewound. This motivates us to adopt the widely used unstructured magnitude pruning to determine the mask $\boldsymbol{m}$. Magnitude pruning prunes the weights with the lowest magnitudes, which often achieves a good trade-off between the model accuracy and the number of zero's weights [29].

Using $a - b \leq |a - b|$, Eq. (3) can be reformulated as $f(y) - f(x) - g(x)^T(y - x) \leq |f(y) - f(x) - g(x)^T(y - x)| \leq L/2 \cdot \|y - x\|_2^2$. Thus, we can bound the relevant loss difference $\ell(\boldsymbol{w} + \delta \boldsymbol{w}^{\mathrm{f}} \odot \boldsymbol{m}) - \ell(\boldsymbol{w}^{\mathrm{f}}) \geq 0$ as

$$\ell(\boldsymbol{w} + \delta \boldsymbol{w}^{\mathrm{f}} \odot \boldsymbol{m}) - \ell(\boldsymbol{w}^{\mathrm{f}}) \leq g(\boldsymbol{w}^{\mathrm{f}})^{\mathrm{T}} \cdot (\delta \boldsymbol{w}^{\mathrm{f}} \odot (\boldsymbol{m} - \boldsymbol{1})) + L/2 \cdot \|\delta \boldsymbol{w}^{\mathrm{f}} \odot (\boldsymbol{m} - \boldsymbol{1})\|_2^2 \quad (4)$$

where $g(\boldsymbol{w}^{\mathrm{f}})$ denotes the loss gradient at $\boldsymbol{w}^{\mathrm{f}}$. As the loss is optimized at $\boldsymbol{w}^{\mathrm{f}}$, $i.e.$, $g(\boldsymbol{w}^{\mathrm{f}}) \approx \boldsymbol{0}$, we can assume that the gradient term is much smaller than the norm of the weight differences in Eq. (4). Therefore, we have

$$\ell(\boldsymbol{w} + \delta \boldsymbol{w}^{\mathrm{f}} \odot \boldsymbol{m}) - \ell(\boldsymbol{w}^{\mathrm{f}}) \lesssim L/2 \cdot \|\delta \boldsymbol{w}^{\mathrm{f}} \odot (\boldsymbol{1} - \boldsymbol{m})\|_2^2 \qquad (5)$$

The right hand side is clearly minimized if $m_i = 1$ for the largest absolute values of $\delta \boldsymbol{w}^{\mathrm{f}}$. As $\boldsymbol{1}^{\mathrm{T}} \cdot (\boldsymbol{c}^{\mathrm{global}} \odot (\boldsymbol{1} - \boldsymbol{m})) = \|\delta \boldsymbol{w}^{\mathrm{f}} \odot (\boldsymbol{1} - \boldsymbol{m})\|_2^2$, this information is captured in the contribution vector

$$\boldsymbol{c}^{\mathrm{global}} = \delta \boldsymbol{w}^{\mathrm{f}} \odot \delta \boldsymbol{w}^{\mathrm{f}} \qquad (6)$$

The $k \cdot I$ weights with the largest values in $\boldsymbol{c}^{\mathrm{global}}$ are assigned to mask values 1 and are further fine-tuned in the second step, whereas all others are rewound to their initial values in $\boldsymbol{w}$. Alg. 2 in Appendix A.1 shows this first approach.

Local Contribution. As experiments show, one can do better when leveraging in addition some gradient-based information gathered during the first step, $i.e.$, optimizing the initial weights $\boldsymbol{w}$ in Q traditional optimization steps, $\boldsymbol{w} = \boldsymbol{w}^0 \to \cdots \to \boldsymbol{w}^{q-1} \to \boldsymbol{w}^q \to \cdots \to \boldsymbol{w}^Q = \boldsymbol{w}^{\mathrm{f}}$.

Using $-a+b \leq |a-b|$, Eq. (3) can be reformulated as $f(x) - f(y) + g(x)^T(y - x) \leq |f(y) - f(x) - g(x)^T(y - x)| \leq L/2 \cdot \|y - x\|_2^2$. Thus, each optimization step is bounded as

$$\ell(\boldsymbol{w}^{q-1}) - \ell(\boldsymbol{w}^q) \leq -\boldsymbol{g}(\boldsymbol{w}^{q-1})^{\mathrm{T}} \cdot \Delta\boldsymbol{w}^q + L/2 \cdot \|\Delta\boldsymbol{w}^q\|_2^2 \qquad (7)$$

where $\Delta\boldsymbol{w}^q = \boldsymbol{w}^q - \boldsymbol{w}^{q-1}$. For a conventional gradient descent optimizer with a small learning rate we can use the approximation $|\boldsymbol{g}(\boldsymbol{w}^{q-1})^{\mathrm{T}} \cdot \Delta\boldsymbol{w}^q| \gg \|\Delta\boldsymbol{w}^q\|_2^2$ and obtain $\ell(\boldsymbol{w}^{q-1}) - \ell(\boldsymbol{w}^q) \lesssim -\boldsymbol{g}(\boldsymbol{w}^{q-1})^{\mathrm{T}} \cdot \Delta\boldsymbol{w}^q$. Summing up over all optimization iterations yields approximately

$$\ell(\boldsymbol{w}^{\mathrm{f}} - \delta\boldsymbol{w}^{\mathrm{f}}) - \ell(\boldsymbol{w}^{\mathrm{f}}) \lesssim -\sum_{q=1}^{Q} \boldsymbol{g}(\boldsymbol{w}^{q-1})^{\mathrm{T}} \cdot \Delta\boldsymbol{w}^q \qquad (8)$$

Note that we have $\boldsymbol{w} = \boldsymbol{w}^{\mathrm{f}} - \delta\boldsymbol{w}^{\mathrm{f}}$ and $\delta\boldsymbol{w}^{\mathrm{f}} = \sum_{q=1}^{Q} \Delta\boldsymbol{w}^q$. Therefore, with a small updating ratio k, $i.e.$, $\boldsymbol{m} \sim \boldsymbol{0}$, we can reformulate Eq. (8) as $\ell(\boldsymbol{w} + \delta\boldsymbol{w}^{\mathrm{f}} \odot \boldsymbol{m}) - \ell(\boldsymbol{w}^{\mathrm{f}}) \lesssim \mathrm{U}(\boldsymbol{m})$ with the upper bound $\mathrm{U}(\boldsymbol{m}) = -\sum_{q=1}^{Q} \boldsymbol{g}(\boldsymbol{w}^{q-1})^{\mathrm{T}} \cdot (\Delta\boldsymbol{w}^q \odot (\boldsymbol{1} - \boldsymbol{m}))$ where we suppose that the gradients are approximately constant for $\boldsymbol{m} \sim \boldsymbol{0}$ ($i.e.$, $\boldsymbol{m}$ has zero entries almost everywhere). Therefore, an approximate incremental contribution of each weight dimension to the upper bound on the loss difference $\ell(\boldsymbol{w} + \delta\boldsymbol{w}^{\mathrm{f}} \odot \boldsymbol{m}) - \ell(\boldsymbol{w}^{\mathrm{f}})$ can be determined by the negative gradient vector at $\boldsymbol{m} = \boldsymbol{0}$, denoted as

$$\boldsymbol{c}^{\mathrm{local}} = -\frac{\partial\mathrm{U}(\boldsymbol{m})}{\partial\boldsymbol{m}} = -\sum_{q=1}^{Q} \boldsymbol{g}(\boldsymbol{w}^{q-1}) \odot \Delta\boldsymbol{w}^q \qquad (9)$$

which models the accumulated contribution to the overall loss reduction.

Combining Global and Local Contribution. So far, we independently calculate the global and local contributions. To avoid the scale impact, we first normalize each contribution by its significance in its own set (either global or local contribution set). We investigated the impacts and the different combinations of both normalized contributions, see the results in Appendix G. Interestingly, the most straightforward combination ($i.e.$, the sum of both normalized metrics) often yields a satisfied and stable performance. Intuitively, local contribution can better identify critical weights w.r.t. the loss during training, while global contribution may be more robust for a highly non-convex loss landscape. Both metrics may be necessary when selecting weights to rewind. Therefore, the combined contribution is computed as

$$\boldsymbol{c} = \frac{1}{\boldsymbol{1}^{\mathrm{T}} \cdot \boldsymbol{c}^{\mathrm{global}}} \boldsymbol{c}^{\mathrm{global}} + \frac{1}{\boldsymbol{1}^{\mathrm{T}} \cdot \boldsymbol{c}^{\mathrm{local}}} \boldsymbol{c}^{\mathrm{local}} \qquad (10)$$

and $m_i = 1$ for the $k \cdot I$ largest values of $\boldsymbol{c}$ and $m_i = 0$ otherwise. The pseudocode of Deep Partial Updating (DPU), $i.e.$, rewinding according to the combined contribution to the loss reduction, is shown in Algorithm 1. The complexity analysis of the corresponding algorithm are shown in Appendix B.

Algorithm 1: Deep Partial Updating

Input: weights w, updating ratio k, learning rate $\{\alpha^q\}_{q=1}^Q$ in Q iterations
Output: weights $\widetilde{w}$
`/* The first step: full updating and rewinding` `*/`
Initiate $w^0 = w$ and $c^{\text{local}} = 0$;
for $q \leftarrow 1$ **to** Q **do**
    | Compute the loss gradient $g(w^{q-1}) = \partial\ell(w^{q-1})/\partial w^{q-1}$;
    | Compute the optimization step with learning rate α^q as Δw^q;
    | Update $w^q = w^{q-1} + \Delta w^q$;
    | Update $c^{\text{local}} = c^{\text{local}} - g(w^{q-1}) \odot \Delta w^q$;
Set $w^{\text{f}} = w^Q$ and get $\delta w^{\text{f}} = w^{\text{f}} - w$;
Compute $c^{\text{global}} = \delta w^{\text{f}} \odot \delta w^{\text{f}}$;
Compute c as Eq.(10) and sort in descending order;
Create a binary mask m with 1 for the Top-$(k \cdot I)$ indices, 0 for others;
`/* The second step: sparse fine-tuning` `*/`
Initiate $\widetilde{\delta w} = \delta w^{\text{f}} \odot m$ and $\widetilde{w} = w + \widetilde{\delta w}$;
for $q \leftarrow 1$ **to** Q **do**
    | Compute the optimization step on $\widetilde{w}$ with learning rate α^q as $\Delta\widetilde{w}^q$;
    | Update $\widetilde{\delta w} = \widetilde{\delta w} + \Delta\widetilde{w}^q \odot m$ and $\widetilde{w} = w + \widetilde{\delta w}$;

4.2 (Re-)Initialization of Weights

In this section, we discuss the initialization of DPU. $\mathcal{D}^1$ denotes the initial dataset used to train the model w^1 from a randomly initialized model w^0. $\mathcal{D}^1$ corresponds to the available dataset before deployment, or collected in the 0-th round if there are no data available before deployment. $\{\delta\mathcal{D}^r\}_{r=2}^R$ denotes newly collected samples in each subsequent round.

Experimental results show (see Appendix D) that training from a randomly initialized model can yield a higher accuracy *after a large number of rounds*, compared to always training from the last round w^{r-1}. As a possible explanation, the optimizer could end in a hard to escape region of the search space if always trained from the last round for a long sequence of rounds. Thus, we propose to re-initialize the weights after a certain number of rounds. In such a case, Algorithm 1 does not start from the weights w^{r-1} but from the randomly initialized weights. The randomly re-initialized model (weights) can be efficiently sent to the edge devices via a single random seed. The device can determine the weights by means of a random generator. This process realizes a random shift in the search space, which is a communication-efficient way in comparison to other alternatives, such as learning to increase the loss or using the (averaged) weights in the previous rounds, as these fully changed weights still need to be sent to each node. Each time the model is randomly re-initialized, the new partially updated model might suffer from an accuracy drop in a few rounds. However, we can simply avoid such an accuracy drop by not updating the model if the validation accuracy does not increase compared to the last round, see Appendix E. Note that the learned knowledge thrown away by re-initialization can be re-learned afterwards, since all collected samples are continuously stored and accumulated in the server. This also makes our setting different from continual learning, that aims at avoiding catastrophic forgetting without accessing old data.

To determine after how many rounds the model should be re-initialized, we conduct extensive experiments on different partial updating settings, see Appendix E. In conclusion, the model is randomly re-initialized as long as the number of total newly collected data samples exceeds the number of samples when the model was re-initialized last time. For example, assume that at round r the model is randomly (re-)initialized and partially updated from this random model on dataset $\mathcal{D}^r$. Then, the model will be re-initialized again at round $r + n$, if $|\mathcal{D}^{r+n}| > 2 \cdot |\mathcal{D}^r|$, where $|.|$ denotes the number of samples in the dataset.

5   Evaluation

In this section, we experimentally show that through updating a small subset of weights, DPU can reach a similar accuracy as full updating while requiring a significantly lower communication cost. We implement DPU with Pytorch [23], and evaluate on public vision datasets, including MNIST [16], CIFAR10 [15], CIFAR100 [15], ImageNet [30], using multilayer perceptron (MLP), VGGNet [5, 26, 28], ResNet56 [10, 25], MobileNetV1 [12], respectively. Particularly, we partition the experiments into multi-round updating and single-round updating.

Multi-round Updating. We consider there are limited (or even zero) samples before the initial deployment, and data samples are continuously collected and sent from edge devices over a long period (the event rate is often low in real cases [21]). The server retrains the model and sends the updates to each device in multiple rounds. Regarding the highly-constrained communication resources, we choose low resolution image datasets (MNIST [16] and CIFAR10/100 [15]) to evaluate multi-round updating. We conduct one-shot rewinding in multi-round DPU, *i.e.*, rewinding is executed only once to achieve the desired updating ratio at each round as in Algorithm 1, which avoids hand-tuning hyperparameters (*e.g.*, updating ratio schedule) frequently over a large number of rounds.

Single-Round Updating. The deployed model is updated once via server-to-edge communication when new data from other sources become available on the server after some time, *e.g.*, releasing a new version of mobile applications based on newly retrieved internet data. Although DPU is elaborated and designed under multi-round updating settings, it can be applied directly in single-round updating. Since transmission from edge devices may be even not necessary, we evaluate single-round DPU on the large scale ImageNet dataset. Iterative rewinding is adopted here due to its better performance. Particularly, we alternatively perform rewinding 20% of the remaining trainable weights according to Eq. (10) and sparse fine-tuning until reaching the desired updating ratio.

Settings for all Experiments. We randomly select 30% of the original test dataset (original validation dataset for ImageNet) as the validation dataset, and

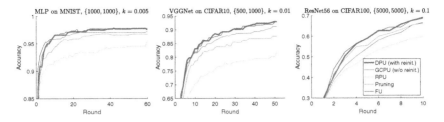

Fig. 2. DPU is compared with other baselines on different benchmarks in terms of the test accuracy during multi-round updating

the remainder as the test dataset. Let $\{|\mathcal{D}^1|, |\delta\mathcal{D}^r|\}$ represent the available data samples along rounds, where $|\delta\mathcal{D}^r|$ is supposed to be constant along rounds. Both $\mathcal{D}^1$ and $\delta\mathcal{D}^r$ are randomly sampled (without replacement) from the original training dataset to simulate the data collection. In each round, the test accuracy is reported, when the validation dataset achieves the highest Top-1 accuracy during retraining. When the validation accuracy does not increase compared to the previous round, the models are not updated to reduce the communication overhead. This strategy is also applied to other baselines to enable a fair comparison. We use the average cross-entropy as the loss function, and use Adam variant of SGD for MLP and VGGNet, Nesterov SGD for ResNet56 and MobileNetV1. More implementation details are provided in Appendix C.

Indexing. DPU generates a sparse tensor. In addition to the updated weights, the indices of these weights also need to be sent to each edge device. A simple implementation is to send the mask m, *i.e.*, a binary vector of I elements. Let S_w denote the bitwidth of each single weight, and S_x denote the bitwidth of each index. Directly sending m yields an overall communication cost of $I \cdot k \cdot S_w + I \cdot S_x$ with $S_x = 1$. To save the communication cost on indexing, we further encode m. Suppose that m is a random binary vector with a probability of k to contain 1. The optimal encoding scheme according to Shannon yields $S_x(k) = k \cdot \log(1/k) + (1-k) \cdot \log(1/(1-k))$. Coding schemes such as Huffman block coding can come close to this bound. We use $S_w \cdot k \cdot I + S_x(k) \cdot I$ to report the size of data transmitted from server to each node at each round, contributed by the partially updated weights plus the encoded indices of these weights.

5.1   Benchmarking Multi-round Updating

Settings. To the best of our knowledge, this is the first work on studying weight-wise partial updating a model using newly collected data in iterative rounds. Therefore, we developed three baselines for comparison, including (i) full updating (FU), where at each round the model is fully updated from a random initialization (*i.e.*, training from scratch, which yields a better performance see Sect. 4.2 and Appendix D); (ii) random partial updating (RPU), where the model is trained from w^{r-1}, while we randomly fix each layer's weights with a ratio of $(1-k)$ and sparsely fine-tune the rest; (iii) global contribution

Table 1. The average accuracy difference over all rounds and the ratio of communication cost over all rounds related to full updating

Method	Average accuracy difference			Ratio of communication cost		
	MLP	VGGNet	ResNet56	MLP	VGGNet	ResNet56
DPU	**−0.17%**	**+0.33%**	**−0.42%**	0.0071	0.0183	0.1147
GCPU	−0.72%	−1.51%	−3.87%	0.0058	0.0198	0.1274
RPU	−4.04%	−11.35%	−7.78%	0.0096	0.0167	0.1274
Pruning [29]	−1.45%	−4.35%	−2.35%	0.0106	0.0141	0.1274

partial updating (GCPU), where the model is trained with Alg. 2 without re-initialization described in Sect. 4.2; (*iv*) a state-of-the-art unstructured pruning method [29], where the model is first trained from a random initialization at each round, then conducts one-shot magnitude pruning, and finally is sparsely fine-tuned with learning rate rewinding. The ratio of nonzero weights in pruning is set to the same as the updating ratio k to ensure the same communication cost. The experiments are conducted on different benchmarks as mentioned earlier.

Results. We report the test accuracy of the model w^r along rounds in Fig. 2. All methods start from the same w^0, an entirely randomly initialized model. As seen in this figure, DPU clearly yields the highest accuracy in comparison to the other partial updating schemes. For example, DPU can yield a final Top-1 accuracy of 92.85% on VGGNet, even exceeds the accuracy (92.73%) of full updating. In addition, we compare three partial updating schemes in terms of the accuracy difference related to full updating averaged over all rounds, and the ratio of the communication cost related to full updating over all rounds in Table 1. As seen in the table, DPU reaches a similar accuracy as full updating, while incurring significantly fewer transmitted data sent from the server to each edge node. Specially, DPU saves around 99.3%, 98.2% and 88.5% of transmitted data on MLP, VGGNet, and ResNet56, respectively (95.3% in average). The communication cost ratios shown in Table 1 differ a little even for the same updating ratio. This is because if the validation accuracy does not increase, the model will not be updated to reduce the communication cost, as discussed earlier.

We further investigate the benefit due to DPU in terms of *the total communication cost reduction*, as DPU has no impact on the edge-to-server communication involving sending newly collected data samples. This experimental setup assumes that all samples in $\delta \mathcal{D}^r$ are collected by N edge nodes during all rounds and sent to the server on a per-round basis. For clarity, let S_d denote the data size of each training sample. During round r, we define per-node communication cost under DPU as $S_d \cdot |\delta \mathcal{D}^r|/N + (S_w \cdot k \cdot I + S_x(k) \cdot I)$. The results are shown in Appendix F.1. We observe that DPU can still achieve a significant reduction on the total communication cost, *e.g.*, reducing up to 88.2% even for the worst case (a single node). Moreover, DPU tends to be more beneficial when the size

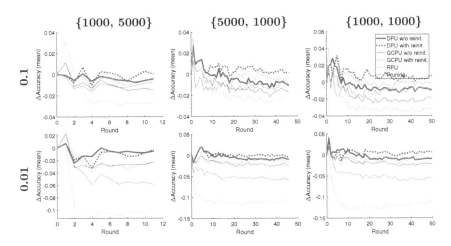

Fig. 3. Comparison w.r.t. the mean accuracy difference (full updating as the reference) under different $\{|\mathcal{D}^1|, |\delta\mathcal{D}^r|\}$ (representing the available data samples along rounds, see in Sect. 5) and updating ratio ($k = 0.1, 0.01$)

of data transmitted by each node to the server becomes smaller. This is intuitive because in this case server-to-edge communication cost (thus the reduction due to DPU) dominates the total communication cost.

5.2 Different Number of Data Samples and Updating Ratios

Settings. In this section, we show that DPU outperforms other baselines under varying number of training samples and updating ratios in multi-round updating. We also conduct ablations concerning the re-initialization of weights discussed in Sect. 4.2. We implement DPU with and without re-initialization, GCPU with and without re-initialization, RPU, and Pruning [29] (see more details in Sect. 5.1) on VGGNet using CIFAR10 dataset. We compare these methods with different amounts of samples $\{|\mathcal{D}^1|, |\delta\mathcal{D}^r|\}$ and different updating ratios k. Without further notations, each experiment runs three times using random data samples.

Results. We compare the difference between the accuracy under each method and that under full updating. The mean accuracy difference over three runs is plotted in Fig. 3. A comprehensive set of results including the absolute accuracy and the standard deviations is provided in Appendix F.2. As seen in Fig. 3, DPU (with re-initialization) always achieves the highest accuracy. DPU also significantly outperforms the pruning method, especially under a small updating ratio. Note that we preferred a smaller updating ratio in our context because it explores the limits of the approach and it indicates that we can improve the deployed model more frequently with the same accumulated server-to-edge communication cost. The dashed curves and the solid curves with the same color can be viewed as the ablation study of our re-initialization scheme. Particularly given a large number of rounds, it is critical to re-initialize the start point $\boldsymbol{w}^{r-1}$ after several rounds (as discussed in Sect. 4.2).

In the first few rounds, partial updating methods almost always yield a higher test accuracy than full updating, *i.e.*, the curves are above zero. This is due to the fact that the amount of available samples is rather small, and partial updating may avoid some co-adaptation in full updating. The partial updating methods perform almost randomly in the first round compared to each other, because the limited data are not sufficient to distinguish critical weights from the random initialization $\boldsymbol{w}^0$. This also motivates us to (partially) update the deployed model when new samples are available.

Pruning Weights vs. Pruning Incremental Weights. One of our chosen baselines, global contribution partial updating (GCPU, see Alg. 2), could be viewed as a counterpart of the pruning method [29], *i.e.*, pruning the incremental weights with the least magnitudes. By comparing GCPU (with or without re-initialization) with "Pruning", we conclude that retaining previous weights yields better performance than zero-outing the weights.

5.3   Benchmarking Single-Round Updating

Settings. To show the versatility of our methods, we test single-round updating for MobileNetV1 [12] on ImageNet [30] with iterative rewinding. Single-round DPU is conducted on different initially deployed models, including a floating-point (FP32) dense model and two compressed models, *i.e.*, a 50%-sparse model and an INT8 quantized model. The sparse model is trained with a state-of-the-art dynamic pruning method [24]; the quantized model is trained with straight-through-estimator with a output-channel-wise floating point scaling factors similar as [28]. To maintain the same on-device inference cost, partial updating is only applied on nonzero values of sparse models; for quantized models, the updated weights are still in INT8 format.

Results. We compare DPU with the vanilla-updates, *i.e.*, the models are trained from a random initialization with the corresponding methods on all available samples. The test accuracy and the ratio of (server-to-edge) communication cost related to full updating on FP32 dense model are reported in Table 2. Results show that DPU often

Table 2. The test accuracy of single-round updating on different initial models (MobileNetV1 on ImageNet). The updating ratio $k = 0.2$. The ratio of communication cost related to full updating is reported in brackets

#Samples	$\{8 \times 10^5, 4.8 \times 10^5\}$		
	Initial	Vanilla-update	DPU
FP32 Dense	68.5%	70.7% (1)	71.1% (0.22)
50%-Sparse	68.1%	70.5% (0.53)	70.8% (0.22)
INT8	68.4%	70.6% (0.25)	70.6% (0.07)

yields a higher accuracy than vanilla updating while requiring substantially lower communication cost.

6   Conclusion

In this paper, we present the weight-wise deep partial updating paradigm, motivated by the fact that continuous full weight updating may be impossible in many edge intelligence scenarios. We present DPU through analytically upper-bounding the loss difference between partial updating and full updating. DPU only updates the weights that make the largest contributions to the upper bound, while reuses the other weights that have less impact on the loss reduction. Extensive experimental results demonstrate the efficacy of DPU which achieves a high inference accuracy while updating a rather small number of weights.

Acknowledgement. Part of Zhongnan Qu and Lothar Thiele's work was supported by the Swiss National Science Foundation in the context of the NCCR Automation. Part of Cong Liu's work was supported by NSF CNS 2135625, CPS 2038727, CNS Career 1750263, and a Darpa Shell grant.

References

1. Ash, J.T., Zhang, C., Krishnamurthy, A., Langford, J., Agarwal, A.: Deep batch active learning by diverse, uncertain gradient lower bounds. In: International Conference on Learning Representations (ICLR) (2020). https://openreview.net/forum?id=ryghZJBKPS

2. Augustin, A., Yi, J., Clausen, T., Townsley, W.M.: A study of lora: long range & low power networks for the internet of things. Sensors **16**(9) (2016). https://doi.org/10.3390/s16091466. https://www.mdpi.com/1424-8220/16/9/1466

3. Bhandare, A., et al.: Efficient 8-bit quantization of transformer neural machine language translation model. In: International Conference on Machine Learning (ICML), Joint Workshop on On-Device Machine Learning & Compact Deep Neural Network Representations (2019)

4. Brown, S., Sreenan, C.: Updating software in wireless sensor networks: a survey. Department of Computer Science, National University of Ireland, Maynooth, Technical report, pp. 1–14 (2006)

5. Courbariaux, M., Bengio, Y., David, J.P.: Binaryconnect: training deep neural networks with binary weights during propagations. In: Annual Conference on Neural Information Processing Systems (NeurIPS) (2015)

6. Evci, U., Gale, T., Menick, J., Castro, P.S., Elsen, E.: Rigging the lottery: making all tickets winners. In: International Conference on Machine Learning (ICML) (2021)

7. Frankle, J., Carbin, M.: The lottery ticket hypothesis: finding sparse, trainable neural networks. In: International Conference on Learning Representations (ICLR) (2019). https://openreview.net/forum?id=rJl-b3RcF7

8. Guo, T.: Cloud-based or on-device: an empirical study of mobile deep inference. In: 2018 IEEE International Conference on Cloud Engineering, IC2E 2018, Orlando, FL, USA, 17–20 April 2018, pp. 184–190. IEEE Computer Society (2018). https://doi.org/10.1109/IC2E.2018.00042

9. Han, S., Mao, H., Dally, W.J.: Deep compression: compressing deep neural networks with pruning, trained quantization and Huffman coding. In: International Conference on Learning Representations (ICLR) (2016)

10. He, K., Zhang, X., Ren, S., Sun, J.: Deep residual learning for image recognition. In: The IEEE/CVF Conference on Computer Vision and Pattern Recognition (CVPR) (2016)

11. Horowitz, M.: 1.1 computing's energy problem (and what we can do about it). In: 2014 IEEE International Solid-State Circuits Conference Digest of Technical Papers (ISSCC) (2014)

12. Howard, A.G., et al.: Mobilenets: efficient convolutional neural networks for mobile vision applications. CoRR (2017)

13. Jung, S., Ahn, H., Cha, S., Moon, T.: Adaptive group sparse regularization for continual learning. CoRR (2020)

14. Kairouz, P., et al.: Advances and open problems in federated learning. CoRR (2019)

15. Krizhevsky, A., Nair, V., Hinton, G.: Cifar10/100 (Canadian institute for advanced research) (2009). http://www.cs.toronto.edu/kriz/cifar.html

16. LeCun, Y., Cortes, C.: MNIST handwritten digit database (2010). http://yann.lecun.com/exdb/mnist/

17. Lee, J., et al.: On-device neural net inference with mobile GPUs. CoRR (2019)

18. Li, A., et al.: LotteryFL: personalized and communication-efficient federated learning with lottery ticket hypothesis on non-IID datasets. CoRR (2020)

19. Lin, Y., Han, S., Mao, H., Wang, Y., Dally, W.J.: Deep gradient compression: Reducing the communication bandwidth for distributed training. In: International Conference on Learning Representations (ICLR) (2018). https://openreview.net/forum?id=SkhQHMW0W

20. Meng, Z., et al.: A two-stage optimized next-view planning framework for 3-D unknown environment exploration, and structural reconstruction. IEEE Robot. Autom. Lett. **2**(3), 1680–1687 (2017)

21. Meyer, M., et al.: Event-triggered natural hazard monitoring with convolutional neural networks on the edge. In: International Conference on Information Processing in Sensor Networks (IPSN), IPSN 2019. Association for Computing Machinery (2019)

22. Nesterov, Y.: Introductory lectures on convex programming volume I: basic course. Lecture Notes **1**, 25 (1998)

23. Paszke, A., et al.: Automatic differentiation in pytorch. In: NIPS Autodiff Workshop: The Future of Gradient-based Machine Learning Software and Techniques (2017)

24. Peste, A., Iofinova, E., Vladu, A., Alistarh, D.: AC/DC: alternating compressed/decompressed training of deep neural networks. In: Annual Conference on Neural Information Processing Systems (NeurIPS) (2021)

25. Pytorch: Pytorch example on resnet (2019). https://github.com/pytorch/vision/blob/master/torchvision/models/resnet.py. Accessed 15 Oct 2019

26. Qu, Z., Zhou, Z., Cheng, Y., Thiele, L.: Adaptive loss-aware quantization for multi-bit networks. In: The IEEE/CVF Conference on Computer Vision and Pattern Recognition (CVPR) (2020)

27. Raghu, A., Raghu, M., Bengio, S., Vinyals, O.: Rapid learning or feature reuse? towards understanding the effectiveness of MAML. In: International Conference on Learning Representations (ICLR) (2020). https://openreview.net/forum?id=rkgMkCEtPB

28. Rastegari, M., Ordonez, V., Redmon, J., Farhadi, A.: XNOR-Net: ImageNet classification using binary convolutional neural networks. In: Leibe, B., Matas, J., Sebe, N., Welling, M. (eds.) ECCV 2016. LNCS, vol. 9908, pp. 525–542. Springer, Cham (2016). https://doi.org/10.1007/978-3-319-46493-0_32

29. Renda, A., Frankle, J., Carbin, M.: Comparing fine-tuning and rewinding in neural network pruning. In: International Conference on Learning Representations (ICLR) (2020). https://openreview.net/forum?id=S1gSj0NKvB

30. Russakovsky, O., et al.: Imagenet large scale visual recognition challenge. Int. J. Comput. Vision **115**(3), 211–252 (2015)

31. Shen, Z., Liu, Z., Qin, J., Savvides, M., Cheng, K.: Partial is better than all: revisiting fine-tuning strategy for few-shot learning. In: The AAAI Conference on Artificial Intelligence (AAAI) (2021)

32. Shokri, R., Shmatikov, V.: Privacy-preserving deep learning. In: Proceedings of the 22nd ACM SIGSAC Conference on Computer and Communications Security, CCS 2015 (2015). https://doi.org/10.1145/2810103.2813687

33. Simonyan, K., Zisserman, A.: Very deep convolutional networks for large-scale image recognition. In: International Conference on Learning Representations (ICLR) (2015)

34. Stowell, D., Wood, M., Pamuła, H., Stylianou, Y., Glotin, H.: Automatic acoustic detection of birds through deep learning: the first bird audio detection challenge. Methods Ecol. Evol. **10**, 368–380 (2018). https://doi.org/10.1111/2041-210X.13103

35. Varshney, A.: Enabling sustainable networked embedded systems. Ph.D. thesis, Uppsala University, Division of Computer Systems, Computer Architecture and Computer Communication (2018)

36. Wang, Z., K, K., Mayhew, S., Roth, D.: Extending multilingual bert to low-resource languages. CoRR (2020)

37. Yoon, J., Yang, E., Lee, J., Hwang, S.J.: Lifelong learning with dynamically expandable networks. In: International Conference on Learning Representations (ICLR) (2018). https://openreview.net/forum?id=Sk7KsfW0-

38. Zhou, H., Lan, J., Liu, R., Yosinski, J.: Deconstructing lottery tickets: zeros, signs, and the supermask. In: Annual Conference on Neural Information Processing Systems (NeurIPS) (2019)

39. Zhuang, F., et al.: A comprehensive survey on transfer learning. CoRR (2019)

Patch Similarity Aware Data-Free Quantization for Vision Transformers

Zhikai Li[1,2], Liping Ma[1], Mengjuan Chen[1], Junrui Xiao[1,2], and Qingyi Gu[1(✉)]

[1] Institute of Automation, Chinese Academy of Sciences, Beijing, China
{lizhikai2020,liping.ma,chenmengjuan2016,xiaojunrui2020,
qingyi.gu}@ia.ac.cn
[2] School of Artificial Intelligence, University of Chinese Academy of Sciences,
Beijing, China

Abstract. Vision transformers have recently gained great success on various computer vision tasks; nevertheless, their high model complexity makes it challenging to deploy on resource-constrained devices. Quantization is an effective approach to reduce model complexity, and data-free quantization, which can address data privacy and security concerns during model deployment, has received widespread interest. Unfortunately, all existing methods, such as BN regularization, were designed for convolutional neural networks and cannot be applied to vision transformers with significantly different model architectures. In this paper, we propose PSAQ-ViT, a Patch Similarity Aware data-free Quantization framework for Vision Transformers, to enable the generation of "realistic" samples based on the vision transformer's unique properties for calibrating the quantization parameters. Specifically, we analyze the self-attention module's properties and reveal a general difference (patch similarity) in its processing of Gaussian noise and real images. The above insights guide us to design a relative value metric to optimize the Gaussian noise to approximate the real images, which are then utilized to calibrate the quantization parameters. Extensive experiments and ablation studies are conducted on various benchmarks to validate the effectiveness of PSAQ-ViT, which can even outperform the real-data-driven methods. Code is available at: https://github.com/zkkli/PSAQ-ViT.

Keywords: Model compression · Data-free quantization · Quantized vision transformer

1 Introduction

With the great success on natural language processing applications, transformer-based models have also demonstrated superior performance on a variety of computer vision tasks [14,19]. However, vision transformers typically employ complicated model architectures with extremely high memory footprints and computational overheads to accomplish the powerful representational capabilities, posing significant challenges for their deployment and real-time inference on resource-constrained edge devices [18,28,32]. Thus, the compression technique for vision transformers is highly desired for real-world applications.

© The Author(s), under exclusive license to Springer Nature Switzerland AG 2022
S. Avidan et al. (Eds.): ECCV 2022, LNCS 13671, pp. 154–170, 2022.
https://doi.org/10.1007/978-3-031-20083-0_10

Model quantization, which converts 32-bit floating-point parameters (weights and activations) to low-precision values, is regarded as a prevalent approach to reduce the complexity of neural networks and accelerate their inference phase [13,20]. To mitigate the accuracy degradation, almost all quantization methods require access to the original dataset for re-training/fine-tuning the model parameters [7,11,16,31,39]. Unfortunately, in scenarios involving sensitive data (*e.g.*, medical and bio-metric data), these methods are no longer applicable due to the unavailability of the original dataset [36,40]. Therefore, data-free quantization is regarded as a potential and practice scheme [3,41].

The main idea of data-free quantization is to generate samples that can match the real-data distribution based on the prior information of the pre-trained full-precision (FP) model, and then utilize these samples to calibrate the quantization parameters. The key issue is how to generate effective and meaningful samples to ensure the calibration accuracy. A notable line of research proposes batch normalization (BN) regularization [3,40], which states that the statistics (*i.e.*, the mean and standard deviation) encoded in the BN layers can represent the distribution of original training data. These methods, however, are only applicable to convolutional neural networks (CNNs) and not to vision transformers, because the latter employs layer normalization (LN), which does not store any previous information like BN. As a result, existing methods cannot be extended and migrated well due to significant differences in model architecture, leaving data-free quantization for vision transformers as a gap.

In this paper, we are motivated to address the above issues, focusing on the following challenge: *how to effectively generate "realistic" samples based on the vision transformer's unique properties?* Since there is no elegant absolute value metric like BN statistics, we intend to investigate the general difference in model inference when the input is Gaussian noise and a real image, and then accordingly design a *relative value* metric to optimize the noise. As stated in [10], in the training phase, the self-attention module is designed to extract the important information from the training data, *i.e.*, to identify the foreground from the background, so that the model can make a good decision. Accordingly, in the inference phase of the pre-trained model, when the input is a real image, the foreground patches and background patches can produce different responses, thus the self-attention module has a diverse patch similarity (*i.e.* the similarity between the responses in the patch dimension); in contrast, the responses to Gaussian noise, whose foreground and background are hard to distinguish, are homogeneous, as shown in Fig. 1.

With the above analysis, we propose PSAQ-ViT, a Patch Similarity Aware data-free Quantization framework for Vision Transformers. Specifically, we utilize the differential entropy of patch similarity to quantify the diversity of responses, which is calculated via kernel density estimation that can ensure gradient back-propagation. Then, the differential entropy is used as the objective function to optimize the Gaussian noise to approximate the real image. Finally, the generated samples are utilized to calibrate the parameters of the quantized vision transformers.

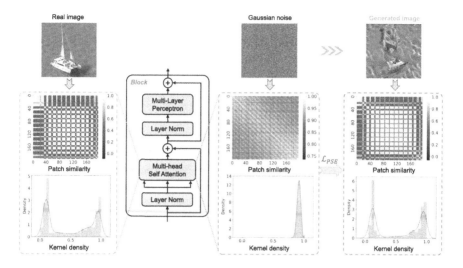

Fig. 1. Illustration of the proposed sample generation approach. When the input is Gaussian noise, patches are grouped into one category (foreground or background), leading to homogeneous patch similarity and a unimodal kernel density curve. Our generated image can potentially represent the real-image features, producing diverse patch similarity and a bimodal kernel density curve, where the left and right peaks describe inter- and intra-category similarity, respectively.

To be specific, our contributions are as follows:

– From an in-depth analysis of the self-attention module, we reveal a general difference in its processing of Gaussian noise and real images, *i.e.*, a substantially distinct diversity of patch similarity. This general difference demonstrates the intrinsic properties of vision transformers' image perception and provides some insights for sample generation.
– With the above insights, we propose PSAQ-ViT, in which we design a relative value metric to optimize the Gaussian image to reduce the general difference and thus approximate the real images, and then utilize them to calibrate the quantization parameters. To the best of our knowledge, this is the first work to quantify vision transformers without any real-world data.
– Extensive experiments on various benchmark models are conducted to demonstrate the effectiveness of PSAQ-ViT, which can generate "realistic" samples and thus enable the outstanding performance of data-free quantization for vision transformers, even outperforming real-data-driven methods.

2   Related Works

2.1   Vision Transformers

Vision transformers, which utilize global information based on self-attention modules, have recently achieved great success on various computer vision tasks

[1,6,15,35,42]. ViT [10] is the first pure vision transformer model, which reshapes the image into a sequence of flattened 2D patches as the input, achieving better performance than CNNs on image classification tasks. Following ViT, DeiT [33] introduces a teacher-student procedure based on a distillation token, which can achieve competitive results on ImageNet with no external data. Swin Transformer [27] presents a hierarchical design with shifting windows for representation, which allows for modeling at various scales and thus boosts the performance of vision transformers. In addition to image classification, transformers have been applied to other computer vision tasks, such as object detection [4,43], semantic segmentation [5], and video recognition [30].

Although these vision transformer models have great potential on computer vision tasks, their powerful representation capabilities are obtained based on the complicated model architectures, which makes them supremely challenging to deploy on resource-constrained devices and execute real-time inference [28,32]. Thus, model compression is a necessary and promising solution to facilitate their real-world applications.

2.2   Model Quantization

Model quantization, which reduces the memory footprint and computational overhead of the models by decreasing the representation precision of the weights and activations, is an effective approach to compressing models in a hardware-friendly manner [13,20]. The mainstream methods exploit quantization-aware training to compensate the accuracy degradation caused by discretization [8,12,23,25,39], and they use the straight-through estimator [2] to approximate the gradient back-propagation of the quantized model. However, these methods rely heavily on the original dataset for re-training/fine-tuning, rendering them inapplicable in many scenarios where the original data is not available [3,40]. In addition, several post-training quantization methods have been proposed to reduce the fine-tuning cost [9,17,22,24,29,34], including schemes for vision transformers [26,28,38], but they still require a small amount of real data for calibration and cannot achieve complete data-free.

Data-Free Quantization. which compresses models without access to any real data, can potentially address the above issues, and thus has received increasing attention. ZeroQ [3] proposes BN regularization to generate samples based on the real-data statistics encoded in the BN layers of the pre-trained FP model, and then use them to fine-tune the model parameters. DSG [40] presents an improved BN regularization scheme that utilizes slack distribution alignment and layerwise sample enhancement to address the homogenization of the generated samples. GDFQ [36] and IntraQ [41] introduce category label information to generate class-conditional samples, further pushing the limit of data-free quantization. However, these methods are only applicable to CNNs, because there is no key structure BN in vision transformers and the LN they employ does not contain any features of the original training data. As a result, there is now a gap in the data-free quantization community of vision transformers.

3   Methodology

In this section, we first introduce the computational process of vision transformers and the uniform quantization strategy in the preliminaries. Our insights
and motivations of the proposed PSAQ-ViT are then presented, followed by a
detailed introduction to the designed patch similarity metric for sample generation. Finally, the overall quantization pipeline for vision transformers is summarized and presented.

3.1   Preliminaries

A standard transformer's input is a one-dimensional sequence of token embeddings. For vision transformers, an image I is reshaped into a sequence of flatted
2D patches, and each patch is then mapped to the hidden size d by a linear
projection to obtain the input vectors $X \in \mathbb{R}^{N \times d}$. Here, N is the number of
patches.

   The vectors X are then input into transformer layers, which are a stack of
blocks composed of a multi-head self-attention (MSA) module and a multi-layer
perceptron (MLP) module apiece. First, MSA calculates the attention between
different patches to extract feature representations with global information as
follows:

$$\mathrm{MSA}(X) = \mathrm{Concat}(\mathrm{head}_1, \cdots, \mathrm{head}_H)W^o$$
$$\text{where head}_i = \mathrm{Attn}(Q_i, K_i, V_i) = \mathrm{softmax}(\frac{Q_i K_i^T}{\sqrt{d}})V_i \quad (1)$$

where H is the number of attention heads. Here, query Q_i, key K_i, and value V_i
are computed by linear projections using matrix multiplication, $i.e.$, $Q_i = XW_i^Q$,
$K_i = XW_i^K$, $V_i = XW_i^V$. Then, the output of MSA is fed into MLP, which
contains two fully-connected layers for feature mapping and information fusion.

   As we can see, in vision transformers, most computational costs are derived
from the large matrix multiplication in MSA and MLP modules. Thus, we intend
to quantize all the parameters in matrix multiplication, including both weights
and activations. In this paper, we perform the uniform quantization strategy,
which is the most popular and hardware-friendly method and is defined as follows:

$$\theta^q = \lfloor \frac{\mathrm{clip}(\theta^p, q_0, q_{2^k-1}) - q_0}{\Delta} \rceil, \text{ where } \Delta = \frac{q_{2^k-1} - q_0}{2^k - 1} \quad (2)$$

where θ^p and θ^q denote the parameters of the FP model and the quantized
model, respectively. Here, q_0 and q_{2^k-1} are clipping values that determine the
quantization scales, $\lfloor \cdot \rceil$ is the round operator, and k is the quantization bit-
precision.

3.2   Our Insights

As mentioned before, the main challenge of data-free quantization for vision
transformers is that they do not have BN layers that store information about

the original training data, resulting in no available absolute value prior information for sample generation and thus no efficient calibration of quantization parameters. Therefore, our interest is to mine deeper into the prior information of the pre-trained vision transformer models and thus explore a reliable *relative value* metric that can well describe the general difference between Gaussian noise and the real image, so that we can reduce this difference to make the Gaussian noise approximate the real image.

Since the self-attention module is the unique structure of vision transformers, its powerful feature extraction capability is believed to contain a certain amount of original data information. Hence, we provide an in-depth analysis of the training process of the self-attention module, and then we observe that the reason the model can make good decisions is that the self-attention module can distinguish the foreground from the background of the training data, thus allocating more attention to the foreground that is more important for the decision. Since the input of vision transformers are independent vectors mapped by 2D patches, the responses of the self-attention module to different patches are significantly different, *i.e.*, the foreground patches receive more attention.

When the pre-trained model executes inference, real images consistently produce the above features, while Gaussian noise, whose foreground is not easily extracted, does not have a similar capability and inevitably leads to homogeneous responses, as shown in Fig. 1. Note that the real images here are only used to verify the general difference (*e.g.*, a certain metric of the real images is always larger than that of Gaussian noise), and they will not be involved in any subsequent process. Therefore, this general difference can indirectly represent the prior information of vision transformers and thus can be used to design the relative value metric to guide the sample generation.

3.3   Patch Similarity Metric

Based on the above insights, we aim to design a reliable metric that can measure the diversity of the self-attention module's responses. For the l-th layer in vision transformers, the output of the MSA module is defined as $O_l \in \mathbb{R}^{B \times H \times N \times d}$ ($l \in \{1, \cdots, L\}$), where each dimension denotes the batch size, number of heads, number of patches, and hidden size, respectively. To simplify the expression, we ignore the batch dimension, *i.e.*, $O_l \in \mathbb{R}^{H \times N \times d}$.

Due to the relative value metric, it is necessary to first normalize O_l to ensure the fairness of the comparison. We accomplish this by calculating the cosine similarity between each subspace vector in the patch dimension, specifying the data range at $[-1, 1]$, as follows:

$$\Gamma_l(u_i, u_j) = \frac{u_i \cdot u_j}{||u_i|| \, ||u_j||} \tag{3}$$

where the numerator represents the inner product of the vectors, and $||\cdot||$ denotes the l_2 norm. Here, $u_i, u_j \in \mathbb{R}^{H \times d}$ ($i, j \in \{1, \cdots, N\}$) is the i-th/j-th vector in the patch dimension of O_l, and $\Gamma_l(u_i, u_j)$ represents the cosine similarity between u_i

and u_j. After pairwise calculations, we obtain the l-th layer's cosine similarity matrix $\boldsymbol{\Gamma}_l = [\Gamma_l(u_i, u_j)]_{N \times N}$, which is a symmetric matrix and is termed as patch similarity. The diversity of patch similarity can potentially represent the diversity of the original data, which not only elegantly achieves data normalization, but also has the additional advantage of achieving reasonable $\frac{Hd}{N}$-fold dimensionality reduction ($\mathbb{R}^{H \times N \times d} \rightarrow \mathbb{R}^{N \times N}$). For instance, for the ViT-B model, the amount of data is reduced by a factor of 3.92, which can greatly improve the subsequent computational efficiency.

Then, the diversity of patch similarity is measured by the information entropy, which can represent the amount of information expressed by the data. To ensure gradient back-propagation, we calculate the differential entropy that has a continuous nature as follows:

$$H_l = -\int \hat{f}_h(x) \cdot \log\left[\hat{f}_h(x)\right] dx \tag{4}$$

where $\hat{f}_h(x)$ is the continuous probability density function of $\boldsymbol{\Gamma}_l$, which is obtained using kernel density estimation as follows:

$$\hat{f}_h(x) = \frac{1}{M}\sum_{m=1}^{M} K_h(x - x_m) = \frac{1}{Mh}\sum_{m=1}^{M} K(\frac{x - x_m}{h}) \tag{5}$$

where $K(\cdot)$ is the kernel (e.g. normal kernel), h is the bandwidth, x_m ($m \in \{1, \cdots, M\}$) is a training point drawn from $\boldsymbol{\Gamma}_l$ and is the center of a kernel, and x is the given test point.

Finally, we sum the differential entropy of each layer to account for the diversity of patch similarity across all layers, and since it is to be maximized, the Patch Similarity Entropy loss is defined as follows:

$$\mathcal{L}_{PSE} = -\sum_{l=1}^{L} H_l \tag{6}$$

3.4   The Overall Pipeline

The whole process of PSAQ-ViT is performed in two stages: first, the Gaussian noise is optimized according to the loss function, which is designed based on the prior information of the pre-trained model, to generate "realistic" samples; second, the generated samples are utilized to calibrate the quantization parameters, thus realizing the vision transformer quantization with no real data participation. These two stages are described in detail below.

Sample Generation: In the sample generation stage, in addition to our proposed patch similarity entropy loss $\mathcal{L}_{PSE}$, which has the greatest contribution to the performance, the optimization objective for sample generation also contains two auxiliary image priors: one-hot loss $\mathcal{L}_{OH}$ and total variance loss $\mathcal{L}_{TV}$, which can ensure more stable convergence to effective images.

Algorithm 1: The PSAQ-ViT Pipeline

Input: A pre-trained FP vision transformer P with parameters θ^p.
Output: A quantized vision transformer Q with parameters θ^q.
Initialize the quantized model Q by Eq. (2);
Randomly produce Gaussian noise $I_G \sim \mathcal{N}(0, 1)$;
# Stage 1: Sample generation
for $t = 1, 2, \ldots$ **do**
    | Input I_G into the pre-trained FP model P;
    | Calculate $\mathcal{L}_{PSE}$ by Eq. (6);
    | Calculate $\mathcal{L}_{OH}$ and $\mathcal{L}_{TV}$ by Eq. (7) and Eq. (8);
    | Combine three losses to obtain $\mathcal{L}_G$ by Eq. (9);
    | Update I_G by back-propagation of $\mathcal{L}_G$;
end
Stage 2: Quantization parameter calibration
Get the generated samples $I = I_G$;
Input I into the quantized model Q;
Determine the clipping values of the activations in Q;

One-hot loss is a popular class prior that describes the class boundary information and motivates the generated images to be predicted to a pre-defined category c [36,41]. Specifically, it encourages to minimize the cross entropy loss as follows:

$$\mathcal{L}_{OH} = CE(P(I), c) \tag{7}$$

where $P(I)$ is the predicted result of the pre-trained model for image I.

Total variance loss is a pixel-level smoothing regularization term for images and can further improve the image quality [37], which is defined as follows:

$$\mathcal{L}_{TV} = \iint |\nabla I(\tau_1, \tau_2)| d\tau_1 d\tau_2 \tag{8}$$

where $\nabla I(\tau_1, \tau_2)$ denotes the gradient of the image I at (τ_1, τ_2).

We combine the above three loss functions to obtain the final objective function for sample generation as follows:

$$\begin{aligned} \mathcal{L}_G &= \mathcal{L}_{PSE} + \alpha \mathcal{L}_{prior} \\ &= \mathcal{L}_{PSE} + \alpha_1 \mathcal{L}_{OH} + \alpha_2 \mathcal{L}_{TV} \end{aligned} \tag{9}$$

where α_1 and α_1 are the balance coefficients.

Quantization Parameter Calibration: In the parameter calibration stage, the weight parameters are fixed and can be calibrated directly, thus the generated samples are only utilized to determine the clipping values (q_0 and q_{2^k-1}) for the activations of each layer to get rid of outliers and to better represent the majority of the given parameters. Note that the calibration process is performed in the form of post-training quantization and does not require resource-consuming fine-tuning. The overall pipeline is summarized in Algorithm 1.

4   Experiments

In this section, PSAQ-ViT is evaluated on various benchmark models for the large-scale image classification task. To the best of our knowledge, there is no published work on data-free quantization of vision transformers, thus the effectiveness of our method is demonstrated by comparing the quantized model calibrated with real images and Gaussian noise at the same settings. Furthermore, ablation studies are conducted to verify the validity of the proposed patch similarity entropy loss.

4.1   Implementation Details

Models and Datasets: We evaluate PSAQ-ViT on various popular vision transformer models, including ViT [10], DeiT [33], and Swin [27]. The dataset we adopt is ImageNet (ILSVRC-2012) [21] for the large-scale image classification task which contains 1000 categories of images (224×224 pixels). The pre-trained models are all obtained from timm[1].

Experimental Settings: All implementations of PSAQ-ViT are done on PyTorch. To demonstrate the validity of our generated images, we employ the most basic quantization parameter calibration method. For weights, symmetric uniform quantization is applied, and the calibration strategy is fixed to Vanilla MinMax; for activations, asymmetric uniform quantization is applied, and the default calibration strategy is Vanilla MinMax if not specifically declared. In all our experiments, the number of images used for calibration is 32. α_1 and α_2 are set to 1.0 and 0.05 after a simple grid search, respectively, and their selection had little effect on the final performance.

4.2   Analysis of Generated Samples

Figure 2 shows the visualization results of the generated images (224×224 pixels), which are obtained based on the ViT-B model pre-trained on ImageNet dataset. Since we use the class prior $\mathcal{L}_{OH}$ in the image generation process, we present them by category, and different images in a category are produced by using different random seeds when initializing the Gaussian noise. It should be highlighted that these images require only a pre-trained model, and not any additional information, especially the original data or any absolute value metrics. Thanks to the proposed optimization objective $\mathcal{L}_{PSE}$, the generated "realistic" images can clearly distinguish the foreground from the background, and the foreground is rich in semantic information. Moreover, according to the subsequent quantization experiments, this excellent property of easily extracting the foreground will have a positive feedback effect on the calibration of the quantization parameters, making the generated images achieve better performance than the real images.

---

[1]   https://github.com/rwightman/pytorch-image-models

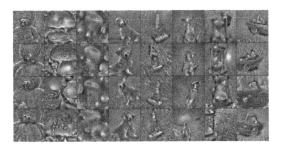

Fig. 2. Class-conditional samples (224×224 pixels) generated by PSAQ-ViT, given only a pre-trained ViT-B model on ImageNet and no additional information.

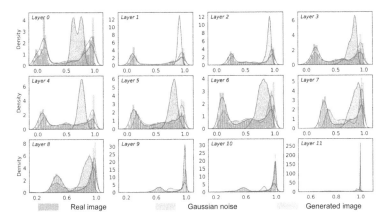

Fig. 3. Comparison of the kernel density curves of the patch similarity for each layer in ViT-B model when the input is the real image, Gaussian noise, and the generated image. The x-axis represents the values of patch similarity. As we can see, the density of each layer corresponding to Gaussian noise shows a concentrated unimodal shape, while the generated image and the real image have similar properties, producing the density with a dispersed bimodal shape.

In addition, since we consider the patch similarity entropy of all layers in Eq. (6), the comparison experiments of the kernel density curves of the patch similarity for each layer in ViT-B model when the input is the real image, Gaussian noise, and the generated image are conducted, as shown in Fig. 3. For the responses to Gaussian noise, the kernel density curves all show a concentrated unimodal shape and the central value of the curve is high, indicating a high degree of similarity between each patch of Gaussian noise and thus a full classification as background or foreground. Fortunately, the kernel density curves corresponding to our generated images are very approximate to those corresponding to the real images. They all show a dispersed bimodal shape, indicating a high diversity of responses, and the left and right peaks of curves describe inter- and intra-category similarity, respectively, which is in line with the expectation that the images can easily be distinguished between foreground and background.

4.3   Quantization Results

Here, we employ the proposed PSAQ-ViT to quantify the ViT-S, ViT-B, DeiT-T, DeiT-S, DeiT-B, Swin-T, and Swin-S models on large-scale ImageNet dataset, and the results are reported in Table 1. Since, to the best of our knowledge, there is no published work on data-free quantization of vision transformers, we set a reasonable baseline for our experiments on our own. *Standard* and *Gaussian noise* denote using real images and Gaussian noise to calibrate the quantization parameters, respectively. Note that all experiments differ only in the images used to calibrate the quantization parameters, and all other settings are the same, such as the calibration strategy and the number of images. Moreover, to demonstrate the robustness of the method, we evaluate different quantization precisions, including W4/A8 and W8/A8.

First, it should be emphasized that PSAQ-ViT can achieve **better performance** than Standard, which requires the real data, on all the aforementioned models, indicating that the generated images are even more effective than the real ones for parameter calibration. The main reason is that the sample generation is based on the prior information in the self-attention module, *i.e.*, facilitating the distinction between foreground and background in images, and then when these samples are utilized to calibrate the quantization parameters, they in turn reinforce the functionality of the self-attention module, thus acting as positive feedback that can reduce the activation outliers to some extent and therefore improve the tolerance to parameter clipping.

The quantization results of each model are discussed in detail below. We begin by discussing the quantization results of ViT-S and ViT-B models. Because we use the vanilla quantization strategy and these models are very sensitive to quantization, different methods can all lead to noticeable accuracy degradation. Despite this, our method achieves the best performance at the same settings, *e.g.* for quantization of ViT-S, our method improves by 0.93% and 1.17% over Standard at W4/A8 and W8/A8 settings, respectively.

DeiT has the same model structure as ViT but with a different training strategy; however, the quantization perturbation on the performance of DeiT is significantly smaller compared to ViT. When the calibration image is Gaussian noise, the representation capability of the quantization model decreases sharply, and its prediction accuracy decreases severely. In comparison, our proposed PSAQ-ViT can achieve very excellent performance. For the W8/A8 quantization of DeiT-T, our method achieves 4-fold compression with almost lossless accuracy (only 0.65% accuracy degradation). PSAQ-ViT is 1.13% and 0.92% higher than real-data-driven Standard in the quantization of W4/A8 and W8/A8 for DeiT-S, respectively. The results of the quantization of DeiT-B, which are similar to those of the previous models, show that our method also achieves the best performance, with an improvement of 0.8% and 0.49% over Standard at W8/A8 and W4/A8 settings, respectively.

The proposed PSAQ-ViT still maintains a high level of robustness to the Swin models. When the quantized Swin-T is calibrated with Gaussian noise, the model performance becomes almost infeasible; nevertheless, our method can guarantee

Table 1. Quantization results on ImageNet dataset. Standard, Gaussian noise, and PSAQ-ViT calibrate the quantization parameters with real images, Gaussian noise, and generated images, respectively, and all other settings are the same. As we can see, PSAQ-ViT can even be more superior than real-data-driven Standard. Here, "No Data" indicates that no real data participate in the quantization process, "Prec." denotes the quantization precision, "Wx/Ay" denotes quantifying the weights and activations to x-bit and y-bit, respectively, and "Top-1" is the Top-1 test accuracy of the quantized vision transformer.

Model	Method	No data	Prec.	Top-1 (%)	Prec.	Top-1 (%)
ViT-S	Standard	×	W4/A8	19.91	W8/A8	30.28
(81.39)	Gaussian noise	✓	W4/A8	15.60	W8/A8	25.22
	PSAQ-ViT (ours)	✓	W4/A8	**20.84**	W8/A8	**31.45**
ViT-B	Standard	×	W4/A8	24.76	W8/A8	36.65
(84.53)	Gaussian noise	✓	W4/A8	19.45	W8/A8	31.63
	PSAQ-ViT (ours)	✓	W4/A8	**25.34**	W8/A8	**37.36**
DeiT-T	Standard	×	W4/A8	65.20	W8/A8	71.27
(72.21)	Gaussian noise	✓	W4/A8	7.80	W8/A8	10.55
	PSAQ-ViT (ours)	✓	W4/A8	**65.57**	W8/A8	**71.56**
DeiT-S	Standard	×	W4/A8	72.10	W8/A8	76.00
(79.85)	Gaussian noise	✓	W4/A8	13.30	W8/A8	18.16
	PSAQ-ViT (ours)	✓	W4/A8	**73.23**	W8/A8	**76.92**
DeiT-B	Standard	×	W4/A8	76.25	W8/A8	78.61
(81.85)	Gaussian noise	✓	W4/A8	11.09	W8/A8	14.72
	PSAQ-ViT (ours)	✓	W4/A8	**77.05**	W8/A8	**79.10**
Swin-T	Standard	×	W4/A8	70.16	W8/A8	74.22
(81.35)	Gaussian noise	✓	W4/A8	0.52	W8/A8	0.62
	PSAQ-ViT (ours)	✓	W4/A8	**71.79**	W8/A8	**75.35**
Swin-S	Standard	×	W4/A8	73.33	W8/A8	75.19
(83.20)	Gaussian noise	✓	W4/A8	5.43	W8/A8	5.66
	PSAQ-ViT (ours)	✓	W4/A8	**75.14**	W8/A8	**76.64**

a small performance degradation. In addition, PSAQ-ViT is very quantization-friendly for Swin-S, achieving substantial performance improvements over Standard, with gains of 1.81% at W4/A8 and 1.45% at W8/A8, respectively.

4.4   Results of Combining with Post-training Quantization

To demonstrate the generality of the proposed method, we evaluate the results of combining PSAQ-ViT with post-training quantization methods, which further improves the performance of quantization. Specifically, instead of using vanilla MinMax, we use EMA [17], Percentile [22], and OMSE [9] to determine the clipping values for activations. Among them, EMA employs a moving average mechanism to smooth the maximum and minimum values of the tensors; Percentile clips the tensors according to the percentile of the parameters (1e−5 percentile is used in the experiments); OMSE minimizes the quantization error to determine the tensors' clipping values. In addition, the experimental results are compared with PTQ-ViT [28], which is the state-of-the-art (SOTA) ranking-aware post-training quantization method for vision transformers. Note that PTQ-ViT has

Table 2. Quantization results of combining with post-training quantization methods on ImageNet dataset. Our PSAQ-ViT combined with simple post-training quantization methods, including EMA [17], Percentile [22], and OMSE [9], can achieve comparable performance to the SOTA ranking-aware post-training method [28] that has high computational complexity and requires the assistance of 1000 real images.

Model	Method	Strategy	No data	Prec.	Top-1 (%)
DeiT-S (79.85)	PTQ-ViT	Ranking-Aware	×	W8/A8	77.47
	PSAQ-ViT (ours)	Vanilla	✓	W8/A8	76.92
		EMA	✓	W8/A8	77.12
		Percentile	✓	W8/A8	**77.31**
		OMSE	✓	W8/A8	76.94
DeiT-B (81.85)	PTQ-ViT	Ranking-Aware	×	W8/A8	80.48
	PSAQ-ViT (ours)	Vanilla	✓	W8/A8	79.10
		EMA	✓	W8/A8	79.99
		Percentile	✓	W8/A8	79.42
		OMSE	✓	W8/A8	**80.26**

higher computational complexity and requires the assistance of 1000 real images, while our method requires only 32 generated images.

The quantization results are reported in Table 2. PSAQ-ViT, when combined with post-training quantization methods, can achieve comparable performance to PTQ-ViT. Meanwhile, different models have different preferences for different calibration strategies. For instance, PSAQ-ViT combined with Percentile shows the best performance on DeiT-S, while DeiT-B achieves the highest accuracy when using OMSE to calibrate the parameters.

4.5   Ablation Studies

We perform ablation studies on DeiT-S and DeiT-B models to demonstrate the effectiveness of different loss functions used for sample generation, and the results are shown in Table 3. We first analyze the experimental results of DeiT-S. Not using any loss function, *i.e.*, calibrating directly with Gaussian noise, certainly leads to an unexpected decrease in accuracy; when only $\mathcal{L}_{OH}$ and $\mathcal{L}_{TV}$ are used to optimize the noise, the accuracy of the quantized model is still far from satisfactory. Using only the proposed patch similarity entropy loss $\mathcal{L}_{PSE}$ can guarantee good quantization performance, and since it is completely decoupled from the other losses, it can be easily combined with them to achieve better results where it has the largest contribution to the final performance. A similar analysis also applies to DeiT-B. It is well demonstrated that our designed $\mathcal{L}_{PSE}$ has an essential driving effect on the quality improvement of the generated images, thus ensuring an effective calibration of the quantization parameters.

Table 3. Ablation study of different loss functions for sample generation. $\mathcal{L}_{PSE}$ has the largest contribution to the final results. In addition, it is fully decoupled from other losses, thus it can further improve performance in combination with other losses.

Model	$\mathcal{L}_{PSE}$	$\mathcal{L}_{OH}$	$\mathcal{L}_{TV}$	Prec	Top-1 (%)
DeiT-S (79.85)	×	×	×	W8/A8	18.16
	×	✓	✓	W8/A8	65.66
	✓	×	×	W8/A8	74.07
	✓	✓	×	W8/A8	75.39
	✓	×	✓	W8/A8	75.28
	✓	✓	✓	W8/A8	**76.92**
DeiT-B (81.85)	×	×	×	W8/A8	14.72
	×	✓	✓	W8/A8	67.95
	✓	×	×	W8/A8	78.07
	✓	✓	×	W8/A8	78.50
	✓	×	✓	W8/A8	78.61
	✓	✓	✓	W8/A8	**79.10**

Table 4. Efficiency analysis of PSAQ-ViT on DeiT-B, which spends less than 4 min on an RTX 3090 GPU, with the majority spent on image generation.

Model	Image generation (s)	Quantization calibration (s)			
DeiT-B	227	Vanilla	0.17	OMSE	0.41

We also perform efficiency analysis of the two stage of PSAQ-ViT divided in Algorithm 1, as shown in Table 4. The whole process takes less than 4 min on an RTX 3090 GPU and most time is spent in the image generation, since the parameter calibration without training produces small overhead.

5   Conclusions

In this paper, we propose PSAQ-ViT, a Patch Similarity Aware data-free Quantization framework for Vision Transformers. First, we perform an in-depth analysis of the unique properties of the self-attention module, revealing a general difference in its processing of Gaussian noise and real images. Based on this insight, we design a relative value metric to optimize the Gaussian noise to approximate the real image. Specifically, we use the differential entropy of patch similarity calculated via kernel density estimation to represent the diversity of the self-attention module's responses, then maximize the entropy to optimize the Gaussian noise, and finally utilize the generated "realistic" samples to efficiently calibrate the quantization parameters. Extensive experiments and ablation studies are conducted to demonstrate the effectiveness of PSAQ-ViT and the proposed patch similarity entropy loss. Thanks to the positive feedback effect of the generated images that are easily distinguished between foreground and background as analyzed in our paper, PSAQ-ViT can even outperform the real-data-driven methods at the same settings.

Acknowledgements. This work was supported in part by the National Natural Science Foundation of China under Grant 62276255; in part by the Scientific Instrument Developing Project of the Chinese Academy of Sciences under Grant YJKYYQ20200045.

References

1. Arnab, A., Dehghani, M., Heigold, G., Sun, C., Lučić, M., Schmid, C.: ViViT: a video vision transformer. In: Proceedings of the IEEE/CVF International Conference on Computer Vision, pp. 6836–6846 (2021)
2. Bengio, Y., Léonard, N., Courville, A.: Estimating or propagating gradients through stochastic neurons for conditional computation. arXiv preprint arXiv:1308.3432 (2013)
3. Cai, Y., Yao, Z., Dong, Z., Gholami, A., Mahoney, M.W., Keutzer, K.: ZeroQ: a novel zero shot quantization framework. In: Proceedings of the IEEE/CVF Conference on Computer Vision and Pattern Recognition, pp. 13169–13178 (2020)
4. Carion, N., Massa, F., Synnaeve, G., Usunier, N., Kirillov, A., Zagoruyko, S.: End-to-end object detection with transformers. In: Vedaldi, A., Bischof, H., Brox, T., Frahm, J.-M. (eds.) ECCV 2020. LNCS, vol. 12346, pp. 213–229. Springer, Cham (2020). https://doi.org/10.1007/978-3-030-58452-8_13
5. Chen, H., et al.: Pre-trained image processing transformer. In: Proceedings of the IEEE/CVF Conference on Computer Vision and Pattern Recognition, pp. 12299–12310 (2021)
6. Chen, X., Yan, B., Zhu, J., Wang, D., Yang, X., Lu, H.: Transformer tracking. In: Proceedings of the IEEE/CVF Conference on Computer Vision and Pattern Recognition, pp. 8126–8135 (2021)
7. Chin, T.-W., Chuang, P.I.-J., Chandra, V., Marculescu, D.: One weight bitwidth to rule them all. In: Bartoli, A., Fusiello, A. (eds.) ECCV 2020. LNCS, vol. 12539, pp. 85–103. Springer, Cham (2020). https://doi.org/10.1007/978-3-030-68238-5_7
8. Choi, J., Wang, Z., Venkataramani, S., Chuang, P.I.J., Srinivasan, V., Gopalakrishnan, K.: PACT: parameterized clipping activation for quantized neural networks. arXiv preprint arXiv:1805.06085 (2018)
9. Choukroun, Y., Kravchik, E., Yang, F., Kisilev, P.: Low-bit quantization of neural networks for efficient inference. In: 2019 IEEE/CVF International Conference on Computer Vision Workshop (ICCVW), pp. 3009–3018. IEEE (2019)
10. Dosovitskiy, A., et al.: An image is worth 16×16 words: transformers for image recognition at scale. arXiv preprint arXiv:2010.11929 (2020)
11. Elthakeb, A.T., Pilligundla, P., Mireshghallah, F., Elgindi, T., Deledalle, C.A., Esmaeilzadeh, H.: Gradient-based deep quantization of neural networks through sinusoidal adaptive regularization. arXiv preprint arXiv:2003.00146 (2020)
12. Esser, S.K., McKinstry, J.L., Bablani, D., Appuswamy, R., Modha, D.S.: Learned step size quantization. arXiv preprint arXiv:1902.08153 (2019)
13. Gholami, A., Kim, S., Dong, Z., Yao, Z., Mahoney, M.W., Keutzer, K.: A survey of quantization methods for efficient neural network inference. arXiv preprint arXiv:2103.13630 (2021)

14. Han, K., et al.: A survey on visual transformer. arXiv e-prints, p. arXiv-2012 (2020)
15. Han, K., Xiao, A., Wu, E., Guo, J., Xu, C., Wang, Y.: Transformer in transformer. Adv. Neural. Inf. Process. Syst. **34**, 15908–15919 (2021)
16. Hubara, I., Courbariaux, M., Soudry, D., El-Yaniv, R., Bengio, Y.: Binarized neural networks. Adv. Neural. Inf. Process. Syst. **29**, 1–9 (2016)
17. Jacob, B., et al.: Quantization and training of neural networks for efficient integer-arithmetic-only inference. In: Proceedings of the IEEE Conference on Computer Vision and Pattern Recognition, pp. 2704–2713 (2018)
18. Jia, D., et al.: Efficient vision transformers via fine-grained manifold distillation. arXiv preprint arXiv:2107.01378 (2021)
19. Khan, S., Naseer, M., Hayat, M., Zamir, S.W., Khan, F.S., Shah, M.: Transformers in vision: a survey. ACM Comput. Surv. (CSUR) **54**, 1–41 (2021)
20. Krishnamoorthi, R.: Quantizing deep convolutional networks for efficient inference: a whitepaper. arXiv preprint arXiv:1806.08342 (2018)
21. Krizhevsky, A., Sutskever, I., Hinton, G.E.: ImageNet classification with deep convolutional neural networks. Adv. Neural Inf. Process. Syst. **25**, 1106–1114 (2012)
22. Li, R., Wang, Y., Liang, F., Qin, H., Yan, J., Fan, R.: Fully quantized network for object detection. In: Proceedings of the IEEE/CVF Conference on Computer Vision and Pattern Recognition, pp. 2810–2819 (2019)
23. Li, Y., Dong, X., Wang, W.: Additive powers-of-two quantization: an efficient non-uniform discretization for neural networks. arXiv preprint arXiv:1909.13144 (2019)
24. Li, Y., et al.: BRECQ: pushing the limit of post-training quantization by block reconstruction. arXiv preprint arXiv:2102.05426 (2021)
25. Li, Z., Gu, Q.: I-ViT: integer-only quantization for efficient vision transformer inference. arXiv preprint arXiv:2207.01405 (2022)
26. Lin, Y., Zhang, T., Sun, P., Li, Z., Zhou, S.: FQ-VIT: fully quantized vision transformer without retraining. arXiv preprint arXiv:2111.13824 (2021)
27. Liu, Z., et al.: Swin transformer: hierarchical vision transformer using shifted windows. In: Proceedings of the IEEE/CVF International Conference on Computer Vision, pp. 10012–10022 (2021)
28. Liu, Z., Wang, Y., Han, K., Zhang, W., Ma, S., Gao, W.: Post-training quantization for vision transformer. Adv. Neural Inf. Process. Syst. **34**, 28092–28103 (2021)
29. Nagel, M., Amjad, R.A., Van Baalen, M., Louizos, C., Blankevoort, T.: Up or down? Adaptive rounding for post-training quantization. In: International Conference on Machine Learning, pp. 7197–7206. PMLR (2020)
30. Neimark, D., Bar, O., Zohar, M., Asselmann, D.: Video transformer network. In: Proceedings of the IEEE/CVF International Conference on Computer Vision, pp. 3163–3172 (2021)
31. Rastegari, M., Ordonez, V., Redmon, J., Farhadi, A.: XNOR-Net: ImageNet classification using binary convolutional neural networks. In: Leibe, B., Matas, J., Sebe, N., Welling, M. (eds.) ECCV 2016. LNCS, vol. 9908, pp. 525–542. Springer, Cham (2016). https://doi.org/10.1007/978-3-319-46493-0_32
32. Tang, Y., et al.: Patch slimming for efficient vision transformers. arXiv preprint arXiv:2106.02852 (2021)
33. Touvron, H., Cord, M., Douze, M., Massa, F., Sablayrolles, A., Jégou, H.: Training data-efficient image transformers and distillation through attention. In: International Conference on Machine Learning, pp. 10347–10357. PMLR (2021)
34. Wu, D., Tang, Q., Zhao, Y., Zhang, M., Fu, Y., Zhang, D.: EasyQuant: post-training quantization via scale optimization. arXiv preprint arXiv:2006.16669 (2020)

35. Wu, K., Peng, H., Chen, M., Fu, J., Chao, H.: Rethinking and improving relative position encoding for vision transformer. In: Proceedings of the IEEE/CVF International Conference on Computer Vision, pp. 10033–10041 (2021)
36. Xu, S., et al.: Generative low-bitwidth data free quantization. In: Vedaldi, A., Bischof, H., Brox, T., Frahm, J.-M. (eds.) ECCV 2020. LNCS, vol. 12357, pp. 1–17. Springer, Cham (2020). https://doi.org/10.1007/978-3-030-58610-2_1
37. Yin, H., et al.: Dreaming to distill: data-free knowledge transfer via deepinversion. In: Proceedings of the IEEE/CVF Conference on Computer Vision and Pattern Recognition, pp. 8715–8724 (2020)
38. Yuan, Z., Xue, C., Chen, Y., Wu, Q., Sun, G.: PTQ4ViT: post-training quantization framework for vision transformers. arXiv preprint arXiv:2111.12293 (2021)
39. Zhang, D., Yang, J., Ye, D., Hua, G.: LQ-Nets: learned quantization for highly accurate and compact deep neural networks. In: Proceedings of the European Conference on Computer Vision (ECCV), pp. 365–382 (2018)
40. Zhang, X., et al.: Diversifying sample generation for accurate data-free quantization. In: Proceedings of the IEEE/CVF Conference on Computer Vision and Pattern Recognition, pp. 15658–15667 (2021)
41. Zhong, Y., et al.: IntraQ: learning synthetic images with intra-class heterogeneity for zero-shot network quantization. arXiv preprint arXiv:2111.09136 (2021)
42. Zhou, D., et al.: DeepViT: towards deeper vision transformer. arXiv preprint arXiv:2103.11886 (2021)
43. Zhu, X., Su, W., Lu, L., Li, B., Wang, X., Dai, J.: Deformable DETR: deformable transformers for end-to-end object detection. arXiv preprint arXiv:2010.04159 (2020)

L3: Accelerator-Friendly Lossless Image Format for High-Resolution, High-Throughput DNN Training

Jonghyun Bae, Woohyeon Baek, Tae Jun Ham, and Jae W. Lee[✉]

Seoul National University, Seoul, Korea
{jonghbae,baneling100,jaewlee}@snu.ac.kr

Abstract. The training process of deep neural networks (DNNs) is usually pipelined with stages for data preparation on CPUs followed by gradient computation on accelerators like GPUs. In an ideal pipeline, the end-to-end training throughput is eventually limited by the throughput of the accelerator, not by that of data preparation. In the past, the DNN training pipeline achieved a near-optimal throughput by utilizing datasets encoded with a lightweight, lossy image format like JPEG. However, as high-resolution, *losslessly*-encoded datasets become more popular for applications requiring high accuracy, a performance problem arises in the data preparation stage due to low-throughput image decoding on the CPU. Thus, we propose L3, a custom lightweight, lossless image format for high-resolution, high-throughput DNN training. The decoding process of L3 is effectively parallelized on the accelerator, thus minimizing CPU intervention for data preparation during DNN training. L3 achieves a 9.29× higher data preparation throughput than PNG, the most popular *lossless* image format, for the Cityscapes dataset on NVIDIA A100 GPU, which leads to 1.71× higher end-to-end training throughput. Compared to JPEG and WebP, two popular *lossy* image formats, L3 provides up to 1.77× and 2.87× higher end-to-end training throughput for ImageNet, respectively, at equivalent metric performance.

Keywords: DNN training · Data preparation · Image processing

1 Introduction

The recent development of deep neural networks (DNNs) has been incited by the large-scale, publicly accessible datasets such as ImageNet [10], CIFAR [19], and SVHN [30]. These datasets are conventionally encoded using a lossy encoding format (e.g., JPEG). However, for emerging application domains requiring high accuracies, such as autonomous driving [6,13,26,58], image generation and restoration [22,23,52,53], denoising [45,55], and medical diagnosis [15,44,57],

Supplementary Information The online version contains supplementary material available at https://doi.org/10.1007/978-3-031-20083-0_11.

the use of lossy image format can potentially result in accuracy loss. Thus, the demands for *losslessly* encoded datasets continue to increase for more accurate pixel-wise segmentation and object representations.

Today's end-to-end DNN training pipeline is composed of the data preparation stage on the CPU, followed by the gradient computation stage on the accelerator (e.g., GPU, TPU). Since the next mini-batch data preparation stage overlaps with the current batch's gradient computation stage, the longest stage limits end-to-end training throughput. In the traditional setting of DNN training pipeline employing lossy-encoded datasets, the gradient computation stage is the primary bottleneck. Thus, most of the research has been conducted to improve the model training throughput on the accelerator by reducing inter-node communication overhead [16,29,41], optimizing GPU memory [25,46,50], and compiler optimization for DNN operators [5,24].

However, data preparation has recently become the critical performance bottleneck, especially for datasets with high-resolution, lossless images, which require a complex decoding process. According to a recent study [38], the throughput ratio of the gradient computation to the data preparation can be as high as $54.9\times$ even with well-optimized data preparation. Balancing the pipeline stages, in this case, would require $50\times$ more CPU cores. There are several recent proposals to address stalls in the data preparation, such as NVIDIA's Data Loading Library (DALI) [33] and TrainBox [38]. However, they only target lossy-encoded datasets with custom hardware support using a dedicated hardware JPEG decoder on NVIDIA A100 GPU [31] and an FPGA accelerator [38].

Thus, we propose L3, a new lightweight, lossless image format, whose decoding can be fully accelerated on data-parallel architectures such as GPUs. L3 eliminates the data preparation stalls by replacing the complex decoding algorithm running on the CPU to achieve the end-to-end training throughput close to the ideal pipeline with zero overhead for data preparation. For standalone execution L3 on NVIDIA A100 GPU delivers a $9.29\times$ higher decoding throughput for the Full HD (1920×1080) Cityscapes dataset than the lossless PNG format on Intel Xeon CPU while maintaining at most 9% loss in the compression ratio. Furthermore, L3 achieves a $1.25\times$ ($1.77\times$) and $1.93\times$ ($2.87\times$) higher geomean (maximum) training throughput than JPEG and lossy WebP, respectively, for seven state-of-the-art object detection and semantic segmentation models at equivalent metric performance.

2   Related Work

Optimized Data Preparation. There exist several recent proposals on improving the data preparation performance for DNN training. DIESEL [51] splits files into two parts—in-memory metadata and on-disk objects, to fully utilize the I/O bandwidth by reducing redundant metadata accesses. CoorDL [27] uses host memory as a cache for the storage and utilizes a new caching policy named MinIO, which does not replace the once-cached dataset elements.

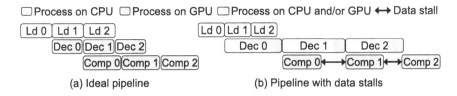

Fig. 1. (a) Ideal DNN training pipeline (b) DNN training Pipeline with data stalls (Ld: Load, Dec: Decode, Comp: Compute)

tf.data [28] proposes an optimized input data pipeline for TensorFlow with optimized parallel I/O, caching, and automatic resource (e.g., CPU, I/O) management to minimize the data preparation latency. These proposals reduce the latency of Load but do not address the Decode which can be the main bottleneck, especially when the input images employ lossless encoding. In contrast, L3 presents a new lossless, accelerator-friendly image format to provide superior Decode throughput, hence effectively eliminating the bottleneck in the Decode stage.

Optimized Lossless Image Decoding. Many proposals attempt to improve the decoding throughput of the lossless image format, such as Huffman decoding [47,54,56], bzip2 [39], and LZSS [36] by utilizing GPUs or specialized hardware. However, there are trade-offs between decoding throughput and compression ratio. So, achieving both high decoding throughput and high compression ratio simultaneously is challenging. Instead, Gompresso [48] utilizes a slightly modified LZ77 algorithm and partition-based Huffman coding to improve the compression/decompression throughput on GPUs. Similarly, Adaptive Loss-Less (ALL) data compression [12] exploits run-length and adaptive dictionary-based coding for GPUs, as well as a partition-based compression/decompression scheme to improve throughput. However, both research target general-purpose applications; hence, they are sub-optimal for DNN training. Instead, L3 is a more specialized format for DNN training to achieve high throughput while maintaining a competitive compression ratio.

3   Background and Motivation

3.1   DNN Training Pipeline

The DNN training operates as a three-stage pipeline: (1) Load image, (2) Decode image, and (3) Compute gradients. For each iteration, a single batch of images are loaded from a storage device to the host main memory (Load), and then those images are decoded with CPU and/or GPU [34,35] (Decode). These decoded images are transferred to the GPU device, and then a gradient computing iteration is performed (Compute). In most DNN frameworks such as PyTorch [43], TensorFlow [1], and MXNet [4], these three steps execute in a pipelined manner.

Since those three stages are pipelined, the overall training throughput is determined by the stage that takes the longest time among Load, Decode, and

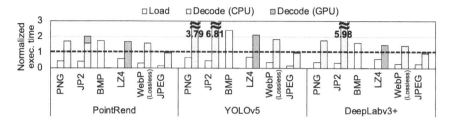

Fig. 2. Load and Decode execution time normalized to the Compute time (in the dotted line) for three representative models in Table 2 and six image formats.

Compute. Ideally, the Load and Decode time should be shorter than Compute time so that the time spent on Load and Decode is completely hidden, as shown in Fig. 1(a). However, as shown in Fig. 1(b), either Decode or Load stage may take longer to bottleneck the pipeline. Such cases are likely to happen when high-resolution images are decoded on CPU.

3.2 Data Preparation Bottleneck

Profiling Data Preparation Bottleneck. Figure 2 compares the Load and Decode time with the Compute time for two segmentation models (PointRend [18], DeepLabv3+ [3]) and one detection model (YOLOv5 [49]). The models run on a platform with Intel Xeon Platinum 8275CL CPU, NVIDIA A100 GPU with 40GB HBM2 DRAM, and 1TB NVMe SSD. The datasets and mini-batch size used for each model are described in Sect. 5.1. The graph shows that the Load and Decode time varies greatly across input image formats, even for the same model. For example, the PNG, JP2, and lossless WebP spend more time on Decode than Compute; this implies that the DNN training pipeline is bottlenecked by Decode. On the other hand, the BMP has a longer Load time than Compute time, indicating that the pipeline is bottlenecked by Load. This is because the uncompressed BMP data tends to be much larger than other compressed image formats. Among those measured, the JPEG is the only one that is not bottlenecked in data preparation (i.e., Load and Decode). However, JPEG is a lossy image format that loses some information in the original data.

Decoding Bottleneck for Lossless Image Formats. JPEG format requires much less time to decode as it utilizes GPU to perform most of the decoding process. Since modern high-end GPUs have substantially higher computation power than CPUs, this greatly helps to prevent Decode from being a bottleneck. However, much less attention has been paid to accelerating *lossless* image formats. NVIDIA proposes the use of LZ4-based format with its library to accelerate the decoding of LZ4-compressed images using GPU [32]; however, as shown in Fig. 2, its decoding speed is much slower than that of JPEG. Also, it is much more challenging to accelerate Decode of other lossless image formats. For example, it is well-known that PNG decoding, especially a process of decoding data

Table 1. Test set accuracy and its standard deviation for seven object detection and semantic segmentation models on PNG (Lossless), JPEG (Lossy), and WebP (Lossy) encoded datasets

Model	Lossless	Lossy	
	PNG (Stdev.)	JPEG (Stdev.)	WebP (Stdev.)
DDRNet23-slim (mIoU)	0.729 (0.001)	0.693 (0.003)	0.689 (0.001)
DeepLabv3+ (mIoU)	0.801 (0.004)	0.801 (0.007)	0.808 (0.009)
MaskFormer (mIoU)	0.685 (0.008)	0.669 (0.004)	0.651 (0.007)
PointRend (mIoU)	0.725 (0.001)	0.711 (0.003)	0.712 (0.002)
EgoNet (PCK@0.1)	0.323 (0.005)	0.300 (0.002)	0.305 (0.002)
PointPillar (mAP)	0.611 (0.006)	0.593 (0.004)	0.582 (0.009)
YOLOv5 (mAP@50)	0.633 (0.004)	0.605 (0.003)	0.614 (0.007)

compressed with dynamic Huffman coding, is difficult to parallelize and thus not well suited for GPU implementation [12,48,56].

3.3 Comparison of Lossless and Lossy Image Formats

Image formats can be lossy or lossless. A lossless image format (e.g., PNG, BMP, JPEG2000, lossless WebP) keeps all information in the raw image. In contrast, lossy image formats (e.g., JPEG, lossy WebP) often require less storage space but lose some information in the raw image. In the context of DNNs, high-resolution, lossless image formats are commonly utilized for domains where model accuracy is critical such as autonomous driving [6,13,26,58], image generation and restoration [22,23,52,53], denoising [45,55], and medical diagnosis [15,44,57]. Table 1 shows the impact of lossy image format on the test set accuracy and its standard deviation of seven object detection and semantic segmentation models, which are often deployed for autonomous driving. We use the default JPEG and lossy WebP quality factor of the Python Pillow package, which is 75. The reported accuracy number is an average of three training runs, and we observe negligible variations in the accuracy. Other than the image format, we use the same hyperparameters taken from the original publication without tuning for both lossless and lossy image formats. The use of the JPEG and lossy WebP results in degradation of the test set accuracy in all models except DeepLabv3+, compared to a case where the lossless PNG format is used for encoding.

4 L3 Design

4.1 Design Goal

L3 is a new image format specialized for a specific use case (i.e., ML/DL training) with a lightweight, lossless encoding/decoding algorithm. Our design goal is to

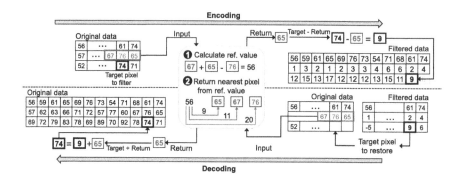

Fig. 3. Example of the custom Paeth filter in L3

eliminate the data preparation bottleneck by (i) maximizing decoding through-put by leveraging the GPU while (ii) providing a good-enough compression ratio not to introduce a new bottleneck in the Load stage. The conventional lossless image formats feature different tradeoffs. If the compression ratio is most impor-tant, WebP would be the best option. If compatibility is most important, PNG would be the one. Instead, we carefully design L3 to efficiently accelerate its decoding process on the GPU, unlike existing lossless image formats that are not suitable for GPU acceleration due to their limited parallelism. With L3, the training system can eliminate the CPU bottleneck to improve the DNN training throughput for datasets with lossless, high-resolution images.

4.2 Encoding/Decoding Algorithm

Encoding and decoding for L3 is a two-stage process: Paeth filtering followed by the row-wise base-delta encoding/decoding. Essentially, L3 utilizes a customized Paeth filter to significantly reduce the number of bits required to represent a pixel, exploiting the spatial redundancy (i.e., nearby pixels have similar val-ues). After Paeth filtering, L3 performs base-delta encoding/decoding to reduce further the number of bits representing each pixel and uses a packed represen-tation to store the image in a compressed format. The compressed image can be decoded by performing the reverse operation of each stage. In what follows, we describe each stage in greater detail.

Custom Paeth Filter. The Paeth filter [37], best known for its usage in PNG encoding, is a popular method to reduce the range of values for each pixel. The original Paeth filter encodes each pixel based on three neighboring pixels (left, top, top-left). However, this selection creates both *row-wise* dependency along the row dimension (left) as well as *column-wise* dependency (top) along the column dimension, thus making it challenging to exploit fine-grained (i.e., pixel-level) parallelism. Therefore, L3 customizes the Paeth filter to make it more amenable to parallel execution by inspecting a different set of neighboring pixels (top-left, top, top-right), thus eliminating column-wise data dependency.

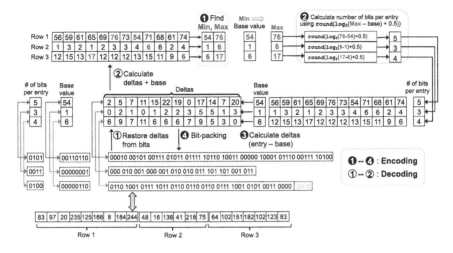

Fig. 4. Example of base-delta encoding/decoding

Figure 3 shows the encoding/decoding process of the custom Paeth filter in L3. In the rest of this paper, we refer to this as the Paeth filter for brevity. First, the Paeth filter calculates the *reference value* using those three neighboring pixels as follows: $Top\_Left + Top\_Right - Top$ (Step ❶). Then, among the three neighboring pixels, the filter selects the one whose value is the closest to the reference value (Step ❷). Finally, the difference between the original pixel value (i.e., 74 in the example) and the selected neighboring pixel value (i.e., 65 in the example) is stored. As an exception, the first row does not have a preceding row, so this row's filter operation is skipped.

The decoding process is similar to the encoding process. Since the first row is stored in a raw data format, we can start decoding from the second row towards the bottom row. The reference value can be computed by inspecting the three neighboring pixels in the preceding row to identify the pixel whose value is the closest to the reference. Then we add the stored residual (i.e., 9 in the example) to the pixel value (i.e., 65 in the example) to recover the original pixel value.

Base-Delta Encoding/Decoding. The outcome of the custom Paeth filter has a reduced value range. L3 applies the base-delta encoding [11,40] to each row to reduce the number of bits representing each pixel. Figure 4 illustrates this base-delta encoding and decoding process. Encoding is a four-step process. First, each row's minimum and maximum values are found (Step ❶). The minimum value is selected as the *base value* of each row. Also, the minimum number of bits required to cover all delta values is computed (Step ❷). Then, the deltas from the base value for all elements are calculated (Step ❸). Finally, a compressed bitstream is generated by representing each delta using the minimum number of bits (Step ❹). Specifically, the first four bits represent the number of bits per delta for each row, and the following eight bits the base value. Then, the deltas of the row, each represented using the minimum number of bits, are appended.

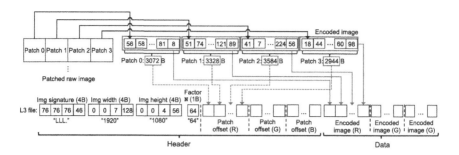

Fig. 5. Example of L3 file formatting

All the rows are concatenated to generate the final encoded stream, as shown at the bottom of Fig. 4.

The decoding process is quite simple. From the compressed stream, the decoder first reads the four bits as well as the following eight bits to identify the number of bits per entry and the base value for the row (Step ①). From this point, the decoder extracts the delta one by one and adds the base value to reconstruct the original value (Step ②). Once the first row is completed, the same process is repeated for the rest of the rows.

4.3 L3 File Format

The L3 encoding/decoding algorithm can work with any large-sized image. In practice, to maximize the parallelism and improve the decoding throughput in GPUs, L3 divides a large image into multiple square *patches* of size $N \times N$, where N is the user-specified parameter. We empirically set $N=32$ for images whose resolution is below 1080×720 (HD), $N=64$ for images whose resolution is between 1080×720 (HD) and 1920×1080 (FHD), and $N=128$ for images whose resolution is between 1920×1080 (FHD) and 3840×2160 (UHD). Once N is set, the image is first separated into three channels (R, G, and B). Then, the image is divided into square-sized patches for each channel, and the encoding algorithm is applied to each patch. The encoded patch is concatenated, and the offset for each patch is recorded and stored in a separate array.

Figure 5 shows the L3 file format. The file consists of the header and the data sections. The header section contains a 4-byte file type signature (also called magic bytes), 4-byte image width, 4-byte image height, 1-byte patch size, and three patch offset arrays corresponding to the three color channels. The data section contains encoded patch data for patches.

4.4 Optimizing L3 Decoder on GPU

Patch-Level Parallelism. Figure 6(a) illustrates patch-level parallelism in L3. Without parallelism, a single thread block, a unit of scheduling for GPU devices, decodes the entire image over multiple iterations, underutilizing GPU resources.

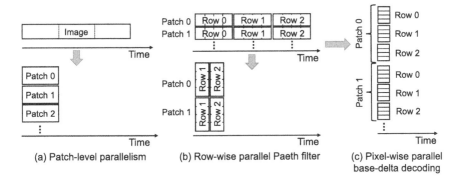

Fig. 6. Parallel execution example of (a) Patch-level parallelism (b) Row-wise parallel Paeth filter (c) Pixel-wise parallel base-delta decoding.

Instead, the L3 decoder first reads the header of each image and then splits it into multiple patches based on the header. Then, the task of decoding a single patch is assigned to a distinct thread block. This enables the GPU to process multiple patches in parallel to exploit the massive parallelism of GPU devices.

Row-Wise Parallel Paeth Filter. Even within a single patch, it is necessary for GPU devices to exploit fine-grained parallelism to maximize performance. For the Paeth filter, we implement the decoder to process a set of pixels within a single row in parallel. Figure 6(b) presents the row-wise parallel processing of the Paeth filter. Because of both row-wise and column-wise data dependencies between three neighbor pixels (Sect. 4.2), the original Paeth filter should decode the pixels in a single patch sequentially from top-left to bottom-right. Our use of the custom Paeth filter makes parallelization easier as it inspects three top pixels in the preceding row for decoding, instead of the one left pixel and two top pixels (i.e., top, top-left) as in the original Paeth filter. Starting from the second row, pixels within a single row are decoded in parallel (i.e., each CUDA thread processes a single pixel in a row). Once the decoding for a row is completed, the decoding for the next row within the same patch begins. This process is repeated for all rows in the patch.

Pixel-Wise Parallel Base-Delta Decoding. Figure 6(c) shows the pixel-level parallel base-delta decoding in L3. The original execution takes the same order with the original Paeth filter–it reconstructs pixels from the first pixel to the last one sequentially in the patch. The L3 decoder assigns a thread for base-delta decoding each pixel to exploit pixel-level parallelism within a row. Basically, each thread extracts the delta corresponding to the pixel and then adds the base value to reconstruct the original value.

Overlapping Decoding and Gradient Computing on GPU. To execute both computing and decoding concurrently on the GPU, we allocate both processes to separate *CUDA streams*. We prioritize the computing stream over the

Table 2. Detailed descriptions about models, datasets, and mini-batch sizes

Dataset (resolution) # of train/val/test	Model (backbone)	Batch size
Cityscapes (1920 × 1080) 2975/500/1525	DDRNet23-slim	36
	DeepLabv3+ (MobileNetv2)	48
	MaskFormer (ResNet50)	32
	PointRend (SemanticFPN)	40
KITTI (1024 × 720) 3519/3462/500	EgoNet	24
	PointPillars	48
	YOLOv5	128
DIV2K (2040 × 1200)	For compression ratio only	
FFHQ (5760 × 3840)	For compression ratio only	
RAISE-1K (4928 × 3264)	For compression ratio only	

decoding stream to prevent the computing (processing the current batch) from being interfered with by the decoding (processing the subsequent batch).

5 Evaluation

5.1 Methodology

Experimental Setup. We implement the decoder of L3 by extending NVIDIA Data Loading Library (DALI) [33] (version 1.6.0) by registering L3 as a new image format. We initiate the training for each model by running DALI-integrated PyTorch (version 1.9.0) [43]. For experiments, we use `p4d.24xlarge` AWS EC2 instance with 8 × NVIDIA A100 GPUs with 40 GB HBM2 DRAM per each, 96 vCPUs on Intel Xeon Platinum 8275CL CPU, and 8 × 1TB NVMe SSD.

Models and Datasets. We compare L3 with various lossless (PNG, JP2, BMP, LZ4, lossless WebP) and lossy (JPEG, lossy WebP) image formats and decoding method. Note that WebP support both lossless and lossy compression. For PNG, JP2, BMP, WebP, and JPEG, we use the implementation of the NVIDIA DALI framework [33]. For LZ4 decoding we use NVIDIA `nvcomp` library [32]. We use seven object detection and semantic segmentation models (DDR: DDRNet23-slim [14], DL: DeepLabv3+ [3], MF: MaskFormer [7], PR: PointRend [18], EN: EgoNet [21], PP: PointPillars [20], and YOLO: YOLOv5 [49]) and two datasets

Table 3. Test set accuracy of L3 (Lossless), JPEG (Lossy), and WebP (Lossy) and its quality factor

Model	L3 acc.	JPEG acc. (Q-factor)	WebP acc. (Q-factor)
DDR	0.729	0.730 (95)	0.726 (95)
DL	0.801	0.801 (75)	0.808 (75)
MF	0.685	0.686 (90)	0.685 (95)
PR	0.725	0.721 (85)	0.722 (85)
EN	0.323	0.323 (80)	0.324 (80)
PP	0.611	0.613 (85)	0.616 (95)
YOLO	0.633	0.635 (85)	0.635 (85)

Table 4. Data compression ratio (compressed size/decompressed size). Lower is better.

	Cityscapes	KITTI	DIV2K	FFHQ	RAISE1K	Random	Black
Raw (MB)	5.93	2.24	7.01	63.28	46.02	5.93	5.93
PNG	0.39×	0.62×	0.67×	0.25×	0.57×	1.01×	0.001×
JP2	0.38×	0.57×	0.59×	0.28×	0.56×	1.09×	0.005×
BMP	1.00×	1.00×	1.00×	1.00×	1.00×	1.00×	1.00×
LZ4	0.79×	0.85×	0.83×	0.44×	0.69×	1.01×	0.026×
WebP	0.29×	0.55×	0.49×	0.15×	0.37×	1.00×	0.000002×
L3	0.44×	0.64×	0.76×	0.33×	0.63×	1.02×	0.13×

(Cityscapes [8] and KITTI [13]) for throughput comparison. For the comparison of compression ratios of the lossless formats, we use three additional high-resolution datasets: DIV2K [2], FFHQ [17], and RAISE-1K [9]. Table 2 summarizes the model, dataset, and batch size used for training. The batch size is chosen to the maximum value for each model without causing an out-of-memory error. For the hyperparameters not shown in the table, we use the default values suggested in the papers or open-source implementations.

Quality Factor of Lossy Image Formats. As we discuss in Sect. 3.3, a lossy image format may yield a lower metric performance than the lossless image format at the default configuration. For fair comparison, we tune the quality factor (Q-factor) of a lossy image format such that it achieves an equivalent metric performance to the lossless image format. Specifically, we set the Q-factor to be the minimum value while its accuracy falls within 1% of that of the lossless image format. Table 3 reports the accuracy of L3, JPEG (Lossy), and WebP (Lossy) with the corresponding quality factors for the latter two. We use these values of Q-factor to compare the training throughput of JPEG and WebP (Lossy) with that of L3.

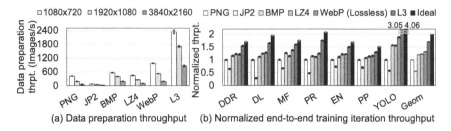

Fig. 7. Throughput comparison with lossless-encoded datasets (a) Data preparation (Load+Decode) throughput (b) Normalized end-to-end training iteration throughput

5.2 Compression Ratio

Table 4 reports the compression ratio of the lossless formats, including L3, where lower is better. For this experiment, we use five image datasets, as well as two additional synthetic images. The random image (Random) selects a random value for each pixel in the range of 0 through 255. The black image (Black) sets all pixel values to zero. Except for the black image, the compression ratio of L3 is worse than PNG and WebP (Lossless), falling within 6–9% of the PNG and 9–30% of the WebP (Lossless) format. However, we reiterate that the design objective of L3 is not to maximize the compression ratio but to balance the DNN training pipeline with more balanced resource utilization. In particular, L3 eliminates the data preparation bottleneck by (i) maximizing decoding throughput while (ii) providing a good-enough compression to not cause data stalls on the Load stage. L3's compression ratio is substantially worse in Black because L3 does not incorporate a recurring patterns compression (e.g., run-length encoding, Huffman coding). However, we find that containing many trivially repeated patterns is not common in lossless, high-resolution images.

5.3 Throughput Comparison with Lossless Decoders

Data Preparation Throughput. Figure 7(a) compares the data preparation throughput (Load+Decode) of L3 and other lossless decoding formats with varying resolutions. We utilize the FHD Cityscapes dataset and its scaled-down/up versions using the Lanczos filter in Python Pillow package, which provides the highest quality for image down/upscaling [42]. Compared to PNG, L3 improves the data preparation throughput by 5.67×, 9.29×, and 15.71× for HD, FHD, and UHD images, respectively. Furthermore, L3 outperforms WebP (Lossless), a state-of-the-art lossless image format, by a factor of 2.41×, 3.25×, and 4.51× for the same datasets. Thus, L3 achieves substantially higher data preparation throughput than all the other lossless formats, regardless of the resolution.

DNN Training Throughput. Figure 7(b) presents the normalized end-to-end training throughput (iterations/sec) of different image formats on various models. The training throughput is normalized to PNG. *Ideal* refers to the case

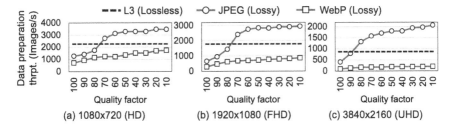

Fig. 8. Data preparation throughput of JPEG (Lossy) and WebP (Lossy)-encoded Cityscapes datasets on various quality factors. (a) 1080×720 (HD) (b) 1920×1080 (FHD) (c) 3840 × 2160 (UHD) resolution

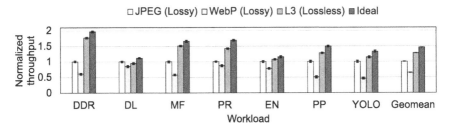

Fig. 9. Normalized training iteration throughput with JPEG (Lossy), WebP (Lossy), and L3 (Lossless)-encoded dataset

when the data preparation stage is completely hidden by the `Compute` stage, thus yielding zero overhead. The choice of the image format has a significant impact on the training throughput as the data preparation time can potentially bottleneck the DNN training pipeline.

Overall, L3 achieves the best end-to-end training throughput with a 1.71× geomean speedup compared to PNG. PNG and JP2 achieve lower throughput mainly because their complex algorithms running on CPU make them slow. The training throughput of BMP is limited by the I/O bandwidth (`Load`). LZ4 decoding utilizes GPUs, but its throughput is much lower than L3, achieving only 1.23× throughput speedup compared to PNG. WebP (Lossless) is the most recently proposed lossless image format, which achieves a 1.32× higher geomean training throughput than PNG. In contrast, L3 achieves a 1.71× higher throughput than PNG, which outperforms WebP (Lossless) by a significant margin.

5.4   Throughput Comparison with Lossy Decoders

Impact of Q-Factor on Data Preparation Throughput. Figure 8 shows the trade-off between quality factor and data preparation throughput of JPEG (Lossy) and WebP (Lossy). We use the scaled-down/up versions of Cityscapes dataset and convert it to JPEG and WebP format. The black dotted line marks the data preparation throughput of L3 (Lossless) for a given image resolution.

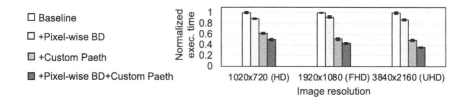

Fig. 10. Normalized L3 decoder execution time on various image resolution (`Baseline`: Seq. original Paeth filter+Seq. base-delta decoding, `+Pixel-wise BD`: Baseline+pixel-wise parallel base-delta decoding, `+Custom Paeth`: Baseline+row-wise parallel custom Paeth filter, `+Pixel-wise BD+Custom Paeth`: Pixel-wise parallel base-delta decoding+Row-wise parallel custom Paeth filter)

WebP (Lossy) decoding is done on CPU, whereas JPEG decoding on the dedicated hardware JPEG accelerator on NVIDIA A100 GPU using nvJPEG library. Even with the lowest quality factor (i.e., highest decoding throughput), WebP (Lossy) has a lower decoding throughput than L3 for all three image resolutions. This leads to a substantially lower end-to-end training throughput of WebP (Lossy) than L3.

In case of JPEG, there is a crossover point of the quality factor, where JPEG starts to outperform L3 in terms of decoding throughput. As we increase the image resolution from HD to UHD, L3 observes diminishing returns in throughput gains (in terms of pixels/sec) as the utilization of CUDA cores becomes high. JPEG performance scales a bit more gracefully to image resolution than L3 as the entire decoder runs on a custom JPEG hardware accelerator which is available only on NVIDIA A100 GPU. However, L3 can provide more robust performance on various data-parallel accelerators without requiring format-specific hardware support. In contrast, JPEG performance depends highly on the existence of custom hardware and enough CPU cores to sustain high throughput.

DNN Training Throughput. Figure 9 shows the end-to-end training iteration throughput for JPEG and WebP (Lossy) at equivalent test set accuracy. For comparable metric performance the quality factor of JPEG and WebP is set to the value in Table 3). The throughput is normalized to that of JPEG. Overall, L3 achieves up to 1.77× and 2.87× higher end-to-end training throughput than JPEG and WebP (Lossy), respectively. The geomean throughput gains are 1.25× for JPEG and 1.93× for WebP. Decreasing the quality factor may boost training throughput, but this comes with the cost of degrading the metric performance.

5.5 Ablation Study: Execution Time of L3 Decoder

L3 customizes the Paeth filter to make it more amenable to parallel execution by inspecting a different set of neighboring pixels (top-left, top, top-right), thus eliminating column-wise data dependency. Also, L3 unleashes pixel-level parallelism within the base-delta decoding step. Finally, L3 processes the decoding tasks in parallel with patch-level parallelism to exploit the massive parallelism of GPU devices. The patch-level parallelism reduces the decoding time by 76.8%,

83.9%, and 86.5% for HD, FHD, and UHD resolutions, respectively. We set the state in which patch-level parallelism is applied as baseline.

Figure 10 shows an ablation study to quantify the benefits of each component. The execution time is normalized by the baseline with sequential base-delta decoding and the original Paeth filter (Baseline). We use the FHD Cityscapes dataset and scale up/down the images with the Lanczos filter in Python Pillow package. The figure shows that pixel-wise parallel base-delta decoding (+Pixel-wise BD) and the custom Paeth filter (+Custom Paeth) reduces the decoding time by an average of 10.8% and 46.0%, respectively, over the baseline for the three image resolutions. With both optimizations applied (+Pixel-wise BD+Custom Paeth), the overall decoding time is reduced by 49.5%, 56.7%, and 59.1% from the Baseline for HD, FHD, and UHD images, respectively.

6   Conclusion

We propose L3, a new lightweight, lossless image format for high-resolution, high-throughput DNN training. The decoding algorithm of L3 is accelerated on data-parallel architectures such as GPUs. Thus, L3 yields much faster decoding than the existing lossless image formats whose decoders are mostly running on the CPU. L3 effectively eliminates data preparation bottlenecks in the DNN training pipeline. L3 can be readily deployed on GPU without requiring any support for specialized hardware. L3 can significantly reduce the end-to-end training time by providing higher throughput, higher accuracy, or both, compared to the existing lossless and lossy image formats at equivalent test set accuracy.

Acknowledgements. This work was supported by SNU-SK Hynix Solution Research Center (S3RC), and National Research Foundation of Korea (NRF) grant funded by Korea government (MSIT) (NRF-2020R1A2C3010663). The source code is available at https://github.com/SNU-ARC/L3.git.

References

1. Abadi, M., et al.: TensorFlow: a system for large-scale machine learning. In: Proceedings of the 12th USENIX Symposium on Operating Systems Design and Implementation, pp. 265–283. USENIX Association (2016)
2. Agustsson, E., Timofte, R.: NTIRE 2017 challenge on single image super-resolution: dataset and study. In: Proceedings of the IEEE Conference on Computer Vision and Pattern Recognition Workshops (2017)
3. Chen, L.-C., Zhu, Y., Papandreou, G., Schroff, F., Adam, H.: Encoder-decoder with atrous separable convolution for semantic image segmentation. In: Ferrari, V., Hebert, M., Sminchisescu, C., Weiss, Y. (eds.) ECCV 2018. LNCS, vol. 11211, pp. 833–851. Springer, Cham (2018). https://doi.org/10.1007/978-3-030-01234-2_49
4. Chen, T., et al.: MXNet: a flexible and efficient machine learning library for heterogeneous distributed systems. arXiv preprint arXiv:1512.01274 (2015)
5. Chen, T., et al.: TVM: an automated end-to-end optimizing compiler for deep learning. In: Proceedings of the 13th USENIX Symposium on Operating Systems Design and Implementation, pp. 578–594. USENIX Association (2018)

6. Chen, X., Ma, H., Wan, J., Li, B., Xia, T.: Multi-view 3D object detection network for autonomous driving. In: Proceedings of the IEEE Conference on Computer Vision and Pattern Recognition (2017)

7. Cheng, B., Schwing, A., Kirillov, A.: Per-pixel classification is not all you need for semantic segmentation. In: Proceedings of the Advances in Neural Information Processing Systems, vol. 34, pp. 17864–17875. Curran Associates, Inc. (2021)

8. Cordts, M., et al.: The cityscapes dataset for semantic urban scene understanding. In: Proceedings of the IEEE Conference on Computer Vision and Pattern Recognition (2016)

9. Dang-Nguyen, D.T., Pasquini, C., Conotter, V., Boato, G.: RAISE: a raw images dataset for digital image forensics. In: Proceedings of the 6th ACM Multimedia Systems Conference, pp. 219–224. Association for Computing Machinery (2015)

10. Deng, J., Dong, W., Socher, R., Li, L.J., Li, K., Fei-fei, L.: ImageNet: a large-scale hierarchical image database. In: Proceedings of the IEEE Conference on Computer Vision and Pattern Recognition. IEEE (2009)

11. Farrens, M., Park, A.: Dynamic base register caching: a technique for reducing address bus width. In: Proceedings of the 18th Annual International Symposium on Computer Architecture, pp. 128–137. Association for Computing Machinery (1991)

12. Funasaka, S., Nakano, K., Ito, Y.: Adaptive loss-less data compression method optimized for GPU decompression. Concurrency Comput. Pract. Experience **29**(24), e4283 (2017)

13. Geiger, A., Lenz, P., Urtasun, R.: Are we ready for autonomous driving? The KITTI vision benchmark suite. In: Proceedings of the IEEE Conference on Computer Vision and Pattern Recognition (2012)

14. Hong, Y., Pan, H., Sun, W., Jia, Y.: Deep dual-resolution networks for real-time and accurate semantic segmentation of road scenes. arXiv preprint arXiv:2101.06085 (2021)

15. Hou, L., et al.: High resolution medical image analysis with spatial partitioning. arXiv preprint arXiv:1909.03108 (2019)

16. Huang, Y., et al.: GPipe: efficient training of giant neural networks using pipeline parallelism. In: Proceedings of the Advances in Neural Information Processing Systems, vol. 32. Curran Associates, Inc. (2019)

17. Karras, T., Laine, S., Aila, T.: A style-based generator architecture for generative adversarial networks. In: Proceedings of the IEEE/CVF Conference on Computer Vision and Pattern Recognition (2019)

18. Kirillov, A., Wu, Y., He, K., Girshick, R.: PointRend: image segmentation as rendering. In: Proceedings of the IEEE/CVF Conference on Computer Vision and Pattern Recognition (2020)

19. Krizhevsky, A., Hinton, G.: Learning multiple layers of features from tiny images. Technical report. Citeseer (2009)

20. Lang, A.H., Vora, S., Caesar, H., Zhou, L., Yang, J., Beijbom, O.: PointPillars: fast encoders for object detection from point clouds. In: Proceedings of the IEEE/CVF Conference on Computer Vision and Pattern Recognition (2019)

21. Li, S., Yan, Z., Li, H., Cheng, K.T.: Exploring intermediate representation for monocular vehicle pose estimation. In: Proceedings of the IEEE/CVF Conference on Computer Vision and Pattern Recognition, pp. 1873–1883 (2021)

22. Liang, J., Cao, J., Sun, G., Zhang, K., Van Gool, L., Timofte, R.: SwinIR: image restoration using swin transformer. In: Proceedings of the IEEE/CVF International Conference on Computer Vision Workshops, pp. 1833–1844 (2021)

23. Lim, B., Son, S., Kim, H., Nah, S., Mu Lee, K.: Enhanced deep residual networks for single image super-resolution. In: Proceedings of the IEEE Conference on Computer Vision and Pattern Recognition Workshops, pp. 136–144 (2017)
24. Ma, L., et al.: Rammer: enabling holistic deep learning compiler optimizations with rTasks. In: Proceedings of the 14th USENIX Symposium on Operating Systems Design and Implementation, pp. 881–897. USENIX Association (2020)
25. Markthub, P., Belviranli, M.E., Lee, S., Vetter, J.S., Matsuoka, S.: DRAGON: breaking GPU memory capacity limits with direct NVM access. In: Proceedings of the International Conference for High Performance Computing, Networking, Storage, and Analysis, pp. 32:1–32:13. IEEE (2018)
26. Menze, M., Geiger, A.: Object scene flow for autonomous vehicles. In: Proceedings of the IEEE Conference on Computer Vision and Pattern Recognition (2015)
27. Mohan, J., Phanishayee, A., Raniwala, A., Chidambaram, V.: Analyzing and mitigating data stalls in DNN training. Proc. VLDB Endowment **14**(5), 771–784 (2021)
28. Murray, D.G., Simsa, J., Klimovic, A., Indyk, I.: tf.data: a machine learning data processing framework. Proceedings of the VLDB Endowment. **14**(12), 2945–2958 (2021)
29. Narayanan, D., et al.: PipeDream: generalized pipeline parallelism for DNN training. In: Proceedings of the 27th ACM Symposium on Operating Systems Principles, pp. 1–15. Association for Computing Machinery (2019)
30. Netzer, Y., Wang, T., Coates, A., Bissacco, A., Wu, B., Ng, A.Y.: Reading digits in natural images with unsupervised feature learning. In: Proceedings of the NIPS Workshop on Deep Learning and Unsupervised Feature Learning (2011)
31. NVIDIA: NVIDIA A100 tensor core GPU architecture (2020). https://images.nvidia.com/aem-dam/en-zz/Solutions/data-center/nvidia-ampere-architecture-whitepaper.pdf
32. NVIDIA: nvcomp: a library for fast lossless compression/decompression on the GPU (2021). https://github.com/NVIDIA/nvcomp
33. NVIDIA: the NVIDIA data loading library (DALI) (2021). https://github.com/NVIDIA/DALI
34. NVIDIA: nvJPEG libraries: GPU-accelerated JPEG decoder, encoder and transcoder (2021). https://developer.nvidia.com/nvjpeg
35. NVIDIA: nvJPEG2000 libraries (2021). https://docs.nvidia.com/cuda/nvjpeg2000
36. Ozsoy, A., Swany, M.: CULZSS: LZSS lossless data compression on CUDA. In: Proceedings of the 2011 IEEE International Conference on Cluster Computing, pp. 403–411 (2011)
37. Paeth, A.W.: II.9 - image file compression made easy. In: Graphics Gems II, pp. 93–100. Morgan Kaufmann (1991)
38. Park, P., Jeong, H., Kim, J.: TrainBox: an extreme-scale neural network training server architecture by systematically balancing operations. In: Proceedings of the 2020 53rd Annual IEEE/ACM International Symposium on Microarchitecture, pp. 825–838 (2020)
39. Patel, R.A., Zhang, Y., Mak, J., Davidson, A., Owens, J.D.: Parallel lossless data compression on the GPU. In: Proceedings of the 2012 Innovative Parallel Computing, pp. 1–9 (2012)
40. Pekhimenko, G., Seshadri, V., Mutlu, O., Gibbons, P.B., Kozuch, M.A., Mowry, T.C.: Base-delta-immediate compression: practical data compression for on-chip caches. In: Proceedings of the 21st International Conference on Parallel Architectures and Compilation Techniques, pp. 377–388. Association for Computing Machinery (2012)

41. Peng, Y., et al.: A generic communication scheduler for distributed DNN training acceleration. In: Proceedings of the 27th ACM Symposium on Operating Systems Principles, pp. 16–29. Association for Computing Machinery (2019)
42. Pillow: python pillow filters (2021). https://pillow.readthedocs.io/en/stable/handbook/concepts.html#filters
43. PyTorch: pyTorch (2021). https://pytorch.org
44. Rebsamen, M., Suter, Y., Wiest, R., Reyes, M., Rummel, C.: Brain morphometry estimation: from hours to seconds using deep learning. Front. Neurol. **11**, 244 (2020)
45. Ren, C., He, X., Wang, C., Zhao, Z.: Adaptive consistency prior based deep network for image denoising. In: Proceedings of the IEEE/CVF Conference on Computer Vision and Pattern Recognition, pp. 8596–8606 (2021)
46. Rhu, M., Gimelshein, N., Clemons, J., Zulfiqar, A., Keckler, S.W.: vDNN: virtualized deep neural networks for scalable, memory-efficient neural network design. In: Proceedings of the 49th Annual IEEE/ACM International Symposium on Microarchitecture, pp. 18:1–18:13. IEEE (2016)
47. Sarangi, S., Baas, B.: Canonical huffman decoder on fine-grain many-core processor arrays. In: Proceedings of the 2021 26th Asia and South Pacific Design Automation Conference, pp. 512–517 (2021)
48. Sitaridi, E., Mueller, R., Kaldewey, T., Lohman, G., Ross, K.A.: Massively-parallel lossless data decompression. In: Proceedings of the 2016 45th International Conference on Parallel Processing, pp. 242–247 (2016)
49. Ultralytics: Yolov5 (2021). https://github.com/ultralytics/yolov5/
50. Wang, L., et al.: SuperNeurons: dynamic GPU memory management for training deep neural networks. In: Proceedings of the 23rd ACM SIGPLAN Symposium on Principles and Practice of Parallel Programming, pp. 41–53. ACM (2018)
51. Wang, L., et al.: DIESEL: a dataset-based distributed storage and caching system for large-scale deep learning training. In: Proceedings of the 49th International Conference on Parallel Processing. Association for Computing Machinery (2020)
52. Wang, X., et al.: ESRGAN: enhanced super-resolution generative adversarial networks. In: Proceedings of the European Conference on Computer Vision Workshops (2018)
53. Wang, Z., Cun, X., Bao, J., Zhou, W., Liu, J., Li, H.: Uformer: a general u-shaped transformer for image restoration. In: Proceedings of the IEEE/CVF Conference on Computer Vision and Pattern Recognition, pp. 17683–17693 (2022)
54. Weißenberger, A., Schmidt, B.: Massively parallel huffman decoding on GPUs. In: Proceedings of the 47th International Conference on Parallel Processing. Association for Computing Machinery (2018)
55. Xu, L., Zhang, J., Cheng, X., Zhang, F., Wei, X., Ren, J.: Efficient deep image denoising via class specific convolution. In: Proceedings of the AAAI Conference on Artificial Intelligence, vol. 35, no. 4, pp. 3039–3046 (2021)
56. Yamamoto, N., Nakano, K., Ito, Y., Takafuji, D., Kasagi, A., Tabaru, T.: Huffman coding with gap arrays for GPU acceleration. In: Proceedings of the 49th International Conference on Parallel Processing. Association for Computing Machinery (2020)
57. Zhou, S., Nie, D., Adeli, E., Yin, J., Lian, J., Shen, D.: High-resolution encoder-decoder networks for low-contrast medical image segmentation. IEEE Trans. Image Process. **29**, 461–475 (2020)
58. Zhou, Y., Tuzel, O.: VoxelNet: end-to-end learning for point cloud based 3D object detection. In: Proceedings of the IEEE Conference on Computer Vision and Pattern Recognition (2018)

Streaming Multiscale Deep Equilibrium Models

Can Ufuk Ertenli$^{(\boxtimes)}$, Emre Akbas , and Ramazan Gokberk Cinbis

Department of Computer Engineering, Middle East Technical University (METU),
Ankara, Turkey
{ufuk.ertenli,eakbas,gcinbis}@metu.edu.tr

Abstract. We present StreamDEQ, a method that infers frame-wise representations on videos with minimal per-frame computation. In contrast to conventional methods where compute time grows at least linearly with the network depth, we aim to update the representations in a continuous manner. For this purpose, we leverage the recently emerging implicit layer models, which infer the representation of an image by solving a fixed-point problem. Our main insight is to leverage the slowly changing nature of videos and use the previous frame representation as an initial condition on each frame. This scheme effectively recycles the recent inference computations and greatly reduces the needed processing time. Through extensive experimental analysis, we show that StreamDEQ is able to recover near-optimal representations in a few frames time, and maintain an up-to-date representation throughout the video duration. Our experiments on video semantic segmentation and video object detection show that StreamDEQ achieves on par accuracy with the baseline (standard MDEQ) while being more than 3× faster. Code and additional results are available at https://ufukertenli.github.io/streamdeq/.

Keywords: Implicit layer models · Video analysis and understanding · Video object detection · Video semantic segmentation

1 Introduction

Modern convolutional deep networks excel at numerous recognition tasks. It is commonly observed that deeper models tend to outperform their shallower counterparts [21,25,48], e.g. the prediction quality tends to increase with network depth using the architectures with residual connections [21]. Due to the sequential nature of the layer-wise calculations, however, increasing the network depth results in longer inference times. While the increase in inference duration can

---

E. Akbas and R. G. Cinbis—Equal contribution for senior authorship.

Supplementary Information The online version contains supplementary material available at https://doi.org/10.1007/978-3-031-20083-0_12.

S. Avidan et al. (Eds.): ECCV 2022, LNCS 13671, pp. 189–205, 2022.
https://doi.org/10.1007/978-3-031-20083-0_12

be acceptable for various offline recognition problems, it is typically of concern for many streaming video analysis tasks. For example, in perception modules of autonomous systems, it is not only necessary to keep up with the frame rate but also desirable to minimize the computational burden of each recognition component to reduce the hardware requirements and/or save resources for additional tasks. Similar concerns arise in large-scale video analysis tasks, e.g. on video sharing platforms, a small increase in per-frame calculations can add up to great increments in total consumption.

Various techniques have been proposed to speed up the inference in deep networks. A widely studied idea is to apply a large model to selected *key-frames* and then either interpolate its features to the intermediate frames [51,53] or apply a smaller model to them [35,47]. However, such approaches come with several potential complications: (i) each time the larger model is applied, the model lags behind, the handling of which demands a complicated system design. (ii) Most methods require optical flow or motion estimates [51,53], which brings in an additional estimation problem and an additional point of failure. In addition, the time cost of the flow estimation naturally reduces the time budget for all dependent steps. (iii) Special techniques need to be developed to maintain the compatibility of the representations and/or confidence scores obtained across the key and intermediate frames. (iv) The training schemes tend to be complicated due to the need for training over video mini-batches. It is also noteworthy that several models, e.g. [29,45], rely on forward **and** backward flow estimates, making them less suitable for streaming recognition problems due to non-causal processing.

A related approach is to select a subset of each frame to process. These methods typically aim to identify the most informative regions in the input [1,14, 38]. For static images, the region selection process can continue until the model becomes confident about its predictions. However, when applied to videos, such subset selection strategies share shortcomings similar to approaches relying on flow-based intra-frame prediction approximations. The inputs change over time, therefore, the models have to choose between relying on optical flow to warp the rest of the features or to omit them entirely, which may result in obsolete representations over time [51].

In this context, the recently introduced *implicit layer models*, pioneered by the work on Deep Equilibrium Model (DEQ) [2] and Multiscale Deep Equilibrium Model (MDEQ) [4], offer a fundamentally different alternative to deep neural networks. DEQ (and MDEQ) shows that by using the fixed-points of a network as the representation, one can gain the representation power of deep models, using a network with only a few layers. The potential of DEQ to eliminate long chains of computations over network layers, therefore, renders it an attractive candidate towards building efficient streaming recognition models.

However, while DEQ provides a way to learn deep representations using shallow networks, the test-time inference process involves iterative fixed-point estimation algorithms, such as root finding or fixed-point iterations. Since each iteration can be interpreted as an increment in the network depth, DEQ effec-

Fig. 1. Our method, StreamDEQ, exploits the temporal smoothness between successive frames and extracts features via a small number of solver iterations, starting from the previous frame's representation as initial solution. StreamDEQ accumulates and transfers the extracted information continuously over successive frames; effectively sharing computations across video frames in a causal manner

tively constructs deep networks for inference, and therefore, can still suffer from the run-time costs as in explicit deep networks.

Our main insight is the potential to speed up the inference process, *i.e.* fixed-point estimation, by exploiting the temporal smoothness across neighboring frames in videos.[1] More specifically, we observe that the fully estimated MDEQ representation of a frame can be used for obtaining the approximate representations of the following frames, using only a few inference iterations. We further develop the idea, and show that even without fully estimating the representation at any time step, the implicit layer representation can be kept up-to-date by running the inference iterations over the iteration steps *and* video time steps, in a continuous manner. The final scheme, starting from scratch, accumulates and transfers the extracted information throughout the video duration. We, therefore, refer to the proposed method as Streaming DEQ, or StreamDEQ for short.

The main difference between standard DEQ and StreamDEQ is illustrated in Fig. 1. While DEQ typically requires a large number of inference iterations, StreamDEQ enables inference with only a few iterations per frame by leveraging the relevance of the most recent frame's representation. On the start of a new stream, or after a major content change (e.g. a shot change), StreamDEQ quickly adapts to the video in a few frames, much like a person adapting her/his focus and attention when watching a new video. In the following frames, it continuously updates the representation to adapt to minor changes (e.g. objects moving, entering or exiting the scene).

Overall, StreamDEQ provides a simple and *lean* solution to the streaming recognition with implicit layer models, where a single model naturally performs cost-effective recognition, without relying on external inputs and heuristics, such as optical flow [22], post-processing methods (Seq-NMS [20] or tubelet re-scoring [29,30]). Our method also maintains the causality of the system, and executes in a continuous manner. We also note that it allows dynamic time budgeting; the duration of the inference process can be tuned on-the-fly by a

---

[1] We do not refer to a mathematical definition of smoothness, but rather emphasize that the changes between neighboring frames are small.

controller, depending on the instantaneous compute system load, which can be a desirable feature in real-world scenarios.

We verify the effectiveness of the proposed method through extensive experiments on video semantic segmentation and video object detection. Our experimental results show that StreamDEQ recovers near-optimal representations at much lower inference costs. More specifically, on the ImageNet-VID video object detection task, StreamDEQ converges to the mAP scores of 50.4 and 54.8 using only 4 and 8 inference iterations per frame, respectively. In comparison, the standard DEQ inference scheme yields only 8.2 and 32.6 mAP scores using 4 and 8 iterations, respectively. Similarly, on the Cityscapes semantic segmentation task, using StreamDEQ instead of the standard DEQ inference scheme improves the converged streaming mIoU score from 42.3 to 71.5 when 4 inference iterations are used per frame, and from 73.2 to 78.2 when 8 iterations are used per frame.

2   Related Work

Here, we summarize efficient video processing methods, video object detection and segmentation models. Furthermore, we discuss saliency based techniques for video processing. Finally, we give an overview of implicit models (DEQs).

Efficient Video Processing and Inference. There have been many efforts to improve video processing efficiency to reach real-time processing speeds. Most of these works take a system-oriented approach [10,31,39]. For example, Carreira et al. [10] develop an efficient parallelization scheme over multiple GPUs and process different parts of a model in separate GPUs to improve efficiency while sacrificing accuracy due to frame delays. Narayanan et al. [39] propose a novel scheduling mechanism that efficiently schedules and divides forward and backward passes over multiple GPUs. In another work, Li et al. [31] use a dynamic scheduler in which the model chooses to skip a frame when the delays build up to the point where it would be impossible to calculate the results of the next frame in the allotted time.

We also note that works on low-cost network designs, such as MobileNets [23, 43] and low-resolution networks [34,49], are also relevant. Such efforts are valuable especially for replacing network components with more compute-friendly counterparts. However, the advantages of such techniques can also be limited due to natural trade-offs between speed and performance [50] as the lower-cost network components tend to have lower expressive power. Nevertheless, one can easily incorporate low-cost model design principles into DEQ or StreamDEQ models, thanks to the architecture-agnostic definition of implicit layer models. While such efforts may bring reductions in inference wall-clock time, they are outside the scope of our work.

Video Semantic Segmentation. Semantic segmentation is a costly, spatially dense prediction task. Its application to videos remains relatively limited. To reduce the computational cost, most works rely on exploiting temporal relations between frames using methods such as feature warping [19,26,28,47], feature propagation [32,44], feature aggregation [24], and knowledge distillation [36].

Gadde et al. [19] propose warping features of the previous frame at different depths based on optical flow. Xu et al. [47] evaluate regions of the input frame and decide whether to warp the features with a cheap flow network or use the large segmentation model based on a confidence score. Huang et al. [26] keep a moving average over time by combining the segmentation maps from the current frame with the warped map from the previous frame. Jain et al. [28] warp high-quality features from the previous key-frame and fuse them with lower quality features calculated on the current frame to make predictions.

Shelhamer et al. [44] propose an adaptive method that schedules updates to the multi-level feature map so that features of layers with smaller changes are carried forward (without any transformation). Li et al. [32] introduce an adaptive key-frame scheduling method based on the deviation of low-level features compared to the previous key-frame and if the deviation is small, the features are propagated with spatially variant convolution.

Hu et al. [24] use a set of shallow networks, each calculating features of consecutive frames starting from scratch. Then, these features are aggregated at the current frame with an attention-based module. Liu et al. [36] propose to use an expensive network during training including optical flow and applies knowledge distillation on a student network to benefit from the high capacity of a teacher network while cutting computational costs thanks to a smaller and more efficient student network which the authors use during inference.

In contrast to all these approaches, the proposed StreamDEQ method directly leverages the similarities across video frames, without requiring any ad-hoc video handling strategies, as a way to adapt the implicit layer inference mechanism to efficient streaming video analysis.

Video Object Detection. Most modern video object detection methods exploit temporal information to improve the accuracy and/or efficiency. To this end, optical flow [29,45,52,53], feature aggregation [8,12,46,52] and post-processing techniques [20,30] are prominently used.

Zhu et al. [53] use optical flow estimates to warp features on selected key-frames to intermediate frames for increased efficiency. Zhu et al. [52] also propose FGFA that uses optical flow to warp features of nearby frames to the current frame and aggregates these features adaptively based on a weight calculated from feature similarity. Kang et al. [29] create links between objects through time (tubelets) from the predictions calculated with optical flow across a video linking objects through time and apply tubelet re-scoring to keep detections of high-confidence. Wang et al. [45] adds an instance level calibration module to FGFA [52] and combines them to generate better predictions.

Bertasius et al. [8] sample features from neighboring support frames via deformable convolution that learns object offsets between frames and aggregates these features over these frames. Wu et al. [46] focus on linking object proposals in a video according to their semantic similarities. Chen et al. [12] propose a model that aggregates local and global information with a long-range memory.

Another common way to improve performance is to apply a post-processing method. For example, Han et al. [20] introduce Seq-NMS to exploit temporal consistency by constructing a temporal graph to link objects in adjacent frames. With a similar idea, Kang et al. [30] generate tubelets by combining single image detections through the video and use a tracker to re-scored the tubelets as a post-processing to improve temporal consistency.

Saliency Based Techniques. To reduce computational cost, another viable approach is to select important regions in an image and process only those small patches [1,14,37,38]. Video extensions of these models also exist [7,18,51].

Mnih et al. [38] and Ba et al. [1] model human eye movements by capturing *glimpses* from images with a recurrent structure and process those glimpses at each step. Cordonnier et al. [14] propose selecting most important regions to process by first processing a downsampled version of the image. Liu et al. [37] stops processing for regions with high-confidence predictions at an earlier stage.

Bazzani et al. [7] and Denil et al. [18] approach video processing in a human-like manner where the model *looks at* a different patch around the objects of interest at each frame and tracks them. Zhu et al. [51] takes a key-frame based approach. At each key-frame they process the full input and at intermediate frames, they update the feature maps partially based on temporal consistency.

Implicit Layer Models. Implicit layer models have seen a recent surge of interest and have been successful at numerous tasks. DEQs [2] are a recent addition to the implicit model family aimed at solving sequence modeling tasks. DEQs pose a fixed-point solving problem as a root finding problem and utilize an iterative root finding algorithm to find a solution. Multiscale Deep Equilibrium Models (MDEQ) [4] are the extension of the base DEQ to image-based models. Bai et al. [6] propose adding a Jacobian regularization term to improve model training.

In addition, Huang et al. [27] propose re-using the fixed-point across training iterations with the drawback of having to stay in full-batch mode for the training. Also, Bai et al. [5] suggests a new initialization scheme that is realized through a small network. Furthermore, inferring information from the last few iterations reduces the number of solver iterations required for convergence.

3 Proposed Method

We build our method on the Deep Equilibrium Model (DEQ). In this section, we first give an overview of DEQ and then present the details of our method.

3.1   DEQ Overview

Weight-tied networks are models where some or all layers share the same weights [3,16]. A DEQ is essentially a weight-tied network with only one shallow block. DEQ leverages the fact that continuously applying the same layer to its output tends to guide the output to an equilibrium point, *i.e.* a fixed-point. Let $\mathbf{x}$ represent the model's input, $\mathbf{z}^*$ the equilibrium point and f_θ the applied shallow block, then a DEQ can be described as

$$\lim_{i \to \infty} \mathbf{z}^{[i+1]} = \lim_{i \to \infty} f_\theta(\mathbf{z}^{[i]}; \mathbf{x}) \equiv f_\theta(\mathbf{z}^*; \mathbf{x}) = \mathbf{z}^*. \tag{1}$$

Since the *depth* of processing is obtained through repeatedly applying the same layer(s), these models are also called *implicit deep models*. DEQ's fundamental difference from a standard weight-tied model is that the fixed-point is found by root finding algorithms in both forward and backward passes as in Eq. 2:

$$g_\theta(\mathbf{z}; \mathbf{x}) = f_\theta(\mathbf{z}; \mathbf{x}) - \mathbf{z} = 0 \implies \mathbf{z}^* = \text{RootFind}(g_\theta; \mathbf{x}). \tag{2}$$

DEQ uses the Broyden's method [9] for this purpose. The accuracy of the solution depends on the number of Broyden iterations [5,6]. While more iterations yield better accuracy, they increase computation cost. DEQs have been successfully adapted to computer vision tasks, too, with the introduction of Multiscale Deep Equilibrium Models (MDEQ) [4]. MDEQ is a multiscale model where each scale is driven to equilibrium together with other scales again by using Broyden's method. Iterations start with $\mathbf{z}^{[0]} = 0$ and continue N times to obtain the final solution, $\mathbf{z}^{[N]}$. N is set to 26 for ImageNet classification and 27 for Cityscapes semantic segmentation in MDEQ [4].

3.2   Streaming DEQ

Let $\mathbf{X}$ be a $H \times W \times 3 \times T$ dimensional tensor representing a video where T is the temporal dimension. We represent the frame at time t with $\mathbf{x}_t$ which is a $H \times W \times 3$ tensor. It should be noted that, we primarily target videos with temporal continuity, without too frequent shot changes. But we also study the effects of shot changes in Sect. 4.

To process a video, DEQ can be applied to each video frame $\mathbf{x}_t$ to obtain $\mathbf{z}_t$, the representation of that frame. This amounts to running the Broyden solver for N iterations starting from $\mathbf{z}_t^{[0]} = \mathbf{0}$ for each frame.

However, we know a priori that transitions between subsequent video frames are typically smooth, *i.e.* $\mathbf{x}_{t+1} \sim \mathbf{x}_t$. From this observation, we hypothesize that the corresponding fixed-points, *i.e.* representations $\mathbf{z}_t^*$ and $\mathbf{z}_{t+1}^*$, are likely to be similar. Therefore, the representation of the previous frame can be used effectively as a starting point for inferring the representation of the current frame. To validate this hypothesis, we run an analysis on the ImageNet-VID [42] dataset using the ImageNet pretrained MDEQ model. We assume that at each frame $\mathbf{x}_t$, we have access to the reference representation, $\mathbf{z}_{t-1}^*$, of the previous frame. Reference representations are obtained by running the MDEQ model until convergence ($N = 26$ iterations). At each frame, we use the reference representation of the previous frame as the starting point of the solver:

$$\mathbf{z}_t^{[0]} = \mathbf{z}_{t-1}^*, \tag{3}$$

and run the solver for various but small numbers of iterations, M. To analyze the amount of change in representations over time, we use an ImageNet-pretrained model (MDEQ-XL), since ImageNet representations are known to be useful in many transfer learning tasks. In Fig. 2, we show the squared Euclidean distance between $\mathbf{z}_t^{[M]}$ and $\mathbf{z}_t^*$ for various t values, when the solver is started with Eq. 3. Dashed lines correspond to the squared Euclidean distance between MDEQ-XL's reference and M-iteration based representations.

From the results presented in Fig. 2, we observe that when the solver is initialized with the preceding frame's fixed-point, the inference process quickly

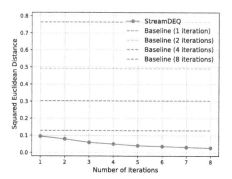

Fig. 2. Squared Euclidean approximation error as a function of inference steps, when the solver is initialized with the reference representation of the preceding frame

converges towards the reference representation. We also observe that after starting from the reference representation of the previous frame and performing only 1 iteration on the current frame, the approximate representation is already more similar to the reference representation than starting from scratch and performing 8 iterations.

Next, we examine the case where the inference method is given access to the reference representations only at certain frames. To simulate this case, at each video clip, we compute the reference representation only at the first frame $\mathbf{x}_0$, *i.e.* $\mathbf{z}_0^*$. In all following ones, we initialize the solver with the estimated representation of the preceding frame and run the solver for M iterations. That is,

$$\mathbf{z}_1^{[0]} = \mathbf{z}_0^* \text{ and } \mathbf{z}_t^{[0]} = \mathbf{z}_{t-1}^{[M]}. \tag{4}$$

We present the results of this scheme for $M \in \{1, 2, 4, 8\}$ in the left hand side of Fig. 3. We observe that starting with the reference representation on the initial frame is still useful but for longer clips its effect diminishes. Still, this scheme helps us maintain a stable performance even after several frames. For example, starting with the reference representation and then applying $M = 2$ iterations per frame throughout the following 20 frames yields a representation that is closer to the reference representation of the final frame than the one given by baseline DEQ inference with 4 solver iterations. This result shows that the M-step inference scheme is able to keep up with the changes in the scene by starting from a good initial point.

While this scheme can provide efficient inference on novel frames, we would still need the reference representations of the initial frames, or key-frame(s),

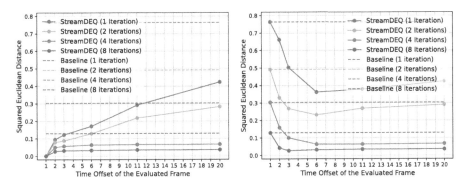

Fig. 3. Distance between the reference representations and StreamDEQ estimations for varying number of iterations, when StreamDEQ is initialized with reference representations (on the left) or just zeros (on the right) on the very first frame

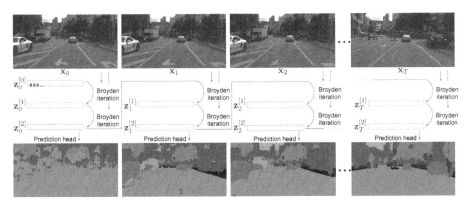

Fig. 4. StreamDEQ applied to a streaming video, performing two iterations per frame. The representation inference process is initialized with zeros in the very first frame ($\mathbf{z}_0^{[0]} = \mathbf{0}$), and with the most recent representation ($\mathbf{z}_t^{[0]} = \mathbf{z}_{t-1}^{[2]}$) in the rest of the stream. This scheme effectively recycles all recent computations for time-efficient inference on a new frame, and therefore, allows approximating a long inference chain (*i.e.* a deep network) by a few inference steps (*i.e.* a few layers) throughout the video stream

which would share the same problems with key-frame based video recognition approaches, e.g. [35,47,53]. To address this problem, we further develop the idea, and hypothesize that we can start from scratch (*i.e.* all zeros), do a limited number of iterations per frame, and pass the representation to the next frame as the starting point. That is,

$$\mathbf{z}_0^{[0]} = 0 \text{ and } \mathbf{z}_t^{[0]} = \mathbf{z}_{t-1}^{[M]}. \tag{5}$$

We present the representation distance results for this final scheme in Fig. 3 (right). The representation distances to the reference representations stabilize in 20 frames. Converged distance values (in 20 frames) are almost the same with those of the previous scheme (Eq. 4). Additionally, the initial representations have relatively large distances but these differences get smaller as new frames arrive. We call this final scheme as StreamDEQ. This scheme avoids any heavy processing in any one of the frames, and completely avoids the concept of key-frames. The number of Broyden iterations can be tuned, which allows easy control over the time-vs-accuracy trade-off. Therefore, the inference iterations can be run as much as the time budget allows. An illustration of the StreamDEQ inference process is given in Fig. 4.

4   Experimental Results

We evaluate our method on video semantic segmentation and video object detection. In the following, we provide technical details regarding training and inference setups, the datasets used, and present our experimental findings. We use the PyTorch [40] framework for all experiments.

4.1   Video Semantic Segmentation

Experimental Setup. We use the Cityscapes semantic segmentation dataset [15], which consists of 5K finely annotated and 20K coarsely annotated images. These finely annotated images are divided into train, val and test set, each containing 2975, 500, and 1525 images, respectively. They correspond to frames extracted from video clips where each annotated image is the 20th frame of its respective clip. To evaluate over videos, we use these clips up to the 20th frame, which has fine annotations, and perform the evaluation on that frame.

We use the pretrained MDEQ-XL segmentation model from the MDEQ paper [4] and do not perform any additional training. We also do not make any changes to its evaluation setup or hyperparameters, perform the evaluation on Cityscapes `val` and report *mean intersection over union* (mIoU) results. For further details, we refer the reader to MDEQ [4].

Results. We present the results of StreamDEQ for two scenarios. The first scenario corresponds to Eq. 4, where we use the reference representations of the first frame to initialize the solver, and apply StreamDEQ then on. Results of this experiment in Fig. 5 (left) show that as the offset of the evaluated frame increases, mIoU starts decreasing, which is expected because the further we move away from the first frame the more irrelevant its representation will become. However, mIoU then stabilizes at a value that is proportional to the number of Broyden iterations (the more iterations, the better the mIoU). This shows that StreamDEQ is able to extract better features over time. StreamDEQ's performance with 8 iterations is still comparable with the baseline (MDEQ) with 27 iterations.

The second scenario corresponds to our final StreamDEQ proposal (*i.e.* Eq. 5), where we initialize the solver from scratch, *i.e.* with all zeros, and apply StreamDEQ. Results of this case are shown in Fig. 5 (right). As the videos

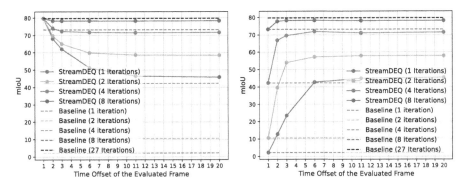

Fig. 5. StreamDEQ semantic segmentation results (in mIoU) on the Cityscapes dataset as a function of solver iterations when the first frame representation is initialized with the reference representation (left) versus zeros (right)

progress, it might be expected that the Broyden solver cannot keep up with the changing scenes. However, we observe that even after 20 frames, the accuracy does not drop. Additionally, the impact of this method is clearer for lower numbers of iterations. For example, performing 1 iteration on every frame without our method would only yield an mIoU score of 2.2. However, StreamDEQ obtains a mIoU score of 44.9 in 10 frames. This is an improvement of over $20\times$. For 8 iterations, StreamDEQ is able to obtain 78.1 mIoU in 10 frames whereas the non-streaming baseline achieves only 73.2 mIoU. Moreover, the converged mIoU values (at larger frame offsets) are similar in Fig. 5 (left) and Fig. 5 (right). Therefore, we conclude that the initial point where we start the solver becomes less important as the video streams and the performance stabilizes at some value higher than the non-streaming case.

We also illustrate these results qualitatively in Fig. 6. For 1 iteration, while the baseline cannot produce any meaningful segmentation, StreamDEQ starts capturing many segments correctly at the 4th frame. With 2 iterations, while the DEQ baseline still produces poor results, StreamDEQ starts to yield accurate predictions in early frames compared to the single iteration case. With 4 iterations, while both models provide rough but relevant predictions in the first frame, StreamDEQ predictions start to become clearly more accurate in the following frames; for example, tree trunks and the sky become visible only when StreamDEQ is applied.

We examine the effects of increasing the number of iterations on inference speed in Table 1. We note that our method does not introduce any computation overhead other than the time it takes to store the fixed-point representation of the previous frame. Therefore, we observe a linear increase in compute times as the number of iterations increases. StreamDEQ with 4 iterations achieves an mIoU score of 71.5 at 530 ms per image. MDEQ with 4 iterations can only achieve 42.3 mIoU. Additional inference results can be found in the videos provided on our project page.

Table 1. Inference time comparisons of the proposed method with differing number of iterations on Cityscapes and ImageNet-VID datasets

		Cityscapes		ImageNet-VID	
Model	# iterations	mIoU	FPS	mAP@50	FPS
StreamDEQ	1	45.5	4.3	9.1	10.3
StreamDEQ	2	57.9	2.9	39.5	9.2
StreamDEQ	4	71.5	1.9	50.4	6.2
StreamDEQ	8	78.2	1.1	54.8	3.5
MDEQ (Baseline)	27/26	79.7	0.3	55.0	1.2

Effect of Shot Changes. We also study the effects of shot changes for the video semantic segmentation task. For this purpose, we initialize the solver with the reference representations from a random frame from the Cityscapes or ImageNet-VID datasets and run StreamDEQ starting from the representations of that frame. Results of ImageNet-VID to Cityscapes shot change experiments can be found in Fig. 7. In this case, even though at first, the obtained scores are lower, following a similar trajectory to the ones in Fig. 5 (right), mIoU scores stabilize to a value close to our original experiment. We conclude that, even with occasional shot changes, our method is able to adapt to the new scene in a few frames. The Cityscapes-to-Cityscapes shot change experiments, with similar results, are provided in the supplementary material.

4.2   Video Object Detection

Experimental Setup. For the video object detection task, we evaluate our method on the ImageNet-VID dataset [42] utilizing the MMDetection [11] and MMTracking [13] frameworks. ImageNet-VID dataset consists of 3862 training and 555 validation videos from 30 classes that are a subset of the 200 classes of the ImageNet-DET dataset. The frames and annotations for each video are available at a rate of 25-30 FPS per video. Note that the ImageNet-DET dataset consists only of images rather than videos. We follow the widely used protocol [12,17,45, 46,52] and train our model on the combination of ImageNet-VID and ImageNet-DET datasets using the 30 overlapping classes. We use a mini-batch size of 4, distributed to 4 NVIDIA A100 GPUs. We resize each image to have a shorter side of 600 pixels and train the model for a total of 12 epochs in 3 stages. We initialize the learning rate to 0.01 and reduce it by a factor of 10 at epochs 7 and 10. We test the model on ImageNet-VID val and report mAP@50 scores following the common practice.

We adopt Faster R-CNN [41] by replacing its ResNet backbone with the MDEQ-XL model. To incorporate multi-level representations, we also use a Feature Pyramid Network (FPN) [33] module after MDEQ-XL. Without any additional modifications, we directly use the model while keeping the model hyperparameters and remaining architecture details same as in Faster R-CNN models

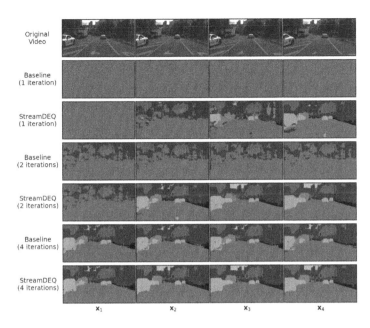

Fig. 6. Qualitative comparison of the baseline with StreamDEQ with different numbers of iterations on the Cityscapes dataset

for ResNet backbones. Exceptionally, we only modify the number of channels for the FPN module to match that of MDEQ-XL. We start training with the ImageNet pretrained MDEQ-XL model from MDEQ [4]. During training, we use 26 solver iterations for both forward and backward passes of the MDEQ, following the ImageNet classification experiments in MDEQ [4]. Unlike the video object detection models [12,17,52], we train our model in the causal single-frame setting, meaning we do not use any temporal information for improved training. Furthermore, we also incorporate the Jacobian regularization loss for MDEQs introduced by Bai et al. [6] to stabilize training.

Results. To the best of our knowledge, this is the first time that an implicit model (MDEQ) is used for an object detection task. We achieve 55.0 mAP@50 on ImageNet-VID val. We are aware that Faster R-CNN with ResNet-18 backbone yields 64.9 mAP@50, however, Faster R-CNN is highly optimized to perform well with ResNet backbones. Yet, we use this same setting with an MDEQ without any parameter optimization, as our focus is not on constructing a MDEQ-based state-of-the-art video object detector. We believe that there is room for improvement in detector design and tuning details, which we leave for future work.

Similar to the video segmentation task, we run StreamDEQ with different numbers of iterations. We present the results of this experiment in Fig. 8. We observe the same trends with the segmentation task. Over time, detection performance increases and stabilizes at a value proportional to the number of Broyden

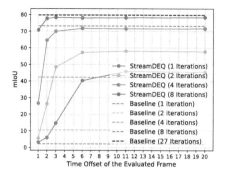

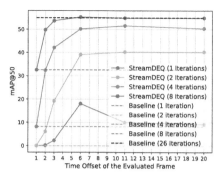

Fig. 7. mIoU results of StreamDEQ with shot changes from ImageNet-VID

Fig. 8. mAP@50 results of StreamDEQ for various number of iterations after initialization with zeros from the beginning of a clip on the ImageNet-VID dataset

iterations. The 1 and 2 iteration cases do not produce any detection results in the non-streaming mode, but if we perform 2 iterations with StreamDEQ, we improve the performance from 0 to 39.5 mAP@50 in 20 frames.

In addition, we also compare the inference speed of StreamDEQ with our baseline for different number of iterations in Table 1. In the 8 iteration case, we obtain a score of 54.8 with StreamDEQ which is only 0.2 lower than our baseline model, while being almost 3 times faster.

5 Conclusions and Future Work

In this paper, we proposed StreamDEQ, an efficient streaming video application of the multiscale implicit deep model, MDEQ. We showed that our model can start from scratch (*i.e.* all zeros) and efficiently update its representations to reach near-optimal representations as the video streams. We validated this claim on video semantic segmentation and video object detection tasks with thorough experiments. StreamDEQ presents a viable approach for both real-time video analysis and off-line large-scale methods. StreamDEQ is not specific to segmentation or object detection, and can be used as a drop-in replacement for most other structured prediction problems on streaming videos as long as the prediction task is performed by an implicit model. In addition, application to time series prediction and event camera based recognition tasks can be interesting future work directions.

Acknowledgments. The numerical calculations were partially performed at TUBITAK ULAKBIM, High Performance and Grid Computing Center (TRUBA) and METU Robotics and AI Technologies Research Center (ROMER) resources. Dr. Cinbis is supported by a Google Faculty Research Award. Dr. Akbas is supported by the BAGEP Award of the Science Academy, Turkey.

References

1. Ba, J., Mnih, V., Kavukcuoglu, K.: Multiple object recognition with visual attention. In: ICLR (2015)
2. Bai, S., Kolter, J.Z., Koltun, V.: Deep equilibrium models. In: Advances in Neural Information Processing Systems (NeurIPS) (2019)
3. Bai, S., Kolter, J.Z., Koltun, V.: Trellis networks for sequence modeling. In: International Conference on Learning Representations (2019). https://openreview.net/forum?id=HyeVtoRqtQ
4. Bai, S., Koltun, V., Kolter, J.Z.: Multiscale deep equilibrium models. In: Advances in Neural Information Processing Systems (NeurIPS) (2020)
5. Bai, S., Koltun, V., Kolter, J.Z.: Neural deep equilibrium solvers. In: International Conference on Learning Representations (2021)
6. Bai, S., Koltun, V., Kolter, J.Z.: Stabilizing equilibrium models by jacobian regularization. In: International Conference on Machine Learning (ICML) (2021)
7. Bazzani, L., de Freitas, N., Larochelle, H., Murino, V., Ting, J.A.: Learning attentional policies for tracking and recognition in video with deep networks. In: ICML (2011)
8. Bertasius, G., Torresani, L., Shi, J.: Object detection in video with spatiotemporal sampling networks. In: Proceedings of the European Conference on Computer Vision (ECCV), pp. 331–346 (2018)
9. Broyden, C.G.: A class of methods for solving nonlinear simultaneous equations. Math. Comput. **19**(92), 577–593 (1965)
10. Carreira, J., Patraucean, V., Mazare, L., Zisserman, A., Osindero, S.: Massively parallel video networks. In: Proceedings of the European Conference on Computer Vision (ECCV), pp. 649–666 (2018)
11. Chen, K., et al.: MMDetection: open MMlab detection toolbox and benchmark. arXiv preprint arXiv:1906.07155 (2019)
12. Chen, Y., Cao, Y., Hu, H., Wang, L.: Memory enhanced global-local aggregation for video object detection. In: Proceedings of the IEEE/CVF Conference on Computer Vision and Pattern Recognition, pp. 10337–10346 (2020)
13. Contributors, M.: MMTracking: OpenMMLab video perception toolbox and benchmark (2020). https://github.com/open-mmlab/mmtracking
14. Cordonnier, J.B., et al.: Differentiable patch selection for image recognition. In: Proceedings of the IEEE/CVF Conference on Computer Vision and Pattern Recognition, pp. 2351–2360 (2021)
15. Cordts, M., et al.: The cityscapes dataset for semantic urban scene understanding. In: Proceedings of the IEEE Conference on Computer Vision and Pattern Recognition (CVPR) (2016)
16. Dehghani, M., Gouws, S., Vinyals, O., Uszkoreit, J., Kaiser, L.: Universal transformers. In: International Conference on Learning Representations (2019). https://openreview.net/forum?id=HyzdRiR9Y7
17. Deng, J., Pan, Y., Yao, T., Zhou, W., Li, H., Mei, T.: Relation distillation networks for video object detection. In: Proceedings of the IEEE/CVF International Conference on Computer Vision, pp. 7023–7032 (2019)
18. Denil, M., Bazzani, L., Larochelle, H., de Freitas, N.: Learning where to attend with deep architectures for image tracking. Neural Comput. **24**(8), 2151–2184 (2012)
19. Gadde, R., Jampani, V., Gehler, P.V.: Semantic video CNNs through representation warping. In: Proceedings of the IEEE International Conference on Computer Vision, pp. 4453–4462 (2017)

20. Han, W., et al.: SEQ-NMS for video object detection. arXiv preprint arXiv:1602.08465 (2016)
21. He, K., Zhang, X., Ren, S., Sun, J.: Deep residual learning for image recognition. In: Proceedings of the IEEE Conference on Computer Vision and Pattern Recognition, pp. 770–778 (2016)
22. Horn, B.K., Schunck, B.G.: Determining optical flow. Artif. Intell. **17**(1–3), 185–203 (1981)
23. Howard, A.G., et al.: MobileNets: efficient convolutional neural networks for mobile vision applications. arXiv preprint arXiv:1704.04861 (2017)
24. Hu, P., Caba, F., Wang, O., Lin, Z., Sclaroff, S., Perazzi, F.: Temporally distributed networks for fast video semantic segmentation. In: Proceedings of the IEEE/CVF Conference on Computer Vision and Pattern Recognition, pp. 8818–8827 (2020)
25. Huang, G., Liu, Z., Van Der Maaten, L., Weinberger, K.Q.: Densely connected convolutional networks. In: Proceedings of the IEEE Conference on Computer Vision and Pattern Recognition, pp. 4700–4708 (2017)
26. Huang, P.Y., Hsu, W.T., Chiu, C.Y., Wu, T.F., Sun, M.: Efficient uncertainty estimation for semantic segmentation in videos. In: Proceedings of the European Conference on Computer Vision (ECCV), pp. 520–535 (2018)
27. Huang, Z., Bai, S., Kolter, J.Z.: (Implicit)^2: Implicit layers for implicit representations. In: Advances in Neural Information Processing Systems (NeurIPS) (2021)
28. Jain, S., Wang, X., Gonzalez, J.E.: Accel: a corrective fusion network for efficient semantic segmentation on video. In: Proceedings of the IEEE/CVF Conference on Computer Vision and Pattern Recognition, pp. 8866–8875 (2019)
29. Kang, K., et al.: T-CNN: tubelets with convolutional neural networks for object detection from videos. IEEE Trans. Circuits Syst. Video Technol. **28**(10), 2896–2907 (2017)
30. Kang, K., Ouyang, W., Li, H., Wang, X.: Object detection from video tubelets with convolutional neural networks. In: Proceedings of the IEEE Conference on Computer Vision and Pattern Recognition, pp. 817–825 (2016)
31. Li, M., Wang, Y.-X., Ramanan, D.: Towards streaming perception. In: Vedaldi, A., Bischof, H., Brox, T., Frahm, J.-M. (eds.) ECCV 2020. LNCS, vol. 12347, pp. 473–488. Springer, Cham (2020). https://doi.org/10.1007/978-3-030-58536-5_28
32. Li, Y., Shi, J., Lin, D.: Low-latency video semantic segmentation. In: Proceedings of the IEEE Conference on Computer Vision and Pattern Recognition, pp. 5997–6005 (2018)
33. Lin, T.Y., Dollár, P., Girshick, R., He, K., Hariharan, B., Belongie, S.: Feature pyramid networks for object detection. In: Proceedings of the IEEE Conference on Computer Vision and Pattern Recognition, pp. 2117–2125 (2017)
34. Liu, M., Zhu, M.: Mobile video object detection with temporally-aware feature maps. In: Proceedings of the IEEE Conference on Computer Vision and Pattern Recognition, pp. 5686–5695 (2018)
35. Liu, M., Zhu, M., White, M., Li, Y., Kalenichenko, D.: Looking fast and slow: memory-guided mobile video object detection. arXiv preprint arXiv:1903.10172 (2019)
36. Liu, Y., Shen, C., Yu, C., Wang, J.: Efficient semantic video segmentation with per-frame inference. In: Vedaldi, A., Bischof, H., Brox, T., Frahm, J.-M. (eds.) ECCV 2020. LNCS, vol. 12355, pp. 352–368. Springer, Cham (2020). https://doi.org/10.1007/978-3-030-58607-2_21
37. Liu, Z., Wang, H.J., Xu, Z., Darrell, T., Shelhamer, E.: Confidence adaptive anytime pixel-level recognition. In: International Conference on Learning Representations (2022). https://openreview.net/forum?id=kNKFOXleuC

38. Mnih, V., Heess, N., Graves, A., Kavukcuoglu, K.: Recurrent models of visual attention. In: Ghahramani, Z., Welling, M., Cortes, C., Lawrence, N., Weinberger, K.Q. (eds.) Advances in Neural Information Processing Systems, vol. 27. Curran Associates, Inc. (2014). https://proceedings.neurips.cc/paper/2014/file/09c6c3783b4a70054da74f2538ed47c6-Paper.pdf

39. Narayanan, D., et al.: PipeDream: generalized pipeline parallelism for DNN training. In: Proceedings of the 27th ACM Symposium on Operating Systems Principles, pp. 1–15 (2019)

40. Paszke, A., et al.: PyTorch: an imperative style, high-performance deep learning library. In: Advances in Neural Information Processing Systems, vol. 32, pp. 8024–8035. Curran Associates, Inc. (2019). https://papers.neurips.cc/paper/9015-pytorch-an-imperative-style-high-performance-deep-learning-library.pdf

41. Ren, S., He, K., Girshick, R., Sun, J.: Faster R-CNN: towards real-time object detection with region proposal networks. Adv. Neural Inf. Process. Syst. **28**, 1–9 (2015)

42. Russakovsky, O., et al.: ImageNet large scale visual recognition challenge. Int. J. Comput. Vision **115**(3), 211–252 (2015). https://doi.org/10.1007/s11263-015-0816-y

43. Sandler, M., Howard, A., Zhu, M., Zhmoginov, A., Chen, L.C.: MobileNetV2: inverted residuals and linear bottlenecks. In: Proceedings of the IEEE Conference on Computer Vision and Pattern Recognition, pp. 4510–4520 (2018)

44. Shelhamer, E., Rakelly, K., Hoffman, J., Darrell, T.: Clockwork convnets for video semantic segmentation. In: Hua, G., Jégou, H. (eds.) ECCV 2016. LNCS, vol. 9915, pp. 852–868. Springer, Cham (2016). https://doi.org/10.1007/978-3-319-49409-8_69

45. Wang, S., Zhou, Y., Yan, J., Deng, Z.: Fully motion-aware network for video object detection. In: Proceedings of the European Conference on Computer Vision (ECCV), pp. 542–557 (2018)

46. Wu, H., Chen, Y., Wang, N., Zhang, Z.: Sequence level semantics aggregation for video object detection. In: Proceedings of the IEEE/CVF International Conference on Computer Vision, pp. 9217–9225 (2019)

47. Xu, Y.S., Fu, T.J., Yang, H.K., Lee, C.Y.: Dynamic video segmentation network. In: Proceedings of the IEEE Conference on Computer Vision and Pattern Recognition, pp. 6556–6565 (2018)

48. Zagoruyko, S., Komodakis, N.: Wide residual networks. arXiv preprint arXiv:1605.07146 (2016)

49. Zhao, B., Zhao, B., Tang, L., Han, Y., Wang, W.: Deep spatial-temporal joint feature representation for video object detection. Sensors **18**(3), 774 (2018)

50. Zhu, H., Wei, H., Li, B., Yuan, X., Kehtarnavaz, N.: A review of video object detection: datasets, metrics and methods. Appl. Sci. **10**(21), 7834 (2020)

51. Zhu, X., Dai, J., Yuan, L., Wei, Y.: Towards high performance video object detection. In: Proceedings of the IEEE Conference on Computer Vision and Pattern Recognition, pp. 7210–7218 (2018)

52. Zhu, X., Wang, Y., Dai, J., Yuan, L., Wei, Y.: Flow-guided feature aggregation for video object detection. In: Proceedings of the IEEE International Conference on Computer Vision, pp. 408–417 (2017)

53. Zhu, X., Xiong, Y., Dai, J., Yuan, L., Wei, Y.: Deep feature flow for video recognition. In: Proceedings of the IEEE Conference on Computer Vision and Pattern Recognition, pp. 2349–2358 (2017)

Symmetry Regularization and Saturating Nonlinearity for Robust Quantization

Sein Park[1], Yeongsang Jang[2], and Eunhyeok Park[1,2]($\boxtimes$)

[1] Graduate School of Artificial Intelligence, POSTECH, Pohang, Korea
[2] Department of Computer Science and Engineering, POSTECH, Pohang, Korea
{seinpark,jangys,eh.park}@postech.ac.kr

Abstract. Robust quantization improves the tolerance of networks for various implementations, allowing reliable output in different bit-widths or fragmented low-precision arithmetic. In this work, we perform extensive analyses to identify the sources of quantization error and present three insights to robustify a network against quantization: reduction of error propagation, range clamping for error minimization, and inherited robustness against quantization. Based on these insights, we propose two novel methods called symmetry regularization (SymReg) and saturating nonlinearity (SatNL). Applying the proposed methods during training can enhance the robustness of arbitrary neural networks against quantization on existing post-training quantization (PTQ) and quantization-aware training (QAT) algorithms and enables us to obtain a single weight flexible enough to maintain the output quality under various conditions. We conduct extensive studies on CIFAR and ImageNet datasets and validate the effectiveness of the proposed methods.

Keywords: Robust quantization · Post-training quantization (PTQ) · Quantization-aware training (QAT)

1 Introduction

Deep learning algorithms have shown excellence in diverse applications, but the increasing memory footprint and computation overhead have become obstacles to utilizing them. To exploit the excellence of deep neural networks (DNNs) in practice, neural network optimization is becoming more and more important. Neural network quantization is a representative optimization technique beneficial to footprint reduction and performance improvement. Due to its practical advantages, advanced hardware is already equipped with low-precision support, such as the well-known float16, bfloat16, and int8-based operations [19,28,36,37,40,42],

---

S. Park and Y. Jang—Equal contribution.

Supplementary Information The online version contains supplementary material available at https://doi.org/10.1007/978-3-031-20083-0_13.

even with 4-bit or lower-precision acceleration [18, 27, 29, 33, 39]. With the aid of a judiciously designed quantization algorithm, we could enjoy the benefit of low-precision computation in reality.

However, quantization has the substantial limitation of accuracy degradation due to the limited representation capability. Many studies have been actively proposed to address this problem, and Quantization-aware training (QAT) is a representative approach where end-to-end training is applied to refine the simulated error of pseudo (or fake-) quantization [6, 9, 20, 22, 30, 47]. QAT is advantageous in the sense of minimal accuracy degradation in the given bit-width. Recently, post-training quantization (PTQ) has emerged as an alternative approach that quantizes the pre-trained network without fine-tuning [2, 3, 5, 10, 16, 22, 25, 26, 41, 46]. PTQ allows us to exploit the benefit of low-precision computation with a minimal number of training datasets, thereby having many more practical use cases compared to QAT.

Nonetheless, both QAT and PTQ have severe drawbacks: QAT requires access to the entire dataset and the expenses of an additional training stage. In addition, the model with QAT is specialized for the target precision and quantization scheme, thereby lacking robustness in different bit-widths. On the other hand, PTQ suffers from notable accuracy degradation to QAT due to the lower degree of freedom and insufficient information to compensate for the errors. Many studies have been proposed to overcome this limitation, but a bit-width of 8- or more is still required for the advanced networks [5, 16, 41].

Recently, alternative approaches have been proposed to enhance the robustness of networks against quantization [12, 14, 34, 45]. Improving the robustness of networks has diverse advantages, including allowing the quantized model to maintain the accuracy in bit-widths other than in which it trained and preserving the quality of output with various quantization algorithms. Practically, these properties help to utilize the pareto-front optimal points of energy consumption and computation, such as exploiting a high-precision model when the resource (e.g., battery) is sufficient and reducing precision dynamically when the resource is scarce. In addition, numerous companies are now designing their own accelerators having divergent and fragmented low-precision implementations. When we need to support multiple accelerators, preparing a single low-precision model robust enough to endure the minor modification of different implementations could be an attractive option for rapid deployment. Robust quantization enables diverse appealing applications, having strong importance in practice.

In this work, we propose two novel methods to increase the robustness of neural networks based on three insights about the error component of quantization. The paper is organized as follows; first, we perform an extensive analyses to identify the source of errors from quantization and indicate three motivations to robustify the network: reduction of error propagation, range clamping for error minimization, and inherited robustness against quantization (Sect. 3). To address those motivations, we introduce two novel ideas: symmetry regularization (SymReg) for the reduction of error propagation and saturating nonlinearity (SatNL) for the others (Sect. 4). According to our extensive experiments, the proposed methods are beneficial for maximizing the robustness of networks after QAT or PTQ, showing state-of-the-art results (Sect. 5). We then clarify the limitation of this study in Sect. 6 and conclude the paper in Sect. 7.

2   Related Work

2.1   Quantization-Aware Training and Post-training Quantization

QAT [6,9,20,22,30,47] shows the potential of low-precision computation, where the milestone networks (e.g., VGG [35], GoogleNet [38], and ResNet [13]) could be quantized into sub-4-bit without accuracy loss [6,9], and advanced light-weight networks (e.g., MobileNet-V2) could be quantized into 4-bit with negligible accuracy loss [30]. Meanwhile, PTQ [2,3,5,10,16,22,25,26,46] applies a conservative bit-width to maintain the quality of output, even though it offers performance benefits with relaxed constraints. In this work, we aim to maximize the benefits of QAT and PTQ through the advantages of robust quantization.

2.2   Robustness of Neural Networks

A line of work closely related to ours is the analysis of the robustness of neural networks, which has attracted attention recently. Currently, the Hessian-aware metric is often used to identify the robustness of neural networks. Previous studies pointed out that the second derivative of loss is a good approximation of network sensitivity [1,8], proposed a way to estimate the approximate Hessian metric efficiently, and showed the potential of sensitivity-aware quantization [7,8,23,41,44]. On the other hand, few studies have focused on easing the sensitivity of networks during the training phase to improve their generalization performance [11,21]. The proposed (adaptive) sharpness-aware minimization, (A)SAM, makes the network have lower Hessian spectra than the networks trained without it. It is expected to be beneficial for enhancing the endurance of the network for quantization, but we observe that the benefit of (A)SAM is degraded in the quantization domain. In this paper, we propose a novel idea, SatNL, to maximize the robustness of quantization with (A)SAM training.

Moreover, few studies have tried to minimize quantization error in the view of robustness. GDRQ [45] and BR [12] attempted to improve the robustness of networks via regularization in QAT tasks. Gradient l_1-regularization [14] lowered the sensitivity of networks for quantization via regularizing l_1-norm of gradient, and KURE [34] regularized the weights in a uniform distribution to have minimal accuracy drop after the quantization. The two studies [14,34] are the most relevant to ours in that sharing the same objective of robust quantization. However, according to our observation, the former becomes unstable in advanced networks (i.e., MobileNet-V2/V3) and the latter is orthogonal to ours. Our methods show results comparable to those of KURE, and we can maximize the endurance of the network by applying ours with KURE jointly, as will be shown in Sect. 5.

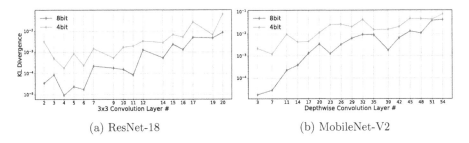

(a) ResNet-18   (b) MobileNet-V2

Fig. 1. KL divergence of 3×3 convolution and depthwise convolution output between original model and model quantized with PTQ [2] (a) ResNet-18 (b) MobileNet-V2.

3   Motivation

In this section, we provide the analysis of the error sources induced by quantization and explain the motivations to enhance the robustness of the networks for each error source.

3.1   Reduction of Error Propagation

Several previous studies have indicated that quantization introduces the distortion of statistics compared to the original distribution [10,24]. Moreover, when we apply quantization over the entire network, each layer introduces additional distortion to the output. As a result, the error continues to propagate and accumulate over the networks, as shown in Fig. 1, resulting in a large amount of accuracy degradation. Many studies related to PTQ have attempted to mitigate this problem by explicitly minimizing the difference in statistics before and after the quantization [4,10]. However, in this paper, we propose an alternative approach to minimize the difference in statistics on any quantization algorithms.

To achieve this goal, we focus on minimizing the biased quantization error problem [4,25]. Consider the linear or convolution operation $y = W \cdot x$, where W is an arbitrary fixed weight and x is an activation assumed i.i.d. variable with $E_i[x_i] = \mu_x$. In this condition, we can estimate the expected value of a single output unit $y_j = \Sigma_i^N W_{j,i} \cdot x_i$:

$$E_j[y_j] = N\mu_x E_i[W_{j,i}], \tag{1}$$

where N is the number of elements and i is the index of input x. When we apply the quantization to the weight, the expected value of output drifts due to the distortion of the weight as follows:

$$E_j[\tilde{y}_j - y_j] = N\mu_x \cdot \left(E_i[Q(W_{j,i})] - E_i[W_{j,i}] \right), \tag{2}$$

where y_j and $\tilde{y}_j$ are the j-th output with the original weight and quantized weight, respectively. To minimize error propagation, we should minimize the difference of averaged weight in the output-channel dimension. However, satisfying

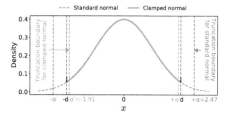
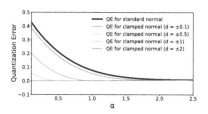

Fig. 2. Histogram of standard normal vs clamped normal ($d = \pm2$), and their truncation boundaries.

Fig. 3. Analysis of quantization error for standard normal and clamped normal.

Eq. (2) strictly for any quantization algorithm is highly challenging. To ease the difficulty of the objective, we adopt the additional condition as given by:

$$E_i[Q(W_{j,i})] = E_i[W_{j,i}] = 0. \tag{3}$$

By forcing the mean of the weights before and after the quantization toward 0, Eq. (2) is satisfied as a sufficient condition. When the full-precision weight is symmetric, the mean of the full-precision weight is zero. In addition, when we apply a symmetric quantization, which is commonly used for the weight quantization due to hardware compatibility [17,40,42], the drift in the positive values could be amortized by the drift in the negative values; thereby, the mean of the quantized weight is also zero.

In summary, if we can force the full-precision weight in a symmetric distribution, the statistics distortion and error propagation after the quantization could be minimized. Unlike the explicit bias correction process [25], weight symmetry inherits the robustness against bias drift, enabling us effortless transition to different quantization policies.

To guide the convergence of weight toward the symmetric distribution, we propose a novel regularization in Sect. 4.1. When we apply this regularization during pre-train the model, the difference in statistics before and after PTQ is reduced. Furthermore, the statistics distortion is also minimized when we utilize the fine-tuned weight after QAT in different bit-widths without an additional fine-tuning stage, helping to maintain the quality of output.

3.2 Range Clamping for Error Minimization

In linear quantization, the quantization levels are evenly distributed in between the truncation boundaries. Applying quantization to the full-precision tensor induces quantization error, which can be decomposed into the truncation error and the rounding error. The truncation and rounding errors are inevitable because of the limited number of quantization levels, but the difference of domain (i.e., the unbounded full-precision and the truncated quantization) enlarges the quantization error. A previous study [31] pointed out that the data with infrequent but large values require a widened truncation boundary, which increases the rounding error significantly. To mitigate the quantization error, the domain of full-precision data should be narrowed to a bounded range.

Motivated by this limitation, we propose a straightforward idea of introducing the range clamping to the full-precision weight[1], as shown in Fig. 2. Assume that the original weight follows a normal distribution whose PDF is $f(x) \sim N(0, 1)$, just as the convention of previous studies [2,34]. When we apply the b-bit symmetric quantization toward minimizing the L2 norm, the quantization error can be estimated as follows [2]:

$$\text{Quantization Error} = E[(W - Q(W))^2]$$

$$\approx 2 \cdot \overbrace{\int_\alpha^\infty f(x) \cdot (x - \alpha)^2 dx}^{\text{truncation error}} + \overbrace{\frac{\alpha^2}{3 \cdot 2^{2b}}}^{\text{rounding error}} , \tag{4}$$

where α is the truncation boundary that minimizes $||W - Q(W)||_2$. On the other hand, the clamping modifies the distribution of weight, whose cumulative distribution function $G(x)$ is expressed as

$$G(x; d) = \begin{cases} 0, & x \le -d \\ F(x) + F(-|d|), & -d < x < d \\ 1, & d \le x, \end{cases} \tag{5}$$

where d is the newly introduced clamping target and $F(x)$ is the cumulative distribution function of $f(x)$. Then, the quantization error of the clamped distribution is expressed as

$$\text{Quantization Error}' = E[(W' - Q(W'))^2]$$

$$\approx 2 \cdot \overbrace{\left(F(-|d|) \cdot (d - \alpha')^2 + \int_{\alpha'}^d f(x) \cdot (x - \alpha')^2 dx \right)}^{\text{truncation error}} + \overbrace{\frac{\alpha'^2}{3 \cdot 2^{2b}}}^{\text{rounding error}} , \tag{6}$$

where α' is the truncation boundary that minimizes $||W' - Q(W')||_2$.

When we compare the errors of Eq. (4) and Eq. (6), Eq. (6) always has a smaller error than Eq. (4), as shown in Fig. 3. The proofs of Eq. (4) to Eq. (6) are provided in the supplementary material.

This analysis indicates that the range clamping of full-precision data is beneficial for minimizing quantization error. Thereby, if we train a network with the range clamping nonlinearity, the network could have a strong endurance for quantization. In addition, the quantization error could be minimized in the different precision, resulting in low accuracy loss other than the bit-width we quantized. Indeed, a similar idea was addressed in MobileNet-V2 [32] in terms of ReLU6 for activation. The range clamping can be seen as an extension of the idea of ReLU6 for weight.

---

[1] Please note that we intentionally use different expressions to distinguish quantization's truncation and the clamping of full-precision data.

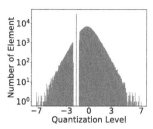
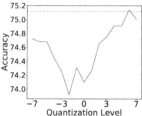
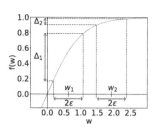

Fig. 4. Example of single-level quantization and corresponding result. The dashed line represents the baseline full-precision accuracy, and the solid line shows that of the quantized network.

Fig. 5. Saturating nonlinearity for weight and the adversarial boundary of ASAM.

One possible drawback of this idea is the accuracy degradation due to the limited degree of freedom. For instance, when we set the clamping target close to zero, the quantization error becomes negligible, but the training from scratch may fail or show poor accuracy. Therefore, it is necessary to find a sweet spot that reduces the quantization error by a large margin while maintaining the quality of output. Empirically, we determine the practical implementation of the truncation that has negligible quality degradation while minimizing quantization loss significantly and provide it in Sect. 4.2.

3.3   Inherited Robustness Against Quantization

In addition to the range clamping for error minimization, we inspect another approach that enhances the inherited robustness of networks against quantization based on Hessian-aware loss sharpness minimization. Recently, several studies have focused on enhancing the generalization of neural networks based on the training pipeline aware of the sharpness of loss surface [11,21]. Those studies have aimed to guide the convergence of networks toward the flat minimum having smaller Hessian values, where the minor distortion of weight could be ignored without affecting the output. From a similar perspective, the Hessian of weight is utilized as an important metric for measuring the sensitivity of networks regarding quantization [7,8,23]. Motivated by these examples, we try to adopt the Hessian-aware training to enhance the robustness of networks for quantization by guiding the convergence of networks into a smooth loss surface. We utilize (A)SAM [11,21], and empirically observe that training with (A)SAM is beneficial for minimizing quantization error.

However, we also observe that there is room for improvement in terms of enhancing robustness for quantization with (A)SAM, because the quantization sensitivity differs depending on the value of quantization levels. Figure 4 shows the accuracy degradation after applying single-level PTQ to the weights of MobileNet-V2 on CIFAR-100, where the weights corresponding to the specific quantization level are quantized while the rest remain in full-precision. As shown in the figure, the accuracy degradation increases when the smaller weights

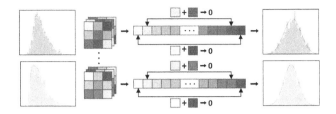

Fig. 6. Visualization of proposed SymReg.

are quantized. This indicates that the weights near zero are more vulnerable to quantization than the weights having large values. We speculate that because the majority of weights are concentrated near zero, the accumulated error of quantization is inversely proportional to the magnitude of the value. Thereby, to maintain the quality of output with quantization, the sharpness minimization should be applied in different strengths depending on the magnitude of weight.

To realize the aforementioned objective, we propose a novel idea that introduces the saturating nonlinearity to the weight, as visualized in Fig. 5. When we apply (A)SAM with the nonlinearity, the robustness boundary of (A)SAM, which is equally assigned in the weight before the nonlinearity, covers a different range in output depending on the slope of the nonlinearity. SatNL has the largest slope near zero; as a result, the effect of the (A)SAM algorithm turns friendly to the quantization.

4   Implementation

Based on the motivations in the previous section, we propose the practical implementations, called symmetry regularization (SymReg) and saturating nonlinearity (SatNL).

4.1   Symmetry Regularization

In Sect. 3.1, we claimed that the symmetric weight could minimize the error propagation. To realize this, we propose an additional regularization called symmetry regularization (SymReg).

The weight symmetry is achievable when every weight has a corresponding mate with an identical magnitude but a different sign. After sorting the weights and assigning the index in ascending order, we can make pairs where one element is selected in ascending order $\tilde{w}_i$ while the other is in descending order $\tilde{w}_{N-i}$. When we calculate the $L1$ norm of the sum of each pair, the expectation should become zero as follows:

$$\frac{2}{N} \sum_{n=1}^{\frac{N}{2}} |\tilde{w}_n + \tilde{w}_{N-n}| = 0. \tag{7}$$

Based on this intuition, we design a layer-wise SymReg that guides the convergence of weight into the symmetric distribution as shown in Fig. 6. The SymReg is defined as:

$$\mathcal{L}_{sym1} = \frac{2}{C \cdot N} \sum_{c=1}^{C} \sum_{i=1}^{\frac{N}{2}} |w_i^c + w_{N-i}^c|, \tag{8}$$

where w_i^c represents the i-th smallest element in the c-th channel.

$\mathcal{L}_{sym1}$ is applied when we train the full-precision network in addition to the conventional loss functions. However, if the restriction of $\mathcal{L}_{sym1}$ is too severe, it could lead to a minor accuracy degradation. To minimize it, we propose the relaxed SymReg, which measures the symmetry in a 2:2 relation with more degree of freedom instead of a 1:1 relation. We empirically observe that more than 3:3 relation degrades the benefit of SymReg.

$$\mathcal{L}_{sym2} = \frac{4}{C \cdot N} \sum_{c=1}^{C} \sum_{i=1}^{\frac{N}{4}} |w_{2i}^c + w_{2i+1}^c + w_{N-2i}^c + w_{N-2i-1}^c|. \tag{9}$$

In the experiment, we combine $\mathcal{L}_{sym1}$ and $\mathcal{L}_{sym2}$ adequately to minimize accuracy degradation while exploiting the benefit of propagation error minimization. The overall loss is expressed as follows:

$$\mathcal{L} = \mathcal{L}_{CE} + \lambda_1 \cdot \mathcal{L}_{sym1} + \lambda_2 \cdot \mathcal{L}_{sym2}. \tag{10}$$

4.2 Saturating Nonlinearity (SatNL)

In Sect. 3.2, we showed that the truncation of full-precision weight could be highly beneficial for minimizing quantization error. In addition, in Sect. 3.3, we showed the empirical analysis indicating that the quantization sensitivity differs depending on the magnitude of the weight. To resolve these two problems simultaneously, we propose applying a specialized nonlinearity function f on top of the weight as follows: $conv(W, X) \rightarrow conv(f(W), X)$.

The nonlinearity should satisfy three properties. 1. It needs to be an odd function. Because we assume that the weight is quantized by the symmetric quantization, it would be better to have an identical range of the negative and positive regions to maximize the quantization level efficiency. 2. The range of output needs to be bounded. To satisfy the criteria of Sect. 3.2, the weight after the nonlinearity should be narrowed to a certain range. 3. The slope is gradually decreased as the input value is increased. To maximize the benefit of (A)SAM, the nonlinearity should be saturated as the value is increased. Empirically, we choose the hyper-tangent (tanh) function as nonlinearity which satisfies all three conditions. Note that the normalized tanh was used in QAT studies [47], but there is a fundamental difference in terms of the intention of using it. In this study, we exploit the tanh function intentionally to maximize the robustness against quantization and turn (A)SAM algorithm friendly to quantization. Other nonlinearity functions could be applicable when the three conditions are met. According to our experiments, the impact on the final accuracy is negligible regardless of the nonlinearity functions, while the desired properties for robustness are valid. Additional analysis is in the supplementary material.

5   Experiments

To show the superiority of the proposed methods, we conduct extensive studies on CIFAR-100 and ImageNet datasets with representative networks (i.e., ResNet-18 [13], MobileNet-V2 [32], and MobileNet-V3 [15]). We use layer-wise asymmetric quantization for activation and output-channel-wise symmetric quantization for weight to enable acceleration on existing hardwares [17,40,42, 43].

For QAT, we apply LSQ [9], which is the advanced differentiable quantization scheme that allows the quantization of ResNet-18 into 3-bit without accuracy loss in the ImageNet. For PTQ, we apply the ACIQ [2], AdaQuant [16], and QDrop [41] algorithms. ACIQ is a well-known PTQ that analytically finds the optimal quantization boundary, and QDrop is a state-of-the-art PTQ algorithm with small calibration sets. By utilizing multiple quantization algorithms with different properties, we aim to validate the universal robustness.

Moreover, we use a common practice that fixes the bit-width of the first and the last layers into 8-bit. All of the other layers are quantized into the given bit-width identically unless explicitly specified otherwise. SymReg is not applied to the depthwise convolution and the linear layer in MobileNet-V2/V3. In 3×3 depthwise convolution, SymReg degrades the expression capability significantly by forcing one out of nine elements to become zero. In ASAM, ρ is fixed as 1 except 0.2 for MobileNet-V3 because ASAM with high ρ becomes unstable. The hyper-parameters of SymReg λ_1/λ_2 are empirically set as 0.1/0.1 for ImageNet and CIFAR-100. The details of the training parameters (i.e., learning rate, decay, epochs, etc.), are provided in the supplementary material.

5.1   Robustness of Bit-Precision for PTQ

Table 1 shows the accuracy degradation after PTQ of ResNet-18, and MobileNet-V2/V3 on the ImageNet dataset. Our/baseline models are trained from scratch with/without the proposed ideas and then quantized into low-precision with PTQ algorithms. Our experiments show that the proposed methods are beneficial for minimizing the accuracy degradation after PTQ in every point regardless of the PTQ details. When we combine ours with the advanced PTQ (i.e., QDrop), we can quantize ResNet-18 into 4-bit with accuracy loss of less than 1 %. Compared to the baseline, we reduce 1.02 % of the top-1 accuracy degradation. In addition, the synergy of the advanced PTQ algorithm and our methods enables sub-8-bit quantization for the advanced networks with minimal accuracy degradation. QDrop with ours achieves 69.87 % in 4-bit MobileNet-V2, which is the highest accuracy after 4-bit PTQ to the best of our knowledge. Notably, when

Table 1. Results of applying PTQ to baseline and network with proposed ideas on the ImageNet dataset. The values in the table represent the top-1 accuracy. The dashed cells represent the points where the PTQ fails to converge.

Model	PTQ	Method	Weight/activation bit-width configuration							
			FP	4/FP	3/FP	2/FP	6/6	5/5	4/4	3/3
ResNet-18	ACIQ [2]	Baseline	70.54	47.44	-	-	68.70	64.87	38.46	-
		Ours	70.92	69.22	49.06	-	70.02	68.99	66.65	42.95
	AdaQuant [16]	Baseline	70.54	69.29	66.18	3.23	70.17	69.55	67.67	57.57
		Ours	70.92	70.36	68.84	48.39	70.75	70.37	69.35	64.04
	QDrop [41]	Baseline	70.54	70.15	69.39	66.40	70.27	69.93	68.91	65.75
		Ours	70.92	70.69	70.06	66.95	70.81	70.57	69.93	67.45
MobileNet-V2	ACIQ [2]	Baseline	72.22	28.68	-	-	69.30	64.20	18.15	-
		Ours	72.87	70.07	40.79	-	71.07	68.66	58.30	6.25
	AdaQuant [16]	Baseline	72.22	70.67	59.80	-	71.52	70.72	63.70	-
		Ours	72.87	72.23	69.03	-	72.27	71.76	68.91	18.36
	QDrop [41]	Baseline	72.22	71.41	68.32	48.68	71.57	70.64	67.08	50.79
		Ours	72.87	72.44	71.18	61.68	72.61	72.05	69.87	62.55
MobileNet-V3	ACIQ [2]	Baseline	74.52	29.65	-	-	-	-	-	-
		Ours	74.43	61.95	1.04	-	-	-	-	-
	AdaQuant [16]	Baseline	74.52	72.92	64.17	-	72.73	68.95	43.88	-
		Ours	74.43	73.51	70.50	2.87	72.69	71.02	62.73	-

we select the layer-wise bit-width depending on the sensitivity of layers, where the depthwise and squeeze-excitation layers are quantized into 8-bit and the rest of the layers are quantized into 4-bit with AdaQuant, MobileNet-V2/V3 shows 2.96 %/4.34 % of accuracy degradation respectively[2]. In this configuration, we can enjoy the benefit of 4-bit computation in 85.3 % of computation in the case of MobileNet-V2. Without ours, the accuracy degradation is 7.25 % and 9.44 % respectively in mixed precision, which is too poor to be used in real applications. The combination of network robustness enhancement and sensitivity-aware quantization could be a good candidate for practical deployment. According to our experiment, SymReg and SatNL extended the training time by 2.53 % when we train MobileNet-V2 with ASAM on the ImageNet dataset. By spending this one-time overhead, a robust network that can minimize accuracy degradation regardless of the PTQ scheme can be achieved. Additional experiments compared with KURE are supported in supplementary materials.

5.2   Ablation Study

Table 2 shows the effect of the proposed methods based on the accuracy degradation with PTQ. When we add an additional component of the proposed methods progressively (+ SymReg, + SatNL, + All), the PTQ error is gradually reduced, showing that the proposed methods enhance the robustness of the network for PTQ in diverse aspects. SymReg and SatNL are beneficial for robustness but introduces a slight accuracy degradation in full-precision. However, the accuracy degradation could be mitigated with the assistance of ASAM. When we

---

[2] Note that the column of mixed-precision results is omitted in Table 1 for brevity.

Table 2. Results of ablation study of proposed methods on MobileNet-V2 at CIFAR-100 dataset. The weight is quantized into the given bit-width with ACIQ. "All" means every method (+ SymReg + SatNL + ASAM). The values in the table represent the mean and standard deviation of top-1 accuracy with eight trials (10 trials except min. and max. results). FP means full-precision.

	FP	4-bit	3-bit	2-bit
Baseline	$74.70_{\pm0.16}$	$73.32_{\pm0.32}$	$66.81_{\pm0.86}$	$6.92_{\pm2.68}$
+ SymReg	$74.66_{\pm0.16}$	$73.49_{\pm0.30}$	$69.69_{\pm0.81}$	$25.01_{\pm4.80}$
+ SatNL	$74.80_{\pm0.11}$	$73.43_{\pm0.36}$	$68.65_{\pm1.23}$	$14.45_{\pm2.93}$
+ SymReg + SatNL	$74.41_{\pm0.10}$	$73.34_{\pm0.31}$	$69.66_{\pm0.62}$	$33.68_{\pm5.87}$
+ ASAM	$75.52_{\pm0.19}$	$74.42_{\pm0.16}$	$69.71_{\pm0.71}$	$11.80_{\pm3.96}$
+ SatNL + ASAM	$75.55_{\pm0.21}$	$74.53_{\pm0.21}$	$70.78_{\pm0.81}$	$25.77_{\pm4.03}$
+ All	$75.33_{\pm0.16}$	$74.55_{\pm0.27}$	$72.17_{\pm0.22}$	$39.27_{\pm2.56}$
+ KURE	$74.97_{\pm0.11}$	$74.22_{\pm0.10}$	$71.26_{\pm0.47}$	$34.90_{\pm4.51}$
+ KURE + ASAM	$75.57_{\pm0.18}$	$74.96_{\pm0.11}$	$72.48_{\pm0.46}$	$42.73_{\pm4.71}$
+ KURE + All	$75.41_{\pm0.3}$	$74.91_{\pm0.29}$	$73.34_{\pm0.25}$	$37.58_{\pm2.99}$

compare + ASAM and + SatNL + ASAM, the latter shows higher accuracy in full-precision and better robustness in lower precision, showing that the benefit of ASAM is boosted with SatNL.

When we compare the performance of the proposed methods with KURE, our methods lower accuracy degradation in the given bit-width than that of KURE. Meanwhile, when we apply ASAM with KURE, the robustness becomes comparable to ours. Moreover, the implementation details of KURE and ours are orthogonal. Thus, we can maximize the robustness by combining KURE and ours (KURE + ALL). This could be a state-of-the-art method for enhancing the robustness of networks, as far as we know.

5.3   Robustness for Quantization Step Size

To validate the effect of the proposed methods, we measure the accuracy changes depending on the step size as shown in Fig. 7. In all cases, including PTQ and QAT, the proposed methods maintain the quality of output in the various step sizes. In the case of MobileNet-V2 for the ImageNet dataset, ours maintains 25.72 % higher accuracy compared to the baseline when the step size is changed by 8 %. In addition, ours shows comparable or better results to the previous best method, KURE, for the optimized network (i.e., MobileNet-V2). As indicated in the previous studies [17,34], the degree of freedom for the quantization step size could be restricted depending on the hardware implementation. For instance, some hardware supports step sizes having predefined values or limited resolution. In such a case, the robustness of the quantization step size is essential to maintain output quality after PTQ or QAT. Because our methods improve the robustness of the step size by a large margin, we expect that our methods could be helpful for the deployment of quantized networks in practice.

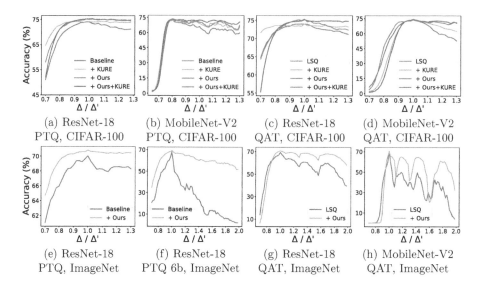

Fig. 7. Robustness of quantized network when we change step size of quantization operator for weight. The networks are optimized for the step size Δ', and the accuracy is measured with the scaled step size Δ. All networks are quantized into 4-bit (except for (f) into 6-bit) with PTQ [16] and QAT [9], including activation and weight. Additional results are included in the supplementary material.

5.4   Robustness of Bit Precision for QAT

Because most of the existing QAT methods specialize the weight fine-tuning for the specific configuration, the accuracy of the quantized network in different bit-widths is reduced significantly. Meanwhile, ours enhance the robustness of the quantized network, allowing stable accuracy in different bit-widths without fine-tuning. Figure 8 shows the accuracy degradation depending on the operation bit-widths other than the one we trained via QAT. With the proposed methods, one can train a generic model that can produce reliable output in multiple bit-widths with existing QAT algorithms.

Table 3 shows the accuracy of QAT for ResNet-18/MobileNet-V2 in different bit-widths with/without the proposed methods. As shown in the table, the proposed fine-tuning scheme does not reduce the accuracy of the quantized net-

Table 3. Effect of our methods for top-1 accuracy of ResNet-18 and MobileNet-V2 quantized by LSQ [9] on the ImageNet dataset (90 epochs of fine-tuning).

Model	FP	Fine-tuning	4/4	3/3	2/2
ResNet-18	70.542	LSQ	69.39	68.80	66.26
		LSQ + Ours	70.74	69.73	66.58
MobileNet-V2	72.24	LSQ	70.46	67.51	44.87
		LSQ + Ours	71.16	66.93	43.41

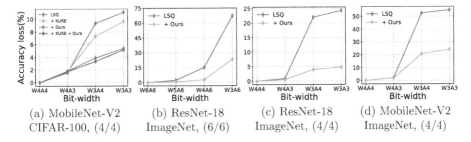

Fig. 8. Robustness of QAT model with and without proposed methods. (W/A) represents the initial bit-width of activation and weight, WXAY indicates the changed bit-width. The accuracy is measured without additional fine-tuning.

work in all cases of ResNet-18 and 4-bit MobileNet-V2. In the case of 3-/2-bit MobileNet-V2, we speculated that the accuracy is slightly degraded due to the overlapped effect of strong regularization and the limited degree of freedom. When applying quantization with 4-bit or higher precision, we can enjoy the benefit of robustness based on the proposed methods without losing accuracy.

6   Discussion

Our proposed methods are advantageous in minimizing quantization error after PTQ and QAT. However, one remaining important topic that we could not address in this paper is the robustness of the activation. In this paper, we rely on the PTQ/QAT algorithms for activation quantization. Unlike weight, activation is input-dependent, and the distribution is diverse depending on the behavior of nonlinear functions. This is a challenging problem and left as future work.

7   Conclusion

Enhancing the robustness of neural networks for quantization maximizes the benefit we can get from the low-precision operations. In this study, we reported three important motivations for minimizing the accuracy degradation after quantization: reduction of error propagation, range clamping for error minimization, and inherited robustness against quantization. Based on these insights, we proposed two novel ideas, symmetry regularization (SymReg) and saturating nonlinearity (SatNL). Our extensive experiments verified the advantages of the proposed methods, which significantly reduce the quantization error of diverse QAT and PTQ algorithms. Enhancing the robustness of quantization is achievable with negligible extra cost, but it enables us to exploit the benefit of low-precision computation with minimal accuracy degradation. We expect that the robustness of networks will minimize the deployment overhead for energy-efficient NPUs, thereby positively affecting the environment.

Acknowledgements. This work was supported by IITP grant funded by the Korea government (MSIT, No. 2019-0-01906, No. 2021-0-00105, and No. 2021-0-00310), SK Hynix Inc. and Google Asia Pacific.

References

1. Alizadeh, M., Behboodi, A., van Baalen, M., Louizos, C., Blankevoort, T., Welling, M.: Gradient l1 regularization for quantization robustness. In: 8th International Conference on Learning Representations, ICLR 2020, Addis Ababa, Ethiopia, 26–30 April 2020. OpenReview.net (2020). https://openreview.net/forum?id=ryxK0JBtPr

2. Banner, R., Nahshan, Y., Hoffer, E., Soudry, D.: ACIQ: analytical clipping for integer quantization of neural networks. CoRR abs/1810.05723 (2018). http://arxiv.org/abs/1810.05723

3. Banner, R., Nahshan, Y., Soudry, D.: Post training 4-bit quantization of convolutional networks for rapid-deployment. In: NeurIPS (2019)

4. Brock, A., De, S., Smith, S.L.: Characterizing signal propagation to close the performance gap in unnormalized ResNets. In: International Conference on Learning Representations (2021). https://openreview.net/forum?id=IX3Nnir2omJ

5. Cai, Y., Yao, Z., Dong, Z., Gholami, A., Mahoney, M.W., Keutzer, K.: ZeroQ: a novel zero shot quantization framework. In: 2020 IEEE/CVF Conference on Computer Vision and Pattern Recognition (CVPR), pp. 13166–13175 (2020)

6. Choi, J., Wang, Z., Venkataramani, S., Chuang, P.I.J., Srinivasan, V., Gopalakrishnan, K.: PACT: parameterized clipping activation for quantized neural networks (2018). https://openreview.net/forum?id=By5ugjyCb

7. Dong, Z., Yao, Z., Arfeen, D., Gholami, A., Mahoney, M.W., Keutzer, K.: HAWQ-V2: hessian aware trace-weighted quantization of neural networks. In: Larochelle, H., Ranzato, M., Hadsell, R., Balcan, M.F., Lin, H. (eds.) Advances in Neural Information Processing Systems, vol. 33, pp. 18518–18529. Curran Associates, Inc. (2020). https://proceedings.neurips.cc/paper/2020/file/d77c703536718b95308130ff2e5cf9ee-Paper.pdf

8. Dong, Z., Yao, Z., Gholami, A., Mahoney, M.W., Keutzer, K.: HAWQ: hessian aware quantization of neural networks with mixed-precision. In; 2019 IEEE/CVF International Conference on Computer Vision (ICCV), pp. 293–302 (2019)

9. Esser, S.K., McKinstry, J.L., Bablani, D., Appuswamy, R., Modha, D.S.: Learned step size quantization. In: International Conference on Learning Representations (2020). https://openreview.net/forum?id=rkgO66VKDS

10. Finkelstein, A., Almog, U., Grobman, M.: Fighting quantization bias with bias. arXiv preprint arXiv:1906.03193 (2019)

11. Foret, P., Kleiner, A., Mobahi, H., Neyshabur, B.: Sharpness-aware minimization for efficiently improving generalization. In: International Conference on Learning Representations (2021). https://openreview.net/forum?id=6Tm1mposlrM

12. Han, T., Li, D., Liu, J., Tian, L., Shan, Y.: Improving low-precision network quantization via bin regularization. In: Proceedings of the IEEE/CVF International Conference on Computer Vision, pp. 5261–5270 (2021)

13. He, K., Zhang, X., Ren, S., Sun, J.: Deep residual learning for image recognition. In: 2016 IEEE Conference on Computer Vision and Pattern Recognition, CVPR 2016, Las Vegas, NV, USA, 27–30 June 2016, pp. 770–778. IEEE Computer Society (2016). https://doi.org/10.1109/CVPR.2016.90

14. Hoffman, J., Roberts, D.A., Yaida, S.: Robust learning with Jacobian regularization (2020). https://openreview.net/forum?id=ryl-RTEYvB

15. Howard, A., et al.: Searching for MobileNetV3. In: Proceedings of the IEEE/CVF International Conference on Computer Vision, pp. 1314–1324 (2019)

16. Hubara, I., Nahshan, Y., Hanani, Y., Banner, R., Soudry, D.: Accurate post training quantization with small calibration sets. In: Meila, M., Zhang, T. (eds.) Proceedings of the 38th International Conference on Machine Learning, ICML 2021,

18–24 July 2021, Virtual Event. Proceedings of Machine Learning Research, vol. 139, pp. 4466–4475. PMLR (2021). http://proceedings.mlr.press/v139/hubara21a. html

17. Jacob, B., et al.: Quantization and training of neural networks for efficient integer-arithmetic-only inference. In: Proceedings of the IEEE Conference on Computer Vision and Pattern Recognition, pp. 2704–2713 (2018)

18. Jang, J., et al.: Sparsity-aware and re-configurable NPU architecture for Samsung flagship mobile SoC. In: 48th ACM/IEEE Annual International Symposium on Computer Architecture, ISCA 2021, Valencia, Spain, 14–18 June 2021, pp. 15–28. IEEE (2021). https://doi.org/10.1109/ISCA52012.2021.00011

19. Jouppi, N.P., et al.: Ten lessons from three generations shaped Google's TPUv4i: industrial product. In: 48th ACM/IEEE Annual International Symposium on Computer Architecture, ISCA 2021, Valencia, Spain, 14–18 June 2021, pp. 1–14. IEEE (2021). https://doi.org/10.1109/ISCA52012.2021.00010

20. Jung, S.H., et al.: Learning to quantize deep networks by optimizing quantization intervals with task loss. In: 2019 IEEE/CVF Conference on Computer Vision and Pattern Recognition (CVPR), pp. 4345–4354 (2019)

21. Kwon, J., Kim, J., Park, H., Choi, I.K.: ASAM: adaptive sharpness-aware minimization for scale-invariant learning of deep neural networks. In: Meila, M., Zhang, T. (eds.) Proceedings of the 38th International Conference on Machine Learning. Proceedings of Machine Learning Research, vol. 139, pp. 5905–5914. PMLR (2021). https://proceedings.mlr.press/v139/kwon21b.html

22. Lee, J.H., Ha, S., Choi, S., Lee, W.J., Lee, S.: Quantization for rapid deployment of deep neural networks (2019). https://openreview.net/forum?id=HkzZBi0cFQ

23. Li, Y., et al.: BRECQ: pushing the limit of post-training quantization by block reconstruction. In: International Conference on Learning Representations (2021). https://openreview.net/forum?id=POWv6hDd9XH

24. Lin, J., Gan, C., Han, S.: Defensive quantization: when efficiency meets robustness. In: International Conference on Learning Representations (2019). https://openreview.net/forum?id=ryetZ20ctX

25. Nagel, M., van Baalen, M., Blankevoort, T., Welling, M.: Data-free quantization through weight equalization and bias correction. In: Proceedings of the IEEE/CVF International Conference on Computer Vision, pp. 1325–1334 (2019)

26. Nahshan, Y., et al.: Loss aware post-training quantization. arXiv abs/1911.07190 (2021)

27. Int4 precision for AI inference (2019). https://devblogs.nvidia.com/int4-for-ai-inference/. Accessed 16 Nov 2021

28. NVIDIA a100 tensor core GPU architecture (2020). https://images.nvidia.com/aem-dam/en-zz/Solutions/data-center/nvidia-ampere-architecture-whitepaper. pdf. Accessed 16 Nov 2021

29. Park, E., Kim, D., Yoo, S.: Energy-efficient neural network accelerator based on outlier-aware low-precision computation. In: International Symposium on Computer Architecture (ISCA) (2018)

30. Park, E., Yoo, S.: PROFIT: a novel training method for sub-4-bit MobileNet models. In: Vedaldi, A., Bischof, H., Brox, T., Frahm, J.-M. (eds.) ECCV 2020. LNCS, vol. 12351, pp. 430–446. Springer, Cham (2020). https://doi.org/10.1007/978-3-030-58539-6_26

31. Park, E., Yoo, S., Vajda, P.: Value-aware quantization for training and inference of neural networks. In: Ferrari, V., Hebert, M., Sminchisescu, C., Weiss, Y. (eds.) ECCV 2018. LNCS, vol. 11208, pp. 608–624. Springer, Cham (2018). https://doi.org/10.1007/978-3-030-01225-0_36

32. Sandler, M., Howard, A., Zhu, M., Zhmoginov, A., Chen, L.C.: MobileNetV2: inverted residuals and linear bottlenecks. In: Proceedings of the IEEE Conference on Computer Vision and Pattern Recognition, pp. 4510–4520 (2018)
33. Sharma, H., et al.: Bit fusion: bit-level dynamically composable architecture for accelerating deep neural networks. In: International Symposium on Computer Architecture (ISCA) (2018)
34. Shkolnik, M., et al.: Robust quantization: one model to rule them all. In: Larochelle, H., Ranzato, M., Hadsell, R., Balcan, M.F., Lin, H.T. (eds.) Advances in Neural Information Processing Systems 33: Annual Conference on Neural Information Processing Systems 2020, NeurIPS 2020, December, pp. 6–12, 2020. Virtual (2020). https://proceedings.neurips.cc/paper/2020/hash/3948ead63a9f2944218de038d8934305-Abstract.html
35. Simonyan, K., Zisserman, A.: Very deep convolutional networks for large-scale image recognition. In: International Conference on Learning Representations (2015)
36. Snapdragon neural processing engine SDK (2017). https://developer.qualcomm.com/docs/snpe/index.html. Accessed 16 Nov 2021
37. Song, J., et al.: 7.1 an 11.5 TOPS/W 1024-MAC butterfly structure dual-core sparsity-aware neural processing unit in 8nm flagship mobile SoC. In: International Solid-State Circuits Conference (ISSCC) (2019)
38. Szegedy, C., et al.: Going deeper with convolutions. In: Computer Vision and Pattern Recognition (CVPR) (2015). http://arxiv.org/abs/1409.4842
39. Tulloch, A., Jia, Y.: High performance ultra-low-precision convolutions on mobile devices. arXiv:1712.02427 (2017)
40. Tulloch, A., Jia, Y.: Quantization and training of neural networks for efficient integer-arithmetic-only inference. In: Conference on Computer Vision and Pattern Recognition (CVPR) (2018)
41. Wei, X., Gong, R., Li, Y., Liu, X., Yu, F.: QDrop: randomly dropping quantization for extremely low-bit post-training quantization. In: International Conference on Learning Representations (2022)
42. Wu, H.: NVIDIA low precision inference on GPU. In: GPU Technology Conference (2019)
43. Wu, H., Judd, P., Zhang, X., Isaev, M., Micikevicius, P.: Integer quantization for deep learning inference: principles and empirical evaluation. CoRR abs/2004.09602 (2020). https://arxiv.org/abs/2004.09602
44. Yao, Z., et al.: HAWQV3: dyadic neural network quantization. In: ICML (2021)
45. Yu, H., Wen, T., Cheng, G., Sun, J., Han, Q., Shi, J.: Low-bit quantization needs good distribution. In: Proceedings of the IEEE/CVF Conference on Computer Vision and Pattern Recognition Workshops, pp. 680–681 (2020)
46. Zhao, R., Hu, Y., Dotzel, J., Sa, C.D., Zhang, Z.: Improving neural network quantization without retraining using outlier channel splitting. In: Chaudhuri, K., Salakhutdinov, R. (eds.) Proceedings of the 36th International Conference on Machine Learning, ICML 2019, Long Beach, California, USA, 9–15 June 2019. Proceedings of Machine Learning Research, vol. 97, pp. 7543–7552. PMLR (2019). http://proceedings.mlr.press/v97/zhao19c.html
47. Zhou, S., Ni, Z., Zhou, X., Wen, H., Wu, Y., Zou, Y.: DoReFa-Net: training low bitwidth convolutional neural networks with low bitwidth gradients. arXiv abs/1606.06160 (2016)

SP-Net: Slowly Progressing Dynamic Inference Networks

Huanyu Wang[1], Wenhu Zhang[2], Shihao Su[1], Hui Wang[1], Zhenwei Miao[3], Xin Zhan[3], and Xi Li[1,4,5(✉)]

[1] College of Computer Science and Technology, Zhejiang University, Hangzhou, China
{huanyuhello,shihaocs,wanghui_17,xilizju}@zju.edu.cn
[2] Polytechnic Institute, Zhejiang University, Hangzhou, China
wenhuzhang@zju.edu.cn
[3] Alibaba Group, Hangzhou, Zhejiang, China
{zhenwei.mzw,zhanxin.zx}@alibaba-inc.com
[4] Shanghai Institute for Advanced Study, Zhejiang University, Hangzhou, China
[5] Shanghai AI Laboratory, Shanghai, China

Abstract. Dynamic inference networks improve computational efficiency by executing a subset of network components, i.e., executing path, conditioned on input sample. Prevalent methods typically assign routers to computational blocks so that a computational block can be skipped or executed. However, such inference mechanisms are prone to suffer instability in the optimization of dynamic inference networks. First, a dynamic inference network is more sensitive to its routers than its computational blocks. Second, the components executed by the network vary with samples, resulting in unstable feature evolution throughout the network. To alleviate the problems above, we propose SP-Nets to slow down the progress from two aspects. First, we design a dynamic auxiliary module to slow down the progress in routers from the perspective of historical information. Moreover, we regularize the feature evolution directions across the network to smoothen the feature extraction in the aspect of information flow. As a result, we conduct extensive experiments on three widely used benchmarks and show that our proposed SP-Nets achieve state-of-the-art performance in terms of efficiency and accuracy.

Keywords: Dynamic inference · Slowly progressing · Executing path regularization · Feature evolution regularization

1 Introduction

Recent years have witnessed a significant development in deep neural networks. The excellent performance of these networks is ascribed not only to the sophisticated design of network modules but also the increasing depth of the network.

H. Wang and W. Zhang—Equal contribution to this work.

Supplementary Information The online version contains supplementary material available at https://doi.org/10.1007/978-3-031-20083-0_14.

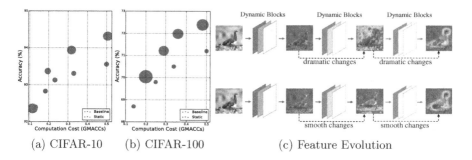

(a) CIFAR-10   (b) CIFAR-100   (c) Feature Evolution

Fig. 1. (a) The stability of static networks (ResNets [12]) and vanilla dynamic inference networks on CIFAR-10. The size of dots is the variance of results. (b) The stability of static networks and vanilla dynamic inference networks on CIFAR-100. The size of dots is the variance of results. (c) Upper: The features evolve unstable throughout the network. Lower: The features evolve stable throughout the network.

However, the merit comes with the price that deep neural networks are highly dependent on computational resources to ensure both efficiency and accuracy, limiting the deployment of these powerful models in real-world scenarios. Hence, to improve the inference efficiency, extensive efforts have been devoted to model compression methods e.g., weight quantization [11,24,36], knowledge distillation [3,13,44], and low-rank approximation [15,16,20]. More often than not, these methods reduce the computational budgets at the expense of a slight drop on performance. Being different from previous streams of methods, dynamic inference methods save computational resources by dynamically assigning adequate computation conditioned on input samples to prevent performance drop. Specifically, dynamic inference networks adaptively execute part of network components and skip the rest when inferring a given sample.

A number of dynamic inference methods have been developed by assigning a router for each convolutional block. At inference time, each router makes a binary executing decision for the current block. Essentially, the binary decision alters the block between a convolution function and an identity mapping. To optimize the discrete executing decisions in an end-to-end fashion, relaxation functions, e.g., Softmax and Gumbel-Softmax, are introduced in the aforementioned methods. However, such an inference mechanism makes the optimization of the dynamic inference network unstable. As shown in Fig. 1(a) and Fig. 1(b), the vanilla dynamic inference methods show larger variance than static inference methods. First, a dynamic inference network is more sensitive to its routers than its convolutional blocks. In essence, the relaxed executing decision in the training process is a coefficient in the linear combination of a convolution function and an identity mapping. Thus, a slight difference in the executing decision results in changes of each element in the output feature map while changes in the input feature impose less effect on the output. Second, since the information flow of the network is calculated in a chain, selectively executing a subset of the network components would cause the interruption of the flow. As a result, features usually evolve unstably across the whole network. For an instance, compared with the features that evolve stable in the lower figure of Fig. 1(c), features in the upper one of Fig. 1(c) show unstable evolution throughout the network.

In this paper, we propose a slowly progressing dynamic inference network to stabilize the optimization via slowing down the progress from the following two aspects. First, we slow down the progress in routers. Considering the unbalance of sensitivity between the parameters in convolutional blocks and routers, we take advantage of the historical information by introducing a dynamic auxiliary module to provide a guidance. Specifically, the auxiliary module is implemented as a momentum-based moving average of the dynamic inference network and gives out pseudo executing paths. Guided by the pseudo executing paths, the routers are more stable against instant changes in paths, and the dynamic inference network reduces gradient variance, thus stabilizing the training procedure. Moreover, we slow down the feature evolution between blocks to ease the interruption of information flow brought by dynamic inference. In this way, the feature evolves smoothly throughout the information flow, and skipping some of the blocks brings in less drastic changes in feature maps. The most straightforward solution to the interruption brought by the varying executing components is minimizing the changes between blocks. However, such a solution cost severe harm to the performance of networks. Thus, we take an alternative strategy that restricts the direction of feature evolution to remedy the interruption of information flow in dynamic inference.

The contributions of this paper are summarized as follows:

- We propose a slowly progressing dynamic inference network, which effectively stabilizes the optimization of dynamic inference networks.
- We slow down the progress in routers by taking advantage of the information from historical iterations to solve the unbalance of sensitivity between parameters in convolutional blocks and routers.
- We slow down the feature evolution by regularizing the direction of the feature evolution, making the feature evolves smoothly throughout the network.
- We conduct experiments on three widely used benchmarks and show that our method obtains state-of-the-art results in terms of performance and efficiency.

2 Relates Work

Dynamic Inference Networks. Dynamic inference networks have emerged as a promising technique to skip blocks or layers at inference time for acceleration [27,31,37,40]. Specifically, ConvNet-AIG [31] proposed a convolutional network that adaptively defines its inference graph conditioned on the input images. It proposes a router to make the execution decision for each convolutional block. SkipNet [40] introduced a method with LSTM gate-ways to determine whether the current block would be skipped or not. Besides, BlockDrop [37] adopted an extra policy network to sample executing paths from the whole routing space to speed up the inference of ResNet. Spatial dynamic convolutions for fast inference were proposed in [28,32,39,43]. Multi-scale networks were introduced in [14,42].

They learn easy samples at low resolutions, while hard samples at high resolutions. Channel-based dynamic inference methods [27] were introduced as well. Recently, various dynamic methods with different kinds of selection have been proposed. Multi-kernel methods [2] select different CNN kernels for better performance. Recursive network [10] are introduced to reuse the networks.

Different from these method who mainly focus on designing dynamic inference mechanisms, we pay attention to the training problem of the dynamic inference network and stabilize the dynamic inference network via slow down the process in router and smoothen feature evolution throughout the network.

Early Prediction Networks. While a dynamic inference network has only one exit, an early prediction network is characterized by multiple exits. In an early prediction network, the network exits once the criterion for a certain sample is satisfied. SACT [5] proposed a halting unit for a recurrent neural network (RNN) to realize early prediction, adopting a stopping unit for each point on feature maps. Since then, early prediction frameworks have been widely used in classification for efficient inference. Considering multi-scale inputs, MSDN [14] introduced early-exit branches. According to the allowed time budget, McIntosh et al. [21] proposed an RNN architecture to dynamically determine the exit. Li *et al.* [19] proposed a self-distillation mechanism to supervise inter-layer outputs with deeper layers. Instead of bypassing residual units, DCP [8] generated decisions to save the computational cost for channels. Hydranets [30] proposed to replace the last residual block with a Mixture-of-Experts layer. Recently, methods have been adopted to other applications, such as action recognition [9,22] and object detection [47].

Our method belongs to dynamic inference networks and does not have multiple exits as early prediction networks do at inference time. However, at training time, we insert several classifiers at different stages of the network.

Knowledge Distillation Methods. Knowledge distillation proposed a concept of distillation, where a lightweight student model is trained to learn the outputs logits of an over-parameterized teacher model. Initially, knowledge distillation [13] is applied to image classification by utilizing the probabilities of each class, generated from the teacher model as soft labels for training a student model [1,13] or learning the intermediate feature maps [25,45]. Moreover, self-distillation is an extension of knowledge distillation, which leverages the network itself as a teacher. Specifically, Born-Again [7] proposes a well-trained model as a teacher to guide a student model from scratch. Furthermore, self ensemble is another way to take advantage of the knowledge distillation. For example, mean teacher [29] proposes an exponential moving average of the model weights. Besides, ensemble in temporal is introduced in [17] to regularized consistent output by a sample wised moving average on the predictions. Recently, self ensemble is also introduced in domain adaptation [41], unsupervised domain adaptation task [6], and medical images segmentation [23].

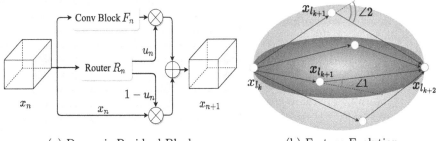

(a) Dynamic Residual Block.   (b) Feature Evolution.

Different from these methods who conduct knowledge distillation to transfer knowledge, we introduce it for stabilizing the training of the dynamic execution decisions in dynamic inference network. To our best knowledge, this is the first work to introduce the distillation in dynamic inference.

3   Method

In this section, we introduce our slowly progressing dynamic inference network (SP-Net) in detail. First, we illustrate the preliminaries of dynamic inference and overview our proposed SP-Net. Second, we introduce the dynamic auxiliary module to provide guidance. Next, we explain the feature evolution regularization. Finally, we summarizes the optimization of the proposed model.

3.1   Preliminaries and Overview

Dynamic Inference Network. Given a N-block dynamic inference network $\mathcal{F}^d$, we term the n-th convolution block F_n, where $n \in \{1, \cdots, N\}$. Generally, a dynamic inference network assigns a router to each convolutional block to decide the execution state, as shown in Fig. 2(a). R_n, the router of F_n, makes a binary executing decision u_n for the current block F_n, where u_n is either 0 or 1, representing skipping the block or executing the block respectively. Combining the execution state, the output feature x_{n+1} of the current block F_n is calculated as

$$
\begin{aligned}
x_{n+1} &= u_n \cdot F_n(x_n) + (1 - u_n) \cdot x_n \\
&= R_n(x_n) \cdot F_n(x_n) + (1 - R_n(x_n)) \cdot x_n.
\end{aligned}
\tag{1}
$$

Let R_n be the router of the n-th convolutional block. To obtain a binary execution decision, R_n usually applies functions that are not continuously differentiable, *e.g.*, argmax, rounding, or sampling. Utilizing such functions hampers the backpropagation of training the network in an end-to-end fashion. Thus, during training, u_n is usually relaxed into a continuous form v_n. Specifically, $v_n \in [0, 1]$ stands for the probability of executing the convolutional block and it is obtained by some relaxation functions involving operations such as Softmax, Gumbel-Softmax, *etc.* Collecting all executing decisions for respective blocks forms an executing path for a given sample $(u_1, u_2, \cdots, u_N)$. Based on the relaxation function, a continuous executing path $(v_1, v_2, \cdots, v_N)$ is obtained accordingly.

The Proposed Framework. In our method, we improve the optimization of the dynamic inference network from two aspects: utilizing the historical information and regularizing the direction of feature evolution. The whole framework of our method is shown in Fig. 2. Our proposed framework contains a dynamic inference network, $\mathcal{F}^d$, a dynamic auxiliary module, $\mathcal{F}^t$, and several attached classifiers.

To analyze the unbalance of sensitivity, we define the parameters of routers R_n as H_n, and the parameters of convolutional blocks F_n as W_n, respectively. In this way, the Eq. (1) is reformulated as,

$$x_{n+1} = R_n(x_n, H_n) \cdot F_n(x_n, W_n) + (1 - R_n(x_n, H_n)) \cdot x_n. \tag{2}$$

The gradient of output feature x_{n+1} with respect to parameters in routers, *i.e.*, H_n and convolutional block, *i.e.*, W_n are computed as,

$$\frac{\partial x_{n+1}}{\partial H_n} = \frac{\partial R_n(\cdot)}{\partial H_n}[F_n(x_n, W_n) - x_n], \tag{3}$$

$$\frac{\partial x_{n+1}}{\partial W_n} = \frac{\partial F_n(\cdot)}{\partial W_n}R_n(x_n, H_n). \tag{4}$$

As shown in Eq. (3), the gradient with respect to router parameters is proportional to $F_n(x_n, W_n) - x_n$, which is the difference between the input feature and the output of convolutional block. Convolutional blocks are able to extract features with large variations from input, resulting in an intense fluctuation in the gradient of routers. In contrast, the gradients with respect to parameters in convolutional blocks are proportional to $R_n(x_n, H_n) \in [0, 1]$, which are a more restricted fluctuation interval. Therefore, the difference between the gradients of routers and convolutional blocks shows that they are of unbalanced sensitivity.

In order to alleviate the unbalance of sensitivity, we introduce a dynamic auxiliary module, of which the network structure is the same as the dynamic inference model. We denote the parameters of $\mathcal{F}^d$ as θ^d and those of $\mathcal{F}^t$ as θ^t. During training, the parameters in the dynamic auxiliary module are progressively updated from the dynamic inference network, in every step by

$$\theta^t = m \cdot \theta^t + (1 - m) \cdot \theta^d, \tag{5}$$

where $m \in [0, 1)$ is a momentum coefficient. In this way, the dynamic auxiliary module $\mathcal{F}^t$ provides historical information including executing decisions and predicted logits. A knowledge distillation loss is employed to transfer the historical information to $\mathcal{F}^d$. It is worth noting that, only the parameters of the dynamic inference network, *i.e.*, θ^d, are updated by back-propagation and the dynamic auxiliary module is not involved in the workflow at inference time.

Second, as defined in Eq. (1), when u_n is zero, the whole information flow would degrade into a skip connection. In this way, the information flow from x_n to x_{n+1} are interrupted, resulting in unstable feature extraction. To solve this problem, we propose to regularize the features of consecutive stages evolve in the same direction. Specifically, we attach several classifiers at stages where down-sampling is applied [46] as shown in Fig. 2. Each classifier is trained with

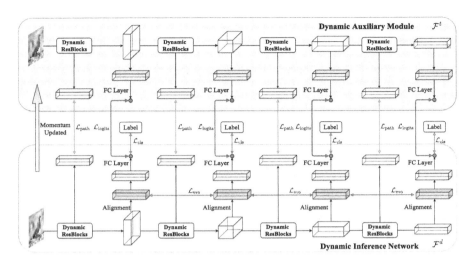

Fig. 2. Illustration of the proposed Slowly Progressing Dynamic Inference Network. The proposed framework consists of a dynamic inference network and a dynamic auxiliary module. We attach several classifiers at different stages of the network.

semantic labels of input. As a result, these classifiers capture the features of the given sample at different resolutions across the network. Then, we regularize the feature evolution angles among the features input to classifiers. These classifiers except the one at the last stage are only utilized in the training period, so there is no additional computation and parameters at inference time.

3.2 Dynamic Auxiliary Module

For convenience, we omit the superscript for dynamic inference network $\mathcal{F}^d$ and the dynamic auxiliary module $\mathcal{F}^t$ in the definition. The set of feature maps extracted by the network is denoted as $\{x_1, x_2, \cdots, x_N\}$ and the feature maps fed to the classifiers is denoted as $\{x_{l_1}, x_{l_2}, \cdots, x_{l_K}\} \subseteq \{x_1, x_2, \cdots, x_N\}$. To be more specific, x_{l_k} is the feature map at the end of the k-th stage of the network. In this way, the output classification results of the dynamic inference network and dynamic auxiliary module are

$$y_1, \cdots, y_K = c_1(x_{l_1}), \cdots, c_K(x_{l_K}), \tag{6}$$

where c_k is the k-th classifier and y_k is the logits output by the k-th classifier. As defined in Sect. 3.1, we term the executing path $p = (v_1, \cdots, v_N)$.

Executing Path Distillation. With the obtained executing path in the dynamic auxiliary module p^t and the executing path in the dynamic inference network p^d, we conduct an executing path distillation, defined as

$$\mathcal{L}_{\text{path}} = \|p^t - p^d\|_2^2 = \sum_{n=1}^{N} (v_n^t - v_n^d)^2, \tag{7}$$

where v_n^t and v_n^d are the executing decision of the n-th dynamic residual block in the dynamic auxiliary module and dynamic inference network respectively.

Logits Distillation. We transfer the knowledge from the outputs logits $\{y_1^t, \cdots, y_K^t\}$ of dynamic auxiliary module to the dynamic inference network by,

$$\mathcal{L}_{\text{logits}} = \sum_{k=1}^{K} \text{KL}(y_k^t \| y_k^d), \tag{8}$$

where $\text{KL}(\cdot \| \cdot)$ is the Kullback-Leibler divergence loss and K is the total number of classifiers in dynamic inference network.

3.3   Feature Evolution Direction

As defined in Eq. (1), feature evolution in a N-block network starts from the input image x_1 to the feature for classification x_N. To make feature evolution from input image to classification feature in a stable process, we propose to regularize the feature evolution in the same direction by minimizing the angles between features from consecutive stages.

Feature Evolution Angle. In most situations, the features from different stages are of different resolutions. Therefore, we introduce a fully connected layer to align them into the same size. Let a_k be the alignment layer before the k-th classifier and x_{l_k} be the input feature to the alignment layer. In this way, the feature evolution angles are calculated by the first-order difference between the aligned features of consecutive stages as follows,

$$\overrightarrow{x_{l_1}, x_{l_2}} = a_2(x_{l_2}) - a_1(x_{l_1}),$$
$$\cdots \tag{9}$$
$$\overrightarrow{x_{l_{K-1}}, x_{l_K}} = a_K(x_{l_K}) - a_{K-1}(x_{l_{K-1}}).$$

Then, we regularize the cosine similarity on the first-order difference of features, *i.e.*, the evolution angles, making the feature evolution in the same direction as

$$\mathcal{L}_{\text{evo}} = \frac{1}{2} \cdot \sum_{m=1}^{M-2} (1 - \cos(\overrightarrow{x_{m+2}, x_{m+1}}, \overrightarrow{x_{m+1}, x_m})), \tag{10}$$

where $\cos(\cdot, \cdot)$ is the cosine similarity. As shown in Fig. 2(b), with $\mathcal{L}_{\text{evo}}$, the feature evolution angle throughout the network would be smaller. Thus, the optimization space is limited and the feature variance throughout the network is effectively smoothened. Therefore, although part of components is skipped at inference time, the interruption brought by dynamic inference is eased.

Multi-stage Classification. Next, to capture the semantic information along feature evolution, all classifiers are supervised by the semantic labels as

$$\mathcal{L}_{\text{cls}} = \sum_{k=1}^{K} \text{CE}(y_k^d, Y), \tag{11}$$

where $\text{CE}(\cdot, \cdot)$ is the Cross-Entropy loss and Y is the semantic label.

3.4 Optimization

Finally, we put all loss together and optimize the dynamic inference network in an end-to-end fashion, the objective function is written as

$$\mathcal{L}_{\text{total}} = \mathcal{L}_{\text{cls}} + \alpha \cdot \mathcal{L}_{\text{path}} + \beta \cdot \mathcal{L}_{\text{logits}} + \gamma \cdot \mathcal{L}_{\text{evo}}, \tag{12}$$

where $\mathcal{L}_{\text{cls}}$ is defined in Eq. (11), $\mathcal{L}_{\text{path}}$ is defined in Eq. (7), $\mathcal{L}_{\text{logits}}$ is defined in Eq. (8), and $\mathcal{L}_{\text{evo}}$ is defined in Eq. (10).

4   Experiment

In this section, we first introduce the experimental settings and implementation details. Second, we compare the results of our proposed method with state-of-the-arts on three benchmarks in terms of efficiency and accuracy. Next, we conduct ablation studies to validate the effectiveness of our design. Finally, we present a qualitative analysis of different designs. Our code is publicly available at https://github.com/huanyuhello/SP-Net.

4.1   Experimental Settings

Datasets and Models. We evaluate the proposed method on three popular classification benchmarks: CIFAR-10, CIFAR-100, and ImageNet. CIFAR-10/100 consists of 50,000 training images and 10,000 testing images with a resolution of 32×32 and annotated by 10/100 classes. ImageNet consists of 1,281,167 training images and 50,000 testing images with a resolution of 224×224 and annotated by 1,000 classes. We conduct extensive experiments based on ResNets, *i.e.*, ResNet-32/110 for CIFAR-10/100 and ResNet-50/101 for ImageNet. For evaluation, we utilize the top-1 metric to measure the classification accuracy and GMACCs (billions of multiply-accumulate operations) to measure the computational cost.

Implementation Details. At training time, we train the whole network for 320 epochs with a batch size of 256 on CIFAR-10/100, and 120 epochs with a batch size of 128 on ImageNet. The initial learning rate of is 0.1 with different schedules. It decreases the learning rate to its 10% at each milestone. Milestones of CIFAR-10/100 are set at epochs 150 and 250, while milestones of ImageNet are set at epochs 30, 60, and 90. Moreover, we employ stochastic gradient descent (SGD) with a momentum of 0.9 and a weight decay of 1e−4 in our method. Besides, the hyper-parameters α, β, and γ, are set to 1, 0.5, and 1e−4 in Eq. (12).

Table 1. Performance comparison with state-of-the-arts on CIFAR-10.

Methods	Backbones	GMACCs	Acc. (%)
ResNet-32 [12]	—	0.14	92.40
ResNet-110 [12]	—	0.50	93.60
Dynamic inference			
SkipNet [40]	ResNet-74	0.09	92.38
BlockDrop [37]	ResNet-110	0.17	93.60
ConvAIG [31]	ResNet-110	0.41	94.24
IamNN [18]	ResNet-101	1.10	94.60
CGap [4]	ResNet-110	0.19	93.43
CoDiNet [34]	ResNet-110	0.29	94.47
RDI-Net [33]	ResNet-110	0.38	95.10
Early prediction			
ACT [5]	ResNet-110	0.38	93.50
SACT [5]	ResNet-110	0.31	93.40
DDI [35]	ResNet-74	0.14	93.88
DG-Net [26]	ResNet-101	3.20	93.99
DG-Net (light)	ResNet-101	2.22	91.99
SP-Net	ResNet-110	0.46	95.22
SP-Net (light)	ResNet-110	0.13	93.79

4.2 Performance Comparison

In this section, we compare the performance of SP-Net with state-of-the-arts on CIFAR and ImageNet datasets w.r.t accuracy and computational cost (Table 1).

Comparison on CIFAR-10. In this section, we compare our method with related method on CIFAR-10. We compare with baselines, dynamic inference methods including: SkipNet [40], BlockDrop [37], ConvAIG [31], CoDiNet [34], IamNN [18], CGap [4], and RDI-Net [33], and early prediction methods including: ACT [5], SACT [5], DDI [35], and DG-Net [26]. Specifically, dynamic inference methods concentrate on skipping the unnecessary blocks with a trainable routing module. Among these methods, RDI-Net achieves 95.10% accuracy with 0.38 GMACCs. CoDiNet achieves 94.47% accuracy with 0.29 GMACCs. Differently, early prediction methods define multiple classifiers and adaptively exit the network. Among these methods, ACT achieves 93.50% with 0.38 GMACCs, while SACT achieves 93.40% with 0.31 GMACCs. In comparison, we provide two versions of model. The light-weight version achieves 93.79% with 0.13 GMACCs, while the normal version achieves 95.22% accuracy with 0.46 GMACCs.

Comparison on ImageNet. In this section, we compare our method with related method on ImageNet including: ConvAIG [31], SkipNet [40], LCNet [38], Block-

Table 2. Performance comparison with state-of-the-arts on ImageNet.

Methods	Backbones	GMACCs	Acc. (%)
ResNet-50 [12]	—	7.72	75.36
ResNet-101 [12]	—	15.26	76.45
Dynamic inference			
ConvAIG [31]	ResNet-50	6.12	76.18
SkipNet [40]	ResNet-101	13.40	77.40
SkipNet (light)	ResNet-101	7.20	75.22
LCNet [38]	ResNet-50	5.78	74.10
BlockDrop [37]	ResNet-101	14.64	76.80
DG-Net [26]	ResNet-101	14.10	76.80
CoDiNet [34]	ResNet-50	6.20	76.63
RDI-Net [33]	ResNet-50	7.42	76.96
Early prediction			
MSDN [14]	DenseNets	4.60	74.24
RA-Net [42]	DenseNets	4.80	75.10
IamNN [18]	ResNet-101	8.00	69.50
ACT [5]	ResNets	13.40	75.30
SACT [5]	ResNets	14.40	75.80
SP-Net	ResNet-50	7.24	77.21
SP-Net (light)	ResNet-50	5.62	76.41

Drop [37], DG-Net [26], CoDiNet [34], RDI-Net [33], MSDN [14], RA-Net [42], IamNN [18], ACT [5], and SACT [5]. As shown in Table 2, the backbone network ResNet-50 achieves 75.36% with 7.72 GMACCs. In dynamic inference methods, ConvAIG achieves 76.18% with 6.12 GMACCs. SkipNet achieves 75.22% with 7.20 GMACCs. Recently, CoDiNet proposes to regularize the consistency and diversity among the executing paths, which achieves 76.63% with 6.20 GMACCs. Similarly, the RDI-Net proposes to ranking the similarity among input samples and their executing paths, which achieves 76.96% with 7.42 GMACCs. In comparison, SP-Net achieves 77.21% with 7.24 GMACCs, which improves 1.75% with 0.32 GMACCs reduction than the static one.

4.3   Ablation Studies

In this part, we discuss the effectiveness of each module in the proposed method. First, we perform the ablation studies on different proposals. Then, we discuss the customizable dynamic routing module to strike the balance.

Evaluation of Proposals. As shown in Table 3, we conduct the ablation studies on our proposed modules. We conduct our method based on ResNet-110 and

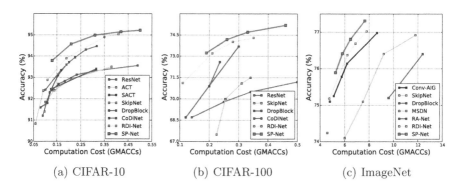

(a) CIFAR-10   (b) CIFAR-100   (c) ImageNet

Fig. 3. The accuracy against computation comparison with state-of-the-art methods.

Table 3. Ablation Studies of our method on CIFAR-10/100. Baseline refers to vanilla dynamic inference network. A refers to the dynamic auxiliary module. B refers to multi-stage classification. C refers to the feature evolution direction regularization.

Method	Components			CIFAR-10		CIFAR-100	
	A	B	C	Acc. (%)	GMACC	Acc. (%)	GMACC
ResNet110	—	—	—	93.61	0.51	71.24	0.51
Baseline	—	—	—	93.83	0.47	70.17	0.46
EXP-1	✓	—	—	94.40	0.48	73.63	0.47
EXP-2	✓	✓	—	95.04	0.47	74.91	0.46
EXP-3	✓	✓	✓	95.22	0.46	75.14	0.46

show a series of experiments. The baseline refers to the vanilla dynamic inference network. Component A refers to dynamic auxiliary module. Component B refers to multi-task regularization. And Component C refers to feature evolution direction regularization. Specifically, the baseline method achieves an accuracy of 93.83% with 0.47% GMACCs and 70.17% with 0.46 GMACCs on CIFAR-10 and CIFAR-100, respectively. With the dynamic auxiliary module, it increases to 94.40% under similar computational cost on CIFAR-10. Moreover, when applying the multi-stage loss, the performance improves by 0.56% on CIFAR-10 and 1.28% on CIFAR-100. Finally, putting all the components together, our method achieves 1.61% improvements with 0.05 GMACCs cost reduction on CIFAR-10 and 3.9% improvements with 0.05 GMACCs cost reduction on CIFAR-100.

Customizable Computation Cost. As shown in Fig. 3, we show the performance of our method compared with the baseline under different computational costs. To enable the computational cost controllable, we apply a cost loss to adjust the executing rates of each block as $\mathcal{L}_{\text{cost}} = \sum_{n=1}^{N} \|v_n^d - t\|_2^2$, where v_n^d is the executing rates, *i.e.*, relaxed executing decision, of the n-th convolution block. t is a hyper-parameter, standing for the target executing rate of the dynamic inference network. We set t to 0.2, 0.4, 0.6, 0.8, and 0.9 on our method and

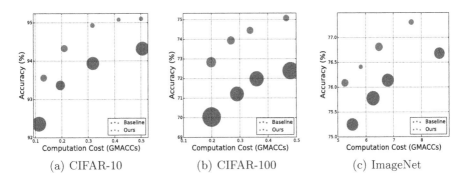

(a) CIFAR-10   (b) CIFAR-100   (c) ImageNet

Fig. 4. The stability and accuracy against computational cost of our method comparing to vanilla dynamic inference networks (Baselines) on CIFAR-10, CIFAR-100, and ImageNet. It is worth noting that the size of dots is the variance of results.

Table 4. Ablation Studies on the momentum coefficient, *i.e.*, m, of SP-Net on CIFAR-10/100. We set momentum from 0.5 to 0.8 under the same setting.

Method	Momentum Coefficient	CIFAR-10		CIFAR-100	
		Acc. (%)	GMACCs	Acc. (%)	GMACCs
EXP-1	$m = 0.5$	94.97	0.47	74.79	0.46
EXP-2	$m = 0.6$	95.03	0.46	75.14	0.46
EXP-3	$m = 0.7$	95.22	0.46	74.86	0.45
EXP-4	$m = 0.8$	94.82	0.47	74.53	0.45

the baseline method. With different target executing rates t, the accuracy varies with the computational cost accordingly. On CIFAR-10, when t is set to 0.2, our method achieves 93.75% at the computational cost of 0.14 GMACCs compared with the baseline method of 90.48% with 0.11 GMACCs. When t is set to 0.4, our method achieves 94.47% accuracy with 0.21 GMACCs, while the baseline method achieves 93.52% with 0.19 GMACCs. Similarly, on CIFAR-100, when t is set to 0.8, our method achieves 75.14% accuracy with 0.46 GMACCs compared with the baseline method of 70.17% accuracy with 0.46 GMACCs. Besides, we train each setting multiple times and demonstrate the variance of the performance as the size of the dots in Fig. 4 on CIFAR-10, CIFAR-100, and ImageNet. The variances of our SP-Nets are smaller than baseline methods, which indicates that our proposed SP-Net not only improves the performance of dynamic inference but also stabilizes the training process accordingly.

4.4   Qualitative Analysis

In this section, we present a qualitative analysis of the key design of our proposed method. First, we conduct a case study on the momentum coefficient m to update

Table 5. Ablation Studies on different feature evolution regularization strategies on CIFAR-10/100. EXP-1 refers to regularization on each adjacent convolutional blocks across the whole network; EXP-2 refers to regularization on each adjacent convolutional blocks within the same resolutions of features; EXP-3 refers to regularization on different resolutions of features across the whole network.

Method	CIFAR-10		CIFAR-100	
	Acc. (%)	GMACCs	Acc. (%)	GMACCs
EXP-1	94.43	0.48	73.57	0.50
EXP-2	94.66	0.49	74.05	0.49
EXP-3	95.22	0.46	74.86	0.46

the parameter in the dynamic auxiliary module. Second, we evaluate different strategies to regularize the feature evolution.

Study on Momentum Coefficient. The performance of different momentum coefficient under a similar computational cost is shown in Table 4. We perform the momentum coefficients m from 0.5 to 0.9. The accuracy increases from 94.97% with 0.47 GMACCs to 95.22% with 0.46 GMACCs with the increasing of the coefficient from 0.5 to 0.7 and then decreases on CIFAR-10. Specifically, on CIFAR-10, when m is set to 0.7, our method achieves the best performance of 95.15% with 0.46 GMACCs. Differently, on CIFAR-100, when m is set to 0.6, our method achieves the best performance of 75.14% with 0.46 GMACCs. Then performance increases from 74.79% with 0.46 GMACCs to 75.14% with the increasing of the coefficient from 0.5 to 0.6 and then decrease to 74.53% with 0.45 GMACCs. Empirically, the coefficient is usually set to a high value in traditional knowledge distillation. However, in the dynamic inference scenarios, since the parameters of routers are much more sensitive than those of convolutional blocks, a high momentum coefficient would lead the optimization collapse.

Study on Feature Evolution Regularization. We study different strategies of feature evolution regularization as shown in Table 5. We design three methods to regularize the feature evolution direction. EXP-1 refers to regularizing on each adjacent convolutional block across the whole network, achieving 94.43% with 0.48 GMACCs on CIFAR-10. EXP-2 refers to regularizing on each adjacent convolutional blocks within the same resolutions of features, achieving 94.66% with 0.49 GMACCs on CIFAR-10. EXP-3 refers to regularizing on different resolutions of features across the whole network, achieving 95.22% with 0.46 GMACCs on CIFAR-10. As a result, regularizing on different resolutions of features achieves the highest accuracy on both CIFAR-10 and CIFAR-100.

5   Conclusion

In this paper, we concentrate on researching an under-explored field in dynamic inference method. We propose a novel strategy called slowly progressing dynamic inference networks via stabilizing executing path and regularizing feature evolution. First, we design a dynamic auxiliary module to solve the imbalance sensitivity between parameters in routers and networks. Besides, we invite a feature evolution regularization making the features evolve smooth throughout the network. As a result, our method achieves state-of-the-art performance on three widely used benchmark in term of accuracy and computational cost reduction.

Acknowledgement. This work was supported by Alibaba Innovative Research (AIR) Program, Alibaba Research Intern Program, National Key Research and Development Program of China under Grant 2020 AAA0107400, Zhejiang Provincial Natural Science Foundation of China under Grant LR19F020004, and National Natural Science Foundation of China under Grant U20A20222.

References

1. Ba, J., Caruana, R.: Do deep nets really need to be deep? In: Proceedings of the Advances in Neural Information Processing Systems, pp. 2654–2662. Curran Associates, Inc. (2014)
2. Chen, Y., Dai, X., Liu, M., Chen, D., Yuan, L., Liu, Z.: Dynamic convolution: attention over convolution kernels. In: Proceedings of the IEEE Conference on Computer Vision and Pattern Recognition (2020)
3. Chen, Z., Zhang, L., Cao, Z., Guo, J.: Distilling the knowledge from handcrafted features for human activity recognition. IEEE Trans. Ind. Inf. **14**, 4334–4342 (2018)
4. Du, X., Li, Z., Ma, Y., Cao, Y.: Efficient network construction through structural plasticity. IEEE J. Emerg. Sel. Top. Circ. Syst. **9**, 453–464 (2019)
5. Figurnov, M., et al.: Spatially adaptive computation time for residual networks. In: Proceedings of the IEEE Conference on Computer Vision and Pattern Recognition (2017)
6. French, G., Mackiewicz, M., Fisher, M.H.: Self-ensembling for visual domain adaptation. In: Proceedings of the International Conference on Learning Representations (2018)
7. Furlanello, T., Lipton, Z.C., Tschannen, M., Itti, L., Anandkumar, A.: Born-again neural networks. In: Proceedings of the International Conference on Machine Learning, pp. 1602–1611 (2018)
8. Gao, X., Zhao, Y., Dudziak, L., Mullins, R., Xu, C.: Dynamic channel pruning: feature boosting and suppression. In: Proceedings of the International Conference on Learning Representations (2019)
9. Ghodrati, A., Bejnordi, B.E., Habibian, A.: FrameExit: conditional early exiting for efficient video recognition. In: Proceedings IEEE Conference on Computer Vision and Pattern Recognition (2021)
10. Guo, Q., Yu, Z., Wu, Y., Liang, D., Qin, H., Yan, J.: Dynamic recursive neural network. In: Proceedings of the IEEE Conference on Computer Vision and Pattern Recognition (2019)
11. Han, S., Mao, H., Dally, W.J.: Deep compression: compressing deep neural networks with pruning, trained quantization and Huffman coding. In: Proceedings of the International Conference on Learning Representations (2016)
12. He, K., Zhang, X., Ren, S., Sun, J.: Deep residual learning for image recognition. In: Proceedings of the IEEE Conference on Computer Vision and Pattern Recognition (2016)

13. Hinton, G., Vinyals, O., Dean, J.: Distilling the knowledge in a neural network. In: Proceedings of the Advances in Neural Information Processing Systems (2015)
14. Huang, G., Chen, D., Li, T., Wu, F., van der Maaten, L., Weinberger, K.: Multi-scale dense networks for resource efficient image classification. In: Proceedings of the International Conference on Learning Representations (2018)
15. Ioannou, Y., Robertson, D., Shotton, J., Cipolla, R., Criminisi, A.: Training CNNs with low-rank filters for efficient image classification. In: Proceedings of the International Conference on Learning Representations (2016)
16. Jaderberg, M., Vedaldi, A., Zisserman, A.: Speeding up convolutional neural networks with low rank expansions. In: Proceedings of the British Machine Vision Conference (2014)
17. Laine, S., Aila, T.: Temporal ensembling for semi-supervised learning. In: Proceedings of the International Conference on Learning Representations (2017)
18. Leroux, S., Molchanov, P., Simoens, P., Dhoedt, B., Breuel, T., Kautz, J.: IamNN: iterative and adaptive mobile neural network for efficient image classification. arXiv:1804.10123 (2018)
19. Li, H., Zhang, H., Qi, X., Yang, R., Huang, G.: Improved techniques for training adaptive deep networks. In: Proceedings of the IEEE International Conference on Computer Vision (2019)
20. McIntosh, L., Maheswaranathan, N., Sussillo, D., Shlens, J.: Convolutional neural networks with low-rank regularization. In: Proceedings of the International Conference on Learning Representations (2016)
21. McIntosh, L., Maheswaranathan, N., Sussillo, D., Shlens, J.: Recurrent segmentation for variable computational budgets. In: Proceedings of the IEEE Conference on Computer Vision and Pattern Recognition (2018)
22. Meng, Y., et al.: AR-Net: adaptive frame resolution for efficient action recognition. In: Vedaldi, A., Bischof, H., Brox, T., Frahm, J.-M. (eds.) ECCV 2020. LNCS, vol. 12352, pp. 86–104. Springer, Cham (2020). https://doi.org/10.1007/978-3-030-58571-6_6
23. Perone, C.S., Ballester, P.L., Barros, R.C., Cohen-Adad, J.: Unsupervised domain adaptation for medical imaging segmentation with self-ensembling. arXiv preprint arXiv:1811.06042 (2018)
24. Polino, A., Pascanu, R., Alistarh, D.: Model compression via distillation and quantization. In: Proceedings of the International Conference on Learning Representations (2018)
25. Romero, A., Ballas, N., Kahou, S.E., Chassang, A., Gatta, C., Bengio, Y.: FitNets: hints for thin deep nets. In: Proceedings of the International Conference on Learning Representations (2015)
26. Shafiee, M.S., Shafiee, M.J., Wong, A.: Dynamic representations toward efficient inference on deep neural networks by decision gates. In: Proceedings of the CVPR Workshop (2019)
27. Su, Z., Fang, L., Kang, W., Hu, D., Pietikäinen, M., Liu, L.: Dynamic group convolution for accelerating convolutional neural networks. In: Vedaldi, A., Bischof, H., Brox, T., Frahm, J.-M. (eds.) ECCV 2020. LNCS, vol. 12351, pp. 138–155. Springer, Cham (2020). https://doi.org/10.1007/978-3-030-58539-6_9
28. Sun, F., Qin, M., Zhang, T., Liu, L., Chen, Y.K., Xie, Y.: Computation on sparse neural networks: an inspiration for future hardware. arXiv:2004.11946 (2020)
29. Tarvainen, A., Valpola, H.: Mean teachers are better role models: weight-averaged consistency targets improve semi-supervised deep learning results. In: Proceedings of the Advances in Neural Information Processing Systems, pp. 1195–1204 (2017)

30. Teja Mullapudi, R., Mark, W.R., Shazeer, N., Fatahalian, K.: HydraNets: specialized dynamic architectures for efficient inference. In: Proceedings of the IEEE Conference on Computer Vision and Pattern Recognition (2018)
31. Veit, A., Belongie, S.: Convolutional networks with adaptive inference graphs. In: Ferrari, V., Hebert, M., Sminchisescu, C., Weiss, Y. (eds.) ECCV 2018. LNCS, vol. 11205, pp. 3–18. Springer, Cham (2018). https://doi.org/10.1007/978-3-030-01246-5_1
32. Verelst, T., Tuytelaars, T.: Dynamic convolutions: exploiting spatial sparsity for faster inference. In: Proceedings of the IEEE Conference on Computer Vision and Pattern Recognition (2020)
33. Wang, H., Li, S., Su, S., Qin, Z., Li, X.: RDI-Net: relational dynamic inference networks. In: Proceedings of the IEEE International Conference on Computer Vision (2021)
34. Wang, H., Qin, Z., Li, S., Li, X.: CoDiNet: path distribution modeling with consistency and diversity for dynamic routing. IEEE Trans. Pattern Anal. Mach. Intell. (2021)
35. Wang, Y., et al.: Dual dynamic inference: enabling more efficient, adaptive and controllable deep inference. IEEE J. Sel. Top. Signal Process. **14**, 623–633 (2020)
36. Wu, J., Leng, C., Wang, Y., Hu, Q., Cheng, J.: Quantized convolutional neural networks for mobile devices. In: Proceedings of the IEEE Conference on Computer Vision and Pattern Recognition (2016)
37. Wu, Z., et al.: BlockDrop: dynamic inference paths in residual networks. In: Proceedings of the IEEE Conference on Computer Vision and Pattern Recognition (2018)
38. Xia, W., Yin, H., Dai, X., Jha, N.K.: Fully dynamic inference with deep neural networks. arXiv:2007.15151 (2020)
39. Xie, Z., Zhang, Z., Zhu, X., Huang, G., Lin, S.: Spatially adaptive inference with stochastic feature sampling and interpolation. In: Vedaldi, A., Bischof, H., Brox, T., Frahm, J.-M. (eds.) ECCV 2020. LNCS, vol. 12346, pp. 531–548. Springer, Cham (2020). https://doi.org/10.1007/978-3-030-58452-8_31
40. Wang, X., Yu, F., Dou, Z.-Y., Darrell, T., Gonzalez, J.E.: SkipNet: learning dynamic routing in convolutional networks. In: Ferrari, V., Hebert, M., Sminchisescu, C., Weiss, Y. (eds.) ECCV 2018. LNCS, vol. 11217, pp. 420–436. Springer, Cham (2018). https://doi.org/10.1007/978-3-030-01261-8_25
41. Xu, Y., Du, B., Zhang, L., Zhang, Q., Wang, G., Zhang, L.: Self-ensembling attention networks: addressing domain shift for semantic segmentation. In: Proceedings of the AAAI Conference on Artificial Intelligence, pp. 5581–5588 (2019)
42. Yang, L., Han, Y., Chen, X., Song, S., Dai, J., Huang, G.: Resolution adaptive networks for efficient inference. In: Proceedings of the IEEE Conference on Computer Vision and Pattern Recognition (2020)
43. Yu, J., Huang, T.S.: Universally slimmable networks and improved training techniques. In: Proceedings of the IEEE International Conference on Computer Vision (2019)
44. Yu, R., et al.: NISP: pruning networks using neuron importance score propagation. In: Proceedings of the IEEE Conference on Computer Vision and Pattern Recognition (2018)
45. Zagoruyko, S., Komodakis, N.: Paying more attention to attention: improving the performance of convolutional neural networks via attention transfer. In: Proceedings of the International Conference on Learning Representations (2017)

46. Zhang, L., Shi, Y., Shi, Z., Ma, K., Bao, C.: Task-oriented feature distillation. In: Larochelle, H., Ranzato, M., Hadsell, R., Balcan, M.F., Lin, H. (eds.) Proceedings of the Advances in Neural Information Processing Systems, pp. 14759–14771 (2020)
47. Zhang, P., Zhong, Y., Li, X.: SlimYOLOv3: narrower, faster and better for real-time UAV applications. In: Proceedings of the IEEE Conference on Computer Vision and Pattern Recognition (2019)

Equivariance and Invariance Inductive Bias for Learning from Insufficient Data

Tan Wad[1]([envelope]), Qianru Sun[2], Sugiri Pranata[3], Karlekar Jayashree[3],
and Hanwang Zhang[1]

[1] Nanyang Technological University, Singapore, Singapore
{tan317,hanwangzhang}@ntu.edu.sg
[2] Singapore Management University, Singapore, Singapore
qianrusun@smu.edu.sg
[3] Panasonic R&D Center Singapore, Singapore, Singapore
{sugiri.pranata,karlekar.jayashree}@sg.panasonic.com

Abstract. We are interested in learning robust models from insufficient data, without the need for any externally pre-trained checkpoints. First, compared to sufficient data, we show why insufficient data renders the model more easily biased to the limited training environments that are usually different from testing. For example, if all the training `swan` samples are "white", the model may wrongly use the "white" environment to represent the intrinsic class `swan`. Then, we justify that **equivariance** inductive bias can retain the class feature while **invariance** inductive bias can remove the environmental feature, leaving the class feature that generalizes to any environmental changes in testing. To impose them on learning, for equivariance, we demonstrate that any off-the-shelf contrastive-based self-supervised feature learning method can be deployed; for invariance, we propose a class-wise invariant risk minimization (IRM) that efficiently tackles the challenge of missing environmental annotation in conventional IRM. State-of-the-art experimental results on real-world benchmarks (VIPriors, ImageNet100 and NICO) validate the great potential of **equivariance** and **invariance** in data-efficient learning. The code is available at https://github.com/Wangt-CN/EqInv.

Keywords: Inductive bias · Equivariance · Invariant risk minimization

1 Introduction

Data is never too big. As illustrated in Fig. 1(a), if we have sufficiently large training sample size of `swan` and `dog`, *e.g.*, dogs and cats in any environment such as different colors, shapes, poses, and backgrounds, by using a conventional softmax cross-entropy based "`swan` vs. `dog`" classifier, we can obtain a "perfect" model that discards the *shared* **environmental** features but retains the *discriminative* **class** features. The underlying common sense is that if the model

---

Supplementary Information The online version contains supplementary material available at https://doi.org/10.1007/978-3-031-20083-0_15.

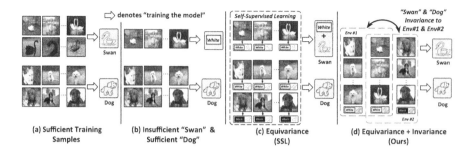

Fig. 1. Illustration of how the proposed equivariance and invariance inductive biases help learning from insufficient data. Cartoon figures such as 🦢 denote the class feature. Boxed words such as [White] denote environmental features. Grey-boxed figures denote the learned model. For simple illustration, we omit the environment as background.

has seen any "case" in training, the testing data is merely a seen IID subset of the training data, yielding testing accuracy as good as training [76].

In this paper, we are interested in learning from insufficient data. Besides the common motivation that collecting data is expensive, we believe that how to narrow the performance gap between insufficient and sufficient data is the key to tackling the non-IID challenge in machine generalization—even if the training data is sufficient, the testing can still be out of the training distribution (OOD) [32,67,70]. After all, we can always frame up exceptional testing samples that fail the trained model [28,37]. Note that **different from few-shot learning** which widely adopts pre-training on large-scale training set [71,73,74], our task does not allow using any externally pre-trained checkpoint and backbone[1].

Figure 1(b) illustrates why insufficient data hurts generalization. Without loss of generality, we conduct a thought experiment that we have limited swan only in "white" color environment while sufficient dog in diverse environments. So, we can expect that the "dog" feature will still be extracted to represent dog model, but the "white" feature will be recklessly learned to represent swan. This is because training swan model by using either "swan" or "white" feature yields the similar training risk: 1) if the former, the training loss is minimized as in the perfect case of Fig. 1(a); 2) if the latter, the only training error possible would be misclassifying "white dog" as swan. However, it can be easily corrected in practice, *e.g.*, by discriminatively training a sample-to-model distance prior that $\|z_{\text{dog}}\| > \|z_{\text{white}}\|$, where z denotes the feature vector[2].

Why, under the same training risk, does the swan model prefer "white" but not "swan" feature? First, feature extraction in deep network follows a bottom-up, low-level to high-level fashion [49]—"simple" features such as color can be easily learned at lower layers, while "complex" features such as object parts will be only emerged in higher layers [69,82,86]. Second, the commonly used cross-

---

[1] See https://vipriors.github.io/ for details.

[2] The distance between the "white dog" sample vector $(z_{\text{white}}, z_{\text{dog}})$ and the swan model vector $(z_{\text{white}}, \mathbf{0})$ is: $\|(z_{\text{white}}, z_{\text{dog}}) - (z_{\text{white}}, \mathbf{0})\| = \|(\mathbf{0}, z_{\text{dog}})\| = \|z_{\text{dog}}\|$; similarly, we have the distance between "white dog" and dog model as $\|z_{\text{white}}\|$.

entropy loss encourages models to stop learning once "simple" features suffice to minimize the loss [25,26]. As a result, "swan" features like "feather", "beak", and "wing" will be lost after training. Such mechanism is also well-known as the shortcut bias [25] or spurious correlation in causality literature [65,78]. We will provide formal justifications in Sect. 3.1.

By comparing the difference between Fig. 1(a) and Fig. 1(b), we can see that the crux of improving the generalization of insufficient data is to recover the missing "swan" class feature while removing the "white" environmental feature. To this end, we propose two inductive biases to guide the learning: *equivariance* for class preservation and *invariance* for environment removal.

Equivariance. This prior requires that the feature representation of a sample should be equivariant to its semantic changes, *e.g.*, any change applied to the sample should be faithfully reflected in the feature change (see Appendix for the mathematical definition). Therefore, if we impose a contrastive loss for each sample feature learning, we can encourage that different samples are mapped into different features (see Sect. 3.2 for a detailed analysis and our choice). As illustrated in Fig. 1(c), equivariance avoids the degenerated case in Fig. 1(b), where all "white swan" samples collapse to the same "white" feature. Thus, for a testing "black swan", the retained "swan" feature can win back some swan scores despite losing the similarity between "black" and "white". It is worth noting that the equivariance prior may theoretically shed light on the recent findings that self-supervised learning features can improve model robustness [35,36,68,79]. We will leave it as future work.

Invariance. Although equivariance preserves all the features, due to the limited environments, the swan model may still be confounded by the "white" environment, that is, a testing "black swan" may still be misclassified as dog, *e.g.*, when $\|(z_{black} - z_{white}, z_{swan} - z_{swan})\| > \|(z_{black} - 0, z_{swan} - z_{dog})\|$. Inspired by invariant risk minimization [4] (IRM) that removes the environmental bias by imposing the environmental invariance prior (Sect. 3.3), as shown in Fig. 1(d), if we split the training data into two environments: "white swan" vs. "white dog" and "white swan" vs. "black dog", we can learn a common classifier (*i.e.*, a feature selector) that focuses on the "swan" and "dog" features, which are the *only invariance* across the two kinds of color environments—one is identical as "white" and the other contains two colors. Yet, conventional IRM requires environment annotation, which is however impractical. To this end, in Sect. 4, we propose **class-wise IRM** based on contrastive objective that works efficiently without the need for the annotation. We term the overall algorithm of using the two inductive biases, *i.e.*, **equivariance** and **invariance**, as EQINV.

We validate the effectiveness of EQINV on three real-world visual classification benchmarks: 1) VIPriors ImageNet classification [12], where we evaluate 10/20/50 samples per class; 2) NICO [32], where the training and testing environmental distributions are severely different; and 3) ImageNet100 [75] which denotes the case of sufficient training data. On all datasets, we observe significant improvements over baseline learners. Our EQINV achieves a new single-model state-of-the-art on test split: 52.27% on VIPriors-50 and 64.14% on NICO.

2   Related Work

Visual Inductive Bias. For a learning problem with many possible solutions, inductive bias is a heuristic prior knowledge that regularizes the learning behavior to find a better solution [57]. It is ubiquitous in any modern deep learning models: from the shallow MLP [55] to the complex deep ResNet [3,10] and Transformers [18,80]. Inductive biases can be generally grouped into two camps: 1) Equivariance: the feature representation should faithfully preserve all the data semantics [19,20,52]. 2) Invariance: generalization is about learning to be invariant to the diverse environments [7,77]. Common practical examples are the pooling/striding in CNN [44], dropout [33], denoising autoencoder [42], batch normalization [23], and data augmentations [11,14].

Data-Efficient Learning. Most existing works re-use existing datasets [13,87] and synthesize artificial training data [22,48]. We work is more related to those that overcome the data dependency by adding prior knowledge to deep nets [9, 27]. Note that data-efficient learning is more general than the popular setting of few-shot learning [71,73,74] which still requires external large pre-training data as initialization or meta-learning. In this work, we offer a theoretical analysis for the difference between learning from insufficient and sufficient data, by posing it in an OOD generalization problem.

OOD Generalization. Conventional machine generalization is based on the Independent and Identically Distributed (IID) assumption for training and testing data [76]. However, this assumption is often challenged in practice—the Out-of-Distribution (OOD) problem degrades the generalization significantly [34,67,84]. Most existing works can be framed into the stable causal effect pursuit [43,65,78] or finding an invariant feature subspace [62,81]. Recently, Invariant Risk Minimization (IRM) takes a different optimization strategy such as convergence speed regularization [4,47] and game theory [1]. Our proposed class-wise IRM makes it more practical by relaxing the restrictions on needing environment annotation.

3   Justifications of the Two Inductive Biases

As we discussed in Sect. 1, given an image $X = x$ with label $Y = y$, our goal is to extract the intrinsic class feature $\phi(x)$ invariant to the environmental changes $z \in Z$. Specifically, Z is defined as all the class-agnostic items in the task of interest. For example, spatial location is the intrinsic class feature in object detection task, but an environmental feature in image classification. This goal can be achieved by using the interventional Empirical Risk Minimization (ERM) [43]. It replaces the observational distribution $P(Y|X)$ with the interventional distribution $P(Y|do(X))$, which removes the environmental effects from the prediction of Y, making $Y = y$ only affected by $X = x$ [63]. The interventional empirical risk $\mathcal{R}$ with classifier f can be written as (See Appendix for the detailed derivation):

$$\mathcal{R} = \mathbb{E}_{x \sim P(X), y \sim P(Y|do(X))} \mathcal{L}(f(\phi(x)), y)$$
$$= \sum_x \sum_y \sum_z \mathcal{L}(f(\phi(x)), y) P(y|x, z) P(z) P(x), \tag{1}$$

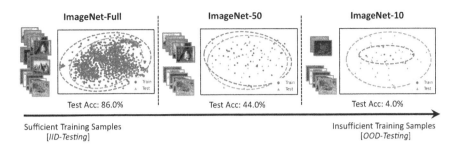

Fig. 2. The t-SNE [56] data visualization of class "hen" on different-scale ImageNet dataset using CLIP [66] pretrained feature extractor. Blue dot and orange triangle represent training and testing samples, respectively. The testing accuracy is evaluated by ResNet-50 [31] trained from scratch on each dataset. See Appendix for details.

where $\mathcal{L}(f(\phi(x)), y)$ is the standard cross-entropy loss. Note that Eq. (1) is hard to implement since the environment Z is unobserved in general.

When the training data is sufficient, X can be almost observed in any environment Z, leading to the approximate independence of Z and X, *i.e.*, $P(Z|X) \approx P(Z)$. Then $\mathcal{R}$ in Eq. (1) approaches to the conventional ERM $\hat{\mathcal{R}}$:

$$\mathcal{R} \approx \hat{\mathcal{R}} = \sum_x \sum_y \mathcal{L}(f(\phi(x)), y) P(y|x) P(x) = \mathbb{E}_{(x,y) \sim P(X,Y)} \mathcal{L}(f(\phi(x)), y), \quad (2)$$

3.1 Model Deficiency in Data Insufficiency

However, when the training data is insufficient, $P(Z|X)$ is no longer approximate to $P(Z)$ and thus $\hat{\mathcal{R}} \not\approx \mathcal{R}$. For example, $P(Z = \boxed{\text{White}} | \text{🐔}) > P(Z = \boxed{\text{Black}} | \text{🐔})$. Then, as we discussed in Sect. 1, some simple environmental semantics Z, *e.g.*, *e.g.*, $Z = \boxed{\text{White}}$, are more likely dominant in minimizing $\hat{\mathcal{R}}$ due to $P(y|x) = P(y|x, z) P(z|x)$ in Eq. (2), resulting the learned ϕ that mainly captures the dominant environment but missing the intrinsic class feature. Empirical results in Fig. 2 also support such analysis. We show the ImageNet classification results of class **hen** using various training sizes. We can observe that with the decreasing of training samples, the accuracy degrades significantly, from 86.0% to 4.0%. After all, when the training size is infinite, any testing data is a subset of training.

3.2 Inductive Bias I: Equivariant Feature Learning

To win back the missing intrinsic class feature, we impose the contrastive-based self-supervised learning (SSL) techniques [15,30,61], without the need for any external data, to achieve the equivariance. In this paper, we follow the definition and implementation in [77] to achieve sample-equivariant by using contrastive learning, *i.e.*, different samples should be respectively mapped to different features. Given an image x, the data augmentation of x constitute the positive

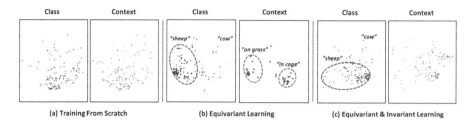

Fig. 3. The t-SNE [56] visualization of learned features *w.r.t* both class and context annotations on NICO dataset with (a) training from scratch; (b) equivariant learning; and (c) equivariant & invariant learning.

example x^+, whereas augmentations of other images constitute N negatives x^-. The key of contrastive loss is to map positive samples closer, while pushing apart negative ones in the feature space:

$$\mathbb{E}_{x,x^+,\{x_i^-\}_{i=1}^N} \left[-\log \frac{\exp(\phi(x)^{\mathrm{T}}\phi(x^+))}{\exp(\phi(x)^{\mathrm{T}}\phi(x^+)) + \sum_{i=1}^N \exp(\phi(x)^{\mathrm{T}}\phi(x_i^-))} \right]. \quad (3)$$

Note that we are open to any SSL choice, which is investigated in Sect. 5.

We visualize the features learned by training from scratch and utilizing the equivariance inductive bias on NICO [32] dataset with both class and context annotations. In Fig. 3(a), it is obvious that there is no clear boundary to distinguish the semantics of class and context in the feature space, while in Fig. 3(b), the features are well clustered corresponding to both class and context.

3.3   Inductive Bias II: Invariant Risk Minimization

Although the equivariance inductive bias preserves all the features, the `swan` model may still be confounded by the "white" feature during the downstream fine-tuning, causing $\hat{\mathcal{R}} \neq \mathcal{R}$. To mitigate such shortcut bias, a straightforward solution is to use Inverse Probability Weighting (IPW) [5,41,53] (also known as reweighting [6,51,60]) to down weight the overwhelmed "white" feature in `swan`. However, they must follow the positivity assumption [39], *i.e.*, all the environmental semantics Z should exist in each class. However, when the training data is insufficient, such assumption no longer holds. For example, how do you down weight "white" over "black" if there is even no "black swan" sample?

Recently, Invariant Risk Minimization (IRM) [4,47] resolves the non-positivity issue by imposing the invariance inductive bias to directly remove the effect of environmental semantics Z. Specifically, IRM first splits the training data into multiple environments $e \in \mathcal{E}$. Then, it regularizes ϕ to be *equally* optimal in different splits, *i.e.*, invariant across environments:

$$\sum_e \mathcal{L}_e(w^{\mathrm{T}}\phi(x),y) + \lambda\|\nabla_{w=1}\mathcal{L}_e(w^{\mathrm{T}}\phi(x),y)\|_2^2, \quad (4)$$

where λ is trade-off hyper-parameter, w stands for a dummy classifier [4] to calculate the gradient penalty across splits—though different environments may induce different losses, the feature ϕ must regularize them optimal at the same time (the lower gradient the better) in the same way (by using the common dummy classifier). Note that each environment should contain a unique mode of environmental feature distribution [2,4,21]: suppose that we have k environmental features that are distributed as $\{p_1, p_2, ..., p_k\}$. If $p_i^{e_1} \neq p_i^{e_2}$, $i = 1$ to k, IRM under the two environments will remove all the k features—the keeping of any one will be penalized by the second term of Eq. (4).

Conventional IRM requires the environment annotations, which are generally impossible in practice. To this end, we propose a novel class-wise IRM to regularize the invariance within each class, without the need for environment supervision. We show the qualitative results of imposing such invariance inductive bias in Fig. 3(c). Compared to Fig. 3(b), we can observe that after applying our proposed class-wise IRM, the equivariance of intrinsic class features are reserved with well-clustered data points while the context labels are no longer responsive—the environment features are removed.

4   Our EqInv Algorithm

Figure 4 depicts the computing flow of EqInv . In the following, we elaborate each of its components.

Input: Insufficient training samples denoted as the pairs $\{(x, y)\}$ of an image x and its label y.

Output: Robust classification model $f \cdot \phi$ with intrinsic class feature $\phi(x)$ and unbiased classifier $f(\phi(x))$.

Step 1: Equivariant Learning via SSL. As introduced in Sect. 3, a wide range of SSL pretext tasks are sufficient for encoding the sample-equivariance. For fair comparison with other methods in VIPriors challenge dataset [12], we use MoCo-v2 [16,30], Simsiam [17], and IP-IRM [77] to learn ϕ in Fig 4(a). We leave the results based on the most recent MAE [29] in Appendix.

Step 2: Environment Construction Based on Adjusted Similarity. Now we are ready to use IRM to remove the environmental features in ϕ. Yet, conventional IRM does not apply as we do not have environment annotations. So, this step aims to automatically construct environments $\mathcal{E}$. However, it is extremely challenging to identify the combinatorial number of unique environmental modes—improper environmental split may contain shared modes, which cannot be removed. To this end, we propose an efficient *class-wise* approximation that seeks *two* environments *w.r.t.* each class. Our key motivation is that, for insufficient training data, the environmental variance within each class is relatively simple and thus we assume that it is single-modal. Therefore, as shown in Fig. 4(b), we propose to use each class (we call anchor class) as an anchor environmental mode to split the samples of the rest of the classes (we call other classes) into two groups: similar to the anchor or not. As a result, for C classes, we will have totally $2C$ approximately unique environments. Intuitively, this

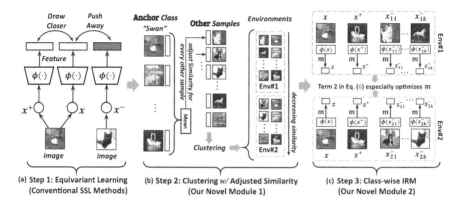

(a) Step 1: Equivariant Learning
(Conventional SSL Methods)

(b) Step 2: Clustering w/ Adjusted Similarity
(Our Novel Module 1)

(c) Step 3: Class-wise IRM
(Our Novel Module 2)

Fig. 4. The flowchart of our proposed EQINV with 3 steps. Rectangle with shading denotes the feature and $\mathcal{E}_j$ represents the generated environment of class j. x_{1k}^- and x_{2k}^- in (c) are the k-th negative samples of subset e_1 and e_2, respectively. We highlight that class-wise IRM optimizes the mask layer m (and an extra MLP g) without gradients flowing back to the feature extract ϕ.

class-wise strategy can effectively remove the severely dominant context bias in a class. For example, if all swan samples are "white", the "white" feature can still be identified as a non-discriminative color feature, thanks to the "black" and "white" samples of dog class.

For each anchor class containing l images, environment Env#1 contains these l samples as positive and the "similar" samples from other classes as negative; environment Env#2 contains the same positive samples while the "dissimilar" samples from other classes as negative. A straightforward way to define the "similarity" between two samples is to use cosine similarity. We compute the cosine similarity between the pair images sampled from anchor class and other classes, respectively. We get the matrix $\mathbf{S} \in \mathbb{R}^{l \times n}$, where n is the number of images in other classes. Then, we average this matrix along the axis of anchor class, as in the pseudocode: $\mathbf{s}^+ = \text{mean}(\mathbf{S}, \dim = 0)$. After ranking $\mathbf{s}^+$, it is easy to get "similar" samples (corresponding to higher half values in $\mathbf{s}^+$) grouped in Env#1 and "dissimilar" samples (corresponding to lower half values in $\mathbf{s}^+$) grouped in Env#2. It is an even split. Figure 5(a) shows the resultant environments for anchor class 0 on the Colored MNIST[3] [60] using the above straightforward cosine similarity. We can see that the digit classes distribute differently in Env#1 and Env#2, indicating that the difference of the two environments also include class information, which will be disastrously removed after applying IRM.

To this end, we propose a similarity adjustment method. It is to adjust every sample-to-class similarity by subtracting a class-to-class similarity, where the sample belongs to the class. First, we calculate the class-to-class similarity $\bar{s}_i$

---

[3] It is modified from MNIST dataset [50] by injecting *color* bias on each digit (class). The non-bias ratio is 0.5%, *e.g.*, 99.5% samples of 0 are in red and only 0.5% in uniform colors.

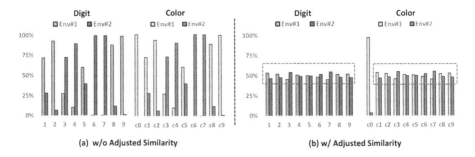

Fig. 5. The obtained environments $\mathcal{E}_0$ for an example `anchor` class 0 on the Colored MNIST [60], by using (a) the vanilla cosine similarity and (b) our adjusted similarity. On X-axis, 1-9 are `other` digit classes, and c0-c9 denote 10 colors used to create this color-bias dataset. On Y-axis, the percentage point denotes the proportion of a digit (or a color) grouped into a specific environment.

between the i-th ($i = 1, ..., C - 1$) `other` class and the `anchor` class: $\bar{s}_i = \text{mean}(\mathbf{s}^+[a_i : b_i])$, where we assume that the image index range of the i-th `other` class is $[a_i : 1 : b_i]$. Such similarity can be viewed as a purer "class effect" to be removed from the total effect of both class and environment—only "environment effect" is then left. Therefore, for any sample x^j from the i-th `other` class, its adjusted similarity to the `anchor` class is: $s = \mathbf{s}^+[j] - \bar{s}_i$. Using this similarity, we obtain new environments and show statistics in Fig. 5(b). It is impressive that the biased color of `anchor` class 0 (*i.e.*, the 0-th color c0 or red) varies between Env#1 and Env#2, but the classes and other colors (red dashed boxes) distribute almost uniformly in these two environments. It means the effects of class and environment are disentangled.

Step 3: Class-wise Invariant Risk Minimization. With the automatically constructed environments, we are ready to remove the environmental feature from ϕ. In particular, we propose a class-wise IRM based on the contrastive objective, which is defined as follows. As shown in Fig. 4(c), given a training image x in environment e of class i, we use a learnable vector mask layer m multiplied on $\phi(x)$ to select the invariant feature. Then, we follow [15] to build a projection layer $g(\cdot)$ to obtain $z = g(m \circ \phi(x))$ for contrastive supervision, where g is a one-hidden-layer MLP with ReLU activation and $\circ$ denotes element-wise production. For each `anchor` class k, we define an environment-based supervised contrastive loss [45]. It is different from the traditional self-supervised contrastive loss. Specifically, our loss is computed within each environment $e \in \mathcal{E}_k$. We take the representations of `anchor` class samples (in e) as positives z^+, and the representations of `other` class samples (in e) as negatives z^-, and we have:

$$\ell(e \in \mathcal{E}_k, w = 1) = \sum_{z \in e} \frac{1}{N^+} \sum_{z^+ \in e} \left[-\log \frac{\exp(z^T z^+ \cdot w)}{\exp(z^T z^+ \cdot w) + \sum_{z^- \in e} \exp(z^T z^- \cdot w)} \right],$$
$$(5)$$

where N^+ denotes the number of the positive samples in the current minibatch and $w = 1$ is a "dummy" classifier to calculate the gradient penalty term [4].

Therefore, the proposed class-wise IRM loss[4] is:

$$\mathcal{L}_k = \sum_{e \in \mathcal{E}_k} \ell(e, w = 1) + \lambda \|\nabla_{w=1} \ell(e, w = 1)\|_2^2, \qquad (6)$$

where λ is the trade-off hyper-parameter. The overall training objective is the combination of minimizing a conventional cross entropy $\mathcal{L}_{ce}$ and the class-wise IRM regularization $\mathcal{L}_k$:

$$\min_{f,g,m,\phi} \mathcal{L}_{ce}(f, m, \phi) + \sum_{k=1}^{C} \mathcal{L}_k(g, m), \qquad (7)$$

where we use $f(m \circ \phi(x))$ for inference. It is worth noting that each loss trains a different set of parameters—ϕ is frozen during the class-wise IRM penalty update. As the equivariance of ϕ is only guaranteed by SSL pretraining, compared to the expensive SSL equivariance regularization in training [77], our frozen strategy is more efficient to mitigate the adversary effect introduced by the invariance bias, which may however discard equivariant features to achieve invariance. We investigate this phenomenon empirically in Sect. 5.4.

5   Experiments

5.1   Datasets and Settings

VIPriors [12] dataset is proposed in VIPrior challenge [12] for data-efficient learning. It contains the same 1,000 classes as in ImageNet [24], and also follows the same splits of `train`, `val` and `test` data. In all splits, each class contains only 50 images, so the total number of samples in the dataset is $150k$. Some related works [8,54] used the merged set (of `train` and `val`) to train the model. We argue that this to some extent violates the protocol of data-efficient learning—using insufficient training data. In this work, our EqInv models as well as comparing models are trained on the standard `train` set and evaluated on `val` and `test` sets. In addition, we propose two more challenging settings to evaluate the models: **VIPriors-20** and **VIPriors-10**. The only difference with VIPriors is they have 20 and 10 images per class in their `train` sets, respectively. There is no change on `val` and `test` sets. We thus call the original **VIPriors-50**. **NICO** [32] is a real-world image dataset proposed for evaluating OOD methods. The key insight of NICO is to provide image labels as well as context labels (annotated by human). On this dataset, it is convenient to "shift" the distribution of the class by "adjusting" the proportions of specific contexts. In our experiments, we follow the "adjusting" settings in the related work [78]. Specifically, this is a challenging OOD setting using the NICO animal set. It mixes three difficulties: 1) Long-Tailed; 2) Zero-Shot and 3) Orthogonal. See Appendix for more details. **ImageNet100** [75] is a subset of original ImageNet [24] with 100 classes and $1k$ images per class. Different with previous OOD datasets, ImageNet100 is to evaluate the performances of our EqInv and comparison methods in sufficient training data settings.

---

[4] Please note that in implementation, we adopt an advanced version [47] of IRM. Please check appendix for details.

Table 1. Recognition accuracies (%) on the VIPriors-50, −20, −10, NICO and ImageNet-100 (IN-100) datasets. "Aug." represents augmentation. Note that due to the effectiveness of "Random Aug.", we set it as a default configuration for the methods trained from SSL. Our results are highlighted.

	Model	VIPriors-50 [12]		VIPriors-20		VIPriors-10		NICO [32]		IN-100 [75]
		Val	Test	Val	Test	Val	Test	Val	Test	Val
	Baseline	32.30	30.60	13.13	12.39	5.02	4.59	43.08	40.77	83.56
Train from Scratch	*Augmentation*									
	Stronger Aug. [15]	36.60	34.72	16.17	15.21	3.49	3.26	42.31	43.31	83.72
	Random Aug. [22]	41.09	39.18	16.71	16.03	3.88	4.01	45.15	44.92	85.12
	Mixup [83]	34.66	32.75	13.35	12.69	2.47	2.31	40.54	38.77	84.52
	Label smoothing [58]	33.77	31.87	12.71	12.05	4.76	4.43	39.77	38.15	85.22
	Debias Learning									
	Lff [60]	35.04	33.29	13.26	12.58	5.20	4.79	41.62	42.54	83.74
	Augment Feat. [51]	35.41	33.63	13.62	12.97	3.43	3.12	42.31	43.27	83.88
	CaaM [78]	36.13	34.24	14.68	13.99	4.88	4.63	46.38	46.62	84.56
Train from SSL	MoCo-v2 [16]	49.47	46.98	30.76	28.83	18.40	16.97	46.45	45.70	86.30
	+EQINV (Ours)	**54.21**	**52.09**	**38.30**	**36.66**	**26.70**	**25.20**	**52.55**	**51.51**	**88.38**
	SimSiam [17]	42.69	40.75	22.09	21.15	6.84	6.68	41.27	42.68	85.28
	+EQINV (Ours)	**52.55**	**50.36**	**37.29**	**35.65**	**24.74**	**23.33**	**45.67**	**44.77**	**86.80**
	IP-IRM [77]	51.45	48.90	38.91	36.26	29.94	27.88	63.60	60.26	86.94
	+EQINV (Ours)	**54.58**	**52.27**	**41.53**	**39.21**	**32.70**	**30.36**	**66.07**	**64.14**	**87.78**

5.2 Implementation Details

We adopted ResNet−50/−18 as model backbones for VIPriors/ImageNet100 and NICO datasets, respectively. We trained the model with 100 epochs for "training from scratch" methods. We initialized the learning rate as 0.1 and decreased it by 10 times at the 60-th and 80-th epochs. We used SGD optimizer and the batch size was set as 256. For equivariant learning (*i.e.*, SSL), we utilized MoCo-v2 [16], SimSiam [17] and IP-IRM [77] to train the model for 800 epochs without using external data, using their default hyper-parameters. We pretrain the model for 200 epochs on ImageNet100 dataset. Then for downstream fine-tuning, We used SGD optimizer and set batch size as 128. We set epochs as 50, initialized learning rate as 0.05, and decreased it at the 30-th and 40-th epochs. Please check appendix for more implementation details. Below we introduce our baselines including augmentation-based methods, debiased learning methods and domain generalization (DG) methods.

Augmentation-Based Methods are quite simple yet effective techniques in the VIPriors challenges as well as for the task of data-efficient learning. We chose four top-performing methods in this category to compare with: stronger augmentation [15], random augmentation [22], mixup [83] and label smoothing [58].

Debias Learning Methods. Data-efficient learning can be regarded as a task for OOD. We thus compared our EQINV with three state-of-the-art (SOTA) debiased learning methods: Lff [60], Augment Feat. [51] and CaaM [78].

Domain Generalization Methods. Domain Generalization (DG) task also tackles the OOD generalization problem, but requires sufficient domain samples and full ImageNet pretraining. In this paper, we select three SOTA DG approaches (SD [64], SelfReg [46] and SagNet [59]) for comparison. These methods do not require domain labels which share the same setting as ours.

5.3   Comparing to SOTAs

Table 1 shows the overall results comparing to baselines on VIPriors−50, −20, −10, NICO and ImageNet100 datasets. Our EQINV achieves the best performance across all settings. In addition, we have another four observations. 1) Incorporating SSL pretraining, vanilla fine-tuning can achieve much higher accuracy than all the methods of "training from scratch". This validates the efficiency of the equivariance inductive bias (learned by SSL) for Etackling the challenge of lacking training data. 2) When decreasing the training size of VIPriors from 50 to 10 images per class, the comparison methods of training from scratch cannot bring performance boosting even hurt the performance. This is because the extremely insufficent data cannot support to establish an equivariant representation, not mention to process samples with harder augmentations. 3) Interestingly, compared to SSL methods, we can see that the improvement margins by our method are larger in the more challenging VIPriors-10, *e.g.*, 8.2% on MoCo-v2 and 16.7% on SimSiam. It validates the invariance inductive bias learned by the class-wise IRM (in our EQINV) helps to disentangle and alleviate the OOD bias effectively. 4) Results on ImageNet100 dataset show the consistent improvements of EQINV due to the additional supervised contrastive loss, indicating the generalizability of our EQINV in a wide range of cases from insufficient to sufficient data.

Table 2. Test accuracy (%) of DG SOTA methods. V−50/−10 denote VIPriors−50/−10.

	Methods	V-50	V-10		Methods	V-50	V-10
Train from Scratch	Boardline	30.60	4.59	Train from SSL	IP-IRM	48.90	26.88
	SD [64]	33.91	4.85		+SD [64]	49.91	28.01
	SelfReg [46]	23.85	3.64		+SelfReg [46]	36.48	22.75
	SagNet [59]	34.92	5.62		+SagNet [59]	47.82	26.17
					+EQINV	**52.27**	**30.36**

In Table 2, we compare our EQINV with DG methods. We try both "train from scratch" and "train from SSL" to meet the pretraining requirement of DG. We can find that our EQINV outperforms DG methods with large margins, showing the weaknesses of existing OOD methods for handling *insufficient* data.

In Table 3, we compare our EQINV with the solutions from other competition teams in the challenge with the same comparable setting: no `val` is used for training, single model w/o ensemble, similar ResNet50/ResNext50 backbones. We can observe the best performance is by our method. It is worth noting that the competitors Zhao *et al.* also used SSL techniques for pretraining. They took the knowledge distillation [38, 40] as their downstream learning method. Our EQINV outperforms their model with a large margin.

Table 3. Accuracy (%) comparison with other competition teams (single model w/o ensemble) on the `val` set of VIPriors−50.

Team	Backbone	Val Acc
Official Baseline	ResNet50	33.16
Zhao *et al.* [85]	ResNet50	44.60
Wang *et al.* [12]	ResNet50	50.70
Sun *et al.* [72]	ResNext50	51.82
EQINV (Ours)	ResNet50	54.58

Table 4. Evaluation of the effectiveness of our three steps in EQINV on VIPriors−20.

Components			Val	Test
Step 1	Step 2	Step 3		
✗	✗	✗	13.13	12.39
✗	✗	✓	13.01	12.41
✗	✓	✓	15.69	14.17
✓	✗	✗	37.61	34.88
✓	✗	✓	38.87	36.34
✓	✓	✓	40.15	37.78

5.4   Ablation Study

Q1: *What are the effects of different components of* EQINV*?*

A1: We traversed different combinations of our proposed three steps to evaluate their effectiveness. The results are shown in Table 4. We can draw the following observations: 1) By focusing on the first three rows, we can find that the improvements are relatively marginal without the SSL equivariance pretraining. This is reasonable as the feature similarity cannot reflect the semantics change exactly without the equivariance property, thus affects the environments construction (Step 2) and class-wise IRM (Step 3); 2) The comparison between row 4 to 6 indicates the significance of our proposed similarity adjustment (Step 2). It is clear that the vanilla cosine similarity results in clear performance drops due to the inaccurate environment construction.

Q2: *What is the optimal λ for* EQINV*? Why does not the class-wise IRM penalty term update feature backbone ϕ?*

A2: Recall that we highlight such elaborate design in Sect. 4 Step 3. In Fig. 6(a), we evaluate the effect of freezing ϕ for Eq. (6) on VIPriors dataset. First, we can see that setting $\lambda = 10$ with freezing ϕ can achieve the best validation and test results. Second, when increasing λ over 10, we can observe a sharp performance drops for updating ϕ, even down to the random guess ($\approx 1\%$). In contrast, the performances are much more robust with freezing ϕ while varying λ, indicating the non-sensitivity of our EQINV . This validate the adversary effect of the equivariance and invariance. Updating ϕ with large λ would destroy the previously learnt equivaraince inductive bias.

Q3: *Does our* EQINV *achieve invariance with the learned environments $\mathcal{E}$ and the proposed class-wise IRM (i.e., Step 3)?*

A3: In Fig. 6(b), we calculate the variance of intra-class feature with training from SSL and our EQINV on VIPriors-10 data. It represents the feature divergence within the class. We can find that: 1) Compared to our EQINV , the

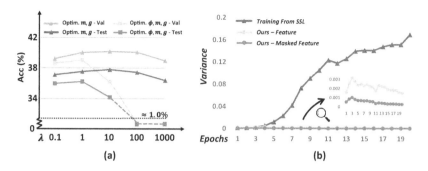

Fig. 6. (a) Accuracies (%) with different optimization schedules and values of λ on the VIPriors-20 `val` and `test` sets. (b) The intra-class feature variance of training from SSL and our EQINV on VIPriors-10 dataset during training process. "Feature" and "Masked Feature" represent $\phi(x)$ and $m \circ \phi(x)$, respectively.

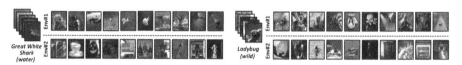

Fig. 7. Visualizations of the top-10 images of generated environments for two classes (*i.e.*, `great white shark` and `ladybug`) on VIPriors-10 dataset. We manually label their main context label (*i.e.*, `water` and `wild`).

variance of training from SSL increases dramatically, indicating that equivariant features are still easily biased to environments without invariance regularizations. 2) The masked feature $m \circ \phi(x)$ of our EQINV achieves continuously lower variance than $\phi(x)$, validates the effectiveness of our learnt mask. See Appendix for more visual attention visualizations.

Q4: *What does the cluster look like for real data with the proposed similarity adjustment (i.e., Step 2)?*

A4: Recall that we have displayed the cluster results on a toy Colored MNIST data in Fig. 5 and validated the superiority of our similarity adjustment. Here we wonder how does it perform on real-world data with much comprehensive semantics? We visualize the top-10 images of Env#1 and Env#2 for two random selected classes in Fig. 7. Interestingly, we can find that images of Env#1 mainly share the context (*e.g.*, `water`) with the anchor class (*e.g.*, `Great White Shark`). In contrast, images of Env#2 have totally different context. More importantly, the classes distribute almost uniformly in both Env#1 and #2, indicating that our adjusted similarity isolate the effect of the class feature.

6   Conclusion

We pointed out the theoretical reasons why learning from insufficient data is inherently more challenging than sufficient data—the latter will be inevitably biased to the limited environmental diversity. To counter such "bad" bias, we proposed to use two "good" inductive biases: equivariance and invariance, which are implemented as the proposed EqInv algorithm. In particular, we used SSL to achieve the equivariant feature learning that wins back the class feature lost by the "bad" bias, and then proposed a class-wise IRM to remove the "bad" bias. For future work, we plan to further narrow down the performance gap by improving the class-muted clustering to construct more unique environments.

Acknowledgement. The authors would like to thank all reviewers for their constructive suggestions. This research is partly supported by the Alibaba-NTU Joint Research Institute, AISG, A*STAR under its AME YIRG Grant (Project No.A20E6c0101).

References

1. Ahuja, K., Shanmugam, K., Varshney, K., Dhurandhar, A.: Invariant risk minimization games. In: ICML, pp. 145–155. PMLR (2020)
2. Ahuja, K., Wang, J., Dhurandhar, A., Shanmugam, K., Varshney, K.R.: Empirical or invariant risk minimization? a sample complexity perspective. arXiv preprint (2020)
3. Allen-Zhu, Z., Li, Y.: What can ResNet learn efficiently, going beyond kernels? In: NeurIPS, vol. 32 (2019)
4. Arjovsky, M., Bottou, L., Gulrajani, I., Lopez-Paz, D.: Invariant risk minimization. arXiv preprint arXiv:1907.02893 (2019)
5. Austin, P.C.: An introduction to propensity score methods for reducing the effects of confounding in observational studies. Multiv. Behav. Res. **46**(3), 399–424 (2011)
6. Bahng, H., Chun, S., Yun, S., Choo, J., Oh, S.J.: Learning de-biased representations with biased representations. In: ICML, pp. 528–539. PMLR (2020)
7. Bardes, A., Ponce, J., LeCun, Y.: VICReg: variance-invariance-covariance regularization for self-supervised learning. arXiv preprint arXiv:2105.04906 (2021)
8. Barz, B., Brigato, L., Iocchi, L., Denzler, J.: A strong baseline for the VIPriors data-efficient image classification challenge. arXiv preprint arXiv:2109.13561 (2021)
9. Battaglia, P.W., et al.: Relational inductive biases, deep learning, and graph networks. arXiv preprint arXiv:1806.01261 (2018)
10. Bietti, A., Mairal, J.: On the inductive bias of neural tangent kernels. In: NeurIPS, vol. 32 (2019)
11. Bouchacourt, D., Ibrahim, M., Morcos, A.: Grounding inductive biases in natural images: invariance stems from variations in data. In: NeurIPS, vol. 34 (2021)
12. Bruintjes, R.J., Lengyel, A., Rios, M.B., Kayhan, O.S., van Gemert, J.: VIPriors 1: visual inductive priors for data-efficient deep learning challenges. arXiv preprint arXiv:2103.03768 (2021)
13. Castro, F.M., Marín-Jiménez, M.J., Guil, N., Schmid, C., Alahari, K.: End-to-end incremental learning. In: ECCV, pp. 233–248 (2018)
14. Chen, S., Dobriban, E., Lee, J.: A group-theoretic framework for data augmentation. NeurIPS **33**, 21321–21333 (2020)
15. Chen, T., Kornblith, S., Norouzi, M., Hinton, G.: A simple framework for contrastive learning of visual representations. In: ICML, pp. 1597–1607. PMLR (2020)
16. Chen, X., Fan, H., Girshick, R., He, K.: Improved baselines with momentum contrastive learning. arXiv preprint arXiv:2003.04297 (2020)

17. Chen, X., He, K.: Exploring simple siamese representation learning. arXiv preprint arXiv:2011.10566 (2020)
18. Chrupała, G.: Symbolic inductive bias for visually grounded learning of spoken language. arXiv preprint arXiv:1812.09244 (2018)
19. Cohen, T., Weiler, M., Kicanaoglu, B., Welling, M.: Gauge equivariant convolutional networks and the icosahedral CNN. In: ICML, pp. 1321–1330. PMLR (2019)
20. Cohen, T., Welling, M.: Group equivariant convolutional networks. In: ICML, pp. 2990–2999. PMLR (2016)
21. Creager, E., Jacobsen, J.H., Zemel, R.: Environment inference for invariant learning. In: ICML, pp. 2189–2200. PMLR (2021)
22. Cubuk, E.D., Zoph, B., Shlens, J., Le, Q.V.: RandAugment: practical automated data augmentation with a reduced search space. In: CVPR Workshops, pp. 702–703 (2020)
23. Daneshmand, H., Joudaki, A., Bach, F.: Batch normalization orthogonalizes representations in deep random networks. In: NeurIPS, vol. 34 (2021)
24. Deng, J., Dong, W., Socher, R., Li, L.J., Li, K., Fei-Fei, L.: ImageNet: a large-scale hierarchical image database. In: CVPR, pp. 248–255. IEEE (2009)
25. Geirhos, R.: Shortcut learning in deep neural networks. Nat. Mach. Intell. **2**(11), 665–673 (2020)
26. Geirhos, R., Rubisch, P., Michaelis, C., Bethge, M., Wichmann, F.A., Brendel, W.: ImageNet-trained CNNs are biased towards texture; increasing shape bias improves accuracy and robustness. ICLR (2018)
27. Gondal, M.W., et al.: On the transfer of inductive bias from simulation to the real world: a new disentanglement dataset. In: NeurIPS, vol. 32 (2019)
28. Goodfellow, I.J., Shlens, J., Szegedy, C.: Explaining and harnessing adversarial examples. arXiv preprint arXiv:1412.6572 (2014)
29. He, K., Chen, X., Xie, S., Li, Y., Dollár, P., Girshick, R.: Masked autoencoders are scalable vision learners. arXiv preprint arXiv:2111.06377 (2021)
30. He, K., Fan, H., Wu, Y., Xie, S., Girshick, R.: Momentum contrast for unsupervised visual representation learning. arXiv preprint arXiv:1911.05722 (2019)
31. He, K., Zhang, X., Ren, S., Sun, J.: Deep residual learning for image recognition. In: CVPR, pp. 770–778 (2016)
32. He, Y., Shen, Z., Cui, P.: Towards non-IID image classification: a dataset and baselines. Pattern Recogn. **110**, 107383 (2021)
33. Helmbold, D.P., Long, P.M.: On the inductive bias of dropout. J. Mach. Learn. Res. **16**(1), 3403–3454 (2015)
34. Hendrycks, D., Dietterich, T.: Benchmarking neural network robustness to common corruptions and perturbations. ICLR (2019)
35. Hendrycks, D., Liu, X., Wallace, E., Dziedzic, A., Krishnan, R., Song, D.: Pretrained transformers improve out-of-distribution robustness. arXiv preprint arXiv:2004.06100 (2020)
36. Hendrycks, D., Mazeika, M., Kadavath, S., Song, D.: Using self-supervised learning can improve model robustness and uncertainty. In: NeurIPS, vol. 32 (2019)
37. Hendrycks, D., Zhao, K., Basart, S., Steinhardt, J., Song, D.: Natural adversarial examples. In: CVPR, pp. 15262–15271 (2021)
38. Heo, B., Kim, J., Yun, S., Park, H., Kwak, N., Choi, J.Y.: A comprehensive overhaul of feature distillation. In: ICCV, pp. 1921–1930 (2019)
39. Hernán, M.A., Robins, J.M.: Causal inference (2010)
40. Hinton, G., Vinyals, O., Dean, J., et al.: Distilling the knowledge in a neural network. arXiv preprint arXiv:1503.02531 2(7) (2015)

41. Imbens, G.W., Rubin, D.B.: Causal Inference in Statistics, Social, and Biomedical Sciences. Cambridge University Press, Cambridge (2015)
42. Jo, Y., Chun, S.Y., Choi, J.: Rethinking deep image prior for denoising. In: ICCV, pp. 5087–5096 (2021)
43. Jung, Y., Tian, J., Bareinboim, E.: Learning causal effects via weighted empirical risk minimization. NeurIPS **33**, 12697–12709 (2020)
44. Kayhan, O.S., Gemert, J.C.v.: On translation invariance in CNNs: convolutional layers can exploit absolute spatial location. In: CVPR, pp. 14274–14285 (2020)
45. Khosla, P., et al.: Supervised contrastive learning. NeurIPS **33**, 18661–18673 (2020)
46. Kim, D., Yoo, Y., Park, S., Kim, J., Lee, J.: SelfReg: self-supervised contrastive regularization for domain generalization. In: ICCV, pp. 9619–9628 (2021)
47. Krueger, D., et al.: Out-of-distribution generalization via risk extrapolation (rex). arXiv preprint (2020)
48. Lahiri, A., Kwatra, V., Frueh, C., Lewis, J., Bregler, C.: LipSync3D: data-efficient learning of personalized 3D talking faces from video using pose and lighting normalization. In: CVPR, pp. 2755–2764 (2021)
49. LeCun, Y., Bengio, Y., Hinton, G.: Deep learning. Nature **521**(7553), 436–444 (2015)
50. LeCun, Y., Cortes, C., Burges, C.: MNIST handwritten digit database (2010)
51. Lee, J., Kim, E., Lee, J., Lee, J., Choo, J.: Learning debiased representation via disentangled feature augmentation. In: NeurIPS, vol. 34 (2021)
52. Lenssen, J.E., Fey, M., Libuschewski, P.: Group equivariant capsule networks. In: NeurIPS, vol. 31 (2018)
53. Little, R.J., Rubin, D.B.: Statistical Analysis with Missing Data, vol. 793. John Wiley & Sons, Hoboken (2019)
54. Liu, Q., Mohamadabadi, B.B., El-Khamy, M., Lee, J.: Diversification is all you need: towards data efficient image understanding (2020)
55. Loshchilov, I., Hutter, F.: Decoupled weight decay regularization. arXiv preprint arXiv:1711.05101 (2017)
56. Van der Maaten, L., Hinton, G.: Visualizing data using t-SNE. J. Mach. Learn. Res. **9**(11), 2579–2605 (2008)
57. Mitchell, T.M.: The need for biases in learning generalizations. Department of Computer Science, Laboratory for Computer Science Research . . . (1980)
58. Müller, R., Kornblith, S., Hinton, G.E.: When does label smoothing help? In: NeurIPS, vol. 32 (2019)
59. Nam, H., Lee, H., Park, J., Yoon, W., Yoo, D.: Reducing domain gap by reducing style bias. In: CVPR, pp. 8690–8699 (2021)
60. Nam, J., Cha, H., Ahn, S., Lee, J., Shin, J.: Learning from failure: de-biasing classifier from biased classifier. NeurIPS **33**, 20673–20684 (2020)
61. Van den Oord, A., Li, Y., Vinyals, O.: Representation learning with contrastive predictive coding. arXiv e-prints pp. arXiv-1807 (2018)
62. Pan, S.J., Tsang, I.W., Kwok, J.T., Yang, Q.: Domain adaptation via transfer component analysis. IEEE Trans. Neural Netw. **22**(2), 199–210 (2010)
63. Pearl, J.: Causality. Cambridge University Press, Cambridge (2009)
64. Pezeshki, M., Kaba, O., Bengio, Y., Courville, A.C., Precup, D., Lajoie, G.: Gradient starvation: a learning proclivity in neural networks. NeurIPS **34**, 1256–1272 (2021)
65. Pfister, N., Bühlmann, P., Peters, J.: Invariant causal prediction for sequential data. J. Am. Stat. Assoc. **114**(527), 1264–1276 (2019)
66. Radford, A., et al.: Learning transferable visual models from natural language supervision. Image **2**, T2 (2021)

67. Recht, B., Roelofs, R., Schmidt, L., Shankar, V.: Do imagenet classifiers generalize to imagenet? In: ICML, pp. 5389–5400. PMLR (2019)
68. Saito, K., Kim, D., Sclaroff, S., Saenko, K.: Universal domain adaptation through self supervision. NeurIPS **33**, 16282–16292 (2020)
69. Selvaraju, R.R., Cogswell, M., Das, A., Vedantam, R., Parikh, D., Batra, D.: Gradcam: visual explanations from deep networks via gradient-based localization. In: Proceedings of the IEEE international conference on computer vision, pp. 618–626 (2017)
70. Shen, Z., Liu, J., He, Y., Zhang, X., Xu, R., Yu, H., Cui, P.: Towards out-of-distribution generalization: a survey. arXiv preprint arXiv:2108.13624 (2021)
71. Snell, J., Swersky, K., Zemel, R.: Prototypical networks for few-shot learning. In: NeurIPS, vol. 30 (2017)
72. Sun, P., Jin, X., Su, W., He, Y., Xue, H., Lu, Q.: A visual inductive priors framework for data-efficient image classification. In: Bartoli, A., Fusiello, A. (eds.) ECCV 2020. LNCS, vol. 12536, pp. 511–520. Springer, Cham (2020). https://doi.org/10.1007/978-3-030-66096-3_35
73. Sun, Q., Liu, Y., Chua, T.S., Schiele, B.: Meta-transfer learning for few-shot learning. In: CVPR, pp. 403–412 (2019)
74. Sung, F., Yang, Y., Zhang, L., Xiang, T., Torr, P.H., Hospedales, T.M.: Learning to compare: Relation network for few-shot learning. In: CVPR, pp. 1199–1208 (2018)
75. Tian, Y., Krishnan, D., Isola, P.: Contrastive multiview coding. arXiv preprint arXiv:1906.05849 (2019)
76. Vapnik, V.: Principles of risk minimization for learning theory. In: NeurIPS (1992)
77. Wang, T., Yue, Z., Huang, J., Sun, Q., Zhang, H.: Self-supervised learning disentangled group representation as feature. In: Conference and Workshop on Neural Information Processing Systems (NeurIPS) (2021)
78. Wang, T., Zhou, C., Sun, Q., Zhang, H.: Causal attention for unbiased visual recognition. In: ICCV, pp. 3091–3100 (2021)
79. Wen, Z., Li, Y.: Toward understanding the feature learning process of self-supervised contrastive learning. In: ICML, pp. 11112–11122. PMLR (2021)
80. Xu, Y., Zhang, Q., Zhang, J., Tao, D.: Vitae: vision transformer advanced by exploring intrinsic inductive bias. In: NeurIPS, vol. 34 (2021)
81. You, K., Long, M., Cao, Z., Wang, J., Jordan, M.I.: Universal domain adaptation. In: CVPR, pp. 2720–2729 (2019)
82. Zeiler, M.D., Fergus, R.: Visualizing and understanding convolutional networks. In: Fleet, D., Pajdla, T., Schiele, B., Tuytelaars, T. (eds.) ECCV 2014. LNCS, vol. 8689, pp. 818–833. Springer, Cham (2014). https://doi.org/10.1007/978-3-319-10590-1_53
83. Zhang, H., Cisse, M., Dauphin, Y.N., Lopez-Paz, D.: Mixup: beyond empirical risk minimization. arXiv preprint arXiv:1710.09412 (2017)
84. Zhang, X., Zhou, L., Xu, R., Cui, P., Shen, Z., Liu, H.: Towards unsupervised domain generalization. In: CVPR, pp. 4910–4920 (2022)
85. Zhao, B., Wen, X.: Distilling visual priors from self-supervised learning. In: Bartoli, A., Fusiello, A. (eds.) ECCV 2020. LNCS, vol. 12536, pp. 422–429. Springer, Cham (2020). https://doi.org/10.1007/978-3-030-66096-3_29
86. Zhou, B., Khosla, A., Lapedriza, A., Oliva, A., Torralba, A.: Learning deep features for discriminative localization. In: CVPR, pp. 2921–2929 (2016)
87. Zhu, F., Cheng, Z., Zhang, X.y., Liu, C.l.: Class-incremental learning via dual augmentation. In: NeurIPS, vol. 34 (2021)

Mixed-Precision Neural Network Quantization via Learned Layer-Wise Importance

Chen Tang[1], Kai Ouyang[1], Zhi Wang[1,4(✉)], Yifei Zhu[2], Wen Ji[3,4],
Yaowei Wang[4], and Wenwu Zhu[1]

[1] Tsinghua University, Beijing, China
{tc20,oyk20}@mails.tsinghua.edu.cn,wangzhi@sz.tsinghua.edu.cn
[2] Shanghai Jiao Tong University, Shanghai, China
[3] Institute of Computing Technology, Chinese Academy of Sciences, Beijing, China
[4] Peng Cheng Laboratory, Shenzhen, China

Abstract. The exponentially large discrete search space in mixed-precision quantization (MPQ) makes it hard to determine the optimal bit-width for each layer. Previous works usually resort to *iterative search* methods on the training set, which consume hundreds or even thousands of GPU-hours. In this study, we reveal that some unique learnable parameters in quantization, namely the scale factors in the quantizer, can serve as importance indicators of a layer, reflecting the contribution of that layer to the final accuracy at certain bit-widths. These importance indicators naturally perceive the numerical transformation during quantization-aware training, which can precisely provide quantization sensitivity metrics of layers. However, a deep network always contains hundreds of such indicators, and training them one by one would lead to an excessive time cost. To overcome this issue, we propose a joint training scheme that can obtain all indicators at once. It considerably speeds up the indicators training process by parallelizing the original sequential training processes. With these learned importance indicators, we formulate the MPQ search problem as a *one-time* integer linear programming (ILP) problem. That avoids the *iterative* search and significantly reduces search time without limiting the bit-width search space. For example, MPQ search on ResNet18 with our indicators takes only 0.06 s, which improves time efficiency exponentially compared to iterative search methods. Also, extensive experiments show our approach can achieve SOTA accuracy on ImageNet for far-ranging models with various constraints (*e.g.*, BitOps, compress rate).

Keywords: Mixed-precision quantization · Model compression

C. Tang and K. Ouyang—Equal contribution.

Supplementary Information The online version contains supplementary material available at https://doi.org/10.1007/978-3-031-20083-0_16.

1   Introduction

Neural network quantization can effectively compress the size and runtime overhead of a network by reducing the bit-width of the network. Using an equal bit-
width for the entire network, a.k.a, fixed-precision quantization, is sub-optimal
because different layers typically exhibit different sensitivities to quantization
[4,25]. It forces the quantization-insensitive layers to work at the same bit-width
as the quantization-sensitive ones, missing the opportunity further to reduce the
average bit-width of the whole network.

Mixed-precision quantization has thus become the focus of network quantization research, with its finer-grained quantization by allowing different bit-widths
for different layers. In this way, the quantization-insensitive layers can use much
lower bit-widths than the quantization-sensitive layers, thus providing more flexible accuracy-efficiency trade-off adjustment than the fixed-precision quantization. Finer-grained quantization also means exponentially larger searching space
to search from. Suppose we have an L-layers network, each layer has n optional
bit-widths for weights and activations, the resulting search space is n^{2L}.

Most of the prior works are search-based. HAQ [25] and AutoQ [21] utilize
deep reinforcement learning (DRL) to search the bit-widths by modeling bit-
width determination problem as a Markov Decision Process. However, due to the
exploration-exploitation dilemma, most existing DRL-based methods require a
significant amount of time to finish the search process. DNAS [26] and SPOS [14]
apply Neural Architecture Search (NAS) algorithms to achieve a differentiable
search process. As a common drawback of NAS, the search space needs to be
greatly and manually limited in order to make the search process feasible, otherwise the search time can be quite high. In a word, the search-based approach
is very time-consuming due to the need to evaluate the searched policy on the
training set for multiple rounds (e.g., 600 rounds in HAQ [25]).

Different from these search-based approaches, some studies aim to define
some "critics" to judge the quantization sensitivity of the layer. HAWQ [12] and
HAWQ-v2 [11] employ second-order information (Hessian eigenvalue or trace)
to measure the sensitivity of layers and leverage them to allocate bit-widths.
MPQCO [6] proposes an efficient approach to compute the Hessian matrix and
formulate a Multiple-Choice Knapsack Problem (MCKP) to determine the bit-
widths assignment. Although these approaches reduce the searching time as
compared to the search-based methods, they have the following defects:

(1) Biased approximation. HAWQ and HAWQv2 approximate the Hessian information on the *full-precision* (unquantized) network to measure the relative sensitivity of layers. This leads to not only an approximation error in these measurements themselves, but more importantly, an inability to perceive the existence
of quantization operations. A full-precision model is a far cry from a quantized
model. We argue that using the information from the full-precision model to
determine the bit-widths assignment of the quantized model is seriously biased
and results in a sub-optimal searched MPQ policy.
(2) Limited search space. MPQCO approximates its objective function with
second-order Taylor expansion. However, the inherent problem in its expan-

Table 1. A comparison of our method and existing works. Iterative search avoiding can significantly reduce the MPQ policy search time. Unlimited search space can provide more potentially excellent MPQ policies. Quantization-aware search can avoid the biased approximation on the full-precision model. Fully automatic bit-width assignment can effectively save human efforts and also reduce the MPQ policy search time. *: MPQCO only can provide the quantization-aware search for weights.

Method	AutoQ	DNAS	HAWQ	HAWQv2	MPQCO	Ours
Iterative search avoiding	No	No	Yes	Yes	Yes	Yes
Unlimited search space	Yes	No	Yes	Yes	No	Yes
Quantization-aware search	Yes	Yes	No	No	Partial yes*	Yes
Fully automatic bit-width assignment	Yes	Yes	No	Yes	No	Yes

sion makes it impossible to quantize the activations with mixed-precision, which significantly limits the search space. A limited search space means that a large number of potentially excellent MPQ policies cannot be accessed during searching, making it more likely to result in sub-optimal performance due to a large number of MPQ policies being abandoned. Moreover, MPQCO needs to assign the bit-witdhs of activations manually, which requires expert involvement and leaves a considerable room for improving search efficiency.

To tackle these problems, we propose to allocate bit-widths for each layer according to the *learned end-to-end importance indicators*. Specifically, we reveal that the learnable scale factors in each layer's quantization function (*i.e.*, quantizer), initially used to adjust the quantization mappings in classic quantization-aware training (QAT) [13,18], can be used as the importance indicators to distinguish whether one layer is more quantization-sensitive than other layers or not. As we will discuss later, they can perceive the numerical error transfer process and capture layers' characteristics in the quantization process (*i.e.*, rounding and clamping) during QAT, resulting in a significant difference in the value of quantization-sensitive and insensitive layers. Since these indicators are learned end-to-end in QAT, errors that might arise from the approximation-based methods are avoided. Moreover, the detached two indicators of each layer for weights and activations allow us to explore the whole search space without limitation.

Besides, an L-layer network with n optional bit-widths for each layer's weights and activations has $M = 2 \times L \times n$ importance indicators. Separately training these M indicators requires M training processes, which is time-prohibitive for deep networks and large-scale datasets. To overcome this bottleneck, we propose a joint scheme to parallelize these M training processes in a once QAT. That considerably reduces the indicators training processes by $M\times$.

Then, based on these obtained layer-wise importance indicators, we transform the original iterative MPQ search problem into a one-time ILP-based mixed-precision search to determine bit-widths for each layer automatically. For example, a sensitive layer (*i.e.*, larger importance) will receive a higher bit-width than an insensitive (*i.e.*, smaller importance) layer. By this means, the time-consuming iterative search is eliminated, since we no longer need to use training data during the search. A concise comparison of our method and existing works is shown in Table 1.

To summarize, our contributions are the following:

- We demonstrate that a small number of learnable parameters (*i.e.*, the scale factors in the quantizer) can act as importance indicators, to reflect the relative contribution of layers to performance in quantization. These indicators are learned end-to-end without performing time-consuming fine-tuning or approximating quantization-unaware second-order information.
- We transform the original *iterative* MPQ search problem into a *one-time* ILP problem by leveraging the learned importance of each layer, increasing time efficiency exponentially without limiting the bit-widths search space. Especially, we achieve about 330× MPQ policy search speedup compared to AutoQ on ResNet50, while preventing 1.7% top-1 accuracy drop.
- Extensive experiments are conducted on a bunch of models to demonstrate the state-of-the-art results of our method. The accuracy gap between full-precision and quantized model of ResNet50 is further narrowed to only 0.6%, while the model size is reduced by 12.2×.

2   Related Work

2.1   Neural Network Quantization

Fixed-Precision Quantization. Fixed-precision quantization [1,3,29,30] focus on using the same bit-width for all (or most of) the layers. In particular, [28] introduces a learnable quantizer, [7] uses the learnable upper bound for activations. [13,18] proposes to use the learnable scale factor (or quantization intervals) instead of the hand-crafted one.

Mixed-Precision Quantization. To achieve a better balance between accuracy and efficiency, many mixed-precision quantization methods which search the optimal bit-width for each layer are proposed.

Search-Based Methods. Search-based methods aim to sample the vast search space of choosing bit-width assignments more effectively and obtain higher performance with fewer evaluation times. [25] and [21] exploit DRL to determine the bit-widths automatically at a layer and kernel level. After that, [24] determines the bit-width by parametrizing the optimal quantizer with the step size and dynamic range. Furthermore, [15] repurposes the Gumbel-Softmax estimator into a smooth estimator of a pair of quantization parameters. In addition, many NAS-based methods have emerged recently [4,14,26,27]. They usually organize the MPQ search problem as a directed acyclic graph (DAG) and make the problem solvable by standard optimization methods (*e.g.*, stochastic gradient descent) through differentiable NAS-based algorithms. DiffQ [9] uses pseudo quantization noise to perform differentiable quantization and search accordingly.

Criterion-Based Methods. Different from exploration approaches, [12] leverages the Hessian eigenvalue to judge the sensitivity of layers, and manually selects the bit width accordingly. [11] further measures the sensitivity by the Hessian trace, and allocates the bit-width based on Pareto frontier automatically. Furthermore, [6] reformulates the problem as a MCKP and proposes a greedy search algorithm to solve it efficiently. The successful achievement of criterion-based methods is that they reduce search costs greatly, but causing a biased approximation or limited search space as we discussed above.

2.2 Indicator-Based Model Compression

Measuring the importance of layers or channels using learned (*e.g.*, scaling factors of batch normalization layers) or approximated indicators are seen as promising work thanks to its excellent efficiency and performance. Early pruning work [19] uses second-derivative information to make a trade-off between network complexity and accuracy. [20] pruning the unimportant channels according to the corresponding BN layer scale factors. [5] sums the scale factors of BN layer to decide which corresponding convolution layer to choose in NAS search process. However, quantization is inherently different from these studies due to the presence of numerical precision transformation.

3   Method

3.1   Quantization Preliminary

Quantization maps the continuous values to discrete values. The uniform quantization function (a.k.a quantizer) under b bits in QAT maps the input $float32$ activations and weights to the homologous quantized values $[0, 2^b - 1]$ and $[-2^{b-1}, 2^{b-1}-1]$. The quantization functions $Q_b(\cdot)$ that quantize the input values v to quantized values v^q can be expressed as follows:

$$v^q = Q_b(v; s) = round(clip(\frac{v}{s}, min_b, max_b)) \times s, \tag{1}$$

where min_b and max_b are the minimum and maximum quantization value [2,13]. For activations, $min_b = 0$ and $max_b = 2^b - 1$. For weights, $min_b = -2^{b-1}$ and $max_b = 2^{b-1}-1$. s is a learnable scalar parameter used to adjust the quantization mappings, called the *step-size scale factor*. For a network, each layer has two distinct scale factors in the weights and activations quantizer, respectively.

To understand the role of the scale factor, we consider a toy quantizer example under b bits and omit the $clip(\cdot)$ function. Namely,

$$v^q = round(\frac{v}{s}) \times s = \hat{v^q} \times s, \tag{2}$$

where $\hat{v^q}$ is the quantized integer value on the discrete domain.

Obviously, for two continuous values v_i and v_j ($v_i \neq v_j$), their quantized integer values $\left|\hat{v_i^q} - \hat{v_j^q}\right| = 0$ if and only if $0 < |v_i - v_j| \leq \frac{s}{2}$. Thus s actually controls the distance between two adjacent quantized values. A larger s means that more different continuous values are mapped to the same quantized value.

3.2   From Accuracy to Layer-Wise Importance

Suppose we have an L-layer network with full-precision parameter tensor $\mathcal{W}$, each layer has n optional bit-widths $\mathcal{B} = \{b_0, ..., b_{n-1}\}$ for activation and weights of each layer, respectively. The bit-width combination of weights and activations $(b_w^{(l)}, b_a^{(l)})$ for layer l is $b_w^{(l)} \in \mathcal{B}$ and $b_a^{(l)} \in \mathcal{B}$. Thus $\mathcal{S} = \{(b_w^{(l)}, b_a^{(l)})\}_{l=0}^{L}$ is the bit-width combination for the whole network, and we use $\mathcal{W}_{\mathcal{S}}$ to denote the quantized parameter tensor. All possible $\mathcal{S}$ construct the search space $\mathcal{A}$. Mixed-precision quantization aims to find the appropriate bit-width combination (*i.e.,* searched MPQ policy) $\mathcal{S}^* \in \mathcal{A}$ for the whole network to maximize the validation accuracy $\mathcal{ACC}_{val}$, under certain constraints C (*e.g.,* model size, BitOps, etc.). The objective can be formalized as follows:

$$\mathcal{S}^* = \arg\max_{\mathcal{S} \sim \Gamma(\mathcal{A})} \mathcal{ACC}_{val}(f(\mathbf{x}; \mathcal{S}, \mathcal{W}_{\mathcal{S}}), \mathbf{y}) \qquad (3)$$

$$\text{s.t.}\quad \mathcal{W}_{\mathcal{S}} = \arg\min_{\mathcal{W}} \mathcal{L}_{train}(f(\mathbf{x}; \mathcal{S}, \mathcal{W}), \mathbf{y}) \qquad (3a)$$

$$BitOps(\mathcal{S}) \leq C \qquad (3b)$$

where $f(\cdot)$ denotes the network, $\mathcal{L}(\cdot)$ is the loss function of task (*e.g.,* cross-entropy), $\mathbf{x}$ and $\mathbf{y}$ are the input data and labels, $\Gamma(\mathcal{A})$ is the prior distribution of $\mathcal{S} \in \mathcal{A}$. For simplicity, we omit the data symbol of training set and validation set, and the parameter tensor of quantizer. This optimization problem is combinatorial and intractable, since it has an extremely large discrete search space $\mathcal{A}$. As above, although it can be solvable by DRL or NAS methods, the time cost is still very expensive. This is due to the need to evaluate the goodness of a specific quantization policy $\mathcal{S}$ on the training set to obtain metrics $\mathcal{L}_{train}$ iteratively to guide the ensuing search. As an example, AutoQ [21] needs more than 1000 GPU-hours to determine a final quantization strategy $\mathcal{S}^*$ [6].

Therefore, we focus on *replacing the iterative evaluation on the training set with some once-obtained importance score of each layer*. In this way, the layer-wise importance score indicates the impact of quantization between and within each layer on the final performance, thus avoiding time-consuming iterative accuracy evaluations. Unlike the approximated Hessian-based approach [11,12], which is imperceptible to quantization operations or limits the search space [6], we propose to *learn the importance in the Quantization-Aware Training*.

3.3   Learned Layer-Wise Importance Indicators

Quantization mapping is critical for a quantized layer since it decides how to use confined quantization levels, and improper mapping is harmful to the performance [18]. As shown in Eq. 1, during QAT, the learnable scale factor of the quantizer in each layer is trained to adjust the corresponding quantization mapping *properly* at a specific bit-width. This means that it can naturally capture

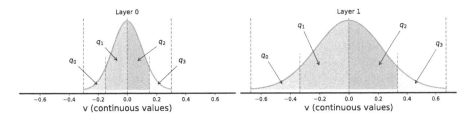

Fig. 1. Illustration of distribution for two example layers, both under 2 bits quantization. The grey dashed line (representing the scale factor of quantizer s) separates the different quantization levels (*i.e.*, $2^2 = 4$ quantized values $\{q_0, q_1, q_2, q_3\}$ for 2 bits). For example, the continuous values in green and red area are quantized to the same quantization level q_1 and q_2, respectively. (Color figure online)

certain quantization characteristics to describe the layers due to its controlled quantization mapping being optimized directly by the task loss.

As we discussed in Sect. 3.1, the continuous values in a uniform range fall into the same quantization level, the specific range is controlled by the scale factor s of this layer. We consider two example layers with *well-trained* s and weights (*i.e.*,, both in a local minimum after a quantization-aware training) and quantized through Eq. 2. As shown in Fig. 1, the continuous distribution of layer 1 is much wider than layer 0. According to the analysis in Sect. 3.1, this results in layer 1 more different continuous values being mapped to the same quantized values (*e.g.*, the values in green area are mapped to q_1), and therefore having a large learned scale factor s. However, while the green area in layer 0 and layer 1 are both quantized to the value q_1, the green area of layer 1 contains a much broader continuous range, and the same goes for other areas. In extreme cases, this extinguishes the inherent differences of original continuous values, thus reducing the expressiveness of the quantized model [23]. To overcome this and maintain the numerical diversity, we should give more available quantization levels to those layers with large scale factors, namely, increasing their bit-width.

Therefore, the numerically significant difference in the learned scale factors of heterogeneous layers can properly assist us to judge the sensitivity of layer. Moreover, the operation involved in the scale factor takes place in the quantizer, which allows it to be directly aware of quantization. Last but not least, there are two quantizers for activations and weights for a layer, respectively, which means that we can obtain the importance of weights and activations separately.

Feasibility Verification. Despite the success of indicator-based methods for model compression [5,19,20] to avoid a time-consuming search process, to the best of our knowledge, there is no literature to demonstrate that the end-to-end learned importance indicators can be used for quantization. To verify the scale factors of quantizer can be used for this purpose, we conduct a contrast experiment for MobileNetv1 [17] on ImageNet [10] as follows.

In the MobileNet family, it is well-known that the depth-wise convolutions (DW-convs) have fewer parameters than the point-wise convolutions (PW-convs); thus, the DW-convs are generally more susceptible to quantization than PW-convs [15,23]. Therefore, we separately quantized the DW-conv and PW-conv for each of the five DW-PW pairs in MobileNetv1 to observe whether the scale factors of the quantizer and accuracy vary. Specifically, we quantized each layer in MobileNetv1 to 2 or 4 bits to observe the accuracy degradation. Each time we quantized only *one layer* to low bits while other layers are not quantized and updated, *i.e.*, we quantized 20 ($5 \times 2 \times 2$) networks independently. If the accuracy of layer l_i degrades higher when the quantization level changes from 4 bits to 2 bits than layer l_j, then l_i is *more sensitive* to quantization than l_j. In addition, the input channels and output channels of these five DW-PW pairs are all 512. Namely, we used the same number of I/O channels to control the variables.

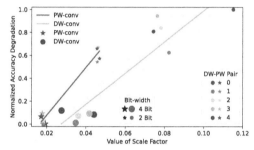

The results of the controlled variable experiment are shown in Fig. 2. Based on the results, we can draw the following conclusions: When the quantization bit-width decreases from 4 to 2 bits, the accuracy degradation of PW-convs is much lower than that of DW-convs, which consists of the prior knowledge that DW-convs are very sensitive. Meanwhile, the values of scale factors of all PW-convs are prominent smaller than those of DW-convs under the

Fig. 2. Results of the contrast experiment of MobileNetv1 on ImageNet. "•" and "⋆" respectively indicate that the DW-conv layer or the PW-conv layer is quantized. Different colors indicate that different layers are quantized. Large labels indicate that the quantization bit-width is set to 4 bits and small labels of 2 bits. (Colr figure online)

same bit-width. That indicates the values of scale factor of whose sensitive layers are bigger than whose insensitive layers, which means the scale factor's value can adequately reflect the quantization sensitivity of the corresponding layer. Namely, the kind of layer with a large scale factor value is more important than the one with a small scale factor.

Initialization of the Importance Indicators. Initializing the scale factors with the statistics [2,13] of each layer results in the different initialization for each layer. We verify whether the factors still show numerical differences by the same initialization value scheme to erase this initialization difference. That is, for each layer, we empirically initialize each importance indicator of bit b by $s_b = 0.1 \times \frac{1}{b}$ since we observed the value of factor is usually quite small (≤ 0.1) and increases as the bit-width decreases.

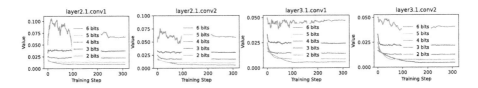

Fig. 3. The importance value of four layers for ResNet18.

As shown in Fig. 3, after training of early instability, the scale factor still showed a significant difference at the end of training. That means the scale factor can still function consistently when using the same initialization for each layer. Nevertheless, we find that, compared to the same initialization value scheme, initialization with statistics [2,13] can speedup and stabilize the training process compared to the same initialization value strategy for each layers, thus we still use the statistics initialization scheme in our experiments.

3.4 One-Time Training for Importance Derivation

Suppose the bit-width options for weights and activations are $\mathcal{B} = \{b_0, ..., b_{n-1}\}$, there are $M = 2 \times L \times n$ importance indicators for an L-layers network. Training these M indicators separately requires M training sessions, which induces huge extra training costs. Therefore, we propose a joint training scheme to obtain importance indicators of all layers corresponding n bit-width options $\mathcal{B}$ at once training.

Specifically, we use a bit-specific importance indicator instead of the original notion s in Eq. 1 for each layer. That is, for the weights and activaions of layer l, we use the notion $s_{w,i}^{(l)}$ and $s_{a,j}^{(l)}$ as the importance indicator for $b_i \in \mathcal{B}$ of weights and $b_j \in \mathcal{B}$ of activations. In this way, n different importance indicators can exist for each layer in a single training session. It is worth noting that the importance indicator parameters are only a tiny percentage of the overall network parameters, thus do not incur too much GPU memory overhead. For example, for ResNet18, if there are 5 bit-width options per layer, we have $M = 2 \times 19 \times 5 = 190$, while the whole network has more than 30 million parameters.

At each training step t, we first perform n times forward and backward propagation corresponding to n bit-width options (i.e., respectively using same bit-width $b_k \in \mathcal{B}$, $k = 0, .., n-1$ for each layer), and inspired by one-shot NAS [8,14] we then introduce one randomly bit-width assignment process for each layer to make sure different bit-widths in different layers can communicate with each other. We define the above procedure as an atomic operation of importance indicators update, in which only the gradients are calculated $n+1$ times, but the importance indicators are not updated during the execution of the operation. After that, we aggregate the above gradients and use them to update the importance indicators. See the Supplementary Material for details and pseudocode.

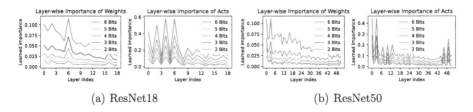

(a) ResNet18   (b) ResNet50

Fig. 4. The importance indicators for ResNet18 and ResNet50.

We show in Fig. 4 all the layer importance indicators obtained by this method in a single training session. We observe that the top layers always show a higher importance value, indicating that these layers require higher bit-widths.

3.5   Mixed-Precision Quantization Search Through Layer-Wise Importance

Now, we consider using these learned importance indicators to allocate bit-widths for each layer automatically. Since these indicators reflect the corresponding layer's contribution to final performance under certain bit-width, we no longer need to use iterative accuracy to evaluate the bit-width combination.

As shown in Fig. 2, the DW-convs always have a higher importance score than PW-convs, and the importance score rise when bit-width reduce, then DW-convs should be quantized to higher bit-width than PW-convs, *e.g.*, 2 bits for PW-convs and 4 bits for DW-convs. For layer l, we use a binary variable $x_{i,j}^{(l)}$ representing the bit-width combination $(b_w^{(l)}, b_a^{(l)}) = (b_i, b_j)$ that b_i bits for weights and b_j bits for activations, whether it is selected or not. Under the given constraint C, our goal is to minimize the summed value of the importance indicator of every layer. Based on that, we reformulate the mixed-precision search into a simple ILP problem as Eq. 4:

$$\underset{\{x_{i,j}^{(l)}\}_{l=0}^{L}}{\arg\min} \sum_{l=0}^{L} (s_{a,j}^{(l)} + \alpha \times s_{w,i}^{(l)}) \times x_{i,j}^{(l)} \tag{4}$$

$$\text{s.t.} \quad \sum_{i}\sum_{j} x_{i,j}^{(l)} = 1 \tag{4a}$$

$$\sum_{l}\sum_{i}\sum_{j} BitOps\left(l, x_{i,j}^{(l)}\right) \leq C \tag{4b}$$

$$\text{vars} \quad x_{i,j}^{(l)} \in \{0,1\} \tag{4c}$$

where Eq. 4a denotes only one bit-width combination selected for layer l, Eq. 4b denotes the summed BitOps of each layer constrains by C. Depending on the deployment scenarios, it can be replaced with other constraints, such as compression rate. α is the hyper-parameter used to form a linear combination of weights and activations importance indicators. Therefore, the final bit-width combination of the whole network $\mathcal{S}^*$ can be obtained by solving Eq. 4.

Please note that, since Eq. 4 do not involve any training data, we no longer need to perform iterative evaluations on the training set as previous works. Thus the MPQ policy search time can be saved exponentially. We solve this ILP by a python library PuLP [22], elapsed time of the solver for ResNet18 is 0.06 s on an 8-core Apple M1 CPU. More details about MPQ policy search efficiency please refer Sect. 4.2.

4   Experiments

In this section, we conduct extensive experiments with the networks ResNet-18/50 [16] and MobileNetv1 [17] on ImageNet [10] classification. We compare our method with the fixed-precision quantization methods including PACT [7], PROFIT [23], LQ-Net [28], and layer-wise MPQ methods HAQ [25], AutoQ [21], SPOS [14], DNAS [26], BP-NAS [27], MPDNN [24], HAWQ [12], HAWQv2 [11], DiffQ [9] and MPQCO [6]. Experimental setups can be found in the Supplementary Material.

4.1   Mixed-Precision Quantization Performance Effectiveness

ResNet18. In Table 2, we show the results of three BitOps (computation cost) constrained MPQ schemes, *i.e.*, 2.5W3A of 19.81G BitOps, 3W3A of 23.07G BitOps and 4W4A of 33.05G BitOps.

Firstly, we observe that in 3-bits level (*i.e.*, 23.07G BitOps) results. We achieve a *least* absolute top-1 accuracy drop than all methods. Please note that the accuracy of our initialization full-precision (FP) model is only 69.6%, which is about 1% lower than some MPQ methods such as SPOS and DNAS. To make a fair comparison, we also provide a result initializing by a higher accuracy FP model (*i.e.*, 70.5%). At this time, the accuracy of the quantized model improves 0.7% and reaches 69.7%, which surpasses all existing methods, especially DNAS 1.0% while DNAS uses a 71.0% FP model as initialization. It is noteworthy that a 2.5W3A (*i.e.*, 19.81G BitOps) result is provided to demonstrate that our method causes less accuracy drop even with a much strict BitOps constraint.

Secondly, in 4-bits level results (*i.e.*, 33.05G BitOps), we also achieve a highest top-1 accuracy than prior arts whether it is fixed-precision quantization method or mixed-precision quantization method. A result initialized by a higher FP precision model is also provided for a fair comparison.

ResNet50. In Table 3, we show the results that not only perform a BitOps constrainted MPQ search but also set a model size constraint (*i.e.*, 12.2 × compression rate). We can observe that our method achieves a much better performance than PACT, LQ-Net, DeepComp, and HAQ, under a much smaller model size (*i.e.*, more than 9MB vs. 7.97MB). In addition, the accuracy degradation of our method is smaller than the criterion-based methods HAWQ, HAWQv2

Table 2. Results for ResNet18 on ImageNet with BitOps constraints. "W-bits" and "A-bits" indicate bit-width of weights and activations respectively. "MP" means mixed-precision quantization. "Top-1/Quant" and "Top-1/FP" indicates the top-1 accuracy of quantized and **F**ull-**P**recision model. "Top-1/Drop" = "Top-1/FP" − "Top-1/Quant".

Method	W-bits	A-bits	Top-1/Quant	Top-1/FP	Top-1/Drop	BitOps (G)
PACT	3	3	68.1	70.4	−2.3	23.09
LQ-Net	3	3	68.2	70.3	−2.1	23.09
Nice	3	3	67.7	69.8	−2.1	23.09
AutoQ	3MP	3MP	67.5	69.9	−2.4	–
SPOS	3MP	3MP	69.4	70.9	−1.5	21.92
DNAS	3MP	3MP	68.7	71.0	−2.3	25.38
Ours	2.5MP	3MP	68.7	69.6	−0.9	**19.81**
Ours	3MP	3MP	69.0	69.6	**−0.6**	23.07
Ours	3MP	3MP	**69.7**	70.5	−0.8	23.07
PACT	4	4	69.2	70.4	−1.2	33.07
LQ-Net	4	4	69.3	70.3	−1.0	33.07
Nice	4	4	69.8	69.8	0	33.07
SPOS	4MP	4MP	70.5	70.9	−0.4	31.81
MPDNN	4MP	4MP	70.0	70.2	−0.2	–
AutoQ	4MP	4MP	68.2	69.9	−1.7	–
DNAS	4MP	4MP	70.6	71.0	−0.4	33.61
MPQCO	4MP	4MP	69.7	69.8	−0.1	–
Ours	4MP	4MP	70.1	69.6	**0.5**	33.05
Ours	4MP	4MP	**70.8**	70.5	0.3	33.05

and MPQCO, which indicates that our quantization-aware search and unlimited search space is necessary for discovering a well performance MPQ policy.

MobileNetv1. In Table 4, we show the results of two BitOps constrainted including a 3-bits level (5.78G BitOps) and a 4-bits level (9.68G BitOps). Especially in the 4-bit level result, we achieve a meaningful accuracy improvement (up to 4.39%) compared to other MPQ methods.

In Table 5, we show the weight only quantization results. We find that the accuracy of our 1.79MB model even surpasses that of the 2.12M HMQ model.

4.2   Mixed-Precision Quantization Policy Search Efficiency

Here, we compare the efficiency of our method to other SOTAs MPQ algorithms with unlimited search space (*i.e.*, MPQ for both weights and activations instead of weights only MPQ, layer-wise MPQ instead of block-wise). Additional results about DiffQ [9] and HAWQ [12] can be found in the Supplementary Material.

Table 3. Results for ResNet50 on ImageNet with BitOps and compression rate constraints. "W-C" means weight compression rate, the size of original full-precision model is 97.28 (MB). "Size" means quantized model size (MB).

Method	W-bits	A-bits	Top-1/Quant	Top-1/Full	Top-1/Drop	W-C	Size (M)
PACT	3	3	75.3	76.9	−1.6	10.67×	9.17
LQ-Net	3	3	74.2	76.0	−1.8	10.67×	9.17
DeepComp	3MP	8	75.1	76.2	−1.1	10.41×	9.36
HAQ	3MP	8	75.3	76.2	−0.9	10.57×	9.22
DiffQ	MP	32	76.3	77.1	−0.8	11.1×	8.8
BP-NAS	4MP	4MP	76.7	77.5	−0.8	11.1×	8.76
AutoQ	4MP	3MP	72.5	74.8	−2.3	−	−
HAWQ	MP	MP	75.5	77.3	−1.8	12.2×	**7.96**
HAWQv2	MP	MP	75.8	77.3	−1.5	12.2×	7.99
MPQCO	2MP	4MP	75.3	76.1	−0.8	12.2×	7.99
Ours	3MP	4MP	**76.9**	77.5	**−0.6**	**12.2×**	7.97

Table 4. Results for MobileNetv1 on ImageNet with BitOps constraints. "W-b" and "A-b" means weight and activation bit-widths. "Top-1" and "Top-5" represent top-1 and top-5 accuracy of quantized model respectively. "B (G)" means BitOps (G).

Method	W-b	A-b	Top-1	Top-5	B (G)
PROFIT	4	4	69.05	88.41	9.68
PACT	6	4	67.51	87.84	14.13
HMQ	3MP	4MP	69.30	−	−
HAQ	4MP	4MP	67.45	87.85	−
HAQ	6MP	4MP	70.40	89.69	−
Ours	3MP	3MP	**69.48**	**89.11**	**5.78**
Ours	4MP	4MP	**71.84**	**90.38**	9.68

Table 5. Weight only quantization results for MobileNetv1 on ImageNet. "W-b" means weight bit-widths. "S (M)" means quantized model size (MB).

Method	W-b	Top-1	Top-5	S (M)
DeepComp	3MP	65.93	86.85	1.60
HAQ	3MP	67.66	88.21	1.58
HMQ	3MP	69.88	−	**1.51**
Ours	3MP	**71.57**	**90.30**	1.79
PACT	8	70.82	89.85	4.01
DeepComp	4MP	71.14	89.84	2.10
HAQ	4MP	71.74	90.36	**2.07**
HMQ	4MP	70.91	−	2.12
Ours	4MP	**72.60**	**90.83**	2.08

The time consumption of our method consists of 3 parts. Namely, *1)* Importance indicators training. *2)* MPQ policy search. *3)* Quantized model fine-tuning. The last part is necessary for all MPQ algorithms while searching the MPQ policy is the biggest bottleneck (*e.g.,* AutoQ needs more than 1000 GPU-hours to determine the final MPQ policy), thus we mainly focus on the first two parts.

Comparison with SOTAs on ResNet50. The time consumption of the first part is to leverage the joint training technique (see Sect. 3.4) to get importance indicators for all layers and their corresponding bit-widths, but it only needs to be done once. It needs to train the network about 50 min (using 50% data of training set) on 4 NVIDIA A100 GPUs (*i.e.*, 3.3 GPU-hours). The time consumption of the second part is to solve the ILP problem. It consumes 0.35 s on a six-core Intel i7-8700 (at 3.2 GHz) CPU, which is negligible.

Hence, suppose we have different z devices with diverse computing capabilities to deploy, our method consumes $50 + 0.35 \times \frac{1}{60} \times z$ minutes to finish the whole MPQ search processes.

Compared with the Search-Based Approach. AutoQ [21] needs 1000 GPU-hours to find the MPQ policy for a single device, which means it needs $1000z$ GPU-hours to search MPQ policies for these z devices. Thus we achieve about **330z×** speedup and obtain a higher accuracy model simultaneously.

Compared with the Criterion-Based Approach. HAWQv2 [11] takes 30 min on 4 GPUs to approximate the Hessian trace. The total time consumption of HAWQv2 for these z devices is $30 + c \times \frac{1}{60} \times z$ minutes, and c is the time consumption for solving a Pareto frontier based MPQ search algorithm with less than 1 min. Thus if z is large enough, our method has almost the same time overhead as HAWQv2. If z is small, *e.g.*, $z = 1$, our method only needs a one-time additional 20-minute investment for the cold start of first part, but resulting in a significant accurate model (*i.e.*, 1.1% top-1 accuracy improvement).

4.3   Ablation Study

In previous analysis, we empirically verify that the layers with bigger scale factor values are more sensitive to quantization when their quantization bit-width is reduced. Motivated by this, we propose our ILP-based MPQ policy search method. However, an intuitive question is *what if we reverse the correlation between scale factors and sensitivity*. Namely, what if we gave the layers with smaller scale factor values more bit-widths instead of fewer bit-widths. And, what if we gave the layers with bigger scale factor values fewer bit-widths instead of more bit-widths.

The result is shown in Table 6, we use "Ours-R" to denote the result of reversed bit-width assignment manner; "Ours" results come from Table 4 directly to represent the routine (not reversed) bit-width assignment manner.

We observe that "Ours-R" has 6.59% top-1 accuracy lower than our routine method under the same BitOps constraint. More seriously, it has 4.23% absolute accuracy gap between "Ours-R" (with 4-bits level constrainted, *i.e.*, 9.68 BitOps) and a 3-bits level (*i.e.*, 5.78G BitOps) routine result. Such a colossal accuracy gap demonstrates that our ILP-based MPQ policy search method is reasonable.

Table 6. Ablation study for MobileNetv1 on ImageNet.

Method	W-bits	A-bits	Top-1/Quant	Top-5/Quant	BitOps
Ours	3MP	3MP	69.48	89.11	5.78
Ours	4MP	4MP	71.84	90.38	9.68
Ours-R	4MP	4MP	65.25	86.15	9.68

5 Conclusion

In this paper, we propose a novel MPQ method that leverages the unique parameters in quantization, namely the scale factors in the quantizer, as the importance indicators to assign the bit-width for each layer. We demonstrate the association between these importance indicators and the quantization sensitivity of layers empirically. We conduct extensive experiments to verify the effectiveness of using these learned importance indicators to represent the contribution of certain layers under specific bit-width to the final performance, as well as to demonstrate the rationality of the bit-width assignment obtained by our method.

Acknowledgements. This work is supported in part by NSFC (Grant No. 61872215 and No. 62072440), the Beijing Natural Science Foundation (Grant No. 4202072), and Shenzhen Science and Technology Program (Grant No. RCYX20200714114523079). Yifei Zhu's work is supported by SJTU Explore-X grant.

References

1. Baskin, C., et al.: Nice: noise injection and clamping estimation for neural network quantization. Mathematics **9**(17), 2144 (2021)
2. Bhalgat, Y., Lee, J., Nagel, M., Blankevoort, T., Kwak, N.: Lsq+: improving low-bit quantization through learnable offsets and better initialization. In: Proceedings of the IEEE/CVF Conference on Computer Vision and Pattern Recognition Workshops, pp. 696–697 (2020)
3. Cai, Z., He, X., Sun, J., Vasconcelos, N.: Deep learning with low precision by half-wave gaussian quantization. In: Proceedings of the IEEE Conference on Computer Vision and Pattern Recognition, pp. 5918–5926 (2017)
4. Cai, Z., Vasconcelos, N.: Rethinking differentiable search for mixed-precision neural networks. In: Proceedings of the IEEE/CVF Conference on Computer Vision and Pattern Recognition, pp. 2349–2358 (2020)
5. Chen, B., et al.: BN-NAS: neural architecture search with batch normalization. In: Proceedings of the IEEE/CVF International Conference on Computer Vision, pp. 307–316 (2021)
6. Chen, W., Wang, P., Cheng, J.: Towards mixed-precision quantization of neural networks via constrained optimization. In: Proceedings of the IEEE/CVF International Conference on Computer Vision, pp. 5350–5359 (2021)
7. Choi, J., Wang, Z., Venkataramani, S., Chuang, P.I.J., Srinivasan, V., Gopalakrishnan, K.: Pact: parameterized clipping activation for quantized neural networks. arXiv preprint arXiv:1805.06085 (2018)

8. Chu, X., Zhang, B., Xu, R.: Fairnas: rethinking evaluation fairness of weight sharing neural architecture search. In: Proceedings of the IEEE/CVF International Conference on Computer Vision, pp. 12239–12248 (2021)

9. Défossez, A., Adi, Y., Synnaeve, G.: Differentiable model compression via pseudo quantization noise. arXiv preprint arXiv:2104.09987 (2021)

10. Deng, J., Dong, W., Socher, R., Li, L.J., Li, K., Fei-Fei, L.: ImageNet: a large-scale hierarchical image database. In: 2009 IEEE Conference on Computer Vision and Pattern Recognition, pp. 248–255. IEEE (2009)

11. Dong, Z., et al.: Hawq-v2: hessian aware trace-weighted quantization of neural networks. In: Advances in Neural Information Processing Systems (2020)

12. Dong, Z., Yao, Z., Gholami, A., Mahoney, M.W., Keutzer, K.: Hawq: hessian aware quantization of neural networks with mixed-precision. In: Proceedings of the IEEE/CVF International Conference on Computer Vision, pp. 293–302 (2019)

13. Esser, S.K., McKinstry, J.L., Bablani, D., Appuswamy, R., Modha, D.S.: Learned step size quantization. In: International Conference on Learning Representations (2020)

14. Guo, Z., et al.: Single path one-shot neural architecture search with uniform sampling. In: Vedaldi, A., Bischof, H., Brox, T., Frahm, J.-M. (eds.) ECCV 2020. LNCS, vol. 12361, pp. 544–560. Springer, Cham (2020). https://doi.org/10.1007/978-3-030-58517-4_32

15. Habi, H.V., Jennings, R.H., Netzer, A.: HMQ: hardware friendly mixed precision quantization block for CNNs. In: Vedaldi, A., Bischof, H., Brox, T., Frahm, J.-M. (eds.) ECCV 2020. LNCS, vol. 12371, pp. 448–463. Springer, Cham (2020). https://doi.org/10.1007/978-3-030-58574-7_27

16. He, K., Zhang, X., Ren, S., Sun, J.: Deep residual learning for image recognition. In: IEEE Conference on Computer Vision and Pattern Recognition (CVPR), pp. 770–778 (2016)

17. Howard, A.G., et al.: Mobilenets: efficient convolutional neural networks for mobile vision applications. arXiv preprint arXiv:1704.04861 (2017)

18. Jung, S., et al.: Learning to quantize deep networks by optimizing quantization intervals with task loss. In: Proceedings of the IEEE/CVF Conference on Computer Vision and Pattern Recognition, pp. 4350–4359 (2019)

19. LeCun, Y., Denker, J.S., Solla, S.A.: Optimal brain damage. In: Advances in Neural Information Processing Systems, pp. 598–605 (1990)

20. Liu, Z., Li, J., Shen, Z., Huang, G., Yan, S., Zhang, C.: Learning efficient convolutional networks through network slimming. In: Proceedings of the IEEE International Conference on Computer Vision, pp. 2736–2744 (2017)

21. Lou, Q., Guo, F., Kim, M., Liu, L., Jiang, L.: AutoQ: automated kernel-wise neural network quantization. In: International Conference on Learning Representations (2020)

22. Mitchell, S., OSullivan, M., Dunning, I.: Pulp: a linear programming toolkit for python. The University of Auckland, Auckland, New Zealand, p. 65 (2011)

23. Park, E., Yoo, S.: PROFIT: a novel training method for sub-4-bit MobileNet models. In: Vedaldi, A., Bischof, H., Brox, T., Frahm, J.-M. (eds.) ECCV 2020. LNCS, vol. 12351, pp. 430–446. Springer, Cham (2020). https://doi.org/10.1007/978-3-030-58539-6_26

24. Uhlich, S., et al.: Mixed precision DNNs: all you need is a good parametrization. In: International Conference on Learning Representations (2020)

25. Wang, K., Liu, Z., Lin, Y., Lin, J., Han, S.: HAQ: hardware-aware automated quantization with mixed precision. In: Proceedings of the IEEE/CVF Conference on Computer Vision and Pattern Recognition, pp. 8612–8620 (2019)

26. Wu, B., Wang, Y., Zhang, P., Tian, Y., Vajda, P., Keutzer, K.: Mixed precision quantization of convnets via differentiable neural architecture search. arXiv preprint arXiv:1812.00090 (2018)
27. Yu, H., Han, Q., Li, J., Shi, J., Cheng, G., Fan, B.: Search what you want: barrier panelty NAS for mixed precision quantization. In: Vedaldi, A., Bischof, H., Brox, T., Frahm, J.-M. (eds.) ECCV 2020. LNCS, vol. 12354, pp. 1–16. Springer, Cham (2020). https://doi.org/10.1007/978-3-030-58545-7_1
28. Zhang, D., Yang, J., Ye, D., Hua, G.: LQ-nets: learned quantization for highly accurate and compact deep neural networks. In: Proceedings of the European Conference on Computer Vision (ECCV), pp. 365–382 (2018)
29. Zhou, A., Yao, A., Guo, Y., Xu, L., Chen, Y.: Incremental network quantization: towards lossless CNNs with low-precision weights. arXiv preprint arXiv:1702.03044 (2017)
30. Zhou, S., Wu, Y., Ni, Z., Zhou, X., Wen, H., Zou, Y.: Dorefa-net: training low bitwidth convolutional neural networks with low bitwidth gradients. arXiv preprint arXiv:1606.06160 (2016)

Event Neural Networks

Matthew Dutson$^{(\boxtimes)}$ (iD), Yin Li (iD), and Mohit Gupta (iD)

University of Wisconsin–Madison, Madison, WI 53715, USA
{dutson,yin.li,mgupta37}@wisc.edu

Abstract. Video data is often repetitive; for example, the contents of adjacent frames are usually strongly correlated. Such redundancy occurs at multiple levels of complexity, from low-level pixel values to textures and high-level semantics. We propose Event Neural Networks (EvNets), which leverage this redundancy to achieve considerable computation savings during video inference. A defining characteristic of EvNets is that each neuron has state variables that provide it with long-term memory, which allows low-cost, high-accuracy inference even in the presence of significant camera motion. We show that it is possible to transform a wide range of neural networks into EvNets without re-training. We demonstrate our method on state-of-the-art architectures for both high- and low-level visual processing, including pose recognition, object detection, optical flow, and image enhancement. We observe roughly an order-of-magnitude reduction in computational costs compared to conventional networks, with minimal reductions in model accuracy.

Keywords: Efficient neural networks · Adaptive inference · Video analysis

1 Introduction

Real-world visual data is repetitive; that is, it has the property of *persistence*. For example, observe the two frames in Fig. 1(a). Despite being separated by one second, they appear quite similar. Human vision relies on the persistent nature of visual data to allocate limited perceptual resources. Instead of ingesting the entire scene at high resolution, the human eye points the fovea (a small region of dense receptor cells) at areas containing motion or detail [51]. This allocation of attention reduces visual processing and eye-to-brain communication.

Processing individual frames using artificial neural networks has proven to be a competitive solution for video inference [55,60]. This paradigm leverages advances in *image* recognition (e.g., pose estimation or object detection) and processes each frame independently without considering temporal continuity, implicitly assuming that adjacent frames are statistically independent. This assumption leads to inefficient use of resources due to the repeated processing of image regions containing little or no new information.

Supplementary Information The online version contains supplementary material available at https://doi.org/10.1007/978-3-031-20083-0_17.

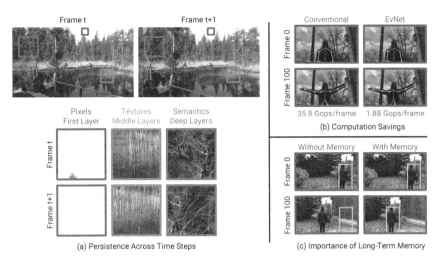

(a) Persistence Across Time Steps

(b) Computation Savings

(c) Importance of Long-Term Memory

Fig. 1. Event Neural Networks. (a) Two frames from a video sequence, separated by 1 s. Video source: [45]. Over this time, some areas of the image maintain consistent pixel values (sky region). However, these areas only represent a small fraction of the frame. In other regions, the pixel values change but the textures (vertical lines) or semantics (tree branches) remain the same. Each type of persistence corresponds to a different depth in the neural hierarchy. EvNets leverage temporal persistence in video streams across multiple levels of complexity. **(b)** EvNets yield significant computation savings while maintaining high accuracy. **(c)** Event neurons have state variables that encode long-term memory, allowing EvNets to perform robust inference even over long video sequences with significant camera motion. A network without long-term memory (left) fails to correctly track the object due to gradual error accumulation.

There has been recent interest in leveraging temporal redundancy for efficient video inference. One simple solution is to skip processing image regions containing few changes in pixel values. However, such methods cannot recognize persistence in textures, patterns, or high-level semantics when it does not coincide with persistent pixel values. See Fig. 1(a).

Because neural networks extract a hierarchy of features from their inputs, they contain a built-in lens for detecting repetition across many levels of visual complexity. Shallow layers detect low-level patterns, and deep layers detect high-level semantics. Temporal repetition at a given level of complexity translates to persistent values at the corresponding depth in the neural hierarchy [14]. Based on this observation, we propose Event Neural Networks (EvNets), a family of neural networks in which neurons transmit (thereby triggering downstream computation) only when there is a significant change in their activation. By applying this strategy over all neurons and layers, we detect and exploit temporal persistence across many levels of complexity.

One of the defining features of EvNets is that each neuron has state variables that provide it with *long-term memory*. Instead of re-computing from scratch for every new input, an EvNet neuron accumulates information over time. Long-term memory allows EvNets to perform robust inference over long video sequences containing significant camera motion. See Fig. 1(c).

We design various structural components for EvNets – both at the individual neuron level (memory state variables) and the network level (layers and transmission policies). We recognize that transmission policies, in particular, are critical for achieving a good accuracy/computation tradeoff, and we describe the policy design space in detail. We show that, with these components, it is possible to transform a broad class of conventional networks into EvNets *without re-training*. We demonstrate our methods on state-of-the-art models for several high- and low-level tasks: pose recognition, object detection, optical flow, and image enhancement. We observe approximately an order-of-magnitude reduction in arithmetic operations with minimal effects on model accuracy.

Scope and Limitations. In this paper, we focus on the theoretical and conceptual properties of EvNets. Although we show results on several video inference tasks, our goal is not to compete with the latest methods for these tasks in terms of accuracy. Instead, we show that, across a range of models and tasks, EvNets can significantly reduce computational costs without decreasing accuracy.

In most of our analyses we do not assume a specific hardware platform or computation model. We mainly report arithmetic and memory operations (a platform-invariant measure of computational cost) instead of wall-clock time (which depends on many situational variables). An important next step is to consider questions relating to the design of hardware-software stacks for EvNets, in order to minimize latency and power consumption.

2   Related Work

Efficient Neural Networks. There are numerous methods for reducing the computational cost of neural networks. Many architectures have been designed to require fewer parameters and arithmetic operations [18,22,29,30,41,59]. Another line of work uses low-precision arithmetic to achieve computation savings [8,21, 40,49]. Our approach is complementary to both architecture- and precision-based efficiency methods. These methods reduce the cost of inference on a single time step, whereas EvNets eliminate repetitive computation between multiple time steps. Pruning algorithms [15,16,25,26] remove redundant neurons or synapses during training to improve efficiency. Instead of pruning universally redundant neurons, an EvNet adaptively ignores temporally redundant neurons.

Adaptive Networks. Adaptive models modify their computation based on the input to suit the difficulty of each inference. Prior approaches consider an ensemble of sub-networks [20,50], equip a network with multiple exits [19,48], select the input resolution at inference time [7,32,57], or dynamically choose the feature resolution [53]. These methods are designed for image recognition tasks and do not explore temporal redundancy. Further, many require custom tailoring or re-training for each task and architecture. In contrast, EvNets can be readily integrated into many existing architectures and do not require re-training.

Temporal Redundancy. Several recent approaches consider temporal redundancy for efficient video inference. Many take a keyframe-oriented approach, computing expensive features on keyframes, then transforming those features for the other frames [6,23,27,44,60,61]. Other methods include using visual trackers [56], skipping redundant frames [13,54], reusing previous frame features [33], distilling results from previous time steps [36], two-stream computation [11], and leveraging video compression [52]. In general, these methods require extensive modifications to the network architecture.

Skip-convolution networks (Skip-Conv) [14] are closely related to EvNets. Skip-Conv reuses activation values that have not changed significantly between frames. However, the algorithm only tracks changes between consecutive frames and thus requires frequent re-initialization to maintain accuracy. Re-initialization leads to reduced efficiency, especially in the presence of camera motion. In contrast, the long-term memory in an EvNet maintains accuracy and efficiency over hundreds of frames, even when there is strong camera motion. See Fig. 1(c).

Sigma-Delta networks [37] exploit temporal redundancy by quantizing the changes in neuron activations. Sigma-Delta networks have been limited so far to simple tasks like digit classification. Unlike Sigma-Delta networks, EvNets do not require quantization (although they do allow it). Compared to Sigma-Delta networks, EvNets achieve superior accuracy/computation tradeoffs (Fig. 5) and generalize better to challenging, real-world tasks (Fig. 7).

DeltaCNN [39] is concurrent work with similar goals to this paper. Like EvNets, DeltaCNN models have mechanisms for integrating long-term changes. They focus on translating theoretical speedups into GPU wall-time savings by enforcing structured sparsity (all channels at a given location transmit together). Despite its practical benefits, this design is inefficient when there is camera motion. In contrast, we emphasize broad conceptual frameworks (e.g., arbitrary sparsity structure) with an eye toward future hardware architectures. (Sect. 6).

Event Sensor Inference. Event sensors [28] generate sparse frames by computing a quantized temporal gradient at each pixel. Many networks designed for inference on event-sensor data have efficient, sparse dynamics [4,34]. However, they make strong assumptions about the mathematical properties of the network (e.g., that it is piecewise linear [34]). EvNets place far fewer constraints on the model and are compatible with a broad range of existing architectures.

Recurrent Neural Networks (RNNs). EvNets use long-term memory to track changes, and are thus loosely connected to RNNs. Long-term memory has been widely adopted in RNNs [17]. Several recent works also propose adaptive inference for RNNs by learning to skip state updates [3] or updating state variables only when a significant change occurs [35,38]. Unlike EvNets, these approaches are tailored for RNNs and generally require re-training.

3   Event Neurons

Consider a neuron in a conventional neural network. Let $\boldsymbol{x} = [x_1, x_2, \ldots, x_n]$ be the vector of input values and y be the output. Suppose the neuron composes a linear function g (e.g., a convolution) with a nonlinear activation f. That is,

$$g(\boldsymbol{x}) = \sum_{i=1}^{n} w_i x_i; \quad y = f(g(\boldsymbol{x})), \tag{1}$$

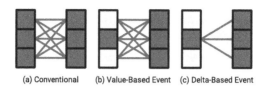

(a) Conventional   (b) Value-Based Event   (c) Delta-Based Event

Fig. 2. Sparse, Delta-Based Transmission. (a) Conventional neurons completely recompute their activations on each time step. **(b)** Value-based event neurons only transmit activations that have changed significantly. However, a value-based transmission can still trigger many computations. **(c)** Delta-based event neurons only transmit differential updates to their activations.

where $\boldsymbol{w} = [w_1, w_2, \ldots, w_n]$ contains the weights of the function g. In a conventional network, every neuron recomputes f and g for every input frame (Fig. 2(a)), resulting in large computational costs over a video sequence.

Inspired by prior methods that exploit persistence in activations [3,14,37], we describe a class of *event neurons* with *sparse, delta-based* transmission.

Sparse, Delta-Based Transmission. An event neuron transmits its output to subsequent layers only when there is a sufficient change between its current activation and the previous transmission. This gating behavior makes the layer output *sparse*. However, a *value* transmission may still trigger many downstream computations (neurons receiving updated input values must recompute their activations from scratch). See Fig. 2(b). Therefore, instead of transmitting an activation *value*, an event neuron transmits a *delta* (differential).

Suppose a neuron receives a vector of incoming differentials $\boldsymbol{\Delta}_{\text{in}}$ (one element per incoming synapse). $\boldsymbol{\Delta}_{\text{in}}$ is sparse, i.e., it only contains nonzero values for upstream neurons that have transmitted. The updated g is given by

$$g(\boldsymbol{x} + \boldsymbol{\Delta}_{\text{in}}) = g(\boldsymbol{x}) + g(\boldsymbol{\Delta}_{\text{in}}). \tag{2}$$

Instead of computing $g(\boldsymbol{x} + \boldsymbol{\Delta}_{\text{in}})$ from scratch, an event neuron stores the value of $g(\boldsymbol{x})$ in a state variable a. When it receives a new input $\boldsymbol{\Delta}_{\text{in}}$, the neuron retrieves the old value of $g(\boldsymbol{x})$ from a, computes $g(\boldsymbol{\Delta}_{\text{in}})$, and saves the value $g(\boldsymbol{x}) + g(\boldsymbol{\Delta}_{\text{in}})$ in a. This process only requires computing products $w_i x_i$ for nonzero elements of $\boldsymbol{\Delta}_{\text{in}}$, i.e., computation scales linearly with the number of transmissions.

The activation function f is nonlinear, so we cannot update it incrementally like g. Whenever a changes, we recompute $f(a)$, then store the updated value in another state variable. f is usually a lightweight function (e.g., ReLU), so the cost of recomputing f is far smaller than computing the products $w_i x_i$.

3.1 Building Event Neurons

An event neuron consists of three state variables, as shown in Fig. 3. The *accumulator* (a) stores the current value of $g(\boldsymbol{x})$. The *best estimate* (b) stores the current value of $f(a)$. The *difference* (d) stores difference between b and the most

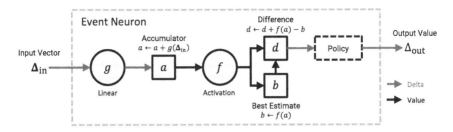

Fig. 3. Building Event Neurons. The state variables and update rules in an event neuron. The incremental updates to a convert from a delta-based representation to value-based. The subtraction $f(a) - b$ returns the output to delta-based.

recent output. When a neuron receives a differential update $\Delta_{\text{in}}^{(t)}$ at time t from one or more of its inputs, it updates these state variables as follows:

$$a^{(t+1)} = a^{(t)} + g(\Delta_{\text{in}}^{(t)}); \quad d^{(t+1)} = d^{(t)} + f(a^{(t+1)}) - b^{(t)}; \quad b^{(t+1)} = f(a^{(t+1)}). \tag{3}$$

A neuron transmits an output Δ_{out} when some condition on d is satisfied. This condition is defined by the *transmission policy*. The transmission policy also gives the relationship between d and Δ_{out}. The policies in this paper simply set $\Delta_{\text{out}} = d$. However, other relationships are possible, and the properties described in Sect. 3.2 hold for other relationships. After a neuron transmits, it sets d to $d - \Delta_{\text{out}}$. See Sect. 4.3 for more details on transmission policies.

3.2 Properties of Event Neurons

Long- and Short-Term Memory. The state variable d accumulates all not-yet-transmitted corrections to the neuron output. It represents the neuron's *long-term memory*, whereas b represents its *short-term memory*. Including a long-term memory keeps the neuron from discarding information when it does not transmit. This *error-retention property* grants certain guarantees on the neuron's behavior, as we demonstrate next.

Error Retention. Consider an event neuron receiving a series of inputs over T time steps, $\Delta_{\text{in}}^{(1)}, \Delta_{\text{in}}^{(2)}, \dots, \Delta_{\text{in}}^{(T)}$. Assume that the state variables a, b, and d have initial values $a^{(0)}$, $f(a^{(0)})$, and zero, respectively. Let the transmitted output values at each time step be $\Delta_{\text{out}}^{(1)}, \Delta_{\text{out}}^{(2)}, \dots, \Delta_{\text{out}}^{(T)}$ (some of these may be zero). By repeatedly applying the neuron update rules, we arrive at the state

$$a^{(T)} = a^{(0)} + g\left(\sum_{t=1}^{T} \Delta_{\text{in}}^{(t)}\right); \quad b^{(T)} = f(a^{(T)});$$
$$d^{(T)} = b^{(T)} - b^{(0)} - \sum_{t=1}^{T} \Delta_{\text{out}}^{(t)}. \tag{4}$$

See the supplementary material for a detailed derivation. Observe that d is equal to the difference between the actual and transmitted changes in the activation.

This is true regardless of the order or temporal distribution of the Δ_{in} and Δ_{out}. Because the neuron stores d, it always has enough information to bring the transmitted activation into exact agreement with the true activation b. We can use this fact to bound the error within an EvNet. For example, we can constrain each neuron's error to the range $[-h, +h]$ by transmitting when $|d| > h$.

The Importance of Long-Term Memory. For comparison, consider a model in which neurons compute the difference between b on adjacent time steps, then either transmit or discard this difference without storing the remainder. This is the model used in Skip-Conv [14]. Under this model, the final state of a neuron depends strongly on the order and temporal distribution of inputs.

For example, suppose a neuron transmits if the frame-to-frame difference exceeds a threshold δ. Consider a scenario where the neuron's activation gradually increases from 0 to 2δ in steps $0.1\delta, 0.2\delta, \ldots, 2\delta$. Gradual changes like this are common in practice (e.g., when panning over a surface with an intensity gradient). Because $0.1\delta < \delta$, the neuron never transmits and ends in a state with error -2δ. The neuron carries this error into all of its future computations. Furthermore, because the neuron discards non-transmitted activations, it has no way to know that this -2δ error exists.

4   Event Networks

4.1   Building Event Networks

So far, we have considered the design and characteristics of individual event neurons. In this section, we broaden our view and consider layers and networks. A "layer" is an atomic tensor operation (e.g., a convolution). By this definition, g and f as defined in Sect. 3.1 correspond to two different layers.

We define three new layer types. An *accumulator layer* consists of a state vector a that contains the a variables for a collection of neurons. A *gate layer* contains state vectors b and d and the transmission policy. A *buffer layer* stores its inputs in a state vector x for future use by the next layer; this is required before non-pointwise, nonlinear layers like max pooling. The state vectors a, b and d are updated using vectorized versions of the rules in Eq. 3. An accumulator layer converts its input from delta-based to value-based, whereas a gate converts from value-based to delta-based.

To create an EvNet, we insert gates and accumulators into a pretrained network such that linear layers receive delta inputs and nonlinear layers receive value inputs (Fig. 4). Note that residual connections do not require any special treatment – in an EvNet, residual connections simply carry deltas instead of values. These deltas are added or concatenated to downstream deltas when the residual branch re-joins the main branch (like in a conventional network).

We place a gate at the beginning of the network and an accumulator at the end. At the input gate, we use pixel values instead of $f(a)$ and update b and d at every timestep. At the output accumulator, we update a sparsely but read all its elements at every frame. Throughout the model, the functions computed by the preexisting layers (the f and g) remain the same.

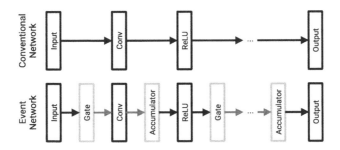

Fig. 4. Building Event Networks. We insert accumulators and gates to make the input to linear layers (e.g., convolutions, fully-connected layers) delta-based and the input to nonlinear layers (e.g., ReLU activations) value-based.

4.2 Network Initialization

The equations in Sect. 3.1 define how to update the neuron state variables, but they do not specify those variables' initial values. Consider a simple initialization strategy where $a = 0$ and $d = 0$ for all neurons. Since the activation function f is nonlinear, the value of the state variable $b = f(a)$ may be nonzero. This nonzero b usually translates to a nonzero value of a in the next layer. However, we initialized $a = 0$ for all neurons. We have an inconsistency.

To address this problem, we define the notion of *internal consistency*. Consider a neuron with state variables a, d, and b. Let $\boldsymbol{b}_{\text{in}}$ and $\boldsymbol{d}_{\text{in}}$ be vectors containing the states of the neurons in the previous layer. We say that a network is in an internally consistent state if, for all neurons,

$$a = g(\boldsymbol{b}_{\text{in}} - \boldsymbol{d}_{\text{in}}); \quad b = f(a). \tag{5}$$

The simplest way to satisfy these criteria is to flush some canonical input through the network. Starting with neurons in the first layer and progressively moving through all subsequent layers, we set $a = g(\boldsymbol{b}_{\text{in}})$, $b = f(a)$, and $d = 0$. In our experiments, we use the first input frame as the canonical input.

4.3 Transmission Policies

A *transmission policy* defines a pair of functions $M(\boldsymbol{d})$ and $P(\boldsymbol{d})$ for each layer. M outputs a binary mask $\boldsymbol{m}$ indicating which neurons should transmit. P outputs the values of $\boldsymbol{\Delta}_{\text{out}}$. In this subsection, we describe the transmission policy design space. The choice of transmission policy is a critical design consideration, strongly influencing the accuracy and efficiency of the final model.

Locality and Granularity. Policies may have different levels of *locality*, defined as the number of elements from $\boldsymbol{d}$ required to compute each element of $\boldsymbol{m}$ and $\boldsymbol{\Delta}_{\text{out}}$. A *global* policy considers all elements of $\boldsymbol{d}$ when computing each value m_i and $\Delta_{\text{out},i}$. A *local* policy considers some strict subset of $\boldsymbol{d}$, and an *isolated* policy considers only the element d_i.

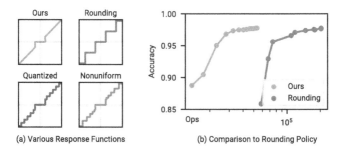

Fig. 5. Policy Design and Quantization. (a) A few sample response functions (assuming an isolated, singular policy). **(b)** A comparison between our policy and a rounding policy (used in Sigma-Delta networks [37]). Results are for a 3-layer fully-connected network on the Temporal MNIST dataset [37].

In addition to its locality, each policy has a *granularity*. The granularity defines how m-values are shared between neurons. A *chunked* policy ties neurons together into local groups, producing one value of m for each group. Neurons in the same group fire in unison. This might be practically desirable for easy parallelization on the hardware. In contrast, a *singular* policy assigns every neuron a separate value of m, so each neuron fires independently.

A Linear-Cost Policy. In this work, we use an isolated, singular policy based on a simple threshold. Specifically,

$$m_i = H(|d_i| - h_i); \quad \Delta_{\text{out},i} = d_i, \tag{6}$$

where H is the Heaviside step function and h_i is the threshold for neuron i. A key advantage of this policy is its low overhead. On receiving an incoming transmission, a neuron evaluates $|d| > h$ (one subtraction) in addition to the usual updates to a, d, and b. Neurons not receiving any updates (e.g., those in a static image region) do not incur any overhead for policy computations. In other words, the policy's cost is linear in the number of updated neurons. Combined with the linear cost of computing the neuron updates, this results in EvNets whose overall *cost scales linearly with the amount of change in the input, not with the quantity of input data received*.

This linear cost has important implications for networks processing data from high-speed sensors (e.g., event sensors [28] or single-photon sensors [10]). Here, the differences between adjacent inputs are often minuscule, and the cost of a policy with fixed per-frame overhead (e.g., a Gumbel gate [14]) could come to dominate the runtime. EvNets with a linear-overhead policy are a natural solution for processing this type of high-speed data.

Policy Design and Quantization. When a policy is both isolated and singular, we can characterize the functions $M(\boldsymbol{d})$ and $P(\boldsymbol{d})$ by scalar functions $M(d_i)$ and $P(d_i)$. Taking the product $M(d_i) \cdot P(d_i)$ gives a *response function* $R(d_i)$ that describes the overall behavior of the neuron. Figure 5(a) illustrates several possible response functions.

Some response functions employ quantization to reduce the cost of computing dot product terms $w_i x_i$ (Eq. 1). Sigma-Delta networks [37] use a rounding policy to quantize neuron outputs; a neuron transmits if this rounded value is nonzero. This rounding policy has significantly worse accuracy-computation tradeoffs (Fig. 5(b)) compared to our proposed policy. This might be caused by coupling the firing threshold with the quantization scale. To increase its output precision a Sigma-Delta network must reduce its firing threshold, possibly resulting in unnecessary transmissions.

5   Experiments and Results

EvNets are widely applicable across architectures and video inference tasks. Any network satisfying a few basic requirements (i.e., frame-based and composing linear functions with nonlinear activations) can be converted to an EvNet *without re-training*. To demonstrate this, we select widespread, representative models for our main experiments: YOLOv3 [41] for video object detection and OpenPose [5] for video pose estimation. Additionally, we conduct ablation experiments and report results on low-level tasks (optical flow and image enhancement).

In the supplement, we include additional results on HRNet [47] for pose estimation. We also include ablations for the effect of layer depth on computation cost, variations in computation over time, the effect of granularity on savings, improved temporal smoothness, and a comparison to simple interpolation.

5.1   Video Pose Estimation

Dataset and Experiment Setup. We conduct experiments on the JHMDB dataset [24] using the widely adopted OpenPose model [5]. We use weights pretrained on the MPII dataset [2] from [5] and evaluate the models on a subset of JHMDB with 319 videos and over 11k frames, following [14]. We report results on the combination of the three JHMDB test splits. We use the PCK metric [58] with a detection threshold of $\alpha = 0.2$, consistent with prior works [14].

Implementation Details. We resize all videos to 320×240, padding as needed to preserve the aspect ratio of the content. The joint definitions in MPII (the training dataset for OpenPose) differ slightly from those in JHMDB. During evaluation, we match the JHMDB "neck", "belly", and "face" joints to the MPII "upper neck," "pelvis," and "head top" joints, respectively.

Baselines. We consider the following baselines, all using the OpenPose model.

- **Conventional:** This is the vanilla OpenPose model without modifications.
- **Skip-Conv:** This is a variant of the Skip-Conv method with norm gates and without periodic state resets.
- **Skip-Conv-8:** This adds state resets to Skip-Conv by re-flushing every 8 frames to reduce the effect of long-term activation drift.

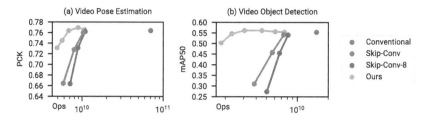

Fig. 6. Pareto Curves. The performance of an EvNet over several different thresholds, with baselines for comparison. The "Skip-Conv-8" model re-flushes the network every 8 frames. EvNets give significant computation savings without sacrificing accuracy. See the supplementary material for a table with this data.

We recognize that Skip-Conv networks can also incorporate a learnable gating function (the Gumbel gate) that uses information from a local window around each neuron. This can also be used for our EvNets (it is local and chunked rather than isolated and singular), but it requires re-training of the network and can incur a higher computational overhead. To keep the analysis fair, we only compare to the Skip-Conv norm gate.

Results. Figure 6(a) presents our results. We vary the policy threshold h to characterize the accuracy/computation Pareto frontier. For both Skip-Conv and EvNets, increasing the threshold reduces the computational cost but increases the error rate. EvNets consistently outperform their direct competitors (Skip-Conv and Skip-Conv reset) on the Pareto frontier, achieving significantly higher accuracy when using a similar amount of computation. Surprisingly, compared to the conventional OpenPose model, EvNets sometimes have slightly better accuracy, even with a large reduction in computation. We hypothesize that this is caused by a weak inter-frame ensembling effect.

Table 1 summarizes the accuracy and computation at the best operating point on the Pareto curve. For each model, we choose the highest threshold that reduces PCK by less than 0.5%. To better understand the accuracy-computation tradeoff, we further report the compute and memory overhead of our EvNets (at the best operating point) in Table 2. We report overhead operations both as a number of operations and as a percentage. This percentage gives the ratio between "extra operations expended" and "number of arithmetic operations saved". For example, an arithmetic overhead of 0.12% indicates that the neuron updates and transmission policy require 0.12 extra arithmetic operations for every 100 operations saved. Overall, EvNets add minimal operation overhead and manageable additional memory.

5.2   Video Object Detection

Dataset, Experiment Setup, and Baselines. We evaluate on the ILSVRC 2015 VID dataset [42] using the popular YOLOv3 model [41] with pre-trained weights from [43]. We report all results on the validation set with 555 videos and over 172k frames, using mean Average Precision (mAP) with an IoU threshold

Table 1. Accuracy and computation. Results on the best threshold for each model. We choose the highest threshold that reduces PCK or mAP by less than 0.5%. See the supplement for a complete list of the points shown in Fig. 6.

Model	Pose estimation			Object detection		
	Thresh.	PCK (%)	Operations	Thresh.	mAP (%)	Operations
Conventional	–	**76.40**	7.055×10^{10}	–	55.38	1.537×10^{10}
Skip-Conv	0.01	76.03	1.027×10^{10}	0.01	54.13	7.340×10^{9}
Skip-Conv-8	0.01	76.21	1.092×10^{10}	0.01	54.06	8.111×10^{9}
EvNet	0.04	76.37	$\mathbf{6.780 \times 10^{9}}$	0.08	**56.19**	$\mathbf{3.061 \times 10^{9}}$

Table 2. Overhead. "Weights" gives the amount of memory required for model weights. "Variables" gives the amount of memory required for the state variables a, b, and d. "Arithmetic" indicates the number of extra arithmetic operations expended for neuron updates (Eq. 3) and policy-related computations. "Load and Store" indicates the number of extra memory access operations. See the text for an explanation of the percentage notation.

Model	Thresh.	Memory costs		Operation overhead	
		Weights	Variables	Arithmetic	Load and store
OpenPose	0.04	206.8 MB	346.2 MB	7.570×10^{7} (0.12 %)	1.342×10^{8} (0.21 %)
YOLO	0.08	248.8 MB	232.2 MB	6.417×10^{7} (0.52 %)	1.040×10^{8} (0.85 %)

of 0.5 (following previous works [6,43,61]). We evaluate the same model variants as in Sect. 5.1 (conventional, EvNet, Skip-Conv, and Skip-Conv reset).

Implementation Details. We resize all videos to 224×384, padding as needed to preserve the aspect ratio. Unlike OpenPose, YOLOv3 includes batch normalization (BN) layers. BN gives us a convenient way to estimate the distribution of activations at each neuron. We use this information to adjust the threshold values. Specifically, we scale the threshold at each neuron by $1/\gamma$ (where γ is the learned BN re-scaling parameter). This scaling makes the policy more sensitive for neurons with a narrower activation distribution, where we would expect equal-sized changes to be more meaningful.

Results. Figure 6 presents our results with varying thresholds. Again, we observe that our EvNets outperform Skip-Conv variants, and sometimes have slightly higher accuracy than the conventional model with greatly reduced compute cost. Table 1 presents the accuracy and computation at the best operating points.

5.3   Low-Level Vision Tasks

We have so far considered only high-level inference tasks. However, EvNets are also an effective strategy for low-level vision. We consider PWC-Net [46] for optical flow computation and HDRNet [12] for video frame enhancement. For brevity, we only show sample results in Fig. 7 and refer the reader to the supplementary material for more details. As with the high-level models, we observe minimal degradation in accuracy and significant computation savings.

Fig. 7. Versatility of EvNets. We demonstrate that EvNets are an effective strategy for many high- and low-level vision tasks. Across tasks, we see significant computation savings while maintaining high-quality output. This frame shows a person mid-jump. The EvNet tracks the subject correctly under rapid motion.

5.4   Ablation and Analysis

Rounding Policy Comparison. Figure 5(b) compares our transmission policy and the rounding policy used in a Sigma-Delta network [37]. We obtain these results by evaluating the fully-connected model from the Sigma-Delta paper (with the authors' original weights) on the Temporal MNIST dataset [37]. We evaluate EvNets with thresholds of the form 10^p, where $p \in \{-1.5, -1.4, \ldots, -0.3, -0.2\}$. We obtain results for the Sigma-Delta network using the original authors' code, which involves training the quantization threshold (the Pareto frontier is a consequence of varying a training penalty scale λ).

Ablation of Long-Term Memory. Figure 8 shows the effect of ablating the long-term memory d (resetting it to zero after each input). We evaluate the OpenPose model on the JHMDB dataset. Other than resetting d, the two models shown are identical. Both models use a threshold of 0.05. We see that long-term memory is critical for maintaining stable accuracy.

Camera Motion. Global camera or scene motion (e.g., camera shake or scene translation) reduces the amount of visual persistence in a video. We would therefore expect camera motion to reduce the savings in an EvNet. To confirm this, we evaluate the OpenPose and YOLO models on a custom-labeled video dataset. We label the camera motion in each video as "none" (perfectly stationary camera), "minor" (slight camera shake), or "major". See the supplement for details. We test OpenPose with a threshold of 0.05 and YOLO with a threshold of 0.06. Because this dataset does not have frame-level labels for pose or object detection, we do not explicitly evaluate task accuracy. However, the thresholds we use here give good accuracy on JHMDB and VID. For OpenPose, the computation savings for "none", "major", and "minor" camera motion are 17.3×, 11.3×, and 8.40×, respectively. For YOLO, the savings are 6.64×, 3.95×, and 2.65×. As expected, we see a reduction in savings when there is strong camera motion, although we still achieve large reductions relative to the conventional model.

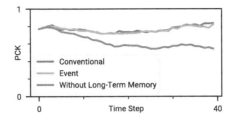

Fig. 8. Ablation of Long-Term Memory. Removing the memory d, as considered in Skip-Conv [14], causes a rapid decay in accuracy. Results using the OpenPose model on the JHMDB dataset. See Sect. 5 for details.

Wall-Time Savings. We now show preliminary results demonstrating wall-time savings in EvNets. We consider the HRNet [47] model (see supplementary) on the JHDMB dataset. We evaluate on an Intel Core i7 8700K CPU.

We implement the model in PyTorch. For the EvNet, we replace the standard convolution with a custom sparse C++ convolution. Our convolution uses an input-stationary design (i.e., an outer loop over input pixels) to skip zero deltas efficiently. In the conventional model, we use a custom C++ convolution with a standard output-stationary design (i.e., an outer loop over output pixels). We use a custom operator in the conventional model to ensure a fair comparison, given the substantial engineering effort invested in the default MKL-DNN library. We implement both operators with standard best practices (e.g., maximizing data-access locality). We compile with GCC 9.4 with the -Ofast flag.

For evaluation, we use an input size of 256×256 and an EvNet threshold of 0.1. The EvNet achieves a PCK of 90.46% and runs in an average of 0.3497 s (7.361×10^8 ops) per frame. The conventional model achieves a PCK of 90.37% and runs in 1.952% (1.019×10^{10} ops) per frame.

6 Discussion

Hardware Platforms. Mainstream GPU hardware is designed for parallel, block-wise computation with coarse control flow. EvNets with neuron-level transmission are inefficient under this computation model. In the long term, we expect to achieve the best performance on specialized hardware designed for extreme parallelism and distributed control. It is important to emphasize that event neurons *do not need to operate by a shared clock*. Each neuron operates independently – consuming new input as it arrives and transmitting output once it is computed. This independence permits an asynchronous, networked execution model in contrast to the ordered, frame-based model in conventional machine learning. Spiking neural networks (SNNs) [31] share this asynchronous computation model and have motivated the development of neuromorphic hardware platforms [1,9] that could be re-purposed for efficient implementation of EvNets.

Acknowledgments. This research was supported by NSF CAREER Award 1943149.

References

1. Akopyan, F., et al.: TrueNorth: design and tool flow of a 65 mW 1 million neuron programmable neurosynaptic chip. IEEE Trans. Comput. Aided Des. Integr. Circ. Syst. **34**(10), 1537–1557 (2015). https://doi.org/10.1109/TCAD.2015.2474396
2. Andriluka, M., Pishchulin, L., Gehler, P., Schiele, B.: 2D human pose estimation: new benchmark and state of the art analysis. In: Proceedings of the IEEE Conference on Computer Vision and Pattern Recognition, pp. 3686–3693 (2014)
3. Campos, V., Jou, B., Giró-i Nieto, X., Torres, J., Chang, S.F.: Skip RNN: learning to skip state updates in recurrent neural networks. In: International Conference on Learning Representations (2018)
4. Cannici, M., Ciccone, M., Romanoni, A., Matteucci, M.: Asynchronous convolutional networks for object detection in neuromorphic cameras. In: 2019 IEEE/CVF Conference on Computer Vision and Pattern Recognition Workshops (CVPRW), pp. 1656–1665 (2019). https://doi.org/10.1109/CVPRW.2019.00209
5. Cao, Z., Simon, T., Wei, S.E., Sheikh, Y.: Realtime multi-person 2D pose estimation using part affinity fields. In: Proceedings of the IEEE Conference on Computer Vision and Pattern Recognition, pp. 7291–7299 (2017)
6. Chen, Y., Cao, Y., Hu, H., Wang, L.: Memory enhanced global-local aggregation for video object detection. In: Proceedings of the IEEE/CVF Conference on Computer Vision and Pattern Recognition, pp. 10337–10346 (2020)
7. Chin, T.W., Ding, R., Marculescu, D.: AdaScale: towards real-time video object detection using adaptive scaling. arXiv:1902.02910 [cs] (2019)
8. Courbariaux, M., Bengio, Y., David, J.P.: BinaryConnect: training deep neural networks with binary weights during propagations. Adv. Neural. Inf. Process. Syst. **28**, 3123–3131 (2015)
9. Davies, M., et al.: Loihi: a neuromorphic manycore processor with on-chip learning. IEEE Micro **38**(1), 82–99 (2018). https://doi.org/10.1109/MM.2018.112130359
10. Dutton, N.A.W., et al.: A SPAD-based QVGA image sensor for single-photon counting and quanta imaging. IEEE Trans. Electron Devices **63**(1), 189–196 (2016). https://doi.org/10.1109/TED.2015.2464682
11. Feichtenhofer, C., Fan, H., Malik, J., He, K.: SlowFast networks for video recognition. In: Proceedings of the IEEE/CVF International Conference on Computer Vision, pp. 6202–6211 (2019)
12. Gharbi, M., Chen, J., Barron, J.T., Hasinoff, S.W., Durand, F.: Deep bilateral learning for real-time image enhancement. ACM Trans. Graph. **36**(4), 118:1–118:12 (2017). https://doi.org/10.1145/3072959.3073592
13. Ghodrati, A., Bejnordi, B.E., Habibian, A.: FrameExit: conditional early exiting for efficient video recognition. In: Proceedings of the IEEE/CVF Conference on Computer Vision and Pattern Recognition, pp. 15608–15618 (2021)
14. Habibian, A., Abati, D., Cohen, T.S., Bejnordi, B.E.: Skip-convolutions for efficient video processing. In: Proceedings of the IEEE/CVF Conference on Computer Vision and Pattern Recognition, pp. 2695–2704 (2021)
15. Han, S., Pool, J., Tran, J., Dally, W.: Learning both weights and connections for efficient neural network. Adv. Neural. Inf. Process. Syst. **28**, 1135–1143 (2015)
16. Hassibi, B., Stork, D.: Second order derivatives for network pruning: Optimal brain surgeon. Adv. Neural. Inf. Process. Syst. **5**, 164–171 (1992)
17. Hochreiter, S., Schmidhuber, J.: Long short-term memory. Neural Comput. **9**(8), 1735–1780 (1997)

18. Howard, A.G., et al.: MobileNets: efficient convolutional neural networks for mobile vision applications. arXiv:1704.04861 [cs] (2017)
19. Huang, G., Chen, D., Li, T., Wu, F., Van Der Maaten, L., Weinberger, K.Q.: Multi-scale dense networks for resource efficient image classification. In: International Conference on Learning Representations (2018)
20. Huang, G., Sun, Yu., Liu, Z., Sedra, D., Weinberger, K.Q.: Deep networks with stochastic depth. In: Leibe, B., Matas, J., Sebe, N., Welling, M. (eds.) ECCV 2016. LNCS, vol. 9908, pp. 646–661. Springer, Cham (2016). https://doi.org/10.1007/978-3-319-46493-0_39
21. Hwang, K., Sung, W.: Fixed-point feedforward deep neural network design using weights +1, 0, and −1. In: Workshop on Signal Processing Systems (SiPS), pp. 1–6 (2014). https://doi.org/10.1109/SiPS.2014.6986082
22. Iandola, F.N., Han, S., Moskewicz, M.W., Ashraf, K., Dally, W.J., Keutzer, K.: SqueezeNet: AlexNet-level accuracy with 50x fewer parameters and <0.5MB model size. arXiv:1602.07360 [cs] (2016)
23. Jain, S., Wang, X., Gonzalez, J.E.: Accel: a corrective fusion network for efficient semantic segmentation on video. In: Proceedings of the IEEE/CVF Conference on Computer Vision and Pattern Recognition, pp. 8866–8875 (2019)
24. Jhuang, H., Gall, J., Zuffi, S., Schmid, C., Black, M.J.: Towards understanding action recognition. In: International Conference on Computer Vision (ICCV), pp. 3192–3199 (2013)
25. LeCun, Y., Denker, J., Solla, S.: Optimal brain damage. Adv. Neural. Inf. Process. Syst. **2**, 598–605 (1989)
26. Li, H., Kadav, A., Durdanovic, I., Samet, H., Graf, H.P.: Pruning filters for efficient ConvNets. arXiv:1608.08710 [cs] (2017)
27. Li, Y., Shi, J., Lin, D.: Low-latency video semantic segmentation. In: Proceedings of the IEEE Conference on Computer Vision and Pattern Recognition, pp. 5997–6005 (2018)
28. Lichtsteiner, P., Posch, C., Delbruck, T.: A 128×128 120 dB 15 Ms latency asynchronous temporal contrast vision sensor. IEEE J. Solid-State Circ. **43**(2), 566–576 (2008). https://doi.org/10.1109/JSSC.2007.914337
29. Lin, T.Y., Goyal, P., Girshick, R., He, K., Dollar, P.: Focal loss for dense object detection. In: International Conference on Computer Vision (ICCV), pp. 2980–2988 (2017)
30. Liu, W., et al.: SSD: single shot multibox detector. In: Leibe, B., Matas, J., Sebe, N., Welling, M. (eds.) ECCV 2016. LNCS, vol. 9905, pp. 21–37. Springer, Cham (2016). https://doi.org/10.1007/978-3-319-46448-0_2
31. Maass, W.: Networks of spiking neurons: the third generation of neural network models. Neural Netw. **10**(9), 1659–1671 (1997). https://doi.org/10.1016/S0893-6080(97)00011-7
32. Meng, Y., et al.: AR-net: adaptive frame resolution for efficient action recognition. In: Vedaldi, A., Bischof, H., Brox, T., Frahm, J.-M. (eds.) ECCV 2020. LNCS, vol. 12352, pp. 86–104. Springer, Cham (2020). https://doi.org/10.1007/978-3-030-58571-6_6
33. Meng, Y., et al.: AdaFuse: adaptive temporal fusion network for efficient action recognition. In: International Conference on Learning Representations (2021)
34. Messikommer, N., Gehrig, D., Loquercio, A., Scaramuzza, D.: Event-based asynchronous sparse convolutional networks. In: Vedaldi, A., Bischof, H., Brox, T., Frahm, J.-M. (eds.) ECCV 2020. LNCS, vol. 12353, pp. 415–431. Springer, Cham (2020). https://doi.org/10.1007/978-3-030-58598-3_25

35. Neil, D., Lee, J.H., Delbruck, T., Liu, S.C.: Delta networks for optimized recurrent network computation. In: Proceedings of the 34th International Conference on Machine Learning, pp. 2584–2593. PMLR (2017)

36. Nie, X., Li, Y., Luo, L., Zhang, N., Feng, J.: Dynamic kernel distillation for efficient pose estimation in videos. In: Proceedings of the IEEE/CVF International Conference on Computer Vision, pp. 6942–6950 (2019)

37. O'Connor, P., Welling, M.: Sigma delta quantized networks. arXiv:1611.02024 [cs] (2016)

38. Pan, B., Lin, W., Fang, X., Huang, C., Zhou, B., Lu, C.: Recurrent residual module for fast inference in videos. In: Proceedings of the IEEE Conference on Computer Vision and Pattern Recognition, pp. 1536–1545 (2018)

39. Parger, M., Tang, C., Twigg, C.D., Keskin, C., Wang, R., Steinberger, M.: DeltaCNN: end-to-end CNN inference of sparse frame differences in videos. In: Proceedings of the IEEE/CVF Conference on Computer Vision and Pattern Recognition, pp. 12497–12506 (2022)

40. Rastegari, M., Ordonez, V., Redmon, J., Farhadi, A.: XNOR-net: ImageNet classification using binary convolutional neural networks. In: Leibe, B., Matas, J., Sebe, N., Welling, M. (eds.) ECCV 2016. LNCS, vol. 9908, pp. 525–542. Springer, Cham (2016). https://doi.org/10.1007/978-3-319-46493-0_32

41. Redmon, J., Divvala, S., Girshick, R., Farhadi, A.: You only look once: unified, real-time object detection. In: Conference on Computer Vision and Pattern Recognition (CVPR), pp. 779–788 (2016)

42. Russakovsky, O., et al.: ImageNet large scale visual recognition challenge. Int. J. Comput. Vis. **115**(3), 211–252 (2015). https://doi.org/10.1007/s11263-015-0816-y

43. Sabater, A., Montesano, L., Murillo, A.C.: Robust and efficient post-processing for video object detection. In: 2020 IEEE/RSJ International Conference on Intelligent Robots and Systems (IROS), pp. 10536–10542 (2020). https://doi.org/10.1109/IROS45743.2020.9341600

44. Shelhamer, E., Rakelly, K., Hoffman, J., Darrell, T.: Clockwork convnets for video semantic segmentation. In: Hua, G., Jégou, H. (eds.) ECCV 2016. LNCS, vol. 9915, pp. 852–868. Springer, Cham (2016). https://doi.org/10.1007/978-3-319-49409-8_69

45. Sillerkiil: Eesti: Aegviidu siniallikad (Aegviidu blue springs in Estonia) (2021)

46. Sun, D., Yang, X., Liu, M.Y., Kautz, J.: PWC-Net: CNNs for optical flow using pyramid, warping, and cost volume. In: Proceedings of the IEEE Conference on Computer Vision and Pattern Recognition, pp. 8934–8943 (2018)

47. Sun, K., Xiao, B., Liu, D., Wang, J.: Deep high-resolution representation learning for human pose estimation. In: Proceedings of the IEEE/CVF Conference on Computer Vision and Pattern Recognition, pp. 5693–5703 (2019)

48. Teerapittayanon, S., McDanel, B., Kung, H.: BranchyNet: fast inference via early exiting from deep neural networks. In: 2016 23rd International Conference on Pattern Recognition (ICPR), pp. 2464–2469 (2016). https://doi.org/10.1109/ICPR.2016.7900006

49. Vanhoucke, V., Senior, A., Mao, M.Z.: Improving the speed of neural networks on CPUs. In: Deep Learning and Unsupervised Feature Learning Workshop, NIPS (2011)

50. Veit, A., Belongie, S.: Convolutional networks with adaptive inference graphs. In: European Conference on Computer Vision (ECCV), pp. 3–18 (2018)

51. Ware, C.: Visual Thinking for Design, 1st edn. Morgan Kaufmann, Burlington, Amsterdam (2008)

52. Wu, C.Y., Zaheer, M., Hu, H., Manmatha, R., Smola, A.J., Krähenbühl, P.: Compressed video action recognition. In: Proceedings of the IEEE Conference on Computer Vision and Pattern Recognition, pp. 6026–6035 (2018)

53. Wu, Z., Xiong, C., Jiang, Y.G., Davis, L.S.: LiteEval: a coarse-to-fine framework for resource efficient video recognition. Adv. Neural Inf. Process. Syst. **32** (2019)

54. Wu, Z., Xiong, C., Ma, C.Y., Socher, R., Davis, L.S.: AdaFrame: adaptive frame selection for fast video recognition. In: Proceedings of the IEEE/CVF Conference on Computer Vision and Pattern Recognition, pp. 1278–1287 (2019)

55. Xiao, B., Wu, H., Wei, Y.: Simple baselines for human pose estimation and tracking. In: Proceedings of the European conference on computer vision (ECCV), pp. 466–481 (2018)

56. Xu, R., et al.: ApproxDet: content and contention-aware approximate object detection for mobiles. In: Proceedings of the 18th Conference on Embedded Networked Sensor Systems, SenSys 2020, pp. 449–462. Association for Computing Machinery, New York (2020). https://doi.org/10.1145/3384419.3431159

57. Yang, L., Han, Y., Chen, X., Song, S., Dai, J., Huang, G.: Resolution adaptive networks for efficient inference. In: Conference on Computer Vision and Pattern Recognition (CVPR), pp. 2369–2378 (2020)

58. Yang, Y., Ramanan, D.: Articulated human detection with flexible mixtures of parts. IEEE Trans. Pattern Anal. Mach. Intell. **35**(12), 2878–2890 (2013). https://doi.org/10.1109/TPAMI.2012.261

59. Zhang, X., Zhou, X., Lin, M., Sun, J.: ShuffleNet: an extremely efficient convolutional neural network for mobile devices. In: Conference on Computer Vision and Pattern Recognition (CVPR), pp. 6848–6856 (2018)

60. Zhu, X., Dai, J., Yuan, L., Wei, Y.: Towards high performance video object detection. In: Proceedings of the IEEE Conference on Computer Vision and Pattern Recognition, pp. 7210–7218 (2018)

61. Zhu, X., Xiong, Y., Dai, J., Yuan, L., Wei, Y.: Deep feature flow for video recognition. In: Proceedings of the IEEE Conference on Computer Vision and Pattern Recognition, pp. 2349–2358 (2017)

EdgeViTs: Competing Light-Weight CNNs on Mobile Devices with Vision Transformers

Junting Pan[1], Adrian Bulat[2], Fuwen Tan[2], Xiatian Zhu[2], Lukasz Dudziak[2], Hongsheng Li[1], Georgios Tzimiropoulos[2,3], and Brais Martinez[2(✉)]

[1] The Chinese University of Hong Kong, Shatin, Hong Kong
[2] Samsung AI, Cambridge, UK
brais.a@samsung.com
[3] Queen Mary University of London, London, UK

Abstract. Self-attention based models such as vision transformers (ViTs) have emerged as a very competitive architecture alternative to convolutional neural networks (CNNs) in computer vision. Despite increasingly stronger variants with ever higher recognition accuracies, due to the quadratic complexity of self-attention, existing ViTs are typically demanding in computation and model size. Although several successful design choices (e.g., the convolutions and hierarchical multi-stage structure) of prior CNNs have been reintroduced into recent ViTs, they are still not sufficient to meet the limited resource requirements of mobile devices. This motivates a very recent attempt to develop light ViTs based on the state-of-the-art MobileNet-v2, but still leaves a performance gap behind. In this work, pushing further along this under-studied direction we introduce **EdgeViTs**, a new family of light-weight ViTs that, for the first time, enable attention based vision models to compete with the best light-weight CNNs in the tradeoff between accuracy and on-device efficiency. This is realized by introducing a highly cost-effective *local-global-local* (LGL) information exchange bottleneck based on optimal integration of self-attention and convolutions. For device-dedicated evaluation, rather than relying on inaccurate proxies like the number of FLOPs or parameters, we adopt a practical approach of focusing directly on on-device latency and, for the first time, energy efficiency. Extensive experiments on image classification, object detection and semantic segmentation validate high efficiency of our EdgeViTs when compared to the state-of-the-art efficient CNNs and ViTs in terms of accuracy-efficiency tradeoff on mobile hardware. Specifically, we show that our models are Pareto-optimal when both accuracy-latency and accuracy-energy tradeoffs are considered, achieving strict dominance over other ViTs in almost all cases and competing with the most efficient CNNs. Code is available at https://github.com/saic-fi/edgevit.

J. Pan—Work done during an internship at Samsung AI Cambridge.

Supplementary Information The online version contains supplementary material available at https://doi.org/10.1007/978-3-031-20083-0_18.

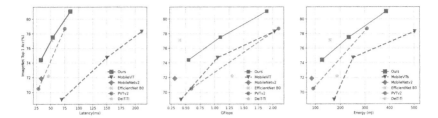

Fig. 1. Our EdgeViTs yield comparable or superior tradeoff between accuracy and efficiency (e.g., run speed, latency) on ImageNet-1K, in comparison to state-of-the-art efficient CNNs [42,47], representative generic ViTs [51,55] and latest Mobile-ViTs [36]. Note that, EdgeViTs outperform MobileViTs, the only specially designed model for mobile device, across all the metrics by a large margin. Whilst our lightest EdgeViT consumes more GFLOPs than EfficientNet B0 [47], it runs faster (33.3 vs. 52.1 ms per image), with a reduced gap in *on-device energy consumption*. *Testing device*: Samsung Galaxy S21 (latency), Snapdragon 888 Hardware Development Kit (energy).

1 Introduction

Vision transformers (ViTs) have rapidly superseded convolutional neural networks (CNNs) on a variety of visual recognition tasks [9,51], particularly when the priors and successful designs of previous CNNs are reintroduced for leveraging the induction bias of visual data such as local grid structures [7,15,30,54]. Due to the quadratic complexity of ViTs and the high-dimension property of visual data, it is indispensable that the computational cost needs to be taken into account in design [54]. Three representative designs to make computationally viable ViTs are (1) the use of a hierarchical architecture with the spatial resolution (i.e., the token sequence length) progressively down-sampled across the stages [7,11,54], (2) the locally-grouped self-attention mechanisms for controlling the length of input token sequences and parameter sharing [15,30], (3) the pooling attention schemes to subsample the key and value by a factor [34,53,54,59]. The general trend has been on designing more complicated and stronger ViTs to challenge the dominance of top-performing CNNs [19,39,47] in computer vision by achieving ever higher accuracies [52,65]. These advances however are still insufficient to satisfy the design requirements and constraints for mobile and edge platforms (e.g., smart phones, robotics, self-driving cars, AR/VR devices), where the vision tasks need to carry out in a timely manner under certain computational budgets. Prior efficient CNNs (e.g., MobileNets [21,22,42], ShuffleNets [35,63], EfficientNets [47,48,50], and etc.) remain the state-of-the-art network architectures for such platforms in the tradeoff between running latency and recognition accuracy (Fig. 1).

In this work, we focus on the development of largely under-studied *efficient ViTs* with the aim to surpass the CNN counterparts on mobile devices. We consider a collection of very practical design requirements for running a ViT model on a target real-world platform as follows: (**1**) *Inference efficiency* needs to be high (e.g., *low latency and energy consumption*) so that the running cost becomes generically affordable and more on-device applications can be supported. This is a direct metric that we really care about in practice. In contrast, the often-used efficiency metric, FLOPs (i.e., the number of multiply-adds), cannot directly translate into the latency and energy consumption on a specific device, with several conditional factors including memory access cost, degree of parallelism, and the platform's characteristics [35]. This is, *not* all operations of a model can be carried out at the same speed and energy cost on a device. Hence, FLOPs is merely an approximate and indirect metric of efficiency. (**2**) *Model size* (i.e., parameter number) is affordable for modern average devices. Given the availability of ever cheaper and larger storage spaces, this constraint has been relaxed significantly. For example, an average smart phone often comes with 32GB or more storage. As a consequence, using it as a threshold metric is no longer valid in most cases. (**3**) *Implementational friendliness* is also critical in real-world applications. For a wider range of deployment, it is necessary that a model can be implemented efficiently using the standard computing operations supported and optimized in the generic deep learning frameworks (e.g., ONNX, TensorRT, and TorchScript), without costly per-framework specialization. Otherwise, the on-device speed of a model might be unsatisfactory even with low FLOPs. For instance, the cyclic shift and its reverse operations introduced in Swin Transformers [30] are rarely supported by the mainstream frameworks, i.e., deployment unfriendly. In the literature, very recent MobileViTs [36] are the only series of ViTs designed for mobile devices. In architecture design, they are a straightforward combination of MobileNetv2 [42] and ViTs [9]. As a very initial attempt in this direction, MobileViTs still lag behind CNN counterparts. Further, its evaluation protocol takes the *model size* (i.e. the parameter number) as the competitor selection criteria (i.e., comparing the accuracy of models only with similar parameter numbers), which however is no longer a hard constraint with modern hardware as discussed above and is hence out of date.

We present a family of light-weight attention based vision models, dubbed as **EdgeViTs**, for the first time, enabling ViTs to compete with the best light-weight CNNs (e.g., MobileNetv2 [42] and EfficientNets [47]) in terms of accuracy-efficiency tradeoff on mobile devices. This sets a milestone in the landscape of light-weight ViTs *vs.* CNNs in the low resource regime. Our EdgeViTs are based on a novel factorization of the standard self-attention for more cost-effective information exchange within every individual layer. This is made possible by introducing a highly light-weight and easy-to-implement *local-global-local* (LGL) information exchange bottleneck characterized with three operations: (**i**) Local information aggregation from neighbor tokens (each corresponding to a specific patch) using efficient depth-wise convolutions; (**ii**) Forming a sparse set of evenly distributed *delegate tokens* for long-range information exchange by self-attention; (**iii**) Diffusing updated information from *delegate tokens* to the *non-delegate tokens* in local neighborhoods via transposed convolutions. As we show in experiments, this design presents a favorable hybrid of self-attention, convolutions, and transposed convolutions, achieving the best accuracy-efficiency tradeoff. It is efficient in that the self-attention is applied to a sparse set of delegate tokens. To support a variety of computational budgets, with our primitive module we establish a family of EdgeViT variants with three computational complexities: small (S), extra-small (XS), extra-extra-small (XXS).

We make the following **contributions: (1)** We investigate the design of light-weight ViTs from the practical on-device deployment and execution perspective. **(2)** For best scalability and deployment, we present a novel family of efficient ViTs, termed as EdgeViTs, designed based on an optimal decomposition of self-attention using standard primitive operations. **(3)** Regarding on-device performance, towards relevance for real-world deployment, we directly consider latency and energy consumption of different models rather than relying on high-level proxies like number of FLOPs or parameters. Our results experimentally verify efficiency of our models in a practical setting and refute some of the claims made in the existing literature. More specifically, extensive experiments on three visual tasks show that our EdgeViTs can match or surpass state-of-the-art light-weight CNNs, whilst consistently outperform the recent MobileViTs in accuracy-efficiency tradeoff, including largely ignored on-device energy evaluation. Importantly, EdgeViTs are consistently Pareto-optimal in terms of both latency and energy efficiency, achieving strict dominance over other ViTs in almost all cases and competing with the most efficient CNNs. On ImageNet classification our EdgeViT-XXS outperforms MobileNetv2 by 2.2% subject to the similar energy-aware efficiency.

2   Related Work

Efficient CNNs. Since the advent of modern CNN architectures [19,45], there has been a steady stream of works focusing on efficient architecture design for on-device deployment. The first widely adopted families bring depthwise separable convolutions in a ResNet-like structure, e.g., MobileNets [22,42], ShuffleNets [35,63]. These works define a space of well-performing efficient architectures, resulting in widespread usage. Successive works further exploit this design space by automating the architectural design choices [21,46,47,49]. As a parallel line of research, net pruning creates efficient architectures by removing spurious parts of a larger network with close-to-zero weights [16,56], or via first training a super-network that is further slimmed to meet a pre-specified computational budget [3,31]. Dynamic computing has also been explored, consisting of the mechanisms that condition the network parameters on the input data [6,61]. Finally, using low bit-width is a very critical technique that can offer different tradeoffs between the accuracy and efficiency [5,16,23].

Vision Transformers. ViTs [9] quickly popularize transformer-based architectures for computer vision. A series of works followed instantly, offering large improvements to the original ViTs in terms of data efficiency [29,51] and architecture design [4,11,30,62]. Among these works, one of the main modifications is to introduce hierarchical designs in multiple stages from convolutional architectures [7,30,54,55]. Several works also focus on improving the positional encoding by using a relative positional embedding [43,44], making it learnable [13], or even replacing it by a attention bias element [14]. All these approaches mostly aim to improve the model performance.

Recently, more efforts have been made towards finding efficient alternatives to the multi-head self-attention (MHSA) module, which is typically the computational bottleneck in the ViT architectures. A particularly effective solution is to reduce the internal spatial dimensions within the MHSA. The MHSA involves projecting the input tensor into key, query and value tensors. Several recent works, e.g. [7,54,55], find that the key and value tensors could be downsampled with a limited loss in accuracy, leading to a better efficiency-accuracy trade-off. Our work extends this idea by also downsampling the query tensors, which further improves the efficiency, as shown in Fig. 2.

There are also alternative approaches reducing the number of tokens dynamically [12,37,40,60]. That is, in the forward pass, tokens deemed to not contain the important information for the target task are pruned or pooled together, reducing the overall complexity thereafter. Finally, encouraged by their potential complementarity, many works have attempted to combine convolutional designs with self-attentions. This ranges from using convolutions at the stem [58], integrating convolutional operations into the MHSA block [25,57], or incorporating the MHSA block into ResNet-like architectures [44]. It is interesting to note that even the original ViTs explored similar tradeoffs [9].

Vision Transformers for Mobile Devices. Whilst the efficiency issue has been taken into account in designing the ViT variants discussed above, they are still not dedicated and satisfactory architectures for on-device applications. There is only one exception, MobileViTs [36], which are introduced very recently. However, compared to the current best light-weight CNNs such as MobileNets [21,42] and EfficicentNets [47], these ViTs are still clearly inferior in terms of the on-device accuracy-efficiency tradeoff. In this work, we present the first family of efficient ViTs that can deliver comparable or even superior tradeoffs in comparison to the best CNNs and ViTs. We also extensively carry out the critical yet largely lacking on-device evaluations with energy consumption analysis.

3   EdgeViTs

3.1   Overview

For designing light-weight ViTs suitable for mobile/edge devices, we adopt a hierarchical pyramid network structure (Fig. 2(a)) used in recent ViT variants [7,8,11,54,55]. A pyramid transformer model typically reduces the spatial resolution but expands the channel dimension across different stages. Each stage consists of multiple transformer-based blocks processing tensors of the same shape, mimicking the ResNet-like networks. The transformer-based blocks heavily rely on the self-attention operations at a quadratic complexity w.r.t the spatial resolution of the visual features. By progressively aggregating the spatial tokens, pyramid vision transformers are potentially more efficient than isotropic models [9].

In this work, we dive deeper into the transformer-based block and introduce a cost-effective bottleneck, *Local-Global-Local* (LGL) (Fig. 2(b)). LGL further reduces the overhead of self-attention with a sparse attention module (Fig. 2(c)), achieving better accuracy-latency balancing.

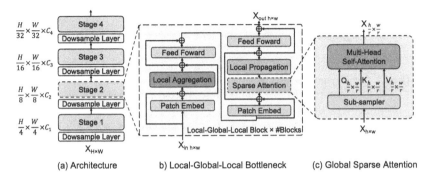

(a) Architecture   b) Local-Global-Local Bottleneck   (c) Global Sparse Attention

Fig. 2. (a) Schematic overview of our four stages EdgeViT architecture, with each stage consisting of a stack of (b) *Local-Global-Local* (LGL) blocks constructed with local aggregation module, sparse-self-attention and local propagation module, patch embedding (PE) and Feed Forward Network (FFN). In this example, h and w refer to input height and width of stage-2: $h = \frac{H}{8}$ and $w = \frac{W}{8}$. C_i refers to the number of channels for stage-i and r denotes the sub-sampling rate.

3.2 *Local-Global-Local* Bottleneck

Self-attention has been shown to be very effective for learning the global context or long-range spatial dependency of an image, which is critical for visual recognition. On the other hand, as images have high spatial redundancy (e.g., nearby patches are semantically similar) [17], applying attention to all the tokens, even in a down-sampled feature map, is inefficient. There is hence an opportunity to reduce the scope of tokens whilst still preserving the underlying information flows that model the global and local contexts. In contrast to previous transformer blocks that perform self-attention at each spatial location, our LGL bottleneck only computes self-attention for a subset of the tokens but enables full spatial interactions, as in the standard multi-head self-attention (MHSA) [9].

To achieve this, we decompose the self-attention into consecutive modules that process the spatial tokens within different ranges (Fig. 2(b)). We introduce three efficient operations: i) *Local aggregation* that integrates signals only from locally proximate tokens; ii) *Global sparse attention* that model long-range relations among a set of delegate tokens where each of them is treated as a representative for a local window; iii) *Local propagation* that diffuses the global contextual information learned by the delegates to the non-delegate tokens with the same window. Combining these, our LGL bottleneck enables information exchanges between any pair of tokens in the same feature map at a low-compute cost. Each of these components is described in detail below:

| (a) Local Aggregation | (b) Global Sparse Attention | (c) Local Propagation |

Fig. 3. Illustration of three key operations involved in the proposed *Local-Global-Local* (LGL) transformer block. In this example, we showcase how *the target token* (the orange square) at the center conducts information exchange with all the others in three sequential steps: (a) Local information from neighbor tokens within the is first aggregated to the target token. (b) Global sparse attention is then computed among the target token and other selected delegates in orange color. (c) Global context information encoded in the target token is last propagated to its neighbor *non-delegate* tokens within the pink area. (Color figure online)

- *Local aggregation*: for each token, we leverage depth-wise and point-wise convolutions to aggregate information in local windows with a size of $k \times k$ (Fig. 3(a)).
- *Global sparse attention*: we sample a sparse set of *delegate tokens* distributed evenly across the space, one *token* for each $r \times r$ window. Here, r denotes the sub-sample rate. We then apply self-attention on these selected tokens only (Fig. 3(b)). This is distinct from all the existing ViTs [7,54,55] where all the spatial tokens are involved as queries in the self-attention computation.
- *Local propagation*: We propagate the global contextual information encoded in the delegate tokens to their neighbor tokens by transposed convolutions (Fig. 3(c)).

Formally, our LGL bottleneck can be formulated as:

$$
\begin{aligned}
X &= \texttt{LocalAgg}(\texttt{Norm}(X_{in})) + X_{in}, \\
Y &= \texttt{FFN}(\texttt{Norm}(X)) + X, \\
Z &= \texttt{LocalProp}(\texttt{GlobalSparseAttn}(\texttt{Norm}(Y))) + Y, \\
X_{out} &= \texttt{FFN}(\texttt{Norm}(Z)) + Z.
\end{aligned}
\tag{1}
$$

Here $X_{in} \in \mathcal{R}^{H \times W \times C}$ indicates the input tensors. $\texttt{Norm}$ is the layer normalization operation [2]. $\texttt{LocalAgg}$ represents the local aggregation operator, $\texttt{FFN}$ is a two-layer perceptron, similar to the position-wise feed-forward network introduced in [9]. $\texttt{GlobalSparseAttn}$ is the global sparse self-attention. $\texttt{LocalProp}$ is the local propagation operator. For simplicity, positional encoding is omitted. Note that, all these operators can be implemented by commonly used and highly optimized operations in the standard deep learning platforms. Hence, our LGL bottleneck is implementation friendly.

Table 1. Configuration of three EdgeViT variants. "#Channels": number of channels per stage. "#Blocks": number of LGL blocks per stage. "#Heads": number of attention heads in MHSA. "#Param": number of parameters.

Model	#Channels	#Blocks	#Heads	FLOPs	#Param
EdgeViT-XXS	[36, 72, 144, 288]	[1,1,3,2]	[1,2,4,8]	0.56G	4.1M
EdgeViT-XS	[48, 96, 240, 384]	[1,1,2,2]	[1,2,4,8]	1.1G	6.7M
EdgeViT-S	[48, 96, 240, 384]	[1,2,3,2]	[1,2,4,8]	1.9G	11.1M

Comparisons to Existing Designs. Our LGL bottleneck shares a similar goal with the recent PVTs [54,55] and Twins-SVTs [7] models that attempt to reduce the self-attention overhead. However, they differ in the core design. PVTs [54,55] perform self-attention where the number of keys and values are reduced by strided-convolutions, whilst the number of queries remains the same. In other words, PVTs still perform self-attention at each grid location. In this work, we question the necessity of positional-wise self-attention and explore to what extent the information exchange enabled by our LGL bottleneck could approximate the standard MHSA (see Sect. 4 for more details). Twins-SVTs [7] combine local-window self-attention [30] with global pooled attention from PvTs [54]. This is different from our hybrid design using both self-attention and convolution operations distributed in a series of *local-global-local* operations. As demonstrated in the experiments (Table 2 and 3), our design achieve a better tradeoff between the model performance and the computation overhead (e.g. latency, energy consumption, etc.).

3.3 Architectures

We build a family of EdgeViTs with the proposed LGL bottleneck at different computational complexities (i.e. 0.5G, 1G, and 2G). The configurations are summarized in Table 1. Following the hierarchical ViTs [7,25,30,54], EdgeViTs consist of four stages with the spatial resolution (*i.e.*, the token sequence length) gradually reduced throughout, and their self-attention module replaced with our LGL bottleneck. For the stage-wise down-sampling, we use a conv-layer with a kernel size of 2×2 and stride 2, except for the first stage where we down-sample the input feature by $\times 4$, and use a 4×4 kernel and a stride of 4. We adopt the conditional positional encoding [8] that has been shown to be superior to the absolute positional encoding. This can be implemented using 2D depth-wise convolutions with a residual connection. In our model, we use 3×3 depth-wise convolutions with zero paddings. It is placed before the local aggregation and global sparse self-attention. The FFN consists of two linear layers with GeLU non-linearity [20] placed in-between. Our local aggregation operator is implemented as a stack of pointwise and depthwise convolutions. The global sparse attention is composed of a spatial uniform sampler with sample rates of $(4, 2, 2, 1)$ for the four stages, and a standard MHSA. The local propagation is implemented with a depthwise separable transposed convolution with the kernel size and stride equal to the sample rate used in the global sparse attention. The exact architecture for the LGL bottleneck is described in the `supplementary material`.

4   Experiments

We benchmark EdgeViTs on visual recognition tasks. We pre-train EdgeViTs on the Imagenet1K recognition task [41], comparing the performances and computation overheads against alternative approaches. We also evaluate the generalization capacity of EdgeViTs on downstream dense prediction tasks: object detection and instance segmentation on the COCO benchmark [28], and semantic segmentation on the ADE20K Scene Parsing benchmark [64]. For on-device execution, we report exucution time (latency) and energy consumption of all relevant models on ImageNet. We do not report on-device measurements on downstream tasks as they reuse ImageNet models.

4.1   Image Classification on ImageNet-1K

Training Settings.   ImageNet-1K [41] provides 1.28 million training images and 50,000 validation images from 1000 categories. We follow the training recipe introduced in DeiT [51]. We optimize the models using AdamW [32] with a batch size of 1024, weight decay of 5×10^{-2}, and momentum of 0.9. The models are trained from *scratch* for 300 epochs with a linear warm-up during the first 5 epochs. Our base learning rate is set as 1×10^{-3}, and decay after the warm-up using a cosine schedule [33]. We apply the same data augmentations as in [7,25,51,55] which include random cropping, random horizontal flipping, mixup, random erasing and label-smoothing. During training, the images are randomly cropped to 224×224. During testing, we use a single center crop of 224×224. We report the top-1 accuracy on the validation set.

Benchmarking Settings.   For latency measurements, we use a Samsung Galaxy S21 mobile phone equipped with a Snapdragon 888 chipset. All relevant models are benchmarked by running a forward pass 50 times using TorchScript lite interpreter via the Android benchmarking app provided by PyTorch [38]. We use CPU implementation, full precision and batch 1 to execute all operations. This choice comes from the fact that this is the only combination that was able to robustly execute all of the models from our paper. In general, more efficient implementations exist, such as those utilizing specialized hardware like Neural Processing Units (NPUs). However, these put more restrictions on what can be executed and many models failed to run in our experiments when trying to use different hardware targets.

For energy measurements, we use a Monsoon High Voltage Power Monitor connected to a Snapdragon 888 Hardware Development Kit (HDK8350) to obtain accurate power readings over the course of running a forward pass of each test model 50 times. The same TorchScript runtime is used as in latency

Table 2. Results on ImageNet-1K validation set. All models are tested on input scale of 224×224, except for MobileViTs [36] that are tested with 256×256 according to their original implementation. * indicates down-scaled architectures beyond original definitions by authors to fit the mobile compute budget. LeViT-384† denotes the LeViT model retrained under the same setting as our EdgeViT.

Model	#Params	FLOPs	CPU (ms)	Acc Top-1 (%)
MobileNet-v2 [42]	3.4M	0.3G	$33.3_{\pm 5.3}$	72.0
MobileNet-v3 0.75 [21]	**4.0M**	**0.16G**	$\mathbf{23.0_{\pm 3.7}}$	**73.3**
EfficientNet-B0 [47]	**5.3M**	**0.4G**	$\mathbf{52.1_{\pm 7.4}}$	**77.1**
MobileViT-XXS [36]	1.3M	0.4G	$69.5_{\pm 5.1}$	69.0
PVT-v2-B0 [55]	3.4M	0.6G	$26.0_{\pm 6.9}$	70.5
Uniformer-Tiny* [25]	3.9M	0.6G	$40.5_{\pm 3.1}$	74.1
Twins-SVT-Tiny* [7]	4.1M	0.6G	$36.9_{\pm 2.3}$	71.2
EdgeViT-XXS	**4.1M**	**0.6G**	$\mathbf{32.8_{\pm 2.7}}$	**74.4**
T2T-ViT-7 [62]	4.3M	1.1G	$48.8_{\pm 6.5}$	71.7
MobileViT-XS [36]	2.4M	1.1G	$150.1_{\pm 6.1}$	74.7
DeiT-Tiny [51]	5.7M	1.3G	$46.2_{\pm 13.6}$	72.2
TNT-Tiny [15]	6.1M	1.4G	$86.4_{\pm 6.0}$	73.9
EdgeViT-XS	**6.7M**	**1.1G**	$\mathbf{54.1_{\pm 2.2}}$	**77.5**
T2T-ViT-12 [62]	6.9M	1.9G	$69.9_{\pm 5.6}$	76.5
PVT-v2-B1 [55]	14M	2.1G	$75.4_{\pm 2.3}$	78.7
MobileViT-S [36]	5.6M	2.0G	$221.3_{\pm 9.3}$	78.3
LeViT-384† [14]	**39.1M**	**2.4G**	$\mathbf{71.3_{\pm 2.3}}$	**79.5**
EdgeViT-S	**11.1M**	**1.9G**	$\mathbf{85.3_{\pm 3.9}}$	**81.0**

measurements. From the power signal reported by the monitor, we derive the average per-inference power and energy consumption by first subtracting background power consumption (i.e., power readings when not running any model) and then identifying 50 continuous regions of significantly higher power draw. Each region like that is considered a single inference and we calculate its total energy as the integral over the individual power samples. Analogously we also calculate average power consumption by averaging over the same set of samples. After energy and power are calculated for each inference, the final statistics of a model are obtained by again averaging over the 50 identified runs. Our methodology follows what can be found in the literature [1].

Results. We compare EdgeViTs to a variety of baseline models, including the classic efficient CNNs, e.g. MobileNetV2 [42], MobileNetV3 [21], EfficientNet [47], and the state-of-the-art ViTs, e.g. MobileViT [36], PVT-v2 [55], DeiT [51], LeViT [14]. As the original LeViT [14] was optimized in a large-scale setting (i.e. 1000 epochs) with knowledge distillation, we perform a comparison by re-training LeViT under the same setting (300 epochs) as EdgeViT without knowledge dis-

Table 3. On-device energy evaluation on ImageNet-1K. All relevant metrics are reported as mean values per forward pass across 50 executions. For facilitating comparison, we define an energy-aware *efficiency* metric as the average gain in top-1 accuracy from each 1 W run for 1 ms (equivalent to consuming 1 mJ of energy). (*Pareto-optimal models* are highlighted in bold in the last column).

Model	Top-1 (%)	CPU (ms)	Energy (mJ)	Power (W)	Efficiency (%/msW)
MobileNet-v2 [42]	72.0	33.3	85.7±7.4	3.31±0.26	0.841
MobileNet-v3 0.75 [21]	**73.3**	**23.0**	**63.0±9.6**	**3.46±0.4**	**1.164**
EfficientNet-B0 [47]	**77.1**	**52.1**	**159.0±26.2**	**3.62±0.45**	**0.485**
PVT-v2-B0 [55]	70.5	26.0	91.7±19.7	3.94±0.68	0.769
PVT-v2-B1 [55]	**78.7**	75.4	309.0±65.8	4.63±0.71	0.255
Twins-SVT-Tiny* [7]	71.2	36.9	114.5±17.3	3.71±0.24	0.622
DeiT-Tiny [51]	72.2	46.2	187.2±7.6	4.77±0.21	0.386
Uniformer-Tiny* [25]	74.1	40.5	134.7±27.3	4.1±0.71	0.55
T2T-ViT-12 [62]	76.5	69.9	266.2±42.6	4.37±0.36	0.287
TNT-Tiny [15]	73.9	86.4	308.7±70.5	3.94±0.63	0.239
LeViT-384† [14]	**79.5**	71.3±2.2	**455.2±125.8**	**6.18±0.74**	**0.173**
MobileViT-XXS [36]	69.0	69.5	175.3±28.7	2.77±0.24	0.394
MobileViT-XS [36]	74.7	150.1	251.5±81.1	2.63±0.61	0.297
MobileViT-S [36]	78.3	221.3	503.6±117.0	2.76±0.21	0.155
EdgeViT-XXS	**74.4**	**32.8**	**127.4±27.3**	**4.27±0.67**	**0.584**
EdgeViT-XS	**77.5**	**54.1**	**234.6±44.0**	**4.77±0.84**	**0.33**
EdgeViT-S	**81.0**	**85.3**	**386.7±43.5**	**4.8±0.26**	**0.209**

tillation. We denote the retrained LeViTs as LeViT-384†. We select the baselines with a complexity of less than 2 GFLOPS as i) in real-world applications, the computational cost remains the top concern; ii) whilst FLOPs is an indirect metric for the latency, it is the most used cost metric in prior works. This selection criterion is different from [36] that instead uses the model size (i.e. the parameter number) which however has become a less restricted facet in mobile devices.

From Table 2, we can learn: i) EdgeViTs significantly outperform other lightweight *ViTs* at a similar level of GFLOPs complexity. Compared to the PVT-v2 family [55], our EdgeViT-XXS/EdgeViT-S achieve 3.9%/2.3% improvements over PVT-v2-B0/PVT-v2-B1. Compared to MobileViTs, EdgeViTs achieve 5.4%, 2.8% and 2.7% gains in the three complexity settings. ii) *ViTs vs. CNNs:* Our EdgeViTs lift the performance of efficient ViTs to approach the level of well-established efficient CNNs. For example, the EdgeViT-XXS performs superior to MobileNet-v2 and MobileNet-v3-0.75 at a similar level of model size, but requires more GFLOPs. However, we observe that the efficient CNNs still surpass efficient ViTs in the accuracy-FLOPs tradeoff by a small margin.

On the other hand, as discuss early, numbers of FLOPs or parameters are merely indicative but do not fully reflect the on-device efficiency [10,35]. We further consider on-device latency and energy consumption directly. Other than the representative ViTs and CNNs, we also compare two recent ViT variants [7,25] with the number of channels and layers re-scaled to fit the complex-

Table 4. Ablation on ImageNet-1K. LA: the local aggregation operator. LP: the local propagation operator. LSA: the Locally-grouped Self-Attention used in [7].

	CPU	Top1		CPU	Top1		CPU	Top1
w/o LA	33.9 ms	72.7	max	34.8 ms	74.3	w/o LP	**32.4 ms**	73.9
LA (LSA)	36.1 ms	74.0	avg	34.5 ms	**74.5**	LP (Bilinear)	34.1 ms	74.1
LA (Ours)	**32.8 ms**	**74.4**	center	**32.8 ms**	74.4	LP (Ours)	32.8 ms	**74.4**
(a) **Local Aggregation**			(b) **Global attention**			(c) **Local Propagation**		

ity need. As presented in Table 2, EdgeViTs demonstrate strong performance with latencies comparable to MobileNets: EdgeViT-XXS achieves a gain of 2.4% over MobileNet-V2 while running slightly faster. EdgeViT-XXS also surpasses MobileNet-V3 by 1.1% but at the cost of being 9.8 ms slower. EdgeViT-XS performs on par with the auto-searched EfficientNet-B0 model. We believe our models could also benefit from the automatic architecture search techniques as use in MobileNet-V3 and EfficientNets. Our models yield clear advantages over alternative ViT models. Compared to MobileViTs in the three GFLOPs settings, EdgeViTs excel by 5.4%, 2.8%, and 2.7% while being ×2, ×2.7, ×2.6 faster.

Energy results are presented in Table 3. In addition to the raw energy and power numbers, for comparison simplicity, we define an energy-aware *efficiency* metric as the average gain in top-1 accuracy (in percentages) from each consumed 1 mJ of energy. We observe that less accurate models tends to be more efficient. This is not a surprise in that improvements in accuracy scale sublinearly with model complexity. However, this also means that comparing efficiency of models with very different top-1 scores might be severely biased by the sole difficulty of achieving certain accuracy levels, which is independent from a model. Therefore, we limit our comparison to identifying *pareto-optimal models*, those upon which no other models can improve in either accuracy or energy efficiency without degrading other metrics. We can see that our EdgeViT family is able to dominate almost all other ViTs, with the only exception being LeViT-384† whose accuracy and efficiency fall between our EdgeViT-S and EdgeViT-XS. When compared to CNNs, our EdgeViTs compete with MobileNet-v3 and EfficientNet-B0 that are more efficient but also less accurate. MobileNet-v2 achieves decent results but is dominated by its newer version, MobileNet-v3. PVT-v2-B0, although high on the efficiency side, is rather inaccurate and hence is favored by highly efficient CNNs. Visibly at the end of the spectrum are the latest MobileViT models which turn out to be neither efficient nor accurate, when compared to the rest. Unlike them, our EdgeViT models, although not as efficient as best CNNs in the absolute sense, exhibit favourable trade-off between efficiency and accuracy by being rather highly accurate while not sacrificing efficiency.

Ablation Study. We conduct detailed ablations to validate our design choices in the LGL bottleneck. We use EdgeViT-XXS as the base model and re-scale the alternative designs to ∼0.5GFLOPs for fair comparison.

Table 5. Comparison to other visual backbones using RetinaNet and Mask-RCNN on COCO `val2017` object detection and instance segmentation. "#Par." refers to number of parameters in million. AP^b and AP^m indicate bounding box AP and mask AP.

Backbone	RetinaNet 1×						Mask R-CNN 1×						
	#Par.	AP	AP_{50}	AP_{75}	AP_S	AP_M	AP_L	#Par.	AP^b	AP^b_{50}	AP^b_{75}	AP^m	AP^m_{50}	AP^m_{75}
PVTv2-B0 [55]	**13.0**	37.2	57.2	39.5	**23.1**	40.4	49.7	**23.5**	38.2	60.5	40.7	36.2	57.8	38.6
EdgeViT-XXS	13.1	**38.7**	**59.0**	**41.0**	22.4	**42.0**	**51.6**	23.8	**39.9**	**62.0**	**43.1**	**36.9**	**59.0**	**39.4**
EdgeViT-XS	16.3	40.6	61.3	43.3	25.2	43.9	54.6	26.5	41.4	63.7	45.0	38.3	60.9	41.3
ResNet18 [19]	**21.3**	31.8	49.6	33.6	16.3	34.3	43.2	**31.2**	34.0	54.0	36.7	31.2	51.0	32.7
PVTv1-Tiny [54]	23.0	36.7	56.9	38.9	22.6	38.8	50.0	32.9	36.7	59.2	39.3	35.1	56.7	37.3
PVTv2-B1 [55]	23.8	41.2	61.9	43.9	25.4	44.5	54.3	33.7	41.8	64.3	45.9	38.8	61.2	41.6
EdgeViT-S	22.6	**43.4**	**64.9**	**46.5**	**26.9**	**47.5**	**58.1**	32.8	**44.8**	**67.4**	**48.9**	**41.0**	**64.2**	**43.8**

Local Aggregation. We compare our local aggregation (LA) operation to the Locally-grouped Self-Attention (LSA) used in [7,30]. It is shown in Table 4a that applying LA consistently improve the performance. Our convolutional LA module performs better than the self-attention based operator (LSA). This validates our choice of using `depth-wise convolutions` in LA for local context learning.

Global Sparse Attention. We explore three options for delegate token sampling: `max`, `avg`, and `center`. All choices perform similarly in terms of accuracy, with our default design `center` being slightly faster.

Local Propagation. We investigate two alternatives to the local propagation operator: i) `w/o LP`: We simply remove the local propagation. Note that EdgeViTsw/o LP has similar complexity to standard EdgeViTs. ii) `Bilinear`: we use the bilinear interpolation, instead of the transposed convolution, to upsample the delegate tokens. Table 4c shows that adding LP improves the top-1 accuracy by 0.5%, with only 0.4 ms overhead.

4.2   Dense Prediction

Following [54,55], we also evaluate the proposed EdgeViTs on COCO Objection Detection/Instance Segmentation [28] and ADE20K Scene Parsing [64]. Here, we use the EdgeViTs as the feature extractor for the main model and initialize it with the ImageNet1K-pretrained weights obtained in our previous experiments.

COCO Object Detection/Instance Segmentation. We demonstrate the performance of our model in main-stream object detection and instance segmentation frameworks: RetinaNet [27] for object detection, Mask R-CNN [18] with the FPN [26] for instance segmentation. Following the training protocol in [7,25,55], we resize the training images to have a shorter side of 800 pixels while keeping the longer side to be smaller than 1333 pixels. During testing, the images are re-scaled to have a shorter size of 800 pixels. The models are fine-tuned with 1× schedule (i.e. 12 epochs) by AdamW [32] using an initial learning rate of 1×10^{-4} and a batch size of 16. We train the models on the COCO 2017 training set and report the mAP@100 score on the COCO 2017 validation set.

Table 6. Semantic segmentation results on the validation set of ADE20K.
Segmentation model: Semantic FPN [24]. *GFLOPs*: Calculated at 512×512 input size.

Backbone	Semantic FPN		
	#Param (M)	GFLOPs	mIoU (%)
PVTv2-B0 [55]	**7.6**	**25.0**	37.2
EdgeViT-XXS	7.9	24.4	**39.7**
EdgeViT-XS	10.6	27.7	41.4
ResNet18 [19]	**15.5**	32.2	32.9
PVTv1-Tiny [54]	17.0	33.2	35.7
PVTv2-B1 [55]	17.8	34.2	42.5
EdgeViT-S	16.9	**32.1**	**45.9**

Results. In Table 5, our EdgeViTs perform consistently better than other visual backbones on RetinaNet [27] and Mask R-CNN [18]. Our smallest variant EdgeViT-XXS, when used on RetinaNet [27], achieves 1.5 higher AP than PVTv2-B0. When used on Mask R-CNN [18], EdgeViT-XXS also surpasses PVTv2-B0 by 1.7 on the bounding box detection task (AP^b), and by 0.7 on the mask segmentation task (AP^m). For EdgeViT-S, we observe even larger gains when comparing to PVTv2-B1: +2.2 on RetinaNet [27], +3.0 AP^b and +1.2 AP^m on Mask R-CNN [18].

ADE20K Scene Parsing. We incorporate the pretrained EdgeViT in the Semantic FPN segmentation model [24]. As in [54,55], we create 512×512 random crops of the images during training and resize the images to have a shorter side of 512 pixels during inference. The models are finetuned by AdamW [32] using an initial learning rate of 1×10^{-4} and a batch size of 16. We train the models for 80K iterations on the ADE20K training set, and report the mean Intersection over Union (mIoU) score on the validation set.

Results. In Table 6, we compare EdgeViTs to both CNN (ResNet-18 [19]) and ViT backbones (PVTs [54,55]) for FPN based Semantic Segmentation [24]. EdgeViTs achieves better performance than all counterparts at similar compute costs. Particularly, EdgeViT-XXS outperforms PVTv2-B0 by 2.5% in mIoU, EdgeViT-S surpasses PVTv2-B1 by a margin of 3.4%.

5 Conclusion

In this work, we investigate the design of efficient ViTs from the on-device deployment perspective. By introducing a novel decomposition of self-attention, we present a family of EdgeViTs that, for the first time, achieve comparable or even superior accuracy-efficiency tradeoff on generic visual recognition tasks, in comparison to a variety of state-of-the-art efficient CNNs and ViTs. We conduct extensive on-device experiments using practically critical and previously underestimated metrics (e.g., energy-aware efficiency) and reveal new insights and observations in the comparison of light-weight CNN and ViT models.

References

1. Almeida, M., Laskaridis, S., Mehrotra, A., Dudziak, L., Leontiadis, I., Lane, N.D.: Smart at what cost? Characterising mobile deep neural networks in the wild. In: ACM Internet Measurement Conference (2021)
2. Ba, J.L., Kiros, J.R., Hinton, G.E.: Layer normalization. arXiv preprint arXiv:1607.06450 (2016)
3. Berman, M., Pishchulin, L., Xu, N., Blaschko, M.B., Medioni, G.: AOWS: adaptive and optimal network width search with latency constraints. In: IEEE Conference on Computer Vision and Pattern Recognition (2020)
4. Bulat, A., Perez Rua, J.M., Sudhakaran, S., Martinez, B., Tzimiropoulos, G.: Space-time mixing attention for video transformer. In: Advances on Neural Information Processing Systems (2021)
5. Bulat, A., Tzimiropoulos, G.: Bit-Mixer: Mixed-precision networks with runtime bit-width selection. In: IEEE International Conference on Computer Vision (2021)
6. Chen, Y., Dai, X., Liu, M., Chen, D., Yuan, L., Liu, Z.: Dynamic convolution: attention over convolution kernels. In: IEEE Conference on Computer Vision and Pattern Recognition (2020)
7. Chu, X., et al.: Twins: revisiting the design of spatial attention in vision transformers. In: Advances on Neural Information Processing Systems (2021)
8. Chu, X., et al.: Conditional positional encodings for vision transformers. arXiv preprint arXiv:2102.10882 (2021)
9. Dosovitskiy, A., et al.: An image is worth 16×16 words: Transformers for image recognition at scale. In: International Conference on Learning Representations (2021)
10. Dudziak, L., Chau, T., Abdelfattah, M.S., Lee, R., Kim, H., Lane, N.D.: BRP-NAS: prediction-based NAS using GCNs. In: Advances on Neural Information Processing Systems (2020)
11. Fan, H., et al.: Multiscale vision transformers. In: IEEE International Conference on Computer Vision (2021)
12. Fayyaz, M., et al.: ATS: adaptive token sampling for efficient vision transformers. arXiv preprint arXiv:2111.15667 (2021)
13. Gehring, J., Auli, M., Grangier, D., Yarats, D., Dauphin, Y.N.: Convolutional sequence to sequence learning. In: International Conference on Machine Learning (2017)
14. Graham, B., et al.: LeViT: a vision transformer in convnet's clothing for faster inference. In: IEEE International Conference on Computer Vision (2021)
15. Han, K., Xiao, A., Wu, E., Guo, J., Xu, C., Wang, Y.: Transformer in transformer. In: Advances on Neural Information Processing Systems (2021)
16. Han, S., Mao, H., Dally, W.J.: Deep compression: compressing deep neural network with pruning, trained quantization and Huffman coding. In: International Conference on Learning Representations (2016)
17. He, K., Chen, X., Xie, S., Li, Y., Dollár, P., Girshick, R.: Masked autoencoders are scalable vision learners. In: IEEE Conference on Computer Vision and Pattern Recognition (2022)

18. He, K., Gkioxari, G., Dollár, P., Girshick, R.: Mask R-CNN. In: IEEE International Conference on Computer Vision (2017)
19. He, K., Zhang, X., Ren, S., Sun, J.: Deep residual learning for image recognition. In: IEEE Conference on Computer Vision and Pattern Recognition (2016)
20. Hendrycks, D., Gimpel, K.: Gaussian error linear units (GELUs). arXiv preprint arXiv:1606.08415 (2016)
21. Howard, A., et al.: Searching for MobileNetV3. In: IEEE International Conference on Computer Vision (2019)
22. Howard, A.G., et al.: MobileNets: efficient convolutional neural networks for mobile vision applications. arXiv preprint arXiv:1704.04861 (2017)
23. Jacob, B., et al.: Quantization and training of neural networks for efficient integer-arithmetic-only inference. In: IEEE Conference on Computer Vision and Pattern Recognition (2018)
24. Kirillov, A., Girshick, R., He, K., Dollár, P.: Panoptic feature pyramid networks. In: IEEE Conference on Computer Vision and Pattern Recognition (2019)
25. Li, K., et al.: UniFormer: unified transformer for efficient spatial-temporal representation learning. In: International Conference on Learning Representations (2022)
26. Lin, T.Y., Dollár, P., Girshick, R., He, K., Hariharan, B., Belongie, S.: Feature pyramid networks for object detection. In: IEEE Conference on Computer Vision and Pattern Recognition (2017)
27. Lin, T.Y., Goyal, P., Girshick, R., He, K., Dollár, P.: Focal loss for dense object detection. In: IEEE International Conference on Computer Vision (2017)
28. Lin, T.-Y., et al.: Microsoft COCO: common objects in context. In: Fleet, D., Pajdla, T., Schiele, B., Tuytelaars, T. (eds.) ECCV 2014. LNCS, vol. 8693, pp. 740–755. Springer, Cham (2014). https://doi.org/10.1007/978-3-319-10602-1_48
29. Liu, Y., Sangineto, E., Bi, W., Sebe, N., Lepri, B., De Nadai, M.: Efficient training of visual transformers with small datasets. In: Advances on Neural Information Processing Systems (2021)
30. Liu, Z., et al.: Swin transformer: hierarchical vision transformer using shifted windows. In: IEEE International Conference on Computer Vision (2021)
31. Liu, Z., Li, J., Shen, Z., Huang, G., Yan, S., Zhang, C.: Learning efficient convolutional networks through network slimming. In: IEEE International Conference on Computer Vision (2017)
32. Loshchilov, I., Hutter, F.: Decoupled weight decay regularization. arXiv preprint arXiv:1711.05101 (2017)
33. Loshchilov, I., Hutter, F.: SGDR: stochastic gradient descent with warm restarts. In: International Conference on Learning Representations (2017)
34. Lu, J., et al.: SOFT: softmax-free transformer with linear complexity. In: Advances on Neural Information Processing Systems (2021)
35. Ma, N., Zhang, X., Zheng, H.-T., Sun, J.: ShuffleNet V2: practical guidelines for efficient CNN architecture design. In: Ferrari, V., Hebert, M., Sminchisescu, C., Weiss, Y. (eds.) Computer Vision – ECCV 2018. LNCS, vol. 11218, pp. 122–138. Springer, Cham (2018). https://doi.org/10.1007/978-3-030-01264-9_8
36. Mehta, S., Rastegari, M.: MobileViT: light-weight, general-purpose, and mobile-friendly vision transformer. In: International Conference on Learning Representations (2022)
37. Pan, B., Jiang, Y., Panda, R., Wang, Z., Feris, R., Oliva, A.: IA-RED^2: Interpretability-aware redundancy reduction for vision transformers. In: Advances on Neural Information Processing Systems (2021)
38. Paszke, A., et al.: PyTorch: an imperative style, high-performance deep learning library. In: Advances on Neural Information Processing Systems (2019)

39. Radosavovic, I., Kosaraju, R.P., Girshick, R., He, K., Dollár, P.: Designing network design spaces. In: IEEE Conference on Computer Vision and Pattern Recognition (2020)
40. Rao, Y., Zhao, W., Liu, B., Lu, J., Zhou, J., Hsieh, C.J.: DynamicViT: efficient vision transformers with dynamic token sparsification. In: Advances on Neural Information Processing Systems (2021)
41. Russakovsky, O., et al.: ImageNet large scale visual recognition challenge. Int. J. Comput. Vis. **115**, 211–252 (2015)
42. Sandler, M., Howard, A., Zhu, M., Zhmoginov, A., Chen, L.C.: MobileNetV2: inverted residuals and linear bottlenecks. In: IEEE Conference on Computer Vision and Pattern Recognition (2018)
43. Shaw, P., Uszkoreit, J., Vaswani, A.: Self-attention with relative position representations. In: North American Chapter of the Association for Computational Linguistics (2018)
44. Srinivas, A., Lin, T.Y., Parmar, N., Shlens, J., Abbeel, P., Vaswani, A.: Bottleneck transformers for visual recognition. In: IEEE Conference on Computer Vision and Pattern Recognition (2021)
45. Szegedy, C., Vanhoucke, V., Ioffe, S., Shlens, J., Wojna, Z.: Rethinking the inception architecture for computer vision. In: IEEE Conference on Computer Vision and Pattern Recognition (2016)
46. Tan, M., et al.: MnasNet: platform-aware neural architecture search for mobile. In: IEEE Conference on Computer Vision and Pattern Recognition (2019)
47. Tan, M., Le, Q.: EfficientNet: rethinking model scaling for convolutional neural networks. In: International Conference on Machine Learning (2019)
48. Tan, M., Le, Q.: EfficientNetV2: smaller models and faster training. In: International Conference on Machine Learning (2021)
49. Tan, M., Le, Q.V.: MixConv: mixed depthwise convolutional kernels. In: British Machine Vision Conference (2019)
50. Tan, M., Pang, R., Le, Q.V.: EfficientDet: scalable and efficient object detection. In: IEEE Conference on Computer Vision and Pattern Recognition (2020)
51. Touvron, H., Cord, M., Douze, M., Massa, F., Sablayrolles, A., Jégou, H.: Training data-efficient image transformers & distillation through attention. In: International Conference on Machine Learning (2021)
52. Touvron, H., Cord, M., Sablayrolles, A., Synnaeve, G., Jégou, H.: Going deeper with image transformers. arXiv preprint arXiv:2103.17239 (2021)
53. Wang, S., Li, B., Khabsa, M., Fang, H., Ma, H.: Linformer: self-attention with linear complexity. arXiv preprint arXiv:2006.04768 (2020)
54. Wang, W., et al.: Pyramid vision transformer: a versatile backbone for dense prediction without convolutions. In: IEEE International Conference on Computer Vision (2021)
55. Wang, W., et al.: PVTv2: improved baselines with pyramid vision transformer. Comput. Vis. Media **8**, 415–424 (2022)
56. Wen, W., Wu, C., Wang, Y., Chen, Y., Li, H.: Learning structured sparsity in deep neural networks. In: Advances on Neural Information Processing Systems (2016)
57. Wu, H., et al.: CvT: introducing convolutions to vision transformers. In: IEEE International Conference on Computer Vision (2021)
58. Xiao, T., Singh, M., Mintun, E., Darrell, T., Dollár, P., Girshick, R.: Early convolutions help transformers see better. In: Advances on Neural Information Processing Systems (2021)
59. Xiong, Y., et al.: Nyströmformer: a nyström-based algorithm for approximating self-attention. In: AAAI Conference on Artificial Intelligence (2021)

60. Xu, Y., et al.: Evo-ViT: slow-fast token evolution for dynamic vision transformer. In: AAAI Conference on Artificial Intelligence (2022)
61. Yang, B., Bender, G., Le, Q.V., Ngiam, J.: CondConv: conditionally parameterized convolutions for efficient inference. In: Advances on Neural Information Processing Systems (2019)
62. Yuan, L., et al.: Tokens-to-token ViT: training vision transformers from scratch on ImageNet. In: IEEE International Conference on Computer Vision (2021)
63. Zhang, X., Zhou, X., Lin, M., Sun, J.: ShuffleNet: an extremely efficient convolutional neural network for mobile devices. In: IEEE Conference on Computer Vision and Pattern Recognition (2018)
64. Zhou, B., Zhao, H., Puig, X., Fidler, S., Barriuso, A., Torralba, A.: Scene parsing through ADE20K dataset. In: IEEE Conference on Computer Vision and Pattern Recognition (2017)
65. Zhou, D., et al.: DeepViT: towards deeper vision transformer. arXiv preprint arXiv:2103.11886 (2021)

PalQuant: Accelerating High-Precision Networks on Low-Precision Accelerators

Qinghao Hu[1] , Gang Li[2] , Qiman Wu[3] , and Jian Cheng[1(✉)]

[1] Institute of Automation, Chinese Academy of Sciences, Beijing, China
huqinghao2014@ia.ac.cn, jcheng@nlpr.ia.ac.cn
[2] Shanghai Jiao Tong University, Shanghai, China
gliaca@sjtu.edu.cn
[3] Baidu Inc., Beijing, China
wuqiman@baidu.com

Abstract. Recently low-precision deep learning accelerators (DLAs) have become popular due to their advantages in chip area and energy consumption, yet the low-precision quantized models on these DLAs bring in severe accuracy degradation. One way to achieve both high accuracy and efficient inference is to deploy high-precision neural networks on low-precision DLAs, which is rarely studied. In this paper, we propose the PArallel Low-precision Quantization (PalQuant) method that approximates high-precision computations via learning parallel low-precision representations from scratch. In addition, we present a novel cyclic shuffle module to boost the cross-group information communication between parallel low-precision groups. Extensive experiments demonstrate that PalQuant has superior performance to state-of-the-art quantization methods in both accuracy and inference speed, e.g., for ResNet-18 network quantization, PalQuant can obtain 0.52% higher accuracy and 1.78× speedup simultaneously over their 4-bit counterpart on a state-of-the-art 2-bit accelerator. Code is available at https://github.com/huqinghao/PalQuant.

Keywords: Quantization · Network acceleration · CNNs

1 Introduction

Recently various model compression techniques have been proposed to deploy deep neural networks on resource-constrained edge devices. Among these, fixed-point quantization [16,30,31] that converts full-precision floating-point operation to low-bit integer counterpart has become the *de facto* method due to its hardware efficiency.

At present, most of the commercial CNN accelerators are designed for high-precision (such as INT16/INT8) arithmetic. One important reason for this is

---

Q. Hu, G. Li and Q. Wu—Equal Contribution.

Supplementary Information The online version contains supplementary material available at https://doi.org/10.1007/978-3-031-20083-0_19.

because the accuracy of a quantized network is hard to retain as the quantization bit-width narrows. Yet low-precision accelerators lead to orders of magnitude decrease in chip-area and energy consumption compared to the high-precision hardware [10]. This motivates plenty of researchers to study how to improve the accuracy of quantized networks on low-precision accelerators [13,15,22,28,31]. While these methods focus on designing low-precision quantization algorithms, the accuracy of quantized networks may be limited by the low computation precision of accelerators.

Different from the above methods, we try to answer the question: *Is it possible to deploy high-precision networks on the existing well-designed low-precision accelerator to capture both model accuracy and inference efficiency?* To achieve this, a naive solution is decomposing a high-precision network at the bit level and conducting inference on low-precision hardware in a nibble iteration manner [1]. For example, to run an 8-bit network on a 2-bit accelerator, we can split each 8-bit operand into four 2-bit operands so that the original 8-bit multiplication can be carried out in $4 \times 4 = 16$ steps using 2-bit multipliers with proper shifting. Although the 2-bit operation is much cheaper than the 8-bit counterpart in terms of chip area and power consumption, there is no gain in inference latency since the total amount of bit operations is unchanged.

In this paper, we investigate the opportunity of fast and accurate inference of high-precision CNN models on low-precision hardware from the algorithm side. We propose PArallel Low-precision Quantization (PalQuant), a hardware-friendly, simple yet effective quantization method for efficient CNN acceleration. Different from the naive bit-level decomposition solution, PalQuant reduces the computational complexity by dividing the expanded low-precision channels into parallel groups.

To encourage information flowing across different groups, we propose a novel cyclic shuffle module that fuses features from two consecutive groups cyclically in a hardware-friendly way. One important property of cyclic shuffle is that it may serve as a complement to channel shuffle. This is mainly due to that cyclic shuffle fuses channel features at group level while the channel shuffle fuses a fraction of channel features from each group. Extensive experiments from both algorithm and hardware sides demonstrate that PalQuant can consistently achieve the highest accuracy and inference speedup than state-of-the-art methods. The contribution of this paper can be summarized as follows:

- We propose the PalQuant algorithm that enables efficient high-precision computation in low-precision accelerators via learning parallel low-precision representations from scratch.
- We propose a novel cyclic shuffle module to help the information flow across parallel low-precision convolution groups.
- Extensive experiments on ImageNet benchmark demonstrate that PalQuant outperforms state-of-the-art quantization methods in terms of both accuracy and computational cost. We also examine the speed-up of PalQuant on two CNN accelerators [14,33], which shows that PalQuant achieves 1.7× speed-ups than state-of-the-arts on ResNet-18.

2 Related Work

2.1 Quantization Methods

Deep Neural Network Quantization methods have become popular in recent years as they can reduce energy consumption and inference latency of deep neural networks. One line of quantization methods aims to reduce the quantization bit-width, while maintaining the network accuracy. Early quantization methods mainly focus on high bit-width fixed-point quantization [16,30], e.g. 8bit or 16bit quantization scheme, which brings in little accuracy degradation.Later, various binary [11,18,31] and ternary quantization methods [25,26,41] are proposed to reduce the multiplication operations in the network inference. Another line of research on quantization methods is to learn good quantization parameters such as quantization step-values, clipping values, and so on. Early quantization methods use fixed quantization parameters [16,19], dynamic parameters based on statistics of the data distribution [5,27], or seek the parameters that minimize the quantization error [31,38]. Recently, researchers propose to use trainable quantization parameters, e.g. clipping values [9,23] and step-size [13,22]. These parameters can be learned by gradient back-propagation which minimizes the task loss.Another related work is WRPN [29] which increases the number of filter maps and reduces the quantization bit-width of feature maps. Our proposed method differs with WRPN [29] in three aspects. First, Our PalQuant and WRPN [29] target at different problems. While WRPN [29] mainly tries to reduce the large memory footprint of high-precision feature maps via wide reduced-precision representations, our PalQuant enables efficient high-precision computations on low-precision accelerators. Second, PalQuant reduces computational complexity and memory access via parallel low-precision computation scheme, and it achieves higher or comparable network accuracy with less computational cost than WRPN [29]. Third, we propose a novel cyclic shuffle module to fuse features across different groups, that further strengthens the model representation.

2.2 Hardware Accelerators for CNN

The extremely high computational complexity of CNN poses a significant challenge to real-world deployment, especially for resource-constraint embedded devices. As a result, energy-efficient FPGA/ASIC-based CNN accelerators have gained increasing popularity recently in both academia and industry. To maintain network accuracy, the computing architecture of early CNN accelerators mainly exploits floating-point and high-precision fixed-point data types (such as 16-bit/8-bit). For example, Zhang *et al.* [37] designs the first FPGA-based accelerator for floating-point CNN inference. DianNao [7], Eyeriss [8], and TPU [21] are ASIC-based accelerators designed for 16-bit quantized CNN.

With the rapid development of quantization methods in the deep learning community, many low-precision CNN accelerators have been proposed to further improve hardware efficiency. With the help of power-of-two quantization [40], Li *et al.* [24] and Tann *et al.* [34] propose multiplier-free architectures for area and power-efficient inference by replacing conventional integer multiplications with shift-based operations. YodaNN [2] is an ASIC-based accelerator tailored for Binary Weight Networks [11]. Bit Fusion [33], Bitblade [32], and BPVeC [14] are precision-scalable CNN accelerators exploiting 2-bit multipliers for arbitrary-precision computation. Umuroglu *et al.* [35] and Zhao *et al.* [39] propose dedicated accelerators for Binarized Neural Networks [20], which can achieve the highest frame-per-second under extremely low area and energy consumption. Lascorz *et al.* [12] accelerates CNN inference through hardware/software co-design, while PalQuant is more general and can be deployed on a variety of accelerators, including bit-parallel and bit-serial accelerators.

3   Preliminaries

3.1   Notation

The input activation and weight of a fully-connected layer in a deep neural network are denoted by $X \in \mathcal{R}^{N \times S}$ and $W \in \mathcal{R}^{T \times S}$, respectively. The layer's output Y can be obtained by:

$$Y = XW^T \tag{1}$$

The quantized input activation and weight are denoted by $\hat{X} \in \mathcal{R}^{N \times S}$ and $\hat{W} \in \mathcal{R}^{T \times S}$, respectively. For a convolutional layer, the output Y can also be obtained by Eq. 1, since the weights and activations can be transformed into matrices.

3.2   Uniform Quantization

Recently uniform quantizer with trainable quantization parameters [15,23] becomes popular. Given a floating-point number x, it first maps x to a range $[0, 1]$ by the following equation:

$$x_n = clip\left(\frac{x - l}{u - l}, 0, 1\right) \tag{2}$$

where $clip(\cdot, \cdot, \cdot)$ denotes the clip function, l and u denotes lower bound and up bound, respectively. Then the integer value x_q can be obtained by the quantization function:$x_q = \lfloor(2^b - 1)x_n\rceil$, where $\lfloor \cdot \rceil$ is rounding-to-nearest operation, and b denotes the quantization bit-width. And the de-quantization function is given by the following formula:

$$\hat{x} = \begin{cases} \frac{x_q}{2^b - 1} & x \in weight \\ 2(\frac{x_q}{2^b - 1} - 0.5) & x \in activation \end{cases} \tag{3}$$

where the formula maps the de-quantized weights to a range $[-1, 1]$, and restricts the de-quantized activations to be non-negative. An additional parameter α [23] is required to multiply the output activations of each layer, which adjusts the output scale of the whole feature map.

3.3   Bit-Level Decomposing

Consider that there is an accelerator that supports only B-bit computation, a naive solution for running a high-precision (M-bit) model on this accelerator is to decompose M-bit representations to multiple B-bit at bit-level, then shift and sum up those B-bit operations results:

$$Y = \hat{X}\hat{W}^T = \sum_{i=0}^{G-1}\sum_{j=0}^{G-1} \hat{X}_i \hat{W}_j^T * 2^{(i+j)*B} \tag{4}$$

where $\hat{X}_i$ and $\hat{W}_j$ are B-bit input and weights that are generated by splitting M-bit input and weights in bit-level, and $G = ceil(M/B)$. Note that the computational cost is unchanged comparing to the original M-bit computation.

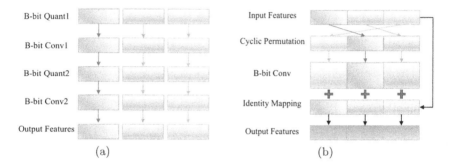

Fig. 1. (a) The proposed parallel low-precision computation scheme. Here the hyper-parameter G is fixed to 3. B-bit Quant and B-bit Conv denote a B-bit quantization layer and B-bit convolutional layer, respectively. (b) The proposed cyclic shuffle module. Here B-bit Conv denotes a B-bit quantized 1×1 group convolutional layer with $G = 3$

4   Proposed Method

4.1   Parallel Low-Precision Quantization

Since the bit-level decomposing method (Eq. 4) has the same amount of bit operations as the original high-precision computation, there is no gain in inference latency. At first, we try to reduce the inference latency by an approximation to Eq. 4:

$$min_{\bar{W}_i} = \|\sum_{i=0}^{G-1}\sum_{j=0}^{G-1} \hat{X}_i \hat{W}_j^T * 2^{(i+j)*B} - \sum_{i=0}^{G-1}(\hat{X}_i\bar{W}_i^T) * 2^{i*B}\|_F^2 \tag{5}$$

where $\hat{W}_j$ and $\hat{X}_i$ are B-bit representations chunked from M-bit $\hat{W}$ and $\hat{X}$, respectively. We found that the B-bit computations in Eq. 5 are naturally in parallel G groups, which provides the potential to be hardware-friendly. In addition, Eq. 5 approximates the standard results with only G parallel B-bit computations while Eq. 4 requires G^2 B-bit computations. This indicates that the proposed approximation only costs $\frac{1}{G}$ computation complexity of the bit-level decomposition method.

Yet there are two problems in Eq. 5. One is that we empirically found that all G solutions of $\bar{W}_i$ degenerate to one similar solution, which results in an inferior network accuracy. We assume that this phenomenon has relations with two aspects: 1. B-bit representations are chunked from M-bit ones, thus they are coupling together with a strong connection[1]; 2. minimizing the feature map reconstruction loss easily causes the over-fitting problem. The other problem in Eq. 5 is that chunking M-bit $\hat{X}$ to multiple B-bit representations tend to be inferior to learning $G\times$ B-bit representations directly by back-propagation, as mentioned in [29].

To tackle the above problems, we propose the parallel low-precision quantization scheme by learning multiple low-precision representations in parallel. It inherits the computation efficiency of parallel low-precision computation from Eq. 5, and cures the solution degeneration problem via training B-bit representations through back-propagation. Specifically, we expand the activation channels of a convolutional layer by $G\times$, and split these activations into G groups along the channel dimension. Then we quantize the activations and weights to B-bit in each group, and there are G groups of B-bit weights in total. Figure 1(a) depicts the proposed parallel low-precision quantization scheme, which can be implemented naturally by quantized group convolution with expanded channels.

4.2   Cyclic Shuffle

Although the proposed parallel low-precision computation scheme enjoys wonderful hardware efficiency, it hinders information flowing across different groups of feature channels. To cure this problem, we propose a novel cyclic shuffle module to enhance information communication across different groups. First, we adopt cyclic permutation [4] to permute the channels in group-level. Given input X with shape $[N \times C \times H \times W]$, it can be reshaped to $\left[N \times G \times \hat{C} \times H \times W\right]$ where $\hat{C} = C/G$. We propose to re-order the channels in group-level via the cyclic permutation. Specifically, let $\mathbf{S} = \{0, 1, 2, 3 \cdots G-1\}$ denotes group indexing set, a cyclic permutation function $\pi(\cdot)$ for $\mathbf{S}$ is

$$\pi(i) \;=\; (i+1) \; mod \; G \qquad (6)$$

A more clear notation of $\pi(\cdot)$ by Cauchy's two-line notation [36] is:

$$\begin{pmatrix} 0 & 1 & 2 & 3 & 4 & \cdots & G-2 & G-1 \\ 1 & 2 & 3 & 4 & 5 & \cdots & G-1 & 0 \end{pmatrix} \qquad (7)$$

---

[1] For example, given the 4-bit number x, the lowest 2-bit number changes from 3 to 0 when x changes from 3 to 4 ($[0\ 0\ 1\ 1] \rightarrow [0\ 1\ 0\ 0]$).

where the first line denotes the channel group indexing set **S** and the second line denotes the corresponding order after permutation. After cyclic permutation, the output Y can be represented by:

$$Y[:, \pi(i), :, :, :] = X[:, i, :, :, :] \tag{8}$$

By using Eq. 8, we can re-order feature channels by permuting the group-level feature cyclically. Under this scheme, the information from each group is received by its following one, which helps the information exchange between feature channels of different groups. Although we have exchanged the information between different groups by the cyclic permutation, each group still holds only one group of feature channels. In order to fuse the feature channels from different groups, we build a novel module, named cyclic shuffle, that composes of the cyclic permutation and a 1×1 group convolution with short-cut connection. Given the input X, the output Z of the cyclic shuffle module can be represented by the following formulas:

$$Y = \text{CyclicPermute}(X) \tag{9}$$
$$Z = X + \text{Conv}_{1 \times 1}(Y) \tag{10}$$

where CyclicPermute$(\cdot)$ denotes the cyclic permutation function (refer to Eq. 8). In the cyclic shuffle module, the cyclic permutation re-orders the feature channels at group level, then the 1×1 group convolution learns a mapping function for channel fusing, finally the short-cut connection together with the element-wise addition operation aggregate feature channels from different groups. Figure 1(b) depicts the proposed cyclic shuffle module.

Our proposed cyclic shuffle module enjoys benefits from three aspects: 1. According to Proposition 1, after passing through G-1 cyclic shuffle modules, each group receives the information flows from all groups of feature channels, that helps cure the side-effect of group convolution. 2. The proposed cyclic shuffle module fuses feature channels at group level, which makes it possible for the intermediate results to stay on chip. As a result, it saves data access to the out-chip memory significantly. 3. Our cyclic shuffle module fuses feature channels in group-level while channel shuffle aggregates a part of feature channels from each group. Therefore, cyclic shuffle can be combined with channel shuffle to achieve higher network accuracy.

Proposition 1. *Suppose the number of groups is G, then G-1 cyclic shuffle modules are enough to assure that each group contains the information flows from all parallel groups.*

Proof. The proof is simple. The cyclic permutation $\pi(\cdot)$ in Eq. 6 is a G-cycle [4], which means the group indexes will be the same as the original order after G times cyclic permutation. As each group fuses a the information flow from its previous group after passing through one cyclic shuffle module, then G-1 cyclic shuffle modules are enough to assure that each group contains all information from G group channels.

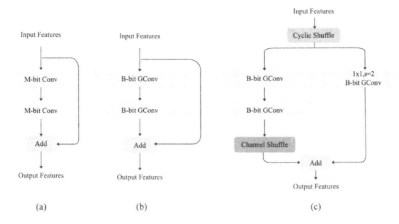

Fig. 2. (a) A typical building block of M-bit quantized ResNet network; (b) A typical building block of PalQuant with $G = 2$; (c) A typical building block of PalQuant with shuffle modules (stride = 2, $G = 2$)

4.3   Overall Framework

In this subsection, we give the overall framework of our proposed method. To approximate M-bit computations on B-bit accelerator, we propose the parallel low-precision quantization scheme which expands the channel dimensions of each layer by G times ($G = ceil(M/B)$) and splits them into G parallel groups. Taking ResNet network quantization for an example, PalQuant makes a simple modification to the standard M-bit basic building block (Fig. 2(a)): enlarging the input channels by $G\times$ and using B-bit quantized group convolution with the number of groups equals to G. The basic building block of PalQuant is depicted in Fig. 2(b).

   To cure the side-effect caused by the group convolution, we propose the cyclic shuffle module to help information communication between different groups. Considering that the cyclic shuffle module contains a 1×1 group convolution layer, we only place the cyclic shuffle module at the beginning of each stage, which brings little extra computational cost. Since cyclic shuffle and channel shuffle fuse information flows in different ways, we use channel shuffle module as a complement. As the channel shuffle in the middle of two consecutive group convolutions will impair the parallel computation flows, we move the channel shuffle module to the end of the convolution block. A typical building block of PalQuant with shuffle modules is depicted in Fig. 2(c).

5   Experiments

In this section, we first give details of experiment settings, then we compare our proposed algorithm PalQuant with state-of-the-art quantization algorithms. To demonstrate the hardware efficacy of PalQuant, we further conduct extensive experiments on CNN accelerators. Finally, we show the ablation study results and the generalization of PalQuant under different hyper-parameter B and G.

Network Architectures and Datasets. Since ResNet [17] is widely used in various computer vision tasks, here we adopt ResNet-18 and ResNet-34 as benchmark networks. All the experiments in this work are conducted on ImageNet dataset which has over 1 million training images and 50,000 validation images.

Training Details. In this work, we use uniform quantization for the input activation and weight of each convolutional or fully-connected layer. In the training stage, a straight-through estimator(STE) [3] is used to approximate the gradient through the rounding function. Based on this gradient estimator, we learn the lower bound l, upper bound u, and the scaling factor α through gradient backpropagation. Following experiment settings in previous work [13,23], the first and the last layer are not quantized unless otherwise specified. We implement our method under the Pytorch framework and train all the networks on a GPU server with eight Nvidia Titan 3090 GPUs. We adopt an SGD optimizer with the momentum set to 0.9. The weight decay is set to 1e−4, and the learning rate is initialized to 0.01 and adjusts with a cosine learning rate decay strategy. Following previous work [13], we train all the models from scratch for 90 epochs with a batch size of 256. In the training stage, random cropping, resizing, and horizontal flipping are adopted while only resizing and center cropping are used in the validation stage.

5.1   Comparison with State-of-the-Arts Algorithms

We first compare our proposed PalQuant with state-of-the-art quantization algorithms. All the accuracy results of other methods, except for LSQ [13] and WRPN [29], are taken from corresponding papers. As LSQ [13] reports the results of pre-activation ResNet, we re-implement LSQ [13] under Pytorch framework and get the quantization results on standard ResNet. And we re-implement WRPN [29] via the same quantization function as ours, and get better results than the reported ones in [29].

Table 1 shows the top-1 accuracy of quantized ResNet-18 and ResNet-34 with different quantization precision. From the table, we can observe that PalQuant outperforms state-of-the-art methods in both accuracy and computational complexity. In detail, PalQuant ($G = 3$) achieves +1.66% (72.66% v.s. 70.7%) higher top-1 accuracy than 4-bit LSQ [13] with 23.4% (22.3G v.s. 29.1G) less computational cost on ResNet-18. With comparable computational budget, PalQuant ($G = 4$) outperforms 4-bit EWGS [23] by 1.06% on ResNet-34. Note that both LSQ [13] and EWGS [23] are finetuned from the pre-trained model, while our method are trained from scratch. In addition, PalQuant($G = 3$) obtains slightly higher top-1 accuracy than WRPN [29] with 23.4% less bitOps on ResNet-18. These results suggest that the proposed PalQuant has strong feature representation power, which leads to the highest accuracy in the experiments under less or equal computational budgets.

Table 1. Comparison Results of Quantized ResNet-18 and ResNet-34 on ImageNet. † Results of WRPN [29] on ResNet34 are taken from the paper

Method	Strategy	Precision	ResNet-18		ResNet-34	
			BitOps	Top1 acc.	BitOps	Top1 acc.
DSQ [15]	Fine-tuned	4b A, 4b W	29.10G	69.60	58.74G	72.80
FAQ [28]	Fine-tuned	4b A, 4b W	29.10G	69.80	58.74G	73.30
QIL [22]	Fine-tuned	4b A, 4b W	29.10G	70.10	58.74G	73.70
EWGS [23]	Fine-tuned	4b A, 4b W	29.10G	70.60	58.74G	73.90
LSQ [13]	Fine-tuned	4b A, 4b W	29.10G	70.70	58.74G	73.50
PalQuant (**Ours**)	Scratch	B = 2, G = 2	14.87G	71.12	**29.68G**	73.90
PalQuant (**Ours**)	Scratch	B = 2, G = 3	22.30G	72.36	44.53G	74.52
PalQuant(**Ours**)	Scratch	B = 2, G = 4	29.73G	72.74	59.37G	74.96
EWGS [23]	Fine-tuned	6b A, 6b W	65.48G	71.01	132.16G	74.14
LSQ [13]	Fine-tuned	6b A, 6b W	65.48G	71.59	132.16G	74.43
WRPN [29]	Scratch	2×, 2b A, 2b W	29.10G	72.31	58.74G	73.32†
PalQuant (**Ours**)	Scratch	B = 2, G = 3	**22.30G**	**72.36**	**44.53G**	**74.52**
FAQ [28]	Fine-tuned	8b A,8b W	116.4G	70.00	234.94G	73.70
LSQ [13]	Fine-tuned	8b A, 8b W	116.4G	71.10	234.94G	73.98
EWGS [23]	Fine-tuned	8b A, 8b W	116.4G	71.14	234.94G	73.97
WRPN [29]	Scratch	2×, 2b A, 2b W	29.10G	72.31	58.74G	73.32†
PalQuant (**Ours**)	Scratch	B = 2, G = 4	**29.73G**	**72.74**	**59.37G**	**74.96**

5.2 Efficacy on Hardware Acceleration

We select the state-of-the-art CNN accelerator Bit Fusion [33] and BPVeC [14] to demonstrate the efficacy of the proposed PalQuant. Bit Fusion [33] is a systolic array-based accelerator designed for low-precision inference, it employs a bit-level decomposable fusion unit for 2/4/8-bit multiplication. To further improve hardware efficiency, BPVeC [14] exploits the Narrow Bit-width Vector Engines (NBVE), an energy and area efficient alternative to conventional MAC that consists of multiple 2-bit multipliers and an adder tree for SIMD-based inner-product computation. The basic building blocks of these two accelerators are shown in Fig. 3. In this experiment, we compare our proposed method against LSQ [13] in terms of inference latency and energy consumption. To mimic resource-constraint computing platforms, we configure both accelerators to 2-bit inference mode with a small on-chip buffer. In this scenario, the on-chip buffer cannot accommodate all weights and activations of a layer, leading to non-negligible data traffic in the memory hierarchy. We develop a cycle-accurate simulator to collect statistics of computation and on-chip buffer access. The power of computational logic and SRAM-based on-chip buffer under 45 nm technology are directly drawn from [33] and [14]. For off-chip data traffic, we assume 15pJ/bit per DDR4 access as in [14].

Table 2 shows the results of hardware acceleration. It is clear to see that our proposed method consistently outperforms the baseline in performance, energy efficiency, and accuracy for the same precision of weight and activation. For example, PalQuant ($G = 2$) obtains 0.52% higher accuracy, 1.78× speedup,

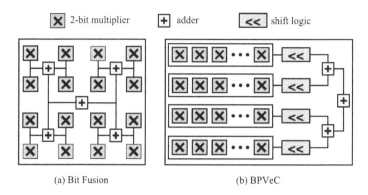

Fig. 3. The basic building blocks for 2-bit inference in Bit Fusion and BPVeC

and $1.91\times$ energy efficiency simultaneously over 4-bit LSQ [13] on Bit Fusion accelerator [33]. The benefit mainly stems from the channel grouping in our method. On the one hand, channel grouping leads to reduced bitOps, which is the source of performance gain. On the other hand, channel grouping eliminates the data dependency between different groups, resulting in reduced on-chip buffer size and off-chip bandwidth requirement, which contributes to the low energy consumption. Besides, the channel shuffle unit at the end, rather than the middle, of a block encourages the intermediate results of group convolution staying on chip, which is another key for saving energy.

In the paper, we evaluate PalQuant on Bit Fusion and BPVeC, yet our algorithm can also witness similar benefits on other bit-parallel accelerators [6,32] which implement a high-precision multiplication as parallel low-precision multiplications followed by reduction. The fundamental reason is that compared to conventional low-precision quantization, our method can always reduce the number of computations and buffer accesses.

5.3   Ablation Study

In this subsection, we ablate the designs of the PalQuant algorithm based on the ResNet-18 network with hyper-parameter $B = 2$ and $G = 2$, and all the experiments follow the same training settings as above.

Ablation: Influence of Different Shuffle Modules. Here we explore the benefits of the proposed cyclic shuffle module. Table 3 provides the comparison of different shuffle modules. The baseline model in this table doesn't contain either cyclic shuffle or channel shuffle. As we can see, using cyclic shuffle or channel shuffle separately achieves similar accuracy improvements. This indicates that both modules help the information communication between different groups. Besides, using cyclic shuffle and channel shuffle jointly demonstrate the highest accuracy gain (up to 3.19%) in terms of top-1 accuracy. This phenomenon verifies our assumption that cyclic shuffle and channel shuffle fuse information flow in different granularities and they may serve as a complement for each other.

Table 2. Comparison of inference latency and energy consumption on two hardware accelerators. We use hyper-parameter $B = 2$

DLA	Method	Precision	ResNet-18			ResNet-34		
			SpeedUp	Energy efficiency	Top1 acc.	SpeedUp	Energy efficiency	Top1 acc.
Bit fusion [33]	LSQ [13]	4b A, 4b W	1x	1x	70.70	1x	1x	73.50
	Ours	G = 2	1.78x	1.91x	71.12	1.78x	1.91x	73.90
	LSQ [13]	6b A, 6b W	1x	1x	71.59	1x	1x	74.43
	Ours	G = 3	2.52x	2.78x	72.36	2.50x	2.78x	74.52
	LSQ [13]	8b A, 8b W	1x	1x	71.14	1x	1x	73.97
	Ours	G = 4	3.21x	3.60x	72.71	3.13x	3.60x	74.96
BPVeC [14]	EWGS [23]	4b A, 4b W	1x	1x	70.60	1x	1x	73.90
	Ours	G = 2	1.77x	1.92x	71.12	1.87x	1.74x	73.90
	EWGS [23]	6b A, 6b W	1x	1x	71.01	1x	1x	74.14
	Ours	G = 3	2.56x	2.84x	72.36	2.46x	2.81x	74.52
	EWGS [23]	8b A, 8b W	1x	1x	71.14	1x	1x	73.97
	Ours	G = 4	3.12x	3.70x	72.71	3.07x	3.67x	74.96

We also explore the way to use channel shuffle. Table 3 shows that using per-stage or per-block channel shuffle has similar performance without cyclic shuffle. Together with cyclic shuffle, per-stage channel shuffle has higher accuracy improvement (+3.19% Top-1 accuracy on ImageNet) than per-block channel shuffle. In this paper, we use per-stage channel shuffle by default unless otherwise specified.

Ablation: Importance of Cyclic Permutation. As the cyclic shuffle module brings in one extra 1×1 group convolution, one may concern that most of the accuracy gains come from the additional computational cost. Experimental results in Table. 4 are against this opinion. PalQuant loses 0.76 % top-1 accuracy after dropping the cyclic permutation and keeping other parts unchanged while dropping the whole cyclic shuffle module causes 0.9 % top-1 accuracy decline. This indicates that the gain brought by the cyclic shuffle module is mainly due to cyclic permutation.

5.4 Generalization

Results With Different Numbers of Parallel Groups. In this section, we explore the generalization of PalQuant under different numbers of parallel groups G. In Table 5, our proposed method outperforms LSQ [13] consistently on a various number of parallel groups, which demonstrates the generalization ability of PalQuant. In addition, PalQuant shows superior performance (nearly 3% higher top-1 accuracy) than LSQ [13] under the same computational budget.

Results with Different Quantization Precision. Next, we explore the performance of PalQuant under different quantization precision B. Table 6 shows that PalQuant outperforms LSQ [13] consistently across different low-precision bit-width, i.e. $B = 2$, 3, and 4. This result means that our proposed scheme can be applied to CNNs accelerators with different quantization precision. We also

Table 3. Influence of different shuffle modules

Cyclic shuffle	Channel shuffle	Top1 acc.	Top5 acc.
		67.94	87.90
	per-block	70.31	89.58
	per-stage	70.24	89.51
✓		70.22	89.55
✓	Per-block	70.62	89.69
✓	**per-stage**	**71.13**	**89.99**
Δ		3.19↑	2.09↑

Table 4. Importance of cyclic permutation

Module	Top1 acc.	Top5 acc.
PalQuant	**71.13**	**89.99**
w.o cyclic permutation	70.37 (−0.76)	89.57 (−0.42)
w.o cyclic shuffle	70.23 (−0.90)	89.51 (−0.48)

quantize the weights with only half of activation bit-width in LSQ [13], which has nearly the same computation complexity as PalQuant. The result shows that our method outperforms LSQ [13] by a large margin under the comparable computational budget. The result of LSQ [13] with 6/3-bit quantization is provided in Appendix.

Table 5. Results with different numbers of parallel groups

Method	Precision	BitOps	Top1 acc.	Top5 acc.
LSQ [13]	4b A, 4b W	29.10G	70.99	89.86
LSQ [13]	4b A, 2b W	14.55G	69.18	88.93
PalQuant (**Ours**)	B = 2, G = 2	14.87G	**71.13**	**89.99**
LSQ [13]	6b A, 6b W	65.48G	71.58	90.24
LSQ [13]	6b A, 2b W	21.83G	69.47	89.06
PalQuant (**Ours**)	B = 2, G = 3	22.30G	**72.36**	**90.63**
LSQ [13]	8b A, 8b W	116.40G	71.10	90.10
LSQ [13]	8b A, 2b W	29.10G	70.71	89.70
PalQuant (**Ours**)	B = 2, G = 4	29.73G	**72.74**	**90.90**

Table 6. Results with different quantization precision

Method	Precision	BitOps	Top1 acc.	Top5 acc.
LSQ [13]	4b A, 4b W	29.10G	70.99	89.86
LSQ [13]	4b A, 2b W	14.55G	69.18	88.93
PalQuant (**Ours**)	B = 2, G = 2	14.87G	**71.13**	**89.99**
LSQ [13]	6b A, 6b W	65.48G	71.58	90.24
LSQ [13]	6b A, 3b W	32.74G	70.896	89.712
PalQuant (**Ours**)	B = 3, G = 2	33.45G	**72.72**	**90.90**
LSQ [13]	8b A, 8b W	116.40G	71.10	90.10
LSQ [13]	8b A, 4b W	58.20G	71.46	90.15
PalQuant (**Ours**)	B = 4, G = 2	59.46G	**73.48**	**91.34**

6   Conclusion

In this paper, we propose the PalQuant, a parallel low-precision quantization method that achieves efficient and accurate high-precision network inference on low-precision accelerators. To help the information flow across parallel groups, we propose the cyclic shuffle module to aggregate parallel information at group level. Extensive experiments demonstrate that PalQuant outperforms state-of-the-art quantization methods in terms of both accuracy and computational complexity.

Acknowledgements. This work was supported in part by National Key Research and Development Program of China (Grant No. 2021ZD0201504), and National Natural Science Foundation of China (Grant No. 62106267).

References

1. Abdel-Aziz, H., Shafiee, A., Shin, J.H., Pedram, A., Hassoun, J.H.: Rethinking floating point overheads for mixed precision DNN accelerators. CoRR abs/2101.11748 (2021). https://arxiv.org/abs/2101.11748
2. Andri, R., Cavigelli, L., Rossi, D., Benini, L.: YodaNN: an architecture for ultralow power binary-weight CNN acceleration. IEEE Trans. Comput. Aided Des. Integr. Circ. Syst. **37**(1), 48–60 (2017)
3. Bengio, Y., Léonard, N., Courville, A.: Estimating or propagating gradients through stochastic neurons for conditional computation. arXiv preprint arXiv:1308.3432 (2013)
4. Bogart, K.P.: Introductory Combinatorics. Saunders College Publishing (1989)
5. Cai, Z., He, X., Sun, J., Vasconcelos, N.: Deep learning with low precision by half-wave gaussian quantization. In: Proceedings of the IEEE Conference on Computer Vision and Pattern Recognition, pp. 5918–5926 (2017)
6. Camus, V., Mei, L., Enz, C., Verhelst, M.: Review and benchmarking of precision-scalable multiply-accumulate unit architectures for embedded neural-network processing. IEEE J. Emerg. Sel. Top. Circ. Syst. **9**(4), 697–711 (2019)

7. Chen, T., et al.: DianNao: a small-footprint high-throughput accelerator for ubiquitous machine-learning. ACM SIGARCH Comput. Archit. News **42**(1), 269–284 (2014)
8. Chen, Y.H., Krishna, T., Emer, J.S., Sze, V.: Eyeriss: an energy-efficient reconfigurable accelerator for deep convolutional neural networks. IEEE J. Solid-State Circ. **52**(1), 127–138 (2016)
9. Choi, J., Wang, Z., Venkataramani, S., Chuang, P.I.J., Srinivasan, V., Gopalakrishnan, K.: PACT: parameterized clipping activation for quantized neural networks. arXiv preprint arXiv:1805.06085 (2018)
10. Conti, F., Schiavone, P.D., Benini, L.: XNOR neural engine: a hardware accelerator IP for 21.6 fJ/op binary neural network inference. CoRR abs/1807.03010 (2018), http://arxiv.org/abs/1807.03010
11. Courbariaux, M., Bengio, Y., David, J.P.: BinaryConnect: training deep neural networks with binary weights during propagations. In: Advances in Neural Information Processing Systems, pp. 3123–3131 (2015)
12. Delmas Lascorz, A., et al.: Bit-tactical: a software/hardware approach to exploiting value and bit sparsity in neural networks. In: ASPLOS, pp. 749–763 (2019)
13. Esser, S.K., McKinstry, J.L., Bablani, D., Appuswamy, R., Modha, D.S.: Learned step size quantization. arXiv preprint arXiv:1902.08153 (2019)
14. Ghodrati, S., Sharma, H., Young, C., Kim, N., Esmaeilzadeh, H.: Bit-parallel vector composability for neural acceleration. In: 2020 57th ACM/IEEE Design Automation Conference (DAC), pp. 1–6 (2020)
15. Gong, R., et al.: Differentiable soft quantization: bridging full-precision and low-bit neural networks. In: Proceedings of the IEEE/CVF International Conference on Computer Vision, pp. 4852–4861 (2019)
16. Gupta, S., Agrawal, A., Gopalakrishnan, K., Narayanan, P.: Deep learning with limited numerical precision. CoRR, abs/1502.02551 392 (2015)
17. He, K., Zhang, X., Ren, S., Sun, J.: Deep residual learning for image recognition. In: IEEE Conference on Computer Vision and Pattern Recognition (CVPR) (2016)
18. Hou, L., Yao, Q., Kwok, J.T.: Loss-aware binarization of deep networks. arXiv preprint arXiv:1611.01600 (2016)
19. Hubara, I., Courbariaux, M., Soudry, D., El-Yaniv, R., Bengio, Y.: Binarized neural networks. In: Advances in Neural Information Processing Systems, vol. 29 (2016)
20. Hubara, I., Courbariaux, M., Soudry, D., El-Yaniv, R., Bengio, Y.: Binarized neural networks. In: Lee, D.D., Sugiyama, M., von Luxburg, U., Guyon, I., Garnett, R. (eds.) Advances in Neural Information Processing Systems 29: Annual Conference on Neural Information Processing Systems, 5–10 December 2016, Barcelona, Spain, pp. 4107–4115 (2016). https://proceedings.neurips.cc/paper/2016/hash/d8330f857a17c53d217014ee776bfd50-Abstract.html
21. Jouppi, N.P., et al.: In-datacenter performance analysis of a tensor processing unit. In: Proceedings of the 44th Annual International Symposium on Computer Architecture, pp. 1–12 (2017)
22. Jung, S., et al.: Learning to quantize deep networks by optimizing quantization intervals with task loss. In: Proceedings of the IEEE/CVF Conference on Computer Vision and Pattern Recognition, pp. 4350–4359 (2019)
23. Lee, J., Kim, D., Ham, B.: Network quantization with element-wise gradient scaling. In: Proceedings of the IEEE/CVF Conference on Computer Vision and Pattern Recognition (2021)
24. Li, F., et al.: A system-level solution for low-power object detection. In: Proceedings of the IEEE/CVF International Conference on Computer Vision Workshops (2019)

25. Li, F., Zhang, B., Liu, B.: Ternary weight networks. arXiv preprint arXiv:1605.04711 (2016)
26. Lin, Z., Courbariaux, M., Memisevic, R., Bengio, Y.: Neural networks with few multiplications. arXiv preprint arXiv:1510.03009 (2015)
27. McKinstry, J.L., et al.: Discovering low-precision networks close to full-precision networks for efficient embedded inference. arXiv preprint arXiv:1809.04191 (2018)
28. McKinstry, J.L., et al.: Discovering low-precision networks close to full-precision networks for efficient inference. In: 2019 Fifth Workshop on Energy Efficient Machine Learning and Cognitive Computing-NeurIPS Edition (EMC2-NIPS), pp. 6–9. IEEE (2019)
29. Mishra, A., Nurvitadhi, E., Cook, J.J., Marr, D.: WRPN: wide reduced-precision networks. arXiv preprint arXiv:1709.01134 (2017)
30. Qiu, J., et al.: Going deeper with embedded fpga platform for convolutional neural network. In: Proceedings of the 2016 ACM/SIGDA International Symposium on Field-Programmable Gate Arrays, pp. 26–35. ACM (2016)
31. Rastegari, M., Ordonez, V., Redmon, J., Farhadi, A.: XNOR-net: ImageNet classification using binary convolutional neural networks. In: Leibe, B., Matas, J., Sebe, N., Welling, M. (eds.) ECCV 2016. LNCS, vol. 9908, pp. 525–542. Springer, Cham (2016). https://doi.org/10.1007/978-3-319-46493-0_32
32. Ryu, S., Kim, H., Yi, W., Kim, J.J.: BitBlade: area and energy-efficient precision-scalable neural network accelerator with bitwise summation. In: Proceedings of the 56th Annual Design Automation Conference 2019, pp. 1–6 (2019)
33. Sharma, H., et al.: Bit fusion: bit-level dynamically composable architecture for accelerating deep neural network. In: 2018 ACM/IEEE 45th Annual International Symposium on Computer Architecture (ISCA), pp. 764–775 (2018). https://doi.org/10.1109/ISCA.2018.00069
34. Tann, H., Hashemi, S., Bahar, R.I., Reda, S.: Hardware-software codesign of accurate, multiplier-free deep neural networks. In: 2017 54th ACM/EDAC/IEEE Design Automation Conference (DAC), pp. 1–6. IEEE (2017)
35. Umuroglu, Y., et al.: FINN: a framework for fast, scalable binarized neural network inference. In: Proceedings of the 2017 ACM/SIGDA International Symposium on Field-Programmable Gate Arrays, pp. 65–74 (2017)
36. Wussing, H.: The Genesis of the Abstract Group Concept: A Contribution to the History of the Origin of Abstract Group Theory. Courier Corporation (2007)
37. Zhang, C., Li, P., Sun, G., Guan, Y., Xiao, B., Cong, J.: Optimizing FPGA-based accelerator design for deep convolutional neural networks. In: Proceedings of the 2015 ACM/SIGDA International Symposium on Field-programmable Gate Arrays, pp. 161–170 (2015)
38. Zhang, D., Yang, J., Ye, D., Hua, G.: LQ-nets: learned quantization for highly accurate and compact deep neural networks. In: Proceedings of the European Conference on Computer Vision (ECCV), pp. 365–382 (2018)
39. Zhao, R., et al.: Accelerating binarized convolutional neural networks with software-programmable FPGAs. In: Proceedings of the 2017 ACM/SIGDA International Symposium on Field-Programmable Gate Arrays, pp. 15–24 (2017)
40. Zhou, A., Yao, A., Guo, Y., Xu, L., Chen, Y.: Incremental network quantization: towards lossless CNNs with low-precision weights. arXiv preprint arXiv:1702.03044 (2017)
41. Zhu, C., Han, S., Mao, H., Dally, W.J.: Trained ternary quantization. arXiv preprint arXiv:1612.01064 (2016)

Disentangled Differentiable Network Pruning

Shangqian Gao⬥, Feihu Huang⬥, Yanfu Zhang⬥, and Heng Huang(✉)⬥

Department of Electrical and Computer Engineering, University of Pittsburgh, Pittsburgh, USA
{shg84,feh23,yaz91,heng.huang}@pitt.edu

Abstract. In this paper, we propose a novel channel pruning method for compression and acceleration of Convolutional Neural Networks (CNNs). Many existing channel pruning works try to discover compact sub-networks by optimizing a regularized loss function through differentiable operations. Usually, a learnable parameter is used to characterize each channel, which entangles the width and channel importance. In this setting, the FLOPs or parameter constraints implicitly restrict the search space of the pruned model. To solve the aforementioned problems, we propose optimizing each layer's width by relaxing the hard equality constraint used in previous works. The relaxation is inspired by the definition of the top-k operation. By doing so, we partially disentangle the learning of width and channel importance, which enables independent parametrization for width and importance and makes pruning more flexible. We also introduce soft top-k to improve the learning of width. Moreover, to make pruning more efficient, we use two neural networks to parameterize the importance and width. The importance generation network considers both inter-channel and inter-layer relationships. The width generation network has similar functions. In addition, our method can be easily optimized by popular SGD methods, which enjoys the benefits of differentiable pruning. Extensive experiments on CIFAR-10 and ImageNet show that our method is competitive with state-of-the-art methods.

Keywords: Model compression · Channel pruning · Differentiable pruning

1 Introduction

Convolutional Neural Networks (CNNs) have accomplished tremendous success in various computer vision tasks [2,28,43,44,47]. To deal with real-world applications, recently, the design of CNNs has become more and more complicated in terms of width, depth, etc. [14,20,28,48]. These complex CNNs can attain better performance on benchmark tasks, but their computational and storage costs

---

Supplementary Information The online version contains supplementary material available at https://doi.org/10.1007/978-3-031-20083-0_20.

S. Avidan et al. (Eds.): ECCV 2022, LNCS 13671, pp. 328–345, 2022.
https://doi.org/10.1007/978-3-031-20083-0_20

increase dramatically. As a result, a typical application based on CNNs can quickly exhaust an embedded or mobile device due to its enormous costs. Given such costs, the application can hardly be deployed on resource-limited platforms. To tackle these problems, many researches [11,12] have been devoted to compressing the original large CNNs into compact models. Among these methods, weight pruning and structural pruning are two popular topics for model compression.

Unlike weight pruning or sparsification, structural pruning, especially channel pruning, is an effective way to truncate the computational cost of a model because it does not require any post-processing steps to achieve actual acceleration and compression. Many existing works [9,24,25,54] try to discover compact sub-networks by optimizing a regularized loss function through differentiable operations. Usually, the width of a layer and the importance of each channel are entangled in this setting since they use one learnable parameter to characterize each channel. Specifically, the FLOPs or parameter constraints implicitly restrict the search space of the pruned model. On the other hand, search based pruning algorithms (through reinforcement learning [17], evolutionary algorithms [33] and so on) can directly generate the width of each layer with flexible importance definition, which leads to competitive performance. However, the costs of search based method is usually more expensive.

Previous differentiable pruning methods [9,24,25,54] entangle width and importance, limiting the potential search space of sub-networks. To tackle this problem, we aim to disentangle the learning of width and importance, and consequently make pruning more flexible. The **disentanglement of pruning** can be understood as using *independent parameterization for channel importance and layer width*. To achieve this, we first observe that the width of a certain layer can be represented by k of the top-k operation. Inspired by the definition of top-k, we then relax the hard equality constraint used in previous works to a soft regularization term, where k and importance scores can be optimized separately. By doing so, we partially disentangle the importance and width of a layer for pruning. Under our setting, the choices of channel importance become more flexible compared to previous works. Additionally, the width k of each layer can be generated directly, which shares similar property of search based algorithms. Following previous works, we also formulate the channel pruning problem as a constrained optimization problem, which can be efficiently optimized through regular SGD methods. Compared to differentiable pruning methods [9,24,25,54], our method disentangles the learning of width and channel importance, which potentially enlarge the search space. Compared to search based algorithms [17,33], our method provides a way to efficiently generate and optimize width without additional costs.

To make the learning efficient, we further parameterize the importance and width by using two neural networks. We use an importance generation network to capture inter-channel and inter-layer relationships. Similarly, a width generation network is used to generate the width of each layer. A soft constraint term is then used to connect importance and width. With these techniques, our method can outperform state-of-the-art pruning methods on CIFAR-10 and ImageNet datasets. Our contributions can be summarized as:

1) We aim to disentangle the learning of width and importance for differentiable channel pruning, which is achieved by relaxing the equality constraint derived by the definition of the top-k operation. By relaxing the equality constraint, width and importance can be parameterized independently. We also extend the discrete top-k masks to soft top-k masks with a smoothstep function allowing custom width for soft windows.
2) To improve the learning efficiency, we parameterize the importance of each channel and width of each layer by using neural networks. The importance generation network is used to capture inter-channel and inter-layer relationships. The width generation network shares similar intuition.
3) Extensive experiments on CIFAR-10 and ImageNet show that our method can outperform existing channel pruning methods on ResNets and MobileNetV2/V3.

2   Related Works

Recently, model compression has drawn a lot of attention from the community. Weight pruning and structural pruning are two popular directions.

2.1   Weight Pruning

Weight pruning eliminates redundant connections without assumptions on the structures of weights. Weight pruning methods can achieve a very high compression rate while they need specially designed sparse matrix libraries to achieve acceleration and compression. As one of the early works, [12] proposes to use L_1 or L_2 magnitude as the criterion to prune weights and connections. SNIP [29] updates the importance of each weight by using gradients from the loss function. Weights with lower importance will be pruned. Lottery ticket hypothesis [7] assumes there exist high-performance sub-networks within the large network at initialization time. Besides one-shot pruning, repeated pruning and fine-tuning can lead to better performance but with larger costs. In rethinking network pruning [35], they challenge the typical model compression process (training, pruning, fine-tuning) and argue that fine-tuning is not necessary. Instead, they show that training the compressed model from scratch with random initialization can obtain better results.

2.2   Structural Pruning

One of the previous works [30] in structural pruning uses the sum of the absolute value of kernel weights as the criterion for filter pruning. Instead of directly pruning filters based on magnitude, structural sparsity learning [52] is proposed to prune redundant structures with Group Lasso regularization. On top of structural sparsity, GrOWL regularization is applied to make similar structures share the same weights [56]. One of the problems when using Group Lasso is that weights with small values could still be important, and it is difficult for structures under

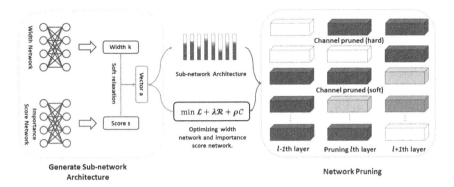

Fig. 1. Flowchart of our proposed method. Importance score generator g_s and width generator g_k are used to generate the importance score and width. We then use them to generate the mask vector **a**, and it is used to produce the sub-network architecture. The network is pruned according to **a**. Finally, g_k and g_s are optimized by minimizing the loss function.

Group Lasso regularization to achieve exact zero values. As a result, [36] proposes to use explicit L_0 regularization to make weights within structures have exact zero values. Besides using the magnitude of structure weights as a criterion, other methods utilize the scaling factor of batchnorm to achieve structure pruning, since batchnorm [22] is widely used in recent neural network designs [14,20]. A straightforward way to achieve channel pruning is to make the scaling factor of batchnorm to be sparse [34]. If the scaling factor falls below a certain threshold, the channel will be removed. Structure sparse selection [21] extends the idea of using scaling factors to more structures, like an entire layer. Another line of research formulates pruning as a constrained optimization problem [8,9,24,25,54,57], and they use learnable parameters (also called gate parameters) to control each channel. These parameters are differentiable in their setting, which enables an efficient end-to-end optimization process. Though these methods have succeeded in channel pruning, the width of each layer and the importance of each channel are entangled, limiting the search space. Besides using gates, Collaborative channel pruning [41] tries to prune channels by using Taylor expansion. Greedy forward selection [53] is proposed to find good sub-networks, which starts from an empty network and greedily adds important channels from the original network. In Automatic Model Compression (AMC) [17], they use policy gradient to update the policy network. This policy network is then used to generate the width of each layer. MetaPruning [33] uses a hypernet to generate parameters for sub-networks, and evolutionary algorithms are utilized to find the best configuration (width) of sub-networks. Our method can generate width directly like MetaPruning and AMC. In addition, our method can be optimized more efficiently through regular stochastic gradient methods.

2.3   Other Methods

Besides weight and channel pruning methods, there are works from other perspectives, including bayesian pruning [37,39], weight quantization [5,42], pruning
for fairness [58], and knowledge distillation [18].

3   Proposed Method

3.1   Preliminary

To better describe our proposed approach, necessary notations are introduced
first. In a CNN, the feature map of lth layer can be represented by $\mathcal{F}_l \in \Re^{C_l \times W_l \times H_l}$, $l = 1, \dots, L$, where C_l is the number of channels, H_l and W_l are
height and width of the current feature map, L is the number of layers. The mini-
batch dimension of feature maps is ignored to simplify notations. sigmoid($\cdot$) is
the sigmoid function. round($\cdot$) rounds inputs to nearest integers. $\mathbb{1}(\cdot)$ is the indicator function.

Usually, differentiable channel pruning algorithms aim to solve the following
problem:

$$\min_{\Theta}\ \mathcal{J}(\Theta) = \mathcal{L}(f(x; \Theta, \mathcal{W}), y) + \lambda \mathcal{R}(\Theta), \tag{1}$$

where x, y are input samples and their labels. $\mathcal{W}$, $\mathcal{L}$ and $\mathcal{R}$ are model weights,
loss functions and regularization functions on parameters or FLOPs. Θ are learn-
able parameters to decide whether to prune the channel. There are many ways
to characterize a channel, such as Gumbel-sigmoid approximation [23], shape
function [25] and so on. $\mathcal{R}$ is the regularization function to control the number
of channels or FLOPs of each layer. Our method aims to disentangle the learning of width and channel importance. As a result, Θ will be reparameterized
into two variables: importance score $\mathbf{s}$ and layer width $\mathbf{k}$. How to achieve such a
disentanglement will be detailed in this section.

3.2   Top-k Operation

In this section, we will introduce how to parameterize the width of a layer. Let us
denote the importance score vector for each channel of a layer as $\mathbf{s} = [s_1, \dots, s_{C_l}]$.
Suppose we need to select k most importance channels out of C_l channels, we
can use a top-k mask vector $\mathbf{a}$, which is given by:

$$a_i = \begin{cases} 1 & \text{if } s_i \text{ is a top-}k \text{ element in } S, \\ 0 & \text{otherwise.} \end{cases} \tag{2}$$

The process of selecting top-k channels is a natural way for channel pruning,
and k represents width of this layer. The relationship between k and a_i can be
represented by the following equation:

$$k = \sum_{i=1}^{C_l} a_i. \tag{3}$$

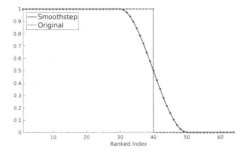

Fig. 2. Soft relaxation vs. naive binary mask. In this figure, we choose $C_l = 64$, $\gamma = 20$, $C_l k = 40$. Solid line represent the continuous function. Square dots represent the actual values taken by the vector $\tilde{\mathbf{a}}$.

Gradients with respect to k through Eq. 3 are not defined. Except Eq. 3, we can use an alternative surrogate to represent k:

$$k = \frac{1}{C_l} \sum_{i=1}^{C_l} \mathbb{1}_{s_i > s_0}(s_i), \tag{4}$$

where s_0 is a value between kth and $k+1$th value, and the indicator function $\mathbb{1}_{s_i > s_0}(\cdot)$ returns 1 if $s_i > s_0$, otherwise it returns 0. Here, to unify the learning of different layers, we abuse the notation k to represent the normalized version of $k \in [0, 1]$. If we enforce the hard equality defined in Eq. 4, it still entangles the learning of importance and width. We then replace it with a regularization term:

$$\mathcal{C}(k, \mathbf{s}) = \left\| k - \frac{1}{C_l} \sum_{i=1}^{C_l} \text{sigmoid}((\bar{s}_i - \bar{s}_0)/t) \right\|^2, \tag{5}$$

which does not enforce a hard constraint and $\bar{s}$ is the unnormalized importance score (outputs before the final activation of g_s defined in Eq. 8). We also relax the indicator function with the sigmoid function of temperature t to facilitate gradient calculations. In practice, we let $\bar{s}_0 = 0$, so that the importance score will match k automatically. The gradients with respect to k can be obtained by utilizing this regularization term, and the width of each layer can be optimized using SGD or other stochastic optimizers. Finally, we achieve pruning by inserting the vector $\mathbf{a}$ to the feature map $\mathcal{F}_l$:

$$\hat{\mathcal{F}}_l = \mathbf{a} \odot \mathcal{F}_l, \tag{6}$$

where $\hat{\mathcal{F}}_l$ is the pruned feature map, $\odot$ is the element-wise product, and $\mathbf{a}$ is first resized to have the same size of $\mathcal{F}_l$.

3.3   Soft Top-k Operation with Smoothstep Function

When performing discrete top-k operation, we place 1 to the first k elements of $\tilde{\mathbf{a}}$. Similarly, we use a smoothstep function [13] to generate values for soft relaxed $\tilde{\mathbf{a}}$:

$$\text{Smoothstep}(x) = \begin{cases} 0, & \text{if } x \leq -\gamma/2 + C_l k, \\ -\frac{2}{\gamma^3}(x - C_l k)^3 + \frac{3}{2\gamma}(x - C_l k) + \frac{1}{2}, \\ \quad \text{if } -\gamma/2 + C_l k \leq x \leq \gamma/2 + C_l k, \\ 1, & \text{if } x \geq \gamma/2 + C_l k. \end{cases} \tag{7}$$

In smoothstep function, γ controls the width of the soft relaxation. $C_l k$ represents the center of the soft relaxation. Outside $[-\gamma/2 + C_l k, \gamma/2 + C_l k]$, smoothstep function performs binary rounding. We provide the comparison between smoothstep function and naive binary masks in Fig. 2. The value of $\tilde{\mathbf{a}}_i = \text{Smoothstep}(i)$. The soft version of $\mathbf{a}$ can be obtained by $\mathbf{a} = P^T \tilde{\mathbf{a}}$. Here, we abuse the notation of $\tilde{\mathbf{a}}$ and $\mathbf{a}$ for the soft relaxed mask vector.

To satisfy Eq. 3, we need $C_l k \approx \sum a_i$. As a result, the center of the soft window should be at $C_l k$. Other functions like sigmoid($\cdot$) can also interpolate between $[0, 1]$. We choose the smoothstep function since it provides a easy way to control the width of soft relaxation. If $C_l k$ is close to C_l (when k is close to 1), the soft range of $\tilde{\mathbf{a}}$ is not symmetric any more on k. We adjust γ to round($C_l - C_l k$) to ensure $C_l k \approx \sum a_i$.

Binary values are often used to control open or close of a channel. However, it is better to use soft relaxed values in certain circumstances. We apply soft relaxation on the mask vector $\mathbf{a}$ for several reasons. In practice, it is hard for us to generate k with discrete values, and discrete constraints on kC_l dramatically increase the difficulty for optimization. Thus, the generated k is within $[0, 1]$. If only binary values are used, then $kC_l = 9.1$ and $kC_l = 8.5$ will produce the same $\mathbf{a}$. Soft relaxation can produce unique $\mathbf{a}$ when $kC_l = 9.1$ or $kC_l = 8.5$. Another benefit of soft relaxation is that we can evaluate more channels compared to the discrete settings. Let us first reindex the vector $\mathbf{s}$ as $\tilde{\mathbf{s}}$ based on the monotone decreasing order of $\mathbf{s}$, then $\tilde{\mathbf{s}} = P\mathbf{s}$, where P is a permutation matrix. Since $\mathbf{a}$ and $\mathbf{s}$ have one-to-one correspondence, sorting $\mathbf{a}$ according to $\mathbf{s}$ can be represented as $\tilde{\mathbf{a}} = P\mathbf{a}$.

3.4   Generating Width and Importance Score

To provide importance score $\mathbf{s}$ for each channel, we use a neural network g_s to learn it from the dataset:

$$\mathcal{S} = g_s(x_s, \Theta_s), \tag{8}$$

where $\mathcal{S} = (\mathbf{s}_1, \cdots, \mathbf{s}_L)$ is the collection of all scores across different layers, Θ_s are learnable parameters of g_s, and x_s is the input of g_s. Before training, we generate x_s randomly, and it is kept fixed during training. We can also use a learnable x_s, which results in similar performance. Previous pruning methods often use a single parameter to control each channel, which can not obtain inter-channel and inter-layer relationships. As a result, g_s is designed to be composed with GRU [4] and fully connected layers. Basically, we use GRU to capture inter-layer relationships, and fully connected layers are for inter-channel relationships. The additional computational costs introduced by g_s is trivial, and it has little impact to the training time. Since $\mathcal{S}$ is not directly involved in the forward computation, we use straight-through gradient estimator [1] to calculate the gradients of it:

Algorithm 1: Disentangled Differentiable Network Pruning

Input: D, p, λ, ρ, E, f,

Freeze $\mathcal{W}$ in f.

Initialization: initialize x_s and x_k for g_s and g_k; randomly initialize Θ_s and Θ_k

for $e := 1$ *to* E **do**

   shuffle(D)

   for *a mini-batch* (x, y) *in* D **do**

      1. generate the width of each layer **k** from g_k by using Eq. 9

      2. generate the importance score of each layer $\mathcal{S}$ from g_s using Eq. 8.

      3. produce the soft mask vector $\tilde{\mathbf{a}}$ with Eq. 7, and obtain $\mathbf{a} = P^T \tilde{\mathbf{a}}$

      4. calculate gradients for Θ_s : $\frac{\partial J}{\partial \Theta_s} = \frac{\partial \mathcal{L}}{\partial \Theta_s} + \rho \frac{\partial \mathcal{C}}{\partial \Theta_s}$ and

      Θ_k : $\frac{\partial J}{\partial \Theta_k} = \lambda \frac{\partial \mathcal{R}}{\partial \Theta_k} + \rho \frac{\partial \mathcal{C}}{\partial \Theta_k}$ separately.

      5. update Θ_k and Θ_s with ADAM.

   end

end

Pruning the model with resulting g_k and g_s, and finetune it.

$\frac{\partial J}{\partial \mathbf{s}} = \frac{\partial J}{\partial \mathbf{a}}$. We also want to emphasize that it's crucial to use simgoid($\cdot$) as the output activation for g_s. Using absolute values [46] or other functions incurs much larger errors when estimating the gradients. This is probably because simgoid($\cdot$) better approximates binary values.

We also use a neural network g_k to generate the width for each layer:

$$\mathbf{k} = g_k(x_k, \Theta_k),\tag{9}$$

where x_k is the input to g_k, Θ_k are parameters for g_k, and $\mathbf{k} = [k_1, \cdots, k_L]$ is a vector contains width of all layers. The output activate function is the sigmoid function again, since we need to restrict the range of $k \in [0, 1]$. g_k is composed with fully connected layers. In addition, like x_s, x_k is also generated randomly, and it is kept fix when training g_k.

3.5 The Proposed Algorithm

With techniques introduced in previous sections, we can start to prune the network. The network pruning problem in our setting can be formulated as the following problem:

$$\min_{\Theta_k, \Theta_s} \; \mathcal{J}(\Theta_k, \Theta_s) = \big\{ \mathcal{L}(f(x; \mathcal{A}, \mathcal{W}), y) + \lambda \mathcal{R}(T(\mathbf{k}), p T_{\text{total}})$$
$$+ \rho \mathcal{C}(\mathbf{k}, \mathcal{S}) \big\}\tag{10}$$

where (x, y) is the input sample and its corresponding label, $f(x; \mathcal{A}, \mathcal{W})$ is a CNN parameterized by $\mathcal{W}$ and controlled by $\mathcal{A} = [\mathbf{a}_1, \cdots, \mathbf{a}_L]$, $\mathcal{R}$ is the FLOPs regularization term, $T(\mathbf{k})$ is the FLOPs of the current sub-network, p is the pruning rate, T_{total} is the total prunable FLOPs, $\mathcal{J}$ is the overall objective function, $\mathcal{C}(\mathbf{k}, \mathcal{S})$ is defined in Eq. 5, and ρ, λ are hyper-parameters for $\mathcal{C}$, $\mathcal{R}$

separately. We let $\mathcal{R}(x,y) = \log(\max(x,y)/y)$, which can quickly push $\mathcal{R}$ to approach 0. Our objective function defined in Eq. 10 can be optimized efficiently by using any stochastic gradient optimizer. Using learnable importance score produces quite strong empirical performance. If better learning mechanism for importance score is designed, it can also be merged into our algorithm.

The overall algorithm is given in Algorithm 1. The input of Algorithm 1 are D: a dataset for pruning, p: the pruning rate defined in Eq. 10, λ and ρ: hyper-parameter for $\mathcal{R}$ and $\mathcal{C}$, f: a neural network to be pruned and E: the number of pruning epochs. In order to facilitate pruning, we usually choose D as a subset of the full training set. In step 4 of Algorithm 1, the gradients of Θ_k and Θ_s are calculated separately because of $\mathcal{C}$. This operation brings marginal computational burden, since $\mathcal{C}$ and $\mathcal{R}$ are not depend on input samples. The fine-tuning process is very time-consuming. As a result, we use the performance of a sub-network within the pre-trained model to represent its quality. This setup is used in many pruning methods, like AMC [17], and we freeze weights $\mathcal{W}$ during pruning. When performing actual pruning, we round $C_l k_l$ to its nearest integer, and soft relaxation is not used. Instead, we use Eq. 2 to generate $\mathbf{a}$, which ensures that $\mathbf{a} \in \{0,1\}$. Like previous differentiable pruning works, our method can be directly applied to pre-trained CNNs, which are flexible to use. The overall flowchart of our method is shown in Fig. 1.

4   Connections to Previous Works

In this section, we will discuss the difference and connections of our methods compared to previous works. To connect our method with previous work, we can use an equality constraint to replace the regularization term in Eq. 10:

$$\min_{\mathbf{k},\mathcal{S}} \mathcal{J}(\mathbf{k},\mathcal{S}) = \mathcal{L}(f(x;\mathbf{a},\mathcal{W}),y) + \lambda \mathcal{R}(T(\mathbf{k}),pT_{\text{total}}),$$

$$s.t.\ k_l = \frac{1}{C_l}\sum_{i=1}^{C_l} \mathbb{1}_{s_i^l > 0.5}(s_i^l),\quad l = 1,\cdots,L. \tag{11}$$

Here, we do not use g_k and g_s to simplify the analysis, and we also let $s_0^l = 0.5$ since we use sigmoid activation functions for g_s. Equation 11 is closely related to Eq. 1. If we insert $\frac{1}{C_l}\sum_{i=1}^{C_l}\mathbb{1}_{s_i^l>0.5}(s_i^l)$ to every k_l in $T(\mathbf{k})$, we almost recover Eq. 1. Compared to Eq. 10, Eq. 11 is more restrictive since it reduces the number of parameters for pruning one layer from $C_l + 1$ to C_l, which is equivalent to saying that disentangled pruning provides an extra degree of freedom compared to previous works. This may explain why using independent parameterization for importance and width achieves better empirical performance than previous works. Also note that Eq. 11 corresponds to set ρ to ∞ in Eq. 10, and $\mathbf{k}$ is no longer a validate variable. If we let $\rho = 0$, we have completely disentangled $\mathbf{k}$ and $\mathcal{S}$. But in this situation, the resulting $\mathbf{k}$ will be a trivial solution because it only depends on $\mathcal{R}$. From this perspective, the proposed method in Eq. 10 actually interpolates between previous differentiable pruning works and the complete disentangled formulation.

Table 1. Comparison results on CIFAR-10 dataset with ResNet-56 and MobileNetV2. Δ-Acc represents the performance changes before and after model pruning. $\pm$ indicates increase or decrease compared to baseline results.

Method	Architecture	Baseline Acc	Pruned Acc	Δ-Acc	↓ FLOPs
AMC [17]	ResNet-56	92.80%	91.90%	−0.90%	50.0%
DCP [59]		93.80%	93.81%	+0.01%	47.0%
CCP [41]		93.50%	93.42%	−0.08%	52.6%
HRank [32]		93.26%	93.17%	−0.09%	50.0%
LeGR [3]		93.90%	93.70%	−0.20%	**53.0%**
DDNP (ours)		93.62%	93.83%	**+0.21%**	51.0%
Uniform [59]	MobileNetV2	94.47%	94.17%	−0.30%	26.0%
DCP [59]		94.47%	94.69%	+0.22%	26.0%
MDP [10]		95.02%	95.14%	+0.12%	28.7%
SCOP [49]		94.48%	94.24%	−0.24%	40.3%
DDNP (ours)		94.58%	94.81%	**+0.23%**	**43.0%**

5 Experiments

5.1 Settings

Similar to many model compression works, CIFAR-10 [27] and ImageNet [6] are used to evaluate the performance of our method. Our method uses p to control the FLOPs budget. The detailed choices of p are listed in the supplementary materials. The architectures of g_s and g_k are also provided in supplementary materials. γ in Eq. 7 is chosen as round($0.1C_l$). γ then depends on layer width C_l, and it empirically works well.

Within the experiment section, our method is called as DDNP (**D**isentangled **D**ifferentiable for **N**etwork **P**runing). For CIFAR-10, we compare with other methods on ResNet-56 and MobileNetV2. For ImageNet, we select ResNet-34, ResNet-50, MobileNetV2 and MobileNetV3 small as our target models. The reason we choose these models is because that ResNet [14], MobileNetV2 [45] and MobileNetV3 [19] are much harder to prune than earlier models like AlexNet [28] and VGG [48].

λ decides the regularization strength in our method. We choose $\lambda = 2$ in all CIFAR-10 experiments and $\lambda = 4$ for all ImageNet experiments. We choose $\rho = 2$ and $t = 0.4$ for both datasets. For CIFAR-10 models, we train ResNet-56 and MobileNet-V2 from scratch following PyTorch examples. After pruning, we finetune the model for 160 epochs using SGD with a start learning rate of 0.1, weight decay 0.0001, and momentum 0.9. For ImageNet models, we directly use the pre-trained models released from pytorch [40]. After pruning, we finetune the model for 100 epochs using SGD with an initial learning rate of 0.1, weight decay 0.0001, and momentum 0.9. For MobileNetV2 on ImageNet, we choose weight decay as 0.00004 and use an initial learning rate of 0.05 with cosine annealing

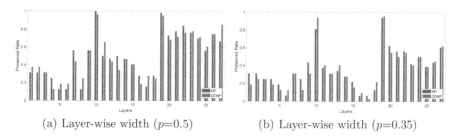

(a) Layer-wise width ($p=0.5$)   (b) Layer-wise width ($p=0.35$)

Fig. 3. (a) and (b): Layer-wise width for two different pruning rates: $p = 0.5/0.35$. We compare DDNP with differentiable pruning (DP) in both figures.

learning rate scheduler, which is the same as the original paper [45]. Most settings for MobileNetV3 small are the same as MobileNetV2. The difference is that weight decay is reduced to 0.00001 following the original setting [19].

For training g_k and g_s, we use ADAM [26] optimizer with a constant learning rate of 0.001 and train them for 200 epochs. We start pruning from the whole network. To achieve this, we add a constant bias to the sigmoid function in g_k, and we set it to 3.0. We randomly sample 2,500 and 25,000 samples from CIFAR-10 and ImageNet, and they are used as the pruning subset D in Algorithm 1. In the experiments, we found that a separate validation set is not necessary. All samples in D come from the original training set. All codes are implemented with pytorch [40].

5.2 CIFAR-10

We present comparison results on CIFAR-10 in Table 1. On ResNet-56, our method achieves the largest performance gain (+0.21% Δ-Acc) compared to other baselines. All methods prune around 50% of FLOPs, and LeGR has the largest pruning rate. At this pruning rate, our method has obvious advantages compared to other methods. Specifically, our method is 0.20% better than DCP regarding Δ-Acc. Although DCP has the second best Δ-Acc, it has the lowest FLOPs reduction rate. CCP and HRank have similar pruning rates and performance, and our method outperforms them by around 0.30% in terms of Δ-Acc. LeGR prunes more FLOPs than our method, but it has a much lower Δ-Acc (−0.20% vs. +0.21%).

For MobileNetV2, our method achieves the best Δ-Acc and prunes most FLOPs (+0.23% Δ-Acc and 43% FLOPs). SCOP prunes slightly less FLOPs, and the performance of SCOP is also lower than our method (−0.24% vs. +0.23% regarding Δ-Acc). Our method and DCP have similar performance, but our method prunes 17% more FLOPs. In summary, the CIFAR-10 results demonstrate that our method is an effective way for network pruning.

5.3 ImageNet Results

The ImageNet results are given in Table 2. We present both base and pruned Top-1/Top-5 accuracy in the table.

Table 2. Comparison results on ImageNet dataset with ResNet-34, ResNet-50, ResNet-101 and MobileNetV2. Δ-Acc represents the performance changes before and after model pruning. ± indicates increase or decrease compared to baseline results.

Method	Architecture	Base/Pruned Top-1	Base/Pruned Top-5	Δ Top-1	Δ Top-5	↓ FLOPs
SFP [15]	ResNet-34	73.93%/71.84%	91.62%/89.70%	−2.09%	−1.92%	41.1%
IE [38]		73.31%/72.83%	-/-	−0.48%	-	24.2%
FPGM [16]		73.92%/72.63%	91.62%/91.08%	−1.29%	−0.54%	41.1%
SCOP [49]		73.31%/72.62%	91.42%/90.98%	−0.69%	−0.44%	**44.8%**
DDNP (ours)		73.31%/**73.03%**	91.42%/**91.23%**	**−0.28%**	**−0.19%**	44.2%
DCP [59]	ResNet-50	76.01%/74.95%	92.93%/92.32%	−1.06%	−0.61%	55.6%
CCP [41]		76.15%/75.21%	92.87%/92.42%	−0.94%	−0.45%	54.1%
MetaPruning [33]		76.60%/75.40%	-/-	−1.20%	-	51.2%
GBN [54]		75.85%/75.18%	92.67%/92.41%	−0.67%	−0.26%	55.1%
HRank [32]		76.15%/74.98%	92.87%/92.33%	−1.17%	−0.54%	43.8%
LeGR [3]		76.10%/75.30%	-/-	−0.80%	-	54.0%
SCOP [49]		76.15%/75.26%	92.87%/92.53%	−0.89%	−0.34%	54.6%
GReg [50]		76.13%/75.36%	-/-	−0.77%	-	**56.7%**
SRR [51]		76.13%/75.11%	92.86%/92.35%	−1.02%	−0.51%	55.1%
CC [31]		76.15%/75.59%	92.87%/92.64%	−0.56%	−0.13%	52.9%
DDNP (ours)		76.13%/**75.89%**	92.86%/**92.90%**	**−0.24%**	**+ 0.04%**	55.0%
Uniform [45]	MobileNetV2	71.80%/69.80%	91.00%/89.60%	−2.00%	−1.40%	30.0%
AMC [17]		71.80%/70.80%	-/-	−1.00%	-	30.0%
AGMC [55]		71.80%/70.87%	-/-	−0.93%	-	30.0%
MetaPruning [33]		72.00%/71.20%	-/-	−0.80%	-	**30.7%**
GFS [53]		72.00%/71.60%	-/-	−0.40%	-	30.0%
DDNP (ours)		72.05%/**72.20%**	90.39%/**90.51%**	**+0.15%**	**+0.12%**	29.5%
Uniform [19]	MobileNetV3 small	67.50%/65.40%	-/-	−2.10%	-	**26.6%**
GFS [53]		67.50%/65.80%	-/-	−1.70%	-	23.5%
DDNP (ours)		67.67%/**67.03%**	87.40%/**86.94%**	**−0.64%**	**−0.46%**	24.5%

ResNet-34. Our method achieves the best Δ Top-1 and Δ Top-5 accuracy with ResNet-34. IE performs the second best regarding Δ Top-1/Δ Top-5, but it prunes much less FLOPs compared to other baselines. SCOP, FPGM, IE, and our method have similar pruning rates. SCOP has the largest FLOPs reduction rate, but the FLOPs gap between our method and SCOP is quite marginal (only 0.6%). Given similar pruning rates, our method outperforms other baselines by at least 0.41% in terms of Δ Top-1 accuracy.

ResNet-50. For ResNet-50, our method achieves the best pruned Top-1/Top-5 accuracy, and the reduction of FLOPs is also obvious. DCP prunes most FLOPs among all comparison baselines. Our method is 0.84% better than DCP regarding Δ Top-1 accuracy while only removing 0.6% less FLOPs than it. The gap between GBN and CC is around 0.09%, and they outperform other baselines. Our method further improves the result of GBN and CC by 0.43% and 0.32% with Δ Top-1 accuracy. CC has the second best performance, but our method prunes 2% more FLOPs than it. Notably, our method achieves no loss on Top-5 accuracy (+0.02%). Also notice that CC considers both channel pruning and weight decomposition, introducing extra performance efficiency trade-off.

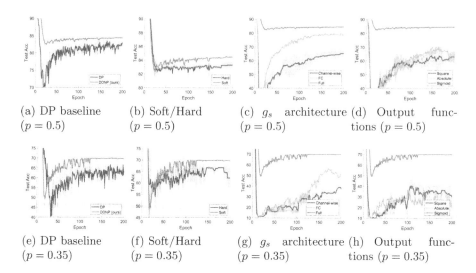

(a) DP baseline ($p = 0.5$)   (b) Soft/Hard ($p = 0.5$)   (c) g_s architecture ($p = 0.5$)   (d) Output functions ($p = 0.5$)

(e) DP baseline ($p = 0.35$)   (f) Soft/Hard ($p = 0.35$)   (g) g_s architecture ($p = 0.35$)   (h) Output functions ($p = 0.35$)

Fig. 4. (a, e): Comparisons of our method and the differentiable pruning (DP) baseline. (b, f): Comparisons of soft and hard setting for the top-k operation. (c, g): Performance during pruning when using different architectures of g_s. (d, h): Performance during pruning when using different output functions of g_s. We run each setting three times and use shaded areas to represent variance. All experiments are done on CIFAR-10 with ResNet-56.

MobileNetV2. MoibleNetV2 is a computationally efficient model by design that is harder to prune than ResNets. With this architecture, all methods prune similar FLOPs within ranges between 29.5% to 30.7%. Our method obtains 72.20%/90.51% Top-1/Top-5 accuracy after pruning, and both of them are better than all the other methods. Compared to the second best method GFS [53], the Top-1/Δ Top-1 accuracy of our method is 0.60%/0.65% higher than it. MetaPruning prunes most FLOPs, but the performance is lower than our method by a large margin. AGMC improves the results of AMC, but the improvement is not very significant.

MobileNetV3 Small. MobileNetV3 small is a tiny model with FLOPs of around 64M. The uniform baseline prunes most FLOPs which is 3.1% and 2.1% higher than GFS and our method, but the absolute FLOPs difference is small (Uniform: 47M; GFS: 49M; Ours: 48.3M). Our method has significant advantages on MobileNet-V3 small, and it is 1.23%/1.06% better than GFS on Top-1/Δ Top-1 accuracy. GFS greedily selects neurons with the largest impact on the loss starting from an empty set, and it performs well across multiple architectures. They argue that forward selection is better than backward elimination with greedy selection. On the contrary, in our setting, disentangled pruning can successfully discover good sub-networks starting from the whole model, especially for compact architectures. The superior performance of our method demonstrates the advantage of disentangled pruning.

5.4   Impacts of Different Settings

In this section, we will demonstrate the effectiveness of different design choices.

We first build a differentiable pruning (DP) baseline by using the Gumbel-sigmoid function. We then compare DP with our method in Fig. 4 (a, e). Our method outperforms DP when $p = 0.5$ and $p = 0.35$. The advantage becomes obvious when the pruning rate is large ($p = 0.35$). This observation suggests that our method can discover better sub-networks than DP across different pruning rates. We also visualize the layer-wise width in Fig. 3. An interesting observation is that, with different p, the relative order of width changes (like the first and second block) with our method.

In Fig. 4 (b, f), we verify the effectiveness of soft top-k defined in Sect. 3.3. The hard setting refers to set $\gamma = 0$ in Eq. 7. From the figure, we can see that soft top-k operation achieves better performance than the hard version. Moreover, when the pruning rate is large, the effect of soft top-k becomes more clear (gap around 5%). These results suggest soft top-k is preferred when disentangling the learning of width and importance.

In Fig. 4 (c, g), we present results by varying the architecture of g_s. We can see that full g_s (GRU+FC) has the best performance, followed by FC and channel-wise importance score. The learning of importance may become difficult when we use disentangled pruning (probably due to gradient calculations with STE), and naive parametrization (one parameter per channel) lacks enough capacity to efficiently capture inter-channel and inter-layer relationships. Using a model with a larger capacity enables fast learning.

Finally, we compare different output functions of g_s in Fig. 4. We compare three different output functions: sigmoid function, absolute function and square function. Recall that we use s and $\bar{s}$ to represent the importance score and unnormalized importance score (outputs before the final activation of g_s). As a result, importance score with sigmoid function, absolute function and square function are defined as: $s = \text{sigmoid}(\bar{s})$, $s^{AF} = | \bar{s} |$ and $s^{SF} = \bar{s}^2$. From the Fig. 4, it is clear that sigmoid($\cdot$) has the best performance, which indicates that better alignment in the forward pass helps improve the quality of gradients when learning importance scores.

6   Conclusion

In previous differentiable pruning works, width and channel importance are entangled during the pruning process. Such a design is straightforward and easy to use, but it restricts the potential search space during the pruning process. To overcome this limitation, we propose to prune neural networks by disentangling width and importance. To achieve such a disentanglement, we propose to relax the hard constraint used in previous methods to a soft regularization term, allowing independent parametrization of width and importance. We also relax hard top-k to soft top-k with the smoothstep function. We further use an importance score generation network and a width network to facilitate the learning process. Moreover, the design choices are empirically verified for our method.

The experimental results on CIFAR-10 and ImageNet demonstrate the strength of our method.

Acknowledgement. This work was partially supported by NSF IIS 1845666, 1852606, 1838627, 1837956, 1956002, 2217003.

References

1. Bengio, Y., Léonard, N., Courville, A.: Estimating or propagating gradients through stochastic neurons for conditional computation. arXiv preprint arXiv:1308.3432 (2013)
2. Bojarski, M., et al.: End to end learning for self-driving cars. arXiv preprint arXiv:1604.07316 (2016)
3. Chin, T.W., Ding, R., Zhang, C., Marculescu, D.: Towards efficient model compression via learned global ranking. In: Proceedings of the IEEE/CVF Conference on Computer Vision and Pattern Recognition, pp. 1518–1528 (2020)
4. Cho, K., van Merrienboer, B., Gulcehre, C., Bougares, F., Schwenk, H., Bengio, Y.: Learning phrase representations using RNN encoder-decoder for statistical machine translation. In: Conference on Empirical Methods in Natural Language Processing (EMNLP 2014) (2014)
5. Courbariaux, M., Bengio, Y., David, J.P.: Binaryconnect: training deep neural networks with binary weights during propagations. In: Advances in Neural Information Processing Systems, pp. 3123–3131 (2015)
6. Deng, J., Dong, W., Socher, R., Li, L.J., Li, K., Fei-Fei, L.: Imagenet: a large-scale hierarchical image database. In: IEEE Conference on Computer Vision and Pattern Recognition, CVPR 2009, pp. 248–255. IEEE (2009)
7. Frankle, J., Carbin, M.: The lottery ticket hypothesis: finding sparse, trainable neural networks. In: International Conference on Learning Representations (2019). https://openreview.net/forum?id=rJl-b3RcF7
8. Ganjdanesh, A., Gao, S., Huang, H.: Interpretations steered network pruning via amortized inferred saliency maps. In: Proceedings of the European Conference on Computer Vision (ECCV) (2022)
9. Gao, S., Huang, F., Pei, J., Huang, H.: Discrete model compression with resource constraint for deep neural networks. In: Proceedings of the IEEE/CVF Conference on Computer Vision and Pattern Recognition, pp. 1899–1908 (2020)
10. Guo, J., Ouyang, W., Xu, D.: Multi-dimensional pruning: a unified framework for model compression. In: Proceedings of the IEEE/CVF Conference on Computer Vision and Pattern Recognition, pp. 1508–1517 (2020)
11. Han, S., Mao, H., Dally, W.J.: Deep compression: compressing deep neural networks with pruning, trained quantization and huffman coding. arXiv preprint arXiv:1510.00149 (2015)
12. Han, S., Pool, J., Tran, J., Dally, W.: Learning both weights and connections for efficient neural network. In: Advances in Neural Information Processing Systems, pp. 1135–1143 (2015)
13. Hazimeh, H., Ponomareva, N., Mol, P., Tan, Z., Mazumder, R.: The tree ensemble layer: differentiability meets conditional computation. In: International Conference on Machine Learning, pp. 4138–4148. PMLR (2020)
14. He, K., Zhang, X., Ren, S., Sun, J.: Deep residual learning for image recognition. In: Proceedings of the IEEE Conference on Computer Vision and Pattern Recognition, pp. 770–778 (2016)

15. He, Y., Kang, G., Dong, X., Fu, Y., Yang, Y.: Soft filter pruning for accelerating deep convolutional neural networks. In: International Joint Conference on Artificial Intelligence (IJCAI), pp. 2234–2240 (2018)

16. He, Y., Liu, P., Wang, Z., Hu, Z., Yang, Y.: Filter pruning via geometric median for deep convolutional neural networks acceleration. In: Proceedings of the IEEE Conference on Computer Vision and Pattern Recognition, pp. 4340–4349 (2019)

17. He, Y., Lin, J., Liu, Z., Wang, H., Li, L.J., Han, S.: AMC: Automl for model compression and acceleration on mobile devices. In: Proceedings of the European Conference on Computer Vision (ECCV), pp. 784–800 (2018)

18. Hinton, G., Vinyals, O., Dean, J.: Distilling the knowledge in a neural network. arXiv preprint arXiv:1503.02531 (2015)

19. Howard, A., et al.: Searching for mobilenetv3. In: Proceedings of the IEEE/CVF International Conference on Computer Vision, pp. 1314–1324 (2019)

20. Huang, G., Liu, Z., Van Der Maaten, L., Weinberger, K.Q.: Densely connected convolutional networks. In: Proceedings of the IEEE Conference on Computer Vision and Pattern Recognition, pp. 4700–4708 (2017)

21. Huang, Z., Wang, N.: Data-driven sparse structure selection for deep neural networks. In: Proceedings of the European Conference on Computer Vision (ECCV), pp. 304–320 (2018)

22. Ioffe, S., Szegedy, C.: Batch normalization: accelerating deep network training by reducing internal covariate shift. In: Proceedings of the 32nd International Conference on International Conference on Machine Learning, ICML, vol. 37, pp. 448–456. JMLR.org (2015). https://dl.acm.org/citation.cfm?id=3045118.3045167

23. Jang, E., Gu, S., Poole, B.: Categorical reparameterization with gumbel-softmax. arXiv preprint arXiv:1611.01144 (2016)

24. Kang, M., Han, B.: Operation-aware soft channel pruning using differentiable masks. In: International Conference on Machine Learning, pp. 5122–5131. PMLR (2020)

25. Kim, J., Park, C., Jung, H., Choe, Y.: Plug-in, trainable gate for streamlining arbitrary neural networks. In: Proceedings of the AAAI Conference on Artificial Intelligence (2020)

26. Kingma, D.P., Ba, J.: Adam: a method for stochastic optimization. arXiv preprint arXiv:1412.6980 (2014)

27. Krizhevsky, A., Hinton, G.: Learning multiple layers of features from tiny images. Technical report, Citeseer (2009)

28. Krizhevsky, A., Sutskever, I., Hinton, G.E.: Imagenet classification with deep convolutional neural networks. In: Advances in Neural Information Processing Systems, pp. 1097–1105 (2012)

29. Lee, N., Ajanthan, T., Torr, P.H.: Snip: single-shot network pruning based on connection sensitivity. In: ICLR (2019)

30. Li, H., Kadav, A., Durdanovic, I., Samet, H., Graf, H.P.: Pruning filters for efficient convnets. In: ICLR (2017)

31. Li, Y., et al.: Towards compact CNNs via collaborative compression. In: Proceedings of the IEEE/CVF Conference on Computer Vision and Pattern Recognition, pp. 6438–6447 (2021)

32. Lin, M., et al.: Hrank: filter pruning using high-rank feature map. In: The IEEE Conference on Computer Vision and Pattern Recognition (CVPR) (2020)

33. Liu, Z., et al.: Metapruning: meta learning for automatic neural network channel pruning. In: Proceedings of the IEEE International Conference on Computer Vision, pp. 3296–3305 (2019)

34. Liu, Z., Li, J., Shen, Z., Huang, G., Yan, S., Zhang, C.: Learning efficient convolutional networks through network slimming. In: ICCV (2017)
35. Liu, Z., Sun, M., Zhou, T., Huang, G., Darrell, T.: Rethinking the value of network pruning. In: International Conference on Learning Representations (2019). https://openreview.net/forum?id=rJlnB3C5Ym
36. Louizos, C., Welling, M., Kingma, D.P.: Learning sparse neural networks through l_0 regularization. In: International Conference on Learning Representations (2018). https://openreview.net/forum?id=H1Y8hhg0b
37. Molchanov, D., Ashukha, A., Vetrov, D.: Variational dropout sparsifies deep neural networks. In: Proceedings of the 34th International Conference on Machine Learning, vol. 70. pp. 2498–2507. JMLR.org (2017)
38. Molchanov, P., Mallya, A., Tyree, S., Frosio, I., Kautz, J.: Importance estimation for neural network pruning. In: Proceedings of the IEEE Conference on Computer Vision and Pattern Recognition, pp. 11264–11272 (2019)
39. Neklyudov, K., Molchanov, D., Ashukha, A., Vetrov, D.P.: Structured Bayesian pruning via log-normal multiplicative noise. In: Advances in Neural Information Processing Systems, pp. 6775–6784 (2017)
40. Paszke, A., et al.: Pytorch: an imperative style, high-performance deep learning library. In: Advances in Neural Information Processing Systems, pp. 8024–8035 (2019)
41. Peng, H., Wu, J., Chen, S., Huang, J.: Collaborative channel pruning for deep networks. In: International Conference on Machine Learning, pp. 5113–5122 (2019)
42. Rastegari, M., Ordonez, V., Redmon, J., Farhadi, A.: XNOR-Net: ImageNet classification using binary convolutional neural networks. In: Leibe, B., Matas, J., Sebe, N., Welling, M. (eds.) ECCV 2016. LNCS, vol. 9908, pp. 525–542. Springer, Cham (2016). https://doi.org/10.1007/978-3-319-46493-0_32
43. Redmon, J., Divvala, S., Girshick, R., Farhadi, A.: You only look once: unified, real-time object detection. In: Proceedings of the IEEE Conference on Computer Vision and Pattern Recognition, pp. 779–788 (2016)
44. Ren, S., He, K., Girshick, R., Sun, J.: Faster R-CNN: towards real-time object detection with region proposal networks. In: Advances in Neural Information Processing Systems, pp. 91–99 (2015)
45. Sandler, M., Howard, A., Zhu, M., Zhmoginov, A., Chen, L.C.: MobileNetV2: inverted residuals and linear bottlenecks. In: Proceedings of the IEEE Conference on Computer Vision and Pattern Recognition, pp. 4510–4520 (2018)
46. Sehwag, V., Wang, S., Mittal, P., Jana, S.: Hydra: pruning adversarially robust neural networks. In: NeurIPS (2020). https://proceedings.neurips.cc/paper/2020/hash/e3a72c791a69f87b05ea7742e04430ed-Abstract.html
47. Simonyan, K., Zisserman, A.: Two-stream convolutional networks for action recognition in videos. In: Advances in Neural Information Processing Systems, pp. 568–576 (2014)
48. Simonyan, K., Zisserman, A.: Very deep convolutional networks for large-scale image recognition. arXiv preprint arXiv:1409.1556 (2014)
49. Tang, Y., et al.: SCOP: scientific control for reliable neural network pruning. In: Advances in Neural Information Processing Systems, vol. 33 (2020)
50. Wang, H., Qin, C., Zhang, Y., Fu, Y.: Neural pruning via growing regularization. In: International Conference on Learning Representations (2021). https://openreview.net/forum?id=o966_Is_nPA
51. Wang, Z., Li, C., Wang, X.: Convolutional neural network pruning with structural redundancy reduction. In: Proceedings of the IEEE/CVF Conference on Computer Vision and Pattern Recognition, pp. 14913–14922 (2021)

52. Wen, W., Wu, C., Wang, Y., Chen, Y., Li, H.: Learning structured sparsity in deep neural networks. In: Advances in Neural Information Processing Systems, pp. 2074–2082 (2016)
53. Ye, M., Gong, C., Nie, L., Zhou, D., Klivans, A., Liu, Q.: Good subnetworks provably exist: pruning via greedy forward selection. In: International Conference on Machine Learning, pp. 10820–10830. PMLR (2020)
54. You, Z., Yan, K., Ye, J., Ma, M., Wang, P.: Gate decorator: global filter pruning method for accelerating deep convolutional neural networks. In: Advances in Neural Information Processing Systems, pp. 2130–2141 (2019)
55. Yu, S., Mazaheri, A., Jannesari, A.: Auto graph encoder-decoder for neural network pruning. In: Proceedings of the IEEE/CVF International Conference on Computer Vision, pp. 6362–6372 (2021)
56. Zhang, D., Wang, H., Figueiredo, M., Balzano, L.: Learning to share: simultaneous parameter tying and sparsification in deep learning (2018)
57. Zhang, Y., Gao, S., Huang, H.: Exploration and estimation for model compression. In: Proceedings of the IEEE/CVF International Conference on Computer Vision, pp. 487–496 (2021)
58. Zhang, Y., Gao, S., Huang, H.: Recover fair deep classification models via altering pre-trained structure. In: Proceedings of the European Conference on Computer Vision (ECCV) (2022)
59. Zhuang, Z., et al.: Discrimination-aware channel pruning for deep neural networks. In: Advances in Neural Information Processing Systems, pp. 875–886 (2018)

IDa-Det: An Information Discrepancy-Aware Distillation for 1-Bit Detectors

Sheng Xu[1], Yanjing Li[1], Bohan Zeng[1], Teli Ma[2], Baochang Zhang[1,3(✉)],
Xianbin Cao[1], Peng Gao[2], and Jinhu Lü[1,3]

[1] Beihang University, Beijing, China
{shengxu,yanjingli,bohanzeng,bczhang}@buaa.edu.cn
[2] Shanghai Artificial Intelligence Laboratory, Shanghai, China
[3] Zhongguancun Laboratory, Beijing, China

Abstract. Knowledge distillation (KD) has been proven to be useful for training compact object detection models. However, we observe that KD is often effective when the teacher model and student counterpart share similar proposal information. This explains why existing KD methods are less effective for 1-bit detectors, caused by a significant information discrepancy between the real-valued teacher and the 1-bit student. This paper presents an Information Discrepancy-aware strategy (IDa-Det) to distill 1-bit detectors that can effectively eliminate information discrepancies and significantly reduce the performance gap between a 1-bit detector and its real-valued counterpart. We formulate the distillation process as a bi-level optimization formulation. At the inner level, we select the representative proposals with maximum information discrepancy. We then introduce a novel entropy distillation loss to reduce the disparity based on the selected proposals. Extensive experiments demonstrate IDa-Det's superiority over state-of-the-art 1-bit detectors and KD methods on both PASCAL VOC and COCO datasets. IDa-Det achieves a 76.9% mAP for a 1-bit Faster-RCNN with ResNet-18 backbone. Our code is open-sourced on https://github.com/SteveTsui/IDa-Det.

Keywords: 1-bit detector · Knowledge distillation · Information discrepancy

1 Introduction

Recently, the object detection task [6,20] has been greatly promoted due to advances in deep convolutional neural networks (DNNs) [8,12]. However, DNN models comprise a large number of parameters and floating-point operations (FLOPs), restricting their deployment on embedded platforms. Techniques such

---

S. Xu, Y. Li and B. Zeng—Equal contribution.

Supplementary Information The online version contains supplementary material available at https://doi.org/10.1007/978-3-031-20083-0_21.

as compact network design [15,24], network pruning [13,16,38], low-rank decomposition [5], and quantization [26,33,36] have been developed to address these restrictions and accomplish an efficient inference on the detection task. Among these, binarized detectors have contributed to object detection by accelerating the CNN feature extracting for real-time bounding box localization and foreground classification [31,34,35]. For example, the 1-bit SSD300 [21] with VGG-16 backbone [28] theoretically achieve the acceleration rate up to 15× with XNOR and Bit-count operations using binarized weights and activations as described in [31]. With extremely high energy-efficiency for embedded devices, they are able to be installed directly on next-generation AI chips. Despite these appealing features, 1-bit detectors' performance often

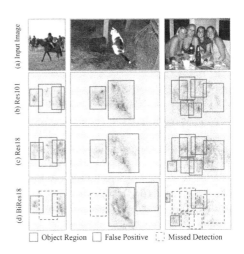

Fig. 1. Input images and saliency maps following [10]. Images are randomly selected from VOC test2007. Each row includes: (a) input images, saliency maps of (b) Faster-RCNN with ResNet-101 backbone (Res101), (c) Faster-RCNN with ResNet-18 backbone (Res18), (d) 1-bit Faster-RCNN with ResNet-18 backbone (BiRes18), respectively.

deteriorates to the point, which explains why they are not widely used in real-world embedded systems.

The recent art [35] employs fine-grained feature imitation (FGFI) [30] to enhance the performance of 1-bit detectors. However, it neglects the intrinsic information discrepancy between 1-bit detectors and real-valued detectors. As shown in Fig. 1, we demonstrate that saliency maps of real-valued Faster-RCNN of the ResNet-101 backbone (often used as the teacher network) and the ResNet-18 backbone, in comparison with 1-bit Faster-RCNN of the ResNet-18 backbone (often used as the student network) from top to bottom. They show that knowledge distillation (KD) methods like [30] are effective for distilling real-valued Faster-RCNNs, only when their teacher model and their student counterpart share small information discrepancy on proposals, as shown in Fig. 1 (b) and (c). This phenomenon does not happen for 1-bit Faster-RCNN, as shown in Fig. 1 (b) and (d). This might explain why existing KD methods are less effective in 1-bit detectors. A statistic on COCO and PASCAL VOC datasets in Fig. 2 show that the discrepancy between proposal saliency maps of Res101 and Res18 (blue) is much smaller than that of Res101 and BiRes18 (orange). That is to say, the smaller the distance is, the less the discrepancy is. Briefly, conventional KD methods show their effectiveness on distilling real-valued detectors but seem to be less effective on distilling 1-bit detectors.

In this paper, we are motivated by the above observation and present an information discrepancy-aware distillation for 1-bit detectors (IDa-Det), which can

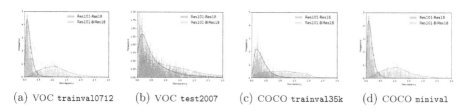

(a) VOC trainval0712   (b) VOC test2007   (c) COCO trainval35k   (d) COCO minival

Fig. 2. The Mahalanobis distance of the gradient in the intermediate neck feature between Res101-Res18 (blue) and Res101-BiRes18 (orange) in various datasets. (Color figure online)

effectively address the information discrepancy problem, leading to an efficient distillation process. As shown in Fig. 3, we introduce a discrepancy-aware method to select proposal pairs and facilitate distilling 1-bit detectors, rather than only using object anchor locations of student models or ground truth as in existing methods [10,30,35]. We further introduce a novel entropy distillation loss to leverage more comprehensive information than the conventional loss functions. By doing so, we achieve a powerful information discrepancy-aware distillation method for 1-bit detectors (IDa-Det). Our contributions are summarized as:

- Unlike existing KD methods, we distill 1-bit detectors by fully considering the information discrepancy into optimization, which is simple yet effective for learning 1-bit detectors.
- We propose an entropy distillation loss further to improve the representation ability of the 1-bit detector and effectively eliminate the information discrepancy.
- We compare our IDa-Det against state-of-the-art 1-bit detectors and KD methods on the VOC and large-scale COCO datasets. Extensive results reveal that our method outperformas the others by a considerable margin. For instance, on VOC test2007, the 1-bit Faster-RCNN with ResNet-18 backbone achieved by IDa-Det obtains 76.9% mAP, achieving a new state-of-the-art.

2   Related Work

1-Bit Detectors. By removing the foreground redundancy, BiDet [31] fully exploits the representational capability of the binarized convolutions. In this way, the information bottleneck is introduced, which limits the amount of data in high-level feature maps, while maximizing the mutual information between feature maps and object detection. The performance of the Faster R-CNN detector is significantly enhanced by the ASDA-FRCNN [34] which suppresses the shared amplitude between the real-value and binary kernels. LWS-Det [35] novelly proposes a layer-wise searching approach, minimizing the angular and amplitude errors for 1-bit detectors. Also, FGFI [30] is used by LWS-Det to distill the backbone feature map further.

Knowledge Distillation. Knowledge distillation (KD), a significant subset of model compression methods, aims to transfer knowledge from a well-trained

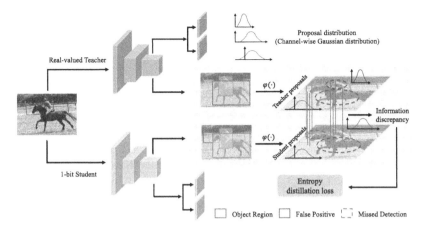

Fig. 3. Overview of the proposed information discrepancy-aware distillation (IDa-Det) framework. We first select representative proposal pairs based on the information discrepancy. Then we propose the entropy distillation loss to eliminate the information discrepancy.

teacher network to a more compact student model. The student is supervised using soft labels created by the teacher, as firstly proposed by [1]. Knowledge distillation is redefined by [14] as training a shallower network after the softmax layer to approximate the teacher's output. Object detectors can be compressed using knowledge distillation, according to numerous recent papers. Chen *et al.* [2] distill the student through all backbone features, regression head, and classification head, but both the imitation of whole feature maps and the distillation in classification head fail to add attention to the important foreground, potentially resulting in a sub-optimal result. Mimicking [17] distills the features from sampled region proposals. However, just replicating the aforementioned regions may lead to misdirection, because the proposals occasionally perform poorly. In order to distill the student, FGFI [30] introduces a unique attention mask to create fine-grained features from foreground object areas. DeFeat [10] balances the background and foreground object regions to efficiently distill the student.

In summary, existing KD frameworks for object detection can only be employed for real-valued students having similar information as their teachers. Thus, they are often ineffective in distilling 1-bit detectors. Unlike prior arts, we identify that the information discrepancy between real-valued teacher and 1-bit students are significant for distillation. We first introduce Mahalanobis distance to identify the information discrepancy and then accordingly distill the features. Meanwhile, we propose a novel entropy distillation loss to prompt the discrimination ability of 1-bit detectors further.

3 The Proposed Method

In this section, we describe our IDa-Det in detail. Firstly, we overview the 1-bit CNNs. We then describe how we employ the information discrepancy method

(IDa) to select representative proposals. Finally, we describe the entropy distillation loss to delicately eliminate the information discrepancy between the real-valued teachers and the 1-bit students.

3.1   Preliminaries

In a specific convolution layer, $\mathbf{w} \in \mathbb{R}^{C_{out} \times C_{in} \times K \times K}$, $\mathbf{a}_{in} \in \mathbb{R}^{C_{in} \times W_{in} \times H_{in}}$, and $\mathbf{a}_{out} \in \mathbb{R}^{C_{out} \times W_{out} \times H_{out}}$ represent its weights and feature maps, where C_{in} and C_{out} represents the number of channels. (H, W) are the height and width of the feature maps, and K denotes the kernel size. We then have

$$\mathbf{a}_{out} = \mathbf{a}_{in} \otimes \mathbf{w}, \tag{1}$$

where $\otimes$ is the convolution operation. We omit the batch normalization (BN) and activation layers for simplicity. The 1-bit model aims to quantize $\mathbf{w}$ and $\mathbf{a}_{in}$ into $\mathbf{b^w} \in \{-1, +1\}^{C_{out} \times C_{in} \times K \times K}$ and $\mathbf{b^{a_{in}}} \in \{-1, +1\}^{C_{in} \times H \times W}$ using the efficient XNOR and Bit-count operations to replace full-precision operations. Following [3,4], the forward process of the 1-bit CNN is

$$\mathbf{a}_{out} = \boldsymbol{\alpha} \circ \mathbf{b^{a_{in}}} \odot \mathbf{b^w}, \tag{2}$$

where $\odot$ is the XNOR, and bit-count operations and $\circ$ denotes the channel-wise multiplication. $\boldsymbol{\alpha} = [\alpha_1, \cdots, \alpha_{C_{out}}] \in \mathbb{R}_+$ is the vector consisting of channel-wise scale factors. $\mathbf{b} = \text{sign}(\cdot)$ denotes the binarized variable using the sign function, which returns 1 if the input is greater than zero, and -1 otherwise. It then enters several non-linear layers, *e.g.*, BN layer, non-linear activation layer, and max-pooling layer. We omit these for simplification. And then, the output $\mathbf{a}_{out}$ is binarized to $\mathbf{b^{a_{out}}}$ via the sign function. The fundamental objective of BNNs is calculating $\mathbf{w}$. We want it to be as close as possible before and after binarization, such that the binarization effect is minimized. Then, we define the reconstruction error as

$$L_R(\mathbf{w}, \boldsymbol{\alpha}) = \mathbf{w} - \boldsymbol{\alpha} \circ \mathbf{b^w}. \tag{3}$$

3.2   Select Proposals with Information Discrepancy

To eliminate the large magnitude scale difference between the real-valued teacher and the 1-bit student, we introduce a channel-wise transformation for the proposals[1] of the intermediate neck. We first apply a transformation $\varphi(\cdot)$ on a proposal $\tilde{R}_n \in \mathbb{R}^{C \times W \times H}$ and have

$$R_{n;c}(x, y) = \varphi(\tilde{R}_{n;c}(x, y)) = \frac{\exp(\frac{\tilde{R}_{n;c}(x,y)}{T})}{\sum_{(x',y') \in (W,H)} \exp(\frac{\tilde{R}_{n;c}(x'y')}{T})}, \tag{4}$$

---

[1] In this paper, the proposal denotes the neck/backbone feature map patched by the region proposal of detectors.

where $(x, y) \in (W, H)$ denote a specific spatial location (x, y) in spatial range (W, H), and $c \in \{1, \cdots, C\}$ is the channel index. $n \in \{1, \cdots, N\}$ is the proposal index. N denotes the number of proposals. $\mathcal{T}$ denotes a hyper-parameter controlling the statistical attributions of the channel-wise alignment operation[2]. After the transformation, features in each channel of a proposal projected into the same feature space [29] and follow a Gaussian distribution as

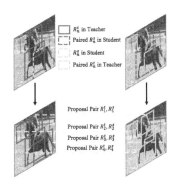

$$p(R_{n;c}) \sim \mathcal{N}(\mu_{n;c}, \sigma_{n;c}^2). \tag{5}$$

Fig. 4. Illustration for the generation of the proposal pairs. Each single proposal in one model generates a counterpart feature map patch in the same location of the other model.

We further evaluate the information discrepancy between the proposals of the teacher and the student. As shown in Fig. 4, the teacher and student have N_T and N_S proposals, respectively. Each single proposal in one model generates a counterpart feature map patch in the same location of the other model, thus total $N_T + N_S$ proposal pairs are considered. To evaluate the information discrepancy, we introduce the Mahalanobis distance of each channel-wise proposal feature and measure the discrepancy as

$$\varepsilon_n = \sum_{c=1}^{C} ||(R_{n;c}^t - R_{n;c}^s)^T \Sigma_{n;c}^{-1}(R_{n;c}^t - R_{n;c}^s)||_2, \tag{6}$$

where $\Sigma_{n;c}$ denotes the covariance matrix of the teacher and student in the c-th channel of the n-th proposal pair. The Mahalanobis distance takes both the pixel-level distance between proposals and the statistical characteristics differences in proposal pairs into account.

To select representative proposals with maximum information discrepancy, we first define a binary distillation mask m_n as

$$m_n = \begin{cases} 1, & \text{if pair}(R_n^t, R_n^s) \text{is selected} \\ 0, & \text{otherwise} \end{cases} \tag{7}$$

where $m_n = 1$ denotes distillation will be applied on such proposal pair, otherwise remain unchanged. For each proposal pair, only when their distribution is quite different, the student model can learn from the teacher counterpart, where a distillation process is needed.

Based on the derivation above, discrepant proposal pairs will be optimized through distillation. For distilling the selected pairs, we resort to maximize conditional probability $p(R_n^s | R_n^t)$. That is to say, after distillation or optimization, feature distributions of teacher proposal and student counterpart become similar

---

[2] In this paper, we set $\mathcal{T} = 4$.

with each other. To this end, we define $p(R_n^s|R_n^t)$ with $m_n, n \in \{1, \cdots, N_T+N_S\}$ in consideration as

$$p(R_n^s|R_n^t; m_n) \sim m_n \mathcal{N}(\mu_n^t, \sigma_n^{t\,2}) + (1 - m_n)\mathcal{N}(\mu_n^s, \sigma_n^{s\,2}). \tag{8}$$

Subsequently, we introduce a bi-level optimization formulation to solve the distillation problem as

$$\max_{R_n^s} p(R_n^s|R_n^t; \mathbf{m}^*), \quad \forall n \in \{0, \cdots, N_T + N_S\},$$
$$\text{s.t.} \ \ \mathbf{m}^* = \arg\max_{\mathbf{m}} \sum_{n=1}^{N_T+N_S} m_n \varepsilon_n, \tag{9}$$

where $\mathbf{m} = [m_1, \cdots, m_{N_T+N_S}]$ and $||\mathbf{m}||_0 = \gamma \cdot (N_T + N_S)$. γ is a hyper-parameter. In this way, we select $\gamma \cdot (N_T + N_S)$ pairs of proposals containing the most representative information discrepancy for distillation. γ controls the proportion of discrepant proposal pairs, further validated in Sect. 4.2.

For each iteration, we first solve the inner-level optimization, $i.e.$, proposal selection, via exhaustive sorting [32]; and then solve the upper-level optimization, distilling the selected pair, based on entropy distillation loss discussed in Sect. 3.3. Considering that there are not too many proposals involved, the process is relatively efficient for the inner-level optimization.

3.3 Entropy Distillation Loss

After selecting a specific number of proposals, we crop the feature based on the proposals we obtained. Most of the SOTA detection models are based on Feature Pyramid Networks (FPN) [19], which can significantly improve the robustness of multi-scale detection. For the Faster-RCNN framework in this paper, we resize the proposals and crop the features from each stage of the neck feature maps. We generate the proposals from the regression layer of the SSD framework and crop the features from the feature map of maximum spatial size. Then we formulate the entropy distillation process as

$$\max_{R_n^s} \ p(R_n^s|R_n^t). \tag{10}$$

Here is **the upper level** of the bi-level optimization, where $\mathbf{m}$ is solved and therefore omitted. We rewrite Eq. 10 and further achieve our entropy distillation loss as

$$L_P(\mathbf{w}, \boldsymbol{\alpha}; \gamma) = (R_n^s - R_n^t) + \text{Cov}(R_n^s, R_n^t)^{-1}(R_n^s - R_n^t)^2 + \log(\text{Cov}(R_n^s, R_n^t)), \tag{11}$$

where $\text{Cov}(R_n^s, R_n^t) = \mathbb{E}(R_n^s R_n^t) - \mathbb{E}(R_n^s)\mathbb{E}(R_n^t)$ denotes the covariance matrix.

Hence, we trained the 1-bit student model end-to-end, the total loss for distilling the student model is defined as

$$L = L_{GT}(\mathbf{w}, \boldsymbol{\alpha}) + \lambda L_P(\mathbf{w}, \boldsymbol{\alpha}; \gamma) + \mu L_R(\mathbf{w}, \boldsymbol{\alpha}), \tag{12}$$

where L_{GT} is the detection loss derived from the ground truth label and L_R is defined in Eq. 3.

4   Experiments

On two mainstream object detection datasets, *i.e.,* PASCAL VOC [6] and COCO [20], extensive experiments are undertaken to test our proposed method. First, we go through the implementation specifics of our IDa-Det. Then, in the ablation studies, we set different hyper-parameters and validate the effectiveness of the components as well as the convergence of our method. Moreover, we illustrate the superiority of our IDa-Det by comparing it to previous state-of-the-art 1-bit CNNs and other KD approaches on 1-bit detectors. Finally, we analyze the deploy efficiency of our IDa-Det on hardware.

4.1   Datasets and Implementation Details

PASCAL VOC. Natural images from 20 different classes are included in the VOC datasets. We use the VOC `trainval2012` and VOC `trainval2007` sets to train our model, which contain around 16k images, and the VOC `test2007` set to evaluate our IDa-Det, which contains 4952 images. We utilize the mean average precision (mAP) as the evaluation matrices, as suggested by [6].

COCO. All our experiments on COCO dataset are conducted on the COCO 2014 [20] object detection track in the training stage, which contains the combination of 80k images and 80 different categories from the COCO `train2014` and 35k images sampled from COCO `val2014`, *i.e.,* COCO `trainval35k`. Then we evaluate our method on the remaining 5k images from the COCO `minival`. We list the average precision (AP) for IoUs$\in [0.5 : 0.05 : 0.95]$, designated as mAP@[.5, .95], using COCO's standard evaluation metric. For further analyzing our method, we also list AP_{50}, AP_{75}, AP_s, AP_m, and AP_l.

Implementation Details. Our IDa-Det is trained with two mainstream object detectors, *i.e.,* two-stage Faster-RCNN[3] [27] and one-stage SSD [21]. In Faster-RCNN, we utilize ResNet-18 and ResNet-34 [12] as backbones. And we utilize VGG-16 [28] as the backbone of SSD framework. PyTorch [25] is used for implementing IDa-Det. We run the experiments on 4 NVIDIA GTX 2080Ti GPUs with 11 GB memory and 128 GB of RAM. We use ImageNet ILSVRC12 to pre-train the backbone of a 1-bit student, following [23]. The SGD optimizer is utilized and the batch size is set as 24 for SSD and 4 for Faster-RCNN, respectively.

We keep the shortcut, first layer, and the last layer (the 1×1 convolution layer of RPN and a FC layer of the bbox head) in the detectors as real-valued on Faster-RCNN framework after implementing 1-bit CNNs following [23]. Following BiDet [31], the extra layer is likewise retained as real-valued for the SSD framework. Following [31] and [9], we modify the network of ResNet-18/34 with an extra shortcut and PReLU [11].

The architecture of VGG-16 is modified with extra residual connections, following [31]. The lateral connection of the FPN [19] neck is replaced with 3×3 1-bit convolution for improving performance. This adjustment is implemented in

---

[3] In this paper, Faster-RCNN denotes the Faster-RCNN implemented with FPN neck.

all of the Faster-RCNN experiments. For Faster-RCNN, we train the model with two stages. Only the backbone is binarized at the first stage. Then we binarize all layers in the second stage. Each stage counts 12 epochs. The learning rate is set as 0.004 and decays by multiplying 0.1 in the 9-th and 11-th epochs. We use a loss coefficient set as 5 and multi-scale training method. For fair comparison, all the methods in this paper are implemented with the same training setup.

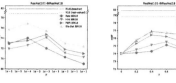

(a) Effect of μ.   (b) Effect of λ and γ.

Fig. 5. On VOC, we (a) select μ on raw detector and different KD methods including Hint [2], FGFI [30], and IDa-Det; (b) select λ and γ on IDa-Det with μ set as $1e{-}4$.

The real-valued counterparts in this paper are also trained for 24 epochs for fair comparison. For SSD, the model is trained for 24 epochs with a learning rate of 0.01, which decays to 0.1 at the 16-th and 22-nd epochs by multiplying 0.1.

We select real-valued Faster-RCNN with ResNet101 backbone (81.9% mAP on VOC and 39.8% mAP on COCO) and real-valued SSD300 with VGG16 backbone (74.5% mAP on VOC and 25.0% mAP on COCO) as teacher network.

4.2   Ablation Study

Selecting the Hyper-parameter. As mentioned above, we select hyper-parameters λ, γ, and μ in this part. We first select the μ, which controls the binarization process. As plotted in Fig. 5 (a), we first fine-tune the hyper-parameter μ controlling the binarization process in four situations: raw BiRes18, and BiRes18 distilled via Hint [2], FGFI [30], and our IDa-Det, respectively. Overall, the performances increase first and then decrease when increasing the value of μ. On raw BiRes18 and IDa-Det BiRes18, the 1-bit student achieves the best performance when μ is set as $1e{-}4$. And μ valued $1e{-}3$ is better for the Hint, and the FGFI distilled 1-bit student. Thus, we set μ as $1e{-}4$ for an extended ablation study. Figure 5 (b) shows that the performances increase first and then decrease with the increase of λ from left to right. In general, the IDa-Det obtains better performances with λ set as 0.4 and 0.6. With varying value of γ, we find $\{\lambda, \gamma\}$ = $\{0.4, 0.6\}$ boost the performance of IDa-Det most, achieving 76.9% mAP on VOC test2007. Based on the ablative study above, we set hyper-parameters λ, γ, and μ as 0.4, 0.6, and $1e{-}4$ for the experiments in this paper.

Effectiveness of Components. We first compare our information discrepancy-aware (IDa) proposal selecting method with other methods to select proposals: Hint [2] (using the neck feature without region mask) and FGFI [30]. We show the effectiveness of IDa on two-stage Faster-RCNN in Table 1. On the Faster-RCNN, the introducing of IDa achieves improvements of the mAP by 2.5%, 2.4%, and 1.8% compared to non-distillation, Hint, and FGFI, under the same student-teacher framework. Then we evaluate the proposed entropy distillation loss against the conventional ℓ_2 loss, inner-product loss, and cosine similarity

Table 1. The effects of different components in IDa-Det with Faster-RCNN model on PASCAL VOC dataset. Hint [2] and FGFI [30] are used to compare with our information discrepancy-aware proposal selection (IDa). IDa and Entropy loss denote main components of the proposed IDa-Det.

Model	Proposal selection	Distillation method	mAP
Res18	✗	✗	78.6
BiRes18	✗	✗	74.0
Res101-BiRes18	Hint	ℓ_2	74.1
Res101-BiRes18	Hint	Entropy loss	74.5
Res101-BiRes18	FGFI	ℓ_2	74.7
Res101-BiRes18	FGFI	Entropy loss	75.0
Res101-BiRes18	IDa	Inner-product	74.8
Res101-BiRes18	IDa	Consine similarity	76.4
Res101-BiRes18	IDa	ℓ_2	76.5
Res101-BiRes18	**IDa**	**Entropy loss**	**76.9**

loss. As depicted in Table 1, our entropy distillation loss improves the distillation performance by 0.4%, 0.3%, and 0.4% with Hint, FGFI, and IDa method compared with ℓ_2 loss. Compared with inner-product and cosine similarity loss, entropy loss outperforms them by 2.1% and 0.5% in mAP on our framework, which further reflects the effectiveness of our method.

4.3   Results on PASCAL VOC

With the same student framework, we compare our IDa-Det with the state-of-the-art 1-bit ReActNet [23] and other KD methods, such as FGFI [30], DeFeat [10], and LWS-Det [35], in the task of object detection with the VOC datasets. The detection results of the multi-bit quantization method DoReFa-Net [37] is also reported. We use the input resolution following [35], *i.e.* 1000×600 for Faster-RCNN and 300×300 for SSD.

Table 2 lists the comparison of several quantization approaches and detection frameworks in terms of computing complexity, storage cost, and the mAP. Our IDa-Det significantly accelerates computation and reduces storage requirements for various detectors. We follow XNOR-Net [26] to calculate memory usage, which is estimated by adding $32\times$ the number of full-precision kernels and $1\times$ the number of binary kernels in the networks. The number of float operations (FLOPs) is calculated in the same way as Bi-Real-Net [22]. The current CPUs can handle both bit-wise XNOR and bit-count operations in parallel. The number of real-valued FLOPs plus $\frac{1}{64}$ of the number of 1-bit multiplications equals the OPs following [22].

Faster-RCNN. We summarize the experimental results on VOC test2007 of 1-bit Faster-RCNNs from lines 2 to 17 in Table 2. Compared with raw ReActNet

Table 2. We report memory usage, FLOPs, and mAP (%) with state-of-the-art 1-bit detectors, other KD methods on VOC `test2007`. The best results are **bold**.

Framework	Backbone	Quantization Method	KDMethod	W/A	Memory Usage (MB)	OPs ($\times 10^9$)	mAP
Faster-RCNN	ResNet-18	Real-valued	✗	32/32	112.88	96.40	78.8
		DoReFa-Net	✗	4/4	21.59	27.15	73.3
		ReActNet	✗	1/1	16.61	18.49	69.6
		Ours	✗				74.0
		LWS-Det					73.2
		Ours	FGFI				74.7
		Ours	DeFeat				74.9
		IDa-Det					**76.9**
	ResNet-34	Real-valued	✗	32/32	145.12	118.80	80.0
		DoReFa-Net	✗	4/4	29.65	32.31	75.6
		ReActNet	✗	1/1	24.68	21.49	72.3
		Ours	✗				75.0
		Ours	FGFI				75.4
		Ours	DeFeat				75.7
		LWS-Det					75.8
		IDa-Det					**78.0**
SSD	VGG-16	Real-valued	✗	32/32	105.16	31.44	74.3
		DoReFa-Net	✗	4/4	29.58	6.67	69.2
		ReActNet	✗	1/1	21.88	2.13	68.4
		Ours	✗				69.5
		LWS-Det					69.5
		Ours	FGFI				70.0
		Ours	DeFeat				71.4
		IDa-Det					**72.5**

[23], our baseline binarization method achieves 4.4%, and 2.7% mAP improvement with ResNet-18/34 backbone respectively. Compared with other KD methods on the 1-bit detector with the same train and test settings, our IDa-Det surpasses FGFI and DeFeat distillation method in a clear margin of 2.2%, 2.0% with ResNet-18 backbone, and 2.6%, 2.3% with ResNet-34 backbone. Our IDa-Det significantly surpasses the prior state-of-the-art, LWS-Det, by 3.7% in ResNet-18 backbone, and 2.2% in ResNet-34 backbone with the same FLOPs and memory usage. All of the improvements have impacts on object detection.

Compared with the raw real-valued detectors, the proposed IDa-Det surpasses real-valued Faster-RCNN with ResNet-18/34 backbone ({76.9%, 78.0%} vs. {76.4%, 77.8%}) by obviously computation acceleration and storage savings by 5.21×/5.53× and 6.80×/5.88×. The above results are of great significance in the real-time inference of object detection.

SSD. On the SSD300 framework with a VGG-16 backbone, our IDa-Det can accelerate computation and save storage by 14.76× and 4.81× faster than

(a) False positives   (b) Missed detection

Fig. 6. Qualitative results on the gain from information discrepancy-aware distilling. The top row shows IDa-Det student's output. The bottom row images are raw student model's output without information discrepancy-aware distilling.

real-valued counterparts, respectively, as illustrated in the bottom section of Table 2. The drop in the performance is relatively minor (72.5% *vs.* 74.3%). Also, our method surpasses other 1-bit networks and KD methods by a sizable margin. Compared to 1-bit ReActNet, our raw 1-bit model can achieve 1.1% higher mAP with the same computation. Compared with FGFI, DeFeat, and LWS-Det, our IDa-Det exceeds them by 3.0%, 2.5%, and 1.1%, respectively.

As shown in Fig. 6, BiRes18 achieved by IDa-Det effectively eliminates the false positives and missed detections compared with raw BiRes18. In summary, we achieve new state-of-the-art performance on PASCAL VOC compared to previous 1-bit detectors and KD methods on various frameworks and backbones. We also achieve competitive results, demonstrating the IDa-Det's superiority.

4.4   Results on COCO

Because of its diversity and scale, the COCO dataset presents a greater challenge in the object detection task, compared with PASCAL VOC. On COCO, we compare our proposed IDa-Det with the state-of-the-art 1-bit ReActNet [23], as well as the KD techniques FGFI [30], DeFeat [10], and LWS-Det [35]. We also report the detection result of the 4-bit quantization method FQN [18] and the DoReFa-Net [37] for reference. Following [35], the input images are scaled to 1333×800 for Faster-RCNN and 300×300 for SSD.

The mAP, AP with different IoU thresholds, and AP of objects with varying scales are all reported in Table 3. Due to the constraints in the width of page, we do not report the FLOPs and memory use in Table 3. We conduct experiments on Faster-RCNN and SSD detectors, the results of which are presented in the folllowing two parts.

Faster-RCNN. Comparing to the state-of-the-art 1-bit ReActNet, our baseline binarized model achieves a 5.7% improvement on mAP@[.5, .95] with the ResNet-18 backbone. Compared to state-of-the-art LWS-Det, FGFI, and DeFeat, our IDa-Det prompts the mAP@[.5, .95] by 2.4%, 1.8%, and 1.4%, respectively. With the ResNet-34 backbone, the proposed IDa-Det surpasses FGFI, DeFeat, and LWS-Det by 1.1%, 0.7%, and 0.6%, respectively. IDa-Det, nevertheless, has substantially reduced FLOPs and memory use.

Table 3. Comparison with state-of-the-art 1-bit detectors and KD methods on COCO `minival`. Optimal results are **bold**.

Framework	Backbone	Quantization Method	KD Method	W/A	mAP @[.5,.95]	AP_{50}	AP_{75}	AP_s	AP_m	AP_l
Faster-RCNN	ResNet-18	Real-valued	✗	32/32	32.2	53.8	34.0	18.0	34.7	41.9
		FQN	✗	4/4	28.1	48.4	29.3	14.5	30.4	38.1
		ReActNet	✗	1/1	21.1	38.5	20.5	9.7	23.5	32.1
		Ours	✗		26.8	46.1	27.9	14.7	28.4	36.0
		LWS-Det			26.9	44.9	27.7	12.9	28.7	38.3
		Ours	FGFI		27.5	46.5	28.8	15.2	28.7	37.5
		Ours	DeFeat		27.9	46.9	29.3	15.8	29.0	38.2
		IDa-Det			**29.3**	**48.7**	**30.9**	**16.7**	**29.8**	**39.9**
	ResNet-34	Real-valued	✗	32/32	35.8	57.6	38.4	21.1	39.0	46.1
		FQN	✗	4/4	31.8	52.9	33.9	17.6	34.4	42.2
		ReActNet	✗	1/1	23.4	43.3	24.4	10.7	25.9	35.5
		Ours	✗		29.0	47.7	30.9	16.6	30.5	39.0
		Ours	FGFI		29.4	48.4	30.3	17.1	30.7	39.7
		Ours	DeFeat		29.8	48.7	30.9	17.6	31.4	40.5
		LWS-Det			29.9	49.2	30.1	15.1	**32.1**	**40.9**
		IDa-Det			**30.5**	**49.2**	**31.8**	**17.7**	31.3	40.6
SSD	VGG-16	Real-valued	✗	32/32	23.2	41.2	23.4	5.3	23.2	39.6
		DoReFa-Net	✗	4/4	19.5	35.0	19.6	5.1	20.5	32.8
		ReActNet	✗	1/1	15.3	30.0	13.2	5.4	16.3	25.0
		Ours	✗		17.2	31.5	16.8	3.2	18.2	31.3
		LWS-Det			17.1	32.9	16.1	**5.5**	17.4	26.7
		Ours	FGFI		17.7	32.3	17.3	3.3	18.9	31.8
		Ours	DeFeat		18.1	32.8	17.9	3.3	19.4	32.6
		IDa-Det			**19.4**	**34.5**	**19.3**	3.7	**21.1**	**35.0**

SSD. On the SSD300 framework, our IDa-Det achieves 19.4% mAP@[.5, .95] with the VGG-16 backbone, surpassing LWS-Det, FGFI, and DeFeat by 2.3%, 1.7%, and 1.3% mAP, respectively.

To summarize, our method outperforms baseline quantization methods and other KD methods in terms of the AP with various IoU thresholds and the AP for objects of varied sizes on COCO, indicating IDa-Det's superiority and generality in many application settings.

4.5 Deployment Efficiency

We implement the 1-bit models achieved by our IDa-Det on ODROID C4, which has a 2.016 GHz 64-bit quad-core ARM Cortex-A55. With evaluating its real speed in practice, the efficiency of our IDa-Det is proved when deployed into real-world mobile devices. We use the SIMD instruction SSHL on ARM NEON, for making inference framework BOLT [7] compatible with our IDa-Det. We compare our IDa-Det to the real-valued backbones in Table 4. We utilize VOC dataset for testing the model. For Faster-RCNN, the input images were scaled

Table 4. Comparison of time cost of real-valued and 1-bit models (Faster-RCNN and SSD) on hardware (single thread).

Framework	Network	Method	W/A	Latency (ms)	Acceleration
Faster-RCNN	ResNet-18	Real-valued	32/32	12043.8	–
		IDa-Det	1/1	2474.4	4.87×
	ResNet-34	Real-valued	32/32	14550.2	–
		IDa-Det	1/1	2971.3	4.72×
SSD	VGG-16	Real-valued	32/32	2788.7	–
		IDa-Det	1/1	200.5	13.91×

to 1000×600 and 300×300 for SSD. We can plainly see that IDa-Det's inference speed is substantially faster with the highly efficient BOLT framework. For example, the acceleration rate achieves about 4.7× on Faster-RCNN, which is slightly lower than the theoretical acceleration rate discussed in Sect. 4.3. Furthermore, IDa-Det achieves 13.91× acceleration with SSD. All deployment results in the object detection are significant on real-world edge devices.

5   Conclusion

This paper presents a novel method for training 1-bit detectors with knowledge distillation to minimize the information discrepancy. IDa-Det employs a maximum entropy model to select the proposals with maximum information discrepancy and proposes a novel entropy distillation loss to supervise the information discrepancy. As a result, our IDa-Det significantly boosts the performance of 1-bit detectors. Extensive experiments show that IDa-Det surpasses state-of-the-art 1-bit detectors and other knowledge distillation methods in object detection.

Acknowledgement. This work was supported in part by the National Natural Science Foundation of China under Grant 62076016, 92067204, 62141604 and the Shanghai Committee of Science and Technology under Grant No. 21DZ1100100.

References

1. Ba, J., Caruana, R.: Do deep nets really need to be deep? In: Proceedings of NeurIPS Workshop (2014)
2. Chen, G., Choi, W., Yu, X., Han, T., Chandraker, M.: Learning efficient object detection models with knowledge distillation. In: Proceedings of NeurIPS (2017)
3. Courbariaux, M., Bengio, Y., David, J.P.: Binaryconnect: training deep neural networks with binary weights during propagations. In: Proceedings of NeurIPS (2015)
4. Courbariaux, M., Hubara, I., Soudry, D., El-Yaniv, R., Bengio, Y.: Binarized neural networks: training deep neural networks with weights and activations constrained to+ 1 or-1. arXiv (2016)

5. Denil, M., Shakibi, B., Dinh, L., Ranzato, M., De Freitas, N.: Predicting parameters in deep learning. In: Proceedings of NeurIPS (2013)
6. Everingham, M., Van Gool, L., Williams, C.K., Winn, J., Zisserman, A.: The pascal visual object classes (VOC) challenge. Int. J. Comput. Vis. **88**(2), 303–338 (2010). https://doi.org/10.1007/s11263-009-0275-4
7. Feng, J.: Bolt. https://github.com/huawei-noah/bolt (2021)
8. Gao, P., Ma, T., Li, H., Dai, J., Qiao, Y.: Convmae: masked convolution meets masked autoencoders. arXiv preprint arXiv:2205.03892 (2022)
9. Gu, J., et al.: Convolutional neural networks for 1-bit CNNs via discrete back propagation. In: Proceedings of AAAI (2019)
10. Guo, J., et al.: Distilling object detectors via decoupled features. In: Proceedings of CVPR (2021)
11. He, K., Zhang, X., Ren, S., Sun, J.: Delving deep into rectifiers: surpassing human-level performance on ImageNet classification. In: Proceedings of ICCV (2015)
12. He, K., Zhang, X., Ren, S., Sun, J.: Deep residual learning for image recognition. In: Proceedings of CVPR (2016)
13. He, Y., Kang, G., Dong, X., Fu, Y., Yang, Y.: Soft filter pruning for accelerating deep convolutional neural networks. In: Proceedings of IJCAI (2018)
14. Hinton, G., Oriol, Dean, J.: Distilling the knowledge in a neural network. In: Proceedings of NeurIPS (2014)
15. Howard, A.G., et al.: MobileNets: efficient convolutional neural networks for mobile vision applications. In: Proceedings of CVPR (2017)
16. Li, H., Kadav, A., Durdanovic, I., Samet, H., Graf, H.P.: Pruning filters for efficient convnets. In: Proceedings of ICLR (2016)
17. Li, Q., Jin, S., Yan, J.: Mimicking very efficient network for object detection. In: Proceedings of CVPR (2017)
18. Li, R., Wang, Y., Liang, F., Qin, H., Yan, J., Fan, R.: Fully quantized network for object detection. In: Proceedings of CVPR (2019)
19. Lin, T.Y., Dollár, P., Girshick, R., He, K., Hariharan, B., Belongie, S.: Feature pyramid networks for object detection. In: Proceedings of CVPR (2017)
20. Lin, T.-Y., et al.: Microsoft COCO: common objects in context. In: Fleet, D., Pajdla, T., Schiele, B., Tuytelaars, T. (eds.) ECCV 2014. LNCS, vol. 8693, pp. 740–755. Springer, Cham (2014). https://doi.org/10.1007/978-3-319-10602-1_48
21. Liu, W., et al.: SSD: single shot multibox detector. In: Leibe, B., Matas, J., Sebe, N., Welling, M. (eds.) ECCV 2016. LNCS, vol. 9905, pp. 21–37. Springer, Cham (2016). https://doi.org/10.1007/978-3-319-46448-0_2
22. Liu, Z., Luo, W., Wu, B., Yang, X., Liu, W., Cheng, K.T.: Bi-real net: binarizing deep network towards real-network performance. Int. J. Comput. Vision **128**(1), 202–219 (2020)
23. Liu, Z., Shen, Z., Savvides, M., Cheng, K.-T.: ReActNet: towards precise binary neural network with generalized activation functions. In: Vedaldi, A., Bischof, H., Brox, T., Frahm, J.-M. (eds.) ECCV 2020. LNCS, vol. 12359, pp. 143–159. Springer, Cham (2020). https://doi.org/10.1007/978-3-030-58568-6_9
24. Ma, N., Zhang, X., Zheng, H.T., Sun, J.: Shufflenet v2: practical guidelines for efficient CNN architecture design. In: Proceedings of ECCV (2018)
25. Paszke, A., et al.: Automatic differentiation in pytorch. In: NeurIPS Workshops (2017)
26. Rastegari, M., Ordonez, V., Redmon, J., Farhadi, A.: XNOR-Net: ImageNet classification using binary convolutional neural networks. In: Leibe, B., Matas, J., Sebe, N., Welling, M. (eds.) ECCV 2016. LNCS, vol. 9908, pp. 525–542. Springer, Cham (2016). https://doi.org/10.1007/978-3-319-46493-0_32

27. Ren, S., He, K., Girshick, R., Sun, J.: Faster R-CNN: towards real-time object detection with region proposal networks. IEEE Trans. Pattern Anal. Mach. Intell. (2016)
28. Simonyan, K., Zisserman, A.: Very deep convolutional networks for large-scale image recognition. In: Proceedings of ICLR (2015)
29. Wang, G.H., Ge, Y., Wu, J.: Distilling knowledge by mimicking features. IEEE Trans. Pattern Anal. Mach. Intell. (2021)
30. Wang, T., Yuan, L., Zhang, X., Feng, J.: Distilling object detectors with fine-grained feature imitation. In: Proceedings of CVPR (2019)
31. Wang, Z., Wu, Z., Lu, J., Zhou, J.: BiDet: an efficient binarized object detector. In: Proceedings of CVPR (2020)
32. Wu, N.: The Maximum Entropy Method, vol. 32. Springer, Heidelberg (2012)
33. Xu, S., Li, Y., Zhao, J., Zhang, B., Guo, G.: Poem: 1-bit point-wise operations based on expectation-maximization for efficient point cloud processing. In: Proceedings of BMVC, pp. 1–10 (2021)
34. Xu, S., Liu, Z., Gong, X., Liu, C., Mao, M., Zhang, B.: Amplitude suppression and direction activation in networks for 1-bit faster R-CNN. In: Proceedings of EMDL (2020)
35. Xu, S., Zhao, J., Lu, J., Zhang, B., Han, S., Doermann, D.: Layer-wise searching for 1-bit detectors. In: Proceedings of CVPR (2021)
36. Zhao, J., Xu, S., Zhang, B., Gu, J., Doermann, D., Guo, G.: Towards compact 1-bit CNNs via Bayesian learning. Int. J. Comput. Vis. **130**(2), 201–225 (2022)
37. Zhou, S., Wu, Y., Ni, Z., Zhou, X., Wen, H., Zou, Y.: DoReFa-Net: training low bitwidth convolutional neural networks with low bitwidth gradients. arXiv (2016)
38. Zhuo, L., et al.: Cogradient descent for bilinear optimization. In: Proceedings of CVPR (2020)

Learning to Weight Samples for Dynamic Early-Exiting Networks

Yizeng Han[1], Yifan Pu[1], Zihang Lai[2], Chaofei Wang[1], Shiji Song[1], Junfeng Cao[3], Wenhui Huang[3], Chao Deng[3], and Gao Huang[1]([✉])

[1] Tsinghua University, Beijing 100084, China
{hanyz18,pyf20,wangcf18}@mails.tsinghua.edu.cn,
{shijis,gaohuang}@tsinghua.edu.cn
[2] Carnegie Mellon University, Pennsylvania 15213, USA
zihangl@andrew.cmu.edu
[3] China Mobile Research Institute, Beijing 100084, China
{caojunfeng,huangwenhui,dengchao}@chinamobile.com

Abstract. Early exiting is an effective paradigm for improving the inference efficiency of deep networks. By constructing classifiers with varying resource demands (the exits), such networks allow *easy* samples to be output at early exits, removing the need for executing deeper layers. While existing works mainly focus on the architectural design of multi-exit networks, the training strategies for such models are largely left unexplored. The current state-of-the-art models treat all samples *the same* during training. However, the *early-exiting* behavior during testing has been ignored, leading to a gap between training and testing. In this paper, we propose to bridge this gap by *sample weighting*. Intuitively, *easy* samples, which generally exit early in the network during inference, should contribute more to training early classifiers. The training of *hard* samples (mostly exit from deeper layers), however, should be emphasized by the late classifiers. Our work proposes to adopt a *weight prediction network* to weight the loss of different training samples at each exit. This weight prediction network and the backbone model are jointly optimized under a *meta-learning* framework with a novel optimization objective. By bringing the adaptive behavior during inference into the training phase, we show that the proposed weighting mechanism consistently improves the trade-off between classification accuracy and inference efficiency. Code is available at https://github.com/LeapLabTHU/L2W-DEN.

Keywords: Sample weighting · Dynamic early exiting · Meta-learning

---

Y. Han and Y. Pu—Equal contribution.
Z. Lai—Work done during an internship at Tsinghua University.

Supplementary Information The online version contains supplementary material available at https://doi.org/10.1007/978-3-031-20083-0_22.

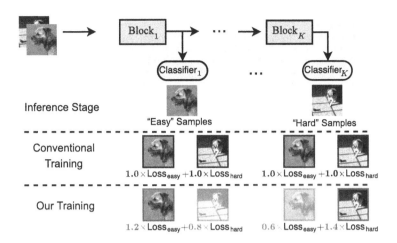

Fig. 1. Our sample weighting strategy. At test time, the classifiers at varying depths handle inputs with different complexity (*top*). However, the conventional training strategy (*middle*) treats all the samples equally at multiple exits. In contrast, our weighting mechanism (*bottom*) enables each classifier to emphasize the samples that it is responsible for.

1   Introduction

Although significant improvements have been achieved by deep neural networks in computer vision [7,13,18,25,33,39,40], the high computational cost of deep models still prevents them from being applied on resource-constrained platforms, such as mobile phones and wearable devices. Improving the inference efficiency of deep learning has become a research trend. Popular solutions include lightweight architecture design [16,49], network pruning [27,28,34,47], weight quantization [10,20], and dynamic neural networks [1,11,12,17,31,36,43,45,46].

Dynamic networks have attracted considerable research interests due to their favorable efficiency and representation power [11]. In particular, they perform a data-dependent inference procedure, and different network components (e.g. layers [45] or channels [1]) could be conditionally skipped based on the input complexity. A typical adaptive inference approach is *early exiting* [17,46], which can be achieved by constructing a deep network with multiple intermediate classifiers (early exits). Once the prediction from an early exit satisfies a certain criterion (e.g. the classification *confidence* exceeds some threshold), the forward propagation is terminated, and the computation of deeper layers will be skipped. Compared to the conventional static models, such an *adaptive* inference mechanism can significantly improve the computational efficiency without sacrificing accuracy. When making predictions with shallow classifiers for canonical (*easy*) samples, a substantial amount of computation will be saved by skipping the calculation of deep layers. Moreover, the network is capable of handling the non-canonical (*hard*) inputs with deep exits (Fig. 1, *Inference Stage*).

Existing works on dynamic networks mainly focus on designing more advanced multi-exit architectures [17,46]. A naive training strategy is generally adopted by summing up the cross-entropy losses obtained from all classifiers. More importantly, the loss of both *easy* and *hard* samples contributes equally to the final optimization objective (Fig. 1, *Conventional Training*), regardless of where a sample may *actually exit*. However, these exits at varying depths have different capabilities, and they are responsible for recognizing samples of varying complexity at test time. Such an adaptive inference behavior has been neglected by the naive training paradigm adopted in previous works [17, 29, 46].

In this paper, we propose to bridge the gap between training and testing by imposing sample-wise weights on the loss of multiple exits. Our motivation is that every classifier is required to handle a *subset* of samples in the adaptive inference scenario. Specifically, the early classifiers only need to recognize some canonical inputs, while the deep classifiers are usually responsible for those non-canonical samples. Therefore, an ideal optimization objective should encourage each exit to emphasize different training samples by weighting their loss. In a nutshell, the challenge for sample weighting in multi-exit models is two-fold: 1) the exiting decisions are made during *inference*, and we have no prior knowledge of where a specific sample exits; 2) setting proper weights is a non-trivial problem.

To address these challenges, we propose to automatically *learn* appropriate weights by leveraging a *weight prediction network* (WPN). The WPN takes the training loss from all exits as input, producing the weights imposed on the samples at every exit. We jointly optimize the backbone model and the WPN in a *meta-learning* manner. A novel optimization objective is constructed to guide the meta-learning procedure. Precisely, we mimic the test-time early-exiting process to find where the samples will exit during inference. The meta objective for each classifier is then defined as the loss only on the samples that *actually* exit at this classifier. Compared to the conventional training strategy, our specialized meta-objective encourages every exit to impose proper weights on different samples for improved performance in dynamic early exiting (Fig. 1, *Our Training*).

We evaluate our method on image recognition tasks in two settings: a class balanced setting and a class imbalanced setting. The experiment results on CIFAR [24], ImageNet [5], and the long-tailed CIFAR [4] demonstrate that the proposed approach consistently improves the trade-off between accuracy and inference efficiency for state-of-the-art early-exiting networks.

2   Related Works

Dynamic Early-Exiting Networks. Early exiting is an effective dynamic inference paradigm, allowing *easy* samples to be output at intermediate classifiers. Existing works mainly focus on designing more advanced architectures. For instance, BranchyNet [41] attaches classifier heads at varying depths in an AlexNet [25]. An alternative option is cascading multiple CNNs (i.e. AlexNet [25], GoogleNet [40] and ResNet [13]) to perform early exiting [2]. It is observed [17] that the classifiers may interfere with each other in these chain-structured or cascaded models. This issue is alleviated via dense connection and multi-scale structure [17]. Resolution adaptive network (RANet) [46] further performs early exiting by conditionally activating high-resolution feature representations.

While considerable efforts have been made on the architecture design, the *training* of these models still follows the routines developed for static networks. The *multi-exit* structural property and the *early-exiting* behavior are usually ignored. Some training techniques are studied to boost the knowledge transfer among exits, yet still neglecting the *adaptive inference* paradigm [29]. In this paper, we put forth a novel optimization *objective*, encouraging each exit to focus on the samples that they would probably handle in dynamic inference.

Sample Wighting. Different training samples have unequal importance. The idea of sample weighting could be traced back to dataset resampling [3] or instance reweighting [48]. These traditional approaches evaluate the sample importance with some prior knowledge. Recent works manage to establish a loss-weight mapping function [9,21,22,26,32,35]. There are mainly two directions: one tends to impose larger weights on *hard* samples to deal with the data imbalance problem (e.g. focal loss [32]); the other focuses more on *easy* samples to reduce the impact brought by noisy labels (e.g. self-paced learning [21,26]).

We study a different case: the test-time data distributions for different exits are divergent. Such distributions *could not be obtained in advance* due to the data-dependent inference procedure. The increased number of classifiers further raises challenges for designing the sample-weight mapping function. With our specially designed optimization objective and the meta-learning algorithm, we effectively produce proper sample weights for training the multi-exit model.

Meta-learning in Sample Weighting. Due to its remarkable progress, meta-learning [8,42] has been extensively studied in sample weighting [37,38,50]. Existing approaches mainly focus on tackling class imbalance or corrupted label problems. In contrast, our goal is to improve the *inference efficiency* of early-exiting networks in a more general setting, without any assumption on the dataset.

3   Method

In this section, we first introduce the preliminaries of dynamic early-exiting networks and their conventional training strategies. Then the sample weighting mechanism and our meta-learning algorithm will be presented.

3.1   Preliminaries

Multi-exit Networks. A typical approach to setting up a K-exit network is attaching $K-1$ intermediate classifiers at varying depths of a deep model [17,46]. For an input sample $\mathbf{x}$, the prediction from the k-th exit can be written as

$$\hat{y}^{(k)} = \arg\max_c p_c^{(k)} = \arg\max_c f_c^{(k)}(\mathbf{x}; \Theta_f^{(k)}), k = 1, 2, \cdots, K, \qquad (1)$$

where $p_c^{(k)}$ is the k-th classifier's output probability for class c, and $f^{(k)}(\cdot; \Theta_f^{(k)})$ represents the k-th sub-network with parameter $\Theta_f^{(k)}$. Note that the parameters

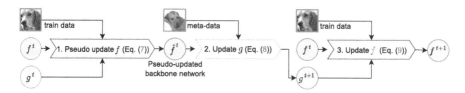

Fig. 2. Our training pipeline in iteration t. It consists of 3 steps: 1) the backbone network f^t is *pseudo-updated* to $\hat{f}^t$ using training samples; 2) a meta objective is computed using $\hat{f}^t$ on meta samples. This objective guides the update of the WPN g^t; 3) f^t is updated using the new weights predicted by g^{t+1}.

of different sub-networks are shared in a nested way except the classifiers. We denote the whole classification model as f and its parameters as Θ_f.

Dynamic Early Exiting. Extensive efforts have been made to perform early exiting based on multi-exit models. A typical approach is terminating the forward propagation *once* the classification *confidence* ($\max_c p_c^{(k)}$) at a certain exit exceeds a given threshold (ϵ_k). The final prediction is obtained by

$$\hat{y} = \hat{y}^{(k)}, \text{ if } \max_c p_c^{(k)} \geq \epsilon_k, \text{ and } \max_c p_c^{(j)} < \epsilon_j, \forall j \leq k, k \leq K - 1. \quad (2)$$

The predictions $\hat{y}^{(k)} (k = 1, 2, \cdots, K)$ are obtained sequentially before satisfying the criterion in Eq. (2) or reaching the last exit. The threshold for exit-k (ϵ_k) can be decided on a validation set according to the computational budget.

Conventional Training Methods. A naive training strategy adopted by existing works [17,46] is directly minimizing a cumulative loss function:

$$\mathcal{L} = \sum_{k=1}^{K} \frac{1}{N} \sum_{i=1}^{N} l_i^{(k)} \triangleq \sum_{k=1}^{K} \frac{1}{N} \sum_{i=1}^{N} \text{CE}(f^{(k)}(\mathbf{x}_i; \Theta_f^{(k)}), y_i), \quad (3)$$

where CE is the cross-entropy loss, and N is the number of training samples.

3.2 Sample-Weighting for Early-Exiting Networks

In this subsection, we first formulate our sample weighting mechanism, and then introduce the proposed meta-learning objective. The optimization method is further presented. See Fig. 2 for an overview of the training pipeline.

Sample Weighting with Weight Prediction Network. We can observe from Eq. (2) that test samples are adaptively allocated to multiple exits according to their prediction confidence during inference. Sub-networks with varying depths are responsible for handling different subsets of samples. Therefore, it is suboptimal to set the same optimization objective for these exits as in Eq. (3). To this end, we propose to ameliorate the training objective by sample weighting:

$$\mathcal{L} = \sum_{k=1}^{K} \frac{1}{N} \sum_{i=1}^{N} w_i^{(k)} l_i^{(k)}. \quad (4)$$

Weight Prediction Network. Since we have no prior knowledge of where a specific sample exits, and the function mapping from input to weight is hard to establish,

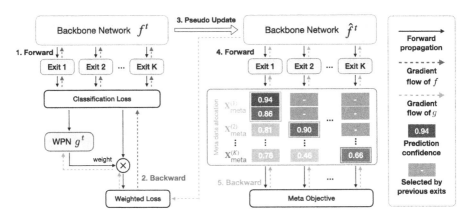

Fig. 3. Updates of the backbone model f and the WPN g in detail. First, we compute the weighted classification loss (Eq. (5)), which guides the pseudo update on the backbone network f^t (Steps 1, 2, and 3). The meta objective (Eq. (6)) is then computed to update the WPN g^t (Steps 4 and 5). (Color figure online)

we propose to automatically *learn* the weight $w_i^{(k)}$ from the input $\mathbf{x}_i$ by a weight prediction network (WPN, denoted by g): $\mathbf{w}_i = [w_i^{(1)}, w_i^{(2)}, \cdots, w_i^{(K)}] = g(\boldsymbol{l}_i; \boldsymbol{\Theta}_g)$, where $\boldsymbol{l}_i = [l_i^{(1)}, l_i^{(2)}, \cdots, l_i^{(K)}]$ is the training loss for sample $\mathbf{x}_i$ at K exits. The WPN g is established as an MLP with one hidden layer, and $\boldsymbol{\Theta}_g$ is the parameters of the WPN. We learn the backbone parameters $\boldsymbol{\Theta}_f$ and the WPN parameters $\boldsymbol{\Theta}_g$ in a meta-learning manner. Note that the WPN is only used for training, and no extra computation will be brought during the inference stage.

The Meta-learning Objective. We construct a novel optimization objective to bring the test-time adaptive behavior into training.

Weighted Classification Loss. With our weighting scheme (given $\boldsymbol{\Theta}_g$), the optimization objective of the backbone model parameter $\boldsymbol{\Theta}_f$ can be written as

$$\boldsymbol{\Theta}_f^\star(\boldsymbol{\Theta}_g) = \arg\min_{\boldsymbol{\Theta}_f} \mathcal{L}_{\mathrm{tr}}(\boldsymbol{\Theta}_f, \boldsymbol{\Theta}_g)$$

$$\triangleq \arg\min_{\boldsymbol{\Theta}_f} \sum_{k=1}^{K} \frac{1}{N} \sum_{i:\mathbf{x}_i \in \mathbf{X}_{\mathrm{tr}}} g^{(k)}(\boldsymbol{l}_i; \boldsymbol{\Theta}_g) \cdot l_i^{(k)}(\boldsymbol{\Theta}_f), \quad (5)$$

where $\mathbf{X}_{\mathrm{tr}}$ is the training set. Following [30], we scale the output of our WPN with $0 < \delta < 1$ to obtain a perturbation $\tilde{\mathbf{w}} \in [-\delta, \delta]^{B \times K}$, where B is the batch size. We further normalize the perturbation $\tilde{\mathbf{w}}$ to ensure the summation of its elements is 0. The final weight imposed on the loss is produced via $\mathbf{w} = 1 + \tilde{\mathbf{w}}$.

Meta Data Allocation. The goal of our weighting mechanism is improving the model performance in the dynamic inference scenario. To construct the optimization objective for the WPN g, we first mimic the early exiting procedure on a meta set $\mathbf{X}_{\mathrm{meta}}$, and obtain the meta samples exiting at different exits

$\mathbf{X}_{\text{meta}}^{(k)}, k = 1, 2, \cdots, K$. Specifically, we define a budget controlling variable q to decide the number of samples that exit at each exit: $N_k = \frac{q^k}{\sum_{k=1}^{K} q^k} \times N_{\text{meta}}, q > 0$, where N_{meta} is the sample number in the meta set $\mathbf{X}_{\text{meta}}$. When $q = 1$, the output numbers for different exits are equal, and $q > 1$ means more samples being output by deeper exits, together with more computation. In *training*, we generally tune the variable q on the validation set and fix it in the meta optimization procedure; in *testing*, we vary q from 0 to 2 to cover a wide range of computational budgets, and plot the curves of accuracy-computation in Sect. 4 (e.g. Fig. 5).

Given q, Fig. 3 (right) illustrates the procedure of obtaining $\mathbf{X}_{\text{meta}}^{(k)}$: each exit selects its most confident N_k samples which have not been output at previous classifiers. Note that for the last exit, $\mathbf{X}_{\text{meta}}^{(K)}$ contains all the unselected samples.

Meta Objective for Dynamic Early Exiting. Instead of correctly recognizing all meta samples with diverse complexity, the meta objective for a specific exit is the classification loss on the samples that *actually exit* at this exit. The overall meta objective is obtained by aggregating the meta objectives from K exits:

$$\boldsymbol{\Theta}_g^\star = \arg\min_{\boldsymbol{\Theta}_g} \mathcal{L}_{\text{meta}}(\boldsymbol{\Theta}_f^\star(\boldsymbol{\Theta}_g))$$

$$\triangleq \arg\min_{\boldsymbol{\Theta}_g} \sum_{k=1}^{K} \frac{1}{N_k} \sum_{j : \mathbf{x}_j \in \mathbf{X}_{\text{meta}}^{(k)}} l^{(k)}(\mathbf{x}_j, y_j; \boldsymbol{\Theta}_f^\star(\boldsymbol{\Theta}_g)). \tag{6}$$

The Optimization Method. We jointly optimize the backbone parameter $\boldsymbol{\Theta}_f$ (Eq. (5)) and the WPN parameter $\boldsymbol{\Theta}_g$ (Eq. (6)) with an online strategy (see Fig. 2 for an overview of the optimization pipeline). Existing meta-learning methods for loss weighting typically require a standalone meta-set, which consists of class-balanced data or clean labels [37,38]. In contrast, we simply reuse the training data as our meta data. Precisely, at iteration t, we first split the input mini-batch into a training batch $\mathbf{X}_{\text{tr}}$ and a meta batch $\mathbf{X}_{\text{meta}}$. The training batch $\mathbf{X}_{\text{tr}}$ is used to construct the classification loss (Eq. (5)) and optimize the model parameters $\boldsymbol{\Theta}_f$. Next, the meta batch $\mathbf{X}_{\text{meta}}$ is leveraged to compute the meta objective (Eq. (6)) and train the WPN parameters $\boldsymbol{\Theta}_g$.

Pseudo update of the backbone network is conducted using $\mathbf{X}_{\text{tr}}$:

$$\hat{\boldsymbol{\Theta}}_f^t(\boldsymbol{\Theta}_g) = \boldsymbol{\Theta}_f^t - \alpha \left. \frac{\partial \mathcal{L}_{\text{tr}}(\boldsymbol{\Theta}_f, \boldsymbol{\Theta}_g)}{\partial \boldsymbol{\Theta}_f} \right|_{\boldsymbol{\Theta}_f^t}, \tag{7}$$

where α is the learning rate. Note that $\hat{\boldsymbol{\Theta}}_f^t$ is a function of $\boldsymbol{\Theta}_g$, and this pseudo update is performed to construct the computational graph for later optimization of $\boldsymbol{\Theta}_g$. See Fig. 3 (gray dashed lines) for the gradient flow of this pseudo update.

Update of the weight prediction network is performed using our meta objective calculated on $\mathbf{X}_{\text{meta}}$. Concretely, we mimic the early exiting procedure on $\mathbf{X}_{\text{meta}}$, and split it into K subsets without intersection $\mathbf{X}_{\text{meta}} = \{\mathbf{X}_{\text{meta}}^{(1)} \cup \mathbf{X}_{\text{meta}}^{(2)} \cup \cdots \cup \mathbf{X}_{\text{meta}}^{(K)}\}$, where $\mathbf{X}_{\text{meta}}^{(k)}$ contains the meta samples which should be output at exit-k according to the criterion in Eq. (2). See also Fig. 3 (right) for the procedure.

Receiving the partition of the meta-data based on the pseudo updated backbone network, the parameters of our weight prediction model can be updated:

$$\mathbf{\Theta}_g^{t+1} = \mathbf{\Theta}_g^t - \beta \frac{\partial \mathcal{L}_{\text{meta}}(\hat{\mathbf{\Theta}}_f^t(\mathbf{\Theta}_g))}{\partial \mathbf{\Theta}_g}\bigg|_{\mathbf{\Theta}_g^t}, \tag{8}$$

where $\mathcal{L}_{\text{meta}}(\hat{\mathbf{\Theta}}^t(\mathbf{\Theta}_g))$ is the aggregation of the classification loss from each exit on its *allocated* meta data (Eq. (6)), and β is the learning rate. The gradient flow for updating $\mathbf{\Theta}_g$ is illustrated in Fig. 3 (the golden dashed lines).

Algorithm 1. The meta-learning Algorithm

Require: Training data $\mathcal{D}$, batch size B, iteration T, interval I, budget controller q.
Ensure: Backbone network parameters $\mathbf{\Theta}_f$.
    for $t = 0$ to $T - 1$ **do**
        $\{\mathbf{X}, \mathbf{y}\} \leftarrow \text{SampleMiniBatch}(\mathcal{D}, B)$.
        Split $\{\mathbf{X}, \mathbf{y}\}$ into $\{\mathbf{X}_{\text{tr}}, \mathbf{X}_{\text{meta}}, \mathbf{y}_{\text{tr}}, \mathbf{y}_{\text{meta}}\}$.
        if $t \mod I = 0$ **then**
            Perform a pseudo update for $\hat{\mathbf{\Theta}}_f^t$ by Eq. (7).
            Perform meta data allocation based on q.
            Update $\mathbf{\Theta}_g^{t+1}$ by Eq. (8).
        end if
        Update $\mathbf{\Theta}_f^{t+1}$ by Eq. (9).
    end for

Update of the backbone network is finally realized based on the updated $\mathbf{\Theta}_g^{t+1}$:

$$\mathbf{\Theta}_f^{t+1} = \mathbf{\Theta}_f^t - \alpha \frac{\partial \mathcal{L}_{\text{tr}}(\mathbf{\Theta}_f, \mathbf{\Theta}_g^{t+1})}{\partial \mathbf{\Theta}_f}\bigg|_{\mathbf{\Theta}_f^t}. \tag{9}$$

We summarize the learning algorithm in Algorithm 1. By mimicking the adaptive inference procedure, our novel objective in Eq. (6) encourages each exit to correctly classify the samples *which are most probably allocated to it* in the early exiting scenario (blue text in Eq. (6)). For example, early exits at shallow layers may focus more on those *easy* samples (see also our visualization results in Sect. 4, Fig. 8). More empirical analysis is presented in Sect. 4.

4   Experiments

In this section, we first conduct ablation studies to validate the design choices made in our weighting mechanism (Sect. 4.1). The main results on CIFAR [24] and ImageNet [5] are then presented in Sect. 4.2 and Sect. 4.3 respectively. Finally, we evaluate our method on the long-tailed CIFAR [4] in a class imbalance setting (Sect. 4.4). A network's performance is measured in terms of the trade-off between the accuracy and the Mul-Adds (multiply-add operations). We apply our training algorithm to two representative dynamic early-exiting architectures, i.e. multi-scale dense network (MSDNet) [17] and resolution adaptive network (RANet) [46]. The experimental setup is provided in the supplementary material.

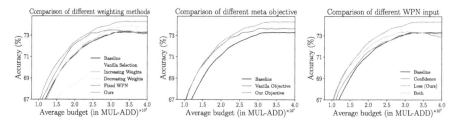

Fig. 4. Ablation studies on CIFAR-100. Left: our meta-learning based approach *v.s.* heuristic weighting mechanisms. Middle: the effectiveness of the proposed meta objective. Right: comparison of different WPN inputs.

4.1   Ablation Studies

We perform ablation studies with a 5-exit MSDNet on CIFAR-100 to validate the effectiveness of different settings and variants in our method.

Meta-learning Algorithm. We first verify the *necessity* of our meta-learning algorithm by comparing it with three variants: 1) the first replaces the weighting scheme with a vanilla selection scheme: the data allocation (Fig. 3) is applied directly to the *training* data, and the loss of each exit is only calculated on its *allocated* samples; 2) the second hand-designs a weighting mechanism with *fixed* weights increasing/decreasing with the exit index; 3) the third uses a *frozen* convergent WPN to weight the samples. For the second variant, the weights are set from 0.6 to 1.4 (or inverse) with uniform steps. Evaluation results on CIFAR-100 are shown in Fig. 4 (left). Several observations can be made:

- The vanilla sample selection strategy results in a drastic drop in accuracy, suggesting the necessity of our sample *weighting* mechanism;
- The gain of hand-crafted weight values is limited, indicating the advantage of our *learning*-based weighting approach;
- Interestingly, our joint optimization outperforms weighting with a frozen WPN. This suggests that it is essential to *dynamically* adjust the weights imposed on the loss at different training stages.

Meta-learning Objective. Our designed meta loss in Eq. (6) encourages every exit to correctly recognize the data *subset* that is most likely handled by the exit in dynamic inference. To clarify the effectiveness of this meta objective, we keep the learning procedure unchanged and substitute Eq. (6) with the classification loss on the whole meta set for each exit. The results are shown in Fig. 4 (middle). While this variant (line Vanilla Objective) outperforms the baseline when the computation is relatively low, our objective achieves higher accuracy when larger computational budget is available. This suggests that emphasizing *hard* samples for deep classifiers is crucial to improving their performance.

The Input of WPN. In addition to the classification loss, the confidence value can also be leveraged to produce the weight perturbations due to its role in making early exiting decisions. We test three types of input: 1) loss only; 2) confidence

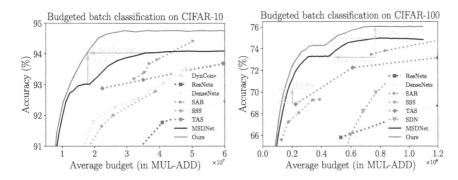

Fig. 5. Dynamic inference results on CIFAR-10 (left) and CIFAR-100 (right)

Table 1. Anytime prediction results of a 7-exit MSDNet on CIFAR-100

Exit index		1	2	3	4	5	6	7
Params ($\times 10^6$)		0.30	0.65	1.11	1.73	2.38	3.05	4.00
Mul-Add ($\times 10^6$)		6.86	14.35	27.29	48.45	76.43	108.90	137.30
Accuracy	MSDNet [17]	61.07	64.55	67.00	69.97	72.55	74.01	74.50
	Ours	**62.47** (↑1.30)	**66.32** (↑1.77)	**68.10** (↑1.10)	**71.29** (↑1.32)	**73.21** (↑0.66)	**74.87** (↑0.86)	**75.81** (↑1.31)
	IMTA [29]	60.29	64.86	69.09	72.71	74.47	75.60	75.19
	Ours + IMTA [29]	**62.26** (↑1.97)	**67.18** (↑2.32)	**70.53** (↑1.44)	**73.10** (↑0.39)	**74.80** (↑0.33)	**76.05** (↑0.45)	**76.31** (↑1.12)

only; 3) the concatenation of loss and confidence. The accuracy-computation curves are shown in Fig. 4 (right). It can be found that the adoption of loss is essential for our method, and the inclusion of confidence could be harmful. We hypothesize that by reflecting the information of *both* network prediction and ground truth, the loss serves as a better candidate for WPN input. The ablation study of the WPN design is presented in the supplementary material.

4.2 CIFAR Results

In this subsection, we report the results on CIFAR in both dynamic inference and anytime prediction settings following previous works [17,29,46]. We first apply our approach to MSDNet [17], and then compare it with the training techniques in [29]. The method is further validated on the RANet architecture [46].

Dynamic Inference Results. We apply our training strategy on MSDNet with 5 and 7 exits and compare with three groups of competitive baseline methods: classic networks (ResNet [13], DenseNet [18]), pruning-based approaches (Sparse Structure Selection (SSS) [19], Transformable Architecture Search (TAS) [6]), and dynamic networks (Shallow-Deep Networks (SDN) [23], Dynamic Convolutions (DynConv) [43], and Spatially Adaptive Feature Refinement (SAR) [12]).

In the dynamic inference scenario, we present the accuracy-computation (measured by Mul-Adds) curve in Fig. 5 (left: CIFAR-10, right: CIFAR-100). From the results, we can observe that the proposed weighting method consistently improves the performance of MSDNet at various computational budgets.

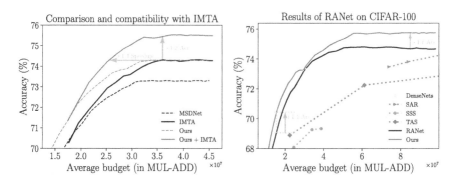

Fig. 6. Dynamic inference results on CIFAR-100. Left: comparison with IMTA [29]. Right: results with two different sized RANets.

Table 2. Anytime prediction results of a 6-exit RANet on CIFAR-100

Exit index		1	2	3	4	5	6
Params ($\times 10^6$)		0.36	0.90	1.30	1.80	2.19	2.62
Mul-Add ($\times 10^6$)		8.37	21.79	32.88	41.57	53.28	58.99
Accuracy	RANet [46]	63.41	67.36	69.62	70.21	71.00	71.78
	Ours	**65.33** (↑1.92)	**68.69** (↑1.30)	**70.36** (↑0.74)	**70.80** (↑0.59)	**72.57** (↑1.57)	**72.45** (↑0.67)

When applied to the 5-exit model, our weighting mechanism obtains significant boosts with the CIFAR-100 Top-1 accuracy increased by about 1.3% when evoking around 15M Mul-Adds. For the 7-exit MSDNet, our model only uses half of the original budget to achieve ∼94.0% Top-1 accuracy on CIFAR-10.

Anytime Prediction Results. We also report the accuracy of each exit on the whole test set in the anytime prediction setting. From the results in Table 1 (a 7-exit MSDNet [17]) and Table 2 (a 6-exit RANet [46]), we surprisingly find that although our meta objective encourages each exit to focus on only a *subset* of samples, the performance on the *whole* test set is consistently increased by a large margin. This phenomenon could bring some insights into the optimization of deep networks: 1) for the shallow exits, emphasizing a small subset of *easy* samples with *high confidence* benefits their generalization performance on the whole dataset [44]; 2) for the deeper exits, our objective forces them to focus on the *hard* samples, which cannot be confidently predicted by previous classifiers. Such a "challenging" goal could further improve their capability of approximating more complex classification boundaries. This coincides with the observation in our ablation studies (Fig. 4 middle): encouraging each exit to correctly classify a *well-selected* subset of training samples is preferable.

Comparison and Compatibility with IMTA. Improved techniques for training adaptive networks (IMTA) [29] are proposed to stabilize training and facilitate the collaboration among exits. These techniques are developed for the *optimization procedure* and ignore the adaptive inference behavior. In contrast, our

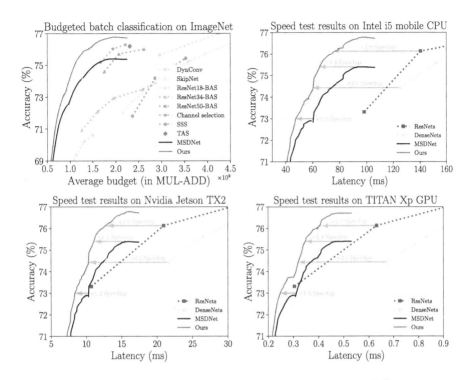

Fig. 7. ImageNet results. Top left: accuracy-computation curves. Others: accuracy-latency curves on different hardware platforms.

meta-learning based method improves the *objective optimization* and takes the inference behavior into account. We empirically compare our approach to IMTA [29], and further combine them to evaluate the compatibility.

The results are presented in Fig. 6 (right, for dynamic inference) and Table 1 (row 6 & 7, for anytime prediction). These results validate that 1) our weighting strategy achieves higher performance, especially at low computational budges; 2) combining our method with IMTA further improves the performance. In particular, note that the accuracy of deep exits can be boosted by a large margin.

Results on RANet. To evaluate the generality of our method, we conduct experiments on another representative multi-exit structure, RANet [46]. The anytime prediction performance of a 6-exit RANet is shown in Table 2, and the dynamic inference results of two different-sized RANets are illustrated in Fig. 6 (right). These results suggest that the proposed approach consistently improves the trade-off between accuracy and computational cost of RANet. Notably, when the Mul-Add is around 15M, our model improves the Top-1 accuracy significantly (∼ 1.4%). The strong performance on MSDNet and RANet indicates that our weighting algorithm is sufficiently *general* and *effective*.

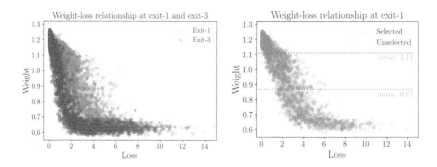

Fig. 8. **Visualization results** of the weight-loss relationship on ImageNet

4.3   ImageNet Results

Next, we evaluate our method on the large-scale ImageNet dataset [5]. We test three different sized MSDNets following [17] and compare them with competitive baselines, including pruning-based approaches (the aforementioned SSS [19] and TAS [6], and Filter Pruning via Geometric Median (Geom) [14]) and dynamic networks (SkipNet [45], Batch-Shaping (BAS) [1], Channel Selection [15] and DynConv [43]). We report the results in the dynamic inference setting here. Anytime prediction results can be found in the supplementary material.

Accuracy-Computation Result. We present the accuracy-computation results on ImageNet in Fig. 7 (top left). The plot shows that the weighting mechanism consistently improves the accuracy-computation trade-off of MSD-Nets. Even though some competing models surpass the baseline MSDNet trained without sample weighting, our method successfully outperforms these competitors.

Accuracy-Latency Result. We benchmark the efficiency of our models on three types of computing devices: Intel i5-1038NG7 mobile CPU, Nvidia Jetson TX2 (a resource-constrained embedded AI computing device) and TITAN Xp GPU. The testing batch sizes are set as 64 for the first two devices and 512 for the last. The testing images have resolutions of 224×224 pixels. The accuracy-latency curves plotted in Fig. 7 demonstrate the significant improvement of our method across all computing platforms. For instance, our model only takes 70% of the original MSDNet's computation time to achieve 75.4% accuracy ($1.4\times$ speed-up).

Visualization. We visualize the weights predicted by WPN toward the end of training in Fig. 8[1]. The left plot presents the weight-loss relationship at exit-1 and exit-3. The right plot shows the weights of the samples selected/not selected by exit-1 in meta-data allocation (Sect. 3.2). Several observations can be made: 1) Since more samples are allocated to early exits, the weights at exit-1 are

---

[1] We set $q = 0.5$ in training, and therefore the proportion of output samples at 5 exits follows an exponential distribution of $[0.52, 0.26, 0.13, 0.06, 0.03]$.

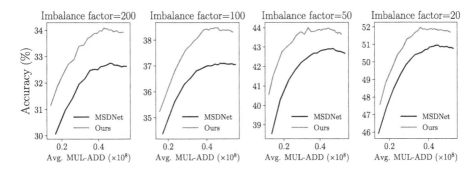

Fig. 9. Results on long-tailed CIFAR-100 with different imbalance factors

generally larger than those at exit-3. This result suggests that the proposed method successfully learns the relationship between exits; 2) Both exits tend to emphasize the samples with a smaller loss. This coincides with the observation in [51] which explains the importance of the well-classified (*easy*) samples in the optimization of deep models; 3) From weights at exit-1, we can observe that the samples with high prediction confidence (which would be selected by exit-1 in meta-data allocation) generally have weights larger than samples with lower confidence (which would be allocated to deeper classifiers). This suggests that the proposed method learns to recognize the samples' exit at different classifiers.

4.4 Class Imbalance Results

The proposed method is finally evaluated in the class imbalance setting. On the long-tailed CIFAR-100 [4] dataset, we test with four imbalance factors (200, 100, 50, and 20), where the imbalance factor is defined as the number of training samples in the largest class divided by the smallest. The budgeted batch classification performance illustrated in Fig. 9 shows that our weighting method consistently outperforms the conventional training scheme by a large margin.

5 Conclusion

In this paper, we propose a meta-learning based sample weighting mechanism for training dynamic early-exiting networks. Our approach aims to bring the test-time adaptive behavior into the training phase by sample weighting. We weight the losses of different samples at each exit by a weight prediction network. This network is jointly optimized with the backbone network guided by a novel meta-learning objective. The proposed weighting scheme can consistently boost the performance of multi-exit models in both anytime prediction and budgeted classification settings. Experiment results validate the effectiveness of our method on the long-tailed image classification task.

Acknowledgement. This work is supported in part by National Key R&D Program of China (2020AAA0105200), the National Natural Science Foundation of China under Grants 62022048, Guoqiang Institute of Tsinghua University and Beijing Academy of Artificial Intelligence. We also appreciate the generous donation of computing resources by High-Flyer AI.

References

1. Bejnordi, B.E., Blankevoort, T., Welling, M.: Batch-shaping for learning conditional channel gated networks. In: ICLR (2020)
2. Bolukbasi, T., Wang, J., Dekel, O., Saligrama, V.: Adaptive neural networks for efficient inference. In: ICML (2017)
3. Chawla, N.V., Bowyer, K.W., Hall, L.O., Kegelmeyer, W.P.: SMOTE: synthetic minority over-sampling technique. JAIR **16**, 321–357 (2002)
4. Cui, Y., Jia, M., Lin, T.Y., Song, Y., Belongie, S.: Class-balanced loss based on effective number of samples. In: CVPR (2019)
5. Deng, J., Dong, W., Socher, R., Li, L.J., Li, K., Fei-Fei, L.: ImageNet: a large-scale hierarchical image database. In: CVPR (2009)
6. Dong, X., Yang, Y.: Network pruning via transformable architecture search. In: NeurIPS (2019)
7. Dosovitskiy, A., et al.: An image is worth 16x16 words: transformers for image recognition at scale. In: ICLR (2021)
8. Finn, C., Abbeel, P., Levine, S.: Model-agnostic meta-learning for fast adaptation of deep networks. In: ICML (2017)
9. Freund, Y., Schapire, R.E.: A decision-theoretic generalization of on-line learning and an application to boosting. JCSS **55**(1), 119–139 (1997)
10. Han, S., Mao, H., Dally, W.J.: Deep compression: compressing deep neural networks with pruning, trained quantization and Huffman coding (2016)
11. Han, Y., Huang, G., Song, S., Yang, L., Wang, H., Wang, Y.: Dynamic neural networks: a survey. TPAMI (2021)
12. Han, Y., Huang, G., Song, S., Yang, L., Zhang, Y., Jiang, H.: Spatially adaptive feature refinement for efficient inference. In: TIP (2021)
13. He, K., Zhang, X., Ren, S., Sun, J.: Deep residual learning for image recognition. In: CVPR (2016)
14. He, Y., Liu, P., Wang, Z., Hu, Z., Yang, Y.: Filter pruning via geometric median for deep convolutional neural networks acceleration. In: CVPR (2019)
15. Herrmann, C., Bowen, R.S., Zabih, R.: Channel selection using gumbel softmax. In: Vedaldi, A., Bischof, H., Brox, T., Frahm, J.-M. (eds.) ECCV 2020. LNCS, vol. 12372, pp. 241–257. Springer, Cham (2020). https://doi.org/10.1007/978-3-030-58583-9_15
16. Howard, A.G., et al.: MobileNets: efficient convolutional neural networks for mobile vision applications. arXiv preprint arXiv:1704.04861 (2017)
17. Huang, G., Chen, D., Li, T., Wu, F., van der Maaten, L., Weinberger, K.: Multi-scale dense networks for resource efficient image classification. In: ICLR (2018)
18. Huang, G., Liu, Z., Van Der Maaten, L., Weinberger, K.Q.: Densely connected convolutional networks. In: CVPR (2017)
19. Huang, Z., Wang, N.: Data-driven sparse structure selection for deep neural networks. In: ECCV (2018)
20. Hubara, I., Courbariaux, M., Soudry, D., El-Yaniv, R., Bengio, Y.: Binarized neural networks. In: NeurIPS (2016)

21. Jiang, L., Meng, D., Mitamura, T., Hauptmann, A.G.: Easy samples first: self-paced reranking for zero-example multimedia search. In: ACM MM (2014)
22. Johnson, J.M., Khoshgoftaar, T.M.: Survey on deep learning with class imbalance. J. Big Data **6**(1), 1–54 (2019). https://doi.org/10.1186/s40537-019-0192-5
23. Kaya, Y., Hong, S., Dumitras, T.: Shallow-deep networks: understanding and mitigating network overthinking. In: ICML (2019)
24. Krizhevsky, A.: Learning multiple layers of features from tiny images. Technical Report (2009)
25. Krizhevsky, A., Sutskever, I., Hinton, G.E.: ImageNet classification with deep convolutional neural networks. In: NeurIPS (2012)
26. Kumar, M., Packer, B., Koller, D.: Self-paced learning for latent variable models. In: NeurIPS (2010)
27. LeCun, Y., Denker, J.S., Solla, S.A.: Optimal brain damage. In: NeurIPS (1990)
28. Li, H., Kadav, A., Durdanovic, I., Samet, H., Graf, H.P.: Pruning filters for efficient convnets. In: ICLR (2017)
29. Li, H., Zhang, H., Qi, X., Yang, R., Huang, G.: Improved techniques for training adaptive deep networks. In: ICCV (2019)
30. Li, S., Ma, W., Zhang, J., Liu, C.H., Liang, J., Wang, G.: Meta-reweighted regularization for unsupervised domain adaptation. TKDE (2021)
31. Lin, J., Rao, Y., Lu, J., Zhou, J.: Runtime neural pruning. In: NeurIPS (2017)
32. Lin, T.Y., Goyal, P., Girshick, R., He, K., Dollár, P.: Focal loss for dense object detection. In: ICCV (2017)
33. Liu, Z., et al.: Swin transformer: hierarchical vision transformer using shifted windows. In: ICCV (2021)
34. Liu, Z., Li, J., Shen, Z., Huang, G., Yan, S., Zhang, C.: Learning efficient convolutional networks through network slimming. In: ICCV (2017)
35. Malisiewicz, T., Gupta, A., Efros, A.A.: Ensemble of exemplar-SVMs for object detection and beyond. In: ICCV (2011)
36. Panda, P., Sengupta, A., Roy, K.: Conditional deep learning for energy-efficient and enhanced pattern recognition. In: DATE (2016)
37. Ren, M., Zeng, W., Yang, B., Urtasun, R.: Learning to reweight examples for robust deep learning. In: ICML (2018)
38. Shu, J., et al.: Meta-weight-net: Learning an explicit mapping for sample weighting. In: NeurIPS (2019)
39. Simonyan, K., Zisserman, A.: Very deep convolutional networks for large-scale image recognition. In: ICLR (2015)
40. Szegedy, C., et al.: Going deeper with convolutions. In: CVPR (2015)
41. Teerapittayanon, S., McDanel, B., Kung, H.T.: Branchynet: fast inference via early exiting from deep neural networks. In: ICPR (2016)
42. THospedales, T., Antoniou, A., Micaelli, P., Storkey, A.: Meta-learning in neural networks: a survey. TPAMI (2021)
43. Verelst, T., Tuytelaars, T.: Dynamic convolutions: exploiting spatial sparsity for faster inference. In: CVPR (2020)
44. Wang, X., Chen, Y., Zhu, W.: A survey on curriculum learning. TPAMI (2021)
45. Wang, X., Yu, F., Dou, Z.Y., Darrell, T., Gonzalez, J.E.: SkipNet: learning dynamic routing in convolutional networks. In: ECCV (2018)
46. Yang, L., Han, Y., Chen, X., Song, S., Dai, J., Huang, G.: Resolution adaptive networks for efficient inference. In: CVPR (2020)
47. Yang, L., et al.: CondenseNet v2: sparse feature reactivation for deep networks. In: CVPR (2021)

48. Zadrozny, B.: Learning and evaluating classifiers under sample selection bias. In: ICML (2004)
49. Zhang, X., Zhou, X., Lin, M., Sun, J.: ShuffleNet: an extremely efficient convolutional neural network for mobile devices. In: CVPR (2018)
50. Zhang, Z., Pfister, T.: Learning fast sample re-weighting without reward data. In: ICCV (2021)
51. Zhao, G., Yang, W., Ren, X., Li, L., Sun, X.: Well-classified examples are underestimated in classification with deep neural networks. In: AAAI (2022)

AdaBin: Improving Binary Neural Networks with Adaptive Binary Sets

Zhijun Tu[1,2], Xinghao Chen[2(✉)], Pengju Ren[1(✉)], and Yunhe Wang[2]

[1] Institute of Artificial Intelligence and Robotics, Xi'an Jiaotong University,
Xi'an, China
tuzhijun123@stu.xjtu.edu.cn, pengjuren@xjtu.edu.cn
[2] Huawei Noah's Ark Lab, Beijing, China
{zhijun.tu,xinghao.chen,yunhe.wang}@huawei.com

Abstract. This paper studies the Binary Neural Networks (BNNs) in which weights and activations are both binarized into 1-bit values, thus greatly reducing the memory usage and computational complexity. Since the modern deep neural networks are of sophisticated design with complex architecture for the accuracy reason, the diversity on distributions of weights and activations is very high. Therefore, the conventional sign function cannot be well used for effectively binarizing full-precision values in BNNs. To this end, we present a simple yet effective approach called AdaBin to adaptively obtain the optimal binary sets $\{b_1, b_2\}$ ($b_1, b_2 \in \mathbb{R}$) of weights and activations for each layer instead of a fixed set (*i.e.*, $\{-1, +1\}$). In this way, the proposed method can better fit different distributions and increase the representation ability of binarized features. In practice, we use the center position and distance of 1-bit values to define a new binary quantization function. For the weights, we propose an equalization method to align the symmetrical center of binary distribution to real-valued distribution, and minimize the Kullback-Leibler divergence of them. Meanwhile, we introduce a gradient-based optimization method to get these two parameters for activations, which are jointly trained in an end-to-end manner. Experimental results on benchmark models and datasets demonstrate that the proposed AdaBin is able to achieve state-of-the-art performance. For instance, we obtain a 66.4% Top-1 accuracy on the ImageNet using ResNet-18 architecture, and a 69.4 mAP on PAS-CAL VOC using SSD300.

Keywords: Binary neural networks · Adaptive binary sets

1 Introduction

Deep Neural Networks (DNNs) have demonstrated powerful learning capacity, and are widely applied in various tasks such as computer vision [27], natural

---

Supplementary Information The online version contains supplementary material available at https://doi.org/10.1007/978-3-031-20083-0_23.

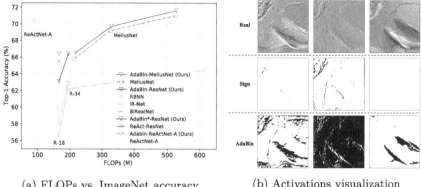

(a) FLOPs vs. ImageNet accuracy   (b) Activations visualization

Fig. 1. (a) Comparisons with state-of-the-art methods. With a little extra computation, the proposed AdaBin achieves better results for various architectures such as ResNet, MeliusNet [5] and ReActNet [33]. (b) Visualization for activations of 2nd layer in ResNet-18 on ImageNet. *Real* denotes real-valued activations, *Sign* and *AdaBin* denote the binary methods of previous BNNs and ours.

language processing [3] and speech recognition [22]. However, the growing complexity of DNNs requires significant storage and computational resources, which makes the deployment of these deep models on embedded devices extremely difficult. Various approaches have been proposed to compress and accelerate DNNs, including low-rank factorization [44], pruning [10,19], quantization [12], knowledge distillation [11,23] and energy-efficient architecture design [9], *etc.* Among these approaches, quantization has attracted great research interests for decades, since the quantized networks with less bit-width require smaller memory footprint, lower energy consumption and shorter calculation delay. Binary Neural Networks (BNNs) are the extreme cases of quantized networks and could obtain the largest compression rate by quantizing the weights and activations into 1-bit values [18,37,41,43]. Different from the floating point matrix operation in traditional DNNs, BNNs replace the multiplication and accumulation with bit-wise operation XNOR and BitCount, which can obtain an about 64× acceleration and 32× memory saving [37]. However, the main drawback of BNNs is the severe accuracy degradation compared to the full-precision model, which also limits its application to more complex tasks, such as detection, segmentation and tracking.

According to the IEEE-754 standard, a 32-bit floating point number has 6.8×10^{38} unique states [1]. In contrast, a 1-bit value only has 2 states $\{b_1, b_2\}$, whose representation ability is very weak compared with that of the full-precision values, since there are only two kinds of the multiplication results of binary values as shown in Table 1a. To achieve a very efficient hardware implementation, the conventional BNN method [13] binarizes both the weights and the activations to either +1 or −1 with sign function. The follow-up approaches on BNNs have made tremendous efforts for enhancing the performance of binary network, but

Table 1. The illustration on the feature representation ability of different binary schemes. The a represents the binarized input and the w represents binarized weights, respectively. $a_{b1}, a_{b2}, w_{b1}, w_{b2} \in \mathbb{R}$, which are not restricted to fixed values for different layers.

a w	-1	+1	a w	0	+1	a w	-1	+1	a w	a_{b1}	a_{b2}
-1	+1	-1	-1	0	-1	0	0	0	w_{b1}	$a_{b1}w_{b1}$	$a_{b2}w_{b1}$
+1	-1	+1	+1	0	+1	+1	-1	+1	w_{b2}	$a_{b1}w_{b2}$	$a_{b2}w_{b2}$
(a) BNN [13]			(b) SiBNN [38]			(c) SiMaN [30]			(d) AdaBin (Ours)		

still restrict the binary values to a fixed set (*i.e.*, $\{-1, +1\}$ or $\{0, +1\}$) for all the layers. Given the fact that the feature distributions in deep neural networks are very diverse, sign function can not provide binary diversity for these different distributions. To this end, we have to rethink the restriction of fixed binary set for further enhancing the capacity of BNNs.

Based on the above observation and analysis, we propose an **Ada**ptive **Bin**ary method (**AdaBin**) to redefine the binary values ($b_1, b_2 \in \mathbb{R}$) with their center position and distance, which aims to obtain the optimal binary set that best matches the real-valued distribution. We propose two corresponding optimization strategies for weights and activations. On one hand, we introduce an equalization method for the weights based on statistical analysis. By aligning the symmetrical center of binary distribution to real-valued distribution and minimizing the Kullback-Leibler divergence (KLD) of them, we can obtain the analytic solutions of center and distance, which makes the weight distribution much balanced. On the other hand, we introduce a gradient-based optimization method for the activations with a loss-aware center and distance, which are initialized in the form of sign function and trained in an end-to-end manner. As shown in Table 1, we present the truth tables of the multiplication results for binary values in different BNNs. Most previous BNNs binarize both the weights and activations into $\{-1, +1\}$ as shown in Table 1b. A few other methods [30,38] attempt to binarize weights and activations into $\{0, +1\}$, as shown in Table 1b and Table 1c. These methods result in 2 or 3 kinds of output representations. Table 1d illustrates the results of our proposed AdaBin method. The activations and weights are not fixed and could provide 4 kinds of output results, which significantly enhances the feature representation of binary networks as shown in Fig. 1b. Meanwhile, we can find that previous binary methods are the special cases of our AdaBin and we extend the binary values from ± 1 to the whole real number domain.

Furthermore, we demonstrate that the proposed AdaBin can also be efficiently implemented by XNOR and BitCount operations with negligible extra calculations and parameters, which could achieve 60.85× acceleration and 31× memory saving in theory. With only minor extra computation, our proposed AdaBin outperforms state-of-the-art methods for various architectures, as shown in Fig. 1a. The contributions of this paper are summarize as follow:

(1) We rethink the limitation of $\{-1, +1\}$ in previous BNNs and propose a simple yet effective binary method called AdaBin, which could seek suitable binary sets by adaptively adjusting the center and distance of 1-bit values.

(2) Two novel strategies are proposed to obtain the optimal binary sets of weights and activations for each layer, which can further close the performance gap between binary neural networks and full-precision networks.

(3) Extensive experiments on CIFAR-10 and ImageNet demonstrate the superior performance of our proposed AdaBin over state-of-the-art methods. Besides, though not tailored for object detection task, AdaBin also outperforms prior task-specific BNN methods by 1.9 mAP on PASCAL VOC dataset.

2   Related Work

Binary neural network was firstly introduced by [13]. They creatively proposed to binarize weights and activations with sign function and replace most arithmetic operations of deep neural networks with bit-wise operations. To reduce the quantization error, XNOR-Net [37] proposed a channel-wise scaling factor to reconstruct the binarized weights, which also becomes one of the most important components of the subsequent BNNs. ABC-Net [32] approximated full-precision weights with the linear combination of multiple binary weight bases and employed multiple binary activations to alleviate information loss. Inspired by the structures of ResNet [21] and DenseNet [25], Bi-Real Net [34] proposed to add shortcuts to minimize the performance gap between the 1-bit and real-valued CNN models, and BinaryDenseNet [6] improved the accuracy of BNNs by increasing the number of concatenate shortcut. IR-Net [36] proposed the Libra-PB, which can minimize the information loss in forward propagation by maximizing the information entropy of the quantized parameters and minimizing the quantization error with the constraint $\{-1, +1\}$. ReActNet [33] proposed to generalize the traditional sign and PReLU functions, denoted as RSign and RPReLU for the respective generalized functions, to enable explicit learning of the distribution reshape and shift at near-zero extra cost.

3   Binarization with Adaptive Binary Sets

In this section, we focus on how to binarize weights and activations respectively, and introduce a new non-linear module to enhance the capacity of BNNs.

We first give a brief introduction on the general binary neural networks. Given an input $a \in \mathbb{R}^{c \times h \times w}$ and weight $\mathrm{w} \in \mathbb{R}^{n \times c \times k \times k}$, then we can get the output $y \in \mathbb{R}^{n \times h' \times w'}$ by convolution operation as Eq. 1.

$$y = \mathrm{Conv}(a, \mathrm{w}). \tag{1}$$

To accelerate the inference process, previous BNNs always partition the input and weight into two clusters, -1 and $+1$ with sign function as Eq. 2.

$$\mathrm{Sign}(x) = \begin{cases} b_1 = -1, & x < 0 \\ b_2 = +1, & x \geq 0 \end{cases}. \tag{2}$$

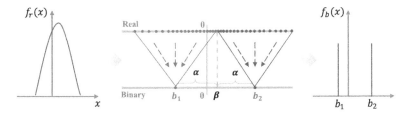

Fig. 2. AdaBin quantizer. The middle represents the mapping from floating point distribution $f_r(x)$ to binary distribution $f_b(x)$. b_1 and b_2 are the two clusters, α and β are the distance and center, respectively.

Then the floating-point multiplication and accumulation could be replaced by bit-wise operation XNOR (denoted as $\odot$) and BitCount as Eq. 3, which will result in much less overhead and latency.

$$y = \text{BitCount}(a_b \odot w_b). \tag{3}$$

In our method, we do not constrain the binarized values to a fixed set like $\{-1, +1\}$. Instead we release b_1 and b_2 to the whole real number domain and utilize the proposed AdaBin quantizer, which could adjust the center position and distance of the two clusters adaptively as Eq. 4. In this way, the binarized distribution can best match the real-valued distribution:

$$\mathcal{B}(x) = \begin{cases} b_1 = \beta - \alpha, & x < \beta \\ b_2 = \beta + \alpha, & x \geq \beta \end{cases}, \tag{4}$$

where the α and β are the half-distance and center of the binary values b_1 and b_2. Figure 2 shows the binarization of AdaBin, as we can see that, the data on the left of the center will be clustered into b_1 and the data on the right of the center will be clustered into b_2. The distance α and center β will change with different distributions, which help partition the floating point data into two optimal clusters adaptively. For the binarization of weights and activations, we exploit the same form of AdaBin but different optimization strategies.

3.1   Weight Equalization

Low-bit quantization greatly weaken the feature extraction ability of filter weights, especially for 1-bit case. Previous BNNs exploit different methods to optimize the binarized weights. XNOR-Net [37] minimizes the mean squared error (MSE) by multiplying a scale factor, and IR-Net [36] obtains the maximum information entropy by weight reshaping and then conduct the same operation as XNOR-Net. However, these methods can not get accurate quantization error between binarized data and real-valued data due to the following limitations. Firstly, the center position of previous binarized values $\{-1, +1\}$ is always 0, which is not aligned with the center of original real-valued weights. Secondly,

MSE is a simple metric to evaluate the quantization error but do not consider the distribution similarity between binarized data and real-valued data. On the contrary, the Kullback-Leibler divergence (KLD) is a measure on probability distributions [28] and is more accurate to evaluate the information loss than MSE. Therefore, we propose to minimize the KLD to achieve a better distribution-match. We apply the AdaBin for weights binarization as Eq. 5:

$$w_b = \mathcal{B}(w) = \begin{cases} w_{b1} = \beta_w - \alpha_w, & w < \beta_w \\ w_{b2} = \beta_w + \alpha_w, & w \geq \beta_w \end{cases}, \tag{5}$$

where α_w and β_w are distance and center of binarized weights, the binary elements of w_b in the forward is $\beta_w - \alpha_w$ and $\beta_w + \alpha_w$. And the KLD of real-valued distribution and binary distribution can be represented as Eq. 6.

$$D_{KL}(P_r \| P_b) = \int_{x \in w \& w_b} P_r(x) \log \frac{P_r(x)}{P_b(x)} dx, \tag{6}$$

where the $P_r(x)$ and $P_b(x)$ denote the distribution probability of real-valued weights and binarized weights. In order to make the binary distribution more balanced, we need to align its symmetrical center (position of mean value) to the real-valued distribution, so that Eq. 7 can be obtained.

$$\beta_w = \mathbb{E}(w) \approx \frac{1}{c \times k \times k} \sum_{m=0}^{c-1} \sum_{j=0}^{k-1} \sum_{i=0}^{k-1} w_{m,j,i}. \tag{7}$$

Therefore, we can further infer that $P_b(w_{b1}) = P_b(w_{b2}) = 0.5$. Since there is no convinced formula of weight distribution for neural networks, it is difficult to calculate the Kullback-Leibler divergence explicitly. However, the weights in such networks typically assume a bell-shaped distribution with tails [2,4,45], and the both sides are symmetrical on the center, then we can get the α_w as Eq. 8, the detailed proof is in the supplementary.

$$\alpha_w = \frac{\|w - \beta_w\|_2}{\sqrt{c \times k \times k}}, \tag{8}$$

where $\| \cdot \|_2$ denotes the ℓ_2-norm. In our method, the distance α_w and center β_w are channel-wise parameters for weight binarization, and updated along the real-valued weights during the training process. As shown in Fig. 3, without distribution reshaping and the constraint that the center of binary values is 0, AdaBin could equalize the weights to make the binarized distribution best match the real-valued distribution. During the inference, we can decompose the binary weights matrix into 1-bit storage format as following:

$$w_b = \alpha_w b_w + \beta_w, b_w \in \{-1, +1\}. \tag{9}$$

So that the same as the previous BNNs, our method can also achieve about 32× memory saving.

3.2   Gradient-Based Activation Binarization

Activation quantization is a challenging task with low bit-width, and has much more impacts to the final performance than weight. HWGQ [8] proposed to address this challenge by applying a half-wave Gaussian quantization method, based on the observation that activation after Batch Normalization tends to have a symmetric, non-sparse distribution, that is close to Gaussian and ReLU is a half-wave rectifier. However, recent BNNs [35] proposed to replace the ReLU with PReLU [20], which could facilitate the training of binary networks. So that HWGQ can not be further applied because of this limitation. Besides, the distribution of real-valued activations is not as stable as weights, which keeps changing for different inputs. Therefore we can not extract the center and distance from the activations as Eq. 7 and Eq. 8, which brings extra cost to calculate them and will greatly weaken the hardware efficiency of binary neural networks during inference. In order to get the optimal binary activation during training, we propose a gradient-based optimization method to minimize the accuracy degradation arising from activation binarization. Firstly, we apply the AdaBin quantizer to activations as Eq. 10.

$$
a_b = \mathcal{B}(a) = \begin{cases} a_{b1} = \beta_a - \alpha_a, & a < \beta_a \\ a_{b2} = \beta_a + \alpha_a, & a \geq \beta_a \end{cases}, \tag{10}
$$

where α_a and β_a are the distance and center of binarized activations, and the binary set of a_b in the forward is $\{\beta_a - \alpha_a, \beta_a + \alpha_a\}$. To make the binary activations adapt to the dataset as much as possible during the training process, we set α_a and β_a as learnable variables, which could be optimized via backward gradient propagation as total loss decreases. In order to ensure that the training process can converge, we need to clip out the gradient of large activation values in the backward as Eq. 11.

$$
\frac{\partial \mathcal{L}}{\partial a} = \frac{\partial \mathcal{L}}{\partial a_b} * \mathbb{1}_{|\frac{a - \beta_a}{\alpha_a}| \leq 1}, \tag{11}
$$

where $\mathcal{L}$ denotes the output loss, a is the real-valued activation and a_b is the binarized activation, $\mathbb{1}_{|x| \leq 1}$ denotes the indicator function that equals to 1 if $|x| \leq 1$ is true and 0 otherwise. This functionality can be achieved by a composite function of hard tanh and sign, thus we rewrite the Eq. 10 as following:

$$
a_b = \alpha_a \times \text{Sign}(\text{Htanh}(\frac{a - \beta_a}{\alpha_a})) + \beta_a. \tag{12}
$$

For simplicity, we denote $g(x) = \text{Sign}(\text{Htanh}(x))$, then we can get the gradient of α_a and β_a as Eq. 13 in the backward:

$$
\begin{aligned}
\frac{\partial \mathcal{L}}{\partial \alpha_a} &= \frac{\partial \mathcal{L}}{\partial a_b} \frac{\partial a_b}{\partial \alpha_a} = \frac{\partial \mathcal{L}}{\partial a_b} (g(\frac{a - \beta_a}{\alpha_a}) - \frac{a}{\alpha_a} g'(\frac{a - \beta_a}{\alpha_a})), \\
\frac{\partial \mathcal{L}}{\partial \beta_a} &= \frac{\partial \mathcal{L}}{\partial a_b} \frac{\partial a_b}{\partial \beta_a} = \frac{\partial \mathcal{L}}{\partial a_b} (1 - g'(\frac{a - \beta_a}{\alpha_a})),
\end{aligned} \tag{13}
$$

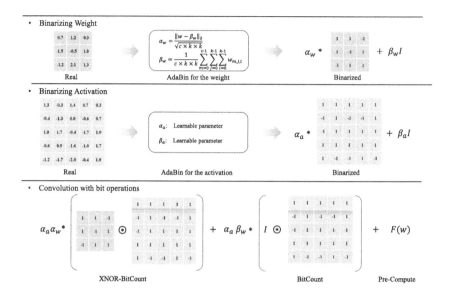

Fig. 3. Binary convolution process. The I represents the identity matrix, and $F(w)$ represents the extra computation with w, which could be pre-computed during the inference.

where $g'(x)$ is the derivative of $g(x)$. We set the initial values of center position β_a and distance α_a to 0 and 1, so that the initial effect of our binary quantizer is equivalent to the sign function [13,34,36]. Then these two parameters of different layers are dynamically updated via gradient descent-based training, and converge to the optimal center and distance values, which is much different from the unified usage of the sign function in the previous BNNs. During inference, the α_a and β_a of all the layers are fixed, then we can binarize the floating point activations into 1-bit as followings:

$$a_b = \alpha_a b_a + \beta_a, b_a \in \{-1, +1\}, \tag{14}$$

where the b_a is the 1-bit storage form and obtained online with input data. Compared with the sign function of previous BNNs, AdaBin will take a little overhead but could significantly improve the feature capacity of activations with the adaptive binary sets for each layer.

3.3 Non-linearity

Prior methods [35] propose to use Parametric Rectified Linear Unit (PReLU) [20] as it is known to facilitate the training of binary networks. PReLU adds an adaptively learnable scaling factor in the negative part and remain unchanged in the positive part. However, we empirically found that the binary values with our proposed AdaBin are almost all positive in very few layers, which invalidate the non-linearity of PReLU. Therefore, to further enhance the representation of

Table 2. Comparisons with state-of-the-art methods on CIFAR-10. W/A denotes the bit width of weights and activations.

Networks	Methods	W/A	Acc. (%)
ResNet-18	Full-precision	32/32	94.8
	RAD [15]	1/1	90.5
	IR-Net [36]		91.5
	RBNN [31]		92.2
	ReCU [42]		92.8
	AdaBin (Ours)		**93.1**
ResNet-20	Full-precision	32/32	91.7
	DoReFa [46]	1/1	79.3
	DSQ [16]		84.1
	IR-Net [36]		86.5
	RBNN [31]		87.8
	AdaBin (Ours)		**88.2**
VGG-Small	Full-precision	32/32	94.1
	LAB [24]	1/1	87.7
	XNOR-Net [37]		89.8
	BNN [13]		89.9
	RAD [15]		90.0
	IR-Net [36]		90.4
	RBNN [31]		91.3
	SLB [43]		92.0
	AdaBin (Ours)		**92.3**

feature maps, we propose to utilize Maxout [17] for the stronger non-linearity in our AdaBin, which is defined as Eq. 15.

$$f_c(x) = \gamma_c^+ \mathrm{ReLU}(x) - \gamma_c^- \mathrm{ReLU}(-x), \qquad (15)$$

where x is the input of the Maxout function, γ_c^+ and γ_c^- are the learnable coefficient for the positive part and negative part of the c-th channel, respectively. Following the setting of PReLU, the initialization of γ_c^+ and γ_c^- are 1 and 0.25.

3.4 Binary Convolution for AdaBin

The goal of BNNs is to replace the computationally expensive multiplication and accumulation with XNOR and BitCount operations. Although the binary sets are not limited to $\{-1, +1\}$, our method can still be accelerated with bit-wise operations by simple linear transformation. As shown in Fig. 3, we can binarize the weights and get the 1-bit matrix b_w offline via Eq. 9, and binarize the activations to get the 1-bit activations b_a online via Eq. 14, then decompose the binary convolution into three items. The first term is the same as the previous BNNs, and the second term only needs to accumulation for one output channel, which can be replaced by BitCount. The third term $F(\mathbf{w})$ could be pre-computed in the inference process. For $n = c = 256$, $k = 3$, $w' = h' = 14$, compared with the binary convolution of IR-Net [37], our method only increases 2.74% operations and 1.37% parameters, which are negligible compared to the total complexity and could achieve 60.85× acceleration and 31× memory saving in theory, the detailed analysis is shown in the supplementary material.

4   Experiments

In this section, we demonstrate the effectiveness of our proposed AdaBin via comparisons with state-of-the-art methods and extensive ablation experiments.

4.1   Results on CIFAR-10

We train AdaBin for 400 epochs with a batch size of 256, where the initial learning rate is set to 0.1 and then decay with CosineAnnealing as IR-Net [36]. We adopt SGD optimizer with a momentum of 0.9, and use the same data augmentation and pre-processing in [21] for training and testing. We compare AdaBin with BNN [13], LAB [24], XNOR-Net [37], DoReFa [46], DSQ [16], RAD [15], IR-Net [36], RBNN [31], ReCU [42] and SLB [43]. Table 2 shows the performance of these methods on CIFAR-10. AdaBin obtains 93.1% accuracy for ResNet-18 architecture, which outperforms the ReCU by 0.3% and reduces the accuracy gap between BNNs and floating-point model to 1.7%. Besides, AdaBin obtains 0.4% accuracy improvement on ResNet-20 compared to the current best method RBNN, and gets 92.3% accuracy while binarizing the weights and activations of VGG-small into 1-bits, which outperforms SLB by 0.3%.

4.2   Results on ImageNet

We train our proposed AdaBin for 120 epochs from scratch and use SGD optimizer with a momentum of 0.9. We set the initial learning rate to 0.1 and then decay with CosineAnnealing following IR-Net [36], and utilize the same data augmentation and pre-processing in [21]. In order to demonstrate the generality of our method, we conduct experiments on two kinds of structures. The first group is the common architectures that are widely used in various computer vision tasks, such as AlexNet [27] and ResNet [21]. Another kind is the binary-specific structures such as BDenseNet [7], MeliusNet [5] and ReActNet [33], which are designed for BNNs and could significantly improve the accuracy with the same amount of parameters as common structures.

Common Structures.   We show the ImageNet performance of AlexNet, ResNet-18 and ResNet-34 on Table 3, and compare AdaBin with recent methods like Bi-Real [34], IR-Net [36], SiBNN [38], RBNN [31], ReCU [42], Real2Bin [35]

Table 3. Comparison with state-of-the-art methods on ImageNet for AlexNet and ResNets. W/A denotes the bit width of weights and activations. * means using the two-step training setting as ReActNet.

Networks	Methods	W/A	Top-1 (%)	Top-5 (%)
AlexNet	Full-precision	32/32	56.6	80.0
	BNN [13]	1/1	27.9	50.4
	DoReFa [46]		43.6	-
	XNOR [37]		44.2	69.2
	SiBNN [38]		50.5	74.6
	AdaBin (Ours)		**53.9**	**77.6**
ResNet-18	Full-precision	32/32	69.6	89.2
	BNN [13]	1/1	42.2	-
	XNOR-Net [37]		51.2	73.2
	Bi-Real [34]		56.4	79.5
	IR-Net [36]		58.1	80.0
	Si-BNN [38]		59.7	81.8
	RBNN [31]		59.9	81.9
	SiMaN [30]		60.1	82.3
	ReCU [42]		61.0	82.6
	AdaBin (Ours)		**63.1**	**84.3**
	IR-Net* [36]	1/1	61.8	83.4
	Real2Bin [35]		65.4	86.2
	ReActNet* [33]		65.5	86.1
	AdaBin* (Ours)		**66.4**	**86.5**
ResNet-34	Full-precision	32/32	73.3	91.3
	ABC-Net [32]	1/1	52.4	76.5
	Bi-Real [34]		62.2	83.9
	IR-Net [36]		62.9	84.1
	SiBNN [38]		63.3	84.4
	RBNN [31]		63.1	84.4
	ReCU [42]		65.1	85.8
	AdaBin (Ours)		**66.4**	**86.6**

and ReActNet [33]. For AlexNet, AdaBin could greatly improve its performance on ImageNet, outperforming the current best method SiBNN by 3.4%, and reducing the accuracy gap between BNNs and floating-point model to only 2.7%. Besides, AdaBin obtains a 63.1% Top-1 accuracy with ResNet-18 structure, which only replaces the binary function and non-linear module of IR-Net [36] with the adaptive quantizer and Maxout but gets 5.0% improvement and outperforms the current best method ReCU by 2.1%. For ResNet-34, AdaBin obtain 1.3% performance improvement compared to the ReCU while binarizing the

Table 4. Comparisons on ImageNet for binary-specific structures.

Networks	Methods	OPs ($\times 10^8$)	Top-1 (%)
BDenseNet28 [7]	Origin	2.09	62.6
	AdaBin	2.11	**63.7** (+1.1)
MeliusNet22 [5]	Origin	2.08	63.6
	AdaBin	2.10	**64.6** (+1.0)
MeliusNet29 [5]	Origin	2.14	65.8
	AdaBin	2.17	**66.5** (+0.7)
MeliusNet42 [5]	Origin	3.25	69.2
	AdaBin	3.28	**69.7** (+0.5)
MeliusNet59 [5]	Origin	5.25	71.0
	AdaBin	5.27	**71.6** (+0.6)
ReActNet-A [33]	Origin	0.87	69.4
	AdaBin	0.88	**70.4** (+1.0)

weights and activations into 1-bits. Besides, we also conduct experiments on ResNet-18 following the training setting as ReActNet. With the two step training strategy, AdaBin could get 66.4% top-1 accuracy, which obtains 0.9% improvement compared to ReActNet.

Binary-Specific Structures. Table 4 shows the performance comparison with BDenseNet, MeliusNet and ReActNet. For BDenseNet28, AdaBin could get 1.1% improvement with the same training setting, which costs negligible extra computational operations. Similarly, when AdaBin is applied to Melius-Net, an advanced version of BDenseNet, it outperforms the original networks by 1.0%, 0.7%, 0.5% and 0.6%, respectively, demonstrating that AdaBin could significantly improve the capacity and quality of binary networks. Besides, we also train the ReActNet-A structure with our AdaBin, following the same training setting with ReActNet [33]. As we can see that, AdaBin could get 1.0% performance improvement with the similar computational operations. Our method could explicitly improve the accuracy of BNNs with a little overhead compared to state-of-the-art methods, as shown in Fig. 1a.

4.3   Results on PASCAL VOC

Table 5 presents the results of object detection on PASCAL VOC dataset for different binary methods. We follow the training strategy as BiDet [34]. The backbone network was pre-trained on ImageNet [14] and then we finetune the whole network for the object detection task. During training, we used the data augmentation techniques in [40], and the Adam optimizer [26] was applied. The learning rate started from 0.001 and decayed twice by multiplying 0.1 at the 160-th and 180-th epoch out of 200 epochs. Following the setting of BiDet [40], we evaluate our proposed AdaBin on both the normal structure and the structure with real-valued shortcut. We compare them with general binary methods BNN [13], XNOR-Net [37] and BiReal-Net [34], and also compare with BiDet [40]

Table 5. The comparison of different methods on PASCAL VOC for object detection. W/A denotes the bit width of weights and activations. * means the the proposed method with extra shortcut for the architectures [40].

Methods	W/A	#Params(M)	FLOPs(M)	mAP
Full-precision	32/32	100.28	31750	72.4
TWN [29]	2/32	24.54	8531	67.8
DoReFa [46]	4/4	29.58	4661	69.2
BNN [13]	1/1	22.06	1275	42.0
XNOR-Net [37]	1/1	22.16	1279	50.2
BiDet [40]	1/1	22.06	1275	52.4
AutoBiDet [39]	1/1	22.06	1275	53.5
AdaBin (Ours)	1/1	22.47	1280	**64.0**
BiReal-Net [34]	1/1	21.88	1277	63.8
BiDet* [40]	1/1	21.88	1277	66.0
AutoBiDet* [39]	1/1	21.88	1277	67.5
AdaBin* (Ours)	1/1	22.47	1282	**69.4**

and AutoBiDet [39], which are specifically designed for high-performance binary detectors. And for reference, we also show the results of the multi-bit quantization method TWN [29] and DoReFa [46] with 4 bit weights and activations. Compared with the previous general BNNs, the proposed AdaBin improves the BNN by 22.0 mAP, XNOR by 13.8 mAP and Bi-Real Net by 5.6 mAP. Even for the task-specific optimization method BiDet, they are 11.6 mAP and 2.6 mAP lower than our method with two structures, and the improved AutoBiDet still lower than AdaBin by 10.5 mAP and 1.9 mAP. Besides, AdaBin with shortcut structure could outperform TWN and DoReFa, which demonstrates that our could significantly enable the binary neural networks to complex tasks.

4.4 Ablation Studies

Effect of AdaBin Quantizer. We conduct the experiments by starting with a vanilla binary neural networks, and then add the AdaBin quantizer of weights and activations gradually. To evaluate different methods fairly, we utilize PReLU for these experiments, which is equal to the Maxout function only with γ^- for negative part. The results are shown in Table 6a, we can see that when combined with existing activation binarization by sign function, our equalization method for binarizing weights could get 0.6% accuracy improvement. Besides, when we free the α_w and β_w to two learnable parameters which are trained in an end-to-end manner as activation, it only get 86.7% accuracy and is much poorer than AdaBin (the last two row). We find that its Kullback-Leibler divergence is also less than AdaBin, which shows the KLD is much important to 1-bit quantization. When keeping the weight binarization as XNOR-Net [37], the proposed

Table 6. (a) Ablation studies of AdaBin for ResNet-20 on CIFAR-10. * means the α_w and β_w are learnable parameters to the binary sets. (b) The ablation studies of Maxout on ImageNet, the scale factor with γ^- equals to PReLU.

W_set	A_set	Non-linearity	Acc.(%)
$\{-\alpha, +\alpha\}$	$\{-1, +1\}$	PReLU	85.7
$\{w_{b1}, w_{b2}\}$	$\{-1, +1\}$	PReLU	86.3
$\{-\alpha, +\alpha\}$	$\{a_{b1}, a_{b2}\}$	PReLU	87.3
$\{w_{b1}, w_{b2}\}$	$\{a_{b1}, a_{b2}\}$	PReLU	87.7
$\{w_{b1}, w_{b2}\}^*$	$\{a_{b1}, a_{b2}\}$	Maxout	86.7
$\{w_{b1}, w_{b2}\}$	$\{a_{b1}, a_{b2}\}$	Maxout	**88.2**

(a) Binary quantizer

Scale factors	Top-1 (%)	Top-5 (%)
None	53.2	77.2
γ^+	62.8	83.9
γ^-	62.9	84.1
γ^-, γ^+	**63.1**	**84.3**

(b) γ in Maxout

gradient-based optimization for binarizing activations could get 1.6% accuracy improvement, as shown in the 3rd row. Combining the proposed weight equalization and activation optimization of AdaBin boosts the accuracy by 2% over vanilla BNN (the 1st vs. 4th row), which shows that AdaBin quantizer could significantly improve the capacity of BNNs.

Effect of γ in Maxout. In addition, we evaluate four activation functions on ImageNet. The first is none, denoting it is an identity connection. The second is Maxout that only with γ^+ for positive part, the third is Maxout only with γ^- for negative part and the last one is the complete Maxout as Eq. 15. As shown in Table 6b, the coefficient of γ^+ and γ^- improve the accuracy by 9.6% and 9.7% individually. The activation function with both coefficients gets the best performance, which justifies the effectiveness of Maxout.

5   Conclusion

In this paper, we propose an adaptive binary method (AdaBin) to binarize weights and activations with optimal value sets, which is the first attempt to relax the constraints of the fixed binary set in prior methods. The proposed AdaBin could make the binary weights best match the real-valued weights and obtain more informative binary activations to enhance the capacity of binary networks. We demonstrate that our method could also be accelerated by XNOR and BitCount operations, achieving 60.85× acceleration and 31× memory saving in theory. Extensive experiments on CIFAR-10 and ImageNet show the superiority of our proposed AdaBin, which outperforms state-of-the-art methods on various architectures, and significantly reduce the performance gap between binary neural networks and real-valued networks. We also present extensive experiments for object detection, which demonstrates that our method can naturally be extended to more complex vision tasks.

Acknowledgments. This work was supported in part by Key-Area Research and Development Program of Guangdong Province No. 2019B010153003, Key Research and Development Program of Shaanxi No. 2022ZDLGY01-08, and Fundamental Research Funds for the Xi'an Jiaotong University No. xhj032021005-05.

References

1. IEEE standard for binary floating-point arithmetic: ANSI/IEEE Std **754–1985**, 1–20 (1985). https://doi.org/10.1109/IEEESTD.1985.82928
2. Anderson, A.G., Berg, C.P.: The high-dimensional geometry of binary neural networks. arXiv preprint arXiv:1705.07199 (2017)
3. Bahdanau, D., Cho, K., Bengio, Y.: Neural machine translation by jointly learning to align and translate. arXiv preprint arXiv:1409.0473 (2014)
4. Baskin, C., et al.: Uniq: uniform noise injection for non-uniform quantization of neural networks. ACM Trans. Comput. Syst. (TOCS) **37**(1–4), 1–15 (2021)
5. Bethge, J., Bartz, C., Yang, H., Chen, Y., Meinel, C.: MeliusNet: an improved network architecture for binary neural networks. In: Proceedings of the IEEE/CVF Winter Conference on Applications of Computer Vision, pp. 1439–1448 (2021)
6. Bethge, J., Yang, H., Bornstein, M., Meinel, C.: Back to simplicity: how to train accurate BNNs from scratch? arXiv preprint arXiv:1906.08637 (2019)
7. Bethge, J., Yang, H., Bornstein, M., Meinel, C.: BinarydenseNet: developing an architecture for binary neural networks. In: Proceedings of the IEEE/CVF International Conference on Computer Vision Workshops (2019)
8. Cai, Z., He, X., Sun, J., Vasconcelos, N.: Deep learning with low precision by half-wave gaussian quantization. In: Proceedings of the IEEE Conference on Computer Vision and Pattern Recognition, pp. 5918–5926 (2017)
9. Chen, H., et al.: AdderNet: do we really need multiplications in deep learning? In: Proceedings of the IEEE/CVF Conference on Computer Vision and Pattern Recognition, pp. 1468–1477 (2020)
10. Chen, X., Zhang, Y., Wang, Y.: MTP: multi-task pruning for efficient semantic segmentation networks. arXiv preprint arXiv:2007.08386 (2020)
11. Chen, X., Zhang, Y., Wang, Y., Shu, H., Xu, C., Xu, C.: Optical flow distillation: towards efficient and stable video style transfer. In: Vedaldi, A., Bischof, H., Brox, T., Frahm, J.-M. (eds.) ECCV 2020. LNCS, vol. 12351, pp. 614–630. Springer, Cham (2020). https://doi.org/10.1007/978-3-030-58539-6_37
12. Choukroun, Y., Kravchik, E., Yang, F., Kisilev, P.: Low-bit quantization of neural networks for efficient inference. In: ICCV Workshops, pp. 3009–3018 (2019)
13. Courbariaux, M., Hubara, I., Soudry, D., El-Yaniv, R., Bengio, Y.: Binarized neural networks: training deep neural networks with weights and activations constrained to+ 1 or-1. arXiv preprint arXiv:1602.02830 (2016)
14. Deng, J., Dong, W., Socher, R., Li, L., Kai Li, Li Fei-Fei: ImageNet: a large-scale hierarchical image database. In: 2009 IEEE Conference on Computer Vision and Pattern Recognition, pp. 248–255 (2009). https://doi.org/10.1109/CVPR.2009.5206848
15. Ding, R., Chin, T.W., Liu, Z., Marculescu, D.: Regularizing activation distribution for training binarized deep networks. In: Proceedings of the IEEE/CVF Conference on Computer Vision and Pattern Recognition, pp. 11408–11417 (2019)
16. Gong, R., et al.: Differentiable soft quantization: Bridging full-precision and low-bit neural networks. In: Proceedings of the IEEE/CVF International Conference on Computer Vision, pp. 4852–4861 (2019)
17. Goodfellow, I., Warde-Farley, D., Mirza, M., Courville, A., Bengio, Y.: Maxout networks. In: International Conference on Machine Learning, pp. 1319–1327. PMLR (2013)
18. Han, K., Wang, Y., Xu, Y., Xu, C., Wu, E., Xu, C.: Training binary neural networks through learning with noisy supervision. In: International Conference on Machine Learning, pp. 4017–4026. PMLR (2020)

19. Han, S., Pool, J., Tran, J., Dally, W.J.: Learning both weights and connections for efficient neural networks. arXiv preprint arXiv:1506.02626 (2015)
20. He, K., Zhang, X., Ren, S., Sun, J.: Delving deep into rectifiers: surpassing human-level performance on imagenet classification. In: Proceedings of the IEEE International Conference on Computer Vision, pp. 1026–1034 (2015)
21. He, K., Zhang, X., Ren, S., Sun, J.: Deep residual learning for image recognition. In: Proceedings of the IEEE Conference on Computer Vision and Pattern Recognition, pp. 770–778 (2016)
22. Hinton, G., et al.: Deep neural networks for acoustic modeling in speech recognition: the shared views of four research groups. IEEE Sig. Process. Mag. **29**(6), 82–97 (2012)
23. Hinton, G., Vinyals, O., Dean, J.: Distilling the knowledge in a neural network. arXiv preprint arXiv:1503.02531 (2015)
24. Hou, L., Yao, Q., Kwok, J.T.: Loss-aware binarization of deep networks. arXiv preprint arXiv:1611.01600 (2016)
25. Huang, G., Liu, Z., Van Der Maaten, L., Weinberger, K.Q.: Densely connected convolutional networks. In: Proceedings of the IEEE Conference on Computer Vision and Pattern Recognition, pp. 4700–4708 (2017)
26. Kingma, D.P., Ba, J.: Adam: a method for stochastic optimization. arXiv preprint arXiv:1412.6980 (2014)
27. Krizhevsky, A., Sutskever, I., Hinton, G.E.: ImageNet classification with deep convolutional neural networks. Adv. Neural. Inf. Process. Syst. **25**, 1097–1105 (2012)
28. Kullback, S., Leibler, R.A.: On information and sufficiency. Ann. Math. Stat. **22**(1), 79–86 (1951)
29. Li, F., Zhang, B., Liu, B.: Ternary weight networks. arXiv preprint arXiv:1605.04711 (2016)
30. Lin, M., et al.: SiMaN: sign-to-magnitude network binarization. arXiv preprint arXiv:2102.07981 (2021)
31. Lin, M., et al.: Rotated binary neural network. In: Advances in Neural Information Processing Systems, vol. 33 (2020)
32. Lin, X., Zhao, C., Pan, W.: Towards accurate binary convolutional neural network. arXiv preprint arXiv:1711.11294 (2017)
33. Liu, Z., Shen, Z., Savvides, M., Cheng, K.-T.: ReActNet: towards precise binary neural network with generalized activation functions. In: Vedaldi, A., Bischof, H., Brox, T., Frahm, J.-M. (eds.) ECCV 2020. LNCS, vol. 12359, pp. 143–159. Springer, Cham (2020). https://doi.org/10.1007/978-3-030-58568-6_9
34. Liu, Z., Wu, B., Luo, W., Yang, X., Liu, W., Cheng, K.T.: Bi-Real Net: enhancing the performance of 1-bit CNNs with improved representational capability and advanced training algorithm. In: Proceedings of the European conference on computer vision (ECCV), pp. 722–737 (2018)
35. Martinez, B., Yang, J., Bulat, A., Tzimiropoulos, G.: Training binary neural networks with real-to-binary convolutions. arXiv preprint arXiv:2003.11535 (2020)
36. Qin, H., et al.: Forward and backward information retention for accurate binary neural networks. In: Proceedings of the IEEE/CVF Conference on Computer Vision and Pattern Recognition, pp. 2250–2259 (2020)
37. Rastegari, M., Ordonez, V., Redmon, J., Farhadi, A.: XNOR-Net: ImageNet classification using binary convolutional neural networks. In: Leibe, B., Matas, J., Sebe, N., Welling, M. (eds.) ECCV 2016. LNCS, vol. 9908, pp. 525–542. Springer, Cham (2016). https://doi.org/10.1007/978-3-319-46493-0_32

38. Wang, P., He, X., Li, G., Zhao, T., Cheng, J.: Sparsity-inducing binarized neural networks. In: Proceedings of the AAAI Conference on Artificial Intelligence, vol. 34, pp. 12192–12199 (2020)
39. Wang, Z., Lu, J., Wu, Z., Zhou, J.: Learning efficient binarized object detectors with information compression. IEEE Trans. Pattern Anal. Mach. Intell. **44**(6), 3082–3095 (2021). https://doi.org/10.1109/TPAMI.2021.3050464
40. Wang, Z., Wu, Z., Lu, J., Zhou, J.: BiDet: an efficient binarized object detector. In: Proceedings of the IEEE/CVF Conference on Computer Vision and Pattern Recognition, pp. 2049–2058 (2020)
41. Xu, Y., Han, K., Xu, C., Tang, Y., Xu, C., Wang, Y.: Learning frequency domain approximation for binary neural networks. In: NeurIPS (2021)
42. Xu, Z., et al.: ReCU: reviving the dead weights in binary neural networks. arXiv preprint arXiv:2103.12369 (2021)
43. Yang, Z., et al.: Searching for low-bit weights in quantized neural networks. arXiv preprint arXiv:2009.08695 (2020)
44. Yu, X., Liu, T., Wang, X., Tao, D.: On compressing deep models by low rank and sparse decomposition. In: Proceedings of the IEEE Conference on Computer Vision and Pattern Recognition, pp. 7370–7379 (2017)
45. Zhang, Z., Shao, W., Gu, J., Wang, X., Luo, P.: Differentiable dynamic quantization with mixed precision and adaptive resolution. In: International Conference on Machine Learning, pp. 12546–12556. PMLR (2021)
46. Zhou, S., Wu, Y., Ni, Z., Zhou, X., Wen, H., Zou, Y.: DoReFa-Net: training low bitwidth convolutional neural networks with low bitwidth gradients. arXiv preprint arXiv:1606.06160 (2016)

Adaptive Token Sampling for Efficient Vision Transformers

Mohsen Fayyaz[1,6]($\boxtimes$), Soroush Abbasi Koohpayegani[2],
Farnoush Rezaei Jafari[3,4], Sunando Sengupta[7], Hamid Reza Vaezi Joze[5],
Eric Sommerlade[7], Hamed Pirsiavash[2], and Jürgen Gall[6]

[1] Microsoft, Berlin, Germany
mohsenfayyaz@microsoft.com
[2] University of California, Davis, Davis, USA
[3] Machine Learning Group, Technische Universität Berlin, Berlin, Germany
[4] Berlin Institute for the Foundations of Learning and Data, Berlin, Germany
[5] Meta Reality Labs, Burlingame, USA
[6] University of Bonn, Bonn, Germany
[7] Microsoft, Reading, UK

Abstract. While state-of-the-art vision transformer models achieve promising results in image classification, they are computationally expensive and require many GFLOPs. Although the GFLOPs of a vision transformer can be decreased by reducing the number of tokens in the network, there is no setting that is optimal for all input images. In this work, we therefore introduce a differentiable parameter-free Adaptive Token Sampler (ATS) module, which can be plugged into any existing vision transformer architecture. ATS empowers vision transformers by scoring and adaptively sampling significant tokens. As a result, the number of tokens is not constant anymore and varies for each input image. By integrating ATS as an additional layer within the current transformer blocks, we can convert them into much more efficient vision transformers with an adaptive number of tokens. Since ATS is a parameter-free module, it can be added to the off-the-shelf pre-trained vision transformers as a plug and play module, thus reducing their GFLOPs without any additional training. Moreover, due to its differentiable design, one can also train a vision transformer equipped with ATS. We evaluate the efficiency of our module in both image and video classification tasks by adding it to multiple SOTA vision transformers. Our proposed module improves the SOTA by reducing their computational costs (GFLOPs) by 2×, while preserving their accuracy on the ImageNet, Kinetics-400, and Kinetics-600 datasets. The code is available at https://adaptivetokensampling.github.io/.

M. Fayyaz, S. A. Koohpayegani, and F. R. Jafari—Equal Contribution
M. Fayyaz and S. A. Koohpayegani—Work has been done during an internship at Microsoft.

Supplementary Information The online version contains supplementary material available at https://doi.org/10.1007/978-3-031-20083-0_24.

1   Introduction

Over the last ten years, there has been a tremendous progress on image and video understanding in the light of new and complex deep learning architectures, which are based on the variants of 2D [23,34,50] and 3D [10,12,17,18,54,56] Convolutional Neural Networks (CNNs). Recently, vision transformers have shown promising results in image classification [13,31,53,63] and action recognition [1,2,39] compared to CNNs. Although vision transformers have a superior representation power, the high computational cost of their transformer blocks make them unsuitable for many edge devices. The computational cost of a vision transformer grows quadratically with respect to the number of tokens it uses. To reduce the number of tokens and thus the computational cost of a vision transformer, DynamicViT [46] proposes a token scoring neural network to predict which tokens are redundant. The approach then keeps a fixed ratio of tokens at each stage. Although DynamicViT reduces the GFLOPs of a given network, its scoring network introduces an additional computational overhead. Furthermore, the scoring network needs to be trained together with the vision transformer and it requires to modify the loss function by adding additional loss terms and hyper-parameters. To alleviate such limitations, EViT [36] employs the attention weights as the tokens' importance scores. A further limitation of both EViT and DynamicViT is that they need to be re-trained if the fixed target ratios need to be changed (*e.g.* due to deployment on a different device). This strongly limits their applications.

In this work, we propose a method to efficiently reduce the number of tokens in any given vision transformer without the mentioned limitations. Our approach is motivated by the observation that in image/action classification, all parts of an input image/video do not contribute equally to the final classification scores and some parts contain irrelevant or redundant information. The amount of relevant information varies depending on the content of an image or video. For instance, in Fig. 7, we can observe examples in which only a few or many patches are required for correct classification. The same holds for the number of tokens used at each stage, as illustrated in Fig. 2. Therefore, we propose an approach that automatically selects an adequate number of tokens at each stage based on the image content, *i.e.* the number of the selected tokens at all network's stages varies for different images, as shown in Fig. 6. It is in contrast to [36,46], where the ratio of the selected tokens needs to be specified for each stage and is constant after training. However, selecting a static number of tokens will on the one hand discard important information for challenging images/videos, which leads to a classification accuracy drop. On the other hand, it will use more tokens than necessary for the easy cases and thus waste computational resources. In this work, we address the question of how a transformer can dynamically adapt its computational resources in a way that not more resources than necessary are used for each input image/video.

To this end, we introduce a novel *Adaptive Token Sampler (ATS)* module. ATS is a differentiable parameter-free module that adaptively down-samples input tokens. To do so, we first assign significance scores to the input tokens by

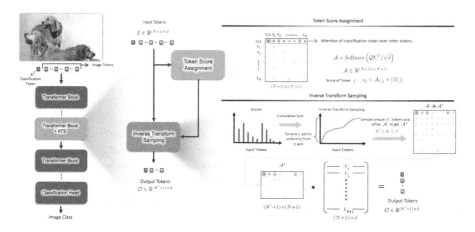

Fig. 1. The Adaptive Token Sampler (ATS) can be integrated into the self-attention layer of any transformer block of a vision transformer model (left). The ATS module takes at each stage a set of input tokens $\mathcal{I}$. The first token is considered as the classification token in each block of the vision transformer. The attention matrix $\mathcal{A}$ is then calculated by the dot product of the queries $\mathcal{Q}$ and keys $\mathcal{K}$, scaled by $\sqrt{d}$. We use the attention weights $\mathcal{A}_{1,2}, \ldots, \mathcal{A}_{1,N+1}$ of the classification token as significance scores $\mathcal{S} \in \mathbb{R}^N$ for pruning the attention matrix $\mathcal{A}$. To reflect the effect of values $\mathcal{V}$ on the output tokens $\mathcal{O}$, we multiply the $\mathcal{A}_{1,j}$ by the magnitude of the corresponding value $\mathcal{V}_j$. We select the significant tokens using inverse transform sampling over the cumulative distribution function of the scores $\mathcal{S}$. Having selected the significant tokens, we then sample the corresponding attention weights (rows of the attention matrix $\mathcal{A}$) to get $\mathcal{A}^s$. Finally, we softly downsample the input tokens $\mathcal{I}$ to output tokens $\mathcal{O}$ using the dot product of $\mathcal{A}^s$ and $\mathcal{V}$.

employing the attention weights of the classification token in the self-attention layer and then select a subset of tokens using inverse transform sampling over the scores. Finally, we softly down-sample the output tokens to remove redundant information with the least amount of information loss. In contrast to [46], our approach does not add any additional learnable parameters to the network. While the ATS module can be added to any off-the-shelf pre-trained vision transformer without any further training, the network equipped with the differentiable ATS module can also be further fine-tuned. Moreover, one may train a model only once and then adjust a maximum limit for the ATS module to adapt it to the resources of different edge devices at the inference time. This eliminates the need of training separate models for different levels of computational resources.

We demonstrate the efficiency of our proposed adaptive token sampler for image classification by integrating it into the current state-of-the-art vision transformers such as DeiT [53], CvT [63], and PS-ViT [68]. As shown in Fig. 4, our approach significantly reduces the GFLOPs of vision transformers of various sizes without significant loss of accuracy. We evaluate the effectiveness of our method by comparing it with other methods designed for reducing the number of tokens, including DynamicViT [46], EViT [36], and Hierarchical Pooling [42]. Extensive experiments on the ImageNet dataset show that our method outperforms existing approaches and provides the best trade-off between computational cost and classification accuracy. We also demonstrate the efficiency of our proposed module for action recognition by adding it to the state-of-the-art video vision transformers such as XViT [2] and TimeSformer [1]. Extensive experiments on the Kinetics-400 and Kinetics-600 datasets show that our method surpasses the performance of existing approaches and leads to the best computational cost/accuracy trade-off. In a nutshell, the adaptive token sampler can significantly scale down the off-the-shelf vision transformers' computational costs and it is therefore very useful for real-world vision-based applications.

2   Related Work

The transformer architecture, which was initially introduced in the NLP community [57], has demonstrated promising performance on various computer vision tasks [3,6,13,39,47,53,66,69–71]. ViT [13] follows the standard transformer architecture to tailor a network that is applicable to images. It splits an input image into a set of non-overlapping patches and produces patch embeddings of lower dimensionality. The network then adds positional embeddings to the patch embeddings and passes them through a number of transformer blocks. An extra learnable class embedding is also added to the patch embeddings to perform classification. Although ViT has shown promising results in image classification, it requires an extensive amount of data to generalize well. DeiT [53] addressed this issue by introducing a distillation token designed to learn from a teacher network. Additionally, it surpassed the performance of ViT. LV-ViT [31] proposed a new objective function for training vision transformers and achieved better performance. TimeSformer [1] proposed a new architecture for video understanding by extending the self-attention mechanism of the standard transformer models to video. The complexity of the TimeSformer's self-attention is $O(T^2S + TS^2)$ where T and S represent temporal and spatial locations respectively. X-ViT [2] reduced this complexity to $O(TS^2)$ by proposing an efficient video transformer.

Besides the accuracy of neural networks, their efficiency plays an important role in deploying them on edge devices. A wide range of techniques have been proposed to speed up the inference of these models. To obtain deep networks that can be deployed on different edge devices, works like [52] proposed more efficient architectures by carefully scaling the depth, width, and resolution of a baseline network based on different resource constraints. [26] aims to meet such resource requirements by introducing hyper-parameters, which can be tuned to build efficient light-weight models. The works [19,59] have adopted quantization techniques to compress and accelerate deep models. Besides quantization techniques, other approaches such as channel pruning [24], run-time neural pruning [45], low-rank matrix decomposition [27,65], and knowledge distillation [25,38] have been used as well to speed up deep networks.

In addition to the works that aim to accelerate the inference of convolutional neural networks, other works aim to improve the efficiency of transformer-based models. In the NLP area, Star-Transformer [21] reduced the number of connections from n^2 to $2n$ by changing the fully-connected topology into a star-shaped structure. TinyBERT [32] improved the network's efficiency by distilling the knowledge of a large teacher BERT into a tiny student network. PoWER-BERT [20] reduced the inference time of the BERT model by identifying and removing redundant and less-informative tokens based on their importance scores estimated from the self-attention weights of the transformer blocks. To reduce the number of FLOPs in character-level language modeling, a new self-attention mechanism with adaptive attention span is proposed in [51]. To enable fast performance in unbatched decoding and improve the scalability of the standard transformers, Scaling Transformers [28] are introduced. These novel transformer architectures are equipped with sparse variants of standard transformer layers.

To improve the efficiency of vision transformers, sparse factorization of the dense attention matrix has been proposed [7], which reduces its complexity to $O(n\sqrt{n})$ for the autoregressive image generation task. [48] tackled this problem by proposing an approach to sparsify the attention matrix. They first cluster all the keys and queries and only consider the similarities of the keys and queries that belong to the same cluster. DynamicViT [46] proposed an additional prediction module that predicts the importance of tokens and discards uninformative tokens for the image classification task. Hierarchical Visual Transformer (HVT) [42] employs token pooling, which is similar to feature map down-sampling in convolutional neural networks, to remove redundant tokens. PS-ViT [68] incorporates a progressive sampling module that iteratively learns to sample distinctive input tokens instead of uniformly sampling input tokens from all over the image. The sampled tokens are then fed into a vision transformer module with fewer transformer encoder layers compared to ViT. TokenLearner [49] introduces a learnable tokenization module that can reduce the computational cost by learning few important tokens conditioned on the input. They have demonstrated that their approach can be applied to both image and video understanding tasks. Token Pooling [40] down-samples tokens by grouping them into a set of clusters and returning the cluster centers. A concurrent work [36] introduces a token reorganization method that first identifies top-k important tokens by computing token attentiveness between the tokens and the classification token and then fuses less informative tokens. IA-RED^2 [41] proposes an interpretability-aware redundancy reduction framework for vision transformers that discards less informative patches in the input data. Most of the mentioned approaches improve the efficiency of vision transformers by introducing architectural changes to the original models or by adding modules that add extra learnable parameters to the networks, while our parameter-free adaptive module can be incorporated into off-the-shelf architectures and reduces their computational complexity without significant accuracy drop and even without requiring any further training.

3   Adaptive Token Sampler

State-of-the-art vision transformers are computationally expensive since their computational costs grow quadratically with respect to the number of tokens, which is static at all stages of the network and corresponds to the number of input patches. Convolutional neural networks deal with the computational cost by reducing the resolution within the network using various pooling operations. It means that the spatial or temporal resolution decreases at the later stages of the network. However, applying such simple strategies, *i.e.* pooling operations with fixed kernels, to vision transformers is not straightforward since the tokens are permutation invariant. Moreover, such static down-sampling approaches are not optimal. On the one hand, a fixed down-sampling method discards important information at some locations of the image or video, like details of the object. On the other hand, it still includes many redundant features that do not contribute to the classification accuracy, for instance, when dealing with an image with a homogeneous background. Therefore, we propose an approach that dynamically adapts the number of tokens at each stage of the network based on the input data such that important information is not discarded and no computational resources are wasted for processing redundant information.

To this end, we propose our novel Adaptive Token Sampler (ATS) module. ATS is a parameter-free differentiable module to sample significant tokens over the input tokens. In our ATS module, we first assign significance scores to the N input tokens and then select a subset of these tokens based on their scores. The upper bound of GFLOPs can be set by defining a maximum limit for the number of tokens sampled, denoted by K. Since the sampling procedure can sample some input tokens several times, we only keep one instance of a token. The number of sampled tokens K' is thus usually lower than K and varies among input images or videos (Fig. 6). Figure 1 gives an overview of our proposed approach.

3.1   Token Scoring

Let $\mathcal{I} \in \mathbb{R}^{(N+1) \times d}$ be the input tokens of a self-attention layer with $N+1$ tokens. Before forwarding the input tokens through the model, ViT concatenates a classification token to the input tokens. The corresponding output token at the final transformer block is then fed to the classification head to get the class probabilities. Practically, this token is placed as the first token in each block and it is considered as a classification token. While we keep the classification token, our goal is to reduce the output tokens $\mathcal{O} \in \mathbb{R}^{(K'+1) \times d}$ such that K' is dynamically adapted based on the input image or video and $K' \leq K \leq N$, where K is a parameter that controls the maximum number of sampled tokens. Figure 6 shows how the number of sampled tokens K' varies for different input data and stages of a network. We first describe how each token is scored.

In a standard self-attention layer [57], the queries $\mathcal{Q} \in \mathbb{R}^{(N+1) \times d}$, keys $\mathcal{K} \in \mathbb{R}^{(N+1) \times d}$, and values $\mathcal{V} \in \mathbb{R}^{(N+1) \times d}$ are computed from the input tokens $\mathcal{I} \in \mathbb{R}^{(N+1) \times d}$. The attention matrix $\mathcal{A}$ is then calculated by the dot product of the queries and keys, scaled by $\sqrt{d}$:

$$\mathcal{A} = \text{Softmax}\left(\mathcal{Q}\mathcal{K}^T / \sqrt{d}\right). \tag{1}$$

Due to the Softmax function, each row of $\mathcal{A} \in \mathbb{R}^{(N+1) \times (N+1)}$ sums up to 1. The output tokens are then calculated using a combination of the values weighted by the attention weights:

$$\mathcal{O} = \mathcal{A}\mathcal{V}. \tag{2}$$

Each row of $\mathcal{A}$ contains the attention weights of an output token. The weights indicate the contributions of all input tokens to the output token. Since $\mathcal{A}_{1,:}$ contains the attention weights of the classification token, $\mathcal{A}_{1,j}$ represents the importance of the input token j for the output classification token. Thus, we use the weights $\mathcal{A}_{1,2}, \ldots, \mathcal{A}_{1,N+1}$ as significance scores for pruning the attention matrix $\mathcal{A}$, as illustrated in Fig. 1. Note that $\mathcal{A}_{1,1}$ is not used since we keep the classification token. As the output tokens $\mathcal{O}$ depend on both $\mathcal{A}$ and $\mathcal{V}$ (2), we also take into account the norm of $\mathcal{V}_j$ for calculating the jth token's significance score. The motivation is that values having a norm close to zero have a low impact and their corresponding tokens are thus less significant. In our experiments, we show that multiplying $\mathcal{A}_{1,j}$ with the norm of $\mathcal{V}_j$ improves the results. The significance score of a token j is thus given by

$$\mathcal{S}_j = \frac{\mathcal{A}_{1,j} \times ||\mathcal{V}_j||}{\sum_{i=2} \mathcal{A}_{1,i} \times ||\mathcal{V}_i||} \tag{3}$$

where $i, j \in \{2 \ldots N\}$. For a multi-head attention layer, we calculate the scores for each head and then sum the scores over all heads.

3.2   Token Sampling

Having computed the significance scores of all tokens, we can prune their corresponding rows from the attention matrix $\mathcal{A}$. To do so, a naive approach is to select K tokens with the highest significance scores (top-K selection). However, this approach does not perform well, as we show in our experiments and it can not adaptively select $K' \leq K$ tokens. The reason why the top-K selection does not work well is that it discards all tokens with lower scores. Some of these tokens, however, can be useful in particular at the earlier stages when the features are less discriminative. For instance, having multiple tokens with similar keys, which may occur in the early stages, will lower their corresponding attention weights due to the Softmax function. Although one of these tokens would be beneficial at the later stages, taking the top-K tokens might discard all of them. Therefore, we suggest sampling tokens based on their significance scores. In this case, the probability of sampling one of the several similar tokens is equal to the sum of their scores. We also observe that the proposed sampling procedure selects more tokens at the earlier stages than the later stages as shown in Fig. 2.

For the sampling step, we suggest using inverse transform sampling to sample tokens based on their significance scores $\mathcal{S}$ (3). Since the scores are normalized, they can be interpreted as probabilities and we can calculate the cumulative distribution function (CDF) of $\mathcal{S}$:

$$\text{CDF}_i = \sum_{j=2}^{j=i} \mathcal{S}_j. \tag{4}$$

Note that we start with the second token since we keep the first token. Having the cumulative distribution function, we obtain the sampling function by taking the inverse of the CDF:

$$\Psi(k) = \text{CDF}^{-1}(k) \tag{5}$$

where $k \in [0, 1]$. In other words, the significance scores are used to calculate the mapping function between the indices of the original tokens and the sampled tokens. To obtain K samples, we can sample K-times from the uniform distribution $U[0, 1]$. While such randomization might be desirable for some applications, deterministic inference is in most cases preferred. Therefore, we use a fixed sampling scheme for training and inference by choosing $k = \{\frac{1}{2K}, \frac{3}{2K} \cdots, \frac{2K-1}{2K}\}$. Since $\Psi(.) \in \mathbb{R}$, we consider the indices of the tokens with the nearest significant scores as the sampling indices.

If a token is sampled more than once, we only keep one instance. As a consequence, the number of unique indices K' is often lower than K as shown in Fig. 6. In fact, $K' < K$ if there is at least one token with a score $S_j \geq 2/K$. In the two extreme cases, either only one dominant token is selected and $K' = 1$ or $K' = K$ if the scores are more or less balanced. Interestingly, more tokens are selected at the earlier stages, where the features are less discriminative and the attention weights are more balanced, and less at the later stages, as shown in Fig. 2. The number and locations of tokens also vary for different input images, as shown in Fig. 7. For images with a homogeneous background that covers a large part of the image, only a few tokens are sampled. In this case, the tokens cover the object in the foreground and are sparsely but uniformly sampled from the background. In cluttered images, many tokens are required. It illustrates the importance of making the token sampling procedure adaptive.

Having indices of the sampled tokens, we refine the attention matrix $\mathcal{A} \in \mathbb{R}^{(N+1)\times(N+1)}$ by selecting the rows that correspond to the sampled $K' + 1$ tokens. We denote the refined attention matrix by $\mathcal{A}^s \in \mathbb{R}^{(K'+1)\times(N+1)}$. To obtain the output tokens $\mathcal{O} \in \mathbb{R}^{(K'+1)\times d}$, we thus replace the attention matrix $\mathcal{A}$ by the refined one $\mathcal{A}^s$ in (2) such that:

$$\mathcal{O} = \mathcal{A}^s \mathcal{V}. \tag{6}$$

These output tokens are then taken as input for the next stage. In our experimental evaluation, we demonstrate the efficiency of the proposed adaptive token sampler, which can be added to any vision transformer.

4   Experiments

In this section, we analyze the performance of our ATS module by adding it to different backbone models and evaluating them on ImageNet [9], Kinetics-400 [33], and Kinetics-600 [4], which are large-scale image and video classification datasets, respectively. In addition, we perform several ablation studies to better analyze our method. For the image classification task, we evaluate our proposed method on the ImageNet [9] dataset with 1.3M images and 1K classes. For the

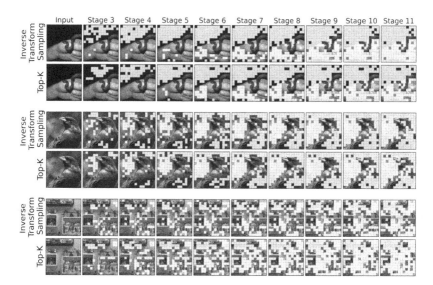

Fig. 2. Visualization of the gradual token sampling procedure in the multi-stage DeiT-S+ATS model. As it can be seen, at each stage, those tokens that are considered to be less significant to the classification are masked and the ones that have contributed the most to the model's prediction are sampled. We also visualize the token sampling results with Top-K selection to have a better comparison to our Inverse Transform Sampling.

action classification task, we evaluate our approach on the Kinetics-400 [33] and Kinetics-600 [4] datasets with 400 and 600 human action classes, respectively. We use the standard training/testing splits and protocols provided by the ImageNet and Kinetics datasets. If not otherwise stated, the number of output tokens of the ATS module are limited by the number of its input tokens. For example, we set $K = 197$ in case of DeiT-S [53]. For the image classification task, we follow the fine-tuning setup of [46] if not mentioned otherwise. The fine-tuned models are initialized by their backbones' pre-trained weights and trained for 30 epochs using PyTorch AdamW optimizer (lr= 5e−4, batch size = 8 × 96). We use the cosine scheduler for training the networks. For more implementation details and also information regarding action classification models, please refer to the supplementary materials.

4.1   Ablation Experiments

First, we analyze different setups for our ATS module. Then, we investigate the efficiency and effects of our ATS module when incorporated in different models. If not otherwise stated, we use the pre-trained DeiT-S [53] model as the backbone and we do not fine-tune the model after adding the ATS module. We integrate the ATS module into stage 3 of the DeiT-S [53] model. We report the results on the ImageNet-1K validation set in all of our ablation studies.

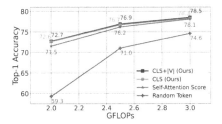

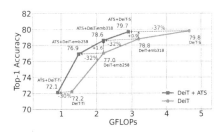

Fig. 3. Impact of different score assignment methods. To achieve different GFLOPs levels, we bound the value of K from above such that the average GFLOPs of our adaptive models over the ImageNet validation set reaches the desired level. For more details, please refer to the supplementary material.

Fig. 4. Performance comparison on the ImageNet validation set. Our proposed adaptive token sampling method achieves a state-of-the-art trade-off between accuracy and GFLOPs. We can reduce the GFLOPs of DeiT-S by 37% while almost maintaining the accuracy.

Significance Scores. As mentioned in Sec. 3.1, we use the attention weights of the classification token as significance scores for selecting our candidate tokens. In this experiment, we evaluate different approaches for calculating significance scores. Instead of directly using the attention weights of the classification token, we sum over the attention weights of all tokens (rows of the attention matrix) to find tokens with highest significance over other tokens. We show the results of this method in Fig. 3 labeled as Self-Attention score. As it can be seen, using the attention weights of the classification token performs better specially in lower FLOPs regimes. The results show that the attention weights of the classification token are a much stronger signal for selecting the candidate tokens. The reason for this is that the classification token will later be used to predict the class probabilities in the final stage of the model. Thus, its corresponding attention weights show which tokens have more impact on the output classification token. Whereas summing over all attention weights only shows us the tokens with highest attention from all other tokens, which may not necessarily be useful for the classification token. To better investigate this observation, we also randomly select another token rather than the classification token and use its attention weights for the score assignment. As shown, this approach performs much worse than the other ones both in high and low FLOPs regimes. We also investigate the impact of using the L_2 norm of the values in (3). As it can be seen in Fig. 3, it improves the results by about 0.2%.

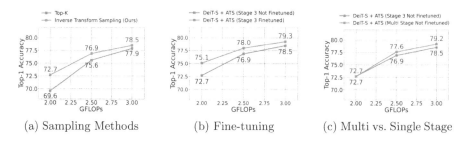

(a) Sampling Methods   (b) Fine-tuning   (c) Multi vs. Single Stage

Fig. 5. For the model with Top-K selection (fixed-rate sampling) (5a), we set K such that the model operates at a desired GFLOPs level. In all three plots, we control the GFLOPs level of our adaptive models as in Fig. 3. We use DeiT-S [53] for these experiments. For more details, please refer to the supplementary material.

Candidate Tokens Selection. As mentioned in Sec. 3.2, we employ the inverse transform sampling approach to softly downsample the input tokens. To better investigate this approach, we also evaluate the model's performance when picking the top K tokens with highest significance scores S. As it can be seen in Fig. 5a, our inverse transform sampling approach outperforms the Top-K selection both in high and low GFLOPs regimes. As discussed earlier, our inverse transform sampling approach based on the CDF of the scores does not hardly discard all tokens with lower significance scores and hence provides a more diverse set of tokens for the following layers. Since earlier transformer blocks are more prone to predict noisier attention weights for the classification token, such a diversified set of tokens can better contribute to the output classification token of the final transformer block. Moreover, the Top-K selection method will result in a fixed token selection rate at every stage that limits the performance of the backbone model. This is shown by the examples in Fig. 2. For a cluttered image (bottom), inverse transform sampling keeps a higher number of tokens across all transformer blocks compared to the Top-K selection and hence preserves the accuracy. On the other hand, for a less detailed image (top), inverse transform sampling will retain less tokens, which results in less computation cost.

Model Scaling. Another common approach for changing the GFLOPs/accuracy trade-off of networks is to change the channel dimension. To demonstrate the efficiency of our adaptive token sampling method, we thus vary the dimensionality. To this end, we first train several DeiT models with different embedding dimensions. Then, we integrate our ATS module into the stages 3 to 11 of these DeiT backbones and fine-tune the networks. In Fig. 4, we can observe that our approach can reduce GFLOPs by 37% while maintaining the DeiT-S backbone's accuracy. We can also observe that the GFLOPs reduction rate gets higher as we increase the embedding dimensions from 192 (DeiT-Ti) to 384 (DeiT-S). The results show that our ATS module can reduce the computation cost of the models with larger embedding dimensions to their variants with smaller embedding dimensions.

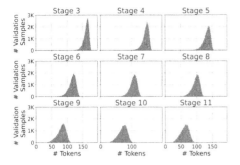

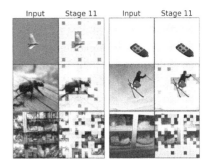

Fig. 6. Histogram of the number of sampled tokens at each ATS stage of our multi-stage DeiT-S+ATS model on the ImageNet validation set. The y-axis corresponds to the number of images and the x-axis to the number of sampled tokens.

Fig. 7. ATS samples less tokens for images with fewer details (top), and a higher number of tokens for more detailed images (bottom). We show the token downsampling results after all ATS stages. For this experiment, we use a multi-stage Deit-S+ATS model.

Visualizations. To better understand the way our ATS module operates, we visualize our token sampling procedure (Inverse Transform Sampling) in Fig. 2. We have incorporated our ATS module in the stages 3 to 11 of the DeiT-S network. The tokens that are discarded at each stage are represented as a mask over the input image. We observe that our DeiT-S+ATS model has gradually removed irrelevant tokens and sampled those tokens which are more significant to the model's prediction. In both examples, our method identified the tokens that are related to the target objects as the most informative tokens.

Adaptive Sampling. In this experiment, we investigate the adaptivity of our token sampling approach. We evaluate our multi-stage DeiT-S+ATS model on the ImageNet validation set. In Fig. 6, we visualize histograms of the number of sampled tokens at each ATS stage. We can observe that the number of selected tokens varies at all stages and for all images. We also qualitatively analyze this nice property of our ATS module in Figs. 2 and 7. We can observe that our ATS module selects a higher number of tokens when it deals with detailed and complex images while it selects a lower number of tokens for less detailed images.

Fine-Tuning. To explore the influence of fine-tuning on the performance of our approach, we fine-tune a DeiT-S+ATS model on the ImageNet training set. We compare the model with and without fine-tuning. As shown in Fig. 5b, fine-tuning can improve the accuracy of the model. In this experiment, we fine-tune the model with $K = 197$ but test it with different K values to reach the desired GFLOPs levels.

ATS Stages. In this experiment, we explore the effect of single-stage and multi-stage integration of the ATS block into vision transformer models. In the single-stage model, we integrate our ATS module into the stage 3 of DeiT-S. In the multi-stage model, we integrate our ATS module into the stages 3 to 11 of DeiT-S. As it can be seen in Fig. 5c, the multi-stage DeiT-S+ATS performs better than the single-stage DeiT-S+ATS. This is due to the fact that a multi-stage DeiT-S+ATS model can gradually decrease the GFLOPs by discarding fewer tokens in earlier stages, while a single-stage DeiT-S+ATS model has to discard more tokens in earlier stages to reach the same GFLOPs level.

4.2   Comparison with State-of-the-Art

We compare the performances of our adaptive models, which are equipped with the ATS module, with state-of-the-art vision transformers for image and video classification on the ImageNet-1K [9] and Kinetics [4,33] datasets, respectively. Tables 1-3 show the results of this comparison. For the image classification task, we incorporate our ATS module into the stages 3 to 11 of the DeiT-S [53] model. We also integrate our ATS module into the 1st to 9th blocks of the 3rd stage of CvT-13 [63] and CvT-21 [63], and into stages 1–9 of the transformer module of PS-ViT [68]. We fine-tune the models on the ImageNet-1K training set. We also evaluate our ATS module for action recognition. To this end, we add our module to the XViT [2] and TimeSformer [1] video vision transformers. For more details, please refer to the supplementary materials.

Image Classification. As it can be seen in Table 1, our ATS module decreases the GFLOPs of all vision transformer models without adding any extra parameters to the backbone models. For the DeiT-S+ATS model, we observe a 37% GFLOPs reduction with only 0.1% reduction of the top-1 accuracy. For the CvT+ATS models, we can also observe a GFLOPs reduction of about 30% with $0.1 - 0.2\%$ reduction of the top-1 accuracy. More details on the efficiency of our ATS module can be found in the supplementary materials (e.g. through-put). Comparing ATS to DynamicViT [46] and HVT [42], which add additional parameters to the model, our approach achieves a better trade-off between accuracy and GFLOPs. Our method also outperforms the EViT-DeiT-S [36] model trained for 30 epochs without adding any extra trainable parameters to the model. We note that the EViT-DeiT-S model can improve its top-1 accuracy by around 0.3% when it is trained for much more training epochs (e.g. 100 epochs). For a fair comparison, we have considered the 30 epochs training setup used by Dynamic-ViT [46]. We have also added our ATS module to the PS-ViT net-work [68]. As it can be seen in Table 1, although PS-ViT has drastically lower GFLOPs compared to its counterparts, its GFLOPs can be further decreased by incorporating ATS in it.

Action Recognition. As it can be seen in Tables 2 and 3, our ATS module dras-tically decreases the GFLOPs of all video vision transformers without adding any extra parameters to the backbone models. For the XViT+ATS model, we observe a 39% GFLOPs reduction with only 0.2% reduction of the top-1 accuracy on Kinetics-400 and a 38.7% GFLOPs reduction with only 0.1% drop of the top-1 accuracy on Kinetics-600. We observe that XViT+ATS achieves a similar accu-racy as TokenLearner [49] on Kinetics-600 while requiring 17.6× less GFLOPs. For TimeSformer-L+ATS, we can observe 50.8% GFLOPs reduction with only

Table 1. Comparison of the multi-stage ATS models with state-of-the-art image classification models with comparable GFLOPs on the ImageNet validation set. We equip DeiT-S [53], PS-ViT [68], and variants of CvT [63] with our ATS module and fine-tune them on the ImageNet training set.

Model	Params (M)	GFLOPs	Resolution	Top-1
ViT-Base/16 [13]	86.6	17.6	224	77.9
HVT-S-1 [42]	22.09	2.4	224	78.0
IA-RED² [41]		2.9	224	78.6
DynamicViT-DeiT-S (30 Epochs) [46]	22.77	2.9	224	79.3
EViT-DeiT-S (30 epochs) [36]	22.1	3.0	224	79.5
DeiT-S+ATS (Ours)	**22.05**	**2.9**	224	79.7
DeiT-S [53]	22.05	4.6	224	79.8
PVT-Small [60]	24.5	3.8	224	79.8
CoaT Mini [64]	10.0	6.8	224	80.8
CrossViT-S [5]	26.7	5.6	224	81.0
PVT-Medium [60]	44.2	6.7	224	81.2
Swin-T [39]	29.0	4.5	766	81.3
T2T-ViT-14 [67]	22.0	5.2	224	81.5
CPVT-Small-GAP [8]	23.0	4.6	817	81.5
CvT-13 [63]	20.0	4.5	224	81.6
CvT-13+ATS (Ours)	20.0	**3.2**	224	81.4
PS-ViT-B/14 [68]	21.3	5.4	224	81.7
PS-ViT-B/14+ATS (Ours)	21.3	**3.7**	224	81.5
RegNetY-8G [44]	39.0	8.0	224	81.7
DeiT-Base/16 [53]	86.6	17.6	224	81.8
CoaT-Lite Small [64]	20.0	4.0	224	81.9
T2T-ViT-19 [67]	39.2	8.9	224	81.9
CrossViT-B [5]	104.7	21.2	224	82.2
T2T-ViT-24 [67]	64.1	14.1	224	82.3
PS-ViT-B/18 [68]	21.3	8.8	224	82.3
PS-ViT-B/18+ATS (Ours)	21.3	**5.6**	224	82.2
CvT-21 [63]	32.0	7.1	224	82.5
CvT-21+ATS (Ours)	32.0	**5.1**	224	82.3
TNT-B [22]	66.0	14.1	224	82.8
RegNetY-16G [44]	84.0	16.0	224	82.9
Swin-S [39]	50.0	8.7	224	83.0
CvT-13₃₈₄ [63]	20.0	16.3	384	83.0
CvT-13₃₈₄+ATS (Ours)	20.0	**11.7**	384	82.9
Swin-B [39]	88.0	15.4	224	83.3
LV-ViT-S [30]	26.2	6.6	224	83.3
CvT-21₃₈₄ [63]	32.0	24.9	384	83.3
CvT-21₃₈₄+ATS (Ours)	32.0	**17.4**	384	83.1

Table 2. Comparison with state-of-the-art on Kinetics-400.

Model	Top-1	Top-5	Views	GFLOPs
STC [10]	68.7	88.5	112	–
bLVNet [15]	73.5	91.2	3 × 3	840
STM [37]	73.7	91.6	–	–
TEA [35]	76.1	92.5	10 × 3	2,100
TSM R50 [29]	74.7	–	10 × 3	650
I3D NL [62]	77.7	93.3	10 × 3	10,800
CorrNet-101 [58]	79.2	–	10 × 3	6,700
ip-CSN-152 [55]	79.2	93.8	10 × 3	3,270
HATNet [11]	79.3	–	–	–
SlowFast 16 × 8 R101+NL [18]	79.8	93.9	10 × 3	7,020
X3D-XXL [17]	80.4	94.6	10 × 3	5,823
TimeSformer-L [1]	80.7	94.7	1 × 3	7,140
TimeSformer-L+ATS (Ours)	80.5	94.6	1 × 3	**3,510**
ViViT-L/16x2 [1]	80.6	94.7	4 × 3	17,352
MViT-B, 64³ [14]	81.2	95.1	3 × 3	4,095
X-ViT (16×) [2]	80.2	94.7	1 × 3	425
X-ViT+ATS (16×) (Ours)	80.0	94.6	1 × 3	**259**
TokenLearner 16at12 (L/16) [49]	82.1	–	4 × 3	4,596

Table 3. Comparison with state-of-the-art on Kinetics-600.

Model	Top-1	Top-5	Views	GFLOPs
AttentionNAS [61]	79.8	94.4	–	1,034
LGD-3D R101 [43]	81.5	95.6	10 × 3	–
HATNET [11]	81.6	–	–	–
SlowFast R101+NL [18]	81.8	95.1	10 × 3	3,480
X3D-XL [17]	81.9	95.5	10 × 3	1,452
X3D-XL+ATFR [16]	82.1	95.6	10 × 3	768
TimeSformer-HR [1]	82.4	96	1 × 3	5,110
TimeSformer-HR+ATS (Ours)	82.2	96	1 × 3	**3,103**
ViViT-L/16x2 [1]	82.5	95.6	4 × 3	17,352
Swin-B [39]	84.0	96.5	4 × 3	3,384
MViT-B-24, 32³ [14]	84.1	96.5	1 × 5	7,080
TokenLearner 16at12(L/16) [49]	84.4	96.0	4 × 3	9,192
X-ViT (16×) [2]	84.5	96.3	1 × 3	850
X-ViT+ATS (16×) (Ours)	84.4	96.2	1 × 3	**521**

0.2% drop of the top-1 accuracy on Kinetics-400. These results demonstrate the generality of our approach that can be applied to both image and video representations.

5 Conclusion

Designing computationally efficient vision transformer models for image and video recognition is a challenging task. In this work, we proposed a novel differentiable parameter-free module called Adaptive Token Sampler (ATS) to increase the efficiency of vision transformers for image and video classification. The new ATS module selects the most informative and distinctive tokens within the stages of a vision transformer model such that as much tokens as needed but not more than necessary are used for each input image or video clip. By integrating our ATS module into the attention layers of current vision transformers, which use a static number of tokens, we can convert them into much more efficient vision transformers with an adaptive number of tokens. We showed that our ATS module can be added to off-the-shelf pre-trained vision transformers as a plug and play module, thus reducing their GFLOPs without any additional training, but it is also possible to train a vision transformer equipped with the ATS module thanks to its differentiable design. We evaluated our approach on the ImageNet-1K image recognition dataset and incorporated our ATS module into three different state-of-the-art vision transformers. We also demonstrated the generality of our approach by incorporating it into different state-of-the-art video vision transformers and evaluating them on the Kinetics-400 and Kinetics-600 datasets. The results show that the ATS module decreases the computation cost (GFLOPs) between 27% and 50.8% with a negligible accuracy drop. Although our experiments are focused on image and video vision transformers, we believe that our approach can also work in other domains such as audio.

Acknowledgments. Farnoush Rezaei Jafari acknowledges support by the Federal Ministry of Education and Research (BMBF) for the Berlin Institute for the Foundations of Learning and Data (BIFOLD) (01IS18037A). Juergen Gall has been supported by the Deutsche Forschungsgemeinschaft (DFG, German Research Foundation) under Germany's Excellence Strategy - EXC 2070 - 390732324, GA1927/4-2 (FOR 2535 Anticipating Human Behavior), and the ERC Consolidator Grant FORHUE (101044724).

References

1. Bertasius, G., Wang, H., Torresani, L.: Is space-time attention all you need for video understanding. In: International Conference on Machine Learning (ICML) (2021)
2. Bulat, A., Perez Rua, J.M., Sudhakaran, S., Martinez, B., Tzimiropoulos, G.: Space-time mixing attention for video transformer. In: Advances in Neural Information Processing Systems (NeurIPS) (2021)
3. Carion, N., Massa, F., Synnaeve, G., Usunier, N., Kirillov, A., Zagoruyko, S.: End-to-End object detection with transformers. In: Vedaldi, A., Bischof, H., Brox, T., Frahm, J.-M. (eds.) ECCV 2020. LNCS, vol. 12346, pp. 213–229. Springer, Cham (2020). https://doi.org/10.1007/978-3-030-58452-8_13
4. Carreira, J., Noland, E., Banki-Horvath, A., Hillier, C., Zisserman, A.: A short note about kinetics-600. In: arXiv preprint. arXiv:1808.01340v1 (2018)
5. Chen, C.F., Fan, Q., Panda, R.: Crossvit: cross-attention multi-scale vision transformer for image classification. In: IEEE/CVF International Conference on Computer Vision (ICCV) (2021)
6. Cheng, B., Schwing, A.G., Kirillov, A.: Per-pixel classification is not all you need for semantic segmentation. In: Advances in Neural Information Processing Systems (NeurIPS) (2021)
7. Child, R., Gray, S., Radford, A., Sutskever, I.: Generating long sequences with sparse transformers. In: arXiv preprint. arXiv:1904.10509 (2019)
8. Chu, X., Tian, Z., Zhang, B., Wang, X., Wei, X., Xia, H., Shen, C.: Conditional positional encodings for vision transformers. In: arXiv preprint. arXiv:2102.10882 (2021)
9. Deng, J., Dong, W., Socher, R., Li, L.J., Li, K., Fei-Fei, L.: Imagenet: a large-scale hierarchical image database. In: IEEE/CVF Conference on Computer Vision and Pattern Recognition (CVPR) (2009)

10. Diba, A., et al.: Spatio-temporal channel correlation networks for action classification. In: Ferrari, V., Hebert, M., Sminchisescu, C., Weiss, Y. (eds.) ECCV 2018. LNCS, vol. 11208, pp. 299–315. Springer, Cham (2018). https://doi.org/10.1007/978-3-030-01225-0_18

11. Diba, A., et al.: Large scale holistic video understanding. In: Vedaldi, A., Bischof, H., Brox, T., Frahm, J.-M. (eds.) ECCV 2020. LNCS, vol. 12350, pp. 593–610. Springer, Cham (2020). https://doi.org/10.1007/978-3-030-58558-7_35

12. Diba, A., Sharma, V., Gool, L.V., Stiefelhagen, R.: Dynamonet: dynamic action and motion network. In: IEEE/CVF International Conference on Computer Vision (ICCV) (2019)

13. Dosovitskiy, A., et al.: An image is worth 16x16 words: transformers for image recognition at scale. In: International Conference on Learning Representations (ICLR) (2021)

14. Fan, H., Xiong, B., Mangalam, K., Li, Y., Yan, Z., Malik, J., Feichtenhofer, C.: Multiscale vision transformers. In: IEEE/CVF International Conference on Computer Vision (ICCV) (2021)

15. Fan, Q., Chen, C.F.R., Kuehne, H., Pistoia, M., Cox, D.: More Is Less: learning Efficient Video Representations by Temporal Aggregation Modules. In: Advances in Neural Information Processing Systems (NeurIPS) (2019)

16. Fayyaz, M., Bahrami, E., Diba, A., Noroozi, M., Adeli, E., Van Gool, L., Gall, J.: 3d cnns with adaptive temporal feature resolutions. In: IEEE/CVF Conference on Computer Vision and Pattern Recognition (CVPR) (2021)

17. Feichtenhofer, C.: X3d: Expanding architectures for efficient video recognition. In: IEEE/CVF Conference on Computer Vision and Pattern Recognition (CVPR) (2020)

18. Feichtenhofer, C., Fan, H., Malik, J., He, K.: Slowfast networks for video recognition. In: IEEE/CVF international conference on computer vision (ICCV) (2019)

19. Gong, Y., Liu, L., Yang, M., Bourdev, L.: Compressing deep convolutional networks using vector quantization. In: arXiv preprint. arXiv:1412.6115 (2014)

20. Goyal, S., Choudhury, A.R., Raje, S.M., Chakaravarthy, V.T., Sabharwal, Y., Verma, A.: Power-bert: accelerating bert inference via progressive word-vector elimination. In: International Conference on Machine Learning (ICML) (2020)

21. Guo, Q., Qiu, X., Liu, P., Shao, Y., Xue, X., Zhang, Z.: Star-transformer. In: arXiv preprint. arXiv:1902.09113 (2019)

22. Han, K., Xiao, A., Wu, E., Guo, J., Xu, C., Wang, Y.: Transformer in transformer. In: Advances in Neural Information Processing Systems (NeurIPS) (2021)

23. He, K., Zhang, X., Ren, S., Sun, J.: Deep residual learning for image recognition. In: IEEE/CVF Conference on Computer Vision and Pattern Recognition (CVPR) (2016)

24. He, Y., Zhang, X., Sun, J.: Channel pruning for accelerating very deep neural networks. In: IEEE/CVF International Conference on Computer Vision (ICCV) (2017)

25. Hinton, G., Vinyals, O., Dean, J.: Distilling the knowledge in a neural network. In: NIPS Deep Learning and Representation Learning Workshop (2015)

26. Howard, A.G., et al.: Mobilenets: efficient convolutional neural networks for mobile vision applications. In: arXiv preprint. arXiv:1704.04861 (2017)

27. Jaderberg, M., Vedaldi, A., Zisserman, A.: Speeding up convolutional neural networks with low rank expansions. In: arXiv preprint. arXiv:1405.3866 (2014)

28. Jaszczur, S., et al.: Sparse is enough in scaling transformers. In: Advances in Neural Information Processing Systems (NeurIPS) (2021)

29. Jiang, B., Wang, M., Gan, W., Wu, W., Yan, J.: Stm: spatiotemporal and motion encoding for action recognition. In: IEEE/CVF International Conference on Computer Vision (ICCV) (2019)
30. Jiang, Z., et al.: Token labeling: training a 85.5% top-1 accuracy vision transformer with 56m parameters on imagenet. In: arXiv preprint. arXiv:2104.10858v2 (2021)
31. Jiang, Z., et al.: All tokens matter: token labeling for training better vision transformers. In: Advances in Neural Information Processing Systems (NeurIPS) (2021)
32. Jiao, X., et al.: Tinybert: distilling bert for natural language understanding. In: arXiv preprint. arXiv:1909.10351 (2020)
33. Kay, W., et al.: The kinetics human action video dataset. In: arXiv preprint. arXiv:1705.06950 (2017)
34. Krizhevsky, A.: One weird trick for parallelizing convolutional neural networks. In: ArXiv preprint. arXiv:1404.5997 (2014)
35. Li, Y., Ji, B., Shi, X., Zhang, J., Kang, B., Wang, L.: Tea: temporal excitation and aggregation for action recognition. In: IEEE/CVF Conference on Computer Vision and Pattern Recognition (CVPR) (2020)
36. Liang, Y., Ge, C., Tong, Z., Song, Y., Wang, J., Xie, P.: Not all patches are what you need: Expediting vision transformers via token reorganizations. In: International Conference on Learning Representations (ICLR) (2022)
37. Lin, J., Gan, C., Han, S.: Tsm: Temporal shift module for efficient video understanding. In: IEEE/CVF International Conference on Computer Vision (ICCV) (2019)
38. Liu, B., Rao, Y., Lu, J., Zhou, J., Hsieh, C.-J.: MetaDistiller: network self-boosting via meta-learned top-down distillation. In: Vedaldi, A., Bischof, H., Brox, T., Frahm, J.-M. (eds.) ECCV 2020. LNCS, vol. 12359, pp. 694–709. Springer, Cham (2020). https://doi.org/10.1007/978-3-030-58568-6_41
39. Liu, Z., et al.: Swin transformer: hierarchical vision transformer using shifted windows. In: Proceedings of the IEEE/CVF International Conference on Computer Vision (ICCV) (2021)
40. Marin, D., Chang, J.H.R., Ranjan, A., Prabhu, A.K., Rastegari, M., Tuzel, O.: Token pooling in vision transformers. arXiv preprint. arXiv:2110.03860 (2021)
41. Pan, B., Panda, R., Jiang, Y., Wang, Z., Feris, R., Oliva, A.: IA-RED^2: interpretability-aware redundancy reduction for vision transformers. In: Advances in Neural Information Processing Systems (NeurIPS) (2021)
42. Pan, Z., Zhuang, B., Liu, J., He, H., Cai, J.: Scalable vision transformers with hierarchical pooling. In: IEEE/CVF International Conference on Computer Vision (ICCV) (2021)
43. Qiu, Z., Yao, T., Ngo, C.W., Tian, X., Mei, T.: Learning spatio-temporal representation with local and global diffusion. In: IEEE/CVF Conference on Computer Vision and Pattern Recognition (CVPR) (2019)
44. Radosavovic, I., Kosaraju, R.P., Girshick, R., He, K., Dollár, P.: Designing network design spaces. In: IEEE/CVF Conference on Computer Vision and Pattern Recognition (CVPR) (2020)
45. Rao, Y., Lu, J., Lin, J., Zhou, J.: Runtime network routing for efficient image classification. In: IEEE Transactions on Pattern Analysis and Machine Intelligence, vol. 41, pp. 2291-2304 (2019)
46. Rao, Y., Zhao, W., Liu, B., Lu, J., Zhou, J., Hsieh, C.J.: Dynamicvit: efficient vision transformers with dynamic token sparsification. In: Advances in Neural Information Processing Systems (NeurIPS) (2021)

47. Rao, Y., Zhao, W., Zhu, Z., Lu, J., Zhou, J.: Global filter networks for image classification. In: Advances in Neural Information Processing Systems (NeurIPS) (2021)
48. Roy, A., Saffar, M., Vaswani, A., Grangier, D.: Efficient content-based sparse attention with routing transformers. In: Transactions of the Association for Computational Linguistics, vol. 9, pp. 53–68 (2021)
49. Ryoo, M.S., Piergiovanni, A., Arnab, A., Dehghani, M., Angelova, A.: Tokenlearner: What can 8 learned tokens do for images and videos? arXiv preprint. arXiv:2106.11297 (2021)
50. Simonyan, K., Zisserman, A.: Very deep convolutional networks for large-scale image recognition. arXiv preprint. arXiv:1409.1556 (2015)
51. Sukhbaatar, S., Grave, E., Bojanowski, P., Joulin, A.: Adaptive attention span in transformers. In: ACL (2019)
52. Tan, M., Le, Q.: EfficientNet: rethinking model scaling for convolutional neural networks. In: International Conference on Machine Learning (ICML) (2019)
53. Touvron, H., Cord, M., Douze, M., Massa, F., Sablayrolles, A., Jegou, H.: Training data-efficient image transformers and distillation through attention. In: International Conference on Machine Learning (ICML) (2021)
54. Tran, D., Bourdev, L., Fergus, R., Torresani, L., Paluri, M.: Learning spatiotemporal features with 3d convolutional networks. In: IEEE International Conference on Computer Vision (ICCV) (2015)
55. Tran, D., Wang, H., Torresani, L., Feiszli, M.: Video classification with channel-separated convolutional networks. In: IEEE/CVF International Conference on Computer Vision (ICCV) (2019)
56. Tran, D., Wang, H., Torresani, L., Ray, J., LeCun, Y., Paluri, M.: A closer look at spatiotemporal convolutions for action recognition. In: IEEE Conference on Computer Vision and Pattern Recognition (CVPR) (2018)
57. Vaswani, A., et al.: Attention is all you need. In: Advances in Neural Information Processing Systems (NeuRIPS) (2017)
58. Wang, H., Tran, D., Torresani, L., Feiszli, M.: Video modeling with correlation networks. In: IEEE/CVF Conference on Computer Vision and Pattern Recognition (CVPR) (2020)
59. Wang, K., Liu, Z., Lin, Y., Lin, J., Han, S.: Haq: Hardware-aware automated quantization with mixed precision. In: IEEE Conference on Computer Vision and Pattern Recognition (CVPR) (2019)
60. Wang, W., Xie, E., Li, X., Fan, D.P., Song, K., Liang, D., Lu, T., Luo, P., Shao, L.: Pyramid vision transformer: a versatile backbone for dense prediction without convolutions. In: IEEE/CVF International Conference on Computer Vision (ICCV) (2021)
61. Wang, X., et al.: AttentionNAS: spatiotemporal attention cell search for video classification. In: Vedaldi, A., Bischof, H., Brox, T., Frahm, J.-M. (eds.) ECCV 2020. LNCS, vol. 12353, pp. 449–465. Springer, Cham (2020). https://doi.org/10.1007/978-3-030-58598-3_27
62. Wang, X., Girshick, R., Gupta, A., He, K.: Non-local neural networks. In: IEEE/CVF Conference on Computer Vision and Pattern Recognition (CVPR) (2018)
63. Wu, H., Xiao, B., Codella, N., Liu, M., Dai, X., Yuan, L., Zhang, L.: Cvt: introducing convolutions to vision transformers. In: IEEE/CVF International Conference on Computer Vision (ICCV) (2021)
64. Xu, W., Xu, Y., Chang, T., Tu, Z.: Co-scale conv-attentional image transformers. In: arXiv preprint. arXiv:2104.06399 (2021)

65. Yu, X., Liu, T., Wang, X., Tao, D.: On compressing deep models by low rank and sparse decomposition. In: IEEE/CVF Conference on Computer Vision and Pattern Recognition (CVPR) (2017)
66. Yu, X., Rao, Y., Wang, Z., Liu, Z., Lu, J., Zhou, J.: Pointr: diverse point cloud completion with geometry-aware transformers. In: IEEE/CVF International Conference on Computer Vision (ICCV) (2021)
67. Yuan, L., et al.: Tokens-to-token vit: training vision transformers from scratch on imagenet. In: arXiv preprint. arXiv:2101.11986 (2021)
68. Yue, X., Sun, S., Kuang, Z., Wei, M., Torr, P., Zhang, W., Lin, D.: Vision transformer with progressive sampling. In: IEEE/CVF International Conference on Computer Vision (ICCV) (2021)
69. Zhao, H., Jiang, L., Jia, J., Torr, P., Koltun, V.: Point transformer. In: IEEE/CVF International Conference on Computer Vision (ICCV) (2021)
70. Zheng, S., et al.: Rethinking semantic segmentation from a sequence-to-sequence perspective with transformers. In: IEEE/CVF Conference on Computer Vision and Pattern Recognition (CVPR) (2021)
71. Zhou, D., et al.: Deepvit: towards deeper vision transformer. arXiv preprint. arXiv:2103.11886 (2021)

Weight Fixing Networks

Christopher Subia-Waud$^{(\boxtimes)}$ and Srinandan Dasmahapatra

University of Southampton, Southampton SO17 1BJ, UK
{cc2u18,sd}@soton.ac.uk

Abstract. Modern iterations of deep learning models contain millions (billions) of unique parameters-each represented by a b-bit number. Popular attempts at compressing neural networks (such as pruning and quantisation) have shown that many of the parameters are superfluous, which we can remove (pruning) or express with $b' < b$ bits (quantisation) without hindering performance. Here we look to go much further in minimising the information content of networks. Rather than a channel or layerwise encoding, we look to lossless whole-network quantisation to minimise the entropy and number of unique parameters in a network. We propose a new method, which we call Weight Fixing Networks (WFN) that we design to realise four model outcome objectives: i) very few unique weights, ii) low-entropy weight encodings, iii) unique weight values which are amenable to energy-saving versions of hardware multiplication, and iv) lossless task-performance. Some of these goals are conflicting. To best balance these conflicts, we combine a few novel (and some well-trodden) tricks; a novel regularisation term, (i, ii) a view of clustering cost as relative distance change (i, ii, iv), and a focus on whole-network re-use of weights (i, iii). Our Imagenet experiments demonstrate lossless compression using 56x fewer unique weights and a 1.9x lower weight-space entropy than SOTA quantisation approaches. Code and model saves can be found at github.com/subiawaud/Weight_Fix_Networks.

Keywords: Compression · Quantization · Minimal description length · Deep learning accelerators

1 Introduction

Deep learning models have a seemingly inexorable trajectory toward growth. Growth in applicability, performance, investment, and optimism. Unfortunately, one area of growth is lamentable - the ever-growing energy and storage costs required to train and make predictions. To fully realise the promise of deep learning methods, work is needed to reduce these costs without hindering task performance.

Here we look to contribute a methodology and refocus towards the goal of reducing both the number of bits to describe a network as well the total number

Supplementary Information The online version contains supplementary material available at https://doi.org/10.1007/978-3-031-20083-0_25.

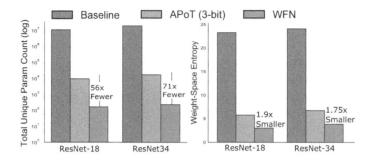

Fig. 1. WFN reduces the total number of weights and the entropy of a network far further than other quantisation works. **Left**: The total number of unique parameters left after quantisation is 56x fewer than APoT for ResNet-18 trained on the Imagenet dataset and 71x for the ResNet-34 model. **Right**: The entropy of the parameters across the network is 1.9x and 1.65x smaller when using the WFN approach over APoT.

of *unique* weights in a network. The motivation to do so is driven both by practical considerations of accelerator designs [1,2], as well as the more theoretical persuasions of the *Minimal Description Length* (MDL) principle [3–5] as a way of determining a *good* model.

The Minimal Description Length. Chaitin's hypothesis captures the thinking behind the MDL principle with the statement "comprehension is compression" [6,7]. That is, to learn is to compress. In this setting, the *best* model minimises the combined cost of describing both the model and the prediction errors. Deep learning models have shown themselves adept at minimising the latter, but it is not clear that we sufficiently compress the model description through unbounded standard gradient descent training. One way to think about MDL in relation to deep learning compression is the following[1]:

Imagine that Bob wants to communicate ground-truth targets to Alice. To achieve this, he can forward both a predictive model and its errors, compressed to as few bits as possible without losing any information. Given these two components, Alice can pass the input data into the model and, given the communicated errors from Bob, make any adjustments to its output to retrieve the desired ground truth. This formulation is the *two-part compression* approach to the problem of learning [6]. The MDL principle says that the *best* model is obtained by minimising the sum of the lengths of the codes for model and errors.

Although the MDL treatment of learning is largely theoretical, it has motivated the design of compressed network architectures [8–10]. We believe that a more direct optimisation to minimise the information theoretic content could bear fruit for downstream hardware translations, but let us start by setting out the description quantities we wish to minimise.

---

[1] Originally posed in 'Keeping neural networks simple by minimizing the description length of the weights' [8].

Describing a Solution. Describing a classification model's error is well captured with the cross-entropy loss. From an information-theoretic perspective, the cross-entropy loss measures the average message length per data point needed to describe the difference in predicted and target output distributions. Essentially, large errors cost more to describe than small errors.

The cost of describing a model is more complex, requiring two elements to be communicated – the weight values, and their arrangement. A metric used to capture both components is the *representation cost*, as outlined in Deep K-means [11] (Eq. 5 below). Here, the cost of describing the model is defined as the summed bit-width of each weight representation times the number of times each weight is used in an inference calculation.

Minimising the Representation Costs in Accelerators. Importantly, this representation cost as a model description can be translated directly into accelerator design savings, as shown by the seminal work in Deep Compression [12] and subsequent accelerator design, EIE [2]. Here the authors cluster neural networks' weights and use Huffman encoding to represent/describe the network cheaply. From an information-theoretic perspective, the motivation for Huffman encoding is simple; this encoding scheme is likely to give us a compressed result closest to our underlying weight-space entropy. However, this work was not focused on the information content, so why was it advantageous to an accelerator design? For this, we need to look at where the computation costs are most concentrated.

The most expensive energy costs in inference calculations lie in memory reads [13,14]. For every off-chip DRAM data read, we pay the equivalent of over two hundred 32-bit multiplications in energy costs[2] [13]. This energy cost concentration has led to the pursuit of data movement minimisation schemes using accelerator dataflow mappings [2,15–17]. These mappings aim to store as much of the network as possible close to computation and maximise weight re-use. From an algorithmic perspective, this makes networks with low entropy content desirable. To make use of a low entropy weight-space, a dataflow mapping can store compressed codewords for each weight which, when decoded, point to the address of the full precision weight, which itself is stored in cheaper access close-to-compute memory. The idea is that the addition of the codeword storage and access costs plus the full weight value access costs can be much smaller than in the unquantised network [11,18]. Several accelerator designs have successfully implemented such a scheme [1,2].

As a simple illustrative example, let us define a filter in a network post-quantisation with the values $[900, 104, 211, 104, 104, 104, 399, 211, 104]$. This network has an entropy of 1.65, meaning each weight can be represented, on average, using a minimum of 1.65 bits, compared to the 9.8bits ($\log(900)$) needed for an uncompressed version of the same network. Using Huffman encoding, we get close to this bound by encoding the network weights as $w \mapsto c(w)$ with:

$$c(104) = 1, c(211) = 01, c(399) = 001, c(900) = 000$$

---

[2] 45 nm process.

and the complete filter can be represented as "000101111001011", totalling just 15 bits, 1.66 bits on average per value. Each decoded codeword points a corresponding weight in the reduced set of unique weights required for computation. These unique weights (since there are very few of them) can be stored close to compute on memory units, processing element scratchpads or SRAM cache depending on the hardware flavour [19,20], all of which have minimal (almost free) access costs. The storage and data movement cost of the encoded weights plus the close-to-compute weight access should be smaller than the storage and movement costs of directly representing the network with the weight values. This link – between minimising the model description and reducing accelerator representational costs – motivates our approach.

Objectives. So we ask ourselves what we could do algorithmically to maximise the benefit of accelerator dataflows and minimise the description length. Since Huffman encoding is used extensively in accelerator designs, we focus on finding networks that reduce the network description when compressed using this scheme. To do this, we first focus on reducing the number of *unique* weights a network uses. Fewer *unique* weights whilst fixing the network topology and the total number of parameters will mean that more weights are re-used more often. Further gains can be achieved if we can concentrate the distribution of weights around a handful of values, enabling frequently used weights to be stored cheaply, close to compute. Finally, we ask what the ideal values of these weights would be. From a computational perspective, not all multiplications are created equal. Powers-of-two, for example, can be implemented as simple bit-shifts. Mapping the weights used most to these values offers potential further optimisation in hardware. Putting these three requirements together: few unique weights; a low-entropy encoding with a distribution of weights highly concentrated around a tiny subset of values; and a focus on powers-of-two values for weights — all motivated to both minimise the MDL as well as the computation costs in accelerator designs — we present our contribution.

Weight Fixing Networks. Our work's overarching objective is to transform a network comprising many weights of any value (limited only by value precision) to one with the same number of weights but just a few unique values and concentrate the weights around an even smaller subset of weights. Rather than selecting the unique weights a priori, we let the optimisation guide the process in an iterative *cluster-then-train* approach. We cluster an ever-increasing subset of weights to one of a few cluster centroids in each iteration. We map the pre-trained network weights to these cluster centroids, which constitute a pool of unique weights. The training stage follows standard gradient descent optimisation to minimise performance loss with two key additions. Firstly, only an ever decreasing subset of the weights are *free* to be updated. We also use a new regularisation term to penalise weights with large relative distances to their nearest clusters. We iteratively cluster subsets of weights to their nearest cluster centre, with the way we determine which subset to move a core component of our contribution.

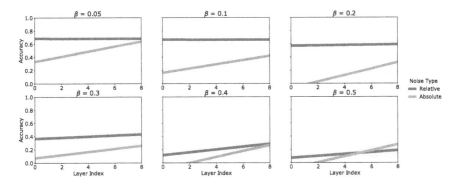

Fig. 2. We explore adding relative vs absolute noise to each of the layers (x-axis). The layer index indicates which layer was selected to have noise added. Each layer index is a separate experiment with the 95% confidence intervals shaded.

Small Relative Distance Change. Rather than selecting subsets with small Euclidean distances to cluster centres, or those that have small magnitude [21], we make the simple observation that the *relative* – as opposed to absolute – weight change matters. We find that the tolerated distance δw_i we can move a weight w_i when quantised depends on the relative distance $|(\delta w_i/w_i)|$. When the new value $w_i + \delta w_i = 0$ — as is the case for pruning methods — then the magnitude of the weight *is* the distance. However, this is not the case more generally. We demonstrate the importance of quantising with small relative changes using simple empirical observations. Using a pre-trained ResNet-18 model, we measure changes to network accuracy when adding relative vs absolute noise to the layers' weights and measure the accuracy change. For relative noise we choose a scale parameter $\beta|w_i^l|$ for each layer-l weight w_i^l, and set $w_i^l \leftarrow w_i^l + \beta|w_i^l|\varepsilon$, $\varepsilon \sim \mathcal{N}(0,1)$. For additive noise perturbations, all weights w_i^l are perturbed by the mean absolute value of weights $\overline{|w^l|}$ in layer l scaled by β: $w_i^l \leftarrow w_i^l + \beta\overline{|w^l|}\varepsilon$.

We run each layer-β combination experiment multiple times – to account for fluctuation in the randomised noise – and present the results in Fig. 2. Even though the mean variation of noise added is the same, noise relative to the original weight value (multiplicative noise) is much better tolerated than absolute (additive noise). Since moving weights to quantisation centres is analogous to adding noise, we translate these results into our approach and prioritise clustering weights with small relative distances first. We find that avoiding significant quantisation errors requires ensuring that $\frac{|\delta w_i|}{|w_i|}$ is small for all weights. If this is not possible, then performance could suffer. In this case, we create an additional cluster centroid in the vicinity of an existing cluster to reduce this relative distance. Our work also challenges the almost universal trend in the literature [22–27] of leaving the first and last layers either at full precision or 8-bit. Instead, we attempt a full network quantisation. The cost of not quantising the first layer – which typically requires the most re-use of weights due to the larger resolution of input maps – and the final linear layer – which often contains the largest number of unique weight values – is too significant to ignore.

With multiple stages of training and clustering, we finish with an appreciably reduced set of unique weights. We introduce a regularisation term that encourages non-uniform, high probability regions in the weight distribution to induce a lower-entropy weight-space. The initial choice of cluster centroids as powers-of-two helps us meet our third objective – energy-saving multiplication. Overall we find four distinct advantages over the works reviewed:

- We assign a cluster value to *all weights* — including the first and last layers.
- We emphasise a low entropy encoding with a regularisation term, achieving entropies smaller than those seen using 3-bit quantisation approaches – over which we report superior performance.
- We require no additional layer-wise scaling, sharing the unique weights across all layers.
- WFN substantially reduces the number of unique parameters in a network when compared to existing SOTA quantisation approaches.

2   Related Work

Clip and Scale Quantisation. Typical quantisation approaches reduce the number of bits used to represent components of the network. Quantisation has been applied to all parts of the network with varying success; the weights, gradients, and activations have all been attempted [23, 25, 28–31]. Primarily, these approaches are motivated by the need to reduce the energy costs of the multiplication of 32-bit floating-point numbers. This form of quantisation maps the weight w_i to $w_i' = s\,\text{round}(w_i)$, where round() is a predetermined rounding function and s a scaling factor. The scaling factor (determined by a clipping range) can be learned channel-wise [24, 32], or more commonly, layerwise in separate formulations. This results in different channels/layers having a diverse pool of codebooks for the network weights/activations/gradients. Quantisation can be performed without training — known as post-training quantisation, or with added training steps – called quantisation-aware training (QAT). Retraining incurs higher initial costs but results in superior performance.

A clipping+scaling quantisation example relevant to our own is the work of [21], where the authors restrict the layerwise rounding of weights to powers-of-two. The use of powers-of-two has the additional benefit of energy-cheap bit-shift multiplication. A follow-up piece of work [22] suggests additive powers-of-two (APoT) instead to capture the pre-quantised distribution of weights better.

Weight Sharing Quantisation. Other formulations of quantisation do not use clipping and scaling factors [11, 33, 39]. Instead, they adopt clustering techniques to cluster the weights and fix the weight values to their assigned group cluster centroid. These weights are stored as codebook indices, allowing for compressed representation methods such as Huffman encoding to squeeze the network further.

We build on these methods, which, unlike clipping+scaling quantisation techniques, share the pool of weights across the entire network. The work by [11] is of particular interest since both the motivation and approach are related to ours. Here the authors use a *spectrally relaxed* k-means regularisation term to encourage the network weights to be more amenable to clustering. In their case, they focus on a filter-row codebook inspired by the row-stationary dataflow used in some accelerator designs [15]. However, their formulation is explored only for convolution, and they restrict clustering to groups of weights (filter rows) rather than individual weights due to computational limitations as recalibrating the k-means regularisation term is expensive during training.

Similarly, [33,35] focus on quantising groups of weights into single codewords rather than the individual weights themselves. Weight-sharing approaches similar to ours include [36]. The authors use the distance from an evolving Gaussian mixture as a regularisation term to prepare the clustering weights. Although it is successful with small dataset-model combinations, the complex optimisation — particularly the additional requirement for Inverse-Gamma priors to lower-bound the mixture variance to prevent mode collapse — limits the method's practical applicability due to the high computational costs of training. In our formulation, the weights already fixed no longer contribute to the regularisation prior, reducing the computational overhead. We further reduce computation by not starting with a complete set of cluster centres but add them iteratively when needed.

3 Method

Algorithm 1: Clustering Np_t weights at the tth iteration.

---

while $|W_{\text{fixed}}^{t+1}| \leq Np_t$ **do**

    $\omega \leftarrow 0$

    $\text{fixed}_{\text{new}} \leftarrow [\ ]$

    while $\text{fixed}_{\text{new}}$ *is empty* **do**

        Increase the order $\omega \leftarrow \omega + 1$

        for each $i = 1 \ldots, |W_{\text{free}}^{t+1}|$

        $c_*^\omega(i) = \min_{c \in \widetilde{C}^\omega} D_{\text{rel}}^+(w_i, c)$

        for each cluster centre $c_k^\omega \in \widetilde{C}^\omega$

        $n_k^\omega = \sum_i \mathbb{I}[c_k^\omega = c_*^\omega(i)]$

        $k^* = \arg\max_k n_k^\omega$

        Sort: $[w_1', \ldots, w_N'] \leftarrow [w_1, \ldots, w_N]$, $w_i' = w_{\pi(i)}$, π permutation

        where $D_{\text{rel}}^+(w_i', c_{k*}^\omega) < D_{\text{rel}}^+(w_{i+1}', c_{k*}^\omega)$

        $i = 1$, mean $= D_{\text{rel}}^+(w_1', c_{k*}^\omega)$

        while mean $\leq \delta^t$ **do**

            $\text{fixed}_{\text{new}} \leftarrow w_i'$

            mean $\leftarrow \frac{i}{i+1} * \text{mean} + \frac{1}{i+1} D_{\text{rel}}^+(w_{i+1}', c_{k*}^\omega)$

            $i \leftarrow i + 1$

    Assign all the weights in $\text{fixed}_{\text{new}}$ to cluster centre $c_*^\omega(i)$, moving them from W_{free}^{t+1} to W_{fixed}^{t+1}

Quantisation. Consider a network $\mathcal{N}$ parameterised by N weights $W = \{w_1, ..., w_N\}$. Quantising a network is the process of reformulating $\mathcal{N}' \leftarrow \mathcal{N}$ where the new network $\mathcal{N}'$ contains weights which all take values from a reduced pool of k cluster centres $C = \{c_1, ..., c_k\}$ where $k \ll N$. After quantisation, each of the connection weights in the original network is replaced by one of the cluster centres $w_i \leftarrow c_j$, $W' = \{w_i' | w_i' \in C, i = 1, \cdots, N\}$, $|W'| = k$, where W' is the set of weights of the new network $\mathcal{N}'$, which has the same topology as the original $\mathcal{N}$.

Method Outline. WFN is comprised of T *fixing iterations* where each iteration $t \in T$ has a training and a clustering stage. The clustering stage is tasked with partitioning the weights into two subsets $W = W_{\text{fixed}}^t \cup W_{\text{free}}^t$. W_{fixed}^t is the set of weights set equal to one of the cluster centre values $c_k \in C$. These *fixed* weights $w_i \in W_{\text{fixed}}^t$ are not updated by gradient decent in this, nor any subsequent training stages. In contrast, the *free-weights* denoted by W_{free}^t remain trainable during the next training stage. With each subsequent iteration t we increase the proportion $p_t = \frac{|W_{\text{fixed}}^t|}{|W|}$ of weights that take on fixed cluster centre values, with $p_0 < p_1 \ldots < p_T = 1$. By iteration T, all weights will be fixed to one of the cluster centres. The training stage combines gradient descent on a cross-entropy classification loss, along with a regularisation term that encourages tight-clusters, in order to maintain lossless performance as we fix more of the weights to cluster centres.

Clustering Stage. In the clustering stage, we work backwards from our goal of minimising the relative distance travelled for each of the weights to determine which values cluster centres $c_i \in C$ should take. For a weight $w_i \in W$ and cluster centre $c_j \in C$ we define a relative distance measure $D_{\text{rel}}(w_i, c_j) = \frac{|w_i - c_j|}{|w_i|}$. To use this in determining the cluster centres, we enforce a threshold δ on this relative distance, $D_{\text{rel}}(w_i, c_j) \leq \delta$ for small δ. We can then define the cluster centres $c_j \in C$ which make this possible using a simple recurrence relation. Assume we have a starting cluster centre value c_j, we want the neighbouring cluster value c_{j+1} to be such that if a network weight w_i is between these clusters $w_i \in [c_j, \frac{c_{j+1}+c_j}{2}]$ then $D_{\text{rel}}(w_i, c_j) \leq \delta$. Plugging in $\frac{c_{j+1}+c_j}{2}$ and c_j into D_{rel} and setting it equal to δ we have:

$$\frac{|\frac{c_{j+1}+c_j}{2} - c_j|}{\frac{c_{j+1}+c_j}{2}} = \delta, \text{ leading to } c_{j+1} = c_j\left(\frac{1+\delta}{1-\delta}\right), \ 0 < \delta < 1, \tag{1}$$

a recurrence relation that provides the next cluster centre value given the previous one. With this, we can generate all the cluster centres given some boundary condition $c_0 = \delta_0$. δ_0 is the lower-bound cluster threshold, and all weights w_i for $|w_i| < \delta_0$ are set to 0 (pruned). This lower bound serves two purposes: firstly, it reduces the number of proposal cluster centres which would otherwise increase exponentially in density around zero, and additionally, the zero-valued weights makes the network more sparse. This will allow sparsity-leveraging hardware to avoid operations that use these weights, reducing the computational overhead. As an upper-bound, we stop the recurrence once a cluster centre is larger than the maximum weight in the network, $\max_j |c_j| \leq \max_i |w_i|$, $w_i \in W, c_j \in C$.

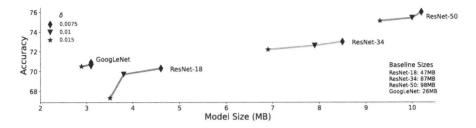

Fig. 3. The accuracy vs model size trade-off can be controlled by the δ parameter. All experiments shown are using the ImageNet dataset, accuracy refers to top-1.

Generating the Proposed Cluster Centres. Putting this together, we have a starting point $c_0 = \delta_0$, a recurrence relation to produce cluster centres given c_0 that maintains a relative distance change when weights are moved to their nearest cluster centre, and a centre generation stopping condition $c_j \leq \max_{i \in W} |w_i|$, $c_j \in C$. We use the δ_0 value as our first proposed cluster centre c_0 with the recurrence relation generating a proposed cluster set of size s. Since all these values will contain only positive values, we join this set with its negated version along with a zero value to create a proposal cluster set $C^S = \{a(\frac{1+\delta}{1-\delta})^j \delta_0 \mid j = 0, 1 \cdots s; \ a = +1, 0, -1\}$ of size $2s + 1$.

To account for the zero threshold δ_0 and for ease of notation as we advance, we make a slight amendment to the definition of the relative distance function $D_{\mathrm{rel}}(w_i, c_j)$:

$$D_{\mathrm{rel}}^{+}(w_i, c_j) = \begin{cases} \frac{|w_i - c_j|}{|w_i|}, & \text{if } |w_i| \geq \delta_0 \\ 0 & \text{otherwise.} \end{cases} \qquad (2)$$

Reducing k with Additive Powers-of-two Approximations. Although using all of the values in C^S as centres to cluster the network weights would meet the requirement for the relative movement of weights to their closest cluster to be less than δ, it would also require a large number of $k = |C^S|$ clusters. In addition, the values in C^S are also of full 16-bit precision, and we would prefer many of the weights to be powers-of-two for ease of multiplication in hardware. With the motivation of reducing k and encouraging powers-of-two clusters whilst maintaining the relative distance movement where possible, we look to a many-to-one mapping of the values of C^S to further cluster the cluster centres. Building on the work of others [21,22], we map each of the values $c_i \in C^S$ to their nearest power-of-two, $\mathrm{round}(c_i) = \mathrm{sgn}(c_i)2^{\lceil \log_2(c_i) \rceil}$ and, for flexibility, we further allow for *additive* powers-of-two rounding. With additive powers-of-two rounding, each cluster value can also come from the sum of powers-of-two values (b-bit) up to order ω where the order represents the number of powers-of-two that can contribute to the approximation.

Minimalist Clustering. We are now ready to present the clustering procedure for a particular iteration t, which we give the pseudo-code for in Algorithm 1. We start the iteration with $\omega = 1$ and a set of weights not yet fixed W_{free}^t.

For the set of cluster centres $\widetilde{C}^\omega$ of order ω, let $c_*^\omega(i) = \min_{c \in \widetilde{C}^\omega} D_{\mathrm{rel}}^+(w_i, c)$ be the one closest to weight w_i. $n_k^\omega = \sum_i \mathbb{I}[c_k^\omega = c_*^\omega(i)]$ counts the number of weights assigned to cluster centre $c_k^\omega \in \widetilde{C}^\omega$, where the indicator function $\mathbb{I}[x]$ is 1 if x is true and 0 otherwise. Let $k^* = \arg\max_k n_k^\omega$ so that $c_{k^*}^\omega$ is the modal cluster. For the cluster k^* let permutation π of $\{1, \ldots, N\}$ that maps $w_i \mapsto w_{\pi(i)}'$, be such that the sequence $(w_1'(k^*), w_2'(k^*), \ldots, w_N'(k^*))$ is arranged in ascending order of relative distance from the cluster $c_{k^*}^\omega$. In other words, $D_{\mathrm{rel}}^+(w_i'(k^*), c_{k^*}^\omega) \leq D_{\mathrm{rel}}^+(w_{i+1}'(k^*), c_{k^*}^\omega)$, for $i = 1, \ldots, (N-1)$. We choose n to be the largest integer such that:

$$\sum_{i=1}^{n} D_{\mathrm{rel}}^+(w_i'(k^*), c_{k^*}^\omega) \leq n\delta, \text{ and } \sum_{i=1}^{n+1} D_{\mathrm{rel}}^+(w_i'(k^*), c_{k^*}^\omega) > (n+1)\delta, \tag{3}$$

and define $\{w_1', w_2', \ldots, w_n'\}$ to be the set of weights to be fixed at this stage of the iteration. These are the weights that can be moved to the cluster centre $c_{k^*}^\omega$ without exceeding the average relative distance δ of the weights from the centre. The corresponding weight indices from the original network $\mathcal{N}$ are in $\{\pi^{-1}(1), \ldots, \pi^{-1}(n)\}$, and called fixed$_{\mathrm{new}}$ in the algorithm. If there are no such weights that can be found, i.e., for some cluster centre l^*, the minimum relative distance $D_{\mathrm{rel}}^+(w_1'(l^*), c_{l^*}) > \delta$, the corresponding set fixed$_{\mathrm{new}}$ is empty. In this case, there are no weights that can move to this cluster centre without breaking the δ constraint and we increase order $\omega \leftarrow \omega + 1$ to compute a new $c_{k^*}^\omega$, repeating the process until $|\text{fixed}_{\mathrm{new}}| > 0$. Once fixed$_{\mathrm{new}}$ is non-empty, we fix the identified weights $\{w_1', w_2', \ldots, w_n'\}$ to their corresponding cluster centre value $c_{k^*}^\omega$ and move them into W_{fixed}^{t+1}. We continue the process of identifying cluster centres and fixing weights to these centres until $|W_{\mathrm{fixed}}^{t+1}| \geq Np_t$, at which point the iteration t is complete and the training stage of iteration $t+1$ begins. Our experiments found that a larger δ has less impact on task performance during early t iterations and so we use a decaying δ value schedule to maximise compression: $\delta^t = \delta(T-t+1)$, $t \in T$. We will show later that, with a small δ, over 75% of the weights can be fixed with $\omega = 1$ and over 95% of weights with $\omega \leq 2$.

Training Stage. Despite the steps taken to minimise the impact of the clustering stage, without retraining, performance would suffer. To negate this, we perform gradient descent to adjust the remaining free weights W_{free}^t. This allows the weights to correct for any loss increase incurred after clustering where training aims to select values W_{free}^t that minimise the cross entropy loss $\mathcal{L}_{\mathrm{cross\text{-}entropy}}$ whilst W_{fixed} remain unchanged.

Cosying up to Clusters. Having the remaining W_{free}^t weights closer to the cluster centroids C post-training makes clustering less damaging to performance. We coerce this situation by adding to the retraining loss a regularisation term

$$\mathcal{L}_{\mathrm{reg}} = \sum_{i \in W_{\mathrm{free}}}^{N} \sum_{j}^{k} D_{\mathrm{reg}}^+(w_i, c_j) p(c_j | w_i), \tag{4}$$

where $p(c_j|w_i) = \frac{e^{-D_{\text{reg}}^+(w_i,c_j)}}{\sum_l^k e^{-D_{\text{reg}}^+(w_i,c_l)}}$. The idea is to penalise the free-weights W_{free}^t in proportion to their distance to the closest cluster. Clusters that are unlikely to be weight w_i's nearest — and therefore final fixed value — do not contribute much to the penalisation term. We update the gradients of the cross-entropy training loss with the regularisation term:

$$\mathbf{w} \leftarrow \mathbf{w} - \eta \left(\nabla_{\mathbf{w}} \mathcal{L}_{\text{cross-entropy}} + \alpha \frac{\mathcal{L}_{\text{cross-entropy}}}{\mathcal{L}_{\text{reg}}} \nabla_{\mathbf{w}} \mathcal{L}_{\text{reg}} \right),$$

with α a hyper-parameter, and η the learning rate schedule. In our implementation we name α times the ratio of loss terms as γ, and we detach γ from the computational graph and treat it as a constant.

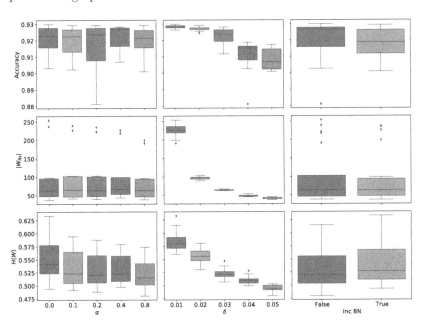

Fig. 4. Exploring the hyper-parameter space with ResNet18 model trained on the CIFAR-10 dataset. Columns; **Left:** varying the regularisation ratio α, **Middle:** varying the distance change value δ, **Right:** whether we fix the batch-norm variables or not. Rows; **Top:** top-1 accuracy test-set CIFAR-10, middle: total number of unique weights, bottom: entropy of the weights.

4 Experiment Details

We apply WFN to fully converged models trained on the CIFAR-10 and ImageNet datasets. Our pre-trained models are all publicly available with strong baseline accuracies[3]: Resnet-(18,34,50) [37] and, GoogLeNet [38]. We run ten

---

[3] CIFAR-10 models: https://github.com/kuangliu/pytorch-cifar, ImageNet models: torchvision.

weight-fixing iterations for three epochs, increasing the percentage of weights fixed until all weights are fixed to a cluster. In total, we train for 30 epochs per experiment using the Adam optimiser [34] with a learning rate 2×10^{-5}. We use grid-search to explore hyper-parameter combinations using ResNet-18 models with the CIFAR-10 dataset and find that the regularisation weighting $\alpha = 0.4$ works well across all experiments reducing the need to further hyper-parameter tuning as we advance. The distance threshold δ gives the practitioner control over the compression-performance trade-off (see Fig. 3), and so we report a range of values. We present the results of a hyper-parameter ablation study using CIFAR-10 in the Fig. 4.

5   Results

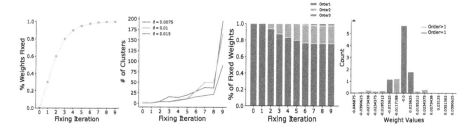

Fig. 5. Far left: We increase the number of weights in the network that are fixed to cluster centres with each fixing iteration. **Middle left:** Decreasing the δ threshold increases the number of cluster centres, but only towards the last few fixing iterations, which helps keep the weight-space entropy down. **Middle right:** The majority of all weights are order 1 (powers-of-two), the increase in order is only needed for outlier weights in the final few fixing iterations. **Far right:** The weight distribution (top-15 most used show) is concentrated around just four values.

We begin by comparing WFN for a range of δ values against a diverse set of quantisation approaches that have comparable compression ratios (CR) in Table 1. The 3-bit quantisation methods we compare include: LSQ [40], LQ-Net [24] and APoT [22]. We additionally compare with the clustering-quantisation methods using the GoogLeNet model: Deep-k-Means [11] whose method is similar to ours, KQ [41], and GreBdec [42]. Whilst the results demonstrate WFN's lossless performance with SOTA CR, this is not the main motivation for the method. Instead, we are interested in how WFN can reduce the number of unique parameters in a network and corresponding weight-space entropy as well as the network representational cost, as defined in [11]. This metric has been tested and verified to linearly correlate with energy estimations deduced using the energy-estimation tool proposed in [43]: $\mathrm{Rep}(\mathcal{N}') = |W| N_w B_w$.

Here, the representation cost is measured as the product of N_w, the number of operations weight w is involved in, B_w its bit-width and $|W|$, the number of unique weights in the network, respectively.

Table 1. A comparison of WFN against other quantisation and weight clustering approaches. The WFN pipeline is able to achieve higher compression ratios than the works compared whilst matching or improving upon baseline accuracies.

Model	Method	Accuracy (%)			Accuracy (%)				
		Top-1	Top-5	CR	Model	Method	Top-1	Top-5	CR
ResNet-18	Baseline	68.9	88.9	1.0	ResNet-34	Baseline	73.3	90.9	1.0
	LQ-Net	68.2	87.9	7.7		LQ-Net	71.9	90.2	8.6
	APoT	69.9	89.2	10.2		APoT	73.4	91.1	10.6
	LSQ	70.2^+	89.4^+	9.0^*		LSQ	73.4^+	91.4^+	9.2^*
	WFN $\delta = 0.015$	67.3	87.6	13.4		WFN $\delta = 0.015$	72.2	90.9	12.6
	WFN $\delta = 0.01$	69.7	89.2	12.3		WFN $\delta = 0.01$	72.6	91.0	11.1
	WFN $\delta = 0.0075$	70.3	89.1	10.2		WFN $\delta = 0.0075$	73.0	91.2	10.3
ResNet-50	Baseline	76.1	92.8	1.0	GoogLeNet	Baseline	69.7	89.6	1.0
	LQ-Net	74.2	91.6	5.9		Deep k-Means	69.4	89.7	3.0
	APoT	75.8	92.7	9.0		GreBdec	67.3	88.9	4.5
	LSQ	75.8^+	92.7^+	8.1^*		KQ	69.2	-	5.8
	WFN $\delta = 0.015$	75.1	92.1	10.3		WFN $\delta = 0.015$	70.5	89.9	9.0
	WFN $\delta = 0.01$	75.4	92.5	9.8		WFN $\delta = 0.01$	70.5	90.0	8.4
	WFN $\delta = 0.0075$	76.0	92.7	9.5		WFN $\delta = 0.0075$	70.9	90.2	8.4

* Estimated from the LSQ paper model size comparison graph, we over-estimate to be as fair as possible.
$^+$ Open-source implementations have so far been unable to replicated the reported results: https://github.com/hustzxd/LSQuantization.

Due to the low weight-space entropy we achieve, we suggest Huffman encoding to represent the network weights (as is used by various accelerator designs [1,2]). Given that the weight-representational bit-width will vary for each weight, we amend the original formulation to account for this, introducing

$$\text{Rep}_{\text{Mixed}}(\mathcal{N}') = \sum_{w_i \in W} N_{w_i} B_{w_i} \tag{5}$$

Here N_{w_i} is the number of times w_i is used in an inference computation, and B_{w_i} its Huffman-encoded representation bit-width of w_i.

The authors of the APoT have released the quantised model weights and code. We use the released model-saves[4] of this SOTA model to compare the entropy, representational cost, unique parameter count, model size and accuracy in Table 2. Our work outperforms APoT in weight-space entropy, unique parameter count and weight representational cost by a large margin. Taking the ResNet-18 experiments as an example, the reduction to just 164 weights compared with APoT's 9237 demonstrates the effectiveness of WFN. This huge reduction is partly due to our full-network quantisation (APoT, as aforementioned, does not quantise the first, last and batch-norm parameters). However, this does not tell the full story; even when we discount these advantages and look at weight subsets ignoring the first, last and batch-norm layers, WFN uses

---

[4] https://github.com/yhhhli/APoT_Quantization.

Table 2. A full metric comparison of WFN Vs. APoT. Params refers to the total number of unique parameter values in the network. No BN-FL refers to the unique parameter count not including the first-last and batch-norm layers. WFN outperforms APoT even when we discount the advantage gained of taking on the challenge of quantising all layers. Model sizes are calculated using LZW compression.

Model	Method	Top-1	Entropy	Params	No BN-FL	$\text{Rep}_{\text{Mixed}}$	Model size
ResNet-18	Baseline	68.9	23.3	10756029	10276369	1.000	46.8 MB
	APoT (3bit)	69.9	5.77	9237	274	0.283	4.56 MB
	WFN $\delta = 0.015$	67.3	2.72	90	81	0.005	3.5 MB
	WFN $\delta = 0.01$	69.7	3.01	164	153	0.007	3.8 MB
	WFN $\delta = 0.0075$	70.3	4.15	193	176	0.018	4.6 MB
ResNet-34	Baseline	73.3	24.1	19014310	18551634	1.000	87.4 MB
	APoT (3bit)	73.4	6.77	16748	389	0.296	8.23 MB
	WFN $\delta = 0.015$	72.2	2.83	117	100	0.002	6.9 MB
	WFN $\delta = 0.01$	72.6	3.48	164	130	0.002	7.9 MB
	WFN $\delta = 0.0075$	73.0	3.87	233	187	0.004	8.5 MB
ResNet-50*	Baseline	76.1	24.2	19915744	18255490	1.000	97.5 MB
	WFN $\delta = 0.015$	75.1	3.55	125	102	0.002	9.3 MB
	WFN $\delta = 0.01$	75.4	4.00	199	163	0.002	10.0 MB
	WFN $\delta = 0.0075$	76.0	4.11	261	217	0.003	10.2 MB

* The APoT model weights for ResNet-50 were not released so we are unable to conduct a comparison for this setting.

many times fewer parameters and half the weight-space entropy — see column 'No BN-FL' in Table 2. Finally, we examine how WFN achieves the reduced weight-space entropy in Fig. 5. Here we see that not only do WFN networks have very few unique weights, but we also observe that the vast majority of all of the weights are a small handful of powers-of-two values (order 1). The other unique weights (outside the top 4) are low frequency and added only in the final fixing iterations.

6   Conclusion

We have presented WFN, a pipeline that can successfully compress whole neural networks. The WFN process produces hardware-friendly representations of networks using just a few unique weights without performance degradation. Our method couples a single network codebook with a focus on reducing the entropy of the weight-space along with the total number of unique weights in the network. The motivation is that this combination of outcomes will offer accelerator designers more scope for weight re-use and the ability to keep most/all weights close to computation to reduce the energy-hungry data movement costs. Additionally, we believe our findings and method offer avenues of further research in understanding the interaction between network compressibility and generalisation, particularly when viewing deep learning through the minimal description length principle lens.

Acknowledgements. This work was supported by the UK Research and Innovation Centre for Doctoral Training in Machine Intelligence for Nano-electronic Devices and Systems [EP/S024298/1]. Thank you to Sulaiman Sadiq for insightful discussions.

References

1. Moons, B., Verhelst, M.: A 0.3–2.6 tops/w precision-scalable processor for real-time large-scale convnets. In: 2016 IEEE Symposium on VLSI Circuits (VLSI-Circuits), pp. 1–2. IEEE (2016)
2. Han, S., et al.: Eie: efficient inference engine on compressed deep neural network. ACM SIGARCH Comput. Archit. News **44**(3), 243–254 (2016)
3. Rissanen, J.: Modeling by shortest data description. Automatica **14**(5), 465–471 (1978)
4. Nannen, V.: A short introduction to model selection, kolmogorov complexity and minimum description length (mdl). arXiv preprint. arXiv:1005.2364 (2010)
5. Barron, A., Rissanen, J., Yu, B.: The minimum description length principle in coding and modeling. IEEE Trans. Inf. Theory **44**(6), 2743–2760 (1998)
6. Blier, L., Ollivier, Y.: The description length of deep learning models. Advances in Neural Information Processing Systems, vol. 31 (2018)
7. Chaitin, G.J., Davies, P.: On the intelligibility of the universe and the notions of simplicity, complexity and irreducibility. Think. about Gödel Turing, pp. 201–225 (2007)
8. Hinton, G.E., Van Camp, D.: Keeping the neural networks simple by minimizing the description length of the weights. In: Proceedings of the Sixth Annual Conference on Computational Learning Theory, pp. 5–13 (1993)
9. Havasi, M., Peharz, R., Hernández-Lobato, J.M.: Minimal random code learning: getting bits back from compressed model parameters. In: 7th International Conference on Learning Representations. ICLR 2019 (2019)
10. Hinton, G.E., Zemel, R.S.: Minimizing description length in an unsupervised neural network (1996)
11. Wu, J., Wang, Y., Wu, Z., Wang, Z., Veeraraghavan, A., Lin, Y.: Deep κ-means: re-training and parameter sharing with harder cluster assignments for compressing deep convolutions. In: 35th International Conference on Machine Learning. ICML 2018, vol. 12, pp. 8523–8532 (2018)
12. Mao, H., Dally, W.J.: deep compression: compressing deep neural. In: 4th International Conference Learning Represent. ICLR (2016)
13. Horowitz, M.: 1.1 Computing's energy problem (and what we can do about it). In: Digest of Technical Papers - IEEE International Solid-State Circuits Conference (2014)
14. Gao, M., Pu, J., Yang, X., Horowitz, M., Kozyrakis, C.: TETRIS: scalable and efficient neural network acceleration with 3D memory. ACM SIGPLAN Not (2017)
15. Chen, Y.H., Krishna, T., Emer, J.S., Sze, V.: Eyeriss: an energy-efficient reconfigurable accelerator for deep convolutional neural networks. IEEE J. Solid-State Circuits **52**(1), 127–138 (2017)
16. Chen, Y., et al.: DaDianNao: a machine-learning supercomputer. In: Proceedings Annual International Symposium on Microarchitecture, MICRO (2015)

17. Chen, Y., Xie, Y., Song, L., Chen, F., Tang, T.: A survey of accelerator architectures for deep neural networks. Engineering **6**(3), 264–274 (2020)
18. Mao, H., Dally, W.J.: Deep compression: compressing deep neural. In: ICLR 2016, pp. 1–14 (2016)
19. Banakar, R., Steinke, S., Lee, B.S., Balakrishnan, M., Marwedel, P.: Comparison of cache-and scratch-pad-based memory systems with respect to performance, area and energy consumption. University of Dortmund, Technical report (762) (2001)
20. Banakar, R., Steinke, S., Lee, B.S., Balakrishnan, M., Marwedel, P.: Scratchpad memory: a design alternative for cache on-chip memory in embedded systems. In: Proceedings of the Tenth International Symposium on Hardware/Software Codesign. CODES 2002 (IEEE Cat. No. 02TH8627). IEEE, pp. 73–78 (2002)
21. Zhou, A., Yao, A., Guo, Y., Xu, L., Chen, Y.: Incremental network quantization: towards lossless cnns with low-precision weights. arXiv preprint. arXiv:1702.03044 (2017)
22. Li, Y., Dong, X., Wang, W.: Additive powers-of-two quantization: an efficient nonuniform discretization for neural networks. In: International Conference on Learning Representations (2019)
23. Jung, S.,et al.: Learning to quantize deep networks by optimizing quantization intervals with task loss. In: Proceedings of the IEEE Conference on Computer Vision and Pattern Recognition (2019)
24. Zhang, D., Yang, J., Ye, D., Hua, G.: Lq-nets: learned quantization for highly accurate and compact deep neural networks. In: Proceedings of the European Conference on Computer Vision (ECCV), pp. 365–382 (2018)
25. Zhou, S., Wu, Y., Ni, Z., Zhou, X., Wen, H., Zou, Y.: DoReFa-Net: training low bitwidth convolutional neural networks with low bitwidth gradients (2016)
26. Yamamoto, K.: Learnable companding quantization for accurate low-bit neural networks. In: Proceedings of the IEEE/CVF Conference on Computer Vision and Pattern Recognition, pp. 5029–5038 (2021)
27. Oh, S., Sim, H., Lee, S., Lee, J.: Automated log-scale quantization for low-cost deep neural networks. In: Proceedings of the IEEE/CVF Conference on Computer Vision and Pattern Recognition (2021)
28. Hubara, I., Courbariaux, M., Soudry, D., El-Yaniv, R., Bengio, Y.: Binarized neural networks. In: Advances in Neural Information Processing Systems (2016)
29. Lee, E.H., Miyashita, D., Chai, E., Murmann, B., Wong, S.S.: LogNet: energy-efficient neural networks using logarithmic computation. In: Proceedings of the IEEE International Conference on Acoustics, Speech and Signal Processing, ICASSP (2017)
30. Yang, J., et al.: Quantization networks. In: Proceedings of the IEEE Conference on Computer Vision and Pattern Recognition (2019)
31. Chmiel, B., Banner, R., Shomron, G., Nahshan, Y., Bronstein, A., Weiser, U., et al.: Robust quantization: One model to rule them all. In: Advances in Neural Information Processing Systems, vol. 33, pp. 5308–5317 (2020)
32. Jacob, B., et al.: Quantization and training of neural networks for efficient integer-arithmetic-only inference. In: Proceedings of the IEEE Conference on Computer Vision and Pattern Recognition, pp. 2704–2713 (2018)
33. Stock, P., Joulin, A., Gribonval, R., Graham, B., Jégou, H.: And the bit goes down: revisiting the quantization of neural networks. In: 8th International Conference on Learning Representations. ICLR 2020, pp. 1–11 (2020)
34. Tartaglione, E., Lathuilière, S., Fiandrotti, A., Cagnazzo, M., Grangetto, M.: HEMP: high-order entropy minimization for neural network compression. Neurocomputing **461**, 244–253 (2021)

35. Fan, A., et al.: Training with quantization noise for extreme model compression. In: 9th International Conference on Learning Representations. ICLR 2021 - Conference on Track Proceedings (2021)
36. Ullrich, K., Meeds, E., Welling, M.: Soft weight-sharing for neural network compression. arXiv, pp. 1–16 (2017)
37. He, K., Zhang, X., Ren, S., Sun, J.: Deep residual learning for image recognition. In: Proceedings of the IEEE Conference on Computer Vision and Pattern Recognition, PP. 770–778 (2016)
38. Chollet, F.: Xception: deep learning with depthwise separable convolutions. In: Proceedings of the 30th IEEE Conference on Computer Vision and Pattern Recognition. CVPR 2017 (2017)
39. Kingma, D.P., Ba, J.L.: Adam: a method for stochastic optimization. In: 3rd International Conference on Learning Representations. ICLR 2015 - Conference on Track Proceedings (2015)
40. Esser, S.K., McKinstry, J.L., Bablani, D., Appuswamy, R., Modha, D.S.: Learned step size quantization. In: 8th International Conference on Learning Representations. ICLR 2020 (2020)
41. Yu, Z., Shi, Y.: Kernel quantization for efficient network compression. IEEE Access 10, 4063–4071 (2022)
42. Yu, X., Liu, T., Wang, X., Tao, D.: On compressing deep models by low rank and sparse decomposition. In: Proceedings of the 30th IEEE conference on computer vision and pattern recognition. CVPR 2017-January, pp. 67–76 (2017)
43. Yang, T.J., Chen, Y.H., Sze, V.: Designing energy-efficient convolutional neural networks using energy-aware pruning. In: Proceedings of the IEEE Conference on Computer Vision and Pattern Recognition, pp. 5687–5695 (2017)

Self-slimmed Vision Transformer

Zhuofan Zong[1], Kunchang Li[3,4], Guanglu Song[2], Yali Wang[3,5], Yu Qiao[3,6], Biao Leng[1], and Yu Liu[2(✉)]

[1] School of Computer Science and Engineering, Beihang University, Beijing, China
[2] SenseTime Research, Hong Kong, China
liuyuisanai@gmail.com
[3] ShenZhen Key Lab of Computer Vision and Pattern Recognition,
SIAT-SenseTime Joint Lab, Shenzhen Institutes of Advanced Technology,
Chinese Academy of Sciences, Shenzhen, China
[4] University of Chinese Academy of Sciences, Beijing, China
[5] SIAT Branch, Shenzhen Institute of Artificial Intelligence and Robotics for Society,
Shenzhen, China
[6] Shanghai AI Laboratory, Shanghai, China

Abstract. Vision transformers (ViTs) have become the popular structures and outperformed convolutional neural networks (CNNs) on various vision tasks. However, such powerful transformers bring a huge computation burden, because of the exhausting token-to-token comparison. The previous works focus on dropping insignificant tokens to reduce the computational cost of ViTs. But when the dropping ratio increases, this hard manner will inevitably discard the vital tokens, which limits its efficiency. To solve the issue, we propose a generic self-slimmed learning approach for vanilla ViTs, namely SiT. Specifically, we first design a novel Token Slimming Module (TSM), which can boost the inference efficiency of ViTs by dynamic token aggregation. As a general method of token hard dropping, our TSM softly integrates redundant tokens into fewer informative ones. It can dynamically zoom visual attention without cutting off discriminative token relations in the images, even with a high slimming ratio. Furthermore, we introduce a concise Feature Recalibration Distillation (FRD) framework, wherein we design a reverse version of TSM (RTSM) to recalibrate the unstructured token in a flexible auto-encoder manner. Due to the similar structure between teacher and student, our FRD can effectively leverage structure knowledge for better convergence. Finally, we conduct extensive experiments to evaluate our SiT. It demonstrates that our method can speed up ViTs by **1.7**× with negligible accuracy drop, and even speed up ViTs by **3.6**× while maintaining **97**% of their performance. Surprisingly, by simply arming LV-ViT with our SiT, we achieve new state-of-the-art performance on ImageNet. Code is available at https://github.com/Sense-X/SiT.

Keywords: Transformer · Token slimming · Knowledge distillation

Z. Zong and K. Li—Contribute equally during their internship at SenseTime.

Supplementary Information The online version contains supplementary material available at https://doi.org/10.1007/978-3-031-20083-0_26.

1 Introduction

Since vision transformer (ViT) [10] started the era of transformer structure in the fundamental computer vision tasks [3,5,35], variant transformers have been designed to challenge the dominance of convolutional neural networks (CNNs). Different from CNNs that stack convolutions to encode local features progressively, ViTs directly capture the long-term token dependencies. However, because of the exhausting token-to-token comparison, current powerful transformers require huge computation, limiting their wide application in reality [12]. Hence, in this paper, we aim to design a generic learning framework for boosting the efficiency of vanilla vision transformers.

Table 1. Comparison to recent model pruning methods for ViT. Our SiT surpasses all the other methods based on structure pruning or hard dropping.

Type	Method	Reference	ImageNet Top-1(%)	Throughput (image/s)	(%)
Baseline	DeiT [29]	ICML21	79.8	1637	0
Structure pruning	S²ViTE [6]	NeurIPS21	79.2(**−0.6**)	2117	29.3
Token hard dropping	PS-ViT [28]	CVPR22	79.4(**−0.5**)	2351	43.6
	IA-RED² [22]	NeurIPS21	79.1(**−0.7**)	2369	44.7
	Dynamic-ViT [24]	NeurIPS21	79.3(**−0.5**)	2575	57.3
	Evo-ViT [36]	AAAI22	79.4(**−0.4**)	2629	60.6
	EViT [20]	ICLR22	79.1(**−0.7**)	2783	70.0
Token soft slimming	Our SiT	ECCV22	79.8(**−0.0**)	2344	43.2
		ECCV22	79.4(**−0.4**)	3308	**102.1**

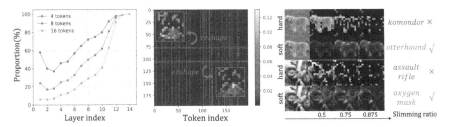

(a) Token similarity becomes higher in deeper layers.
(b) All tokens tend to focus on the same tokens in deeper layers.
(c) Our *soft slimming* can automatically zoom the attention scope based on the object size.

Fig. 1. Our motivation. In Fig (a), we calculate the correlation coefficients among tokens and count the proportion that is at least similar (≥0.7) to 4/8/16 tokens in different layers. As for Fig (b), we randomly select two tokens in the tenth layer to show their attention. Moreover, we compare different token pruning methods in Fig (c). Darker tokens get less attention.

To make ViTs more efficient, we tried to explore the inherent properties of the token-to-token comparison. We conduct a series of experiments based on LV-ViT, which reveals that sparse attention with high token similarity exists in ViTs. Figure 1a shows that token similarity becomes higher in deeper layers, especially in the last three layers. Besides, the attention tends to focus on the specific tokens in the deeper layers (Fig. 1b), which indicates the number of decision-relevant tokens becomes fewer. These observations demonstrate that only a few token candidates indicate meaningful information. It inspires us a feasible structure-agnostic dimension, token number, to reduce the computational cost. Intuitively, we can progressively drop the redundant tokens as the network deepens.

Recent studies have tried to compress tokens via data-independent dropping with minimizing reconstruction error [28], or data-dependent dropping with differentiable scoring [24]. However, data-independent dropping requires layer-by-layer optimization, which is hard to generalize. Moreover, the token hard dropping will inevitably discard the vital tokens as the dropping ratio increases, e.g., the shape of the otterhound is destroyed in the deep layer (Fig. 1c), thus limiting its performance as shown in Table 1.

Can we design a flexible method of token slimming, thus decision-relevant information can be dynamically aggregated into a slimmed token set? To answer this question, we propose token soft slimming. It contains a concise Token Slimming Module (TSM), which generates decision-relevant tokens via a data-dependent weight matrix. As shown in Fig. 1c, by simply inserting multiple TSMs in LV-ViT, our network can learn to localize the key object tokens. More importantly, the attention scope can be zoomed automatically without cutting off the discriminative token relations, e.g., our network can adaptively concentrate on the most informative parts of the otterhound in *softer* manner, while the oxygen mask in *harder* manner.

In DynamicViT [24], self-distillation is introduced in the last layer to minimize the performance drop brought by token sparsification. However, it ignores hint knowledge in the intermediate layers, leading to inevitable knowledge loss. To solve this issue, we introduce a concise Feature Recalibration Distillation (FRD) to achieve stable and efficient model slimming optimization. Note that previous hint knowledge distillation methods [17,25,38,41] are designed for spatial structured tokens. Since the neighbor token information is contiguous, they can apply contiguous upsampling (e.g., deconvolution and interpolation) to find the correspondence between tokens of teacher and student. In contrast, our TSM generates *unstructured* token set, which can not be allocated corresponding supervision directly. To align the token relations among unstructured tokens, we design a reverse version of the token slimming module (RTSM) in a flexible auto-encoder manner. Such an effective way helps our FRD densely transfer all the token information block to block. Benefiting from the innate knowledge inheritance (structure knowledge), our FRD is more suitable for teaching itself. As shown in Table 1, our SiT is able to improve the throughput by 43.2% without any performance drop, and accelerate the inference speed by over 100% with negligible accuracy decrease.

Our self-slimmed learning method is flexible and easy to generalize to all vanilla vision transformers (SiT), e.g., DeiT [29], LV-ViT [16] etc. We conduct extensive experiments on ImageNet [8] to verify the effectiveness and efficiency. Interestingly, our method without self-distillation can perform even better than

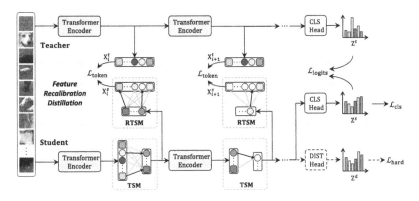

Fig. 2. The framework of our self-slimmed learning. We insert our token slimming modules (TSM) into vanilla vision transformer. To reduce decision-relevant information loss, we apply Feature Recalibration Distillation (FRD), wherein the reverse version of TSM (RTSM) is utilized to recalibrate unstructured token. The dash lines indicate the prediction supervision from an extra CNN teacher is optional and complementary to our method.

DynamicViT [24] with distillation. Besides, the SiT-XS achieves 81.8% top-1 accuracy with **3.6×** inference speed and SiT-L achieves competitive 85.6% top-1 accuracy while running **1.7×** faster. More importantly, our SiT based on LV-ViT achieves the new state-of-the-art performance on ImageNet, surpassing recent CNNs and ViTs.

2   Related Works

2.1   Vision Transformers

Transformer architecture [31] was first proposed for machine translation. The success of transformer in NLP inspires the application of transformers in various vision tasks, for example, DETR [3] for object detection and ViT [10] for image recognition. ViT is the first pure transformer that achieves the state-of-the-art performance on ImageNet [8]. Recent ViT variants mainly focus on better optimization and more powerful performance [4, 7, 9, 11, 13, 15, 18, 19, 33, 34, 37, 39, 40, 42]. However, few of them explore to improve the efficiency of vision transformers [12]. In this paper, we aim to design a general optimization framework named self-slimmed learning to promote the efficiency of ViTs.

2.2   Transformer Slimming

The large computation of self-attention hinders the wide application of ViTs, such as detection and segmentation with the high-resolution input image. To solve this problem, several prior works concentrate on designing sparse attention [21,32] or structure pruning [6]. S^2ViTE [6] dynamically extracts and trains sparse subnetworks of ViTs, while sticking to a fixed small parameter budget. However, model structure pruning struggles to trim down the inference latency. Other works try to reduce the token redundancy [22,24,28,36] by entirely dropping the unimportant tokens, which brings more improvements on throughput compared to structure pruning. Different from the above works, our SiT aggregates all tokens into fewer informative tokens in a soft manner by a concise slimming module. It can automatically zoom the attention scope to localize the key object for better recognition.

3   Method

In this section, we describe our self-slimmed learning for vision transformer (SiT) in detail. First, we introduce the overall architecture of SiT. Then, we explain the vital design of our SiT, *i.e.*, token slimming module (TSM) and feature recalibration distillation (FRD). Finally, we thoroughly compare our TSM and FRD with other counterparts.

3.1   Overview of Self-slimmed Learning

The overall framework is illustrated in Fig. 2. We first design a lightweight Token Slimming Module (TSM) for conventional ViTs to perform token slimming, and its reverse version (RTSM) to recalibrate unstructured tokens for hint knowledge distillation. We divide the slimmed vision transformer into multiple stages (*e.g.*, 4 stages as in prior works [12,21]), where different numbers of tokens participate in feed-forward computation. To decrease the information loss, we propose a block-to-block feature recalibration distillation (FRD), wherein the original vision transformer can serve as a teacher to minimize the difference between itself and the slimmed student. Finally, we integrate TSM and FRD to form a general self-slimmed learning method for all vanilla ViTs.

3.2   Token Slimming Module

Given a sequence of N input tokens with C channels $\mathbf{X} = [\mathbf{x}_1; \mathbf{x}_2; \cdots ; \mathbf{x}_N] \in \mathbb{R}^{N \times C}$, (class token is omitted as it will never be pruned), token slimming aims to dynamically aggregate the redundant tokens to generate $\hat{N}$ informative tokens $\hat{\mathbf{X}} = [\hat{\mathbf{x}}_1; \hat{\mathbf{x}}_2; \cdots ; \hat{\mathbf{x}}_{\hat{N}}]$:

$$\hat{\mathbf{X}} = \hat{\mathbf{A}}\mathbf{X}, \tag{1}$$

where $\hat{\mathbf{A}} \in \mathbb{R}^{\hat{N} \times N}$ is a normalized weight matrix:

$$\sum_{i=1}^{\hat{N}} \hat{\mathbf{A}}_{i,j} = 1, \quad \text{where} \ \ j = 1 \dots N. \tag{2}$$

Such operation is differentiable and friendly to end-to-end training. We follow the design paradigm of self-attention [31] and propose a lightweight token slimming module (TSM) shown in Fig. 3a:

$$\hat{\mathbf{A}} = \text{Softmax}(\frac{W_q \sigma(\mathbf{X} W_k)^T}{\tau}), \tag{3}$$

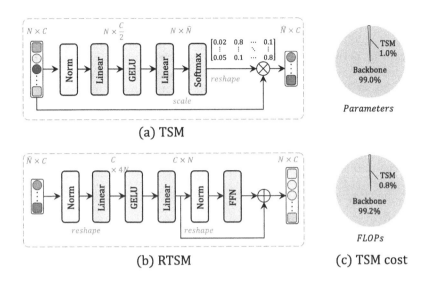

(a) TSM

(b) RTSM

(c) TSM cost

Fig. 3. The token slimming module (TSM) and its reverse version (RTSM).

where $W_k \in \mathbb{R}^{C \times \frac{C}{2}}$ and $W_q \in \mathbb{R}^{\hat{N} \times \frac{C}{2}}$ are both learnable parameters. σ and τ represents the nonlinear function (GELU) and scaling factor respectively. Similar to self-attention, TSM generates a global attention matrix, but it requires much fewer overhead in terms of throughput and memory usage during both training and inference. Figure 3c shows that TSM blocks only require negligible cost. Thanks to the learnable scaling factor τ, the attention tends to be sparse in our experiments, which means it learns to focus on the most informative tokens.

Hard Dropping vs. Soft Slimming. The prior works have tried to compress tokens via hard dropping [24,28], in which the slimming weight $\hat{\mathbf{A}}_{i,j} \in \{0,1\}$ is a binary decision matrix, *i.e.*, dropping or keeping the corresponding token. However, this approach with binary decision leads to severe information loss if numerous tokens are discarded. Such weakness limits their high efficiency on ImageNet [8], wherein the objects often occupy a large part in the pictures. On the contrary, we design soft slimming with a learnable normalized weight $\hat{\mathbf{A}}_{i,j} \in (0,1)$, which is able to discriminate the meaningful tokens in a global view. As shown in Fig. 1c, our soft slimming can dynamically zoom the attention scope to cover the significant regions for classification. It reveals that $\hat{\mathbf{A}}$ can adaptively become a one-hot matrix to help our SiT focus on the most informative part.

3.3 Feature Recalibration Distillation

Feature Recalibration. Though token slimming significantly reduces the inference latency, when using a large slimming rate, it inevitably decreases the accuracy because of the information loss. Hint knowledge distillation is the common method to maintain meaningful information in intermediate layers, wherein the challenge is to align the feature semantics between student and teacher. Previous works [25,38] adopt spatial deconvolution or interpolation to cope with this misalignment between spatial contiguous features. However, it is not suitable for slimmed unstructured tokens with spatially discrete semantics. To solve this problem, we design a reverse version of the token slimming module (RTSM) to recalibrate the unstructured tokens in a flexible auto-encoder manner (Fig. 3b). Therefore, all the token information can be seamlessly transferred from the teacher. Note that we only perform RTSM when training, thus no extra computation is introduced during inference. We first linearly transform the informative tokens into plenty of token candidates, thus utilizing a non-linear function (GELU) to filter the vital representations. Finally, another linear transformation is performed to compress the token candidates:

$$\hat{\mathbf{X}}' = \mathbf{A}_2(\sigma(\mathbf{A}_1\hat{\mathbf{X}})), \tag{4}$$

where $\mathbf{A}_1 \in \mathbb{R}^{4N \times \hat{N}}$ and $\mathbf{A}_2 \in \mathbb{R}^{N \times 4N}$ in our experiments. To further enhance the token representations, we introduce an extra multi-layer perception (MLP) block [31] with residual connection [14]:

$$\mathbf{X}' = \hat{\mathbf{X}}' + \mathrm{MLP}(\hat{\mathbf{X}}'). \tag{5}$$

The recalibrated features $\mathbf{X}'$ will be forced to be consistent with the teacher features in FRD, ensuring the sufficient information of the slimmed tokens $\hat{\mathbf{X}}$.

Knowledge Distillation. Due to the invariant model structure, we design a block-to-block knowledge distillation for the recalibrated features:

$$\mathcal{L}_{\mathrm{token}} = \frac{1}{LN} \sum_{i=1}^{L} \sum_{j=1}^{N} (\mathbf{X}_{i,j}^s - \mathbf{X}_{i,j}^t)^2, \tag{6}$$

where $\mathbf{X}_{i,j}^s$ and $\mathbf{X}_{i,j}^t$ refer to the j-th token embedding at the i-th layer of the student and teacher, respectively. L means the layer number. Note that $\mathbf{X}^s$ refers to the recalibrated tokens $\mathbf{X}'$ in Eq. 5. With such reconstruction loss, the student model will be forced to maintain as much knowledge in the informative tokens $\hat{\mathbf{X}}$. Besides, to further alleviate the classification performance deterioration caused by token slimming, we introduce the logits distillation to minimize the predictions difference between the student and teacher:

$$\mathcal{L}_{\mathrm{logits}} = \mathrm{KL}(\psi(\mathbf{Z}^s), \psi(\mathbf{Z}^t)), \tag{7}$$

where KL denotes Kullback-Leibler divergence loss and ψ is the softmax function. $\mathbf{Z}^s$ and $\mathbf{Z}^t$ are respectively the predictions of the student and teacher model. Moreover, the above FRD is complementary to the hard distillation in [29]:

$$\mathcal{L}_{\mathrm{hard}} = \mathrm{CrossEntropy}(\psi(\mathbf{Z}^d), y^c), \tag{8}$$

where $\mathbf{Z}^d$ indicates the prediction of distillation head and y^c is a hard decision of the extra CNN teacher. It can further improve the performance with longer training epochs. Our final objective of distillation for self-slimmed learning is:

$$\mathcal{L}_{\mathrm{dist}} = \lambda_{\mathrm{token}}\mathcal{L}_{\mathrm{token}} + \lambda_{\mathrm{logits}}\mathcal{L}_{\mathrm{logits}} + \lambda_{\mathrm{hard}}\mathcal{L}_{\mathrm{hard}}, \tag{9}$$

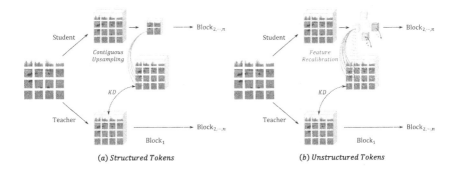

(a) *Structured Tokens*   (b) *Unstructured Tokens*

Fig. 4. Hint knowledge distillation for structured and unstructured tokens.

where λ is the coefficient balancing the three distillation losses. We set $\lambda_{\text{logits}} = 2, \lambda_{\text{token}} = 2$ by default. λ_{hard} is set to 1 when the CNN teacher is involved. As for the training objective of self-slimmed learning, we treat the classification task and the distillation task equally:

$$\mathcal{L}_{\text{cls}} = \text{CrossEntropy}(\psi(Z^s), \overline{y}), \tag{10}$$

$$\mathcal{L}_{\text{global}} = \mathcal{L}_{\text{cls}} + \mathcal{L}_{\text{dist}}, \tag{11}$$

where $\overline{y}$ means the ground truth, *i.e.*, one-hot label.

FRD vs. Other Knowledge Distillation. Firstly, current vision transformers [29,30] simply select a strong teacher network with totally different architectures, such as RegNet for DeiT. Only a few knowledge (*e.g.*, final predictions) can be inherited, thus the semantic information in the intermediate layer is ignored. In FRD, thanks to the consistency between the teacher and student, we naturally conduct densely token-level supervision for each block, which greatly improves the stability of the model mimicking. Besides, the popular hint knowledge distillation method [25,38] are mainly designed for spatial structured tokens. As shown in Fig. 4a, they can simply apply local and contiguous upsampling to reconstruct tokens. However, as shown in Fig. 4b, the token slimming generates an unstructured token set. Each token contains partial information of previous tokens. To recalibrate the unstructured features, we design a concise RTSM in a flexible auto-encoder manner. Thus via reconstruction loss, our FRD can force the student model to maintain sufficient knowledge in the informative tokens.

4   Experiments

4.1   Implementation Details

In this section, we conduct comprehensive experiments to empirically analyze the effectiveness of our proposed self-slimmed learning for vision transformer (SiT). All the models are evaluated on the ImageNet dataset [8]. For our teacher models, we train LV-ViTs [16] following the original settings, but we replace the patch

Table 2. Main results on ImageNet. We apply our self-slimming learning on the state-of-the-art vanilla vision transformer LV-ViT [16]. ‡ means we adopt an extra CNN teacher. Our SiT can speed up LV-ViT **1.7×** with a slight accuracy drop. For fast inference, our SiT can maintain 97% of the performance while speeding up the original transformers by **3.6×**.

Model	Depth	Stage	Embed Dim	Heads	Resolution	#Params (M)	FLOPs (G)
SiT-Ti	14	{1,1,1,11}	320	5	224^2	15.9	1.0
SiT-XS	16	{1,1,1,13}	384	6	224^2	25.6	1.5
SiT-S	16	{9,3,2,2}	384	6	224^2	25.6	4.0
SiT-M	20	{10,4,3,3}	512	8	224^2	55.6	8.1
SiT-L	24	{10,4,3,7}	768	12	288^2	148.2	34.4

(a) Model architecture settings

Model	Student			Teacher	
	Throughput (image/s)	Top-1 (%)	Top-1‡ (%)	Throughput (image/s)	Top-1 (%)
SiT-Ti	5896 (**3.2×**)	80.1 (−2.0)	80.6 (−1.5)	1827	82.1
SiT-XS	4839 (**3.6×**)	81.1 (−2.2)	81.8 (−1.5)	1360	83.3
SiT-S	1892 (**1.4×**)	83.2 (−0.1)	83.4 (+0.1)	1360	83.3
SiT-M	1197 (**1.5×**)	84.1 (−0.1)	84.3 (+0.1)	804	84.2
SiT-L	346 (**1.7×**)	85.6 (−0.1)	-	204	85.7

(b) Efficiency comparisons

embedding module with lightweight stacked convolutions inspired by LeViT [12]. As for student models, all the training hyper-parameters are the same as DeiT [29] by defaults. For initialization, we load all the weights from the corresponding teacher models to accelerate the convergence and train them for 125 epochs. If utilizing an extra CNN teacher, we extend the training time to 300 epochs for better improvements. Moreover, we set different initial learning rates for the backbone and the feature recalibration branch, which are $0.0002 \times \frac{batch\ size}{1024}$ and $0.001 \times \frac{batch\ size}{1024}$ respectively. For token slimming, we insert TSM three times, thus there are four stages in SiT. The default keeping ratio $\hat{N}/N$ is set to 0.5, which means the token number is halved after slimming.

4.2   Main Results

We conduct our self-slimmed learning for LV-ViT [16], which is the state-of-the-art vanilla vision transformer. Table 2 shows our detailed settings for different SiT variants. For SiT-Ti and SiT-XS, we explore their capacity for fast inference, thus we insert TSMs in the early layers. It demonstrates that our self-slimmed method is able to speed up the original vision transformers by **3.6×**, while maintaining at least 97% of their accuracy. Besides, we adopt another CNN teacher to provide the hard label as in DeiT [29]. The results show that complementary

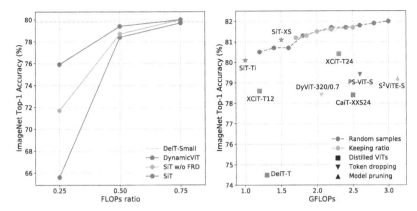

(a) Our SiT achieves much higher accuracy than DynamicViT even without the distillation.

(b) Our randomly sampled models consistently outperform other distilled and pruned ViTs.

Fig. 5. Effectiveness and robustness study. We compare our SiT with the DynamicViT in Fig (a). To verify the robustness of our method, we randomly change the numbers of blocks in each stage (*i.e.*, TSM location) and adjust the keeping ratio of TSM from 0.3 to 0.7 to sample a series of SiT-Ti models in Fig (b). All the models are trained for 125 epochs without a CNN teacher.

prediction supervision can further improve performance. As for other variants, we insert TSMs in the deeper layers. Surprisingly, with negligible accuracy drop, our SiTs are up to **1.7×** faster than their teacher models. It is worth mentioning that, extra CNN distillation brings little improvement, mainly because the CNN teacher is inferior to the original transformer teacher (82.9% *vs.* 83.3%/84.2%).

4.3   Effectiveness and Robustness

Comparison to DynamicViT. In Fig. 5a, we compare our method with DynamicViT [24] on DeiT-S [29]. When dropping too many tokens, the performance of DynamicViT deteriorates dramatically. Though it utilizes knowledge distillation to minimize the gap, our SiT without the distillation consistently surpasses it under different FLOPs ratios, especially under the smallest ratio. Besides, when armed with FRD, our SiT can maintain performance better.

TSM Locations and Keeping Ratio. To verify the robustness of our method, we conduct experiments on SiT-Ti as shown in Fig. 5b. It clearly shows that all of the randomly sampled models outperform popular ViTs with knowledge distillation, *e.g.*, DeiT [29] and XCiT [1]. Besides, compared with other counterparts based on token hard dropping [24,28] and structure pruning [6], our models surpass them by a large margin. These results demonstrate our SiT is insensitive to the setting of TSM locations and keeping ratio. To make a fair comparison with the state-of-the-art ViTs, we set these hyper-parameters according to the FLOPs.

Table 3. Comparison to the state-of-the-art on ImageNet. The models marked in *gray* color are trained with distillation supervision from a powerful CNN for 300 epochs. Our SiT achieves the best performance trade-off.

Model	Resolution	#Params (M)	FLOPs (G)	Throughput (image/s)	ImageNet Top-1(%)
EfficientNet-B1 [26]	240^2	7.8	0.7	2559	79.1
EfficientNet-B2 [26]	260^2	9.1	1.1	1808	80.1
DeiT-T [29]	224^2	5.9	1.3	3346	74.5
LeViT-256 [12]	224^2	18.9	1.1	5802	80.1
SiT-Ti	224^2	15.9	1.0	**5896**	80.1
SiT-Ti	224^2	16.2	1.0	5833	**80.6**
EfficientNet-B3 [26]	300^2	12.2	1.9	1062	81.6
Swin-T [21]	224^2	28.3	4.5	1023	81.3
DeiT-S [29]	224^2	22.4	4.6	1598	81.2
LeViT-384 [12]	224^2	39.1	2.4	3876	81.6
SiT-XS	224^2	25.6	1.5	**4839**	81.1
SiT-XS	224^2	26.0	1.5	4798	**81.8**
EfficientNet-B4 [26]	380^2	19.3	4.6	545	82.9
Swin-B [21]	224^2	87.8	15.5	474	83.3
DeiT-B [29]	224^2	87.3	17.7	718	83.4
LV-ViT-S [16]	224^2	26.2	6.6	1270	83.3
SiT-S	224^2	25.6	4.0	**1892**	83.2
SiT-S	224^2	26.0	4.0	1873	**83.4**
EfficientNet-B6 [26]	528^2	43.0	19.9	153	84.0
EfficientNetV2-S [27]	384^2	21.5	8.5	742	83.9
CaiT-S36 [30]	224^2	68.2	13.9	233	83.9
LV-ViT-M [16]	224^2	55.8	12.7	768	84.1
SiT-M	224^2	55.6	8.1	**1197**	84.1
SiT-M	224^2	56.2	8.1	1185	**84.3**
EfficientNetV2-M [27]	480^2	54.1	25.0	271	85.1
NFNet-F1 [2]	320^2	132.6	36.0	128	84.7
CaiT-M36 [30]	224^2	270.1	53.4	130	85.1
LV-ViT-L [16]	288^2	150.1	58.8	208	85.3
SiT-L	288^2	148.2	34.4	**346**	**85.6**

4.4   Comparison to State-of-the-Art

In Table 3, we compare SiT with other competitive CNNs and ViTs. For a fair comparison, we group these methods according to their top-1 accuracies. The throughput is measured on a single 16GB V100 GPU under the same setting as LeViT [12]. Our SiT-Ti is competitive with LeViT, while the throughput is **3.2×** than that of EfficientNet [26]. Note that EfficientNet is designed via extensive neural architecture search and LeViT is elaborately designed for fast inference. For our larger model variants, they perform better than EfficientNetV2 [27] with simple training strategies. Compared with the original LV-ViT [16], our SiT is **1.5×** faster than those with similar accuracy.

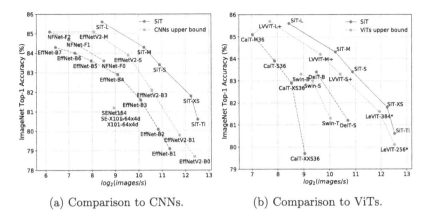

(a) Comparison to CNNs.   (b) Comparison to ViTs.

Fig. 6. Speed *vs.* Accuracy. We compare our SiT with the previous state-of-the-art CNNs and ViTs in Fig (a) and Fig (b), respectively. "LV-ViT+" denotes our improved LV-ViT teacher. Our SiT surpasses the SOTA methods by a large margin, even the efficient models EfficientNetV2 [27] and LeViT [12].

Table 4. Efficiency comparison.

Method	Top-1	Throughput
Structure-width	76.3	2947
Structure-depth	69.4	5652
DynamicViT [24]	75.7	5762
SiT w/o FRD	**77.7**	**5896**

Table 5. Inherited knowledge.

Knowledge	Self	CaiT	RegNet
	83.3	83.5	82.9
Scratch	**80.1**	79.9	79.2
Fine-tuning	**80.5**	80.2	80.0
Fine-tuning+Structure	**81.1**	80.6	80.2

We further visualize the comparisons to the upper bounds of CNNs and ViTs in Fig. 6a and 6b. It clearly shows that our SiT achieves the best balance between throughput and accuracy, surpassing the recent state-of-the-art CNNs and ViTs.

4.5 Ablation Studies

If not otherwise specified, all experiments for ablations are conducted on SiT-Ti and run with only 125 training epochs under the supervision of the original teacher model. "Token-MLP" refers double linear layers along the token dimension.

Does Token Slimming Outperform Model Scaling Down? In Table 4, we compare token slimming with model scaling down rules under the same computation limit. For model scaling down, we adapt the channel and depth individually. Note that the above two models are trained from scratch for 300 epochs with token labeling [16]. For token slimming, we simply insert TSMs without FRD. We also drop tokens and train it with extra distillation as in DynamicViT [24]. It shows that scaling along the channel achieves higher accuracy

Table 6. Robustness of slimming ratios.

Ratio	$\mathcal{L}_{logits}+\mathcal{L}_{token}$	$\mathcal{L}_{hard}$
1	82.1	82.1
0.75	82.0	82.0
0.5	81.6	81.3
0.25	80.1	78.4

Table 7. Slimming.

Method	GFLOPs	Top-1
Baseline	3.5	82.1
3×3 AvgPool	1.0	77.4
3×3 Conv	1.0	79.3
Token-Mixer	1.1	79.3
Our TSM	**1.0**	**80.1**

Table 8. Recalibration.

Method	Top-1
Baseline	79.0
Interpolation	78.3
Deconvolution	78.4
Token-MLP	79.0
Our RTSM	**80.1**

Table 9. Knowledge distillation.

Method	Top-1
Baseline	77.7
$+\mathcal{L}_{logits}$	79.0(+**1.3**)
$+\mathcal{L}_{logits}+\mathcal{L}_{token}$	80.1(+**2.4**)
$+\mathcal{L}_{logits}+\mathcal{L}_{token}+\mathcal{L}_{hard}$	80.2(+**2.5**)
$+\mathcal{L}_{logits}+\mathcal{L}_{token}+\mathcal{L}_{hard}$+Longer training	80.6(+**2.9**)

Table 10. Loss weights.

$\lambda_{token}{:}\lambda_{logit}{:}\lambda_{hard}$	Top-1
1:1:1	79.3
1:2:1	79.4
1:2:2	79.5
2:1:1	**79.6**
2:2:1	**79.6**

than scaling along the depth but with lower throughput. Besides, token slimming can largely improve the throughput with higher performance. However, DynamicViT performs worse than our SiT without distillation, since token hard dropping loses much discriminative information with a large slimming ratio. Such results demonstrate simply inserting our TSM into vanilla ViT is able to achieve great performance.

Does Structure Knowledge Matter to Self-slimmed Learning? We further investigate whether the structure knowledge benefits the performance as shown in Table 5. For the teacher models, we adopt different architectures (LV-ViT-S [16], CaiT-S24 [30], and RegNetY-16GF [23]) but similar accuracies for a fair comparison. It shows that training with the pre-trained weights for 125 epochs converges to higher results than those trained from scratch for 300 epochs. Moreover, we utilize structure knowledge via block-to-block mimicking, which can further boost the performance. It also reveals that higher similarity between students and teachers can bring greater improvements.

Is Self-slimmed Learning Robust to Different FLOPs Ratios? In Table 6, we empirically train models with different FLOPs ratios. When the ratio is large than 0.5, our FRD and CNN distillation are both helpful for maintaining performance. However, when the ratio is small, CNN distillation leads to a higher performance drop, while our FRD only drops the accuracy by 2.0%. These results demonstrate that our method is robust to different FLOPs ratios.

Dynamic vs. Static: Which Aggregation Manner Works Better for Token Slimming? To explore whether dynamic aggregation is better for token slimming, we perform ablation experiments as shown in Table 7. For static aggregation, we choose different data-independent operations and maintain similar

| Stage 1 | Stage 2 | Stage 3 | Stage 4 | Stage 1 | Stage 2 | Stage 3 | Stage 4 | Stage 1 | Stage 2 | Stage 3 | Stage 4 |

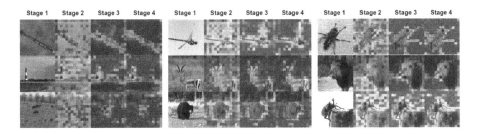

Fig. 7. Visualizations of our progressive token slimming. The blue/red tokens contribute less/more to the final informative tokens. Our method can zoom the attention scope to cover the key object, even with only 12.5% of tokens.

computation: 3×3 average pooling/convolution with stride 2×2, and double linear layers with GELU function ("Token-MLP"). It shows that learnable parameters are vital for token slimming since average pooling leads to a severe accuracy drop. Besides, the static aggregation methods with data-independent weights yield similar but inferior performance to our TSM (79.3% *vs.* 80.1%). Such comparisons prove that our TSM can generate more informative tokens.

Can Contiguous Upsampling Recalibrate the Features? We first recalibrate the original tokens by contiguous upsampling methods, *e.g.*, bilinear interpolation and deconvolution. As shown in Table 8, these two spatial contiguous methods misalign the token relations and hurt the capacity compared with the baseline (without block-to-block mimicking). In contrast, "Token-MLP" does not hurt the token representation, and its accuracy can be further boosted to 80.1% by the insertion of an MLP.

Does Each Distillation Supervision Help? Table 9 presents that the soft logits supervision $\mathcal{L}_{\text{logits}}$ brings 1.4% accuracy gain. When further introducing block-to-block knowledge supervision, our model improves the accuracy by 1.1%. Finally, combining complementary hard label supervision, the accuracy reaches 80.6% with longer training epochs.

What are the Appropriate Loss Weights? Table 10 shows the settings of loss wights are robust in our SiT (trained for 100 epochs). In fact, we simply choose the weight of 2:2:1 to ensure different loss values are close in the early training.

4.6   Visualization

Qualitative Token Slimming Visualization. Figure 7 shows the original images and the token slimming procedure of our SiT-Ti. We observe that the tokens of higher scores, *i.e.*, brighter tokens, are concentrated and tend to cover the key objects in the image. It demonstrates our proposed TSM is able to localize the significant regions and predict accurate scores for the most informative tokens.

5   Conclusions

In this paper, we propose a generic self-slimmed learning method for vanilla vision transformers (SiT), which can speed up the ViTs with negligible accuracy drop. Our concise TSM softly integrates redundant tokens into fewer informative ones. For stable and efficient training, we introduce a novel FRD framework to leverage structure knowledge, which can densely transfer token information in a flexible auto-encoder manner. Extensive experiments demonstrate the effectiveness of our SiT. By simply arming LV-ViT with our SiT, we achieve new state-of-the-art performance on ImageNet, surpassing recent CNNs and ViTs.

Acknowledgements. This work is partially supported by National Key R&D Program of China under Grant 2019YFB2102400, National Natural Science Foundation of China (61876176), the Joint Lab of CAS-HK, Shenzhen Institute of Artificial Intelligence and Robotics for Society, the Shanghai Committee of Science and Technology (Grant No. 21DZ1100100).

References

1. Ali, A., et al.: Xcit: cross-covariance image transformers. In: Advances in Neural Information Processing Systems (2021)
2. Brock, A., De, S., Smith, S.L., Simonyan, K.: High-performance large-scale image recognition without normalization. In: International Conference on Machine Learning (2021)
3. Carion, N., Massa, F., Synnaeve, G., Usunier, N., Kirillov, A., Zagoruyko, S.: End-to-End object detection with transformers. In: Vedaldi, A., Bischof, H., Brox, T., Frahm, J.-M. (eds.) ECCV 2020. LNCS, vol. 12346, pp. 213–229. Springer, Cham (2020). https://doi.org/10.1007/978-3-030-58452-8_13
4. Chen, C.F.R., Fan, Q., Panda, R.: Crossvit: cross-attention multi-scale vision transformer for image classification. In: Proceedings of the IEEE/CVF International Conference on Computer Vision (2021)
5. Chen, H., et al.: Pre-trained image processing transformer. In: Proceedings of the IEEE/CVF Conference on Computer Vision and Pattern Recognition (2021)
6. Chen, T., Cheng, Y., Gan, Z., Yuan, L., Zhang, L., Wang, Z.: Chasing sparsity in vision transformers: an end-to-end exploration. In: Advances in Neural Information Processing Systems (2021)
7. Chu, X., et al.: Twins: revisiting the design of spatial attention in vision transformers. In: Advances in Neural Information Processing Systems (2021)
8. Deng, J., Dong, W., Socher, R., Li, L.J., Li, K., Fei-Fei, L.: Imagenet: a large-scale hierarchical image database. In: Proceedings of the IEEE/CVF Conference on Computer Vision and Pattern Recognition (2009)
9. Dong, X., Bao, J., et al.: Cswin transformer: a general vision transformer backbone with cross-shaped windows. ArXiv abs/2107.00652 (2021)
10. Dosovitskiy, A., Beyer, et al.: An image is worth 16x16 words: transformers for image recognition at scale. In: International Conference on Learning Representations (2020)
11. d'Ascoli, S., Touvron, H., Leavitt, M.L., Morcos, A.S., Biroli, G., Sagun, L.: Convit: improving vision transformers with soft convolutional inductive biases. In: International Conference on Machine Learning (2021)

12. Graham, B., et al.: Levit: a vision transformer in convnet's clothing for faster inference. In: Proceedings of the IEEE/CVF International Conference on Computer Vision (2021)
13. Han, K., Xiao, A., Wu, E., Guo, J., Xu, C., Wang, Y.: Transformer in transformer. In: Advances in Neural Information Processing Systems (2021)
14. He, K., Zhang, X., Ren, S., Sun, J.: Deep residual learning for image recognition. In: IEEE Conference on Computer Vision and Pattern Recognition (2016)
15. Heo, B., Yun, S., Han, D., Chun, S., Choe, J., Oh, S.J.: Rethinking spatial dimensions of vision transformers. In: Proceedings of the IEEE/CVF International Conference on Computer Vision (2021)
16. Jiang, Z.H., et al.: All tokens matter: token labeling for training better vision transformers. In: Advances in Neural Information Processing Systems (2021)
17. Kim, J., Park, S., Kwak, N.: Paraphrasing complex network: network compression via factor transfer. In: Advances in Neural Information Processing Systems (2018)
18. Li, K., et al.: Uniformer: unifying convolution and self-attention for visual recognition. ArXiv abs/2201.09450 (2022)
19. Li, Y., Zhang, K., Cao, J., Timofte, R., Gool, L.V.: Localvit: Bringing locality to vision transformers. ArXiv abs/2104.05707 (2021)
20. Liang, Y., GE, C., Tong, Z., Song, Y., Wang, J., Xie, P.: EVit: expediting vision transformers via token reorganizations. In: International Conference on Learning Representations (2022)
21. Liu, Z., et al.: Swin transformer: hierarchical vision transformer using shifted windows. In: Proceedings of the IEEE/CVF International Conference on Computer Vision (2021)
22. Pan, B., Panda, R., Jiang, Y., Wang, Z., Feris, R., Oliva, A.: Ia-red2: interpretability-aware redundancy reduction for vision transformers. In: Advances in Neural Information Processing Systems (2021)
23. Radosavovic, I., Kosaraju, R.P., Girshick, R.B., He, K., Dollár, P.: Designing network design spaces. In: 2020 IEEE/CVF Conference on Computer Vision and Pattern Recognition (2020)
24. Rao, Y., Zhao, W., Liu, B., Lu, J., Zhou, J., Hsieh, C.J.: Dynamicvit: efficient vision transformers with dynamic token sparsification. In: Advances in Neural Information Processing Systems (2021)
25. Romero, A., Ballas, N., Kahou, S.E., Chassang, A., Gatta, C., Bengio, Y.: Fitnets: hints for thin deep nets. CoRR abs/1412.6550 (2015)
26. Tan, M., Le, Q.: Efficientnet: rethinking model scaling for convolutional neural networks. In: International Conference on Machine Learning (2019)
27. Tan, M., Le, Q.: Efficientnetv2: smaller models and faster training. In: International Conference on Machine Learning (2021)
28. Tang, Y., et al.: Patch slimming for efficient vision transformers. In: Proceedings of the IEEE/CVF Conference on Computer Vision and Pattern Recognition, pp. 12165–12174 (2022)
29. Touvron, H., Cord, M., Douze, M., Massa, F., Sablayrolles, A., Jégou, H.: Training data-efficient image transformers & distillation through attention. In: International Conference on Machine Learning (2021)
30. Touvron, H., Cord, M., Sablayrolles, A., Synnaeve, G., Jégou, H.: Going deeper with image transformers. In: Proceedings of the IEEE/CVF International Conference on Computer Vision (2021)
31. Vaswani, A., et al.: Attention is all you need. In: Advances in Neural Information Processing Systems (2017)

32. Wang, P., et al.: Kvt: k-nn attention for boosting vision transformers. ArXiv abs/2106.00515 (2021)
33. Wang, W., et al.: Pyramid vision transformer: a versatile backbone for dense prediction without convolutions. In: Proceedings of the IEEE/CVF International Conference on Computer Vision (2021)
34. Wu, H., et al.: Cvt: introducing convolutions to vision transformers. In: Proceedings of the IEEE/CVF International Conference on Computer Vision (2021)
35. Xie, E., Wang, W., Yu, Z., Anandkumar, A., Alvarez, J.M., Luo, P.: Segformer: simple and efficient design for semantic segmentation with transformers. In: Advances in Neural Information Processing Systems (2021)
36. Xu, Y., et al.: Evo-vit: slow-fast token evolution for dynamic vision transformer. In: Proceedings of the AAAI Conference on Artificial Intelligence (2022)
37. Yang, J., et al.: Focal self-attention for local-global interactions in vision transformers. ArXiv abs/2107.00641 (2021)
38. Yim, J., Joo, D., Bae, J.H., Kim, J.: A gift from knowledge distillation: Fast optimization, network minimization and transfer learning. In: 2017 IEEE Conference on Computer Vision and Pattern Recognition (CVPR) (2017)
39. Yuan, L., et al.: Tokens-to-token vit: training vision transformers from scratch on imagenet. In: Proceedings of the IEEE/CVF International Conference on Computer Vision (2021)
40. Yuan, L., Hou, Q., Jiang, Z., Feng, J., Yan, S.: Volo: vision outlooker for visual recognition. ArXiv abs/2106.13112 (2021)
41. Zagoruyko, S., Komodakis, N.: Paying more attention to attention: improving the performance of convolutional neural networks via attention transfer. ArXiv abs/1612.03928 (2017)
42. Zhou, D., et al.: Deepvit: towards deeper vision transformer. ArXiv abs/2103.11886 (2021)

Switchable Online Knowledge Distillation

Biao Qian[1], Yang Wang[1(✉)], Hongzhi Yin[2], Richang Hong[1], and Meng Wang[1]

[1] Key Laboratory of Knowledge Engineering with Big Data,
Ministry of Education, School of Computer Science and Information Engineering,
Hefei University of Technology, Hefei, China
yangwang@hfut.edu.cn
[2] The University of Queensland, Brisbane, Australia
h.yin1@uq.edu.au

Abstract. Online Knowledge Distillation (OKD) improves the involved models by reciprocally exploiting the difference between teacher and student. Several crucial bottlenecks over the gap between them — e.g., Why and when does a large gap harm the performance, especially for student? How to quantify the gap between teacher and student? — have received limited formal study. In this paper, we propose **Swit**chable **O**nline **K**nowledge **D**istillation (SwitOKD), to answer these questions. Instead of focusing on the accuracy gap at test phase by the existing arts, the core idea of SwitOKD is to adaptively calibrate the gap at training phase, namely distillation gap, via a switching strategy between two modes — expert mode (pause the teacher while keep the student learning) and learning mode (restart the teacher). To possess an appropriate distillation gap, we further devise an adaptive switching threshold, which provides a formal criterion as to when to switch to learning mode or expert mode, and thus improves the student's performance. Meanwhile, the teacher benefits from our adaptive switching threshold and keeps basically on a par with other online arts. We further extend SwitOKD to multiple networks with two basis topologies. Finally, extensive experiments and analysis validate the merits of SwitOKD for classification over the state-of-the-arts. Our code is available at https://github.com/hfutqian/SwitOKD.

1 Introduction

The essential purpose of Knowledge Distillation (KD) [7,13,14,16,24–26,28,33] is to improve the performance of a *low-capacity student network* (small size, compact) for model compression by distilling the knowledge from a high-capacity teacher network (large size, over parameterized)[1]. The conventional knowledge distillation [2,3,7,9,10,15,27,32] requires a pre-trained teacher to serve as the *expert* network in advance, to be able to provide better supervision for the student in place of one-hot labels. However, it is usually a two-stage offline process, which is inflexible and requires extra computational cost.

Unlike offline fashion, the goal of recently popular online knowledge distillation is to reciprocally train teacher and student from scratch, where they

---

[1] Throughout the rest of the paper, we regard high-capacity network as teacher and low-capacity network as student for simplicity.

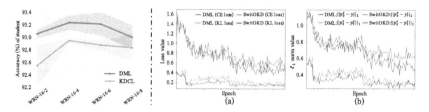

Fig. 1. Left: Illustration of how the large accuracy gap between teacher (WRN-16-2 to WRN-16-8) and student (ResNet-20) affects online distillation process on CIFAR-100 [11]. **Right:** DML [31] bears the emergency of escaping online KD under (a) large *accuracy gap* and (b) large *distillation gap*, whereas SwitOKD extends online KD's lifetime to avoid the degeneration.

learn extra knowledge from each other, and thus improve themselves simultaneously [1,4,22,31]. Typically, Deep Mutual Learning (DML) [31] encourages each network to mutually learn from each other by mimicking their predictions via Kullback Leibler (KL) divergence. Chen *et al.* [1] presents to improve the effectiveness of online distillation by assigning weights to each network with the same architecture. Further, Chung *et al.* [4] proposes to exchange the knowledge of feature map distribution among the networks via an adversarial means. Most of these approaches tend to *equally* train the same or different networks with *small accuracy gap*, where they usually lack richer knowledge from a powerful teacher. In other words, online fashion still fails to resolve the problem of student's performance impairment caused by a large accuracy gap [3,10,15] (see Fig. 1 Left), thus somehow violating the principle of KD. As inspired, we revisit such long-standing issue, and find the fundamental *bottlenecks* below: 1) when and how the gap has negative effect on online distillation process. For example, as the teacher turns from WRN-16-4 to WRN-16-8 (larger gap), the student accuracy rapidly declines (see Fig. 1 Left), while KL loss for the student degenerates into Cross-Entropy (CE) loss (see Fig. 1 Right(a)) as per loss functions in Table 1. To mitigate that, we raise 2) how to quantify the gap and automatically adapt to various accuracy gap, particularly large accuracy gap.

One attempt derives from Guo *et al.* [5], who studied the effect of large accuracy gap on distillation process and found that a large accuracy gap constitutes a certain harm to the performance of teacher. To this end, they propose KDCL, to admit the accuracy improvement of teacher by generating a high-quality soft target, so as to benefit the student. Unfortunately, KDCL pays more attention to teacher, which deviates from the essential purpose of KD; see Table 1.

To sum up, the above online fashions overlook the principle of KD. Meanwhile, they focus on the accuracy gap that is merely obtained at *test* phase, which is not competent for quantifying the gap since it offers no guidance for the distribution alignment of distillation process performed at *training* phase. For instance, the accuracy just depends on the class with maximum probability given a 10-class output, while distillation usually takes into account all of the 10 classes. As opposed to them, we study the gap (the difference in class distribution

Table 1. The varied loss functions for typical distillation methods and the common form of their gradients. τ is set to 1 for theoretical and experimental analysis. *ensemble* is used to generate a soft target by combining the outputs of teacher and student. The gradient for KL divergence loss exactly reflects the difference between the predictions of student and teacher.

Method	Loss function of the networks	The common form of the gradient	Focus on student or not
KD [7] (NeurIPS 2015)	$\mathcal{L} = \alpha\mathcal{L}_{CE}(y,p_s^1) + (1-\alpha)\tau^2\mathcal{L}_{KL}(p_t^\tau,p_s^\tau)$	$(p_s^1 - y) + (p_s^\tau - p_t^\tau)$	✓
KDCL [5] (CVPR 2020)	$\mathcal{L} = \sum_i \mathcal{L}_{CE}(y,p_i^1) + \tau^2\mathcal{L}_{KL}(p_m,p_i^\tau),$ $p_m = ensemble(p_s^\tau, p_t^\tau)$	$(p_i^1 - y) + (p_i^\tau - p_m),$ $i = s,t$	✗
DML [31] (CVPR 2018)	$\mathcal{L}_s = \mathcal{L}_{CE}(y,p_s^1) + \mathcal{L}_{KL}(p_t^\tau,p_s^\tau),$ $\mathcal{L}_t = \mathcal{L}_{CE}(y,p_t^1) + \mathcal{L}_{KL}(p_s^\tau,p_t^\tau)$	$(p_s^1 - y) + (p_s^\tau - p_t^\tau),$ $(p_t^1 - y) + (p_t^\tau - p_s^\tau)$	✗
SwitOKD **(Ours)**	$\mathcal{L}_s = \mathcal{L}_{CE}(y,p_s^1) + \alpha\tau^2\mathcal{L}_{KL}(p_t^\tau,p_s^\tau),$ $p_t^\tau = p_t^{\tau,l} \Leftrightarrow p_t^\tau = p_t^{\tau,e}$	$(p_s^1 - y) + (p_s^\tau - p_t^{\tau,l})$ $\Updownarrow$ $(p_s^1 - y) + (p_s^\tau - p_t^{\tau,e})$	✓

between teacher and student) at *training* phase, namely *distillation gap*, which is quantified by ℓ_1 norm of the gradient (see Sect. 2.1), and how it affects online distillation process from student's perspective. Taking DML [31] as an example, we observe that the gradient for KL loss $||p_s^\tau - p_t^\tau||_1$ increasingly degenerates into that for CE loss $||p_s^\tau - y||_1$ given a large gap; see Fig. 1 Right(b). In such case, the student suffers from *the emergency of escaping online KD process*.

In this paper, we study online knowledge distillation and come up with a novel framework, namely **Swit**chable **O**nline **K**nowledge **D**istillation(SwitOKD), as illustrated in Fig. 2, which stands out new ways to mitigate the adversarial impact of large distillation gap on student. The basic idea of SwitOKD is to calibrate the distillation gap by adaptively pausing the teacher to wait for the learning of student during the *training* phase. Technically, we specify it via an adaptive switching strategy between two types of training modes: namely *learning mode* that is equivalent to reciprocal training from scratch and *expert mode* that freezes teacher's weights while keeps the student learning. Notably, we devise an adaptive switching threshold to endow SwitOKD with the capacity to yield an appropriate distillation gap that is conducive for knowledge transfer from teacher to student. Concurrently, it is nontrivial to devise an "ideal" switching threshold (see Sect. 2.5) due to: 1) not too large — a large threshold aggressively pushes *learning mode* and enlarges the distillation gap, resulting the student into the emergency of escaping online KD process; such fact, as expanded in Sect. 2.5, will further trap teacher to be paused constantly; as opposed to 2) not too small — the teacher constantly performs *expert mode* and receives poor accuracy improvement, suffering from no effective knowledge distilled from teacher to student. The above two conditions lead to 3) adaptiveness — the threshold is adaptively calibrated to balance learning mode and expert mode for extending online KD's lifetime. *Following* SwitOKD, we further establish two fundamental basis topologies to admit the extension of multi-network setting. The extensive experiments on typical datasets demonstrate the superiority of SwitOKD.

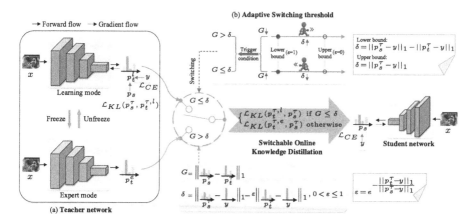

Fig. 2. Illustration of the proposed SwitOKD framework. Our basic idea is to adaptively pause the training of teacher while make the student continuously learn from teacher, to mitigate the adversarial impact of large distillation gap on student. Our framework is achieved by an adaptive switching strategy between two training modes: *learning mode* that is equivalent to training two networks reciprocally and *expert mode* that freezes teacher's parameters while keeps the student learning. Notably, we devise an adaptive switching threshold (b) to admit automatic switching between learning mode and expert mode for an appropriate distillation gap (quantified by G). See Sect. 2.6 for the detailed switching process.

2 Switchable Online Knowledge Distillation

Central to our method are three aspects: (i) quantifying the distillation gap between teacher and student, and analyzing it for online distillation (Sect. 2.1 and 2.2); (ii) an adaptive switching threshold to mitigate the adversarial impact of large distillation gap from student's perspective (Sect. 2.5); and, (iii) extending SwitOKD to multiple networks (Sect. 2.7).

2.1 How to Quantify the Distillation Gap Between Teacher and Student?

Thanks to varied random starts and differences in network structure (*e.g.*, layer, channel, etc.), the prediction difference between teacher and student always exists, which is actually exploited to benefit online distillation. Since the accuracy obtained at *test* phase is not competent to quantify the gap, we propose to quantify the gap at *training* phase, namely *distillation gap*, by computing ℓ_1 norm of the gradient for KL divergence loss (see Table 1), denoted as G, which is more suitable for capturing the same elements (0 entries) and the element-wise difference between the predictions of student and teacher, owing to the *sparsity* property of ℓ_1 norm. Concretely, given a sample x, let $p_t^\mathcal{T}$ and $p_s^\mathcal{T}$ represent the softened outputs of a teacher network $\mathcal{T}(x, \theta_t)$ and a student network $\mathcal{S}(x, \theta_s)$, respectively, then G is formulated as

$$G = ||p_s^\tau - p_t^\tau||_1 = \frac{1}{K} \sum_{k=1}^{K} |p_s^\tau(k) - p_t^\tau(k)|, G \in [0, 2], \tag{1}$$

where $|.|$ denotes the absolute value and τ is the temperature parameter. The k-th element of the softened output p_f^τ is denoted as $p_f^\tau(k) = \frac{exp(z_f(k)/\tau)}{\sum_j^K exp(z_f(j)/\tau)}, f = s, t$; $z_f(k)$ is the k-th value of the logit vector z_f. K is the number of classes. Prior work observes that a great prediction difference between teacher and student has a negative effect on distillation process [3,5,10]. Next, we discuss how the distillation gap affects online distillation process from student's perspective.

2.2 Why is an Appropriate Distillation Gap Crucial?

It is well-accepted that knowledge distillation loss for student is the KL divergence of the soften outputs of teacher p_t^τ and student p_s^τ [7], defined as

$$\mathcal{L}_{KL}(p_t^\tau, p_s^\tau) = \frac{1}{K} \sum_{k=1}^{K} p_t^\tau(k)log\frac{p_t^\tau(k)}{p_s^\tau(k)} = \mathcal{L}_{CE}(p_t^\tau, p_s^\tau) - H(p_t^\tau), \tag{2}$$

where $p_t^\tau(k)$ and $p_s^\tau(k)$ are the k-th element of the output vector p_t^τ and p_s^τ, respectively. $\mathcal{L}_{CE}(.,.)$ represents the Cross-Entropy loss and $H(\cdot)$ means the entropy value. Notably, when p_t^τ stays away from p_s^τ (large distillation gap appears), p_t^τ goes to y, then $\mathcal{L}_{KL}(p_t^\tau, p_s^\tau)$ will degenerate into $\mathcal{L}_{CE}(y, p_s^\tau)$ below:

$$\lim_{p_t^\tau \to y} \mathcal{L}_{KL}(p_t^\tau, p_s^\tau) = \lim_{p_t^\tau \to y} (\mathcal{L}_{CE}(p_t^\tau, p_s^\tau) - H(p_t^\tau)) = \mathcal{L}_{CE}(y, p_s^\tau), \tag{3}$$

where $H(y)$ is a constant (*i.e.*, 0) since y is the one-hot label. The gradient of $\mathcal{L}_{KL}$ w.r.t. z_s also has

$$\lim_{p_t^\tau \to y} \frac{\partial \mathcal{L}_{KL}}{\partial z_s} = \lim_{p_t^\tau \to y} \frac{1}{\tau}(p_s^\tau - p_t^\tau) = \frac{1}{\tau}(p_s^\tau - y), \tag{4}$$

where the gradient for KL loss increasingly degenerates into that for CE loss, resulting student into the emergency of escaping online KD process. The results in Fig. 1 Right also confirm the above analysis. As opposed to that, when p_t^τ goes to p_s^τ (the distillation gap becomes small), $\lim_{p_t^\tau \to p_s^\tau} \mathcal{L}_{KL}(p_t^\tau, p_s^\tau) = 0$, therefore no effective knowledge will be distilled from teacher to student.

How to Yield an Appropriate Gap? Inspired by the above, we need to yield an appropriate distillation gap G to ensure that student can always learn effective knowledge from the teacher throughout the training. In other words, the learning pace of student should continuously keep consistent with that of teacher. Otherwise, the online KD process will terminate. To this end, we propose to maintain an appropriate G. When G is larger than a threshold δ, namely *switching threshold*, we terminate teacher and keep only student learn from teacher, such training status is called *expert mode*. When expert mode progresses, G will decrease until less than δ, it will switch to the other training status of mutually learning between teacher and student, namely *learning mode*. The above two modes alternatively switch under an appropriate δ to keep improving the student's performance. Next, we will offer the details for learning mode (see Sect. 2.3) and expert mode (see Sect. 2.4), which pave the way to our proposed adaptive switching threshold δ (see Sect. 2.5).

2.3   Learning Mode: Independent *vs* Reciprocal

Unlike [3, 7, 10] that pre-train a teacher network in advance, the goal of learning mode is to reduce the distillation gap by training teacher and student network from scratch. Naturally, one naive strategy is to train the teacher independently with the supervision of one-hot label. Then the loss function of teacher and student is given as

$$\mathcal{L}_s^l = \mathcal{L}_{CE}(y, p_s^1) + \alpha\tau^2 \mathcal{L}_{KL}(p_t^{\tau,l}, p_s^\tau), \quad \mathcal{L}_t^l = \mathcal{L}_{CE}(y, p_t^{1,l}), \tag{5}$$

where $p_t^{\tau,l}$ and p_s^τ are the predictions of teacher and student, respectively. α is a balancing hyperparameter. Unfortunately, the independently trained teacher provides poor improvement for student (see Sect. 3.3). Inspired by the fact that the teacher can benefit from reciprocal training [5, 31] and, in turn, admit better guidance for student, we propose to reciprocally train student and teacher, therefore $\mathcal{L}_t^l$ in Eq. (5) can be upgraded to

$$\mathcal{L}_t^l = \mathcal{L}_{CE}(y, p_t^{1,l}) + \beta\tau^2 \mathcal{L}_{KL}(p_s^\tau, p_t^{\tau,l}), \tag{6}$$

where β is a balancing hyperparameter. Thus we can compute the gradient of $\mathcal{L}_s^l$ and $\mathcal{L}_t^l$ w.r.t. z_s and z_t, i.e.,

$$\partial \mathcal{L}_s^l / \partial z_s = (p_s^1 - y) + \alpha\tau(p_s^\tau - p_t^{\tau,l}), \quad \partial \mathcal{L}_t^l / \partial z_t = (p_t^{1,l} - y) + \beta\tau(p_t^{\tau,l} - p_s^\tau). \tag{7}$$

In learning mode, the teacher usually converges faster (yield higher accuracy), owing to its superior learning ability, therefore *the distillation gap will increasingly grow as the training progresses. Meanwhile, for the student, KL loss exhibits a trend to be functionally equivalent to CE loss, causing the effect of knowledge distillation to be weakened.* In this case, SwitOKD will switch to expert mode.

2.4   Expert Mode: Turn to Wait for Student

To mitigate the adversarial impact of large distillation gap on student, SwitOKD attempts to pause the training of teacher while make student continuously learn from teacher, to keep the learning pace consistent, that sets it apart from previous online distillation methods [5, 31]. Indeed, a teacher that is suitable for student rather than one who perfectly imitates one-hot label, can often improve student's performance, in line with our view of an appropriate distillation gap. Accordingly, the loss function for student is similar in spirit to that of Eq. (5):

$$\mathcal{L}_s^e = \mathcal{L}_{CE}(y, p_s^1) + \alpha\tau^2 \mathcal{L}_{KL}(p_t^{\tau,e}, p_s^\tau), \tag{8}$$

where $p_t^{\tau,e}$ is the prediction of teacher network under expert mode. Thus the gradient of $\mathcal{L}_s^e$ w.r.t. z_s is computed as

$$\partial \mathcal{L}_s^e / \partial z_s = (p_s^1 - y) + \alpha\tau(p_s^\tau - p_t^{\tau,e}). \tag{9}$$

In such mode, the student will catch up or even surpass teacher as the training progresses, resulting into no effective knowledge distilled from teacher to student. Then SwitOKD will switch back to learning mode based on our adaptive switching threshold. We discuss that in the next section.

2.5   Adaptive Switching Threshold: Extending Online Knowledge Distillation's Lifetime

Intuitively, a naive strategy is to *manually* select a *fixed* value of δ, which, however, is inflexible and difficult to yield an appropriate distillation gap for improving the student (see Sect. 3.3). We propose an adaptive switching threshold for δ, which offer insights into how to *automatically* switch between learning mode and expert mode. First, observing that the distillation gap $G = ||p_s^{\tau} - p_t^{\tau}||_1 < ||p_s^{\tau} - y||_1$ on average because the teacher is superior to student, and

$$||p_s^{\tau} - p_t^{\tau}||_1 = ||(p_s^{\tau} - y) - (p_t^{\tau} - y)||_1 \geq ||p_s^{\tau} - y||_1 - ||p_t^{\tau} - y||_1, \qquad (10)$$

which further yields $||p_s^{\tau} - y||_1 - ||p_t^{\tau} - y||_1 \leq G < ||p_s^{\tau} - y||_1$, leading to

$$\underbrace{||p_s^{\tau} - y||_1 - ||p_t^{\tau} - y||_1}_{\text{lower bound}} \leq \delta < \underbrace{||p_s^{\tau} - y||_1}_{\text{upper bound}}, \qquad (11)$$

which, as aforementioned in Sect. 1, ought to be neither too large nor too small. To this end, we propose to adaptively adjust δ. Based on Eq. (11), we can further reformulate δ to be:

$$\delta = ||p_s^{\tau} - y||_1 - \varepsilon ||p_t^{\tau} - y||_1, 0 < \varepsilon \leq 1. \qquad (12)$$

It is apparent that ε approaching either 1 or 0 is equivalent to lower or upper bound of Eq. (11). Unpacking Eq. (12), the effect of ε is expected to be: when G becomes large, δ will be decreased towards $||p_s^{\tau} - y||_1 - ||p_t^{\tau} - y||_1$ provided ε approaching 1, then $G > \delta$ holds, which naturally enters into expert mode, and switches back into learning mode vice versa; see Fig. 2(b).

Discussion on ε. As per Eq. (12), the value of δ closely relies on ε, which actually plays the role of tracking the changing trend of G. Intuitively, once the teacher learns faster than student, G will be larger, while $||p_t - y||_1 < ||p_s - y||_1$ holds from Eq. (12). Under such case, small value of δ is expected, leading to a larger value of ε, and vice versa. Hence, ε is inversely proportional to $r = \frac{||p_t^{\tau} - y||_1}{||p_s^{\tau} - y||_1}$. However, if G is very large, the student cannot catch up with the teacher; worse still, the teacher is constantly paused (trapped in expert mode) and cannot improve itself to distill knowledge to student, *making the online KD process terminated*. Hence, we decrease δ, so that, observing that p_t and p_s are very close during the early training time, the teacher can pause more times initially to make student to be in line with teacher, to avoid being largely fall behind at the later training stage (see Sect. 3.2 for detailed validations). Following this, we further decrease the value of r, such that $r = \frac{||p_t^{\tau} - y||_1}{||p_s^{\tau} - y||_1 + ||p_t^{\tau} - y||_1}$, to balance learning mode and expert mode. For normalization issue, we reformulate $\varepsilon = e^{-r}$, leading to the final adaptive switching threshold δ to be:

$$\boxed{\delta = ||p_s^{\tau} - y||_1 - e^{-\frac{||p_t^{\tau} - y||_1}{||p_s^{\tau} - y||_1 + ||p_t^{\tau} - y||_1}} ||p_t^{\tau} - y||_1.} \qquad (13)$$

Algorithm 1. SwitOKD: Switchable Online Knowledge Distillation

---

Input: learning rate η_1, η_2, student network $\mathcal{S}$ parameterized by θ_s, teacher network $\mathcal{T}$ parameterized by θ_t

Output: Trained $\mathcal{S}$, $\mathcal{T}$

1: Randomly initialize $\mathcal{S}$ and $\mathcal{T}$.
2: **for** number of training iterations **do**
3:　Compute $G = ||p_s^\tau - p_t^\tau||_1$.
4:　Compute δ by Eqn. (13).
5:　**if** $G \leq \delta$ **then**
6:　　# **Learning Mode**
7:　　Estimate $\mathcal{L}_s^l$, $\mathcal{L}_t^l$ with Eqn. (6).
8:　Update θ_s, θ_t:
9:　　$\theta_s \leftarrow \theta_s - \eta_1 \frac{\partial \mathcal{L}_s^l}{\partial z_s} \frac{\partial z_s}{\partial \theta_s}$
10:　　$\theta_t \leftarrow \theta_t - \eta_2 \frac{\partial \mathcal{L}_t^l}{\partial z_t} \frac{\partial z_t}{\partial \theta_t}$
11:　**else**
12:　　# **Expert Mode**
13:　　Estimate $\mathcal{L}_s^e$ with Eqn. (8).
14:　　Freeze θ_t, update θ_s:
15:　　$\theta_s \leftarrow \theta_s - \eta_1 \frac{\partial \mathcal{L}_s^e}{\partial z_s} \frac{\partial z_s}{\partial \theta_s}$
16:　**end if**
17: **end for**

Unlike the existing arts [5,31], where they fail to focus on student to follow the principle of KD, SwitOKD can *extend online knowledge distillation's lifetime*, and thus largely improve student's performance, while keep our teacher be basically on par with theirs, thanks to Eq. (13); see Sect. 3.4 for validations.

2.6 Optimization

The above specifies the adaptive switching strategy between two training modes. Specifically, we kick off SwitOKD with learning mode to minimize $\mathcal{L}_s^l$ and $\mathcal{L}_t^l$, then the training mode is switched into expert mode to minimize $\mathcal{L}_s^e$ when $G > \delta$. Following that, SwitOKD switches back to learning mode when $G \leq \delta$. The whole training process is summarized in Algorithm 1.

2.7 Multi-network Learning Framework

To endow SwitOKD with the extendibility to multi-network setting with large distillation gap, we divide these networks into multiple teachers and students, involving switchable online distillation between teachers and students, which is built by two types of fundamental basis topologies below: multiple teachers *vs* one student and one teacher *vs* multiple students. For ease of understanding, we take 3 networks as an example and denote the basis topologies as **2T1S** and **1T2S**, respectively; see Fig. 3. Notably, the training between each teacher-student pair directly follows SwitOKD, while two teachers for **2T1S** (or two students for **1T2S**) mutually transfer knowledge in a conventional two-way manner. Note that, for **1T2S**, only when the switching conditions between teacher and both students are triggered, will the teacher be completely suspended. The detailed validation results are reported in Sect. 3.8.

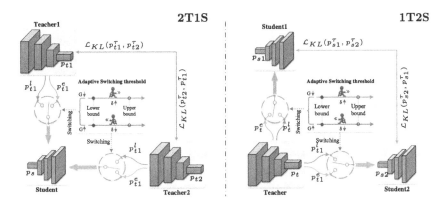

Fig. 3. The multi-network framework for training 3 networks simultaneously, including two fundamental basis topologies: **2T1S** (**Left**) and **1T2S** (**Right**).

3   Experiment

To validate the effectiveness of SwitOKD, we experimentally evaluate various state-of-the-art backbone networks via student-teacher pair below: MobileNet [8], MobileNetV2 [20] (sMobileNetV2 means the width multiplier is s), ResNet [6] and Wide ResNet (WRN) [30] over the typical image classification datasets: **CIFAR-10** and **CIFAR-100** [11] are natural image datasets, including 32×32 RGB images containing 10 and 100 classes. Both of them are split into a training set with 50 k images and a test set with 10k images. **Tiny-ImageNet** [12] consists of 64×64 color images from 200 classes. Each class has 500 training images, 50 validation images, and 50 test images. **ImageNet** [19] contains 1 k object classes with about 1.2 million images for training and 50 k images for validation.

3.1   Experimental Setup

We implement all networks and training procedures with pytorch [17] on an NVIDIA GeForce GTX 1080 Ti GPU and an Intel(R) Core(TM) i7-6950X CPU @ 3.00 GHz. For **CIFAR-10/100**, we use Adam optimizer with the momentum of 0.9, weight decay of 1e−4 and set batch size to 128. The initial learning rate is 0.01 and then divided by 10 at 140, 200 and 250 of the total 300 epochs. For **Tiny-ImageNet** and **ImageNet**, we adopt SGD as the optimizer, and set momentum to 0.9 and weight decay to 5e−4. Specifically, for Tiny-ImageNet, we set batch size to 128, the initial learning rate to 0.01, and the learning rate is dropped by 0.1 at 100, 140 and 180 of the total 200 epochs. For ImageNet, batch size is 512, while the initial learning rate is 0.1 (dropped by 0.1 every 30 epochs and trained for 120 epochs). As for the hyperparameter, we set α, β and τ to 1, and $\tau = \{2, 3\}$ for the classic distillation [5, 7].

Previous sections (Sects. 2.3, 2.4, 2.5 and 2.7) explicate how adaptive switching strategy benefits a student. We offer practical insights into why SwitOKD

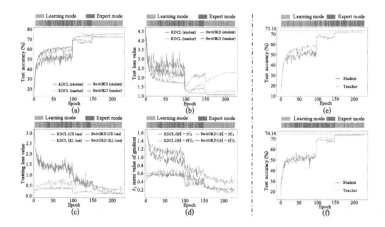

Fig. 4. Left: Illustration of test accuracy (a) and loss (b) for SwitOKD and KDCL [5]. From the perspective of student, (c) shows the comparison of CE loss and KL loss, while (d) is ℓ_1 norm value of the gradient for CE loss and KL loss. **Right:** Illustration of why the parameter r should be decreased from $r = \frac{||p_t^\tau - y||_1}{||p_s^\tau - y||_1}$ (e) to $r = \frac{||p_t^\tau - y||_1}{||p_s^\tau - y||_1 + ||p_t^\tau - y||_1}$ (f). The color bar shows the switching process of SwitOKD, where the cyan and the magenta denote learning mode and expert mode, respectively. (Color figure online)

works well, including ablation study and comparison with the state-of-the-arts, as well as extendibility to multiple networks.

3.2   Why Does SwitOKD Work Well?

One of our aims is to confirm that the core idea of our SwitOKD — using an adaptive switching threshold to achieve adaptive switching strategy — can possess an appropriate distillation gap for the improvement of student. The other is to verify why the parameter r (Sect. 2.5) should be decreased. We perform online distillation with a compact student network (ResNet-32) and a powerful teacher network (WRN-16-8 and WRN-16-2) on CIFAR-100.

Figure 4(a) (b) illustrate that the performance of student is continuously improved with smaller accuracy gap (gray area) compared to KDCL, confirming that our switching strategy can effectively calibrate the distillation gap to extend online KD's lifetime, in keeping with the core idea of SwitOKD. As an extension of Fig. 1, Fig. 4(c) reveals that KL loss for SwitOKD keeps far away from CE loss throughout the training unlike KDCL. Akin to that, the gradient for KL loss $||p_s^\tau - p_t^\tau||_1$ (refer to Eq. (1)) keeps divergent from that for CE loss $||p_s^\tau - y||_1$; see Fig. 4(d). Especially, *the color bar* illustrates the process of switching two modes on top of each other: when to pause the teacher — expert mode (magenta), or restart training — learning mode (cyan), reflecting that an appropriate gap holds with adaptive switching operation.

Figure 4(e)(f) validate the findings below: when $r = \frac{||p_t^\tau - y||_1}{||p_s^\tau - y||_1}$ (e), the teacher is rarely paused at the early stage of training, then the student largely falls behind at the later stage, leading to poor teacher (73.10% vs 74.14%) and student (71.89% vs 73.47%), confirming our analysis in Sect. 2.5 — $r = \frac{||p_t^\tau - y||_1}{||p_s^\tau - y||_1 + ||p_t^\tau - y||_1}$ (f) is desirable to balance learning mode and expert mode.

3.3 Ablation Studies

Is Each Component of SwitOKD Essential? To verify the effectiveness of several components constituting SwitOKD — *switching strategy, adaptive switching threshold* and *teacher's training strategy*, we construct ablation experiments with ResNet-32 (student) and WRN-16-2 (teacher) from the following cases: **A**: SwitOKD without switching; **B**: SwitOKD with fixed δ (*i.e.,* $\delta \in \{0.2, 0.6, 0.8\}$); **C**: teacher's loss $\mathcal{L}_t^l$ (Eq. (6) vs Eq. (5)); **D**: the proposed SwitOKD. Table 2

Table 2. Ablation study about the effectiveness of each component, of which constitutes SwitOKD. The best results are reported with **boldface**.

Case	Threshold δ	Switching or not	Teacher's loss $\mathcal{L}_t^l$	CIFAR-100
A	-	✗	Eqn.(6)	72.91
B	$\delta = 0.2$	✓	Eqn.(6)	72.92
	$\delta = 0.6$			72.83
	$\delta = 0.8$			73.00
C	Eqn.(13)	✓	Eqn.(5)	72.72
D	Eqn.(13)	✓	Eqn.(6)	**73.47**

summarizes our findings, which suggests that SwitOKD shows great superiority (73.47%) to other cases. Especially for case **B**, the manual δ fails to yield an appropriate distillation gap for improving the performance of student, confirming the importance of adaptive switching threshold, subject to our analysis (Sect. 2.5). Notably, the student for case **C** suffers from a large accuracy loss, verifying the benefits of reciprocal training on improving the performance of student (Sect. 2.3).

Why does the Temperature τ Benefit SwitOKD? The temperature τ [7] usually serves as a factor to smooth the predictions of student and teacher. Empirically, temperature parameter enables the output of student to be closer to that of teacher (and thus reduce the distillation gap), improving the performance, in line

Table 3. Ablation study about the effectiveness of varied temperature τ on CIFAR-100. The best results are reported with **boldface**.

τ	0.5	1	2	5	8	10
SwitOKD	64.95	67.24	**67.80**	66.30	66.00	65.64

with our perspective in Sect. 2.4. To highlight the effectiveness of SwitOKD, we simply set $\tau = 1$ for our experiments. To further verify the effectiveness of varied $\tau \in \{0.5, 1, 2, 5, 8, 10\}$, we perform the ablation experiments with MobileNetV2 (student) and WRN-16-2 (teacher). Table 3 summarizes the findings. The optimal student (67.80%) is achieved with a slightly higher $\tau^* = 2$, implying that τ contributes to the calibration of the distillation gap. Note that when $\tau = 10$, the accuracy of student rapidly declines, implying that excessive smoothness can make the gap beyond an optimal range and, in turn, harm the performance of student, consistent with our view of an appropriate gap in Sect. 2.2.

Table 4. Accuracy (%) comparison on Tiny-ImageNet and CIFAR-10/100. (.M) denotes the number of parameters. All the values are measured by computing mean and standard deviation across 3 trials with random seeds. The best results are reported with **boldface**.

	Backbone	Vanilla	DML [31]	KDCL [5]	SwitOKD
Tiny-ImageNet					
Student	1.4MobileNetV2(4.7 M)	50.98 ± 0.32	55.70 ± 0.61	57.79 ± 0.30	**58.71** ± 0.11
Teacher	ResNet-34(21.3 M)	63.18 ± 0.37	64.49 ± 0.43	**65.47** ± 0.32	63.31 ± 0.04
Student	ResNet-20(0.28M)	52.35 ± 0.15	53.98 ± 0.26	53.74 ± 0.39	**55.03** ± 0.19
Teacher	WRN-16-2(0.72 M)	56.59 ± 0.22	57.45 ± 0.19	**57.71** ± 0.30	57.41 ± 0.06
CIFAR-10					
Student	WRN-16-1(0.18 M)	91.45 ± 0.06	91.96 ± 0.08	91.86 ± 0.11	**92.50** ± 0.17
Teacher	WRN-16-8(11.0 M)	95.21 ± 0.12	95.06 ± 0.05	**95.33** ± 0.17	94.76 ± 0.12
CIFAR-100					
Student	0.5MobileNetV2(0.81 M)	60.07 ± 0.40	66.23 ± 0.36	66.83 ± 0.05	**67.24** ± 0.04
Teacher	WRN-16-2(0.70 M)	72.90 ± 0.09	73.85 ± 0.21	73.75 ± 0.26	**73.90** ± 0.40

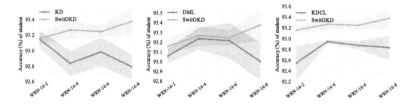

Fig. 5. Illustration of how an appropriate distillation gap yields better student. For KD, DML and KDCL, the accuracy of student (ResNet-20) rapidly declines as the teacher turns to higher capacity (WRN-16-2 to WRN-16-8). As opposed to that, SwitOKD grows steadily, owing to an appropriate distillation gap.

3.4   Comparison with Other Approaches

To verify the superiority of SwitOKD, we first compare with typical online KD methods, including: 1) DML [31] is equivalent to learning mode for SwitOKD; 2) KDCL [5] studies the effect of large accuracy gap at the test phase on online distillation process, but they pay more attention to teacher instead of student. For brevity, "vanilla" refers to the backbone network trained from scratch with classification loss alone. A compact student and a powerful teacher constitute the student-teacher network pair with *large distillation gap* at the training phase.

Table 4 and Fig. 5 summarize our findings below: *First*, switchable online distillation offers a significant and consistent performance improvement over the baseline (vanilla) and the state-of-the-arts for *student*, in line with the principle of KD process. Impressively, SwitOKD achieves 1.05% accuracy improvement to vanilla on CIFAR-10 (7.17% on CIFAR-100). Besides, SwitOKD also shows 0.54% and 0.54% (WRN-16-1/WRN-16-8) accuracy gain over DML and KDCL on CIFAR-100, respectively. Especially with 1.4MobileNetV2/ResNet-34, SwitOKD still obtains significant performance gains of 7.73%, 3.01% and 0.92% (the gains are substantial for Tiny-ImageNet) over vanilla, DML and KDCL.

Table 5. Accuracy (%) comparison of student network with offline KD methods (seen as expert mode of SwitOKD) on CIFAR-100. (.M) denotes the number of parameters. The best results are reported with **boldface**.

Backbone		Vanilla	KD [7]	FitNet [18]	AT [29]	CRD [23]	RCO [10]	SwitOKD
Student	WRN-16-2(0.70 M)	72.79	74.49	73.44	73.35	75.01	75.36	**75.95**
Teacher	WRN-40-2(2.26 M)	76.17	–	–	–	–	–	**76.54**

Table 6. Top-1 accuracy (%) on ImageNet dataset. (.M) denotes the number of parameters. The best results are reported with **boldface**.

	Backbone	Vanilla	DML [31]	KDCL [5]	SwitOKD
Student	ResNet-18(11.7 M)	69.76	70.81	70.91	**71.75**
Teacher	ResNet-34(21.8 M)	73.27	73.47	**73.70**	73.65
Student	0.5MobileNetV2(1.97 M)	63.54	64.22	63.92	**65.11**
Teacher	ResNet-18(11.7 M)	69.76	68.30	**70.60**	68.08

Second, our teachers still benefit from SwitOKD and obtain accuracy improvement *basically on a par* with DML and KDCL, confirming our analysis about the adaptive switching threshold δ (see Eq. (13)) — balance of learning mode and expert mode. Note that, with 0.5MobileNetV2 and WRN-16-2 on CIFAR-100, our teacher (73.90%) upgrades beyond the vanilla (72.90%), even yields comparable accuracy gain (0.05% and 0.15%) over DML and KDCL. By contrast, KDCL has most of the best teachers, but with poor students, owing to its concentration on teacher only.

Finally, we also validate the effectiveness of SwitOKD for student even under a small distillation gap on CIFAR-10 (see Fig. 5), where the students (ResNet-20) still possess significant performance advantages, confirming the necessity of adaptively calibrating an appropriate gap with adaptive switching threshold δ in Sect. 2.5. Especially for Fig. 5 (b)(c), as the teacher turns to higher capacity (WRN-16-2 to WRN-16-8), students' accuracy from DML and KDCL rises at the beginning, then rapidly declines, and reaches the best results when the teacher is WRN-16-4. This, in turn, keeps consistent with our analysis (Sect. 2.2) — an appropriate distillation gap admits student's improvement.

To further validate the switching strategy between two modes, we also compare SwitOKD with offline knowledge distillation approaches (seen as expert mode of SwitOKD) including KD [7], FitNet [18], AT [29] and CRD [23] that require a fixed and pre-trained teacher. Especially, RCO [10] is similar to our approach, which maintains a reasonable performance gap by manually selecting a series of pre-trained intermediate teachers. Table 5 reveals that SwitOKD achieves superior performance over offline fashions, while exceeds the second best results from RCO by 0.59%, implying that SwitOKD strictly follows the essential principle of KD with the adaptive switching strategy.

Table 7. Accuracy (%) comparison with 3 networks on CIFAR-100. WRN-16-2 serves as either teacher (**T**) or student (**S**) for DML and KDCL, while is treated as   for **1T2S** and T for **2T1S**. The best results are reported with **boldface**.

Backbone	Vanilla	DML [31]	KDCL [5]	SwitOKD (1T2S)	SwitOKD (2T1S)
MobileNet	58.65(**S**)	63.75(**S**)	62.13(**S**)	**64.62(S)**	**65.03(S)**
WRN-16-2	73.37(/**T**)	74.30(/**T**)	73.94(/**T**)	**75.02( )**	71.73(**T**)
WRN-16-10	79.45(**T**)	77.82(**T**)	**80.71(T)**	77.33(**T**)	77.07(**T**)

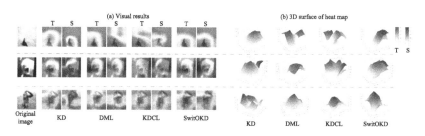

Fig. 6. Visual analysis of why SwitOKD works on CIFAR-100. **Left:**(a) The visual results by superimposing the heat map onto corresponding original image. **Right:**(b) 3D surface of heat maps for teacher and student (the more the peak overlaps, the better the student mimics teacher), where x and y axis denote the width and height of an image, while z axis represents the gray value of the heat map. T: ResNet-34 (teacher); S: ResNet-18 (student).

3.5   Extension to Large-Scale Dataset

Akin to [5,31], as a by-product, SwitOKD can effectively be extended to the large-scale dataset (*i.e.*, ImageNet), benefiting from its good generalization ability; see Table 6. It is observed that the students' accuracy is improved by 1.99% and 1.57% upon the vanilla, which are substantial for ImageNet, validating the scalability of SwitOKD. Particularly, for ResNet-34, our teacher (73.65%) outperforms the vanilla (73.27%) and DML (73.47%), highlighting the importance of our adaptive switching strategy upon δ to balance the teacher and student. Another evidence is shown for 0.5MobileNetV2 and ResNet-18 with larger distillation gap, our student outperforms DML and KDCL by 0.89% and 1.19%, while the teacher also yields comparable performance with DML, keeping consistent with our analysis in Sect. 2.5.

3.6   How About SwitOKD from Visualization Perspective?

To shed more light on why SwitOKD works in Sect. 3.2, we further perform a visual analysis with Grad-cam [21] visualization of image classification via a heat map (red/blue region corresponds to high/low value) that localizes the class-discriminative regions, to confirm that our adaptive switching strategy enables student to mimic teacher well, and thus improves the classification accuracy.

Figure 6(a) illustrates the visual results by superimposing the heat map onto corresponding original image, to indicate whether the object regions of the image is focused (the red area denotes more focus); Fig. 6(b) shows 3D surface of the heat map to reflect the overlap range of heat maps for teacher and student (the more the peak overlaps, the better the student mimics teacher). Combining Fig. 6(a) and (b), it suggests that SwitOKD focuses on the regions of student, which not only keep consistent with that of the teacher — mimic the teacher well (KL loss), but correctly cover the object regions — yield high precision (CE loss), in line with our analysis (Sect. 1): keep the gradient for KL loss divergent from that for CE loss. *The above further confirms the adversarial impact of large distillation gap — the emergency of escaping online KD process* (Sects. 2.2 and 2.3).

3.7   What Improves the Training Efficiency of SwitOKD?

Interestingly, SwitOKD has considerably raised the training efficiency of online distillation process beyond [5,31] since the training of teacher is *paused* (merely involve inference process) under expert mode (Sect. 2.4). We perform efficiency analysis on a single GPU (GTX 1080 Ti), where SwitOKD is compared with other online distillation methods, *e.g.*, DML [31] and KDCL [5]. Figure 7 shows that the time per iteration for SwitOKD (green line) varies greatly, owing to adaptive switching operation. Notably, the total

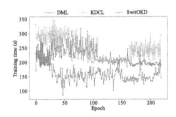

Fig. 7. Efficiency analysis with Mobile NetV2 (student) and ResNet-18 (teacher) on Tiny-ImageNet. (Color figure online)

training time is significantly reduced by 27.3% (9.77h *vs* 13.43h) compared to DML (blue line), while 34.8% (9.77h *vs* 14.99h) compared to KDCL (orange line).

3.8   Extension to Multiple Networks

To show our approach's extendibility for training multiple networks, we conduct the experiments based on three networks with large distillation gap, see Table 7. As can be seen, the students for **1T2S** and **2T1S** achieve significant accuracy gains (5.97%, 1.65%, and 6.38%) over vanilla and outperform other online distillation approaches (*i.e.*, DML [31] and KDCL [5]) with significant margins, while our teachers (WRN-16-10) are basically on a par with DML, consistent with the tendency of performance gain for SwitOKD in Table 4. By contrast, KDCL receives the best teacher (WRN-16-10), but a poor student (MobileNet), in that it pays more attention to teacher instead of student. Notably, **1T2S** achieves a better teacher (77.33% *vs* 77.07% for WRN-16-10) than **2T1S**; the reason is that the teacher for **1T2S** will be completely suspended when the switching conditions between teacher and both students are triggered (Sect. 2.7).

4 Conclusion

In this paper, we propose Switchable Online Knowledge Distillation (SwitOKD), to mitigate the adversarial impact of large distillation gap between teacher and student, where our basic idea is to calibrate the distillation gap by adaptively pausing the teacher to wait for the learning of student. We foster it throughout an adaptive switching strategy between learning mode and expert mode. Notably, an adaptive switching threshold is devised to endow SwitOKD with the capacity to automatically yield an appropriate distillation gap, so that the performance of student and teacher can be improved. Further, we verify SwitOKD's extendibility to multiple networks. The extensive experiments on typical classification datasets validate the effectiveness of SwitOKD.

Acknowledgements. This work is supported by the National Natural Science Foundation of China under grant no U21A20470, 62172136, 61725203, U1936217. Key Research and Technology Development Projects of Anhui Province (no. 202004a5020043).

References

1. Chen, D., Mei, J.P., Wang, C., Feng, Y., Chen, C.: Online knowledge distillation with diverse peers. In: Proceedings of the AAAI Conference on Artificial Intelligence, vol. 34, pp. 3430–3437 (2020)
2. Chen, P., Liu, S., Zhao, H., Jia, J.: Distilling knowledge via knowledge review. In: Proceedings of the IEEE/CVF Conference on Computer Vision and Pattern Recognition, pp. 5008–5017 (2021)
3. Cho, J.H., Hariharan, B.: On the efficacy of knowledge distillation. In: Proceedings of the IEEE/CVF International Conference on Computer Vision (ICCV) (2019)
4. Chung, I., Park, S., Kim, J., Kwak, N.: Feature-map-level online adversarial knowledge distillation. In: International Conference on Machine Learning, pp. 2006–2015. PMLR (2020)
5. Guo, Q., et al.: Online knowledge distillation via collaborative learning. In: IEEE/CVF Conference on Computer Vision and Pattern Recognition (CVPR) (2020)
6. He, K., Zhang, X., Ren, S., Sun, J.: Deep residual learning for image recognition. In: Proceedings of the IEEE Conference on Computer Vision and Pattern Recognition, pp. 770–778 (2016)
7. Hinton, G., Vinyals, O., Dean, J.: Distilling the knowledge in a neural network. In: NIPS (2015)
8. Howard, A.G., et al.: Mobilenets: efficient convolutional neural networks for mobile vision applications. arXiv preprint arXiv:1704.04861 (2017)
9. Huang, Z., et al.: Revisiting knowledge distillation: an inheritance and exploration framework. In: Proceedings of the IEEE/CVF Conference on Computer Vision and Pattern Recognition, pp. 3579–3588 (2021)
10. Jin, X., et al.: Knowledge distillation via route constrained optimization. In: Proceedings of the IEEE International Conference on Computer Vision, pp. 1345–1354 (2019)

11. Krizhevsky, A., Hinton, G., et al.: Learning multiple layers of features from tiny images (2009)
12. Le, Y., Yang, X.: Tiny imagenet visual recognition challenge. CS 231N, **7N**(7), 3 (2015)
13. Li, T., Li, J., Liu, Z., Zhang, C.: Few sample knowledge distillation for efficient network compression. In: Proceedings of the IEEE/CVF Conference on Computer Vision and Pattern Recognition, pp. 14639–14647 (2020)
14. Menon, A.K., Rawat, A.S., Reddi, S., Kim, S., Kumar, S.: A statistical perspective on distillation. In: International Conference on Machine Learning, pp. 7632–7642. PMLR (2021)
15. Mirzadeh, S.I., Farajtabar, M., Li, A., Levine, N., Matsukawa, A., Ghasemzadeh, H.: Improved knowledge distillation via teacher assistant. In: Proceedings of the AAAI Conference on Artificial Intelligence, vol. 34, pp. 5191–5198 (2020)
16. Passalis, N., Tzelepi, M., Tefas, A.: Heterogeneous knowledge distillation using information flow modeling. In: Proceedings of the IEEE/CVF Conference on Computer Vision and Pattern Recognition, pp. 2339–2348 (2020)
17. Paszke, A., et al.: Automatic differentiation in pytorch (2017)
18. Romero, A., Ballas, N., Kahou, S.E., Chassang, A., Gatta, C., Bengio, Y.: Fitnets: Hints for thin deep nets. arXiv preprint arXiv:1412.6550 (2014)
19. Russakovsky, O., et al.: Imagenet large scale visual recognition challenge. Int. J. Comput. Vis. **115**(3), 211–252 (2015). https://doi.org/10.1007/s11263-015-0816-y
20. Sandler, M., Howard, A., Zhu, M., Zhmoginov, A., Chen, L.C.: Mobilenetv 2: inverted residuals and linear bottlenecks. In: Proceedings of the IEEE Conference on Computer Vision and Pattern Recognition, pp. 4510–4520 (2018)
21. Selvaraju, R.R., Cogswell, M., Das, A., Vedantam, R., Parikh, D., Batra, D.: Gradcam: visual explanations from deep networks via gradient-based localization. In: Proceedings of the IEEE International Conference on Computer Vision, pp. 618–626 (2017)
22. Song, G., Chai, W.: Collaborative learning for deep neural networks. In: Advances in Neural Information Processing Systems, pp. 1832–1841 (2018)
23. Tian, Y., Krishnan, D., Isola, P.: Contrastive representation distillation. arXiv preprint arXiv:1910.10699 (2019)
24. Tung, F., Mori, G.: Similarity-preserving knowledge distillation. In: Proceedings of the IEEE International Conference on Computer Vision, pp. 1365–1374 (2019)
25. Wang, X., Zhang, R., Sun, Y., Qi, J.: Kdgan: knowledge distillation with generative adversarial networks. In: Advances in Neural Information Processing Systems, pp. 775–786 (2018)
26. Wang, Y.: Survey on deep multi-modal data analytics: collaboration, rivalry, and fusion. ACM Trans. Multimedia Comput. Commun. Appl. (TOMM) **17**(1s), 1–25 (2021)
27. Xu, G., Liu, Z., Li, X., Loy, C.C.: Knowledge distillation meets self-supervision. In: Vedaldi, A., Bischof, H., Brox, T., Frahm, J.-M. (eds.) ECCV 2020. LNCS, vol. 12354, pp. 588–604. Springer, Cham (2020). https://doi.org/10.1007/978-3-030-58545-7_34
28. Yim, J., Joo, D., Bae, J., Kim, J.: A gift from knowledge distillation: fast optimization, network minimization and transfer learning. In: Proceedings of the IEEE Conference on Computer Vision and Pattern Recognition, pp. 4133–4141 (2017)
29. Zagoruyko, S., Komodakis, N.: Paying more attention to attention: Improving the performance of convolutional neural networks via attention transfer. arXiv preprint arXiv:1612.03928 (2016)

30. Zagoruyko, S., Komodakis, N.: Wide residual networks. arXiv preprint arXiv:1605.07146 (2016)
31. Zhang, Y., Xiang, T., Hospedales, T.M., Lu, H.: Deep mutual learning. In: Proceedings of the IEEE Conference on Computer Vision and Pattern Recognition, pp. 4320–4328 (2018)
32. Zhu, J., et al.: Complementary relation contrastive distillation. In: Proceedings of the IEEE/CVF Conference on Computer Vision and Pattern Recognition, pp. 9260–9269 (2021)
33. Zhu, Y., Wang, Y.: Student customized knowledge distillation: bridging the gap between student and teacher. In: Proceedings of the IEEE/CVF International Conference on Computer Vision, pp. 5057–5066 (2021)

ℓ∞-Robustness and Beyond: Unleashing Efficient Adversarial Training

Hadi M. Dolatabadi$^{(\boxtimes)}$ (ID), Sarah Erfani (ID), and Christopher Leckie (ID)

School of Computing and Information Systems, The University of Melbourne,
Parkville, VIC, Australia
hadi.mohagheghdolatabadi@student.unimelb.edu.au

Abstract. Neural networks are vulnerable to adversarial attacks: adding well-crafted, imperceptible perturbations to their input can modify their output. Adversarial training is one of the most effective approaches in training robust models against such attacks. However, it is much slower than vanilla training of neural networks since it needs to construct adversarial examples for the entire training data at every iteration, hampering its effectiveness. Recently, *Fast Adversarial Training* (FAT) was proposed that can obtain robust models efficiently. However, the reasons behind its success are not fully understood, and more importantly, it can only train robust models for ℓ∞-bounded attacks as it uses FGSM during training. In this paper, by leveraging the theory of coreset selection, we show how selecting a small subset of training data provides a *general*, more principled approach toward reducing the time complexity of robust training. Unlike existing methods, our approach can be adapted to a wide variety of training objectives, including TRADES, ℓ_p-PGD, and Perceptual Adversarial Training (PAT). Our experimental results indicate that our approach speeds up adversarial training by 2–3 times while experiencing a slight reduction in the clean and robust accuracy.

Keywords: Adversarial training · Coreset selection · Efficient training

1 Introduction

Neural networks have achieved great success in the past decade. Today, they are one of the primary candidates in solving a wide variety of machine learning tasks, from object detection and classification [12,42] to photo-realistic image generation [14,38] and beyond. Despite their impressive performance, neural networks are vulnerable to adversarial attacks [3,35]: adding well-crafted, imperceptible perturbations to their input can change their output. This unexpected behavior of neural networks prevents their widespread deployment in safety-critical applications, including autonomous driving [8] and medical diagnosis [24]. As such, training robust neural networks against adversarial attacks is of paramount importance and has gained lots of attention.

Supplementary Information The online version contains supplementary material available at https://doi.org/10.1007/978-3-031-20083-0_28.

Adversarial training is one of the most successful approaches in defending neural networks against adversarial attacks[1]. This approach first constructs a perturbed version of the training data. Then, the neural network is optimized on these perturbed inputs instead of the clean samples. This procedure must be done iteratively as the perturbations depend on the neural network weights. Since the weights are optimized during training, the perturbations must also be adjusted for each data sample in every iteration.

Various adversarial training methods primarily differ in how they define and find the perturbed version of the input [22,25,44]. However, they all require repetitive construction of these perturbations during training which is often cast as another non-linear optimization problem. As such, the time and computational complexity of adversarial training is massively higher than vanilla training. In practice, neural networks require massive amounts of training data [1] and need to be trained multiple times with various hyper-parameters to get their best performance [16]. Thus, reducing the time/computational complexity of adversarial training is critical in enabling the environmentally efficient application of robust neural networks in real-world scenarios [33,34].

Fast Adversarial Training (FAT) [41] is a successful approach proposed for efficient training of robust neural networks. Contrary to the common belief that building the perturbed versions of the inputs using *Fast Gradient Sign Method* (FGSM) [10] does not help in training arbitrary robust models [25,36], Wong *et al.* [41] show that by carefully applying uniformly random initialization before the FGSM step one can make this training approach work. Using FGSM to generate the perturbed input in a single step combined with implementation tricks such as mixed precision and cyclic learning rate, FAT can significantly reduce the training time of robust neural networks.

Despite its success, FAT may exhibit unexpected behavior in different settings. For instance, it was shown that FAT suffers from *catastrophic overfitting* where the robust accuracy during training suddenly drops to 0% [2,41]. A more fundamental issue with FAT and its variations such as `GradAlign` [2] is that they are specifically designed and implemented for ℓ_∞ adversarial training. This is because FGSM, particularly an ℓ_∞ perturbation generator, is at the heart of these methods. As a result, the quest for a unified, systematic approach that can reduce the time complexity of all types of adversarial training is not over.

Motivated by the limited scope of FAT, in this paper we take an important step towards finding a general yet principled approach for reducing the time complexity of adversarial training. We notice that repetitive construction of adversarial examples for each data point is the main bottleneck of robust training. While this process needs to be done iteratively, we speculate that per-

---

[1] Note that adversarial training in the literature generally refers to a particular approach proposed by Madry *et al.* [25]. For the purposes of this paper, we refer to any method that builds adversarial attacks around the training data and incorporates them into the training of the neural network as adversarial training. Using this taxonomy, methods such as TRADES [44], ℓ_p-PGD [25] or Perceptual Adversarial Training (PAT) [22] are all considered different versions of adversarial training.

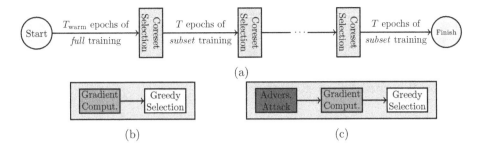

Fig. 1. Overview of neural network training using coreset selection. (a) Selection is done every T epochs. During the next episodes, the network is only trained on this subset. (b) Coreset selection module for vanilla training. (c) Coreset selection module for adversarial training.

haps we can find a subset of the training data that is more important to robust network optimization than the rest. Specifically, we ask the following research question: *Can we train an adversarially robust neural network using a subset of the entire training data without sacrificing clean or robust accuracy?*

In this paper, we show that the answer to this question is affirmative: by selecting a *weighted* subset of the data based on the neural network state, we run *weighted* adversarial training only on this selected subset. We draw an elegant connection between adversarial training and adaptive coreset selection algorithms to achieve this goal. In particular, we use Danskin's theorem and demonstrate how the entire training data can effectively be approximated with an informative weighted subset. To conduct this selection, our study shows that one needs to build adversarial examples for the entire training data and solve a respective subset selection objective. Afterward, training can be performed on this selected subset of the training data. In our approach, shown in Fig. 1, adversarial coreset selection is only required every few epochs, effectively reducing the training time of robust learning algorithms. We demonstrate how our proposed method can be used as a general framework in conjunction with different adversarial training objectives, opening the door to a more principled approach for efficient training of robust neural networks in a general setting. Our experimental results show that one can reduce the training time of various robust training objectives by 2–3 times without sacrificing too much clean or robust accuracy. In summary, we make the following contributions:

- We propose a practical yet principled algorithm for efficient training of robust neural networks based on adaptive coreset selection. To the best of our knowledge, we are the first to use coreset selection in adversarial training.
- We show that our approach can be applied to a variety of robust learning objectives, including TRADES [44], ℓ_p-PGD [25] and Perceptual [22] Adversarial Training. Our approach encompasses a broader range of robust models compared to the limited scope of the existing methods.

– Through extensive experiments, we show that the proposed approach can result in a 2–3 fold reduction of the training time, with only a slight reduction in the clean and robust accuracy.

2   Background and Related Work

2.1   Adversarial Training

Let $\mathcal{D} = \{(\boldsymbol{x}_i, y_i)\}_{i=1}^n \subset \mathbb{X} \times \mathbb{C}$ denote a training dataset consisting of n i.i.d. samples. Each data point contains an input data $\boldsymbol{x}_i$ from domain $\mathbb{X}$ and an associated label y_i taking one of k possible values $\mathbb{C} = [k] = \{1, 2, \ldots, k\}$. Without loss of generality, in this paper we focus on the image domain $\mathbb{X}$. Furthermore, assume that $f_{\boldsymbol{\theta}} : \mathbb{X} \to \mathbb{R}^k$ denotes a neural network classifier with parameters $\boldsymbol{\theta}$ that takes $\boldsymbol{x} \in \mathbb{X}$ as input and maps it to a logit value $f_{\boldsymbol{\theta}}(\boldsymbol{x}) \in \mathbb{R}^k$. Then, training a neural network in its most general format can be written as the following minimization problem:

$$\min_{\boldsymbol{\theta}} \sum_{i \in V} \boldsymbol{\Phi}(\boldsymbol{x}_i, y_i; f_{\boldsymbol{\theta}}), \tag{1}$$

Here, $\boldsymbol{\Phi}(\boldsymbol{x}, y; f_{\boldsymbol{\theta}})$ is a function that takes a data point $(\boldsymbol{x}, y)$ and a function $f_{\boldsymbol{\theta}}$ as its inputs, and its output is a measure of discrepancy between the input $\boldsymbol{x}$ and its ground-truth label y. Also, $V = [n] = \{1, 2, \ldots, n\}$ denotes the entire training data. By writing the training objective in this format, we can denote both vanilla and adversarial training using the same notation. Below we show how various choices of the function $\boldsymbol{\Phi}$ amount to different training objectives.

Vanilla Training. In case of vanilla training, the function $\boldsymbol{\Phi}$ is a simple evaluation of an appropriate loss function over the neural network output $f_{\boldsymbol{\theta}}(\boldsymbol{x})$ and the ground-truth label y. For instance, for vanilla training we can have:

$$\boldsymbol{\Phi}(\boldsymbol{x}, y; f_{\boldsymbol{\theta}}) = \mathcal{L}_{\mathrm{CE}}(f_{\boldsymbol{\theta}}(\boldsymbol{x}), y), \tag{2}$$

where $\mathcal{L}_{\mathrm{CE}}(\cdot, \cdot)$ is the cross-entropy loss.

FGSM, ℓ_p-PGD, and Perceptual Adversarial Training. In these cases, the training objective is itself an optimization problem:

$$\boldsymbol{\Phi}(\boldsymbol{x}, y; f_{\boldsymbol{\theta}}) = \max_{\tilde{\boldsymbol{x}}} \mathcal{L}_{\mathrm{CE}}(f_{\boldsymbol{\theta}}(\tilde{\boldsymbol{x}}), y) \text{ s.t. } \mathrm{d}(\tilde{\boldsymbol{x}}, \boldsymbol{x}) \leq \varepsilon \tag{3}$$

where $\mathrm{d}(\cdot, \cdot)$ is an appropriate distance measure over image domain $\mathbb{X}$, and ε denotes a scalar. The constraint over $\mathrm{d}(\tilde{\boldsymbol{x}}, \boldsymbol{x})$ is used to ensure visual similarity between $\tilde{\boldsymbol{x}}$ and $\boldsymbol{x}$. Solving Eq. (3) amounts to finding an adversarial example $\tilde{\boldsymbol{x}}$ for the clean sample $\boldsymbol{x}$ [25]. Different choices of the visual similarity measure $\mathrm{d}(\cdot, \cdot)$ and solvers for Eq. (3) result in different adversarial training objectives.

– FGSM [10] assumes that $\mathrm{d}(\tilde{\boldsymbol{x}}, \boldsymbol{x}) = \|\tilde{\boldsymbol{x}} - \boldsymbol{x}\|_\infty$. Using this ℓ_∞ assumption, the solution to Eq. (3) is computed using one iteration of gradient ascent.

– ℓ_p-PGD [25] utilizes ℓ_p norms as a proxy for visual similarity $d(\cdot,\cdot)$. Then, several steps of projected gradient ascent is taken to solve Eq. (3).
– Perceptual Adversarial Training (PAT) [22] replaces $d(\cdot,\cdot)$ with *Learned Perceptual Image Patch Similarity* (LPIPS) distance [45]. Then, Laidlaw *et al.*[22] propose to solve this maximization objective using either projected gradient ascent or Lagrangian relaxation.

TRADES Adversarial Training. This approach uses a combination of Eqs. (2) and (3). The intuition behind TRADES [44] is to create a trade-off between clean and robust accuracy. In particular, the objective is written as:

$$\Phi(x, y; f_\theta) = \mathcal{L}_{CE}(f_\theta(x), y) + \max_{\tilde{x}} \mathcal{L}_{CE}(f_\theta(\tilde{x}), f_\theta(x))/\lambda, \qquad (4)$$

such that $d(\tilde{x}, x) \leq \varepsilon$. Here, λ is a coefficient that controls the trade-off.

2.2 Coreset Selection

Adaptive data subset selection, and *coreset selection* in general, is concerned with finding a weighted subset of the data that can approximate specific attributes of the entire population [9]. Traditionally, coreset selection has been used for different machine learning tasks such as k-means and k-medians [11], Naïve Bayes and nearest neighbor classifiers [39], and Bayesian inference [4].

Recently, coreset selection algorithms are being developed for neural network training [16,17,27,28]. The main idea behind such methods is to approximate the full gradient using a weighted subset of the training data. These algorithms start with computing the gradient of the loss function with respect to the neural network weights. This gradient is computed for *every* data sample in the training set. Then, a selection criterion is formed. This criterion aims to find a *weighted subset* of the training data that can approximate the full gradient. In Sect. 3 we provide a detailed account of these methods.

Existing coreset selection algorithms can only be used for the vanilla training of neural networks. As such, they still suffer from adversarial vulnerability. This paper extends coreset selection algorithms to robust neural network training and shows how they can be adopted to various robust training objectives.

3 Proposed Method

As discussed in Sect. 1, the main bottleneck in the time/computational complexity of adversarial training stems from constructing adversarial examples for the entire training set at each epoch. FAT [41] tries to eliminate this issue by using FGSM as its adversarial example generator. However, this simplification 1) may lead to catastrophic overfitting [2,41], and 2) is not easy to generalize to all types of adversarial training as FGSM is designed explicitly for ℓ_∞ attacks.

Instead of using a faster adversarial example generator, here we take a different, *orthogonal* path and try to reduce the training set size effectively. This way, the original adversarial training algorithm can still be used on this smaller subset of training data. This approach can reduce the training time while optimizing a similar objective as the original training. In this sense, it leads to a more *unified* method that can be used along with various types of adversarial training objectives, including the ones that already exist and the ones that will be proposed in the future.

The main hurdle in materializing this idea is the following question: *How should we select this subset of the training data without hurting either the clean or robust accuracy?* To answer this question, we propose to use coreset selection on the training data to reduce the sample size and improve training efficiency.

3.1   Problem Statement

Using our general notation from Sect. 2.1, we write both vanilla and adversarial training using the same objective:

$$\min_{\theta} \sum_{i \in V} \Phi\left(x_i, y_i; f_\theta\right), \tag{5}$$

where V denotes the entire training data, and depending on the training task, $\Phi\left(x_i, y_i; f_\theta\right)$ takes any of the Eqs. (2) to (4) forms. We adopt this notation to make our analysis more accessible.

As discussed in Sect. 2.2, coreset selection can be seen as a two-step process. First, the gradient of the loss function with respect to the neural network weights is computed for each training sample. Then, based on the gradients obtained in step one, a weighted subset (a.k.a. the coreset) of the training data is formed (see Fig. 1b). This subset is obtained such that the weighted gradients of the samples inside the coreset can provide a good approximation of the full gradient.

Specifically, using our universal notation in Eq. (5), we write coreset selection for both vanilla and adversarial training as:

$$\min_{S \subseteq V, \gamma} \left\| \sum_{i \in V} \nabla_\theta \Phi\left(x_i, y_i; f_\theta\right) - \sum_{j \in S} \gamma_j \nabla_\theta \Phi\left(x_j, y_j; f_\theta\right) \right\|, \tag{6}$$

where $S \subseteq V$ is the coreset, and γ_j's are the weights of each sample in the coreset. Once the coreset S is found, instead of training the neural network using Eq. (5), we can optimize its parameters using a weighted objective over the coreset:

$$\min_{\theta} \sum_{j \in S} \gamma_j \Phi\left(x_j, y_j; f_\theta\right). \tag{7}$$

It can be shown that solving Eq. (6) is NP-hard [27,28]. Roughly, various coreset selection methods differ in how they approximate the solution of the aforementioned objective. For instance, CRAIG [27] casts this objective as a *submodular set cover problem* and uses existing greedy solvers to get an approximate solution. As another example, GRADMATCH [16] analyzes the convergence of stochastic gradient descent using adaptive data subset selection. Based on this study, Killamsetty *et al.* [16] propose to use Orthogonal Matching Pursuit (OMP) [7,31] as a greedy solver of the data selection objective. More information about these methods is provided in Appendix A.

The issue with the aforementioned coreset selection methods is that they are designed explicitly for vanilla training of neural networks (see Fig. 1b), and they do not reflect the requirements of adversarial training. As such, we should modify these methods to make them suitable for our purpose of robust neural network training. Meanwhile, we should also consider the fact that the field of coreset selection is still evolving. Thus, we aim to find a general modification that can later be used alongside newer versions of greedy coreset selection algorithms.

We notice that various coreset selection methods proposed for vanilla neural network training only differ in their choice of greedy solvers. Therefore, we narrow down the changes we want to make to the first step of coreset selection: gradient computation. Then, existing greedy solvers can be used to find the subset of training data that we are looking for. To this end, we draw a connection between coreset selection methods and adversarial training using Danskin's theorem, as outlined next. Our analysis shows that for adversarial coreset selection, one needs to add a pre-processing step where adversarial attacks for the raw training data need to be computed (see Fig. 1c).

3.2   Coreset Selection for Efficient Adversarial Training

As discussed above, to construct the Eq. (6) objective, we need to compute the loss gradient with respect to the neural network weights. Once done, we can use existing greedy solvers to find the solution. The gradient computation needs to be performed for the entire training set. In particular, using our notation from Sect. 2.1, this step can be written as:

$$\nabla_\theta \Phi\left(x_i, y_i; f_\theta\right) \quad \forall \quad i \in V, \tag{8}$$

where V denotes the training set.

For vanilla neural network training (see Sect. 2.1) the above gradient is simply equal to $\nabla_\theta \mathcal{L}_{\mathrm{CE}}\left(f_\theta(x_i), y_i\right)$ which can be computed using standard backpropagation. In contrast, for the adversarial training objectives in Eqs. (3) and (4), this gradient requires taking partial derivative of a maximization objective. To this end, we use the famous Dasnkin's theorem [6] as stated below.

Theorem 1 (Theorem A.1 [25]). *Let $\mathcal{S}$ be a nonempty compact topological space, $\ell : \mathbb{R}^m \times \mathcal{S} \to \mathbb{R}$ be such that $\ell(\cdot, \delta)$ is differentiable for every $\delta \in \mathcal{S}$, and $\nabla_\theta \ell(\theta, \delta)$ is continuous on $\mathbb{R}^m \times \mathcal{S}$. Also, let $\delta^*(\theta) = \{\delta \in \arg\max_{\delta \in \mathcal{S}} \ell(\theta, \delta)\}$. Then, the corresponding max-function $\phi(\theta) = \max_{\delta \in \mathcal{S}} \ell(\theta, \delta)$ is locally Lipschitz continuous, directionally differentiable, and its directional derivatives along vector h satisfy:*

$$\phi'(\theta, h) = \sup_{\delta \in \delta^*(\theta)} h^\top \nabla_\theta \ell(\theta, \delta).$$

In particular, if for some $\boldsymbol{\theta} \in \mathbb{R}^m$ the set $\boldsymbol{\delta}^*(\boldsymbol{\theta}) = \{\boldsymbol{\delta}_{\boldsymbol{\theta}}^*\}$ is a singleton, then the max-function is differentiable at $\boldsymbol{\theta}$ and

$$\nabla \phi(\boldsymbol{\theta}) = \nabla_{\boldsymbol{\theta}} \ell\left(\boldsymbol{\theta}, \boldsymbol{\delta}_{\boldsymbol{\theta}}^*\right).$$

In summary, Theorem 1 indicates how to take the gradient of a max-function. To this end, it suffices to 1) find the maximizer, and 2) evaluate the normal gradient at this point.

Now that we have stated Danskin's theorem, we are ready to show how it can provide the connection between coreset selection and the adversarial training objectives of Eqs. (3) and (4). We do this for the two cases of adversarial training and TRADES as outlined next.

Case 1. (ℓ_p-PGD and Perceptual Adversarial Training) Going back to Eq. (8), we know that to perform coreset selection, we need to compute this gradient term for our objective in Eq. (3). In other words, we need to compute:

$$\nabla_{\boldsymbol{\theta}} \boldsymbol{\Phi}\left(\boldsymbol{x}, y; f_{\boldsymbol{\theta}}\right) = \nabla_{\boldsymbol{\theta}} \max_{\tilde{\boldsymbol{x}}} \mathcal{L}_{\mathrm{CE}}\left(f_{\boldsymbol{\theta}}(\tilde{\boldsymbol{x}}), y\right) \tag{9}$$

under the constraint $\mathrm{d}\left(\tilde{\boldsymbol{x}}, \boldsymbol{x}\right) \leq \varepsilon$ for every training sample. Based on Danskin's theorem, we can deduce:

$$\nabla_{\boldsymbol{\theta}} \boldsymbol{\Phi}\left(\boldsymbol{x}, y; f_{\boldsymbol{\theta}}\right) = \nabla_{\boldsymbol{\theta}} \mathcal{L}_{\mathrm{CE}}\left(f_{\boldsymbol{\theta}}(\boldsymbol{x}^*), y\right), \tag{10}$$

where $\boldsymbol{x}^*$ is the solution to:

$$\arg\max_{\tilde{\boldsymbol{x}}} \mathcal{L}_{\mathrm{CE}}\left(f_{\boldsymbol{\theta}}(\tilde{\boldsymbol{x}}), y\right) \quad \text{s.t.} \quad \mathrm{d}\left(\tilde{\boldsymbol{x}}, \boldsymbol{x}\right) \leq \varepsilon. \tag{11}$$

The conditions under which Danskin's theorem hold might not be satisfied for neural networks in general. This is due to the presence of functions with discontinuous gradients, such as ReLU activation, in neural networks. More importantly, finding the exact solution of Eq. (11) is not straightforward as neural networks are highly non-convex. Usually, the exact solution $\boldsymbol{x}^*$ is replaced with its approximation, which is an adversarial example generated under the Eq. (11) objective [18]. Based on this approximation, we can re-write Eq. (10) as:

$$\nabla_{\boldsymbol{\theta}} \boldsymbol{\Phi}\left(\boldsymbol{x}, y; f_{\boldsymbol{\theta}}\right) \approx \nabla_{\boldsymbol{\theta}} \mathcal{L}_{\mathrm{CE}}\left(f_{\boldsymbol{\theta}}(\boldsymbol{x}_{\mathrm{adv}}), y\right). \tag{12}$$

In other words, to perform coreset selection for ℓ_p-PGD [25] and Perceptual [22] Adversarial Training, one needs to add a pre-processing step to the gradient computation. At this step, adversarial examples for the entire training set must be constructed. Then, the coresets can be built as in vanilla neural networks.

Case 2. (TRADES Adversarial Training) For TRADES [44], the gradient computation is slightly different as the objective in Eq. (4) consists of two terms. In this case, the gradient can be written as:

$$\nabla_{\boldsymbol{\theta}} \boldsymbol{\Phi}\left(\boldsymbol{x}, y; f_{\boldsymbol{\theta}}\right) = \nabla_{\boldsymbol{\theta}} \mathcal{L}_{\mathrm{CE}}\left(f_{\boldsymbol{\theta}}(\boldsymbol{x}), y\right) + \nabla_{\boldsymbol{\theta}} \max_{\tilde{\boldsymbol{x}}} \mathcal{L}_{\mathrm{CE}}\left(f_{\boldsymbol{\theta}}(\tilde{\boldsymbol{x}}), f_{\boldsymbol{\theta}}(\boldsymbol{x})\right) / \lambda, \tag{13}$$

where $d(\tilde{x}, x) \leq \varepsilon$. The first term is the normal gradient of the neural network. For the second term, we apply Danskin's theorem to obtain:

$$\nabla_\theta \Phi(x, y; f_\theta) \approx \nabla_\theta \mathcal{L}_{CE}(f_\theta(x), y) + \nabla_\theta \mathcal{L}_{CE}(f_\theta(x_{adv}), f_\theta(x))/\lambda, \qquad (14)$$

where x_{adv} is an approximate solution to:

$$\arg\max_{\tilde{x}} \mathcal{L}_{CE}(f_\theta(\tilde{x}), f_\theta(x))/\lambda \quad \text{s.t.} \quad d(\tilde{x}, x) \leq \varepsilon. \qquad (15)$$

Having found the loss gradients $\nabla_\theta \Phi(x_i, y_i; f_\theta)$ for ℓ_p-PGD, PAT (Case 1), and TRADES (Case 2), we can construct Eq. (6) and use existing greedy solvers like CRAIG [27] or GRADMATCH [16] to find the coreset. As we saw, adversarial coreset selection requires adding a pre-processing step where we need to build perturbed versions of the training data using their respective objectives in Eqs. (11) and (15). Then, the gradients are computed using Eqs. (12) and (14). Afterward, greedy subset selection algorithms are used to construct the coresets based on the value of the gradients. Finally, having selected the coreset data, one can run *weighted* adversarial training only on the data that remains in the coreset. As can be seen, we are not changing the essence of the training objective in this process. We are just reducing the dataset size to enhance our proposed solution's computational efficiency; as such, we can use it along with any adversarial training objective.

3.3   Practical Considerations

Since coreset selection depends on the current values of the neural network weights, it is important to update the coresets as the training evolves. Prior work [16,17] has shown that this selection needs to be done every T epochs, where T is usually greater than 15. Also, we employ small yet crucial practical changes while using coreset selection to increase efficiency. We summarize these practical tweaks below. Further detail can be found in [16,27].

Gradient Approximation. As we saw, both Eqs. (12) and (14) require computation of the loss gradient with respect to the neural network weights. This is equal to backpropagation through the entire neural network, which is not very efficient. Instead, it is common to replace the exact gradients in Eqs. (12) and (14) with their last-layer approximation [15,16,27]. In other words, instead of backpropagating through the entire network, one can backpropagate up until the penultimate layer. This estimate has an approximate complexity equal to forwardpropagation, and it has been shown to work well in practice [16,17,27,28].

Batch-Wise Coreset Selection. As discussed in Sect. 3.2, data selection is usually done in a *sample-wise* fashion where each data sample is separately considered to be selected. This way, one must find the data candidates from the entire training set. To increase efficiency, Killamsetty *et al.* [16] proposed the *batch-wise* variant. In this type of coreset selection, the data is first split into several batches. Then, the algorithm makes a selection out of these batches. Intuitively, this change increases efficiency as the sample size is reduced from the number of data points to the number of batches.

Warm-Start with the Entire Data. Finally, we warm-start the training using the entire dataset. Afterward, coreset selection is activated, and training is only performed using the data in the coreset.

Final Algorithm. Figure 1 and Algorithm 1 in Appendix B.1 summarize our coreset selection approach for adversarial training. As can be seen, our proposed method is a generic and principled approach in contrast to existing methods such as FAT [41]. In particular, our approach provides the following advantages compared to existing methods:

1. The proposed approach does not involve algorithmic level manipulations and dependency on specific training attributes such as ℓ_∞ bound or cyclic learning rate. Also, it controls the training speed through coreset size, which can be specified solely based on available computational resources.
2. The simplicity of our method makes it compatible with any existing/future adversarial training objectives. Furthermore, as we will see in Sect. 4, our approach can be combined with any greedy coreset selection algorithms to deliver robust neural networks.

These characteristics increase the likelihood of applying our proposed method for robust neural network training no matter the training objective. This contrasts with existing methods that solely focus on a particular training objective.

4   Experimental Results

In this section, we present our experimental results[2]. We show how our proposed approach can efficiently reduce the training time of various robust objectives in different settings. To this end, we train neural networks using TRADES [44], ℓ_p-PGD [25] and PAT [22] on CIFAR-10 [19], SVHN [30], and a subset of ImageNet [32] with 12 classes. For TRADES and ℓ_p-PGD training, we use ResNet-18 [12] classifiers, while for PAT we use ResNet-50 architectures.

4.1   TRADES and ℓ_p-PGD Robust Training

In our first experiments, we train ResNet-18 classifiers on CIFAR-10 and SVHN datasets using TRADES, ℓ_∞ and ℓ_2-PGD adversarial training objectives. In each case, we set the training hyper-parameters such as the learning rate, the number of epochs, and attack parameters. Then, we train the network using the entire training data and our adversarial coreset selection approach. For our approach, we use batch-wise versions of CRAIG [27] and GRADMATCH [16] with warm-start.

---

[2] Our implementation can be found in this repository.

Table 1. Clean (ACC) and robust (RACC) accuracy, and total training time (T) of different adversarial training methods. For each objective, all the hyper-parameters were kept the same as full training. For our proposed approach, the difference with full training is shown in parentheses. The results are averaged over 5 runs. More detail can be found in Appendix C.

Objec.	Data	Training method	Performance measures		
			↑ ACC (%)	↑ RACC (%)	↓ T (mins)
TRADES	CIFAR-10	Adv. CRAIG (Ours)	83.03 (−2.38)	41.45 (−2.74)	179.20 (−165.09)
		Adv. GRADMATCH (Ours)	83.07 (−2.34)	41.52 (−2.67)	178.73 (−165.56)
		Full Adv. training	85.41	44.19	344.29
ℓ_∞-PGD	CIFAR-10	Adv. CRAIG (Ours)	80.37 (−2.77)	45.07 (+3.68)	148.01 (−144.86)
		Adv. GRADMATCH (Ours)	80.67 (−2.47)	45.23 (+3.84)	148.03 (−144.84)
		Full Adv. training	83.14	41.39	292.87
ℓ_2-PGD	SVHN	Adv. CRAIG (Ours)	95.42 (+0.10)	49.68 (−3.34)	130.04 (−259.42)
		Adv. GRADMATCH (Ours)	95.57 (+0.25)	50.41 (−2.61)	125.53 (−263.93)
		Full Adv. training	95.32	53.02	389.46

We set the *coreset size* (the percentage of training data to be selected) to *50%* for CIFAR-10 and *30%* for SVHN to get a reasonable balance between accuracy and training time. We report the clean and robust accuracy (in %) as well as the total training time (in minutes) in Table 1. For our approach, we also report the difference with full training in parentheses. In each case, we evaluate the robust accuracy using an attack with similar attributes as the training objective (for more information, see Appendix C).

As seen, in all cases, we reduce the training time by more than a factor of two while keeping the clean and robust accuracy almost intact. Note that in these experiments, all the training attributes such as the hyper-parameters, learning rate scheduler, etc. are the same among different training schemes. This is important since we want to clearly show the relative boost in performance that one can achieve just by using coreset selection. Nonetheless, it is likely that by tweaking the hyper-parameters of our approach, one can obtain even better results in terms of clean and robust accuracy.

4.2   Perceptual Adversarial Training vs. Unseen Attacks

As discussed in Sect. 2, PAT [22] replaces the visual similarity measure d($\cdot$, $\cdot$) in Eq. (3) with LPIPS [45] distance. The logic behind this choice is that ℓ_p norms can only capture a small portion of images similar to the clean one, limiting the search space of adversarial attacks. Motivated by this reason, Laidlaw *et al.* [22] propose two different ways of finding the solution to Eq. (3) when d($\cdot$, $\cdot$) is the LPIPS distance. The first version uses PGD, and the second is a relaxation of

Table 2. Clean (ACC) and robust (RACC) accuracy and total training time (T) of Perceptual adversarial training for CIFAR-10 and ImageNet-12 datasets. At inference, the networks are evaluated against five attacks that were not seen during training (Unseen RACC) and different versions of Perceptual Adversarial Attack (Seen RACC). In each case, the average is reported. For more information and details about the experiment, please see the Appendices C and D.

Data	Training method	↑ ACC (%)	↑ RACC (%)		↓ T (mins)
			Unseen	Seen	
CIFAR-10	Adv. CRAIG (Ours)	83.21 (−2.81)	46.55 (−1.49)	13.49 (−1.83)	767.34 (−915.60)
	Adv. GRADMATCH (Ours)	83.14 (−2.88)	46.11 (−1.93)	13.74 (−1.54)	787.26 (−895.68)
	Full PAT (Fast-LPA)	86.02	48.04	15.32	1682.94
ImageNet	Adv. CRAIG (Ours)	86.99 (−4.23)	53.05 (−0.18)	22.56 (−0.77)	2817.06 (−2796.06)
	Adv. GRADMATCH (Ours)	87.08 (−4.14)	53.17 (−0.06)	20.74 (−2.59)	2865.72 (−2747.40)
	Full PAT (Fast-LPA)	91.22	53.23	23.33	5613.12

the original problem using the Lagrangian form. We refer to these two versions as PPGD (Perceptual PGD) and LPA (Lagrangian Perceptual Attack), respectively. Then, Laidlaw *et al.* [22] proposed to utilize a fast version of LPA to enable its efficient usage in adversarial training.

For our next set of experiments, we show how our approach can be adapted to this unusual training objective. This is done to showcase the compatibility of our proposed method with different training objectives as opposed to existing methods that are carefully tuned for a particular training objective. To this end, we train ResNet-50 classifiers using Fast-LPA. We train the classifiers on CIFAR-10 and ImageNet-12 datasets. Like our previous experiments, we set the hyper-parameters of the training to be fixed and then train the models using the entire training data and our adversarial coreset selection method. For our method, we use batch-wise versions of CRAIG [27] and GRADMATCH [16] with warm-start. The *coreset size* for CIFAR-10 and ImageNet-12 were set to *40%* and *50%*, respectively. We measure the performance of the trained models against unseen attacks during training and the two variants of perceptual attacks as in [22]. The unseen attacks for each dataset were selected similarly to [22]. We also record the total training time taken by each method.

Table 2 summarizes our results on PAT using Fast-LPA (full results can be found in Appendix D). As seen, our adversarial coreset selection approach can deliver a competitive performance in terms of clean and average unseen attack accuracy while reducing the training time by at least a factor of two. These results indicate the flexibility of our adversarial coreset selection that can be combined with various objectives. This is due to the orthogonality of the proposed approach with the existing efficient adversarial training methods. In this case, we see that we can make Fast-LPA even faster using our approach.

Table 3. Clean (ACC) and robust (RACC) accuracy, and average training speed (S_{avg}) of Fast Adversarial Training [41] without and with our adversarial coreset selection on CIFAR-10. The difference with full training is shown in parentheses for our proposed approach.

Training method	Performance measures			
	↑ ACC (%)	↑ RACC (%)	↓ S_{avg} (min/epoch)	↓ T (min)
Fast Adv. training	86.20	47.54	0.5178	31.068
+ Adv. CRAIG (Ours)	82.56 (−3.64)	47.77 (+0.23)	0.2783	16.695 (−14.373)
+ Adv. GRADMATCH (Ours)	82.53 (−3.67)	47.88 (+0.34)	0.2737	16.419 (−14.649)

4.3 Compatibility with Existing Methods

To showcase that our adversarial coreset selection approach is complementary to existing methods, we integrate it with a stable version of Fast Adversarial Training (FAT) [41] that does not use a cyclic learning rate. Specifically, we train a neural network using FAT [41], and then add adversarial coreset selection to this approach and record the training time and clean/robust accuracy. We run the experiments on the CIFAR-10 dataset and train a ResNet-18 for each case. We set the *coreset size* to *50%* for our methods. The results are shown in Table 3. As can be seen, our approach can be easily combined with existing methods to deliver faster training. This is due to the orthogonality of our approach that we discussed previously.

Moreover, we show that adversarial coreset selection gives a better approximation to ℓ_∞-PGD adversarial training compared to using FGSM [10] as done in FAT [41]. To this end, we use our adversarial GRADMATCH to train neural networks with the original ℓ_∞-PGD objective. We also train these networks using FAT [41] that uses FGSM. We train neural networks with a perturbation norm of $\|\varepsilon\|_\infty \leq 8$. Then, we evaluate the trained networks against PGD-50 adversarial attacks with different attack strengths to see how each network generalizes to unseen perturbations. As seen in Fig. 2, adversarial coreset selection is a closer approximation to ℓ_∞-PGD compared to FAT [41]. This indicates the success of the proposed approach in retaining the characteristics of the original objective as opposed to existing methods.

4.4 Ablation Studies

In this section, we perform a few ablation studies to examine the effectiveness of our adversarial coreset selection method. First, we compare a random data selection with adversarial GRADMATCH. Figure 3 shows that for any given coreset size, our adversarial coreset selection method results in a lower robust error. Furthermore, we modify the warm-start epochs for a fixed coreset size of 50%. As seen, the proposed method is not very sensitive to the number of warm-start epochs, although a longer warm-start is generally beneficial. More experiments on the accuracy vs. speed-up trade-off and the importance of warm-start and batch-wise adversarial coreset selection can be found in Appendix D.

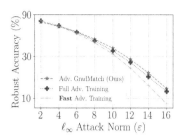

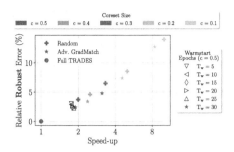

Fig. 2. Robust accuracy as a function of ℓ_∞ attack norm. We train neural networks with a perturbation norm of $\|\varepsilon\|_\infty \leq 8$ on CIFAR-10. At inference, we evaluate the robust accuracy against PGD-50 with various attack strengths.

Fig. 3. Relative robust error vs. speed up for TRADES. We compare our adversarial coreset selection (GRADMATCH) for a given subset size against random data selection. Furthermore, we show our results for a selection of different warmstart settings.

5   Conclusion

In this paper, we proposed a general yet principled approach for efficient adversarial training based on the theory of coreset selection. We discussed how repetitive computation of adversarial attacks for the entire training data could impede the training speed. Unlike previous methods that try to solve this issue by making the adversarial attack more straightforward, here, we took an orthogonal path to reduce the training set size without modifying the attacker. We drew a connection between greedy coreset selection algorithms and adversarial training using Danskin's theorem. We then showed the flexibility of our adversarial coreset selection method by utilizing it for TRADES, ℓ_p-PGD, and Perceptual Adversarial Training. Our experimental results indicate that adversarial coreset selection can reduce the training time by more than 2–3 times with only a slight reduction in the clean and robust accuracy.

Acknowledgements. This research was undertaken using the LIEF HPC-GPGPU Facility hosted at the University of Melbourne. This Facility was established with the assistance of LIEF Grant LE170100200.

References

1. Adadi, A.: A survey on data-efficient algorithms in big data era. J. Big Data **8**(1), 1–54 (2021). https://doi.org/10.1186/s40537-021-00419-9
2. Andriushchenko, M., Flammarion, N.: Understanding and improving fast adversarial training. In: Proceedings of the Advances in Neural Information Processing Systems 33: Annual Conference on Neural Information Processing Systems (NeurIPS) (2020)

3. Biggio, B., Corona, I., Maiorca, D., Nelson, B., Šrndić, N., Laskov, P., Giacinto, G., Roli, F.: Evasion attacks against machine learning at test time. In: Blockeel, H., Kersting, K., Nijssen, S., Železný, F. (eds.) ECML PKDD 2013. LNCS (LNAI), vol. 8190, pp. 387–402. Springer, Heidelberg (2013). https://doi.org/10.1007/978-3-642-40994-3_25

4. Campbell, T., Broderick, T.: Bayesian coreset construction via greedy iterative geodesic ascent. In: Proceedings of the 35th International Conference on Machine Learning (ICML), pp. 697–705 (2018)

5. Croce, F., Hein, M.: Reliable evaluation of adversarial robustness with an ensemble of diverse parameter-free attacks. In: Proceedings of the 37th International Conference on Machine Learning (ICML), pp. 2206–2216 (2020)

6. Danskin, J.M.: The Theory of Max-min and its Application to Weapons Allocation Problems, vol. 5. Springer Science & Business Media (1967). https://doi.org/10.1007/978-3-642-46092-0

7. Elenberg, E.R., Khanna, R., Dimakis, A.G., Negahban, S.N.: Restricted strong convexity implies weak submodularity. CoRR abs/1612.00804 (2016)

8. Eykholt, K., et al.: Robust physical-world attacks on deep learning visual classification. In: Proceeding of the IEEE Conference on Computer Vision and Pattern Recognition (CVPR), pp. 1625–1634 (2018)

9. Feldman, D.: Introduction to core-sets: an updated survey. CoRR abs/2011.09384 (2020)

10. Goodfellow, I.J., Shlens, J., Szegedy, C.: Explaining and harnessing adversarial examples. In: Proceedings of the 3rd International Conference on Learning Representations (ICLR) (2015)

11. Har-Peled, S., Mazumdar, S.: On coresets for k-means and k-median clustering. In: Proceedings of the 36th Annual ACM Symposium on Theory of Computing (STOC), pp. 291–300 (2004)

12. He, K., Zhang, X., Ren, S., Sun, J.: Deep residual learning for image recognition. In: Proceedings of the IEEE Conference on Computer Vision and Pattern Recognition (CVPR), pp. 770–778 (2016)

13. Kang, D., Sun, Y., Hendrycks, D., Brown, T., Steinhardt, J.: Testing robustness against unforeseen adversaries. CoRR abs/1908.08016 (2019)

14. Karras, T., Laine, S., Aittala, M., Hellsten, J., Lehtinen, J., Aila, T.: Analyzing and improving the image quality of stylegan. In: Proceedings of the IEEE Conference on Computer Vision and Pattern Recognition (CVPR), pp. 8107–8116 (2020)

15. Katharopoulos, A., Fleuret, F.: Not all samples are created equal: Deep learning with importance sampling. In: Proceedings of the 35th International Conference on Machine Learning (ICML), pp. 2530–2539 (2018)

16. Killamsetty, K., Sivasubramanian, D., Ramakrishnan, G., De, A., Iyer, R.K.: GRAD-MATCH: gradient matching based data subset selection for efficient deep model training. In: Proceedings of the 38th International Conference on Machine Learning (ICML), pp. 5464–5474 (2021)

17. Killamsetty, K., Sivasubramanian, D., Ramakrishnan, G., Iyer, R.K.: GLISTER: generalization based data subset selection for efficient and robust learning. In: Proceedings of the 35th AAAI Conference on Artificial Intelligence, pp. 8110–8118 (2021)

18. Kolter, Z., Madry, A.: Adversarial robustness: theory and practice. In: Tutorial in the Advances in Neural Information Processing Systems 31: Annual Conference on Neural Information Processing Systems (NeurIPS) (2018). https://adversarial-ml-tutorial.org/

19. Krizhevsky, A., Hinton, G.: Learning multiple layers of features from tiny images. Master's thesis, Department of Computer Science, University of Toronto (2009)
20. Krizhevsky, A., Sutskever, I., Hinton, G.E.: ImageNet classification with deep convolutional neural networks. In: Proceedings of the Advances in Neural Information Processing Systems 25: Annual Conference on Neural Information Processing Systems (NeurIPS), pp. 1106–1114 (2012)
21. Laidlaw, C., Feizi, S.: Functional adversarial attacks. In: Proceedings of the Advances in Neural Information Processing Systems 32: Annual Conference on Neural Information Processing Systems (NeurIPS), pp. 10408–10418 (2019)
22. Laidlaw, C., Singla, S., Feizi, S.: Perceptual adversarial robustness: defense against unseen threat models. In: Proceedings of the 9th International Conference on Learning Representations (ICLR) (2021)
23. Liu, Y., Ma, X., Bailey, J., Lu, F.: Reflection backdoor: a natural backdoor attack on deep neural networks. In: Proceedings of the 16th European Conference on Computer Vision (ECCV), pp. 182–199 (2020). https://doi.org/10.1007/978-3-030-58607-2_11
24. Ma, X., et al.: Understanding adversarial attacks on deep learning based medical image analysis systems. Pattern Recogn. **110**, 107332 (2021)
25. Madry, A., Makelov, A., Schmidt, L., Tsipras, D., Vladu, A.: Towards deep learning models resistant to adversarial attacks. In: Proceedings of the 6th International Conference on Learning Representations (ICLR) (2018)
26. Minoux, M.: Accelerated greedy algorithms for maximizing submodular set functions. In: Optimization Techniques, pp. 234–243. Springer (1978). https://doi.org/10.1007/BFb0006528
27. Mirzasoleiman, B., Bilmes, J.A., Leskovec, J.: Coresets for data-efficient training of machine learning models. In: Proceedings of the 37th International Conference on Machine Learning (ICML), pp. 6950–6960 (2020)
28. Mirzasoleiman, B., Cao, K., Leskovec, J.: Coresets for robust training of deep neural networks against noisy labels. In: Proceedings of the Advances in Neural Information Processing Systems 33: Annual Conference on Neural Information Processing Systems (NeurIPS) (2020)
29. Nemhauser, G.L., Wolsey, L.A., Fisher, M.L.: An analysis of approximations for maximizing submodular set functions - I. Math. Program. **14**(1), 265–294 (1978). https://doi.org/10.1007/BF01588971
30. Netzer, Y., Wang, T., Coates, A., Bissacco, A., Wu, B., Ng, A.Y.: Reading digits in natural images with unsupervised feature learning. In: NeurIPS Workshop on Deep Learning and Unsupervised Feature Learning (2011)
31. Pati, Y.C., Rezaiifar, R., Krishnaprasad, P.S.: Orthogonal matching pursuit: recursive function approximation with applications to wavelet decomposition. In: Proceedings of 27th Asilomar Conference on Signals, Systems and Computers, vol. 1, pp. 40–44 (1993)
32. Russakovsky, O., et al.: ImageNet large scale visual recognition challenge. Int. J. Comput. Vis. (IJCV) **115**(3), 211–252 (2015). https://doi.org/10.1007/s11263-015-0816-y
33. Schwartz, R., Dodge, J., Smith, N.A., Etzioni, O.: Green AI. Commun. ACM, **63**(12), 54–63 (2020)
34. Strubell, E., Ganesh, A., McCallum, A.: Energy and policy considerations for deep learning in NLP. In: Korhonen, A., Traum, D.R., Màrquez, L. (eds.) Proceedings of the 57th Conference of the Association for Computational Linguistics (ACL), pp. 3645–3650 (2019)

35. Szegedy, C., et al.: Intriguing properties of neural networks. In: Proceedings of the 2nd International Conference on Learning Representations (ICLR) (2014)

36. Tramèr, F., Kurakin, A., Papernot, N., Goodfellow, I.J., Boneh, D., McDaniel, P.D.: Ensemble adversarial training: Attacks and defenses. In: Proceedings of the 6th International Conference on Learning Representations (ICLR) (2018)

37. Tsipras, D., Santurkar, S., Engstrom, L., Turner, A., Madry, A.: Robustness may be at odds with accuracy. In: Proceedings of the 7th International Conference on Learning Representations (ICLR) (2019)

38. Vahdat, A., Kautz, J.: NVAE: a deep hierarchical variational autoencoder. In: Proceedings of the Advances in Neural Information Processing Systems 33: Annual Conference on Neural Information Processing Systems (NeurIPS) (2020)

39. Wei, K., Iyer, R., Bilmes, J.: Submodularity in data subset selection and active learning. In: Proceedings of the 32nd International Conference on Machine Learning (ICML), pp. 1954–1963 (2015)

40. Wolsey, L.A.: An analysis of the greedy algorithm for the submodular set covering problem. Combinatorica **2**(4), 385–393 (1982). https://doi.org/10.1007/BF02579435

41. Wong, E., Rice, L., Kolter, J.Z.: Fast is better than free: revisiting adversarial training. In: Proceedings of the 8th International Conference on Learning Representations (ICLR) (2020)

42. Wu, Y., Kirillov, A., Massa, F., Lo, W.Y., Girshick, R.: Detectron2 (2019). https://github.com/facebookresearch/detectron2

43. Xiao, C., Zhu, J., Li, B., He, W., Liu, M., Song, D.: Spatially transformed adversarial examples. In: Proceedings of the 6th International Conference on Learning Representations (ICLR) (2018)

44. Zhang, H., Yu, Y., Jiao, J., Xing, E.P., Ghaoui, L.E., Jordan, M.I.: Theoretically principled trade-off between robustness and accuracy. In: Proceedings of the 36th International Conference on Machine Learning (ICML), pp. 7472–7482 (2019)

45. Zhang, R., Isola, P., Efros, A.A., Shechtman, E., Wang, O.: The unreasonable effectiveness of deep features as a perceptual metric. In: Proceedings of the IEEE Conference on Computer Vision and Pattern Recognition (CVPR), pp. 586–595 (2018)

Multi-granularity Pruning for Model Acceleration on Mobile Devices

Tianli Zhao[1,2,3,4], Xi Sheryl Zhang[2,3], Wentao Zhu[5], Jiaxing Wang[6],
Sen Yang[7], Ji Liu[8], and Jian Cheng[2,3(✉)]

[1] School of Artificial Intelligence, University of Chinese Academy of Sciences,
Beijing, China
[2] Institute of Automation, Chinese Academy of Sciences, Beijing, China
jcheng@nlpr.ia.ac.cn
[3] AIRIA, Mumbai, India
[4] Maicro.ai, Texas, USA
[5] Amazon Video Prime, Seattle, USA
[6] JD.com, Beijing, China
[7] Snap Inc., Santa Monica, USA
[8] Kwai Inc., Beijing, China

Abstract. For practical deep neural network design on mobile devices,
it is essential to consider the constraints incurred by the computational
resources and the inference latency in various applications. Among deep
network acceleration approaches, pruning is a widely adopted practice
to balance the computational resource consumption and the accuracy,
where unimportant connections can be removed either channel-wisely or
randomly with a minimal impact on model accuracy. The coarse-grained
channel pruning instantly results in a significant latency reduction, while
the fine-grained weight pruning is more flexible to retain accuracy. In this
paper, we present a unified framework for the Joint Channel pruning and
Weight pruning, named JCW, which achieves a better pruning proportion between channel and weight pruning. To fully optimize the trade-
off between latency and accuracy, we further develop a tailored multi-
objective evolutionary algorithm in the JCW framework, which enables
one single round search to obtain the accurate candidate architectures for
various deployment requirements. Extensive experiments demonstrate
that the JCW achieves a better trade-off between the latency and accuracy against previous state-of-the-art pruning methods on the ImageNet
classification dataset.

1 Introduction

Recently, deep learning has prevailed in many machine learning tasks. However, the substantial computational overhead limits its applications on resource-
constrained platforms, e.g., mobile devices. To design a deep network deployable

---

Supplementary Information The online version contains supplementary material
available at https://doi.org/10.1007/978-3-031-20083-0_29.

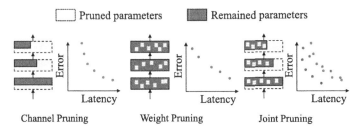

Fig. 1. Illustration of JCW (right), channel pruning (   ) and weight pruning (middle). The JCW conducts joint channel and weight pruning, which achieves a better Pareto-frontier between the accuracy and latency with one single search. (Color figure online)

to the aforementioned platforms, it is necessary to consider the constraint incurred by the available computational resource and reduce the inference latency while maximizing the accuracy.

Pruning has been one of the predominant approaches to accelerating large deep neural networks. The pruning methods can be roughly divided into two categories, channel pruning which removes parameters in a channel-wise manner [20,23,60], and weight pruning which prunes parameters randomly [17,38]. The two mainstream pruning methods mainly focus on accelerating neural networks on one single dimension (e.g. either channel wisely or element wisely), while they may have different impacts on the latency and accuracy. For instance, the results in Table 1 show that the channel pruning method [15] offers better accuracy than the weight pruning method [10] when the inference latency is high. In contrast, the weight pruning [10] yields better accuracy than the channel pruning [15] under a low latency requirement. Inspired by this observation, this work attempts to unveil an important problem overlooked before, *is it possible to achieve a better latency-accuracy trade-off by designing a subtle network pruning method that enjoys benefits from both of the two pruning methods?*

However, it is non-trivial to determine the channel numbers and layer-wise weight sparsity in the joint channel and weight pruning because of the exponentially increasing decision space. Besides, simply applying channel width and weight sparsity search sequentially often leads to a sub-optimal solution because the two search steps are in fact entangled and it is difficult to determine a better balance between channel and weight pruning for a wide range of latency constraints.

To this end, we build an efficient model acceleration paradigm, named as JCW (**J**oint **C**hannel and **W**eight pruning) which accelerates the models by applying both channel and weight pruning jointly and finds the a better balance between channel and weight pruning automatically. Considering both the latency and accuracy, we formulate the model acceleration into a multi-objective optimization (MOO) problem. Specifically, given a predefined base model with L layers, denoting the number of channels of a certain compressed model for channel pruning by $C = \{c^{(l)}\}_{l=1}^{L}$, and weight sparsity for weight pruning by $S = \{s^{(l)}\}_{l=1}^{L}$, we search for a sequence of (C, S) lying on the Pareto-frontier between latency and accuracy:

486   T. Zhao et al.

Table 1. Latency & accuracy of MobileNetV2 models accelerated by fast sparse convolution [10] and DMCP [15]. The best acceleration strategy differs under various latency budgets.

Method	Type	Latency	Accuracy
Fast	Weight pruning	88.94 ms	72.0%
DMCP	Channel pruning	82.95 ms	72.4%
Fast	Weight pruning	35.89 ms	65.2%
DMCP	Channel pruning	33.50 ms	62.7%

$$(C, S)^* = \arg\min_{C,S} \{\mathcal{T}(C, S), \mathcal{E}(C, S)\}, \qquad (1)$$

where $\mathcal{T}(C, S)$ and $\mathcal{E}(C, S)$ denote the latency and error rate of the compressed model, respectively. We further propose a uniform non-dominated sorting selection based on an enhanced evolutionary algorithm, NSGA-II [4], to generate accurate candidate architectures with a wide range of latency in one single round search. To alleviate the search cost, we construct an accuracy predictor based on parameter sharing [16,42], and a latency predictor based on tri-linear interpolation.

We conduct extensive experiments to validate the effectiveness of JCW on the ImageNet dataset [5]. The JCW outperforms previous state-of-the-art model compression approaches by a large margin. Without loss of accuracy, the JCW yields 2.74×, 2.42× and 1.93× speedup over ResNet18 [19], MobileNetV1 [24] and MobileNetV2 [44], respectively.

Our major contributions are summarized as follows,

- We build a general model acceleration framework by the joint channel and weight pruning, JCW, as shown in the right of Fig. 1, which finds a better balance between channel and weight pruning automatically and obtains a much better trade-off between model accuracy and latency on mobile devices. To our best knowledge, we are the first to investigate on achieving the balance between channel and weight pruning automatically for better model acceleration.
- We enhance the Pareto multi-objective optimization with a uniform non-dominated sorting selection. The enhanced search algorithm can find multiple accurate candidate architectures for various computational budgets through one single round search.
- Extensive experiments demonstrate the effectiveness of the joint channel pruning and weight pruning. The JCW outperforms previous state-of-the-art model compression and acceleration approaches by a large margin on the ImageNet classification dataset.

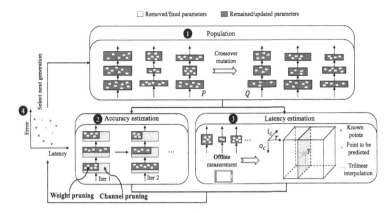

Fig. 2. Illustration of the framework of JCW.

2   Methodology

2.1   Motivation

In this section, we review the two categories of existing network pruning methods – coarse-grained channel pruning and fine-grained weight pruning, and analyze their pros and cons, which give rise to our joint design method.

Channel pruning reduces the width of feature maps by pruning filters channel-wisely [15,48]. As a result, the original network is shrunk into a thinner one. The channel pruning is well-structured and thus conventionally believed to be more convenient for model acceleration than random weight pruning [20,53]. However, a well-known drawback of channel pruning is its difficulty in attaining accuracy because of its strong structural constraints.

In random weight pruning, each individual element of parameters can be freely pruned [17,18]. It is more flexible and generally known to be able to get theoretically smaller models than channel pruning. More recently, Elsken *et al.* [10] made it more practical by arguing that despite of its irregular memory access, it can also be efficiently accelerated on mobile CPUs if implemented properly. However, the acceleration ratio achieved by pure weight pruning is still limited because the problem of irregular memory access still exists. For example, their method can only accelerate the computation by $\sim 3\times$ even when 90% of parameters are removed.

Based on our analysis, we present JCW, a unified framework that combines the advantages of both channel pruning and weight pruning for better model acceleration by applying the two jointly. JCW searches the number of channels and layer-wise weight sparsity jointly and automatically finds the balance between channel and weight pruning for a wide range of latency budgets, thus it is able to achieve a much better accuracy-latency trade-off.

2.2   Problem Formulation

Formally, for some compressed model $\mathcal{A}$ with L layers, we denote the number of channels of each layer by $C_{\mathcal{A}} = \{c_{\mathcal{A}}^{(l)}\}_{l=1}^{L}$, and the weight sparsity[1] of each layer by $S_{\mathcal{A}} = \{s_{\mathcal{A}}^{(l)}\}_{l=1}^{L}$. In this way, each sub-network $\mathcal{A}$ can be represented by a pair of vectors: $\mathcal{A} = \{C_{\mathcal{A}}, S_{\mathcal{A}}\}$. Our goal is to accelerate the inference of networks by *applying channel pruning and weight pruning simultaneously*, while at the same time minimizing the accuracy loss:

$$\mathcal{A}^{*} = \arg\min_{\mathcal{A}}\{\mathcal{T}(C_{\mathcal{A}}, S_{\mathcal{A}}), \mathcal{E}(C_{\mathcal{A}}, S_{\mathcal{A}})\}, \tag{2}$$

where $\mathcal{T}(C_{\mathcal{A}}, S_{\mathcal{A}})$ and $\mathcal{E}(C_{\mathcal{A}}, S_{\mathcal{A}})$ denote the inference latency and task specific error of the model, respectively. For simplicity, we will abbreviate them as $\mathcal{T}(\mathcal{A})$ and $\mathcal{E}(\mathcal{A})$ in the remaining of this paper under the clear context.

A crucial question for solving the problem in Eq. (2) is: *How to determine the number of channels and weight sparsity for each layer?* One alternative way is to first prune the channels of the original model in an automated way, then prune the weights of the channel-pruned model for further acceleration. However, this separated optimization may lead to a sub-optimal solution, because it is difficult to find the optimal balance between channel pruning and weight pruning by hand. Concretely speaking, the optimal architecture for channel pruning may be sub-optimal when further applying weight pruning for acceleration. Therefore, we instead optimize the number of channels and weight sparsity simultaneously in one single optimization run and determine the balance of acceleration between channel pruning and weight pruning automatically.

Another difficulty in solving the problem in Eq. (2) is that there are more than one objective (the latency and accuracy) to be optimized, yielding a multi-objective optimization (MOO) problem naturally. It is challenging to design one model that achieves the best values for both of the two objectives since these two objectives are generally conflict with each other. Therefore, the optimal solutions for the problem in Eq. (2) are not unique, and we need to find a sequence of models lying on the Pareto-frontier between the accuracy and latency[2]. We solve this problem based on the multi-objective evolutionary algorithm.

2.3   Unified Framework

Before going into the details, we first introduce the overview of the whole framework, which is illustrated in Fig. 2. The JCW works in an iterative way, it maintains a sequence of well-performing models $P = \{(C_i, S_i)\}_{i=1}^{n}$ with different number of channels and weight sparsity, here n is the population size. ❶ In each iteration, a new set of models $Q = \{(C_i^{new}, S_i^{new})\}_{i=1}^{n}$ are generated from P

---

[1] We use weight sparsity to denote the ratio of non-zero parameters of remaining channels across the whole paper.

[2] In MOO, the Pareto-frontier is a set of solutions that for each solution, it is not possible to further improve some objectives without degrading other objectives.

through crossover and mutation operators. Then, we estimate the accuracy (❷) and latency (❸) of all the models in $P \cup Q$. ❹ Based on the estimations, we select the models with various latency and relatively low error rates to form the next generation of P. The proposed components are integrated under the learning problem of Eq. (2), and the iteration continues until the qualified models are found. In the sequel, we instantiate different components of the framework, i.e. the accuracy estimator in Sect. 2.4, the latency estimator in Sect. 2.5, and the uniform non-dominated sorting selection of models in Sect. 2.6.

2.4 Accuracy Estimation

In JCW, the accuracy prediction is done efficiently by training a super-net with parameter sharing [16]. Parameter sharing has been widely used in previous one shot NAS methdos [3,16,45,59]. However, the accuracy predictor in JCW is different from theirs in that we support not only coarse-grained channel pruning but also fine-grained weight pruning. More importantly, these methods often train a super-net containing all the models in the whole search space, the training target can be formulated by:

$$\min_{w} \mathbb{E}_{(x,y)\sim\mathcal{D},\mathcal{A}\sim\Omega}[\mathcal{L}(x|y;W_{\mathcal{A}})], \tag{3}$$

where Ω is the search space (e.g. in the scenario of joint channel and weight pruning in this paper, the search space is all the models with different channel numbers and layer-wise sparsity), $\mathcal{D}$ is the training dataset, W is the parameter of the super-net, and $W_{\mathcal{A}}$ are a part of parameters from W to form the model with architectural configuration $\mathcal{A}$. In this case, the large number of models in the super-net are largely coupled with each other [48].

Considering that in our framework, we only need to determine the accuracy rank of a finite number of architectures (i.e. 2 times of the population size) each time, at the beginning of each evolutionary generation, we reconstruct the super-net to contain only models to be evaluated and retrain it to determine the accuracy rank. Formally, denote $\tilde{P} = P \cup Q = \{(C_i, S_i)\}_{i=1}^{2n}$ to be the sequence of models whose accuracy rank are to be estimated, we train the super-net with the following simplified target:

$$\min_{W} \mathbb{E}_{(x,y)\sim\mathcal{D},\mathcal{A}\sim\tilde{P}}[\mathcal{L}(x|y;W_{\mathcal{A}})], \tag{4}$$

and $W_{\mathcal{A}}$ is constructed by selecting the parameters with top channel indices and then applying norm-based weight pruning. Note that the number of models in $\tilde{P}$ is far less than the whole search space Ω. In this way, the coupling between different models in the super-net can be reduced.

2.5 Latency Estimation

Previously, latency prediction is conducted by constructing a look up Table [49, 50,55] or training an estimation model [2,54]. The former one is limited to a small number of candidates, while the latter one requires a large number of architecture-latency pairs to train the estimation model, which is laborious to collect. In contrast, in the JCW, we propose to estimate the latency of models with trilinear interpolation. The latency of one model can be represented by the summation of the latency of each layer:

$$T(C, S) = \sum_{l=1}^{L} T^{(l)}(C, S), \tag{5}$$

where $T^{(l)}(C, S)$ is the latency of the l-th layer of the model represented by $\{C, S\}$. In the context of joint channel pruning and weight pruning, the latency of each layer depends on the number of input/output channels and the weight sparsity of that layer:

$$T^{(l)}(C, S) = \hat{T}^{(l)}(c^{(l-1)}, c^{(l)}, s^{(l)}). \tag{6}$$

For efficient latency estimation, we build a layer-wise latency predictor $\hat{T}^{(l)}(\cdot)$ with trilinear interpolation. This is based on the observation that the latency is locally linear with respect to the layer width and weight sparsity[3]. Specifically, we denote $C_{max} = \{c_{max}^{(l)}\}_{l=1}^{L}$ as the maximum number of channels of each layer. For layer l with maximum input channels of $c_{max}^{(l-1)}$ and maximum output channels of $c_{max}^{(l)}$, we first measure the real runtime on the target device with channel width $0, \frac{1}{N}, \frac{2}{N}, \cdots, 1.0$ of both input and output channels, and weight sparsity of $0, \frac{1}{M}, \frac{2}{M}, \cdots, 1.0$, respectively, generating a 3-D array of architecture-latency samples. We denote $T_{ijk}^{(l)}$ to be the latency of the lth layer's convolution with input, output channels and weight sparsity of $\frac{i}{N} c_{max}^{(l-1)}, \frac{j}{N} c_{max}^{(l)}$, and $\frac{k}{M}$, respectively, and define:

$$c_i^{(l)} = N \frac{c^{(l)}}{c_{max}^{(l)}} - i, \qquad s_i^{(l)} = M s^{(l)} - i \tag{7}$$

to be the normalized array index. Given any input/output channels and weight sparsity, we can easily approximate the latency through trilinear interpolation of the 3-D array[4]:

$$\hat{T}^{(l)}(c^{(l-1)}, c^{(l)}, s^{(l)}) = \sum_{i,j,k} \tau(c_i^{(l-1)}) \tau(c_j^{(l)}) \tau(s_k^{(l)}) T_{ijk}^{(l)}, \tag{8}$$

where:
$$\tau(x) = \max(0, 1 - |x|).$$

In practice, we find that $M = 10, N = 8$ is sufficient for approximating latency with high efficiency and accuracy as illustrated in Fig. 5. Compared to other latency estimation methods [2,55] requiring tens of thousands of architecture-latency pairs, the proposed trilinear interpolation-based latency predictor can be efficiently constructed with as less as 700 data points.

---

[3] Please refer to the Appendix for more results about this observation.
[4] More detailed derivation is given in Appendix.

2.6   Uniform Non-dominated Sorting Selection

An important role of individual selection is to search for well-performed models while keeping model diversity in terms of latency. However, this cannot be fulfilled by the standard selection scheme used in previous multi-objective evolutionary algorithms [4] because in JCW, the accuracy estimation with parameter sharing is not as accurate as latency estimation. As a result, the evolver will put too much emphasis on the latency minimization, and large models will be gradually and incorrectly removed during evolution as shown in the left of Fig. 7.

To handle this obstacle, we propose a uniform non-dominated sorting selection to generate diverse architectures of various latency. We first uniformly sample N points from the interval $[T_{\min}, T_{\max}]$:

$$T_i = T_{\min} + i \times \frac{T_{\max} - T_{\min}}{N - 1}, i = 0, 1, 2, \cdots, N - 1, \tag{9}$$

where $T_{\min}$, $T_{\max}$ are the minimal and maximal latency, respectively. For each T_i, we sort the individuals in the merged population $P \cup Q$ with objectives $\{|T_i - T|, \mathcal{E}\}$ with non-dominated sort [4], where T and $\mathcal{E}$ are latency and error rate of the network architecture.

Let $F_i = \{F_i^{(s)}\}_{s=0}^{n_i-1}$ be frontier stages after sorting architectures with objectives $\{|T_i - T|, \mathcal{E}\}$. We select candidates stage by stage in the order $F_0^{(0)}, F_1^{(0)}, \cdots$ until the number of selected candidates reaches the evolutionary population size. When we select individuals from F_i, architectures with latency close to T_i and relatively low error rate will be selected. In the right of Fig. 7, it demonstrates that our uniform non-dominated sorting selection generates diverse architectures of various latency.

3   Experimental Results

To validate the efficiency of our joint channel and weight pruning (JCW), we conduct extensive experiments, including ablation studies, based on ResNet18 [19], MobileNetV1 [24], and MobileNetV2 [44] on the ImageNet classification dataset [5].

3.1   Implementation Details

We use a similar technique as fast sparse ConvNets [10] for efficient computation of sparse convolution, with two slight improvements. (i) For weight pruning, we group the parameters along the output channels and remove the parameters group-wisely. Specifically, four parameters at the same location of adjacent output channels are grouped together, and parameters in the same group are removed or retained simultaneously. The grouping strategy is beneficial for efficient data reuse [10]. In all of our experiments, the group size is set to 4. (ii) We extend their computation algorithm to support not only matrix multiplication but also regular convolution. We implement an efficient algorithm for the

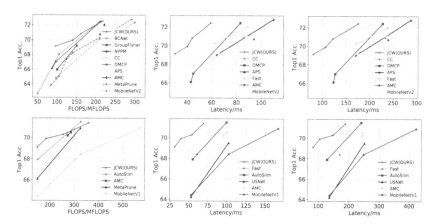

Fig. 3. Comparison of JCW with state of the art model compression/acceleration methods. Top: the results on MobileNetV2. Bottom: the results on MobileNetV1. Left: the comparison of FLOPS. Middle: the comparison of latency on one single Cortex-A72 CPU. Right: the comparison on one single Cortex-A53 CPU.

computation of sparse convolution and utilize it to measure the latency of our searched sparse models.

For evolutionary search, we set the population size to 64 and the number of search steps to 128. We sample a subset of the ImageNet dataset for the supernet training. Specifically, we randomly sample 100 classes from ImageNet, 500 images per class to construct the train set, and randomly sample 50 images from the rest images per class to construct the validation set. We train the supernet for 30 epochs with batch size of 256, where the first 10 epochs are used for parameter sharing warming up. The learning rate and weight decay are set to 0.1 and 0.00004 in all the experiments, respectively. For each model in supernet, the batch normalization (BN) statistics are recalculated with $1,280$ images.

After we complete the search stage, the generated architectures are re-trained on the whole training set and validated on the validation set of the ImageNet dataset. We train the compressed models with ADMM, details about the hyperparameters of different models are given in Appendix.

If not specified, the latency of all the models are measured on one single ARM Cortex-A72 CPU. The latency of models for channel pruning methods is measured with TFLite [14], which is a commonly used mobile-oriented deep learning inference framework for dense deep neural networks. The latency of models with weight sparsity is measured with our implemented high-performance sparse convolution algorithm. We run each model for 20 times and report the average of the runtime.

3.2   Comparison with State-of-the-art Methods

We compare JCW with multiple state-of-the-art model compression and acceleration methods, including BCNet [45], NPPM [13], GroupFisher [34], CC [1],

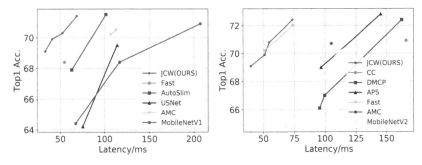

Fig. 4. Comparison in terms of accuracy and latency on 4× Cortex-A53 CPUs.

DMCP [15], APS [48], AMC [23], AutoSlim [57], USNet [58], and Fast [10] in terms of model FLOPS, accuracy, and latency on multiple types of devices. The main results are shown in Fig. 3 and Table 2, and more detailed results are shown in Appendix. The middle of the column shows the comparison in terms of accuracy and latency on a single ARM Cortex-A72 CPU. We see from the figure that JCW outperforms all the baseline methods by a large margin. This proves that JCW achieves a better latency-accuracy trade-off.

Table 2. Experimental results on the Resnet18 model.

Method	Latency A72	Latency A53	Latency A53×4	Acc@1
Uniform 1×	537 ms	1167 ms	402 ms	69.8%
DMCP	341 ms	893 ms	291 ms	69.7%
APS	363 ms	921 ms	322 ms	70.2%
JCW(OURS)	**194 ms**	**518 ms**	**158 ms**	69.7%
	224 ms	**694 ms**	**179 ms**	70.2%

Generalization to Other Deployment Platforms. The latency of models can be largely depended on deployment platforms. Thus a common problem about JCW may be: Whether it can generalize well to other deployment platforms? To resolve this, we directly deploy the searched models on Cortex-A53 CPUs. The latency-accuracy comparisons are shown in the right column of the Fig. 3, we see from the figure that, although not targeting Cortex-A53 during the search, JCW still achieves excellent latency-accuracy trade-off on this platform, outperforming all the other methods by a large margin. In Fig. 4, we further show the comparison of accuracy and latency on multi-device platforms. The latency is further measured on 4× Cortex-A53 CPUs. JCW still outperforms all the other baselines in most cases, except on MobileNetV2 under the latency constraint of 50 ms, where JCW performs very similar to Fast [10].

Generalization to Other Metrics. Besides latency, FLOPs is also an important evaluation index when deploying deep learning models to mobile devices. We thus further compare JCW with other methods in terms of FLOPs and accuracy in the left column of Fig. 3. We see that although targeting latency, JCW still achieves competitive trade-off between FLOPs and accuracy. This is not surprising because JCW includes flexible weight pruning.

To sum up, the proposed JCW can achieve excellent trade-off between latency and accuracy, and generalizes well to other deployment platforms and metrics.

3.3 Ablation Study

Accuracy of Latency Estimation. To evaluate the accuracy of the proposed latency estimation, we compare the real latency and predicted latency of 100 randomly generated MobileNetV1 models with various number of channels and weight sparsity. Specifically, we run each model for 20 times, calculate the average runtime and compare it with the predicted runtime. Results are shown in the left of Fig. 5. From the figure, we can observe that with trilinear interpolation, the predicted latency is highly correlated to the real latency of deep networks. The right of Fig. 5 shows the latency and the number of arithmetic operations of different deep networks. We can see that the number of arithmetic operations (MFLOPS) is positively correlated with latency generally, while the latency does not monotonically increase with the number of operations (MFLOPS). This is mainly because that the model's latency on real hardware platform can be impacted by both computation intensity and other factors such as memory access time. This phenomenon motivates us to design deep networks based on latency instead of FLOPs for model pruning. When deploying a deep network into a practical hardware, we consider the latency of runtime, not the FLOPs.

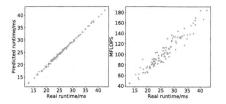

Fig. 5. Left: the real runtime & the predicted runtime with proposed trilinear interpolation. Right: the real runtime & FLOPs of models.

Correlation of Accuracy Predictor. In JCW, the accuracy predictor is designed to support both channel and weight pruning, which has not been studied before. Recall in Sect. 2.4 that in each training, the super-net will be re-initialized and only contains a finite number of architectures. We further evaluate the correlation between predicted accuracy and true accuracy. It is time consuming to train a number of models on the large-scale ImageNet dataset, so we evaluate it on the CIFAR-10 dataset. We train a super-net containing 128 ResNet20 models with various channel widths and weight sparsity, and all the hyper-parameters are the same as described in Sect. 3.1. Figure 6 shows a high correlation between predicted and true accuracy.

Table 3. Comparisons among different variants of JCW for accelerating ResNet18 on ImageNet. JCW consistently outperforms all of its variants.

Method	CP	WP	Optim	Latency	Accuracy
WSO	✗	✓	-	161.72 ms	68.45%
				196.82 ms	69.54%
CWO	✓	✗	-	161.14 ms	66.78%
				205.03 ms	67.75%
SCW	✓	✓	Seq.	197.62 ms	69.29%
				221.65 ms	69.58%
JCW	✓	✓	**Joint**	**160.37 ms**	**69.16%**
				196.44 ms	**69.90%**
				223.73 ms	**70.19%**

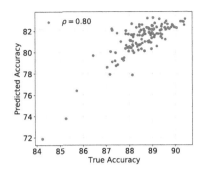

Fig. 6. Correlation between predicted accuracy and true accuracy of ResNet20 models on CIFAR-10 dataset.

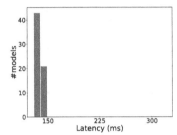
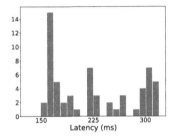

Fig. 7. Left: latency distribution after 50 evolutionary steps with standard non-dominated sorting selection. Right: latency distribution after 50 evolutionary steps with uniform non-dominated sorting selection. The searches are conducted with ResNet18 on ImageNet. The proposed selection generates models of various latency, while the standard selection gives priority to models of low latency.

Effect of Joint Channel and Weight Pruning

The core idea of JCW is to apply joint channel and weight pruning jointly for better latency-accuracy trade-off. To prove its effectiveness, we compare JCW with three of its variants, *i.e.* (i) WSO which only searches for the layer-wise weight sparsity, while the number of channels for each layer remains the maximum value; (ii) CWO which only searches for the number of channels while keeps the full parameters of remaining channels; (iii) SCW which first searches for number of channels and then searches for weight sparsity with fixed channel widths.

Table 3 shows that our JCW outperforms all the other variants, and CWO performs the worst. In particular, under the latency of ∼160 ms, the accuracy of JCW is 0.61% and 2.38% higher than WSO and CWO, respectively. Under the latency of ∼196 ms, the JCW achieves 0.46% and 0.61% higher accuracy than WSO and SCW, respectively. Moreover, we see from the table that SCW, which simply applies channel width and weight sparsity search sequentially, does not achieve a better accuracy-latency trade-off than JCW. This is reasonable because simply applying channel and weight pruning search sequentially will lead to a sub-optimal balance between channel and weight pruning. In contrast, JCW optimizes channel width and weight sparsity jointly in one single search and tries to find the optimal balance between channel and weight pruning, thus achieving the best accuracy-latency trade-off.

Effect of Uniform Non-Dominated Sorting Selection. Considering the search efficiency, we predict the accuracy of models with different architectures using parameter sharing at the cost of inaccurate accuracy prediction. Because of inaccurate model accuracy prediction, the evolver tends to focus on minimizing the other objective, the latency. For instance, let $\mathcal{A}$, $\mathcal{B}$ be two architectures in the combined population to be selected. We assume that $\mathcal{T}(\mathcal{A}) < \mathcal{T}(\mathcal{B})$, $\mathcal{E}(\mathcal{A}) > \mathcal{E}(\mathcal{B})$. Here, $\mathcal{T}$ and $\mathcal{E}$ are real latency and error rate of architectures, respectively. In terms of multi-objective optimization, there is no priority relation between $\mathcal{A}$ and $\mathcal{B}$. In other words, both architectures $\mathcal{A}$ and $\mathcal{B}$ should be selected in the new population with an equal chance. If the accuracy estimation is inaccurate, it is likely that the predicted error rate of $\mathcal{A}$ is smaller than $\mathcal{B}$. In this case, the standard non-dominated sorting selection may remove the architecture $\mathcal{B}$ from the population incorrectly. This motivates us to develop the uniform non-dominated sort scheme, which explicitly selects architectures with a wide range of latency. The uniform non-dominated sorting selection generates diverse architectures of various latency.

To further validate the effectiveness of the proposed selection method, we show the latency distribution of models searched with the above two different selection methods in Fig. 7. The experiments are conducted with ResNet18 on the ImageNet dataset. From the left of Fig. 7, we can observe that the latency of models searched with the original selection scheme are small after 50 evolutionary steps. In contrast, from the right of Fig. 7 we can observe that with the proposed uniform non-dominated sorting selection, models with relatively large latency are also preserved during search. So we conclude that the enhanced evolutionary algorithm with the proposed uniform non-dominated sorting selection is able to generate diverse models with various latency.

4   Related Works

Pruning has long been one of the primary techniques for network compression and acceleration [17,18,20,32]. These methods remove unimportant parameters from the original network and optimize the remaining parts of the networks to retain accuracy. According to the granularity of pruning, these methods can be categorized into fine-grained weight pruning and coarse-grained filter pruning. In weight pruning, parameters are removed in weight-level [7,12,17,18,28]. Weight pruning is flexible to achieve theoretically smaller models and can also be efficiently accelerated on mobile CPUs thanks to the recent work of fast sparse convolution [10]. In contrast, channel pruning compresses networks by removing parameters at the filter level. Most of the early channel pruning methods are based on an filter importance scoring scheme, *e.g.*, the filter norm [21,31], the percentage of zero-activation [26], the reconstruction error of outputs [6,20,37], the increase of loss after pruning [34,39–41], the geometric properties of filters [22,27]. Besides, sparse regularization-based methods [43,53] have also been intensively explored. Channel pruning methods are well-structured, thus it can be directly accelerated without extra implementation efforts.

Apart from the aforementioned pruning methods, many recently emerging pruning methods formulate the network pruning as an architecture search, taking benefits from the automated process in composing architectures to avoid the labor-prohibitive model design. He *et al.* [23] propose to determine the number of channels with reinforcement learning and outperforms human-designed pruning methods. Lin *et al.* [33] search for the channel numbers with population-based algorithm. Wang *et al.* [52] model the problem of channel number search as structural redundancy reduction. Gao *et al.* [13], Wang *et al.* [51] train parameterized accuracy predictors to guide the pruning process. Liu *et al.* [35] train a meta network to predict the weights of compressed models, then conduct evolutionary search for pruning. There is also a vast body of work utilizing the parameter sharing technique to train a supernet for accuracy evaluation and conduct the pruning with evolutionary search [3,16,45], greedy slimming [57], or reinforcement learning [48]. Besides, differentiable channel number search approaches [9,15] have also been investigated.

Besides the above static pruning methods, there is also a vast body of work pruning networks dynamically. These methods decide online inference which connections should be pruned according to the input image [29,30,47]. Dynamic pruning is a promising method to reduce the average computational complexity over a large number of testing samples, while it is not convenient for deployment on mobile devices for real-time applications because its computation complexity depends on the input image and is unpredictable. Our method lies in the family of static pruning.

Efficient model design often involves multiple objectives, *e.g.*, the accuracy, the latency, the model size, *etc.* In this perspective, it is more desired to search for a sequence of Pareto optimal models. Many works have been proposed to deal with multi-objective model design. Hsu *et al.* [25], Tan *et al.* [46] integrate multiple objectives into one correlated reward function and conduct the search with reinforcement learning. However, they need trial and error to design the form and related hyper-parameters for the correlated reward function, which is prohibitively laborious. Some recent works [8,11,36,56] search for Pareto optimal architectures directly with evolutionary algorithm and a selection criterion based on non-dominated sorting [4]. Our work focuses on model pruning, which is orthogonal to these general NAS methods.

To our best knowledge, the JCW is the first work investigating the essential part of pruning: is it possible to absorb both benefits of channel and weight pruning and achieve a better accuracy-latency trade-off by applying the two jointly? We conduct extensive experiments and ablation studies, which demonstrate that the joint channel and weight pruning achieves a better accuracy-latency Pareto frontier than previous pruning approaches.

5 Conclusion

In this work, we propose a joint channel and weight pruning, named JCW, which achieves a better pruning proportion between channel and weight pruning. We further construct a multi-objective optimization considering both model accuracy and inference latency in the JCW, which can be solved by a tailored Pareto-optimization evolutionary algorithm. Extensive experiments demonstrate that the effectiveness of each component of JCW.

Acknowledgements. This work was supported in part by the National Key Research and Development Program of China under Grant 2021ZD0201504, in part by the Strategic Priority Research Program of Chinese Academy of Sciences under Grant XDA27040300.

References

1. Towards compact cnns via collaborative compression. In: IEEE Conference on Computer Vision and Pattern Recognition (CVPR)
2. Berman, M., Pishchulin, L., Xu, N., B.Blaschko, M., Medioni, G.: Aows: adaptive and optimal network width search with latency constraints. In: 2020 IEEE Conference on Computer Vision and Pattern Recognition (CVPR) (2020)
3. Cai, H., Gan, C., Wang, T., Zhang, Z., Han, S.: Once for all: train one network and specialize it for efficient deployment. In: International Conference on Learning Representations (2020)
4. Deb, K., Pratap, A., Agarwal, S., Meyarivan, T.: A fast and elitist multiobjective genetic algorithm: Nsga-ii. IEEE Trans. Evol. Comput. **6**(2), 182–197 (2002). https://doi.org/10.1109/4235.996017
5. Deng, J., Dong, W., Socher, R., Li, L.J., Li, K., Fei-Fei, L.: Imagenet: A large-scale hierarchical image database. In: 2009 IEEE Conference on Computer Vision and Pattern Recognition, pp. 248–255. IEEE (2009)
6. Ding, X., Ding, G., Zhou, X., Guo, Y., Liu, J., Han, J.: Approximated oracle filter pruning for destructive cnn width optimization. In: IEEE Conference on Machine Learning (ICML) (2019)
7. Ding, X., Ding, G., Zhou, X., Guo, Y., Liu, J., Han, J.: Global sparse momentum sgd for pruning very deep neural networks. In: Advances in Neural Information Processing Systems, (NeurIPS) (2019)
8. Dong, J.D., Cheng, A.C., Juan, D.C., Wei, W., Sun, M.: Dpp-net: device-aware progressive search for pareto-optimal neural architectures. In: Proceedings of the European Conference on Computer Vision (ECCV), pp. 517–531 (2018)
9. Dong, X., Yang, Y.: Network pruning via transformable architecture search. In: Advances in Neural Information Processing Systems, pp. 760–771 (2019)
10. Elsen, E., Dukhan, M., Gale, T., Simonyan, K.: Fast sparse convnets. In: Proceedings of the IEEE/CVF Conference on Computer Vision and Pattern Recognition (CVPR) (2020)
11. Elsken, T., Metzen, J.H., Hutter, F.: Efficient multi-objective neural architecture search via lamarckian evolution. In: International Conference on Learning Representations (ICLR) (2019)

12. Frankle, J., Carbin, M.: The lottery ticket hypothesis: finding sparse, trainable neural networks. In: International Conference on Learning Representations (ICLR) (2019)

13. Gao, S., Huang, F., Cai, W., Huang, H.: Network pruning via performance maximization. In: IEEE Conference on Computer Vision and Pattern Recognition (CVPR) (2021)

14. Google-Inc.: machine learning for mobile devices: Tenworflow lite (2020). https://www.tensorflow.org/lite

15. Guo, S., Wang, Y., Li, Q., Yan, J.: Dmcp: differentiable markov channel pruning for neural networks. In: Proceedings of the IEEE/CVF Conference on Computer Vision and Pattern Recognition, pp. 1539–1547 (2020)

16. Guo, Z., et al.: Single path one-shot neural architecture search with uniform sampling. In: Vedaldi, A., Bischof, H., Brox, T., Frahm, J.-M. (eds.) ECCV 2020. LNCS, vol. 12361, pp. 544–560. Springer, Cham (2020). https://doi.org/10.1007/978-3-030-58517-4_32

17. Han, S., Mao, H., Dally, W.J.: Deep compression: compressing deep neural network with pruning, trained quantization and huffman coding. In: International Conference on Learning Representations (ICLR) (2016)

18. Han, S., Pool, J., Tran, J., Dally, W.J.: Learning both weights and connections for efficient neural networks. In: Advances in Neural Information Processing Systems (NIPS) (2015)

19. He, K., Zhang, X., Ren, S., Sun, J.: Deep residual learning for image recognition. In: IEEE Conference on Computer Vision and Pattern Recognition (CVPR) (2015)

20. He, Y., Zhang, X., Sun, J.: Channel pruning for accelerating very deep neural networks. In: 2017 IEEE International Conference on Computer Vision (ICCV) (2017)

21. He, Y., Kang, G., Dong, X., Fu, Y., Yang, Y.: Soft filter pruning for accelerating deep convolutional neural networks. In: Proceedings of International Joint Conference on Artificial Intelligence (IJCAI) (2018)

22. He, Y., Liu, P., Wang, Z., Hu, Z., Yang, Y.: Filter pruning via geometric median for deep convolutional neural networks acceleration. In: IEEE Conference on Computer Vision and Pattern Recognition (CVPR) (2019)

23. He, Y., Lin, J., Liu, Z., Wang, H., Li, L.J., Han, S.: Amc: automl for model compression and acceleration on mobile devices. In: Proceedings of the European Conference on Computer Vision (ECCV) (2018)

24. Howard, A.G., et al.: Mobilenets: efficient convolutional neural networks for mobile vision applications. CoRR abs/1704.04861 (2017). https://arxiv.org/abs/1704.04861

25. Hsu, C.H., et al: Monas: multi-objective neural architecture search using reinforcement learning. arXiv preprint arXiv:1806.10332 (2018)

26. Hu, H., Peng, R., Tai, Y.W., Tang, C.K.: Network trimming: a data-driven neuron pruning approach towards efficient deep architectures. arXiv:1607.03250 (2016)

27. Joo, D., Yi, E., Baek, S., Kim, J.: Linearly replaceable filters for deep network channel pruning. In: The 34th AAAI Conference on Artificial Intelligence (AAAI) (2021)

28. LeCun, Y., Denker, J.S., Solla, S.A.: Optimal brain damage. In: Advances in Neural Information Processing Systems, pp. 598–605 (1990)

29. Li, C., Wang, G., Wang, B., Liang, X., Li, Z., Chang, X.: Dynamic slimmable network. In: Proceedings of the IEEE/CVF Conference on Computer Vision and Pattern Recognition (CVPR), pp. 8607–8617 (2021)

30. Li, F., Li, G., He, X., Cheng, J.: Dynamic dual gating neural networks. In: Proceedings of the IEEE/CVF International Conference on Computer Vision (ICCV), pp. 5330–5339 (2021)
31. Li, H., Kadav, A., Durdanovic, I.: Pruning filters for efficient convnets. In: International Conference on Learning Representation (ICLR) (2017)
32. Li, H., Kadav, A., Durdanovic, I., Samet, H., Graf, H.P.: Pruning filters for efficient convnets. arXiv preprint arXiv:1608.08710 (2016)
33. Lin, M., Ji, R., Zhang, Y., Zhang, B., Wu, Y., Tian, Y.: Channel pruning via automatic structure search. In: Proceedings of International Joint Conference on Artificial Intelligence (IJCAI) (2020)
34. Liu, L., et al.: Group fisher pruning for practical network compression. In: International Conference on Machine Learning (ICML) (2021)
35. Liu, Z., et al.: Metapruning: meta learning for automatic neural network channel pruning. In: Proceedings of the IEEE/CVF International Conference on Computer Vision, pp. 3296–3305 (2019)
36. Lu, Z., et al.: Nsga-net: neural architecture search using multi-objective genetic algorithm. In: Proceedings of the Genetic and Evolutionary Computation Conference, pp. 419–427 (2019)
37. Luo, J., Zhang, H., Zhou, H., Xie, C., Wu, J., Lin, W.: Thinet: pruning cnn filters for a thinner net. IEEE Trans. Pattern Anal. Mach. Intell. TPAMI (2018)
38. Molchanov, D., Ashukha, A., Vetrov, D.: Variational dropout sparsifies deep neural networks. In: Proceedings of the International Conference on Machine Learning (2017)
39. Molchanov, P., Tyree, S., Karras, T., Aila, T., Kautz, J.: Pruning convolutional neural networks for resource efficient inference. In: International Conference on Learning Representations (ICLR) (2017)
40. Molchanov, P., Mallya, A., Tyree, S., Frosio, I., Kautz, J.: Importance estimation for neural network pruning. In: IEEE Conference on Computer Vision and Pattern Recognition (CVPR) (2019)
41. Peng, H., Wu, J., Chen, S., Huang, J.: Collaborative channel pruning for deep neural networks. In: International Conference on Machine Learning (ICML) (2019)
42. Pham, H., Guan, M.Y., Zoph, B., Le, Q.V., Dean, J.: Efficient neural architecture search via parameter sharing. In: International Conference on Machine Learning (2018)
43. Ruan, X., Liu, Y., Li, B., Yuan, C., Hu, W.: Dpfps: dynamic and progressive filter pruning for compressing convolutional neural networks from scratch. In: The 34th AAAI Conference on Artificial Intelligence (AAAI) (2021)
44. Sandler, M., Howard, A.G., Zhu, M., Zhmoginov, A., Chen, L.C.: Mobilenetv 2: inverted residuals and linear bottlenecks. In: IEEE Conference on Computer Vision and Pattern Recognition (CVPR) (2018)
45. Su, X., You, S., Wang, F., Qian, C., Zhang, C., Xu, C.: Bcnet: searching for network width with bilaterally coupled network. In: IEEE Conference on Computer Vision and Pattern Recognition (CVPR) (2021)
46. Tan, M., et al.: Mnasnet: platform-aware neural architecture search for mobile. In: Proceedings of the IEEE/CVF Conference on Computer Vision and Pattern Recognition, pp. 2820–2828 (2019)
47. Tang, Y., et al.: Manifold regularized dynamic network pruning. In: Proceedings of the IEEE/CVF Conference on Computer Vision and Pattern Recognition (CVPR), pp. 5018–5028 (2021)
48. Wang, J., et al.: Revisiting parameter sharing for automatic neural channel number search. In: Advances in Neural Information Processing Systems 33 (2020)

49. Wang, K., Liu, Z., Lin, Y., Lin, J., Han, S.: Haq: Hardware-aware automated quantization with mixed precision. In: IEEE Conference on Computer Vision and Pattern Recognition (CVPR) (2019)
50. Wang, T., et al.: Apq: joint search for network architecture, pruning and quantization policy. In: Proceedings of the IEEE/CVF Conference on Computer Vision and Pattern Recognition, pp. 2078–2087 (2020)
51. Wang, W., et al.: Accelerate cnns from three dimensions: a comprehensive pruning framework. In: International Conference on Machine Learning (ICML) (2021)
52. Wang, Z., Li, C., Wang, X.: Convolutional neural network pruning with structural redundancy reduction. In: IEEE Conference on Computer Vision and Pattern Recognition (CVPR) (2021)
53. Wen, W., Wu, C., Wang, Y., Chen, Y., Li, H.: Learning structured sparsity in deep neural networks. In: Advances in Neural Information Processing Systems (NeurIPS) (2016)
54. Yang, H., Zhu, Y., Liu, J.: Ecc: Platform-independent energy-constrained deep neural network compression via a bilinear regression model. In: IEEE Conference on Computer Vision and Pattern Recognition (CVPR) (2019)
55. Yang, T., et al.: Netadapt: Platform-aware neural network adaption for mobile applications. In: European Conference on Computer Vision (ECCV) (2018)
56. Yang, Z., et al.: Cars: continuous evolution for efficient neural architecture search. In: IEEE Conference on Computer Vision and Pattern Recognition (CVPR) (2020)
57. Yu, J., Huang, T.S.: Autoslim: towards one-shot architecture search for channel numbers. CoRR abs/1903.11728 (2019). https://arxiv.org/abs/1903.11728
58. Yu, J., Huang, T.S.: Universally slimmable networks and improved training techniques. In: Proceedings of the IEEE/CVF International Conference on Computer Vision, pp. 1803–1811 (2019)
59. Yu, J., et al.: BigNAS: scaling up neural architecture search with big single-stage models. In: Vedaldi, A., Bischof, H., Brox, T., Frahm, J.-M. (eds.) ECCV 2020. LNCS, vol. 12352, pp. 702–717. Springer, Cham (2020). https://doi.org/10.1007/978-3-030-58571-6_41
60. Zhuang, Z., et al.: Discrimination-aware channel pruning for deep neural networks. In: Advances in Neural Information Processing Systems (NeurIPS) (2018)

Deep Ensemble Learning by Diverse Knowledge Distillation for Fine-Grained Object Classification

Naoki Okamoto$^{(\boxtimes)}$, Tsubasa Hirakawa , Takayoshi Yamashita ,
and Hironobu Fujiyoshi

Chubu University, Kasugai, Aichi, Japan
{naok,hirakawa}@mprg.cs.chubu.ac.jp,
{takayoshi,fujiyoshi}@isc.chubu.ac.jp

Abstract. Ensemble of networks with bidirectional knowledge distillation does not significantly improve on the performance of ensemble of networks without bidirectional knowledge distillation. We think that this is because there is a relationship between the knowledge in knowledge distillation and the individuality of networks in the ensemble. In this paper, we propose a knowledge distillation for ensemble by optimizing the elements of knowledge distillation as hyperparameters. The proposed method uses graphs to represent diverse knowledge distillations. It automatically designs the knowledge distillation for the optimal ensemble by optimizing the graph structure to maximize the ensemble accuracy. Graph optimization and evaluation experiments using Stanford Dogs, Stanford Cars, CUB-200-2011, CIFAR-10, and CIFAR-100 show that the proposed method achieves higher ensemble accuracy than conventional ensembles.

Keywords: Ensemble learning · Knowledge distillation

1 Introduction

Deep learning models trained under the same conditions, such as network architecture and dataset, produce variations in accuracy and different errors due to random factors such as network initial values and mini-batches. Ensemble and knowledge distillation improve the performance by using multiple networks with different weight parameters for training and inference.

Ensemble performs inference using multiple trained networks. It performs inference on the basis of the average of the output of each network for the input samples and thus improves the performance compared with an inference using a single network. It is also effective against problems such as adversarial attack

Supplementary Information The online version contains supplementary material available at https://doi.org/10.1007/978-3-031-20083-0_30.

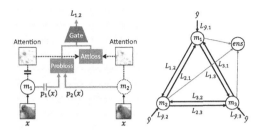

Fig. 1. Ensemble learning with diverse knowledge distillation by graph representation. Loss calculation shows knowledge distillation from m_1 to m_2.

and out-of-distribution detection due to the nature of using multiple networks [5,6,15,21,26]. It is computationally more expensive than a single network, so methods for constructing parameter-efficient ensembles have been proposed [16, 26,28,33,34].

Knowledge distillation is a training method where a network shares the knowledge acquired through training with other networks to reduce parameters or improve network performance. There are two types of knowledge distillation: unidirectional [10] and bidirectional [38]. A typical method for unidirectional knowledge distillation is knowledge distillation (KD) [10]. KD uses probability distributions as knowledge of the network and trains an untrained network with shallow layers using a trained network with deep layers. A typical method for bidirectional knowledge distillation is deep mutual learning (DML) [38]. DML is a method of mutual distillation using multiple untrained networks. Various distillation methods have been proposed depending on the combination of the networks and the type of knowledge [1,3,20,25,27,30,31,35–37]. Minami et al. [22] introduced a graph representation for knowledge distillation to unify the existing methods.

In this paper, we propose knowledge distillation for ensemble. The critical factors of the proposed method are knowledge distillation that promotes diversity among networks and automatic design of the distillation method. We consider knowledge distillation that separates knowledge between networks, in addition to conventional knowledge distillation. The direct separation of probability distributions as knowledge may degrade the performance of the network. Therefore, we perform diverse knowledge distillation from two types of knowledge: probability distributions and attention maps. The proposed method represents the diverse knowledge distillation in a graph [22], as shown in Fig. 1. We define the combination of loss design of knowledge distillation as a hyperparameter and automatically design complex knowledge distillation, which is difficult to design manually by hyperparameter search.

Our contributions are as follows.

- We investigate the relationship between the difference of probability distributions and the effect of ensemble and find a positive correlation.
- We perform knowledge distillation to promote diversity among networks for ensembles. We design a loss of proximity and loss of separation of knowledge

using probability distribution and attention map and weight the loss values using gates to achieve diverse knowledge distillation.
– We use graphs to represent diverse knowledge distillation and automatically design appropriate knowledge distillations by optimizing the graph structure. The networks of the optimized graph improve the ensemble accuracy, and each network is a diverse model with a different attention map.

2   Related Work

In this section, we introduce ensemble and knowledge distillation, which are methods for using multiple networks.

2.1   Ensemble

Ensemble is one of the oldest machine learning methods. Ensemble in deep learning is a simple method of averaging probability distributions or logits, which are the outputs of networks with different weight parameters. It is known that the ensemble accuracy improves depending on the number of networks and that the ensemble accuracy ceases to improve after exceeding a certain number of networks. Ensemble is also effective against problems such as adversarial attack and out-of-distribution detection due to it using multiple networks [5,6,15,21,26].

The training and inference cost of ensemble increases with the number of networks. In knowledge distillation [16,28], a single network can achieve the same performance as ensemble by training the network to approach the probability distribution by the ensemble. Batch ensemble [33] and hyperparameter ensemble [34] prevent increasing the parameters by sharing some of them and reduce the training and inference costs.

2.2   Knowledge Distillation

Knowledge distillation is a training method where a network shares the knowledge acquired through training with other networks to reduce parameters or improve network performance. There are two types of knowledge distillation: unidirectional and bidirectional.

Unidirectional knowledge distillation uses a teacher network, which is a trained network, and a student network, which is an untrained network. The student network trains the outputs of the teacher network as pseudo-labels in addition to the labels. Hinton et al. [10] proposed KD, which trains the student network with small parameters using the probability distribution of the teacher network with large parameters. KD is effective even for teacher and student networks with the same number of parameters [8]. There is also a two-stage knowledge distillation using three networks [23].

Bidirectional knowledge distillation trains multiple student networks at the same time, using the probability distributions of the student networks as pseudo-labels. The first bidirectional knowledge distillation, DML, was proposed by

Table 1. Correlation coefficient for each dataset.

Dataset	Correlation coefficient
Stanford Dogs	0.237
Stanford Cars	0.499
CUB-200-2011	0.322
CIFAR-10	0.386
CIFAR-100	0.325

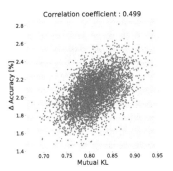

Fig. 2. Relationship on Stanford Cars.

Zhang et al. [38]. In DML, the accuracy of the network increases with the number of networks.

On-the-Fly Native Ensemble (ONE) [16] is knowledge distillation using ensemble. ONE reduces the number of parameters by using a multi-branch network and produces a training effect similar to that of DML by using an ensemble of multi-branches as pseudo-labels.

A variety of knowledge has been proposed, such as probability distributions, feature maps, attention maps, and relationships between samples [1,3,20,25,27, 30,31,35–37]. Minami et al. [22] introduced a graph representation for knowledge distillation to unify the existing methods. Knowledge distillation is also effective in a variety of problem settings [4,19,26,29].

3   Investigating the Relationship Between Ensemble and Knowledge Distillation

We think that there is a correlation between the differences of probability distributions and ensemble accuracy because of the relationship between bidirectional distillation and ensemble. In this section, we investigate the relationship between the difference in probability distributions and ensemble accuracy and verify the change in ensemble accuracy caused by knowledge distillation on the Stanford Dogs dataset [12].

3.1   Relationship Between Bidirectional Knowledge Distillation and Ensemble

We investigate the relationship between KL-divergence, a loss design of DML [38], and the accuracy improvement by ensemble of two models. KL-divergence is a measure of the difference in distribution. However, it is an asymmetric measure, so the analysis defines the measure of difference as

$$\text{Mutual KL} = \frac{1}{2}(KL(p_1 \parallel p_2) + KL(p_2 \parallel p_1)), \tag{1}$$

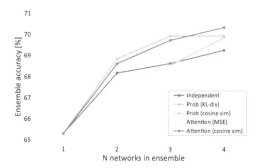

Fig. 3. Ensemble accuracy of loss design for various numbers of networks.

where p_1 and p_2 are the probability distributions of networks 1 and 2, respectively. The accuracy improvement by ensemble is defined as

$$\Delta\text{Accuracy} = \text{ACC}_{ens} - \frac{1}{2}(\text{ACC}_1 + \text{ACC}_2), \tag{2}$$

where ACC_{ens} is the ensemble accuracy, ACC_1 is the accuracy of network 1, and ACC_2 is the accuracy of network 2. The datasets are Stanford Dogs [12], Stanford Cars [13], Caltech-UCSD Birds-200-2011 (CUB-200-2011) [32], CIFAR-10 [14], and CIFAR-100 [14]. The network is ResNet [9], and 100 trained networks are prepared. ResNet-20 is used for CIFAR datasets, and ResNet-18 is used for other datasets. Ensemble is constructed by selecting two networks out of 100, and evaluation is performed on all combinations (4,950 pairs).

Table 1 shows the correlation coefficients for each dataset, and Fig. 2 shows the evaluation results for Stanford Cars. There is a weak positive correlation between the difference of probability distributions and the improvement of accuracy by ensemble. Therefore, it is expected that the ensemble effect can be improved by training to have diversity in the probability distributions between networks.

3.2   Bidirectional Knowledge Distillation to Promote Diversity for Ensemble

On the basis of analysis trends, we consider knowledge distillation that separates knowledge to improve the ensemble accuracy. Therefore, we investigate the effect on the ensemble by training to bring knowledge closer and training to separate knowledge. We use two types of knowledge: the probability distribution that is the final output of the network, and the attention map that represents the information in the middle layer. The loss design for probability distributions uses KL-divergence to bring the probability distributions closer together and cosine similarity to separate the probability distributions. The loss design for the attention map uses mean square error to bring the attention map closer together and cosine similarity to separate the attention maps. We use ResNet-18 [9] as the network and Stanford Dogs [12] as the dataset. The attention map is created from the output of ResBlock4 using Attention Transfer [37].

Figure 3 shows the results of the evaluation of each loss design with the number of networks used in the ensemble from 1 to 4. Here, Independent is the ensemble accuracy without knowledge distillation. The ensemble accuracy is improved by training to bring the probability distributions closer together when the number of networks is small and training to separate the attention maps when the number of networks is large.

4   Proposed Method

We propose an ensemble learning method using knowledge distillation. From the trend in Sect. 3, it is difficult to intentionally design a knowledge distillation method for each number of networks. Therefore, we propose to automatically design an effective ensemble learning method. We use the graph representation in the knowledge transfer graph [22] and optimize the loss design of diverse knowledge distillation as a hyperparameter of the graph by hyperparameter search. We consider various ensemble learning methods by optimizing the structure of the graph to maximize ensemble accuracy. We show that using the automatically designed ensemble learning methods improves the ensemble accuracy and that each network is prompted to a specific attention strategy by the combination of the selected knowledge distillations regardless of the dataset.

4.1   Designing for Loss of Knowledge Distillation to Promote Diversity

We perform knowledge distillation to promote diversity among networks for ensemble. In this paper, we use probability distributions and attention maps as knowledge, and design loss of bringing knowledge closer and loss of separating knowledge. To train as a minimization problem, we use different loss designs for bringing knowledge closer and separating knowledge. We refer to the destination of knowledge distillation as the target network t and the source of knowledge as the source network s.

Loss Design for the Probability Distribution. When the probability distribution is brought closer together, KL-divergence is used, and when it is separated, cosine similarity is used. The loss function using KL-divergence is defined as

$$KL(p_s(x) \parallel p_t(x)) = \sum_{c=1}^{C} p_s^c(x) \log \frac{p_s^c(x)}{p_t^c(x)}, \tag{3}$$

$$L_p = KL(p_s(x) \parallel p_t(x)), \tag{4}$$

where C is the number of the classes, x is the input sample, $p_s(x)$ is the probability distribution of the source network, and $p_t(x)$ is the probability distribution of the target network. The loss function using cosine similarity is defined as

$$L_p = \frac{p_s(x)}{\parallel p_s(x) \parallel_2} \cdot \frac{p_t(x)}{\parallel p_t(x) \parallel_2}. \tag{5}$$

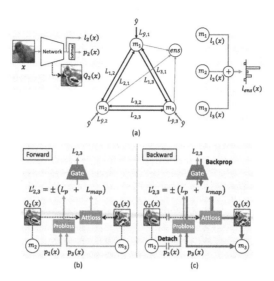

Fig. 4. (a) Ensemble learning with diverse knowledge distillation by graph representation. (b) Loss calculation shows knowledge transfer from m_2 to m_3. (c) Calculated loss gradient information is only propagated in m_3.

Loss Design for the Attention Map. The attention map responds strongly to regions in the input sample that is useful for training. The size of the target object varies from sample to sample, so the similarity may be high even though the map responds strongly to different parts of the target object. Therefore, we crop the attention map. The attention map of the source network is cropped to a square centered on the position with the highest value, and the attention map of the target network is cropped to the same position as the source network. Cropping is performed at multiple sizes, and the average of the similarities at each size is used as the similarity of the attention map. When the attention map is brought closer together, the mean square error is used, and when it is separated, cosine similarity is used. The loss function using mean squared error is defined as

$$L_{map} = \frac{1}{K} \sum_{k=1}^{K} (\frac{Q_s^k(x)}{\| Q_s^k(x) \|_2} - \frac{Q_t^k(x)}{\| Q_t^k(x) \|_2})^2, \tag{6}$$

where K is the number of crops, Q_s is the attention map of the source network and Q_t is the attention map of the target network. The loss function using cosine similarity is defined as

$$L_{map} = \frac{1}{K} \sum_{k=1}^{K} \frac{Q_s^k(x)}{\| Q_s^k(x) \|_2} \cdot \frac{Q_t^k(x)}{\| Q_t^k(x) \|_2}. \tag{7}$$

Introducing Gate. We control knowledge distillation by weighting the above loss values of the probability distribution and the attention map using gates. In this paper, we consider four types of gates: through, cutoff, linear, and correct.

The through gate passes through the loss value of each input sample as it is and is defined as

$$G_{s,t}^{Though}(a) = a. \tag{8}$$

The cutoff gate does not execute loss calculation and is defined as

$$G_{s,t}^{Cutoff}(a) = 0. \tag{9}$$

The linear gate changes the weights linearly with training time and is defined as

$$G_{s,t}^{Linear}(a) = \frac{k}{k_{end}}a, \tag{10}$$

where k is the number of the current iterations and k_{end} is the total number of iterations at the end of the training. The correct gate passes only the samples that the source network answered correctly and is defined as

$$G_{s,t}^{Correct}(a) = \begin{cases} a \ y_s = \hat{y} \\ 0 \ y_s \neq \hat{y} \end{cases}, \tag{11}$$

where y_s is the output of the source network and $\hat{y}$ is a label.

4.2   Graph Representation and Optimization of Graph Structures

A diverse knowledge distillation by the losses in Sect. 4.1 is represented by a graph [22], and an appropriate knowledge distillation is automatically designed by optimizing the graph structure.

Graph Representation for Ensemble. We use a graph representation [22] in a knowledge transfer graph to automatically design knowledge distillation. The ensemble learning using knowledge distillation by the graph representation is shown in Fig. 4. The graph consists of nodes and edges. Nodes define the network node that represents the network and the ensemble node that performs ensemble. Edges represent loss calculation. Edges between network nodes represent knowledge distillation. Edges between the network node and the label represent cross-entropy loss using the output of the node and the label.

The ensemble node performs ensemble by using the outputs of all the network nodes. The process in an ensemble node is defined as

$$l_{ens} = \frac{1}{M} \sum_{m=1}^{M} l_m(x). \tag{12}$$

where M is the number of network nodes, l_m is the logits of the network node, and x is the input sample.

Knowledge Distillation Between Nodes. Edges between network nodes perform knowledge distillation between nodes. Figure 4b and 4c show the process of loss processing at the edge from node m_2 to node m_3. First, the edge computes

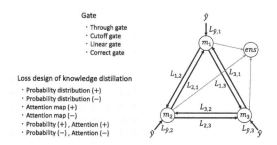

Fig. 5. Hyperparameters in graph structures for ensemble.

the loss of knowledge distillation of the probability distribution and attention map as shown in Fig. 4b. The loss calculation of knowledge distillation at the edge is defined as

$$L'_{s,t} = L_p(x) + L_{map}(x). \tag{13}$$

The final loss of knowledge distillation is then applied to the gate. The loss of knowledge distillation applied to the gate is defined as

$$L_{s,t} = \frac{1}{N} \sum_{n=1}^{N} G_{s,t}(L'_{s,t}(x_n)), \tag{14}$$

where N is the number of the input sample, and $G_{s,t}(\cdot)$ is one of the four types of gate. The gradient of the loss of knowledge distillation changes the network that propagates the gradient depending on the edge direction. Figure 4c shows the gradient flow at the edge from node m_2 to node m_3. In this case, knowledge distillation from node m_2 to node m_3 is performed by cutting the computational graph of node m_2 to propagate the gradient only to node m_3.

The loss calculation is performed for each edge, and the final loss of the network node is defined as

$$L_t = G_{hard,t}(L_{hard}) + \sum_{s=1,s\neq t}^{M} L_{s,t}, \tag{15}$$

where $G_{hard,t}$ is the gate, and L_{hard} is cross-entropy loss using the output of the network node and the label $\hat{y}$.

Hyperparameter Search. Figure 5 shows the hyperparameters of the graph structure. The hyperparameters of the graph are the loss design of the edges between the network nodes and the gate of each edge. There are six loss designs: bring the probability distribution closer to that of the other edge (Eq. 4), separate the probability distribution (Eq. 5), bring the attention map closer to that of the other edge (Eq. 6), separate the attention map (Eq. 7), bring the probability distribution and attention map closer to those of the other edge at the same

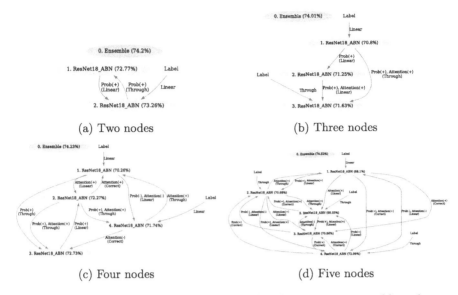

(a) Two nodes   (b) Three nodes

(c) Four nodes   (d) Five nodes

Fig. 6. Graph optimized on Stanford Dogs. Red node represents ensemble node, gray node represents network node, and "Label" represents supervised labels. At each edge, selected loss design and gate are shown, exclusive of cutoff gate. Accuracy in parentheses is the result of one of five trials. (Color figure online)

time (Eqs. 4 and 6), and separate the probability distribution and attention map at the same time (Eqs. 5 and 7). The network to be used as the network node is fixed to that determined before optimization.

The optimization of the graph structure uses random search and the asynchronous successive halving algorithm (ASHA) [18]. The combination of hyperparameters is determined randomly, and the graph evaluates the ensemble node at $1, 2, 4, 8 \cdots 2^k$ epochs. If the accuracy of the ensemble node is less than the median accuracy at the same epoch in the past, the training is terminated and the next graph is trained.

5 Experiments

We evaluate the proposed method. In Sect. 5.2, we visualize the optimized graph structure. In Sect. 5.3, we compare the proposed method with the conventional method. In Sect. 5.4, we evaluate the generalizability of the graph structure on various datasets. In Sect. 5.5, we evaluate the performance of knowledge distillation from the optimized ensemble graph into a single network.

5.1 Experimental Setting

Datasets. We used Stanford Dogs [12], Stanford Cars [13], Caltech-UCSD Birds-200-2011 (CUB-200-2011) [32], CIFAR-10 [14], and CIFAR-100 [14]. Stanford

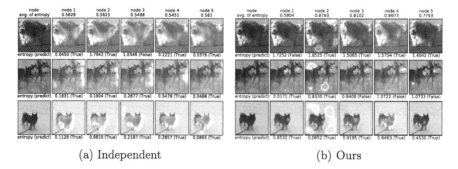

(a) Independent   (b) Ours

Fig. 7. Attention map of ABN in individually training and the optimized graph with five nodes (Fig. 6d). Bottom of map shows prediction results and entropy of probability distribution.

Dogs, Stanford Cars, and CUB-200-2011 belong to the fine-grained object classification task. CIFAR-10 and CIFAR-100 belong to the general object classification task. When optimizing the graphs, we used part of the training data for training and the rest for evaluation. We used 40,000 images for CIFAR and half of the training data for other datasets. For the comparative evaluation discussed in Sect. 5.3 and 5.4, the original training data and testing data were used.

Networks. We used ResNet [9] and attention branch network (ABN) [7] based on ResNet. When training the CIFAR dataset, we used ResNet-20 and ABN based on ResNet-20. When training the other dataset, we used ResNet-18 and ABN based on ResNet-18. The attention map of ResNet is created from the output of ResBlock4 by Attention Transfer [37]. ABN creates an attention map on the basis of the class activation map [39] and weights the attention map to the feature map by using the attention mechanism.

Implementation Details. The training conditions were the same for all experiments. The optimization algorithms were stochastic gradient descent (SGD) and momentum. The initial learning rate was 0.1, momentum was 0.9, coefficient of weight decay was 0.0001, batch size was 16, and number of epochs was 300. The learning rate was decayed by a factor of 10 at 150 and 225 epochs. The attention map of ResNet is cropped to 3×3, 5×5, and 7×7 for loss calculation. The attention map of ABN is cropped to 3×3, 7×7, and 11×11 for loss calculation. In the optimization of the graph, we tried 6,000 combinations of hyperparameters. We used PyTorch [24] as a framework for deep learning and Optuna [2] as a framework for hyperparameter search. For the optimization of a graph, we used 90 Quadro P5000 servers. Each result represents the mean and standard deviation of five trials.

5.2 Visualization of Optimized Graphs

Figure 6 shows the graphs of two to five nodes optimized on Stanford Dogs. With two nodes, we obtained a graph that is an extension of DML. With three

Table 2. Comparison of the accuracy on Stanford Dogs [%].

Method	No. of nodes	ResNet-18		ABN	
		Node	Ensemble	Node	Ensemble
Independent	2	65.31 ± 0.16	68.19 ± 0.20	68.13 ± 0.16	70.90 ± 0.19
DML	2	67.55 ± 0.27	68.86 ± 0.25	69.91 ± 0.46	71.45 ± 0.52
ONE(B × 2)	1	67.96 ± 0.41	68.53 ± 0.39	69.38 ± 0.38	69.81 ± 0.33
Ours	2	**71.38 ± 0.08**	**72.41 ± 0.20**	**72.77 ± 0.23**	**73.86 ± 0.26**
Independent	3	65.08 ± 0.23	68.64 ± 0.38	68.04 ± 0.28	71.41 ± 0.34
DML	3	68.66 ± 0.34	69.95 ± 0.39	70.50 ± 0.26	72.08 ± 0.42
ONE(B × 3)	1	68.49 ± 0.60	68.94 ± 0.56	69.96 ± 0.47	70.44 ± 0.44
Ours	3	**69.58 ± 0.15**	**71.87 ± 0.33**	**70.95 ± 0.16**	**73.41 ± 0.30**
Independent	4	65.29 ± 0.35	69.27 ± 0.49	68.30 ± 0.27	72.06 ± 0.53
DML	4	68.83 ± 0.44	69.95 ± 0.58	**71.50 ± 0.31**	72.87 ± 0.29
ONE(B × 4)	1	68.48 ± 0.32	68.85 ± 0.37	70.16 ± 0.47	70.54 ± 0.54
Ours	4	**70.34 ± 0.12**	**72.71 ± 0.13**	71.46 ± 0.22	**74.16 ± 0.22**
Independent	5	65.00 ± 0.24	69.47 ± 0.13	68.24 ± 0.26	72.32 ± 0.18
DML	5	**68.77 ± 0.17**	69.94 ± 0.20	**71.15 ± 0.28**	72.50 ± 0.16
ONE(B × 5)	1	68.51 ± 0.18	68.95 ± 0.24	70.59 ± 0.28	70.89 ± 0.14
Ours	5	52.28 ± 0.87	**71.35 ± 0.48**	70.23 ± 0.33	**74.14 ± 0.50**

nodes, we obtained a graph that combines the conventional knowledge distillation methods of KD and TA. With four and five nodes, we obtained graphs with a mixture of loss designs that are brought closer together and loss designs that are separated.

Figure 7b shows the attention map of ABN with the five nodes. Each node has a different focus of attention. Looking at the average entropy, nodes 1 and 5, which focus on a single point on the dog's head, have low entropy. Nodes 2, 3, and 4, which focus on the whole image or background, have higher entropy than nodes 1 and 5. This means that inferences are made on the basis of the importance of different locations and the state of attention affects probability distribution.

Figure 7a shows the attention map in the ensemble method using individually trained networks. Compared with the optimized graph, the average entropy of the ensemble method using individually trained networks is lower. This is because the attention regions are almost the same among the networks even though they are trained individually.

5.3 Comparison with Conventional Methods

Table 2 shows the average and ensemble accuracy of the nodes of the proposed and conventional methods on Stanford Dogs. "Ours" is the result of the optimized graph, "Independent" is the result of the individually trained network,

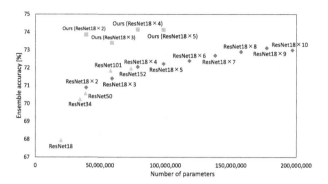

Fig. 8. Relationship between number of parameters and accuracy in Stanford Dogs. Green shows single network, blue shows "Independent," and light blue shows "Ours". (Color figure online)

Table 3. Ensemble accuracy of reused two-node graphs optimized on another dataset [%].

Method	Training Graph	Optimizing Graph	Ensemble
Independent	CUB-200-2011	–	65.26
Ours	CUB-200-2011	Stanford Dogs	**72.06**
Ours	CUB-200-2011	CUB-200-2011	69.81
Independent	Stanford Cars	–	88.49
Ours	Stanford Cars	Stanford Dogs	**89.76**
Ours	Stanford Cars	Stanford Cars	89.44
Independent	CIFAR-100	–	73.16
Ours	CIFAR-100	Stanford Dogs	72.19
Ours	CIFAR-100	CIFAR-100	**74.18**
Independent	CIFAR-10	–	93.99
Ours	CIFAR-10	Stanford Dogs	93.87
Ours	CIFAR-10	CIFAR-100	**94.37**
Ours	CIFAR-10	CIFAR-10	94.15

"DML" is the result of the network with DML [38], and "ONE" is the result of the multi-branch network with ONE [16]. "ONE(B × 2)" is the result of the two-branch network. The ensemble accuracy of "Ours" was higher than those of "Independent," "DML," and "ONE." Comparing "Independent" and "DML," we can see that the improvement in ensemble accuracy was smaller than the improvement in node accuracy. With "Ours," compared with "DML," ensemble accuracy also improved as network accuracy improved. Therefore, we can say that "Ours" obtained the graph that generates more diversity by training.

Figure 8 shows the comparison results with ABN and different base networks. The vertical axis is accuracy, and the horizontal axis is the total number of

Table 4. Accuracy of single network by knowledge distillation [%].

Method	Teacher	Studet
DML	67.97	69.68
KTG	71.71	72.71
SLA	–	69.36
FRSKD	–	71.42
Ours	72.60	**72.94**

parameters. In Stanford Dogs, the accuracy of the single network and "Independent" varied with the number of parameters. "Ours" shows that ensemble with high parameter efficiency can be constructed by mutual learning with diversity without changing the network structure. When the number of networks is increased, ensemble accuracy reaches a ceiling of around 73%. This shows that the proposed method achieved an accuracy that exceeds the limit of a conventional method.

5.4   Generalizability of Graphs

We evaluate the optimized graph in Stanford Dogs on a variety of datasets. Table 3 shows the ensemble accuracy of the two-node graph. On the dataset of the fine-grained object classification, the graph optimized by Stanford Dogs has better accuracy than Independent. On CIFAR-10, the graph optimized by CIFAR-100 has better accuracy than Independent. We believe that there is generalizability in the graph structure when the problem set is the same, and that optimization has resulted in a graph structure that corresponds to the problem set.

5.5   Knowledge Distillation from Ensemble Learning

We evaluate the performance of knowledge distillation from the optimized ensemble graph into a single network. We use ResNet-18 as a student network and DML [38] KTG [22] as a teacher network for knowledge distillation on the Stanford Dogs dataset. We also compare with the state of the art of knowledge distillation, such as SLA [17] and FRSKD [11], which are self-distillation methods. The Table 4 shows the accuracy of teachers and students trained with each method. "Ours" means knowledge distillation using the ensemble of two networks trained by the graph of Fig. 6a as a teacher. From the Table 4, we see that the accuracy of the student network by "Ours" is higher than that of the conventional methods. This is because the ensemble of two networks trained by the graph has diversity for representing dark knowledge to make suitable knowledge transfer.

6   Conclusion and Future Work

This paper proposed a knowledge distillation for ensemble. We investigated loss design for ensemble to promote diversity among the networks and automatically designed knowledge distillation for ensemble by graph representation. Experimental results on five different datasets showed that the proposed method increased the accuracy. The optimization of the graph structure was evaluated on 6,000 randomly determined pairs using the asynchronous successive halving algorithm (ASHA). The number of combinations of graph structures increases in proportion to the number of nodes. Therefore, increasing the number of combinations to be evaluated may result in a better graph structure. Our future work will include introducing Bayesian optimization and fine-tuning graph structures.

Acknowledgements. This paper is based on results obtained from a project, JPNP18002, commissioned by the New Energy and Industrial Technology Development Organization (NEDO).

References

1. Ahn, S., Hu, S.X., Damianou, A., Lawrence, N.D., Dai, Z.: Variational information distillation for knowledge transfer. In: IEEE/CVF Conference on Computer Vision and Pattern Recognition. pp. 9163–9171 (2019)
2. Akiba, T., Sano, S., Yanase, T., Ohta, T., Koyama, M.: Optuna: a next-generation hyperparameter optimization framework. In: Proceedings of the 25rd ACM SIGKDD International Conference on Knowledge Discovery and Data Mining (2019)
3. Chattopadhay, A., Sarkar, A., Howlader, P., Balasubramanian, V.N.: Grad-cam++: generalized gradient-based visual explanations for deep convolutional networks. In: IEEE Winter Conference on Applications of Computer Vision (2018)
4. Chen, G., Choi, W., Yu, X., Han, T., Chandraker, M.: Learning efficient object detection models with knowledge distillation. In: Advances in Neural Information Processing Systems, pp. 742–751 (2017)
5. Dabouei, A., Soleymani, S., Taherkhani, F., Dawson, J., Nasrabadi, N.M.: Exploiting joint robustness to adversarial perturbations. In: IEEE/CVF Conference on Computer Vision and Pattern Recognition (2020)
6. Dvornik, N., Schmid, C., Mairal, J.: Diversity with cooperation: ensemble methods for few-shot classification. In: IEEE/CVF International Conference on Computer Vision (2019)
7. Fukui, H., Hirakawa, T., Yamashita, T., Fujiyoshi, H.: Attention branch network: learning of attention mechanism for visual explanation. In: IEEE/CVF Conference on Computer Vision and Pattern Recognition (2019)
8. Furlanello, T., Lipton, Z., Tschannen, M., Itti, L., Anandkumar, A.: Born again neural networks. In: International Conference on Machine Learning. Proceedings of Machine Learning Research, vol. 80, pp. 1607–1616 (2018)
9. He, K., Zhang, X., Ren, S., Sun, J.: Deep residual learning for image recognition. In: IEEE Conference on Computer Vision and Pattern Recognition, pp. 770–778 (2016)

10. Hinton, G., Vinyals, O., Dean, J.: Distilling the knowledge in a neural network. In: Neural Information Processing Systems Deep Learning and Representation Learning Workshop (2015)
11. Ji, M., Shin, S., Hwang, S., Park, G., Moon, I.C.: Refine myself by teaching myself: feature refinement via self-knowledge distillation. In: IEEE/CVF Conference on Computer Vision and Pattern Recognition, pp. 10664–10673 (2021)
12. Khosla, A., Jayadevaprakash, N., Yao, B., Fei-Fei, L.: Novel dataset for fine-grained image categorization. In: First Workshop on Fine-Grained Visual Categorization, IEEE Conference on Computer Vision and Pattern Recognition (2011)
13. Krause, J., Stark, M., Deng, J., Fei-Fei, L.: 3D object representations for fine-grained categorization. In: 4th International IEEE Workshop on 3D Representation and Recognition. Sydney, Australia (2013)
14. Krizhevsky, A., Hinton, G.: Learning multiple layers of features from tiny images. Technical report, Citeseer (2009)
15. Laine, S., Aila, T.: Temporal ensembling for semi-supervised learning. In: International Conference on Learning Representations (2017)
16. Lan, X., Zhu, X., Gong, S.: Knowledge distillation by on-the-fly native ensemble. In: Advances in Neural Information Processing Systems, pp. 7527–7537 (2018)
17. Lee, H., Hwang, S.J., Shin, J.: Self-supervised label augmentation via input transformations. In: International Conference on Machine Learning, pp. 5714–5724 (2020)
18. Li, L., et al.: A system for massively parallel hyperparameter tuning. In: Dhillon, I.S., Papailiopoulos, D.S., Sze, V. (eds.) Proceedings of Machine Learning and Systems, vol. 2, pp. 230–246 (2020)
19. Liu, Y., Chen, K., Liu, C., Qin, Z., Luo, Z., Wang, J.: Structured knowledge distillation for semantic segmentation. In: IEEE/CVF Conference on Computer Vision and Pattern Recognition (2019)
20. Liu, Y., Cao, J., Li, B., Yuan, C., Hu, W., Li, Y., Duan, Y.: Knowledge distillation via instance relationship graph. In: IEEE/CVF Conference on Computer Vision and Pattern Recognition (2019)
21. Malinin, A., Mlodozeniec, B., Gales, M.: Ensemble distribution distillation. In: International Conference on Learning Representations (2020)
22. Minami, S., Hirakawa, T., Yamashita, T., Fujiyoshi, H.: Knowledge transfer graph for deep collaborative learning. In: Asian Conference on Computer Vision (2020)
23. Mirzadeh, S.I., Farajtabar, M., Li, A., Ghasemzadeh, H.: Improved knowledge distillation via teacher assistant: Bridging the gap between student and teacher. In: Association for the Advancement of Artificial Intelligence (2020)
24. Paszke, A., et al.: Pytorch: An imperative style, high-performance deep learning library. In: Wallach, H., Larochelle, H., Beygelzimer, A., d'Alché-Buc, F., Fox, E., Garnett, R. (eds.) Advances in Neural Information Processing Systems, vol. 32. Curran Associates, Inc. (2019)
25. Peng, B., et al.: Correlation congruence for knowledge distillation. In: IEEE/CVF International Conference on Computer Vision (2019)
26. Radosavovic, I., Dollár, P., Girshick, R., Gkioxari, G., He, K.: Data distillation: towards omni-supervised learning. In: IEEE Conference on Computer Vision and Pattern Recognition (2018)
27. Romero, A., Ballas, N., Kahou, S.E., Chassang, A., Gatta, C., Bengio, Y.: Fitnets: hints for thin deep nets. In: International Conference on Learning Representations (2015)
28. Song, G., Chai, W.: Collaborative learning for deep neural networks. In: Advances in Neural Information Processing Systems, pp. 1837–1846 (2018)

29. Tarvainen, A., Valpola, H.: Mean teachers are better role models: weight-averaged consistency targets improve semi-supervised deep learning results. In: Advances in Neural Information Processing Systems (2017)
30. Tian, Y., Krishnan, D., Isola, P.: Contrastive representation distillation. In: International Conference on Learning Representations (2020)
31. Tung, F., Mori, G.: Similarity-preserving knowledge distillation. In: IEEE/CVF International Conference on Computer Vision (2019)
32. Wah, C., Branson, S., Welinder, P., Perona, P., Belongie, S.: The caltech-ucsd birds-200-2011 dataset. Technical report CNS-TR-2011-001, California Institute of Technology (2011)
33. Wen, Y., Tran, D., Ba, J.: Batchensemble: an alternative approach to efficient ensemble and lifelong learning. In: International Conference on Learning Representations (2020)
34. Wenzel, F., Snoek, J., Tran, D., Jenatton, R.: Hyperparameter ensembles for robustness and uncertainty quantification. In: Larochelle, H., Ranzato, M., Hadsell, R., Balcan, M.F., Lin, H. (eds.) Advances in Neural Information Processing Systems, vol. 33, pp. 6514–6527. Curran Associates, Inc. (2020)
35. Yim, J., Joo, D., Bae, J., Kim, J.: A gift from knowledge distillation: fast optimization, network minimization and transfer learning. In: IEEE Conference on Computer Vision and Pattern Recognition, pp. 4133–4141 (2017)
36. Yu, L., Yazici, V.O., Liu, X., Weijer, J.V.D., Cheng, Y., Ramisa, A.: Learning metrics from teachers: compact networks for image embedding. In: IEEE/CVF Conference on Computer Vision and Pattern Recognition (2019)
37. Zagoruyko, S., Komodakis, N.: Paying more attention to attention: improving the performance of convolutional neural networks via attention transfer. In: International Conference on Learning Representations (2017)
38. Zhang, Y., Xiang, T., Hospedales, T.M., Lu, H.: Deep mutual learning. In: IEEE Conference on Computer Vision and Pattern Recognition (2018)
39. Zhou, B., Khosla, A., Lapedriza, A., Oliva, A., Torralba, A.: Learning deep features for discriminative localization. In: IEEE Conference on Computer Vision and Pattern Recognition (2016)

Helpful or Harmful: Inter-task Association in Continual Learning

Hyundong Jin and Eunwoo Kim[(⊠)]

School of Computer Science and Engineering, Chung-Ang University,
Seoul, South Korea
{jude0316,eunwoo}@cau.ac.kr

Abstract. When optimizing sequentially incoming tasks, deep neural networks generally suffer from catastrophic forgetting due to their lack of ability to maintain knowledge from old tasks. This may lead to a significant performance drop of the previously learned tasks. To alleviate this problem, studies on continual learning have been conducted as a countermeasure. Nevertheless, it suffers from an increase in computational cost due to the expansion of the network size or a change in knowledge that is favorably linked to previous tasks. In this work, we propose a novel approach to differentiate helpful and harmful information for old tasks using a model search to learn a current task effectively. Given a new task, the proposed method discovers an underlying association knowledge from old tasks, which can provide additional support in acquiring the new task knowledge. In addition, by introducing a sensitivity measure to the loss of the current task from the associated tasks, we find cooperative relations between tasks while alleviating harmful interference. We apply the proposed approach to both task- and class-incremental scenarios in continual learning, using a wide range of datasets from small to large scales. Experimental results show that the proposed method outperforms a large variety of continual learning approaches for the experiments while effectively alleviating catastrophic forgetting.

Keywords: Continual learning · Task association · Model search

1 Introduction

Deep learning algorithms are generally optimized for a single task or multiple different tasks. However, it is difficult for them to apply to a more challenging scenario; learning sequentially incoming tasks without accessing the old ones. This is because a single deep network lacks the ability to accommodate both new and old knowledge using a limited set of shared parameters. Specifically, when a network learned from old tasks is further trained with a new task, it can easily forget the previously learned tasks. This phenomenon is called catastrophic forgetting [17,28], a major hindrance in learning sequential tasks with a deep neural network.

Many approaches have been recently proposed as a remedy for memorizing knowledge. They are generally classified into four categories: regularization [2, 5, 17, 21, 42], replay [6, 10, 15, 25, 30, 31, 37, 40], dynamic expansion [3, 32, 33, 41], and structural allocation [1, 26, 27, 36] approaches. The regularization approaches add a new penalty to consolidate old knowledge while training on a new task. However, this will not be effective when a long sequence of tasks is involved because the penalty does not sufficiently prevent the change of contributable parameters. Replay methods retain a small number of instances from old tasks and learn them with new task samples in a joint manner. For a large number of tasks, the number of retained instances of each task decreases due to the size limit to hold the old instances, causing performance degradation. Dynamic expansion methods generally expand the network when a new task comes in or when performance does not meet a predetermined criterion. As a result, they require a large amount of computation costs, constraining their applicability to real-world problems. The last category, structural allocation, constructs disjoint sets of parameters for tasks, respectively, and learns a new task using the older sets. However, it will not be helpful when we use unwanted old parameters that can negatively affect the new task.

The proposed method falls within the last category. Unlike other categories, it encourages to use a disjoint set of parameters for each task, preventing the network from containing the mixture of old and new knowledge. Despite its benefit, the strategy will not be promising if we use adversarial parameters that negatively affect a new task when using old parameter sets altogether [27]. From this, a question arises can we discover helpful knowledge from old tasks to enrich the knowledge of a new task for structural allocation based continual learning.

In this paper, we reformulate the forgetting problem into a task interference problem in continual learning and solve it using model (task) selection to discover cooperative tasks. To this end, we propose a novel approach to differentiate helpful and harmful knowledge from old tasks in a continual learning framework. Specifically, the proposed framework is based upon an architecture consisting of a global feature extractor and multiple heads (classifiers) corresponding to tasks. The proposed approach is based on structural allocation using a train-prune-retrain learning paradigm [27] to address sequential tasks so that the feature extractor consists of disjoint sets of parameters corresponding to tasks, respectively. To find older tasks and exploit their knowledge given a task, we present a model search framework and apply it to the disjoint sets of parameters learned so far (at a coarser level). The proposed method further discovers critical parameters in element-wise by measuring sensitivity to the loss of a new task from the searched old tasks (at a finer level). The discovered knowledge is leveraged when learning the new task. Due to the search mechanism, we select an optimal subnetwork that can accelerate the model inference as well as discover a cooperative relationship among tasks.

We validate our method for both task-incremental and class-incremental learning scenarios. A number of large-scale datasets, including ImageNet [8] and diverse fine-grained datasets, are used in task-incremental learning. In class-incremental learning, ImageNet-50 [30], Split CIFAR-10 [19], and Split MNIST

[20] are used. Experimental results show that the proposed method achieves higher performance than state-of-the-art continual learning methods for both learning scenarios while minimizing the computation cost. The contributions of this work are mainly four-fold:

- We redesign the forgetting problem into a task association problem by taking task interference into account in the structural allocation strategy.
- We propose a novel continual learning approach that can effectively learn a new task by differentiating and absorbing helpful knowledge from old tasks.
- The proposed approach elaborately selects contributable parameters from older tasks using a gradient-based model search at a coarser level and a sensitivity measure at a finer level.
- Experimental results show that the proposal achieves remarkable performance improvement over diverse continual learning competitors while minimizing the computation cost.

2   Related Work

Regularization Method. Regularization approaches [2,5,17,21,42] in continual learning typically append a regularization term to suppress updating of parameters that are important to the previously learned tasks. The importance can be calculated by the Fisher information [17], the change of loss [42], or the derivative of the network output [2]. This line of methods is memory-efficient because it does not require instances of previous tasks and does not expand the size of the network. However, the regularization term does not effectively prevent the change of previous knowledge for many sequential tasks, resulting in drastic performance drop [12].

Replay Method. The approaches rely on some raw samples [6,10,23,25,31,40] or generated instances by a generative model [30,37]. [6] and [25] mitigate forgetting by controlling gradient produced from a few replayed samples. [31] picks some nearest neighbor samples to the average sample per task, and [24] parameterizes exemplar samples which can be optimized. However, they repeatedly train a small number of old instances, potentially leading to an overfitting problem. Also, since generated instances may not exactly imitate the distribution of real instances, we may encounter a performance drop [4].

Dynamic Expansion. Methods falling into this category dynamically expand the network structure when a new task comes in [3,32,33,38,41]. Progressive neural network [33] produces an additional network to perform a new task, which inherits the knowledge of old tasks by connecting to previously learned networks. However, this reveals a drawback that the computation cost increases proportionally to the number of tasks. Dynamically expandable network [41] expands the network architecture by adding a predetermined number of parameters and learning with sparse regularization for size optimization. However, this method is not free from the forgetting issue as many parameters of old tasks are updated for the new task.

Structural Allocation. This family of approaches [1,26,27,36] does not suffer from updating old knowledge with a new one. The key is to assign a disjoint subset of parameters to each task. The disjoint sets of old tasks are fixed [1,26,27] or rarely [36] updated while training a new task. This is usually achieved by pruning parameters [27], learning a mask [26], or attention through gradient descent [36]. However, these methods do not discover relationships between tasks, which may cause negative task interference. Recently, [1] has proposed to select an internal network structure with the channel-gating module, but it produces a non-negligible amount of additional parameters unlike ours.

Contributions. We reformulate the catastrophic forgetting problem into a task interference problem under the structural allocation framework. To solve the problem, we introduce a task association strategy through a model search that effectively differentiates helpful and harmful information from old tasks. Since we explore an optimally shrunk subnetwork from an architecture, it is faster than the structural allocation baseline [27] that leverages all previous knowledge (i.e., using the entire architecture). The proposed task selection approach results in performance improvement over its strong competitors [26,27] while consuming a small number of model parameters.

3   Methodology

3.1   Framework

The proposed method aims to enrich the knowledge of a new task from old tasks without absorbing negative information. To this end, we propose a novel approach to whether each old task and its parameters help learn the new task or not. The proposal is a structural allocation method based on a train-prune-retrain framework [27], where a disjoint set of parameters is assigned to each task in a network. Let us denote incoming tasks as $T^1, ..., T^t$, where the t-th task $T^t = \{\mathcal{X}^t, \mathcal{Y}^t\}$ contains data $\mathcal{X}^t = \{x_j^t\}_{j=1}^n$ and corresponding labels $\mathcal{Y}^t = \{y_j^t\}_{j=1}^n$. The first task T^1 is learned using the feature extractor f_{θ^1} and the classifier $g_{w^1}^1$ parameterized by θ^1 and w^1, respectively. For T^1, we allocate the set of task-specific parameters θ^1 to Θ^1 in the feature extractor, i.e., $\Theta^1 = [\theta^1]$. After learning T^1, redundant parameters in θ^1 are pruned away to reserve parameters for the next task, and the set of survived parameters $\tilde{\theta}^1$ squeezed from θ^1 performs T^1.

For the i-th task T^i, we allocate the set of parameters θ^i to the pruned locations in the feature extractor after learning up to T^{i-1}. One way to train parameters θ^i is to utilize the entire set of parameters $\Theta^i = [\tilde{\theta}^1, \ldots, \tilde{\theta}^{i-1}, \theta^i]$ in the forward pass and then the set of parameters θ^i assigned to the ith task is updated in the backward pass [27]:

$$\text{Forward: } f_{\theta_{fwd}^i}(x^i) \text{ where } \theta_{fwd}^i = \Theta^i,$$
$$\text{Backward: } f_{\theta_{bwd}^i}(x^i) \text{ where } \theta_{bwd}^i = \theta^i. \tag{1}$$

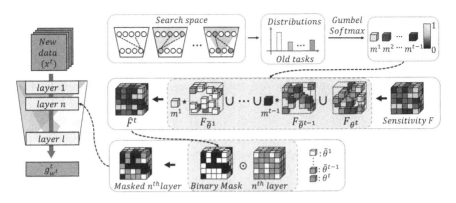

Fig. 1. A graphical illustration of the training process of the proposed method for the t-th task. It searches the old parameters to help optimize the t-th task. The architecture is obtained by the sampled variable m^i. The information from the proposed measure, $\hat{F}^t$, is computed by combining m^i and the sensitivity F. Helpful and harmful parameters are finally identified through a binary mask that is obtained from $\hat{F}^t$. From this, we get a masked layer obtained through the element-wise multiplication of the parameters in each layer and the corresponding mask. Best viewed in color.

After learning θ^i, we apply a prune-retrain approach to get $\tilde{\theta}^i$ and the set of reusable parameters θ^{i+1} for the next task. In summary, we train incoming tasks and sequentially produce disjoint sets of parameters, $\tilde{\theta}^1, \cdots, \tilde{\theta}^{t-1}, \theta^t$ for t tasks.

To discover parameter sets that are critical for the new task t in the search space containing $\{\tilde{\theta}^1, \cdots, \tilde{\theta}^{t-1}\}$, in this work, we discover an association knowledge between tasks based on how each old task affects the new one. Note here that Eq. (1) may not avoid harmful interference between tasks because the forward pass takes the entire parameters from all previous tasks. By selectively using parameters of the previous tasks, which are positively cooperative to the new task, it can mitigate negative task interference. To discover helpful tasks with the corresponding parameters, we first search for an optimal subnetwork by a model search approach at a coarser level. Then, we further explore the parameters within the searched network that are sensitive to the loss of the new task at a finer level. Here, sensitive parameters can be understood as crucial old ones for the new task. By the search mechanism and the sensitivity measure, we can find the most promising architecture for the current task. The entire procedure of the proposal is illustrated in Fig. 1.

3.2 Search

We present a gradient-based search method to find previous tasks that are cooperative to the new task. We first define the search problem for the t-th task in continual learning as

$$\min_{a^t \in \mathcal{A}} \min_{\theta^t} \mathcal{L}(a^t, \theta^t). \tag{2}$$

Given the search space $\mathcal{A}$ containing a pool of candidate networks (corresponding to different combinations of parameters sets), we find an optimal network structure a^t and update the set of parameters θ^t to minimize the loss. Since we aim to find cooperative old tasks for T^t, the search space is 2^{t-1}. As a naïve approach, during training task t, a previous parameter set $\tilde{\theta}^i$ is sampled and leveraged with the probability of

$$P_\alpha(\tilde{\theta}^i) = \mathrm{softmax}(\alpha^i) = \frac{\exp(\alpha^i)}{\sum_{j=1}^{t-1}\exp(\alpha^j)}, \tag{3}$$

where $\alpha = (\alpha^1, \cdots, \alpha^{t-1})$ denotes learnable parameters that determine the sampling probability. Therefore, the output based on the probability over all previous sets becomes

$$\hat{\Theta}^t = \bigcup_{i=1}^{t-1}\left(m^i \cdot \tilde{\theta}^i\right) \cup \theta^t, \tag{4}$$

and it defines the structure a^t. Here, $m^i \in \{0,1\}$ becomes 1 if $\mathrm{softmax}(\alpha^i)$ is greater than pre-determined threshold and 0 otherwise.

However, we may not solve the problem in Eq. (2) using a gradient descent method due to (i) the discrete search space $\mathcal{A}$ and (ii) discontinuous relation between the sampling parameter α^i and m^i. For the discrete optimization problem, we relax it as the following problem:

$$\min_\alpha \min_{\theta^t} \mathbb{E}_{a^t \sim P_\alpha}[\,\mathcal{L}(a^t, \theta^t)\,]. \tag{5}$$

To alleviate the latter problem, we present the Gumbel softmax trick [9,16] that produces the continuous random variable m^i.

$$m^i = \mathrm{GumbelSoftmax}(\alpha^i|\alpha) = \frac{exp((\alpha^i + g^i)/\tau)}{\sum_{j=1}^{t-1} exp((\alpha^j + g^j)/\tau)}, \tag{6}$$

where g^i follows the Gumbel distribution [16] and τ is the control variable. While training, we discover the optimal architecture a^t of the feature extractor for T^t using Eqs. (5) and (6). We optimize the current set of parameters θ^t followed by updating α to select the contributable old parameters.

3.3   Sensitivity Measure

To further explore useful parameters learned from the selected old tasks, we additionally introduce a promising measure that gives a sensitivity of each parameter to the loss of a new task, which can be calculated by the Fisher information [14] as

$$F(\theta) = \sum_n \mathbb{E}_{p_\theta(y|x_n)}[\nabla_\theta \log p_\theta(y|x_n)\nabla_\theta \log p_\theta(y|x_n)^\top], \tag{7}$$

where $\nabla_\theta \log p_\theta(y|x_n)$ denotes the gradient of the log-likelihood at all parameters θ. Fisher information captures the variability of the gradient and tells us which old parameters are sensitive (important) for the current task [17]. Note that while [17] applies Fisher to prevent the update of sensitive old parameters, ours uses them to select helpful parameters.

Let us first express the sensitivity for the ith task. $F_{\tilde{\theta}^i}$ denotes the element-wise Fisher values corresponding to $\tilde{\theta}^i$. For T^t, we gather all the Fisher information up to the task t as follows:

$$F^t(\theta) = \bigcup_{i=1}^{t-1} F_{\tilde{\theta}^i} \cup F_{\theta^t}. \tag{8}$$

Finally, θ^t_{fwd}, exploiting both sampled variable in Eq. (6) and the sensitivity in Eq. (8), is obtained as

$$\theta^t_{fwd} = \bigcup_{i=1}^{t-1} \left(\sigma_\phi(m^i \cdot F_{\tilde{\theta}^i}) \odot \tilde{\theta}^i \right) \cup \theta^t, \tag{9}$$

where $\odot$ is the element-wise multiplication. By applying the threshold function $\sigma_\phi(\cdot)$ to $m^i \cdot F_{\tilde{\theta}^i}$, we discover a final architecture a^t which is used in the forward pass.

3.4 Meta Classifier

To perform the challenging class-incremental learning scenario, where task identity is unknown, we present a meta classifier to predict the task label of each sample and allocate the corresponding network. The meta classifier $h_{\theta^{mc}}(\cdot)$ has a few learnable parameters θ^{mc} but the total parameter increase is negligible. Features are extracted from the exemplar, x^t_e, corresponding to each task. By concatenating the features for t tasks, we compose data $\tilde{x}^t$ to learn the meta classifier, which is obtained by

$$\tilde{x}^t = \bigoplus_{i=1}^t flatten(\hat{a}^i(x^t_e)), \tag{10}$$

where $\bigoplus$ the concatenation operation and $\hat{a}^i$ is the architecture a^i excluding the head classifier $g^i_{w^i}$. Using the collected data, we learn the meta classifier and predict the task identity l to perform the chosen task l with the head classifier $g^l_{w^l}(\cdot)$, where

$$l \leftarrow h_{\theta^{mc}}(\tilde{x}^t), \; l \leq t. \tag{11}$$

To develop the meta classifier that learns discriminative task-wise features, we optimize it with the contrastive loss [7]. Unlike the contrastive loss used in self-supervised learning, which allows two different transformations to apply to an image as a positive pair, we treat samples in the same task as a positive pair. Formally, we compute the loss using positive pairs identified by their task IDs as

$$\mathcal{L}_{con} = \sum_i \sum_{j \neq i} \pi(z_i, z_j)$$

$$\text{where } \pi(z_i, z_j) = -\log \frac{\exp(sim(z_i, z_j)/\gamma)}{\sum_k \mathbb{1}_{[k \neq i]} \exp(sim(z_i, z_k)/\gamma)}, \quad (12)$$

where $\mathbb{1}$, $sim(\cdot)$ and γ are the indicator function, cosine similarity, and temperature parameter, respectively. z is the extracted feature of $h_{\theta^{mc}}$, where z_i and z_j are extracted features from different tasks. The total loss with the cross-entropy loss $\mathcal{L}_{ce}$ for training $h_{\theta^{mc}}$ is

$$\mathcal{L}_{total} = \mathcal{L}_{con} + \lambda \mathcal{L}_{ce}, \quad (13)$$

where λ is a balancing factor between two losses.

4   Experiments

4.1   Datasets

We applied the proposed approach to discover **H**elpful or **H**armful parameters, named $\mathbf{H^2}$, to task-incremental and class-incremental learning scenarios. The datasets used in the task-incremental scenarios include ImageNet, diverse fine-grained datasets, CIFAR-10, and MNIST, where the fine-grained datasets are CUBS [39], Stanford Cars [18], Flowers [29], Wikiart [34], and Sketch [11]. ImageNet and the fine-grained datasets were used altogether in a task-incremental scenario and resized to 224×224 pixels. CIFAR-10 [19] contains $50,000$ training and $10,000$ test samples of the 32×32 pixels. This dataset was divided into five tasks, referred to as Split CIFAR-10, and each task is composed of two classes. MNIST [20] consists of $60,000$ training and $10,000$ test handwritten images of the 28×28 size. The dataset was divided into five tasks, referred to as Split MNIST, similar to Split CIFAR-10.

For class-incremental learning scenarios, we used three datasets: ImageNet-50 [8], Split CIFAR-10, and Split MNIST. The summary of all datasets used in the experiments is shown in Table 1. To construct ImageNet-50, we selected 50 classes randomly from the ImageNet dataset and resized to 32×32, following the practice in [30]. It contains five tasks, each of which has 10 classes and 13,000 images. We also have the same datasets, Split CIFAR-10 and Split MNIST, to task-incremental learning, but their task identities are unknown at test time in class-incremental learning.

Table 1. Datasets used in this work.

Dataset	ImageNet	CUBS	Stanford Cars	Flowers	Wikiart	Sketch	ImageNet-50	CIFAR-10	MNIST
# Train	1,287,167	5,994	8,144	2,040	42,129	16,000	65,000	50,000	60,000
# Test	50,000	5,794	8,041	6,149	10,628	40,000	2,500	10,000	10,000
# Class	1,000	200	196	102	195	250	50	10	10

Table 2. Classification accuracy (Acc, %) and the number of required parameters (Param, $\times 10^6$) on the ImageNet and find-grained datasets. Best results are indicated in bold font.

Method	ImageNet		CUBS		Stanford Cars		Flowers		Wikiart		Sketch		Avg	Avg
	Acc	Param	Acc	Param	Acc	Param	Acc	Param	Acc	Param	Acc	Param	Acc	Param
EWC-On	54.5	23.4	64.1	23.4	52.1	23.4	78.7	23.4	51.4	23.4	30.7	23.4	55.3	23.4
LwF	64.5	23.4	51.7	23.4	43.9	23.4	72.9	23.4	42.7	23.4	45.5	23.4	53.6	23.4
PackNet	75.7	**12.5**	80.4	15.6	86.1	18.0	93.0	19.7	69.4	21.0	76.7	22.0	80.2	18.1
Piggyback	**76.1**	23.4	81.5	20.5	89.6	19.7	94.7	22.3	71.3	16.3	**79.9**	18.0	82.2	20.0
H^2 (Ours)	75.7	**12.5**	**84.1**	**14.5**	**90.6**	**15.9**	**94.9**	**15.2**	**75.1**	**9.4**	76.2	**4.1**	**82.8**	**11.9**

4.2 Implementation Details

We used the ResNet-50 [13] backbone architecture for the task-incremental learning scenario using the ImageNet and fine-grained datasets. The backbone models on Split MNIST and Split CIFAR-10 were a three-layer CNN and ResNet-18, respectively, for both task-incremental and class-incremental learning scenarios. All layers except the last convolutional layer are accompanied by the max pooling operation in the three-layer CNN. We used ResNet-18 for ImageNet-50. The meta classifier used in all class-incremental scenarios consists of three fully connected layers with the hyperparameters γ and λ of 0.5 and 1.6, respectively. The feature z is obtained by passing through two linear layers of the meta-classifier. For large-scale datasets, we resized every image into the size of 224×224 and applied a center crop. We applied random cropping and horizontal flip to augment ImageNet-50 and Split CIFAR-10 except for the Split MNIST dataset. We set the Gumbel softmax parameter τ for all experiments to 1.0. We set the threshold ϕ in Eq. (9) to 0.01 for all the experiments. We trained the backbone and the meta classifier using stochastic gradient descent. We report the average results over five independent runs for all the experiments.

4.3 Task-Incremental Learning

Large-Scale Datasets. We first demonstrated the task association strategy of the proposed method compared with EWC-On [35], LwF [22], PackNet [27] and Piggyback [26]. We applied the compared approaches for six large-scale datasets in order of ImageNet, CUBS, Stanford Cars, Flowers, Wikiart, and Sketch, where we regard each dataset as a task. The scenario is challenging because the datasets are from different domains and rarely share many common classes. For a fair comparison, ours has the same pruning ratio to PackNet for each task.

Table 2 shows the experimental results of the methods on the six datasets. The proposed method, H^2, outperforms the compared approaches for most of the tasks. The regularization approaches, EWC-On and LwF, do not effectively address the large-scale tasks. PackNet does not give satisfactory results, especially for CUBS, Stanford Cars, and Wikiart. This is probably due to negative interference between tasks as the tasks may be irrelevant to one another. Piggyback performs better than PackNet but performs poorer than ours. The results

Table 3. Task-incremental learning results on the Split MNIST and Split CIFAR-10 datasets. We provide the accuracy for each task after training the last task, T^5, where T^i denotes the i-th task. Best results are indicated in bold font.

Method	Split MNIST						Split CIFAR-10					
	T^1	T^2	T^3	T^4	T^5	Avg	T^1	T^2	T^3	T^4	T^5	Avg
Joint	99.9	99.9	99.9	100.0	99.5	99.9	99.6	96.4	97.9	99.8	98.3	98.3
EWC-On	97.1	99.4	93.4	98.2	93.2	96.3	75.8	80.4	80.3	95.2	96.0	85.5
LwF	99.8	97.9	99.7	99.9	98.5	99.2	94.8	87.3	67.1	50.5	51.4	70.2
PackNet	99.7	99.4	99.8	99.8	98.2	99.3	98.8	93.4	95.7	98.7	97.8	96.8
Piggyback	**100.0**	99.6	**100.0**	99.7	97.0	99.2	96.1	85.4	91.8	96.6	96.4	93.3
HAT	99.9	99.6	99.9	99.8	99.0	99.7	98.8	91.1	95.3	98.5	97.8	96.3
TAC	**100.0**	99.4	**100.0**	99.9	99.3	99.7	**99.4**	91.7	95.0	98.3	97.8	96.4
H^2 (Ours)	**100.0**	**99.8**	**100.0**	**100.0**	**99.6**	**99.9**	98.8	**94.1**	**96.5**	**98.8**	**98.3**	**97.3**

show that finding helpful parameters by the proposed measure can alleviate destructive task interference. When it comes to the number of parameters, ours takes a much smaller number of parameters than the compared methods. Overall, the proposed method selects fewer parameters while allowing better performance than the existing approaches, showing its excellence.

Split CIFAR-10 and Split MNIST. We conducted additional task-incremental learning experiments on Split MNIST and Split CIFAR-10, where we construct five incremental tasks for each dataset. We compared with EWC-On [35] and LwF [22] in the regularization category, PackNet [27], Piggyback [26], HAT [36], and TAC [1] in the structural allocation category. We also compared with a joint learning method that learns multiple tasks simultaneously. Note that since Piggyback requires a pretrained backbone, we pretrained a three-layer CNN by collecting 10K data samples from the original MNIST data. For Split CIFAR-10, we used the ImageNet pretrained ResNet-18 backbone for Piggyback.

The results for the Split MNIST are summarized on the left of Table 3. From the table, we can observe that H^2 performs better than other continual learning approaches on average and is almost similar to the joint learning results. The results for Split CIFAR-10 are shown on the right of Table 3. For both experiments, the structural allocation methods consistently perform better than the regularization methods. The proposed method gives higher accuracy than the regularization and other structural allocation methods with a larger margin. Compared to the recently proposed structural allocation method, TAC, ours achieves meaningful performance improvement with a margin of 0.9% for Split CIFAR-10.

4.4   Class-Incremental Learning

ImageNet-50. We evaluated H^2 to the class-incremental learning scenario using ImageNet-50 to compare with two class-incremental learning methods [31,40].

Table 4. Class-incremental learning results on ImageNet-50 in terms of the exemplar size.

Exemplar size	65,000	2,000					5,000				
Method	Joint	iCaRL	BiC	TAC	PackNet	H^2	iCaRL	BiC	TAC	PackNet	H^2
Acc (%)	58.7	43.6	21.2	40.4	48.0	**49.7**	47.3	35.2	42.9	48.2	**55.5**

While task-incremental learning knows task ID, class-incremental learning predicts it through the meta classifier described in Sect. 3.4. The backbone architecture for this scenario was ResNet-18. The dataset consists of five tasks of 10 classes each. For the scenario, the proposed method utilizes exemplars to learn the meta classifier and is compared with other replay-based methods, iCaRL [31] and BiC [40]. We also compared with two structural allocatioin methods, PackNet [27] and TAC [1]. Because PackNet does not address class-incremental learning, we applied the same meta classifier. Even if we make use of exemplars in class-incremental learning, we use them when we train the meta-classifier. Fine-tuning with a small number of exemplars can distort the knowledge learned on complete data of the previous tasks. We also compared H^2 with a joint learning method that has full accessibility to the previous data. We report the average accuracy of the learned tasks after learning the last task.

Table 4 shows the experimental results under two exemplar sizes. The proposed method yields better results than the replay-based methods. The proposed method achieves 6.1%, 28.5%, 1.7% and 9.3% higher accuracy than iCaRL, BiC, PackNet, and TAC, respectively, for the exemplar size of 2,000. For the exemplar size of 5,000, ours achieves more significant performance improvement over those methods with the margins of 8.2%, 20.3%, 5.8%, and 12.6%, respectively. The results show that ours is more efficient in terms of accuracy under the same memory sizes than the competitors for class-incremental learning.

Split CIFAR-10 and Split MNIST. We also conducted additional class-incremental learning scenarios on the Split CIFAR-10 and Split MNIST datasets, respectively. In this scenario, we compared H^2 with the same approaches in the previous experiment. A three-layer CNN was used for Split MNIST, and ResNet-18 was used for Split CIFAR-10. We show the results under different exemplar sizes, from small to large, obtained after learning all tasks.

Figure 2 (left) shows the results of the compared methods for Split MNIST. H^2 gives higher accuracy than other approaches for most exemplar sizes. The structural allocation methods, TAC and PackNet, perform better than the replay methods, iCaRL and BiC, because they do not update old parameter sets. However, these methods perform poorer than the proposed approach using the search mechanism for task association. The performance gap between the proposed method and the joint learning baseline decreases as the number of exemplars increases. Figure 2 (right) shows the results of the methods on Split CIFAR-10. Notably, ours outperforms other compared approaches for all of the exemplar sizes. As the number of stored exemplars increases, the gap between H^2 and

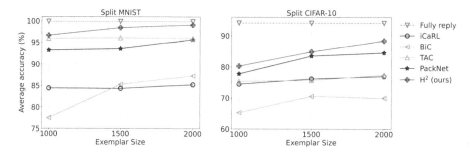

Fig. 2. Class-incremental learning results on the Split MNIST and Split CIFAR-10 datasets under different exemplar sizes.

TAC increases (4.9%, 9.2%, and 10.9% for 1,000, 1,500, and 2,000 exemplars, respectively). Similarly, ours outperforms PackNet equipped with the presented meta classifier for all exemplar sizes.

4.5   Analysis

Ablation Study. To demonstrate the efficiency of the proposed approach, we conducted an ablation study. We compared ours by removing the search method, the sensitivity measure, and both strategies, i.e., using all parameters [27] under the same learning framework as ours. In addition, we compared ours by removing the contrastive loss to show the effect on training the meta classifier. We also provide the results of the random selection, which chooses the parameters of the previous tasks randomly with the same ratio as the proposed method for a fair comparison. For the experiment, we conducted both task-incremental and class-incremental learning scenarios using Split CIFAR-10. We report the average accuracy of the tasks learned each time step.

Table 5 summarizes the ablation study. The proposed method of searching cooperative tasks and parameters gives higher accuracy than other strategies for both learning scenarios. Random selection performs poorly for most of the tasks in the scenarios, indicating that selectively incorporating helpful parameters from old tasks improves performance and arbitrarily selecting old parameters worsens the performance. The method using all parameters without search and sensitivity shows disappointing results (0.5% and 2.9% margins for task-incremental and class-incremental learning, respectively) compared to ours. This reveals that some parameters influence the current task negatively. We find that applying both approaches is the most promising way and is better than applying the individual method. Note that the performance gap is clearly shown in class-incremental learning that is more challenging than task-incremental learning.

We also show the parameter consumption, FLOPs, and memory requirement of the proposed method (except the meta classifier) in Fig. 3. We compute parameter consumption and FLOPs following the practice of *ShrinkBench*[1]. Since the

---

[1] https://github.com/jjgo/shrinkbench.

Table 5. Ablation study of the proposed method for Split CIFAR-10.

Method	Task-incremental					Class-incremental				
	T^1	T^2	T^3	T^4	T^5	T^1	T^2	T^3	T^4	T^5
Ours	**98.8**	**96.4**	**96.4**	**97.1**	**97.3**	**98.8**	**91.2**	**87.4**	**86.1**	**80.3**
w/o search	**98.8**	96.1	96.0	96.7	97.0	**98.8**	89.5	84.9	84.5	79.6
w/o sensitivity	**98.8**	96.2	96.1	96.7	96.9	**98.8**	90.0	83.8	82.6	78.7
w/o search and sensitivity [27]	**98.8**	96.1	95.9	96.6	96.8	**98.8**	89.3	83.5	81.1	77.4
w/o contrastive loss			N/A			**98.8**	90.7	86.7	83.2	77.1
Random search	**98.8**	95.1	95.1	95.9	96.3	**98.8**	86.3	80.8	77.9	68.7

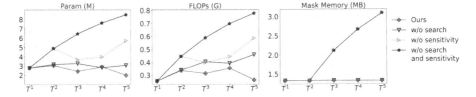

Fig. 3. Ablation study on parameter consumption, FLOPs, and memory of the proposed method for Split CIFAR-10. Best viewed in color.

random selection method has the same number of parameters as the proposal, it was excluded from the experimental results. The proposed method takes the smallest number of parameters and FLOPs to perform tasks on average under the same threshold. The proposed methods without search and sensitivity consume a larger number of parameters than the proposal. Especially, the amount of parameters consumed to perform a task continues to increase ($4.2\times$ higher than ours for T^5) when we exclude the search and sensitivity approach, as shown on the left of the figure. The result shows that ours containing search and sensitivity makes it highly efficient for parameter consumption. In addition, we compute the memory of the mask to perform each task. Since the proposed method and the method without search use a binary mask, the memory for performing each task is constant. For example, when the backbone network has n parameters, the binary mask requires n bits. However, the memory of the mask to perform each task generally continues to increase for the other methods.

Task Order. To analyze the performance of the proposed method with respect to different task orders, we conducted an additional experiment using the ImageNet-50 (I) and Split CIFAR-10 (C) datasets, respectively. Specifically, the datasets were divided into five tasks as done in Sects. 4.3 and 4.4, respectively. In this study, we performed both task- and class-incremental learning scenarios. We produced ten random sequences for each scenario and evaluated the compared methods using the average result. We used 1,000 and 2,000 exemplars for Split CIFAR-10 and ImageNet-50, respectively, in class-incremental learning.

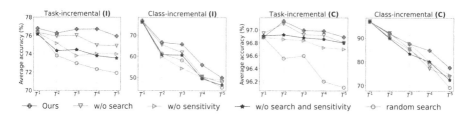

Fig. 4. Ablation study of the proposed method with respect to different orders of tasks for ImageNet-50 (I) and Split CIFAR-10 (C). Best viewed in color.

Figure 4 reports the average accuracies of the tasks trained up to each time step. Overall, the results on ImageNet-50 and Split CIFAR-10 show a similar trend to those in Table 5. In Fig. 4 (left two subfigures), the proposed method outperforms other strategies without search, sensitivity, or both, by large margins of 1.06%, 1.97%, and 2.42%, respectively, in the task-incremental scenario. Similarly, the proposed method outperforms other methods for all time steps on average in the class-incremental scenario. The experiments using Split CIFAR-10 are shown in Fig. 4 (right two subfigures). The results show a similar trend to those on ImageNet-50. Note that compared to the results for task-incremental learning, the class-incremental learning performance degrades significantly due to its difficulty without knowing the task oracle.

5   Conclusions

We have proposed a novel approach to find an optimal architecture that exploits useful knowledge from old tasks in structural allocation-based continual learning. The proposed method simultaneously associates cooperative tasks and explores the optimal architecture by a model search with continuous relaxation. We have also presented element-wise parameter selection from the sensitivity measure incorporated with the model search. The proposed method takes fewer parameters than its competitors while avoiding negative interference. In experiments, we have compared with existing methods for task- and class-incremental learning scenarios using small to large-scale datasets. Experimental results show that the proposed method achieves excellent performance over diverse continual learning competitors while minimizing the computation cost.

Acknowledgement. This work was supported in part by Samsung Research Funding & Incubation Center of Samsung Electronics under Project Number SRFC-IT2002-05, and in part by Institute of Information & communications Technology Planning & Evaluation (IITP) grant funded by the Korea government(MSIT) (No. 2022-0-00124, Development of Artificial Intelligence Technology for Self-Improving Competency-Aware Learning Capabilities and No. 2021-0-01341, Artificial Intelligence Graduate School Program (Chung-Ang University)).

References

1. Abati, D., Tomczak, J., Blankevoort, T., Calderara, S., Cucchiara, R., Bejnordi, B.E.: Conditional channel gated networks for task-aware continual learning. In: Proceedings of the IEEE/CVF Conference on Computer Vision and Pattern Recognition, pp. 3931–3940 (2020)
2. Aljundi, R., Babiloni, F., Elhoseiny, M., Rohrbach, M., Tuytelaars, T.: Memory aware synapses: learning what (not) to forget. In: Ferrari, V., Hebert, M., Sminchisescu, C., Weiss, Y. (eds.) ECCV 2018. LNCS, vol. 11207, pp. 144–161. Springer, Cham (2018). https://doi.org/10.1007/978-3-030-01219-9_9
3. Aljundi, R., Chakravarty, P., Tuytelaars, T.: Expert gate: lifelong learning with a network of experts. In: Proceedings of the IEEE Conference on Computer Vision and Pattern Recognition, pp. 3366–3375 (2017)
4. Belouadah, E., Popescu, A., Kanellos, I.: A comprehensive study of class incremental learning algorithms for visual tasks. Neural Netw. **135**, 38–54 (2020)
5. Chaudhry, A., Dokania, P.K., Ajanthan, T., Torr, P.H.S.: Riemannian walk for incremental learning: understanding forgetting and intransigence. In: Ferrari, V., Hebert, M., Sminchisescu, C., Weiss, Y. (eds.) ECCV 2018. LNCS, vol. 11215, pp. 556–572. Springer, Cham (2018). https://doi.org/10.1007/978-3-030-01252-6_33
6. Chaudhry, A., Marc'Aurelio, R., Rohrbach, M., Elhoseiny, M.: Efficient lifelong learning with a-gem. In: 7th International Conference on Learning Representations, ICLR 2019. International Conference on Learning Representations, ICLR (2019)
7. Chen, T., Kornblith, S., Norouzi, M., Hinton, G.: A simple framework for contrastive learning of visual representations. In: International conference on machine learning, pp. 1597–1607. PMLR (2020)
8. Deng, J., Dong, W., Socher, R., Li, L.J., Li, K., Fei-Fei, L.: ImageNet: a large-scale hierarchical image database. In: 2009 IEEE Conference on Computer Vision and Pattern Recognition, pp. 248–255 (2009)
9. Dong, X., Yang, Y.: Network pruning via transformable architecture search. Adv. Neural Inf. Process. Syst. **32** (2019)
10. Douillard, A., Cord, M., Ollion, C., Robert, T., Valle, E.: PODNet: pooled outputs distillation for small-tasks incremental learning. In: Vedaldi, A., Bischof, H., Brox, T., Frahm, J.-M. (eds.) ECCV 2020. LNCS, vol. 12365, pp. 86–102. Springer, Cham (2020). https://doi.org/10.1007/978-3-030-58565-5_6
11. Eitz, M., Hays, J., Alexa, M.: How do humans sketch objects? ACM Trans. Graph. **31**(4), 1–10 (2012)
12. Farquhar, S., Gal, Y.: Towards robust evaluations of continual learning. arXiv preprint arXiv:1805.09733 (2018)
13. He, K., Zhang, X., Ren, S., Sun, J.: Deep residual learning for image recognition. In: 2016 IEEE Conference on Computer Vision and Pattern Recognition, pp. 770–778 (2016)
14. Hecht-Nielsen, R.: Theory of the backpropagation neural network. In: Neural Networks for Perception, pp. 65–93. Elsevier (1992)
15. Hou, S., Pan, X., Loy, C.C., Wang, Z., Lin, D.: Learning a unified classifier incrementally via rebalancing. In: Proceedings of the IEEE Conference on Computer Vision and Pattern Recognition, pp. 831–839 (2019)
16. Jang, E., Gu, S., Poole, B.: Categorical reparameterization with Gumbel-Softmax. In: 5th International Conference on Learning Representations, ICLR (2017)
17. Kirkpatrick, J., et al.: Overcoming catastrophic forgetting in neural networks. Proc. Natl. Acad. Sci. U.S.A. **114**(13), 3521–3526 (2017)

18. Krause, J., Stark, M., Deng, J., Fei-Fei, L.: 3D object representations for fine-grained categorization. In: 2013 IEEE International Conference on Computer Vision Workshops, pp. 554–561. IEEE (2013)
19. Krizhevsky, A., Hinton, G.: Learning multiple layers of features from tiny images (2009)
20. LeCun, Y.: The mnist database of handwritten digits (1998). http://yann.lecun.com/exdb/mnist/
21. Lee, S.W., Kim, J.H., Jun, J., Ha, J.W., Zhang, B.T.: Overcoming catastrophic forgetting by incremental moment matching. Adv. Neural. Inf. Process. Syst. **30**, 4652–4662 (2017)
22. Li, Z., Hoiem, D.: Learning without forgetting. In: Leibe, B., Matas, J., Sebe, N., Welling, M. (eds.) ECCV 2016. LNCS, vol. 9908, pp. 614–629. Springer, Cham (2016). https://doi.org/10.1007/978-3-319-46493-0_37
23. Liu, Y., Schiele, B., Sun, Q.: Adaptive aggregation networks for class-incremental learning. In: Proceedings of the IEEE/CVF Conference on Computer Vision and Pattern Recognition, pp. 2544–2553 (2021)
24. Liu, Y., Su, Y., Liu, A.A., Schiele, B., Sun, Q.: Mnemonics training: multi-class incremental learning without forgetting. In: Proceedings of the IEEE/CVF conference on Computer Vision and Pattern Recognition, pp. 12245–12254 (2020)
25. Lopez-Paz, D., Ranzato, M.: Gradient episodic memory for continual learning. In: Proceedings of the 31st International Conference on Neural Information Processing Systems, pp. 6470–6479 (2017)
26. Mallya, A., Davis, D., Lazebnik, S.: Piggyback: adapting a single network to multiple tasks by learning to mask weights. In: Proceedings of the European Conference on Computer Vision, pp. 67–82 (2018)
27. Mallya, A., Lazebnik, S.: Packnet: adding multiple tasks to a single network by iterative pruning. In: Proceedings of the IEEE Conference on Computer Vision and Pattern Recognition, pp. 7765–7773 (2018)
28. McCloskey, M., Cohen, N.J.: Catastrophic interference in connectionist networks: the sequential learning problem. In: Psychology of learning and motivation, vol. 24, pp. 109–165. Elsevier (1989)
29. Nilsback, M.E., Zisserman, A.: Automated flower classification over a large number of classes. In: 2008 Sixth Indian Conference on Computer Vision, Graphics & Image Processing, pp. 722–729. IEEE (2008)
30. Ostapenko, O., Puscas, M., Klein, T., Jahnichen, P., Nabi, M.: Learning to remember: a synaptic plasticity driven framework for continual learning. In: Proceedings of the IEEE Conference on Computer Vision and Pattern Recognition, pp. 11321–11329 (2019)
31. Rebuffi, S.A., Kolesnikov, A., Sperl, G., Lampert, C.H.: iCaRL: incremental classifier and representation learning. In: Proceedings of the IEEE conference on Computer Vision and Pattern Recognition, pp. 2001–2010 (2017)
32. Rosenfeld, A., Tsotsos, J.: Incremental learning through deep adaptation. IEEE Trans. Pattern Anal. Mach. Intell. **42**(3), 651–663 (2018)
33. Rusu, A.A., et al.: Progressive neural networks. arXiv preprint arXiv:1606.04671 (2016)
34. Saleh, B., Elgammal, A.: Large-scale classification of fine-art paintings: learning the right metric on the right feature. Int. J. Digital Art Hist. (2) (2016)
35. Schwarz, J., et al.: Progress & compress: a scalable framework for continual learning. In: International Conference on Machine Learning, pp. 4528–4537. PMLR (2018)

36. Serra, J., Suris, D., Miron, M., Karatzoglou, A.: Overcoming catastrophic forgetting with hard attention to the task. In: International Conference on Machine Learning, pp. 4548–4557. PMLR (2018)
37. Shin, H., Lee, J.K., Kim, J., Kim, J.: Continual learning with deep generative replay. In: Proceedings of the 31st International Conference on Neural Information Processing Systems, pp. 2994–3003 (2017)
38. Tu, C.H., Wu, C.E., Chen, C.S.: Extending conditional convolution structures for enhancing multitasking continual learning. In: 2020 Asia-Pacific Signal and Information Processing Association Annual Summit and Conference (APSIPA ASC), pp. 1605–1610. IEEE (2020)
39. Wah, C., Branson, S., Welinder, P., Perona, P., Belongie, S.: The caltech-ucsd birds-200-2011 dataset (2011)
40. Wu, Y., et al.: Large scale incremental learning. In: Proceedings of the IEEE Conference on Computer Vision and Pattern Recognition, pp. 374–382 (2019)
41. Yoon, J., Yang, E., Lee, J., Hwang, S.J.: Lifelong learning with dynamically expandable networks. In: International Conference on Learning Representations (2018)
42. Zenke, F., Poole, B., Ganguli, S.: Continual learning through synaptic intelligence. In: Proceedings of the 34th International Conference on Machine Learning-Volume 70, pp. 3987–3995 (2017)

Towards Accurate Binary Neural Networks via Modeling Contextual Dependencies

Xingrun Xing[1], Yangguang Li[2], Wei Li[3], Wenrui Ding[1], Yalong Jiang[1(✉)],
Yufeng Wang[1(✉)], Jing Shao[2], Chunlei Liu[1], and Xianglong Liu[1]

[1] Beihang University, Beijing, China
{sy2002215,ding,allenyljiang,wyfeng,liuchunlei,xlliu}@buaa.edu.cn
[2] SenseTime Group, Hong Kong, China
liyangguang@sensetime.com, shaojing@senseauto.com
[3] Nanyang Technological University, Jurong West, Singapore
wei.l@ntu.edu.sg

Abstract. Existing Binary Neural Networks (BNNs) mainly operate on local convolutions with binarization function. However, such simple bit operations lack the ability of modeling contextual dependencies, which is critical for learning discriminative deep representations in vision models. In this work, we tackle this issue by presenting new designs of binary neural modules, which enables BNNs to learn effective contextual dependencies. First, we propose a binary multi-layer perceptron (MLP) block as an alternative to binary convolution blocks to directly model contextual dependencies. Both short-range and long-range feature dependencies are modeled by binary MLPs, where the former provides local inductive bias and the latter breaks limited receptive field in binary convolutions. Second, to improve the robustness of binary models with contextual dependencies, we compute the contextual dynamic embeddings to determine the binarization thresholds in general binary convolutional blocks. Armed with our binary MLP blocks and improved binary convolution, we build the BNNs with explicit Contextual Dependency modeling, termed as BCD-Net. On the standard ImageNet-1K classification benchmark, the BCD-Net achieves 72.3% Top-1 accuracy and outperforms leading binary methods by a large margin. In particular, the proposed BCDNet exceeds the state-of-the-art ReActNet-A by 2.9% Top-1 accuracy with similar operations. Our code is available at https://github.com/Sense-GVT/BCDNet.

Keywords: Binary neural network · Contextual dependency · Binary MLP

1 Introduction

Over the last decade, deep learning methods have shown impressive results for a multitude of computer vision tasks. However, these models require massive

Supplementary Information The online version contains supplementary material available at https://doi.org/10.1007/978-3-031-20083-0_32.

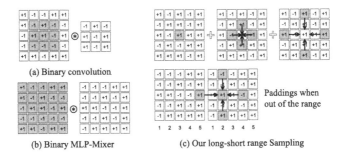

(a) Binary convolution

(b) Binary MLP-Mixer

(c) Our long-short range Sampling

Paddings when
out of the range

Fig. 1. Comparisons of binary operations: (a) The binary convolutions have inductive bias but limited local perceptions; (b) The binary token-mixing MLPs are sharing contextual perceptions but difficult to optimize; (c) Our proposed binary MLPs achieve inductive bias in short-range and explore long-range dependencies concurrently.

parameters and computation to achieve strong performance, hindering their application in practical embedded devices with limited storage and computing resources. Aiming at reduce computational cost in deep models, model compression has drawn growing interests and developed into various methods, such as quantization [14,19,46], pruning [11,41], knowledge distillation [9,33], etc. Binary neural networks (BNNs) [10,22] can be viewed as an extreme case of low-bit quantization and become one of the most prevailing approaches to save computational cost at an extremely high rate. BNNs binarize the weights and activations to save at most $64\times$ operations and $32\times$ memory. However, current BNNs for vision tasks are built upon simple local convolutions, contributing to limited perception field in each layer. As such, it is of significant interest to explore more efficient contextual binary operations for BNNs.

Despite the advantage of inductive bias (e.g. locality) of convolutions for vision tasks, recent studies have shown improving interests in contextual operations, including the vision transformers (ViTs) and MLPs, that achieves state-of-the-art performance. Different from local convolutions, ViTs and MLPs treat images as tokens and model the dependencies across all tokens at once. To explore the binary contextual interaction in BNNs, we start by implementing a vanilla binary MLP-Mixer model, which we find is difficult to optimize compared with traditional binary convolutions. The model also fails to improve BNN accuracy with the help of contextual dependencies. This suggests local inductive bias is more crucial for BNNs compared with real-valued models.

Motivated by this, we aim to take advantage of contextual information while maintaining local inductive bias concurrently. A long-short range binary MLP operation is first proposed as an alternative to binary convolutions, where the short-range and long-range dependencies are introduced upon the standard binary convolutions, as shown in Fig. 1(c). We indicate the original location modeling as pointwise branch. The pointwise and short-range dependencies provide inductive bias similar to convolutions, meanwhile we break the local perceptions with long-range modeling. In summary, the three modeling patterns jointly

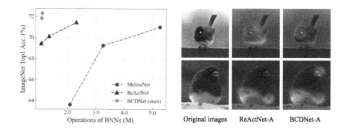

Fig. 2. The performance of binary sparse MLPs express state-of-the-art binary CNNs; class activation maps (CAM) suggest that contextual binary MLPs are better to detect salient features, for example the 'beak' for the 'bird' class.

perform a contextual interaction at each location. In this way, the proposed long-short range binary MLPs take the essence of convolutions and MLP-Mixer while alleviate their problems (Fig. 1 (a) and (b)).

Furthermore, we leverage the proposed long-short-range binary MLP blocks to replace some of the last convolutional blocks to improve the contextual perception field of BNNs while maintaining almost the same calculation. In addition, to access global perceptions in general binary convolutions, we further propose to compute the contextual dynamic embeddings to determine binarization thresholds. In binary convolutional blocks, we first aggregate global features with global average pooling, and then transform pooled features as global embeddings, which adapts binarization at inference. With the binary contextual MLPs and dynamic convolutions, we present a new binary networks named BCDNet to enhance the **B**inary **C**ontextual **D**ependencies that is deficient in previous binary CNNs. As shown in Fig. 2, with similar architecture and computational cost, BCDNets achieves 2.9% accuracy gain against the state-of-the-art ReActNet-A model [22]. It also detects more discriminative features in the contextual range.

Our contributions are summarized as follows:

- We make the first attempt to break local binary convolutions and explore more efficient contextual binary operations.
- We design a long-short range binary MLP module to enhance modeling contextual dependencies in BNNs.
- To our best knowledge, the proposed BCDNet reports the best BNN performance on the large-scale ImageNet dataset currently.

2   Related Work

2.1   Binary Neural Networks

1-bit neural networks are first introduced by BNN [4] but encounters severe quantity error, especially in large scale datasets. To improve performance, some works focus to optimize binary weights. For example, XNOR-Net [30] reduce binarization error with scaling factors; RBNN [16] learns rotation matrixes to reduce angular bias. SA-BNN [18] reduces flips in training; AdamBNN [21] explores influence of optimizers. Others try to modify architectures to improve expression ability. For instance, Bi-Real Net [23] adds more skip connections and achieves better performance; ReActNet [22] employs channel shift and the MobileNet-V1 [8] architecture; CP-NAS [24] adopts architecture search to explore connections. However, previous works only focus on CNN-based BNNs and ignore the performance upper-bound of binary convolutions. This work provides MLP based BNNs and exceed state of the arts.

2.2 Vision Transformers and MLPs

Recently, transformers [5,15,47] are widely used to explore contextual dependencies. ViTs [6] first define to represent an image as 16×16 tokens using an embedding layer, and then stack self-attention and MLP layers. To improve performance of ViTs, DeiTs [36] propose a distillation token which benefits transformers a lot. More recently, Swin transformers [20], Twins [3] and NesT [45] explore various self-attentions to save computational cost, which makes transformers become the first choice in vision tasks. At the same time, works such as MLP-Mixers [34], ResMLP [35] simplify transformers using spatial and channel MLPs, but encounter overfeating. To this end, works such as S^2MLPs [43], CycleMLPs [2] adopt surrounding shift operations to replace token-mixing MLPs. However, these works only focus on shifts and full precision circumstances. For BNNs, there still lacks in efficient contextual operations.

3 Background

In this section, we first review the binary convolution in recent binary convolutional neural networks [4,22,30]. Then, we follow the design in the binary convolution and make an initial attempt to binarize the MLP layer in vision MLP models [2,34,35,43]. We further conduct pilot studies to evaluate the performance of a binarized MLP-Mixer [34] and a ResNet+MLP architecture on ImageNet.

3.1 Binary Operations

Binary Convolution. By replacing arithmetic operations with more efficient bitwise operations, binary convolutions can drastically reduce memory allocation and accesses. Typically, full precision weights $\boldsymbol{W}$ and activations $\boldsymbol{A}$ are first binarized as $\boldsymbol{W}_b$ and $\boldsymbol{A}_b$ using the quantization function $Q_b(\cdot)$. As such, convolutions can be implemented as XNOR-Bitcounting operations [4,30]:

$$\texttt{BiConv}(\boldsymbol{A}) = \alpha Q_b(\boldsymbol{W})^T Q_b(\boldsymbol{A}) = \frac{\|\boldsymbol{W}\|_{\ell_1}}{k \times k \times c_{in}} \texttt{bitcount}(\boldsymbol{W}_b{}^T \oplus \boldsymbol{A}_b), \quad (1)$$

where k and c_{in} indicate the kernel size and input channels respectively, and α serves as the scale factor $\frac{\|\boldsymbol{W}\|_{\ell_1}}{k \times k \times c_{in}}$. Following [23], the quantization function $Q_b(\cdot)$ can be implemented in the form of a differentiable polynomial function:

$$Q_b(x) = \begin{cases} -1 & \text{if } x < -1 \\ 2x + x^2 & \text{if } -1 \leqslant x < 0 \\ 2x - x^2 & \text{if } 0 \leqslant x < 1 \\ 1 & \text{otherwise} \end{cases}, \quad \frac{\partial Q_b(x)}{\partial x} = \begin{cases} 2 + 2x & \text{if } -1 \leqslant x < 0 \\ 2 - 2x & \text{if } 0 \leqslant x < 1 \\ 0 & \text{otherwise} \end{cases}, \quad (2)$$

where gradients are approximated during back-propagation to optimize model weights W. Compared with real-valued convolutions with the 32-bit weight parameters, binary convolutions output 1-bit representations and can obtain up to 32× memory and 64× operation savings, leading to much faster test-time inference and lower power consumption.

Binary MLP. The MLP layer conducts regular matrix multiplications, which can be treated as 1×1 convolution that are repeatedly applied across either spatial locations or feature channels. For spatial dimensions, A single token-mixing MLP layer can be binarized in the form of:

$$\texttt{TM-BiMLP}(A) = \frac{\|W\|_{\ell_1}}{h \times w} \texttt{bitcount}(W_b^T \oplus A_b), \quad (3)$$

where h and w indicate the height and width of input features. Next, we binarize MLP-Mixer-S as shown in Fig. 3 (left). For binary MLP-Mixer, one MLP block is formulated by 2 binarized token-mixing MLP layers and 2 binarized token-wise MLP layers, and the expending ratio is 4. Later, we further explore a unified architecture composed of binary convolutions and MLPs. As shown in Fig. 3 (middle), our initial attempt is to replace the last 3 binary convolution blocks as binarized MLP-Mixer blocks according to similar operations, which aims at modeling contextual dependencies in high level. Because MLP-Mixer blocks have large expending ratio and multiple feed-forward layers, the binarized MLP-Mixer block presents similar or larger model size than the binary convolution.

3.2   Analysis

In order to evaluate the performance of token-mixing binary MLP layers introduced above, we binarize operations in MLP-Mixer-S [34] except the first and last layers. For comparison, we also evaluate binary ResNet-18 implemented by ReActNet and a binary ResNet+MLP architecture mentioned above, where the convolutional and MLP blocks are implemented by the ReActNet and binarized MLP-Mixer ones respectively. We train binary models on the public ILSVRC2021 ImageNet dataset [31] for 100 epochs. Complete training configurations appear in the Supplementary A.

Summary. As shown in Fig. 3, although MLP-Mixers achieves larger perception field than CNNs, the binary version cannot even achieve 40% top1 accuracy, which is worse than the first BNN model [4]. Compared with the binary MLPs that directly omit inductive bias, binary convolutions achieve best accuracy with the help of local perceptions and transformation invariance. Moreover, in the unified architecture of binary convolutions and MLPs, binary token-mixing layers

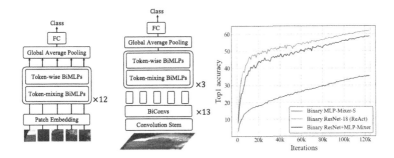

Fig. 3. Initial implementations of binary MLP-Mixer (left) and a binary ResNet adding MLP architecture (middle). The comparison of performance with binary MLP-Mixer-S, ResNet-18 and ResNet+MLP is shown on the right.

still can not benefit from contextual perceptions, but much easier to train than stand-alone binary MLPs. Here we draw the following analysis:

- In the design of contextual binary operations, one need consider the inductive bias sufficiently. However, only preventing local dependencies in binary convolutions also barriers BNNs to capture long-range interactions. To this end, comprehensive design of local and contextual modeling is desired.
- In the design of BNN architectures, binary convolution stage is necessary, as stand-alone binary MLP-Mixers are difficult to optimize.

4   Method

Our proposed binary model is built on modeling contextual dependencies. An overview of the BCDNet architecture is illustrate in Fig. 4. We first describe the fundamentals of binary MLP block. Then, we introduce the contextual design of binary convolutions with dynamic embeddings. Finally, we present the complete BCDNet architecture with modeling contextual dependencies.

4.1   Binary MLP Block with Long-Short Range Dependencies

We first introduce three basic components in our binary MLP blocks, including a long-short range sampling scheme, a token reconstruction function and binary MLP operations, and then indicate the overall binary MLP block.

Long-Short Range Sampling. As shown in Fig. 1(c), we design a comprehensive sampling function which consists of three sampling patterns: the pointwise, short-range and long-range samplings. First, pointwise and short-range tokens provide enough local inductive bias like convolutions. Second, long-range tokens are responding to explore contextual perceptions like self-attentions and token-mixing MLPs. In every block, one pointwise, four short-range, and four

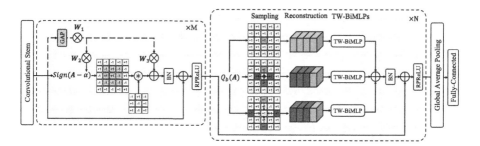

Fig. 4. Overall architecture of BCDNet, which consists of M binary convolution blocks and N binary MLP blocks. In practice, $M = 11$ and $N = 9$.

long-range tokens are sampled at each location, which compose a sparse contextual sampling. Given a sampling distance (r_1, r_2) in spatial, we use a sampling function to indicate the index of a token shifted from a location (x_1, x_2):

$$S(r_1, r_2) = \{y : y = (x_1 + r_1, x_2 + r_2)\}. \tag{4}$$

In each location, we donate sampled long-range, short-range and the pointwise tokens as $\{S(0, \pm h/2), S(\pm w/2, 0)\}$, $\{S(0, \pm 1), S(\pm 1, 0)\}$ and $S(0, 0)$ respectively. Note that, when an index comes out of range, we just pad features from the other side and regard the index as recurrent (shown in Fig. 1 (c)).

Token Reconstruction Across Samplings. Based on sampled tokens above, one simple way to aggregate them is concatenation. However, this will increase the channel numbers and calculations. To save calculations, we select and concatenate parts of channels of sampled tokens similar to [38], which keeps channels unchanged in reconstructed tokens. We choose the spatial shift operation [38] to reconstruct four sampled long-range tokens or four short-range tokens respectively, which is zero parameters and FLOPs. In each location, we can reconstruct a new short-range token A_b^S from parts of four sampled short-range tokens:

$$A_b^S = Cat\{A_b[0 : c/4]_{S(-1,0)}, A_b[c/4 : c/2]_{S(1,0)},$$
$$A_b[c/2 : 3c/4]_{S(0,-1)}, A_b[3c/4 : c]_{S(0,1)}\}, \tag{5}$$

where A_b is a binarized input token, $S(r_1, r_2)$ indexes its location, and $[:]$ indexes its channels. Similarly, a new long-range token A_b^L is reconstructed from parts of four long-range tokens:

$$A_b^L = Cat\{A_b[0 : c/4]_{S(-h/2,0)}, A_b[c/4 : c/2]_{S(h/2,0)},$$
$$A_b[c/2 : 3c/4]_{S(0,-w/2)}, A_b[3c/4, c]_{S(0,w/2)}\}. \tag{6}$$

Here, we set the short-range interaction distance as one token and long-range as h/2 or w/2 tokens in space. In summary, we can obtain an pointwise token, a reconstructed short-range token and a long-range token (A_b, A_b^S, A_b^L) for each location, so that a sparse contextual token sampling is formulated.

Token-Wise Binary MLP. In the following, we model three kinds of sampled tokens (A_b, A_b^S, A_b^L) independently using three token-wise binary MLPs. In MLP blocks, We follow ReActNets [22] and apply $Sign(.)$ and $RSign(.)$ to binarize weights and activations respectively. Given binarized weights W_b and activations A_b, the token-wise binary MLP is defined as a binarized fully connected layer across the channel dimension:

$$\text{TW-BiMLP}(A) = \frac{\|W\|_{\ell_1}}{c_{in}} \text{popcount}(W_b^T \oplus A_b), \tag{7}$$

where $\|W\|_{\ell_1}$ indicates the L1 norm of real valued weights across the channel dimension. Similar to XNOR-Net [30], we compute a scaling factor $\frac{\|W_b\|_{\ell_1}}{c_{in}}$ to minimize binarization error in weights. Next, the binary matrix product can be implemented by XNOR and bit-counting. The difference between Eq. 3 and 7 is fully connected dimensions.

Binary MLP Block. To explore block-wise contextual interactions, we propose the long-short range binary MLP block as an alternative to binary convolutional residual blocks in a network. Different from real-valued MLP-Mixers that aggregate all tokens indistinctively, our primary design is independently sampling pointwise, short-range, and long-range tokens according to a long-short sparse sampling rule. After a token reconstruction function to downsample channels, we use three token-wise binary MLPs to model three kinds of dependencies comprehensively. As shown in Fig. 4, we first binarize activations using RSign(). In every location, three binary MLPs $\{P, S, L\}$ independently model three kinds of dependencies and are then added together:

$$A^i = P\left(A_b^{i-1}\right) + S\left(A_b^S\right) + L\left(A_b^L\right), \tag{8}$$

where $\{A_b, A_b^S, A_b^L\}$ indicates pointwise binary tokens, reconstructed short-range tokens and long-range tokens respectively. Like residual blocks, batch norm, skip connection and activation functions are also attached as in Fig. 4.

When the channel size is large (e.g., 512 and 1024), a binary MLP block consumes about 1/3 operations compared with a typical 3×3 binary convolution with the same input and output channels. Our binary MLPs can be easily adopted as stand-alone or plug-in building blocks in BNNs With similar configurations of regular layers.

4.2   Binary Convolutions with Dynamic Contextual Embeddings

To enhance model robustness with contextual dependencies, we improve the binary convolutions by introducing dynamic embedding to determine binarization thresholds $\alpha \in \mathbb{R}^c$ before the sign functions. Unlike ReActNets [22] that learns overall thresholds from the whole dataset without representing image-specific characteristics, we propose to infer binarization thresholds according to the input images dynamically. Specifically, in each residual block, we first obtain the mean value of each channel using the global average pooling (GAP), followed by the matrix product operation:

$$\alpha(\boldsymbol{A}) = \texttt{GAP}(\boldsymbol{A})\boldsymbol{W}_1 + \boldsymbol{b}_\alpha, \tag{9}$$

where $\boldsymbol{A} \in \mathbb{R}^{c \times h \times w}$ is an input tensor, and $\{\boldsymbol{W}_1 \in \mathbb{R}^{c \times \frac{c}{4}}, \boldsymbol{b}_\alpha \in \mathbb{R}^{\frac{c}{4}}\}$ are learnable parameters. Note that, this matrix product increases negligible calculation due to the GAP function. Based on the contextual feature $\alpha(\boldsymbol{A})$, the channel-wise thresholds $\beta(\boldsymbol{A}) \in \mathbb{R}^c$ is obtained by a matrix product operation:

$$\beta(\boldsymbol{A}) = \alpha(\boldsymbol{A})\boldsymbol{W}_2 + \boldsymbol{b}_\beta, \tag{10}$$

with learnable parameters $\{\boldsymbol{W}_2 \in \mathbb{R}^{\frac{c}{4} \times c}, \boldsymbol{b}_\beta \in \mathbb{R}^c\}$. After merging $\beta(\boldsymbol{A})$ with the following sign function, we quantize inputs using instance specific thresholds:

$$Q_b(x) = \texttt{sign}(x - \beta(\boldsymbol{A})) = \begin{cases} +1, x > \beta(\boldsymbol{A}) \\ -1, x \le \beta(\boldsymbol{A}). \end{cases} \tag{11}$$

To compensate the activation distribution variance caused by subtracting binarization thresholds, we learn additional embedding $\gamma(\boldsymbol{A})$ after binary convolutions:

$$\gamma(\boldsymbol{A}) = \alpha(\boldsymbol{A})\boldsymbol{W}_3 + \boldsymbol{b}_\gamma, \tag{12}$$

where dynamic embeddings $\gamma(\boldsymbol{A}) \in \mathbb{R}^{c'}$ are also transformed by a matrix product with parameters $\boldsymbol{W}_3 \in \mathbb{R}^{\frac{c}{4} \times c'}$, $\boldsymbol{b}_\gamma \in \mathbb{R}^{c'}$. As such, the output tensor $\boldsymbol{A}'$ is given by:

$$\boldsymbol{A}' = \boldsymbol{A}' + \gamma(\boldsymbol{A}), \tag{13}$$

In summary, two dynamic embeddings that model contextual dependencies are attached before and after the binary convolution respectively, formulating a dynamic binarization strategy for given inputs.

4.3   Overall Network

In Fig. 4, we outline the overall architecture of BCDNet. BCDNet is composed of two stages: the binary CNN-embedding stage with M binary convolutional blocks, and the binary MLP stage with N proposed binary MLP blocks. In a standard ViT or MLP architecture, the first embedding layer are implemented by a non-overlapping stride convolution. However, as suggested by [39], this patchify stem layer leads to optimization problems and can be sensitive to training strategies. In Sect. 3.2, we find stand-alone contextual binary models are difficult to optimize. Instead, we adopt our binary MLP modules to enhance high level features. First, we replace several binary convolution blocks with our binary MLPs based on ReActNet-A, which we indicate as BCDNet-A. To exploit the trade-off between convolution and MLP blocks, we perform a grid search under the same computational cost, and determine 11 convolutional blocks and 9 MLP blocks in Fig. 5. Second, we apply both improved binary convolution and MLP blocks and introduce a BCDNet-B model to further improve performance.

5   Experiments

5.1   Implementation Details

We train and evaluate the proposed BNNs on the large-scale ImageNet-1K dataset [31], which contains 1.28M images for training and 50K images for validation. We use 224×224 resolution images with standard data augmentations similar to Real-To-Binary [27]. Following ReActNet [22] and Real-to-Binary [27], we conduct a two-step training strategy with knowledge distillation and weight decay. We first train binary activation and real-valed weight networks for 300K iterations. Then, we initialize model weights from the first step, and train binary weight and activation models for 600K iterations. We use the AdamW [25] optimizer with a cosine learning rate schedule and batch size of 256. In two steps, we set the learning rate 1×10^{-4} and 5×10^{-5} and the weight decay 1×10^{-5} and 0, accordingly. For supervision, we use real labels with smooth rate 0.1 and distill from a full precision ResNet50. Detailed evaluation of distillation affect is shown in Supplementary B. For dynamic embeddings, we initialize W_2, W_3 and all biases as zeros and finetune another 300K iterations in Table 1.

Considering one binary MLP block has almost 1/3 operations of a 3×3 binary convolution block, BCDNet-A simply replace the last three 3×3 convolution blocks in ReActNet-A with 9 binary MLP blocks and keeps almost the same computational cost (Table 1); 1×1 binary convolutions in ReActNet-A remain unchanged. The replacing number is determined by a grid search in Fig. 5. Based on BCDNet-A, we further attach contextual embeddings in the CNN stage to improve binary convolutions and formulate BCDNet-B. We follow previous works [1] and use BOPs, FLOPs and OPs to measure the computational cost, where OPs=1/64BOPs + FLOPs. Note that the FLOPs indicate full precision MACs in BNNs as [23].

5.2   Comparisons with State of the Arts

In Table 1, we present comparisons of BCDNets with state-of-the-art binary or low-bit neural networks. To our best of knowledge, all BNNs are based on convolutional networks. BCDNets achieve the highest accuracy, which indicates the necessity to improve binary convolutions with contextual interactions. With similar architecture and computational cost, we observe that BCDNet-A is able to exceed ReActNet-A by a large margin (+2.4% top1 accuracy). Moreover, with additional dynamic embeddings, we find BCDNet-B further improves 0.5% top-1 accuracy with a little computational overhead, which validates the effectiveness of contextual dynamic embeddings. Also, BCDNet-B outperforms the full-precision MobileNet-V2 for the first time with a binary model.

Generalization to ResNet-18. The proposed binary MLP block can be easily used as plug-in blocks and improve contextual modeling ability in many BNNs. In Table 2, we compare results in the ResNet-18 (R18) architecture. We replace the last three convolutional layers by 9 binary MLP blocks so as to keep the same

Table 1. Comparisons with state of the arts. "W/A" is bit-width of weights and activations. We underline ReActNet-A with consideration of our models share the similar architecture and operations. "†" indicates operations reported by ReActNets which may omit some small operations. We also report operations of BCDNets only considering conv. and fc. in "(.)".

Methods	W/A	BOPs ($\times 10^9$)	FLOPs ($\times 10^8$)	OPs ($\times 10^8$)	Top1 (%)	Top5 (%)
Mobile-V1 [8]	32/32	0	5.69	5.69	70.6	–
Mobile-V2 [32]	32/32	0	3.00	3.00	72.0	–
ResNet-18 [7]	32/32	0	18.14	18.14	69.3	89.2
BWN [30]	1/32	–	–	–	60.8	83.0
LQ-Net [44]	1/2	–	–	–	62.6	84.3
DoReFa [48]	2/2	–	–	–	62.6	84.4
SLB [42]	1/8	–	–	–	66.2	86.5
ABC-Net [17]	(1/1)×5	–	–	–	65.0	85.9
Bi-Real-34 [23]	1/1	3.53	1.39	1.93	62.2	83.9
Real-to-Bin [27]	1/1	1.68	1.56	1.83	65.4	86.2
FDA-BNN* [40]	1/1	–	–	–	66.0	86.4
SA-BNN-50 [18]	1/1	–	–	–	68.7	87.4
MeliusNet-22 [1]	1/1	4.62	1.35	2.08	63.6	84.7
MeliusNet-42 [1]	1/1	9.69	1.74	3.25	69.2	88.3
MeliusNet-59 [1]	1/1	18.3	2.45	5.25	71.0	89.7
ReActNet-A [22]	1/1	<u>4.82</u>	0.31 (0.12†)	1.06 (0.87†)	<u>69.4</u>	–
ReActNet-B [22]	1/1	4.69	0.61 (0.44†)	1.34 (1.17†)	70.1	–
ReActNet-C [22]	1/1	4.69	1.57 (1.40†)	2.30 (2.14†)	71.4	–
BCDNet-A	**1/1**	**<u>4.82</u>**	**0.32 (0.12)**	**1.08 (0.87)**	**71.8 (+2.4)**	**90.3**
BCDNet-B	**1/1**	**<u>4.82</u>**	**0.34 (0.14)**	**1.09 (0.89)**	**72.3 (+2.9)**	**90.5**

Table 2. Comparison with ResNet-18 architectures on the ImageNet dataset. W/A indicates bit-width of weights and activations respectively.

Methods	W/A	Top1 Acc.	Methods	W/A	Top1 Acc.
ResNet-18	32/32	69.3	RBNN [16]	1/1	59.9
XNOR-Net [30]	1/1	51.2	SA-BNN [18]	1/1	61.7
Bi-Real [23]	1/1	56.4	ReActNet (R18) [22]	1/1	65.5
Real-to-Bin [27]	1/1	65.4	FDA-BNN [40]	1/1	60.2
IR-Net [29]	1/1	58.1	FDA-BNN* [40]	1/1	66.0
SLB [42]	1/1	61.3	**BCDNet-A (R18)**	**1/1**	**66.9**
SLB [42]	1/8	66.2	**BCDNet-B (R18)**	**1/1**	**67.9**

operations. Compared with ReActNet (R18), without changing the first three stages, BCDNet-A (R18) improves 1.4% top-1 accuracy with binary MLPs. For BCDNet-B (R18), significant improvement (+1.0%) is obtained with the help of dynamic embeddings.

Table 3. Evaluation on fine-grained classification benchmarks. "†" indicates the comparison with ImageNet pretraining.

Method	W/A	CUB-200	Flowers	Aircraft	Cars	Dogs	Avg
ResNet18	32/32	52.3	44.2	66.8	39.2	47.1	49.9
BiRealNet18	1/1	19.5	27.0	18.6	14.8	17.9	19.6
ReActNet-A	1/1	29.9	29.8	18.4	19.4	18.7	23.2
BCDNet-A	1/1	34.3	25.0	18.1	25.5	20.7	24.7
ReActNet-A†	1/1	83.7	80.8	74.2	85.4	70.3	78.9
BCDNet-A†	1/1	90.6	83.0	81.1	91.1	75.2	84.2

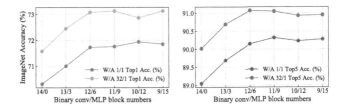

Fig. 5. Performance of replacing binary convolutional blocks with binary MLP blocks. We start from full convolutional ReActNet-A.

Generalization to Fine-grained Datasets. As shown in Table 3, we also report results on 5 fine-grained datasets: CUB-200-2011 [37], Oxford-flowers102 [28], Aircraft [26], Stanford-cars [13], Stanford-dogs [12]. Detailed training strategies are reported in Supplementary C. These datasets have much less images than ImageNet but more concentrate on detailed differences of attributions. As shown in Table 3, BCDNet-A exceeds ReActNet-A 1.5% average accuracy, and are able to work well in most cases. We also report results with ImageNet pretraining. Pretraining helps BCDNet-A more than pure binary convolution based ReActNet-A. Modeling contextual dependencies helps BCDNet-A to exceed ReActNet-A by 5.3% on average, which shows BCDNet is more friendly to pretraining.

5.3 Ablation Studies

Table 4. Replacement stage settings

Method	Stages	Top1 Acc
ReActNet-18	–	61.89%
BCDNet-18	4	63.58%
BCDNet-18	3, 4	63.47%
BCDNet-18	1, 2, 3, 4	63.10%

Table 5. Replacement number settings

Setting	Block OPs ($\times10^6$)	Top1 Acc
BiConv	3.813 ($\times1.00$)	61.89%
BiMLP$\times1$	1.505 ($\times0.39$)	61.32%
BiMLP$\times2$	3.011 ($\times0.79$)	62.64%
BiMLP$\times3$	4.516 ($\times1.18$)	63.58%

Table 6. Evaluation of different range modeling. We report checkpoint results for different combinations of long- and short-range branches.

Training time	Stage1	Stage1+200K	Stage1+400K	Stage1+600K
Bitwidth W/A	32/1	1/1	1/1	1/1
Convolution	71.58/90.02	66.30/86.61	68.27/87.84	70.31/89.05
P-L-L	72.85/90.82	67.06/87.32	69.14/88.42	70.79/89.39
P-S-S	71.98/90.56	68.24/88.10	69.87/89.07	71.72/89.97
P-S-L	**73.13/91.05**	**68.63/88.45**	**70.46/89.37**	**71.76/90.32**

Efficiency of Binary MLPs. We first study where to replace binary convolutional blocks as proposed binary MLP blocks (BiMLPs) in a ReActNet (RN18), and then report the influence of different configuration of replacements. Training settings in Table 4, 5 are reported in Supplementary A. In Table 4, we gradually replace all 3×3 binary conv stages of a ReActNet RN18 (except downsampling layers). In practice, we use the same MLP sampling range settings as the last stage MLPs. Results exceed original ReActNet in each drop-in replacement. We find only replacing the last stage is the most efficient. In Table 5, we replace every 3×3 binary convolutions with 1, 2 or 3 BiMLPs in the last stage of a ReActNet RN18 (except the downsampling layer). One BiMLP has 0.39 times single block OPs, while drops accuracy slightly; three BiMLPs have 1.18 times block OPs while significantly improve accuracy. In the MobileNet-V1 backbone, we simply replace one convolutional block by three binary MLP blocks once at a time in the ReActNet-A. Figure 5 presents comparisons between MLPs and convolutions, where overall OPs are basically the same ($1.06 \times 10^8 \sim 1.08 \times 10^8$). Due to the influence of randomness, 9~12 MLP blocks make out the best for a ReActNet-A architecture. For simplicity, we just choose the setting of 11 (1,1,2,2,4,1 blocks for each resolution) convolutional and 9 MLP blocks in other experiments.

Evaluation of Contextual Dependencies. Contextual dependencies come from long-short range sampling and the MLP architecture. In Table 6, we keep three branches in every binary MLP block and evaluate different sampling range combinations. 'P', 'S' and 'L' indicate the pointwise (original location) branch, the short-range branch and the long-range branch, with token sampling ranges of 0, 1 and half of the resolution respectively. We also report results of 200k, 400k and 600k training iterations. In Table 6, combination of long-short ranges ("P-S-L") with different receptive field achieves better performance. Compared with only long ranges ("P-L-L"), "P-S-L" has inductive bias and achieve better performance. Compared with only short ranges ("P-S-S"), "P-S-L" places long-range modeling and can be easier to converge in early training times (e.g., 200k, 400k iterations). In the MLP architecture, even though without long-ranges, "P-S-S" still exceeds binary CNNs because it decouples a 3×3 binary convolution to three BiMLPs and expands reception field in each layer (from 3×3 to 7×7).

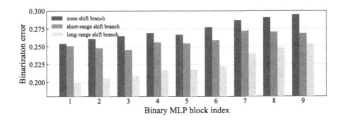

Fig. 6. Binarization error of different branches in MLP layers.

Binarization Error Analysis. We define the average binarization error as:

$$error = \frac{1}{n} \sum |\frac{\|sign(\boldsymbol{W})\|_{\ell_1}}{c_{in}} sign(\boldsymbol{W}) - \boldsymbol{W}| \tag{14}$$

where $\frac{\|sign(\boldsymbol{W})\|_{\ell_1}}{c_{in}}$ follows Eq. 14; $\boldsymbol{W}$ is the weight matrix in a binary MLP layer with n numbers. For BCDNet-A, we calculate the average binarization error of each MLP layer in Fig. 6. Note that, when binarization error increases, gradient estimation is inaccurate when backpropagation. We find in almost every block, errors in 3 branches follows a order of: `pointwise>short-range>long-range`. Binarization error is improved by decreasing of exchange distance, which indicates the long-range branch is easier to optimize than the short-range branch.

6   Conclusion

This work improves binary neural networks by modeling contextual dependencies. A long-short range binary MLP module is proposed to explore comprehensive contextual perceptions, which can be an efficient alternative to traditional binary convolutions. Equipped with our binary MLPs and improved binary convolutions, BCDNets exceed state of the arts significantly.

Acknowledgements. This work was supported by the National Natural Science Foundation of China (U20B2042), the National Natural Science Foundation of China (62076019) and Aeronautical Science Fund (ASF) of China (2020Z071051001).

References

1. Bethge, J., Bartz, C., Yang, H., Chen, Y., Meinel, C.: MeliusNet: can binary neural networks achieve mobileNet-level accuracy? arXiv preprint arXiv:2001.05936 (2020)
2. Chen, S., Xie, E., Ge, C., Liang, D., Luo, P.: CycleMLP: a MLP-like architecture for dense prediction. arXiv preprint arXiv:2107.10224 (2021)
3. Chu, X., et al.: Twins: revisiting spatial attention design in vision transformers (2021)

4. Courbariaux, M., Hubara, I., Soudry, D., El-Yaniv, R., Bengio, Y.: Binarized neural networks: training deep neural networks with weights and activations constrained to+ 1 or-1. arXiv preprint arXiv:1602.02830 (2016)
5. Dai, Z., Cai, B., Lin, Y., Chen, J.: UP-DETR: unsupervised pre-training for object detection with transformers. In: Proceedings of the IEEE/CVF Conference on Computer Vision and Pattern Recognition, pp. 1601–1610 (2021)
6. Dosovitskiy, A., et al.: An image is worth 16x16 words: transformers for image recognition at scale. In: International Conference on Learning Representations (2020)
7. He, K., Zhang, X., Ren, S., Sun, J.: Deep residual learning for image recognition. In: Proceedings of the IEEE Conference on Computer Vision and Pattern Recognition, pp. 770–778 (2016)
8. Howard, A.G., et al.: MobileNets: efficient convolutional neural networks for mobile vision applications. arXiv preprint arXiv:1704.04861 (2017)
9. Ji, M., Shin, S., Hwang, S., Park, G., Moon, I.C.: Refine myself by teaching myself: feature refinement via self-knowledge distillation. In: Proceedings of the IEEE/CVF Conference on Computer Vision and Pattern Recognition, pp. 10664–10673 (2021)
10. Jiang, X., Wang, N., Xin, J., Li, K., Yang, X., Gao, X.: Training binary neural network without batch normalization for image super-resolution. In: Proceedings of the AAAI Conference on Artificial Intelligence, vol. 35, pp. 1700–1707 (2021)
11. Joo, D., Yi, E., Baek, S., Kim, J.: Linearly replaceable filters for deep network channel pruning. In: Proceedings of the AAAI Conference on Artificial Intelligence, vol. 35, pp. 8021–8029 (2021)
12. Khosla, A., Jayadevaprakash, N., Yao, B., Li, F.F.: Novel dataset for fine-grained image categorization: stanford dogs. In: Proceedings CVPR Workshop on Fine-Grained Visual Categorization (FGVC), vol. 2. Citeseer (2011)
13. Krause, J., Stark, M., Deng, J., Fei-Fei, L.: 3D object representations for fine-grained categorization. In: Proceedings of the IEEE International Conference on Computer Vision Workshops, pp. 554–561 (2013)
14. Lee, J., Kim, D., Ham, B.: Network quantization with element-wise gradient scaling. In: Proceedings of the IEEE/CVF Conference on Computer Vision and Pattern Recognition, pp. 6448–6457 (2021)
15. Lin, K., Wang, L., Liu, Z.: End-to-end human pose and mesh reconstruction with transformers. In: Proceedings of the IEEE/CVF Conference on Computer Vision and Pattern Recognition, pp. 1954–1963 (2021)
16. Lin, M., et al.: Rotated binary neural network. In: Advances in Neural Information Processing Systems, vol. 33 (2020)
17. Lin, X., Zhao, C., Pan, W.: Towards accurate binary convolutional neural network. In: Advances in Neural Information Processing Systems, vol. 30 (2017)
18. Liu, C., Chen, P., Zhuang, B., Shen, C., Zhang, B., Ding, W.: SA-BNN: state-aware binary neural network. In: Proceedings of the AAAI Conference on Artificial Intelligence, vol. 35, pp. 2091–2099 (2021)
19. Liu, X., Ye, M., Zhou, D., Liu, Q.: Post-training quantization with multiple points: mixed precision without mixed precision. In: Proceedings of the AAAI Conference on Artificial Intelligence, vol. 35, pp. 8697–8705 (2021)
20. Liu, Z., et al.: Swin transformer: hierarchical vision transformer using shifted windows. arXiv preprint arXiv:2103.14030 (2021)
21. Liu, Z., Shen, Z., Li, S., Helwegen, K., Huang, D., Cheng, K.T.: How do Adam and training strategies help BNNs optimization? In: International Conference on Machine Learning. PMLR (2021)

22. Liu, Z., Shen, Z., Savvides, M., Cheng, K.-T.: ReActNet: towards precise binary neural network with generalized activation functions. In: Vedaldi, A., Bischof, H., Brox, T., Frahm, J.-M. (eds.) ECCV 2020. LNCS, vol. 12359, pp. 143–159. Springer, Cham (2020). https://doi.org/10.1007/978-3-030-58568-6_9

23. Liu, Z., Wu, B., Luo, W., Yang, X., Liu, W., Cheng, K.T.: Bi-Real Net: enhancing the performance of 1-bit CNNs with improved representational capability and advanced training algorithm. In: Proceedings of the European conference on computer vision (ECCV), pp. 722–737 (2018)

24. Li'an Zhuo, B.Z., Chen, H., Yang, L., Chen, C., Zhu, Y., Doermann, D.: CP-NAS: child-parent neural architecture search for 1-bit CNNs. In: IJCAI (2020)

25. Loshchilov, I., Hutter, F.: Decoupled weight decay regularization. arXiv preprint arXiv:1711.05101 (2017)

26. Maji, S., Rahtu, E., Kannala, J., Blaschko, M., Vedaldi, A.: Fine-grained visual classification of aircraft. arXiv preprint arXiv:1306.5151 (2013)

27. Martinez, B., Yang, J., Bulat, A., Tzimiropoulos, G.: Training binary neural networks with real-to-binary convolutions. arXiv preprint arXiv:2003.11535 (2020)

28. Nilsback, M.E., Zisserman, A.: Automated flower classification over a large number of classes. In: 2008 Sixth Indian Conference on Computer Vision, Graphics & Image Processing, pp. 722–729. IEEE (2008)

29. Qin, H., et al.: Forward and backward information retention for accurate binary neural networks. In: Proceedings of the IEEE/CVF Conference on Computer Vision and Pattern Recognition, pp. 2250–2259 (2020)

30. Rastegari, M., Ordonez, V., Redmon, J., Farhadi, A.: XNOR-Net: ImageNet classification using binary convolutional neural networks. In: Leibe, B., Matas, J., Sebe, N., Welling, M. (eds.) ECCV 2016. LNCS, vol. 9908, pp. 525–542. Springer, Cham (2016). https://doi.org/10.1007/978-3-319-46493-0_32

31. Russakovsky, O., et al.: ImageNet large scale visual recognition challenge. Int. J. Comput. Vis. **115**(3), 211–252 (2015). https://doi.org/10.1007/s11263-015-0816-y

32. Sandler, M., Howard, A., Zhu, M., Zhmoginov, A., Chen, L.C.: MobileNetv 2: inverted residuals and linear bottlenecks. In: Proceedings of the IEEE Conference on Computer Vision and Pattern Recognition, pp. 4510–4520 (2018)

33. Song, L., Wu, J., Yang, M., Zhang, Q., Li, Y., Yuan, J.: Robust knowledge transfer via hybrid forward on the teacher-student model. In: Proceedings of the AAAI Conference on Artificial Intelligence, vol. 35, pp. 2558–2566 (2021)

34. Tolstikhin, I., et al.: MLP-mixer: an all-MLP architecture for vision. arXiv preprint arXiv:2105.01601 (2021)

35. Touvron, H., et al.: ResMLP: feedforward networks for image classification with data-efficient training. ArXiv abs/2105.03404 (2021)

36. Touvron, H., Cord, M., Douze, M., Massa, F., Sablayrolles, A., Jegou, H.: Training data-efficient image transformers & distillation through attention. In: International Conference on Machine Learning, vol. 139, pp. 10347–10357 (2021)

37. Wah, C., Branson, S., Welinder, P., Perona, P., Belongie, S.: The caltech-UCSD birds-200-2011 dataset (2011)

38. Wu, B., et al.: Shift: a zero flop, zero parameter alternative to spatial convolutions. In: Proceedings of the IEEE Conference on Computer Vision and Pattern Recognition, pp. 9127–9135 (2018)

39. Xiao, T., Dollar, P., Singh, M., Mintun, E., Darrell, T., Girshick, R.: Early convolutions help transformers see better. In: Advances in Neural Information Processing Systems, vol. 34 (2021)

40. Xu, Y., Han, K., Xu, C., Tang, Y., Xu, C., Wang, Y.: Learning frequency domain approximation for binary neural networks. Adv. Neural. Inf. Process. Syst. **34**, 25553–25565 (2021)
41. Yamamoto, K.: Learnable companding quantization for accurate low-bit neural networks. In: Proceedings of the IEEE/CVF Conference on Computer Vision and Pattern Recognition, pp. 5029–5038 (2021)
42. Yang, Z., et al.: Searching for low-bit weights in quantized neural networks. Adv. Neural. Inf. Process. Syst. **33**, 4091–4102 (2020)
43. Yu, T., Li, X., Cai, Y., Sun, M., Li, P.: S2-MLP: spatial-shift MLP architecture for vision. arXiv preprint arXiv:2106.07477 (2021)
44. Zhang, D., Yang, J., Ye, D., Hua, G.: LQ-Nets: learned quantization for highly accurate and compact deep neural networks. In: Proceedings of the European conference on computer vision (ECCV), pp. 365–382 (2018)
45. Zhang, Z., Zhang, H., Zhao, L., Chen, T., Pfister, T.: Aggregating nested transformers. arXiv preprint arXiv:2105.12723 (2021)
46. Zhao, K., et al.: Distribution adaptive int8 quantization for training CNNs. In: Proceedings of the Thirty-Fifth AAAI Conference on Artificial Intelligence (2021)
47. Zheng, S., et al.: Rethinking semantic segmentation from a sequence-to-sequence perspective with transformers. In: Proceedings of the IEEE/CVF Conference on Computer Vision and Pattern Recognition, pp. 6881–6890 (2021)
48. Zhou, S., Wu, Y., Ni, Z., Zhou, X., Wen, H., Zou, Y.: DoReFa-Net: training low bitwidth convolutional neural networks with low bitwidth gradients. arXiv preprint arXiv:1606.06160 (2016)

SPIN: An Empirical Evaluation on Sharing Parameters of Isotropic Networks

Chien-Yu Lin[1]($\boxtimes$), Anish Prabhu[2], Thomas Merth[2], Sachin Mehta[2], Anurag Ranjan[2], Maxwell Horton[2], and Mohammad Rastegari[2]

[1] University of Washington, Seattle, USA
cylinbao@gmail.com
[2] Apple, Inc., Cupertino, USA

Abstract. Recent isotropic networks, such as ConvMixer and Vision Transformers, have found significant success across visual recognition tasks, matching or outperforming non-isotropic Convolutional Neural Networks. Isotropic architectures are particularly well-suited to cross-layer weight sharing, an effective neural network compression technique. In this paper, we perform an empirical evaluation on methods for sharing parameters in isotropic networks (SPIN). We present a framework to formalize major weight sharing design decisions and perform a comprehensive empirical evaluation of this design space. Guided by our experimental results, we propose a weight sharing strategy to generate a family of models with better overall efficiency, in terms of FLOPs and parameters versus accuracy, compared to traditional scaling methods alone, for example compressing ConvMixer by 1.9× while improving accuracy on ImageNet. Finally, we perform a qualitative study to further understand the behavior of weight sharing in isotropic architectures. The code is available at https://github.com/apple/ml-spin.

Keywords: Parameter sharing · Isotropic networks · Efficient CNNs

1 Introduction

Isotropic neural networks have the property that all of the weights and intermediate features have identical dimensionality, respectively (see Fig. 1). Some notable convolutional neural networks (CNNs) with isotropic structure [15,25] have been proposed recently in the computer vision domain, and have been applied to different visual recognition tasks, including image classification, object detection,

---

C-Y. Lin and A. Prabhu—Equal contribution.
C.-Y. Lin—Work done while interning at Apple.

Supplementary Information The online version contains supplementary material available at https://doi.org/10.1007/978-3-031-20083-0_33.

S. Avidan et al. (Eds.): ECCV 2022, LNCS 13671, pp. 553–568, 2022.
https://doi.org/10.1007/978-3-031-20083-0_33

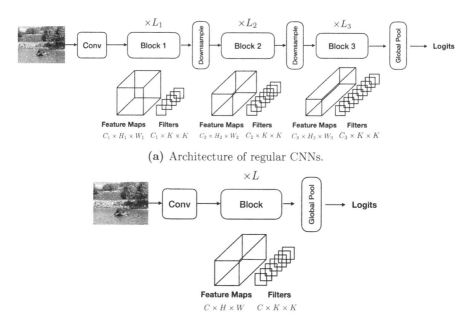

(a) Architecture of regular CNNs.

(b) Architecture of isotropic CNNs.

Fig. 1. Basic architectures of regular and isotropic CNNs. (a) Regular CNNs vary the shape of intermediate features and weight tensors in the network while (b) isotropic CNNs fix the shape of all intermediate features and weight tensors in the network.

and action recognition. These isotropic CNNs contrast with the typical "hierarchical" design paradigm, in which spatial resolution and channel depth are varied throughout the network (e.g., VGG [19] and ResNet [6]).

The Vision Transformer (ViT) [3] architecture also exhibits this isotropic property, although softmax self-attention and linear projections are used for feature extraction instead of spatial convolutions. Follow-up works have experimented with various modifications to ViT models (e.g. replacing softmax self-attention with linear projections [24], factorized attention [27], and non-learned transformations [26]); however, the isotropic nature of the network is usually retained.

Recent isotropic models (e.g., ViT [3], ConvMixer [25], and ConvNext [15]) attain state-of-the-art performance for visual recognition tasks, but are computationally expensive to deploy in resource constrained inference scenarios. In some cases, the parameter footprint of these models can introduce memory transfer bottlenecks in hardware that is not well equipped to handle large amounts of data (e.g. microcontrollers, FPGAs, and mobile phones) [13]. Furthermore, "over-the-air" updates of these large models can become impractical for continuous deployment scenarios with limited internet bandwidth. Parameter (or weight) sharing[1], is one approach which compresses neural networks, potentially enabling the deployment of large models in these constrained environments.

Isotropic DNNs, as shown Fig. 1, are constructed such that a layer's weight tensor has identical dimensionality to that of other layers. Thus, cross-layer parameter sharing becomes a straightforward technique to apply, as shown in ALBERT [13]. On the other hand, weight tensors within non-isotropic networks cannot be shared in this straightforward fashion without intermediate weight transformations (to coerce the weights to the appropriate dimensionality). In Appendix A, we show that the search space of possible topologies for straightforward cross-layer parameter sharing is significantly larger for isotropic networks, compared to "multi-staged" networks (an abstraction of traditional, non-isotropic networks). This rich search space requires a comprehensive exploration. Therefore, in this paper, we focus on isotropic networks, with the goal of finding practical parameter sharing techniques that enable high-performing, low-parameter neural networks for visual understanding tasks. To extensively explore the weight sharing design space for isotropic networks, we experiment with different orthogonal design choices (Sect. 3.2). Specifically, we explore (1) different sharing topologies, (2) dynamic transformations, and (3) weight fusion initialization strategies from pretrained non-sharing networks. Our results show that parameter sharing is a simple and effective method for compressing large neural networks versus standard architectural scaling approaches (e.g. reduction of input image size, channel size, and model depth). Using a weight sharing strategy discovered from our design space exploration, we achieve nearly identical accuracy (to non-parameter sharing, iso-FLOP baselines) with significantly reduced parameter counts. Beyond the empirical accuracy versus efficiency experiments, we also investigate network representation analysis (Sect. 5) and model generalization (Appendix F) for parameter sharing isotropic models.

2   Related Works

Cross-Layer Parameter Sharing. Cross-layer parameter sharing has been explored for both CNN- and Transformer-based models [2,9,11–13,20,21]. For instance, Kim et al. [9] applies cross-layer parameter sharing across an entire heterogeneous CNN. However, they share weights at the granularity of filters, whereas we share weights at the granularity of layers. In terms of our framework, Kubilius et al. [11] experiments with Uniform-Strided, proposing a heterogeneous network based off of the human visual cortex. With isotropic networks, we can decouple parameter sharing methods from the constraints imposed by heterogeneous networks. Thus, we expand the scope of weight sharing structures from their work to isotropic networks.

Cross-layer parameter sharing is explored for isotropic Transformer models for the task of neural language modeling [2,13] and vision [21]. Lan et al.

---

[1] We interchangeably use the terms parameter and weight sharing throughout this paper.

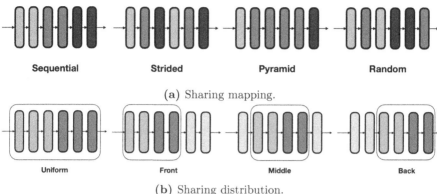

(a) Sharing mapping.

(b) Sharing distribution.

Fig. 2. Sharing topologies. In (a), sharing mapping determines which layers share the same weights while in (b), sharing distribution determines how the weight sharing layers are distributed in the network. Layers with the same color share weights. Layers outside of the sharing section do not share weights. Best viewed in color.

[13] experiments with Uniform-Sequential, and Dehghain et al. [2] experiments with universal sharing (i.e. all layers are shared). Takase et al. [21] experiments with 3 strategies, namely Uniform-Sequential, Uniform-Strided, and Cycle. In this paper, we extend these works by decomposing the sharing topology into combinations of different sharing mappings (Fig. 2a) and sharing distributions (Fig. 2b).

Dynamic Recurrence for Sharing Parameters. Several works [1,4,14,18] explore parameter sharing through the lens of dynamically repeating layers. However, each technique is applied to a different model architecture, and evaluated in different ways. Thus, without a common framework, it's difficult to get a comprehensive understanding of how these techniques compare. While this work focuses only on static weight sharing, we outline a framework that may encompass even these dynamic sharing schemes. In general, we view this work as complementary to explorations on dynamic parameter sharing, since our analysis and results could be used to help design new dynamic sharing schemes.

3   Sharing Parameters in Isotropic Networks

In this section, we first motivate why we focus on isotropic networks for weight sharing (Sect. 3.1), followed by a comprehensive design space exploration of methods for weight sharing, including empirical results (Sect. 3.2).

3.1   Why Isotropic Networks?

Isotropic networks, shown in Fig. 1b, are simple by design, easy to analyze, and enable flexible weight sharing, as compared to heterogeneous networks.

Simplicity of Design. Standard CNN architectural design, whether manual [6,17]) or automated through methods like neural architecture search [7,22]), require searching a complex search space, including what blocks to use, where and when to downsample the input, and how the number of channels should vary throughout the architecture. On the other hand, isotropic architectures form a much simpler design space, where just a single block (e.g., attention block in Vision Transformers or convolutional block in ConvMixer) along with network's depth and width must be chosen. The simplicity of implementation for these architectures enables us to more easily design generic weight sharing methods across various isotropic architectures. The architecture search space of these networks is also relatively smaller than non-isotropic networks, which makes them a convenient choice for large scale empirical studies.

Increased Weight Sharing Flexibility. Isotropic architectures provide significantly more flexibility for designing a weight sharing strategy than traditional networks.

We define the *sharing topology* to be the underlying structure of how weight tensors are shared throughout the network. Suppose we have an isotropic network with $L \geq 1$ layers and a weight tensor "budget" of $1 \leq P \leq L$. The problem of determining the optimal sharing topology can be seen as a variant of the set cover problem; we seek a set cover with no more than B disjoint subsets, which maximizes the accuracy of the resulting network. More formally, a possible sharing topology is an ordered collection of disjoint subsets $\mathcal{T} = (\mathcal{S}_1, \mathcal{S}_2, ..., \mathcal{S}_P)$, where $\cup_{i=1}^{B} \mathcal{S}_i = \{1, 2, ..., L\}$ for some $1 \leq P \leq L$. We define $\frac{L}{P}$ to be the *share rate*.

We characterize the search space in Appendix A, showing that isotropic networks support significantly more weight sharing topologies than heterogeneous networks (when sharing at the granularity of weight tensors). This substantially increased search space may yield more effective weight sharing strategies in isotropic networks than non-isotropic DNNs, a reason why we are particularly interested in isotropic networks.

Cross-Layer Representation Analysis. To better understand if the weights of isotropic architectures are amenable to compression through weight sharing, we study the representation of these networks across layers. We hypothesize that layers with similar output representations will be more compressible via weight sharing. To build intuition, we use Centered Kernel Alignment (CKA) [10], a method that allows us to effectively measure similarity across layers.

Figure 3 shows the pairwise analysis of CKA across layers within the ConvMixer network. We find significant representational similarity for nearby layers. This is not unexpected, given the analysis of prior works on iterative refinement in residual networks [8]. Interestingly, we find that CKA generally peaks in the middle of the network for different configurations of ConvMixer. Overall, these findings suggest that isotropic architectures may be amenable to weight sharing, and we use this analysis to guide our experiments exploring various sharing topologies in Sect. 3.2.

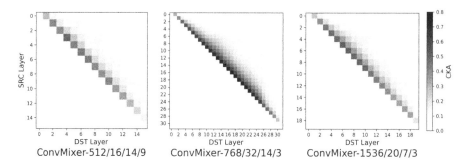

ConvMixer-512/16/14/9   ConvMixer-768/32/14/3   ConvMixer-1536/20/7/3

Fig. 3. CKA similarity analysis on ConvMixer's intermediate feature maps shows that the output feature maps of neighboring layers and especially the middle layers have the highest similarity. Here, we compute the CKA similarity of each layer's output feature maps. The diagonal line and the lower triangle part are masked out for clarity. The CKA for the diagonal line is 1 since they are identical. The CKA for the lower triangle is the mirror of the upper triangle. Best viewed on screen.

3.2 Weight Sharing Design Space Exploration

When considering approaches to sharing weights within a neural network, there is an expansive design space to consider. This section provides insights as well as empirical evaluation to help navigate this design space. We first consider the weight sharing topology. Then, we introduce lightweight dynamic transformations on the weights to increase the representational power of the weight-shared networks. Finally, we explore how to use the trained weights of an uncompressed network to further improve accuracy in weight-sharing isotropic networks. All experiments done in this section are based on a ConvMixer model with 768 channels, depth of 32, patch extraction kernel size of 14, and convolutional kernel size of 3.

Weight Sharing Topologies. Isotropic networks provide a vast design space for sharing topologies. We perform an empirical study of various sharing topologies for the ConvMixer architecture, evaluated on the ImageNet dataset. We characterize these topologies by the (1) *sharing mapping* (shown in Fig. 2a), which describes the structure of shared layers, and (2) the *sharing distribution* (shown in Fig. 2b), which describes which subset of layers sharing is applied to. We study the following sharing mappings:

1. **Sequential:** Neighboring layers are shared in this topology. There is motivated by our cross layer similarity analysis in Sect. 3.1 and Fig. 3, which suggest that local structures of recurrence may be promising.
2. **Strided:** This topology defines the recurrence on the network level rather than locally. If we consider having P blocks with unique weights, we first run all of the layers sequentially, then we repeat this whole structure L/P times.
3. **Pyramid:** This topology is an extension of Sequential, which has increasingly more shared sequential layers as you approach the center of the network. This

Table 1. Effect of different sharing distributions and mappings on the performance of weight-shared (WS) ConvMixer with a share rate of 2. In order to maintain the fixed share rate 2 for non-uniform sharing distributions (i.e., Middle, Front and Back), we apply sharing to 8 layers with share rate $3\times$ and have 16 independent layers. For Middle-Pyramid, the network is defined as $[4 \times 1, 1 \times 2, 2 \times 3, 2 \times 4, 2 \times 3, 1 \times 2, 4 \times 1]$, where for each element $N \times S$, N stands for the number of sharing layers and S the share rate for the layer. All experiments were done with a ConvMixer with 768 channels, depth of 32, patch extraction kernel size of 14, and convolutional kernel size of 3.

Network	Sharing Distribution	Sharing Mapping	Params (M)	FLOPs (G)	Top-1 Acc (%)
ConvMixer	–	–	20.46	5.03	75.71
WS-ConvMixer	Uniform	Sequential	11.02	5.03	**73.29**
		Strided			72.80
		Random			Diverged
WS-ConvMixer	Middle	Sequential	11.02	5.03	73.14
		Pyramid			**73.22**
WS-ConvMixer	Front	Sequential	11.02	5.03	**73.31**
	Back				72.35

is inspired by (1) empirical results in Fig. 3 that show a similar structure in the layer-wise similarity and (2) neural network compression methods (e.g. quantization and sparsity methods), which leave the beginning and end of the network uncompressed [5,16].

4. **Random:** We randomly select which layers are shared within the network, allowing us to understand how much the choice of topology actually matters.

For the sharing distribution, we consider applying (1) **Uniform**, where sharing mapping is applied to all layers, (2) **Front**, where sharing mapping is applied to the front of the network, (3) **Middle**, where sharing mapping is applied to the middle of the network, and (4) **Back**, where sharing mapping is applied to the back of the network. Note that front, middle and back sharing distributions results in a non-uniform distribution of share rates across layers.

Figure 2 visualizes different sharing topologies while Table 1 shows the results of these sharing methods on the ImageNet dataset. When share rate is 2, ConvMixer with uniform-sequential, middle-pyramid, and front-sequential sharing topology result in similar accuracy (2.5% less than the non-shared model) while other combinations result in lower accuracy. These results are consistent with the layer-wise similarity study in Sect. 3.1, and suggests that layer-wise similarity may be a reasonable metric for determining which layers to share. Because of the simplicity and flexibility of *uniform-sequential* sharing topology, we use it in the following experiments unless otherwise stated explicitly.

Lightweight Dynamic Transformations on Shared Weights. To improve the performance of a weight shared network, we introduce lightweight dynamic transformations on top of the shared weights for each individual layer. With this, we potentially improve the representational power of the weight sharing network without increasing the parameter count significantly.

To introduce the lightweight dynamic transformation used in this study, we consider a set of N layers to be shared, with a shared weight tensor W_s. In the absence of dynamic transforms, the weight tensor W_s would simply be shared among all N layers. We consider $W_i \in \mathbb{R}^{C \times C \times K \times K}$ to be the weights of the i-th layer, where C is the channel size and K is the kernel size. With a dynamic weight transformation function f_i, the weights W_i at the i-th layer becomes

$$W_i = f_i(W_s) \tag{1}$$

The choose f_i to be a learnable lightweight affine transformation that allows us to transform the weights without introducing heavy computation and parameter overhead. Specifically, $f_i(W) = \mathbf{a} * W + \mathbf{b}$ applies a grouped point-wise convolution with weights $\mathbf{a} \in \mathbb{R}^{C \times G}$ and bias $\mathbf{b} \in \mathbb{R}^C$ to W, where G is the number of groups. The number of groups, $G \in [1, C]$, can be varied to modulate the amount of inter-channel mixing.

Table 2 shows the effect of different number of groups in the dynamic weight transformation on the performance and efficiency (in terms of parameters and FLOPs) of ConvMixer on the ImageNet dataset. As Table 2 shows, using $G = 64$, the dynamic weight transformation slightly improves accuracy by 0.07% (from 73.29% to 73.36%) with 7% more parameters (from 11.02 M to 11.8 M) and 11.9% more FLOPs (from 5.03 G to 5.63 G). Despite having stronger expressive power, dynamic weight transformation does not provide significant accuracy improvement with under 10% of overhead on number of weights and FLOPs and sometimes even degrading accuracy.

Initializing Weights from Pretrained Non-sharing Networks. Here we consider how we can use the weights of a pretrained, uncompressed network to improve the parameter shared version of an isotropic network. To this end, we introduce transformations on the original weights to generate the weights of the shared network for a given sharing topology. We define $V_j \in \mathbb{R}^{C \times C \times K \times K}$ to be the j-th pretrained weight in the original network, and $u_j \in \mathbb{R}^C$ to be the corresponding pretrained bias. The chosen sharing topology defines a disjoint set cover of the original network's layers, where each disjoint subset maps a group of layers from the original network to a single shared weight layer. Concretely, if the weight W_i is shared among S_i layers $\{i_1, i_2, ..., i_{S_i}\}$ in the compressed network, then we define $W_i = F_i(V_{i_1}, V_{i_2}, ..., V_{i_{S_i}})$, where we can design each F_i. We refer to F as the *fusion strategy*. In all experiments we propagate the gradient back to the original, underlying V_j weights. Importantly, F does not incur a cost at inference-time, since we can constant-fold this function once we finish training.

Table 2. Effect of affine transformations on the performance of Weight Shared ConvMixer model with a sharing rate of 2. All experiments were done with a ConvMixer with 768 channels, depth of 32, patch extraction kernel size of 14, and convolutional kernel size of 3.

Network	Weight Transformation?	Group Rate	Params (M)	FLOPs (G)	Top-1 Acc (%)
ConvMixer	–	–	20.46	5.03	75.71
WS-ConvMixer	✗	–	11.02	5.03	73.29
	✓	1	11.05	5.04	72.87
	✓	16	11.20	5.17	73.20
	✓	32	11.40	5.31	73.14
	✓	64	11.80	5.63	**73.36**

Table 3. Effect of different fusion strategies (Sect. 3.2) on the performance of ConvMixer. All experiments were done with a ConvMixer with 768 channels, depth of 32, patch extraction kernel size of 14, and convolutional kernel size of 3. All weight sharing ConvMixer models share groups of 2 sequential layers.

Network	Fusion Strategy	Params (M)	FLOPs (G)	Top-1 Acc (%)
ConvMixer	–	20.5	5.03	75.71
WS-ConvMixer	–	10.84	5.03	73.23
	Choose First			74.81
	Mean			74.91
	Scalar Weighted Mean			**75.15**
	Channel Weighted Mean			**75.15**
	Pointwise Convoulution			Diverged

One simple fusion strategy would be to randomly initialize a single weight tensor for this layer. Note that this is the approach we have used in all previous experiments. We empirically explore the following fusion strategies:

- **Choose First:** In this setup we take the first of the set of weights within the set: $W_i = F_i(V_{i_1}, V_{i_2}, ..., V_{i_S}) = V_{i_1}$. The choice of the first weight (V_{i_1}), rather than any other weight, is arbitrary. Training this method from scratch is equivalent to our vanilla weight sharing strategy.
- **Mean:** We take the average of all the weight tensors within the set, $W_i = \frac{1}{S_i} \sum_{k=1}^{S_i} V_{i_k}$ and $b_i = \frac{1}{S_i} \sum_{k=1}^{S_i} u_{i_k}$.
- **Scalar Weighted Mean**: Same as the average, except each weight tensor gets a learned scalar weighting, $W_i = \frac{1}{S_i} \sum_{k=1}^{S_i} \alpha_{i_k} V_{i_k}$, $\alpha_i \in \mathbb{R}$. We take a

simple mean of the bias, just as in the Mean strategy. The idea here is to provide the ability to learn more complex fusions, of which Choose First strategy, and Mean are special cases.

- **Channel Weighted Mean:** Rather than a scalar per layer, each weight tensor has a learned scalar for every filter, $W_i = \frac{1}{S_i} \sum_{k=1}^{S_i} \vec{\alpha_i} V_{i_k}$, $\vec{\alpha_i} \in \mathbb{R}^C$. Again, we take a simple mean of the bias, just as the Mean strategy. This strategy should allow the model to choose filters from specific weight tensors, or learn linear combinations.
- **Pointwise Convolution:** In this transformation, a pointwise convolution is applied to each layers weights, that maps to the same size filter, $W_i = \frac{1}{S_i} \sum_{k=1}^{S_i} A_i * V_{i_k}$, $A_i \in \mathbb{R}^{C \times C}$. This should allow arbitrary mixing and permutations of the kernels of each layer.

Table 3 shows that the Channel Weighted Mean fusion strategy allows us to compress the model by $2\times$ while maintaining the performance of original network. Furthermore, in Sect. 5, we show that weight fusion strategies allow us to learn representations similar to the original network.

4   Effect of Parameter Sharing on Different Isotropic Networks on the ImageNet Dataset

We evaluate the performance of the parameter sharing methods introduced in Sect. 3.2 on a variety of isotropic architectures. For more information on the training set-up and details, see Appendix C.

4.1   Parameter Sharing for ConvMixer

Typically, when considering model scaling, practitioners often vary parameters including the network depth, width, and image resolution, which scale the performance characteristics of the model [23]. In Table 4, we show that weight sharing models can significantly outperform baselines with the same FLOPs and parameters generated through traditional scaling alone, for example improving accuracy by roughly 10% Top-1 in some cases. We also show a full family of weight sharing ConvMixer models across multiple architectures in Table 5, and find that weight sharing can reduce parameters by over $2\times$ in many architectures while maintaining similar accuracy. These results show that weight sharing, in addition to typical scaling methods, is an effective axis for model scaling.

4.2   Parameter Sharing for Other Isotropic Networks

Although our evaluations have focused on ConvMixer, the methods discussed in Sect. 3 are generic and can be applied to any isotropic model. Here, we show results of applying parameter sharing to ConvNeXt [15] and the Vision Transformer (ViT) architecture.

Table 4. Weight sharing vs. model scaling for the ConvMixer model on ImageNet. For a fair comparison, we generate models with similar FLOPs and network parameters to our family of weight sharing models using traditional model scaling methods. Weight sharing methods achieve significantly better performance than traditional model scaling. See Table 5 for more details on the weight sharing model.

Network (C/D/P/K)	Resolution	Weight Sharing?	Share Rate	Params (M)	FLOPs (G)	Top-1 Acc(%)
768/32/14/3	224	✗	–	20.5	5.03	75.71
576/32/14/3	322	✗	–	11.8	5.92	70.326
768/16/14/3	322	✗	–	10.84	5.32	74.20
768/32/14/3	224	✓	2	11.02	5.03	**75.14**
384/32/14/3	448	✗	–	5.5	5.23	58.83
768/8/14/3	448	✗	–	6.04	5.38	68.31
768/32/14/3	224	✓	4	6.3	5.03	**71.91**
288/32/14/3	644	✗	–	3.25	6.23	40.46
768/4/14/3	644	✗	–	3.63	6.04	57.75
768/32/14/3	224	✓	8	3.95	5.03	**67.19**

ConvNeXt. Table 6 shows the results of parameter sharing on the ConvNeXt isotropic architecture. With parameter sharing, we are able to compress the model by 2× while maintaining similar accuracy on the ImageNet dataset.

Vision Transformer (ViT). We also apply our weight sharing method to a Vision Transformer, a self-attention based isotropic network. Due to space limit, we report accuracy numbers in Appendix B. Furthermore, we discuss the differences between applying weight sharing methods to CNNs versus transformers.

4.3 Comparison with State-of-the-Art Weight Sharing Methods

Table 7 compares the performance of weight sharing methods discussed in Sect. 3.2 with existing methods [4,9,14] on ImageNet. Compared to existing methods, our weight sharing schemes are effective; achieving higher compression rate while maintaining accuracy. For example, ConvMixer-768/32, ConvMixer-156/20, and ConvNeXt-18 with weight sharing and weight fusion achieve 1.86x, 1.91x and 1.92 share rate while having a similar accuracy. Existing weight sharing techniques [4,9] can only achieve at most 1.58x and 1.45x share rate at while maintaining accuracy. Although [14] can achieve 12x share rate, it results in a 8.8% accuracy drop.

These results show that isotropic networks can achieve a high share rate while maintaining accuracy with simple weight sharing methods. The traditional pyramid style networks, while using complicated sharing schemes [4,9,14], the

Table 5. Weight sharing family of ConvMixer model on ImageNet. Significant compression rates can be achieved without loss in accuracy across multiple isotropic ConvMixer models. We also generate a full family of weight sharing models by varying the *share rate*, which is the reduction factor in number of unique layers for the weight shared model compared to the original. C/D/P/K represents the dimension of channel, depth, patch and kernel of the model. If *weight fusion* is specified, the channel weighted mean strategy described in Sect. 3.2 is used.

Network (C/D/P/K)	Weight Sharing?	Share Rate	Weight Fusion?	Params (M)	FLOPs (G)	Top-1 Acc(%)
1536/20/7/3	✗	–	–	49.4	48.96	78.03
	✓	2	✓	25.8		**78.47**
	✓	4	✓	14		75.76
	✓	10	✗	6.9		72.27
768/32/14/3	✗	–	–	20.5	5.03	75.71
	✓	2	✓	11.02		**75.14**
	✓	4	✓	6.3		71.91
	✓	8	✓	3.95		67.19
512/16/14/9	✗	–	–	5.7	1.33	67.48
	✓	2	✓	3.63		**65.04**
	✓	4	✓	2.58		59.34
	✓	8	✗	2.05		54.25

Table 6. Effect of weight sharing on the ConvNeXt model on ImageNet. WS-ConvNeXxt has 2x less number of parameters but still achieves similar accuracy to the original ConvNeXt model.

Network	Depth	Share Rate	Params (M)	FLOPs (G)	Top-1 Acc(%)
ConvNeXt	18	–	22.3	4.3	78.7
	9	–	11.5	2.2	75.3
WS-ConvNeXt	18	2	11.5	4.3	**78.07**
		4	6.7		76.11
		6	4.3		72.07
		9	3.1		68.75

share rate is usually limited. Note that although our sharing schemes can achieve higher share rates, existing methods like [9,14] are able to directly reduce FLOPs, which our method does not address.

Table 7. Share rate and ImageNet accuracy comparison with existing weight sharing methods.

Network	Share Rate	Params (M)	FLOPs (G)	Top-1 Acc(%)
ConvMixer-768/32 [25] (baseline)	–	20.5	5.03	75.71
WS-ConvMixer-768/32-S2 (ours)	**1.86**	**11.02**	5.03	**75.14**
ConvMixer-1536/20 [25] (baseline)	–	49.4	48.96	78.03
WS-ConvMixer-1536/20-S2 (ours)	**1.91**	**25.8**	48.96	**78.47**
ConvNeXt-18 [15] (baseline)	–	22.3	4.3	78.7
WS-ConvNeXt-18-S2 (ours)	**1.92**	**11.5**	4.3	**78.07**
WS-ConvNeXt-18-S4 (ours)	3.33	6.7	4.3	76.11
ResNet-152 [6] (baseline)	–	60	11.5	78.3
IamNN [14]	12	5	2.5–9	69.5
ResNet-101 [6] (baseline)	–	44.54	7.6	77.95
DR-ResNet-65 [4]	1.58	28.12	5.49	78.12
DR-ResNet-44 [4]	2.2	20.21	4.25	77.27
ResNet-50 [6] (baseline)	–	25.56	3.8	76.45
DR-ResNet-35 [4]	1.45	17.61	3.12	76.48
ResNet50-OrthoReg [9]	1.25	20.51	4.11	76.36
ResNet50-OrthoReg-SharedAll [9]	1.6	16.02	4.11	75.65

5   Representation Analysis

In this sections, we perform qualitative analysis of our weight sharing models to better understand why they lead to improved performance and how they change model behavior. To do this, we first analyze the representations learned by the original network, compared to one trained with weight sharing. We follow a similar set-up to Sect. 3.1. We use CKA as a metric for representational similarity and compute pairwise similarity across all layers in both the networks we aim to compare. In Fig. 4(a) we first compare the representations learned by a vanilla weight sharing method to the representations of the original network. We find that there is no clear relationship between the representations learned. Once we introduce the weight fusion initialization strategy (Sect. 3.2), we find significant similarity in representations learned, as shown in Fig. 4(b). This suggests that our weight fusion initialization can guide the weight shared models to learn similar features to the original network. In Appendix F, we further analyze the weight shared models and characterize their robustness compared to standard networks.

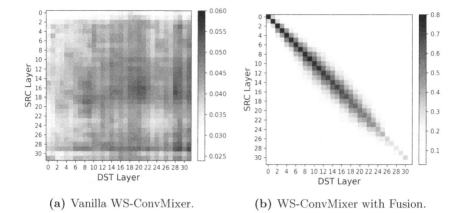

(a) Vanilla WS-ConvMixer.   (b) WS-ConvMixer with Fusion.

Fig. 4. (a) The CKA similarity analysis of a standard ConvMixer's intermediate feature maps compared to a vanilla weight shared ConvMixer, with share rate of 2. (b) The same analysis but compare to a weight shared ConvMixer initialized with weight fusion. The channel weighted mean fusion strategy is used (see Sect. 3.2).

6 Conclusion

Isotropic networks have the unique property in which all layers in the model have the same structure, which naturally enables parameter sharing. In this paper, we perform a comprehensive design space exploration of shared parameters in isotropic networks (SPIN), including the weight sharing topology, dynamic transformations and weight fusion strategies. Our experiments show that, when applying these techniques, we can compress state-of-the-art isotropic networks by up to 2 times without losing any accuracy across many isotropic architectures. Finally, we analyze the representations learned by weight shared networks and qualitatively show that the techniques we introduced, specifically fusion strategies, guide the weight shared model to learn similar representations to the original network. These results suggest that parameters sharing is an effective axis to consider when designing efficient isotropic neural networks.

References

1. Battash, B., Wolf, L.: Adaptive and iteratively improving recurrent lateral connections. CoRR abs/1910.11105 (2019). http://arxiv.org/abs/1910.11105
2. Dehghani, M., Gouws, S., Vinyals, O., Uszkoreit, J., Kaiser, L.: Universal transformers. ArXiv abs/1807.03819 (2019)
3. Dosovitskiy, A., et al.: An image is worth 16x16 words: transformers for image recognition at scale. arXiv preprint arXiv:2010.11929 (2020)
4. Guo, Q., Yu, Z., Wu, Y., Liang, D., Qin, H., Yan, J.: Dynamic recursive neural network. In: 2019 IEEE/CVF Conference on Computer Vision and Pattern Recognition (CVPR), pp. 5142–5151 (2019)

5. Han, S., Pool, J., Tran, J., Dally, W.J.: Learning both weights and connections for efficient neural network. ArXiv abs/1506.02626 (2015)

6. He, K., Zhang, X., Ren, S., Sun, J.: Deep residual learning for image recognition (2015)

7. Howard, A.G., et al.: Searching for mobilenetv3. In: 2019 IEEE/CVF International Conference on Computer Vision (ICCV), pp. 1314–1324 (2019)

8. Jastrzebski, S., Arpit, D., Ballas, N., Verma, V., Che, T., Bengio, Y.: Residual connections encourage iterative inference. CoRR abs/1710.04773 (2017). http://arxiv.org/abs/1710.04773

9. Kim, D., Kang, W.: Learning shared filter bases for efficient convnets. CoRR abs/2006.05066 (2020). https://arxiv.org/abs/2006.05066

10. Kornblith, S., Norouzi, M., Lee, H., Hinton, G.E.: Similarity of neural network representations revisited. ArXiv abs/1905.00414 (2019)

11. Kubilius, J., et al.: Aligning artificial neural networks to the brain yields shallow recurrent architectures (2018)

12. Kubilius, J., et al.: Brain-like object recognition with high-performing shallow recurrent ANNs. CoRR abs/1909.06161 (2019). http://arxiv.org/abs/1909.06161

13. Lan, Z., Chen, M., Goodman, S., Gimpel, K., Sharma, P., Soricut, R.: Albert: a lite BERT for self-supervised learning of language representations. In: ICLR. OpenReview.net (2020). http://dblp.uni-trier.de/db/conf/iclr/iclr2020.htmlLanCGGSS20

14. Leroux, S., Molchanov, P., Simoens, P., Dhoedt, B., Breuel, T.M., Kautz, J.: IamNN: iterative and adaptive mobile neural network for efficient image classification. CoRR abs/1804.10123 (2018). http://arxiv.org/abs/1804.10123

15. Liu, Z., Mao, H., Wu, C.Y., Feichtenhofer, C., Darrell, T., Xie, S.: A convnet for the 2020s. arXiv preprint arXiv:2201.03545 (2022)

16. Rastegari, M., Ordonez, V., Redmon, J., Farhadi, A.: XNOR-Net: ImageNet classification using binary convolutional neural networks. In: Leibe, B., Matas, J., Sebe, N., Welling, M. (eds.) ECCV 2016. LNCS, vol. 9908, pp. 525–542. Springer, Cham (2016). https://doi.org/10.1007/978-3-319-46493-0_32

17. Sandler, M., Howard, A.G., Zhu, M., Zhmoginov, A., Chen, L.C.: MobileNetV 2: inverted residuals and linear bottlenecks. In: 2018 IEEE/CVF Conference on Computer Vision and Pattern Recognition, pp. 4510–4520 (2018)

18. Shen, Z., Liu, Z., Xing, E.P.: Sliced recursive transformer. CoRR abs/2111.05297 (2021). https://arxiv.org/abs/2111.05297

19. Simonyan, K., Zisserman, A.: Very deep convolutional networks for large-scale image recognition (2015)

20. Spoerer, C.J., Kietzmann, T.C., Mehrer, J., Charest, I., Kriegeskorte, N.: Recurrent networks can recycle neural resources to flexibly trade speed for accuracy in visual recognition. bioRxiv (2020). https://doi.org/10.1101/677237, https://www.biorxiv.org/content/early/2020/03/26/677237

21. Takase, S., Kiyono, S.: Lessons on parameter sharing across layers in transformers. CoRR abs/2104.06022 (2021). https://arxiv.org/abs/2104.06022

22. Tan, M., Chen, B., Pang, R., Vasudevan, V., Le, Q.V.: MnasNet: platform-aware neural architecture search for mobile. In: 2019 IEEE/CVF Conference on Computer Vision and Pattern Recognition (CVPR), pp. 2815–2823 (2019)

23. Tan, M., Le, Q.: EfficientNet: rethinking model scaling for convolutional neural networks. In: Chaudhuri, K., Salakhutdinov, R. (eds.) Proceedings of the 36th International Conference on Machine Learning. Proceedings of Machine Learning Research, vol. 97, pp. 6105–6114. PMLR (2019)

24. Tolstikhin, I.O., et al.: MLP-mixer: An all-MLP architecture for vision. CoRR abs/2105.01601 (2021). https://arxiv.org/abs/2105.01601

25. Trockman, A., Kolter, J.Z.: Patches are all you need? CoRR abs/2201.09792 (2022). https://arxiv.org/abs/2201.09792
26. Wang, G., Zhao, Y., Tang, C., Luo, C., Zeng, W.: When shift operation meets vision transformer: an extremely simple alternative to attention mechanism. CoRR abs/2201.10801 (2022). https://arxiv.org/abs/2201.10801
27. Zhai, S., Talbott, W., Srivastava, N., Huang, C., Goh, H., Zhang, R., Susskind, J.M.: An attention free transformer. CoRR abs/2105.14103 (2021). https://arxiv.org/abs/2105.14103

Ensemble Knowledge Guided Sub-network Search and Fine-Tuning for Filter Pruning

Seunghyun Lee[iD] and Byung Cheol Song[(⊠)][iD]

Inha University, Incheon, Republic of Korea
bcsong@inha.ac.kr

Abstract. Conventional NAS-based pruning algorithms aim to find the sub-network with the best validation performance. However, validation performance does not successfully represent test performance, i.e., potential performance. Also, although fine-tuning the pruned network to restore the performance drop is an inevitable process, few studies have handled this issue. This paper provides a novel Ensemble Knowledge Guidance (EKG) to solve both problems at once. First, we experimentally prove that the fluctuation of loss landscape can be an effective metric to evaluate the potential performance. In order to search a sub-network with the smoothest loss landscape at a low cost, we employ EKG as a search reward. EKG utilized for the following search iteration is composed of the ensemble knowledge of interim sub-networks, i.e., the by-products of the sub-network evaluation. Next, we reuse EKG to provide a gentle and informative guidance to the pruned network while fine-tuning the pruned network. Since EKG is implemented as a memory bank in both phases, it requires a negligible cost. For example, in the case of ResNet-50, just 315 GPU hours are required to remove around 45.04% of FLOPS without any performance degradation, which can operate even on a low-spec workstation. the source code is available at here.

1 Introduction

Network pruning is attracting a lot of attention as a lightweight technique to reduce computation and memory cost by directly removing parameters of deep neural networks (DNNs). In particular, filter pruning is advantageous in accelerating using the basic linear algebra subprograms (BLAS) library because it eliminates parameters in units of filters. Recently, filter pruning has been regarded as a kind of neural architecture search (NAS), that is, the sub-network search process. Actually, some methods using the existing NAS algorithms succeeded in finding a high-performance pruned network [6,24,25,39]. Based on 'supernet' capable of dynamic inference, the other methods significantly reduced

Supplementary Information The online version contains supplementary material available at https://doi.org/10.1007/978-3-031-20083-0_34.

computational complexity while maintaining high performance of NAS-based pruning algorithm [2,11,35,44].

In general, NAS-based pruning algorithms consist of a search phase and a fine-tuning phase. We paid attention to the fact that prior arts seldom dealt with critical factors in each phase. First, note that rewards used in the search phase so far are not accurate enough to find the optimal sub-network. NAS-based pruning usually samples the valida-

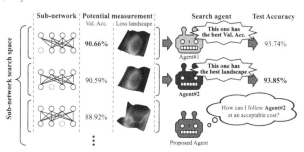

Fig. 1. Conceptual visualization showing the performance change of the search algorithm according to reward and the intrinsic goal of the proposed agent

tion set from the training dataset and uses the validation performance as a reward. In other words, it is assumed that validation performance has a high correlation with test performance, i.e., potential performance. Although this approach improves the search speed greatly, it implies the possibility of overfitting the sub-network to the validation set. Despite such a risk, most previous works adopted validation performance as the reward without any doubt. Second, there were no prior arts considering the characteristics of the pruned network in the fine-tuning phase. During the filter pruning procedure, some information loss is unavoidable. To recover the information loss, we should fine-tune the pruned network. Since the pruned network still possesses the information of the pre-trained network, an approach differentiated from randomly-initialized networks is required. However, prior arts employed general learning strategies used when learning pre-trained networks as they are or adopted primitive knowledge distillation (KD).

To solve two problems mentioned above, this paper presents a single solution, i.e., Ensemble Knowledge Guidance (EKG). In general, a student network constrained by teacher knowledge has a flatter local minima [3,26,37] and faster convergence speed [38], compared to ordinary optimization algorithms. Based on these characteristics, we formulate a way to improve the search and fine-tuning phases by using EKG as follows: First, we show that loss landscape fluctuation is a useful tool to evaluate the potential performance of sub-network. Figure 1 visualizes the concept. The sub-network with the highest potential performance is expected to be sufficiently generalized as well as to have a fast learning speed. Since the smoothness of loss landscape can estimate both factors, i.e., generalization and learning speed [9,20,33], we employ it to evaluate sub-networks. However, since loss landscape-based reward requires massive computational cost, it is impractical to utilize the smoothness of loss landscape as a search reward. So, we use EKG to find an optimal sub-network with the smoothest loss landscape at an acceptable cost. Based on the property of KD, EKG implicitly selects a sub-network with a smoother loss landscape and discovers the optimal sub-network more precisely than previous works that solely use the validation loss.

Here, as the source of knowledge, the output features of interim sub-networks, i.e., by-products of each search iteration, are stored and ensembled in the memory bank. Therefore, EKG incurs only a negligible cost because we don't have to infer the numerous genuine teacher networks in every training iteration.

Furthermore, EKG is applied once again to fine-tune the pruned network. As mentioned above, KD is an effective way to improve a small network. However, as the pruning rate increases, the gap between the pre-trained and pruned networks also increases, which makes it difficult to transfer knowledge [26]. To bridge the performance gap, we adopt interim sub-networks once again as teacher-assistant networks and build a memory bank with their knowledge. Then, according to the learning status of the pruned network, the knowledge of the memory bank is ensembled and transferred. Since the knowledge always maintains an appropriate gap with the pruned network, over-constraints can be avoided.

As a result, the proposed EKG can improve the performance of the pruned network while affecting both phases with a marginal cost increase. For example, when pruning ResNet-50, 45.04% FLOPS can be removed without any performance drop in only 315 GPU hours, which is an impressive result considering that our experiments are performed on a low spec workstation. Finally, the contributions of this paper are summarized as follows:

- As a new tool to measure the potential performance of sub-network in NAS-based pruning, the smoothness of loss landscape is presented. Also, the experimental evidence that the loss landscape fluctuation has a higher correlation with the test performance than the validation performance is provided.
- EKG is proposed to find a high potential sub-network and boost the fine-tuning process without complexity increase, which makes a high-performance light-weighted neural network democratized.
- To our knowledge, this paper provides the world-first approach to store the information of the search phase in a memory bank and to reuse it in the fine-tuning phase of the pruned network. The proposed memory bank contributes to greatly improving the performance of the pruned network.

2   Related Work

This section reviews the research trend of filter pruning. Existing filter pruning algorithms can be categorized into two groups. The first approach is filter importance scoring (FIS). FIS scores the importance of each filter according to a specific criterion and removes lower-score filters. Early FIS techniques adopted filter norms as criteria [13]. However, they faced with a problem of 'smaller-norm-less-importance criteria.' So, various scoring techniques were proposed to solve this problem, e.g., gradient-norm [27,43], geometric relation [16,39], and graph structure [19,40]. Although the latest FIS techniques improved the performance of the pruned network, a question still remains as to whether the existing filter scores are closely related to the performance degradation.

The second is a NAS-based approach and is more intuitive than the first. NAS-based methods define the number of filters as a search space and search for the optimal sub-network by learning and evaluating each sub-network [6, 24,25]. Here, the performance itself is considered a score. However, the NAS-based methodology inevitably requires a huge cost because each sub-network must be actually trained. Thus, supernet was proposed to reduce this massive cost. A supernet in which a low index filter has higher importance than a high index filter is a network trained to enable dynamic inference [11,44,46]. Once a supernet is trained, the search agent can explore the search space by free, so the search cost can be greatly reduced. Thus, supernet-based pruning has become the most dominant methodology. For example, [35,36] achieved state-of-the-art (SOTA) performance by adopting a more sophisticated search algorithm, and [44,45] presented a few effective methods for training supernets. However, since a supernet is not publicly available in general, it must be re-trained every time. In other words, supernet-based methods have a fatal drawback in that they cannot use well-trained networks [10,42].

The most important virtue of the pruning algorithm is to find a sub-network that minimizes information loss, but the process of restoring it is just as important. Surprisingly, how to fine-tune the pruned network has been rarely studied. For example, most pruning algorithms adopted a naive way of training the pruned network according to a general training strategy. If following such a naive way, a huge cost is required for fine-tuning. As another example, a method to re-initialize the pruned network according to lottery-ticket-hypothesis was proposed [7]. This method can guide the pruned network fallen into the local minima back to the global minima, but its learning cost is still huge. The last example is using KD [17]. In general, the pre-trained network has a similar structure to the pruned network, but has a higher complexity and is more informative. So, we can intuitively configure the pre-trained and pruned networks as a teacher-student pair. Recently, a few methods have been proposed to improve the performance of the pruned network with the pre-trained network as a teacher [6]. However, if the pruning rate increases, that is, if the performance gap between teacher and student increases, the information of the teacher network may not be transferred well [26]. Also, since the pruned network has information about the target dataset to some extent, unlike general student networks, fine-tuning for the pruning algorithm is required.

3   Method: Ensemble Knowledge Guidance

3.1   Accurate Potential Performance Evaluation

Conventional NAS-based pruning iteratively eliminates redundant filter set θ from the pre-trained network's filter set Θ_0 until the target floating point operations (FLOPs) are reached. This process is expressed by

$$\Theta_{i+1} = \Theta_i \setminus \theta_i^* \tag{1}$$

$$\theta_i^* = \underset{\theta_i}{\operatorname{argmax}} \; \mathcal{R}(\Theta_i, \theta_i) \tag{2}$$

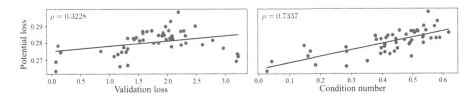

Fig. 2. (Left) Potential loss vs. validation loss (Right) Potential loss vs. condition number. 50 sub-networks of ResNet-56 trained on CIFAR10 were used for this experiment

where i is the search step, $\setminus$ is the set subtraction, and $\mathcal{R}$ indicates the reward of the search agent. It is costly to handle all filter combinations, so the agent usually divides the search space into intra-layer and inter-layer spaces. Then, the redundant filter set candidate $\phi_{i,l}$ is determined in each layer, and the candidate with the highest reward is selected. If a supernet is used as a pre-trained network, a candidate of each layer can be determined at no cost. As a result, the search process is regarded as a task that evaluates candidates, and is re-written by

$$\Theta_{i+1} = \Theta_i \setminus \phi_{i,l^*} \tag{3}$$

$$l^* = \operatorname*{argmax}_{l} \mathcal{R}(\Theta_i, \phi_{i,l})$$
$$\text{s.t. } \Phi_i = \{\phi_{i,l} | 1 \leq l \leq L\} \tag{4}$$

where L stands for the number of layers in DNN.

Among prior arts, some concentrated on how to construct candidates while maintaining the above-mentioned framework [2,44–46], and others presented new search agents to select the best candidate [35,36]. On the other hand, most of the previous studies employed sub-network's performance as a search reward $\mathcal{R}$ without any doubt. Since it is too heavy to use the entire training dataset for validation, the validation set $\mathcal{D}^{val}$ is sampled from the training dataset. Therefore, assuming that the loss function $\mathcal{L}$ is used as a metric to measure performance and $\phi_{i,l}$ is removed from Θ_i by a certain NAS-based pruning algorithm, the search reward is defined by

$$\mathcal{R}(\Theta_i, \phi_{i,l}) = -\mathcal{L}(\mathcal{D}^{val}; \Theta_i \setminus \phi_{i,l}) \tag{5}$$

The reason to search for the sub-network with the highest validation performance is that its potential performance after fine-tuning is expected to be high. In other words, it is assumed that validation performance is highly correlated with potential performance. To examine whether this assumption holds, we randomly sampled sub-networks of ResNet-56 [14] trained on CIFAR10 [18] and measured the validation performance and test performance after fine-tuning. Specifically, Pearson Correlation Coefficient [34] (PCC) $\rho_{X,Y}$ of the validation loss (X) and potential loss (Y) was calculated. For the detailed training configuration of this experiment, please refer to the supplementary material. The left plot of Fig. 2 shows that the correlation between the two losses is not so high. Especially, note that the validation loss has high variance in the low potential loss region of the most interest. This phenomenon makes it difficult to adopt the validation loss as a reliable reward. Since the search process is a sort of optimization, exploring the validation set makes the sub-network fit the validation set. In other words, there is a risk of overfitting. Therefore, instead of simply measuring validation performance, we need another indicator to measure generality.

We introduce loss landscape [20] as a means to analyze the generality of sub-networks. The more the loss landscape is smooth and close to convex, the more the network is robust and easy to be trained. Therefore, quantitatively measuring fluctuations in the loss landscape can determine how well the network is generalized, which provides higher-order information than performance. Because a network with fewer filters has relatively large fluctuations in the loss landscape, the loss landscape of the sub-network becomes more complex as pruning progresses. However, if the information of a certain filter is redundant, the generality error as well as fluctuations in the loss landscape do not increase. Based on this insight, we assume that a sub-network can be evaluated through the fluctuation of the loss landscape. To verify whether this assumption is valid, we examine the PCC of loss landscape fluctuation (X) and potential loss (Y). As an index representing the loss landscape fluctuation, we employed the condition number (CN), i.e., the ratio of minimum and maximum eigenvalues of the Hessian [20,30], which is defined by

$$\text{Condition number} = \left| \lambda^{\min} / \lambda^{\max} \right| \tag{6}$$

The right plot of Fig. 2 shows that CN has a higher correlation with potential loss than validation loss. In particular, note that the variance of CN is very small in the low potential loss region. This proves that CN is a more reliable indicator to evaluate sub-networks. Based on this experimental result, the next subsection designs a search process to select a sub-network with a smoother loss landscape.

3.2   Search with Ensemble Knowledge Guidance

CN must be a better indicator to evaluate potential performance than validation performance. However, since the Hessian matrix required to calculate CN causes a huge computation cost, it is burdensome to directly use CN as a search reward.

To design a new sub-network search process of a reasonable cost, we adopt EKG. KD is widely used to enhance the performance of a small-size network (student) with a large-size network (teacher). Many researchers have pointed out that the performance of a student is improved because the loss landscape of the student gets smoother by receiving the teacher's knowledge, that is, the generality error of the student is reduced [3,26,37]. Student networks are saturated with a smoother loss landscape under multi-directional constraints due to knowledge and target loss. However, if the pruning rate gets higher, a performance gap between the pre-trained network and the sub-network gets larger, making teacher knowledge not transferred effectively [26]. Fortunately, interim sub-networks have suitable properties as teacher-assistant networks [26]. In detail, interim sub-networks have intermediate performance between the pre-trained

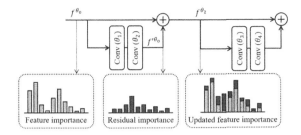

Fig. 3. An example of a position sensing a feature map f^{θ_2} for scoring a filter θ_2 in the proposed method. Compared with f'^{θ_2}, which only grasps residual importance, f^{θ_2} can observe that importance is well measured regardless of architecture characteristics

and pruned networks and are computationally efficient because there is no additional training cost. So, we build a memory bank and ensemble the knowledge of interim sub-networks at every step to keep the knowledge guidance at the middle point of the pre-trained network and the current sub-network. Finally, $\mathcal{R}$ is re-defined by

$$\mathcal{R}(\Theta_i, \phi_{i,l}) = -\mathcal{L}(\mathcal{D}^{val}; \Theta_i \setminus \phi_{i,l}) - \mathcal{L}(\mathcal{T}_i; \Theta_i \setminus \phi_{i,l}) \tag{7}$$

$$\mathcal{T}_i = \frac{1}{i+1} \sum_{j=0}^{i} \mathcal{O}(\mathcal{D}^{val}; \Theta_j) \tag{8}$$

where $\mathcal{O}(\cdot; \cdot)$ indicates the inference result.

Based on the proposed reward, we reconfigure the existing search process as follows. First, a score is given to each filter according to Eq. (7), and filters of low scores are selected as candidates. Here, in order to more accurately evaluate the importance of θ, the input feature map of the next layer f^θ is used as in Fig. 3. Since f^θ has both the information generated from θ and the characteristics of the network architecture, it can more accurately represent the importance of each filter. Using the well-known Taylor expansion [27,43], we propose the following scoring function:

$$\mathcal{S}(\theta) = \left| \frac{\partial \mathcal{R}(\Theta_i, \cdot)}{\partial f^\theta} f^\theta \right| \tag{9}$$

The detailed derivation process of the above formula is depicted in the supplementary material. Next, filters of lower scores as much as a specific ratio r for each layer become candidates, and a candidate with the highest reward is selected through a greedy search. Finally, the optimal sub-network Θ^* is searched iteratively.

In fact, the concept of transferring knowledge of a large sub-network to a small sub-network is already widely used for learning supernets [2,11,35,44,46]. We can say that the proposed reward based on EKG inherits this concept. Here, scoring based on Taylor expansion can compute the scores of all filters at once,

---

Algorithm 1: The proposed pruning algorithm

Input : $\Theta_0, \mathcal{D}^{train}, r$ **Output** : Fine-tuned network

1: Sample $\mathcal{D}^{subset}$ and $\mathcal{D}^{val}$ in $\mathcal{D}^{train}$ and fine-tune Θ_0 in an epoch on $\mathcal{D}^{subset}$

2: Store initial ensemble knowledge $\mathcal{T}_0$

3: **Repeat**

4:    Compute filter importance scores $\mathcal{S}(\theta)$ by Eq. (9)

5:    Sample candidates Φ_i in each layers with r.

6:    Select $\phi_{i,l}^*$ by Eq. (4) that maximizes Eq. (7).

7:    Get next pruned network Θ_{i+1} by Eq. (3).

8:    Update ensemble knowledge $\mathcal{T}_{i+1}$ by Eq. (8)

9:     $i = i + 1$

10: **Until** FLOPs reduction rate reaches the goal

11: $\Theta^* = \Theta_i$

12: Build memory bank by Eq. (10)

13: **Repeat**

14:    Get training sample in $\mathcal{D}^{train}$ and apply two augmentation functions.

15:    Minimize loss function in Eq. (12).

16: **Until** Training is done

---

so it not only requires a much lower cost than learning the supernet itself but also has the advantage of using well-trained networks as they are.

3.3 Fine-Tuning with Ensemble Knowledge Guidance

Even though a sub-network with high potential is available, the performance of the sub-network can be maximized through fine-tuning. So, we once again utilize EKG. As mentioned in Sect. 3.2, the pruning process generates many qualified teacher networks, i.e., interim sub-networks. If this knowledge is transferred to the pruned network again, the effect of inverse-tracking the information loss of the pruning process can be accomplished. However, acquiring knowledge by inferencing of all teacher networks at every iteration requires a huge computational cost. To transfer teacher knowledge efficiently, we propose to sample interim sub-networks with a uniform performance gap and store their knowledge in a memory bank. This entire procedure is expressed by

$$\mathcal{M} = \{\mathbf{M}_k = \mathcal{O}(\mathcal{D}^{train}, \mathcal{T}_k) | 1 \leq k \leq K\} \tag{10}$$

$$\mathcal{T}_k = \operatorname*{argmin}_{\Theta_i} \left| \frac{K-k}{K}\mathcal{L}(\Theta^*) + \frac{k}{K}\mathcal{L}(\Theta_0) - \mathcal{L}(\Theta_i) \right| \tag{11}$$

where K indicates the number of interim sub-networks to be sampled.

Memory bank knowledge smoothly bridges the performance gap between pretrained and pruned networks, but fixed knowledge can often cause overfitting. To resolve this side-effect, we employ contrastive learning. Contrastive learning is to minimize the gap between two data representations to which different augmentations are applied [4,10]. So, it allows DNNs to learn more generalized representations. We set memory bank knowledge as the center of the augmented data distribution and transfer it into the sub-network. As in the search process, we select the teacher network from $\mathcal{M}$ according to the performance of the pruned network, and then the ensemble knowledge is transferred. Therefore, the loss function $\mathcal{L}^{ft}$ for fine-tuning is defined by

$$\mathcal{L}^{ft} = \sum_{a=1}^{2} \mathcal{L}(\mathcal{D}^{train}, \mathcal{A}_a; \Theta^*) + \mathcal{L}(\overline{\mathbf{M}}; \Theta_i \setminus \phi_{i,l}) \tag{12}$$

$$\overline{\mathbf{M}} = \mathbb{E}\left\{ \mathbf{M}_k \;\middle|\; \begin{array}{l} 1 \leq k \leq K \\ \mathcal{L}(\mathbf{M}_k) \leq \mathcal{L}(\mathcal{D}^{train}; \Theta^*) \end{array} \right\} \tag{13}$$

where $\mathcal{A}$ stands for augmentation function. The proposed fine-tuning strategy injects extra information into the pruned network at almost no cost. Also, it not only improves the learning speed but also has a regularization effect so that the pruned network has higher performance. Algorithm 1 summarizes the proposed method. The proposed search process has the following differentiation points from the existing supernet-based search algorithm.

- Because of the ease of using the pre-trained network and the accurate potential performance evaluation, a better sub-network can be obtained.
- The potential of sub-network is maximized by transferring the information lost in the search phase back to the fine-tuning phase, resulting in high performance.
- Since the proposed search process has almost no additional cost compared to prior arts, it can operate even on a low-performance computer system.

4   Experimental Results

This section experimentally proves the performance of EKG. First, we verify that CN can be a metric that evaluates sub-networks more accurately than validation performance. Second, an ablation study is given. Third, we prove that the proposed method accomplishes more effective pruning than conventional methods in various configurations and achieves SOTA performance at a reasonable cost. The datasets for experiments were CIFAR10, CIFAR100 [18], and ImageNet [31]. ResNet family [14] and MobileNet-v2 [32] were used as network architectures. The proposed method was implemented in Tensorflow [1]. The CIFAR pre-trained network was trained by ourselves, and the ImageNet pre-trained network is a version released in Tensorpack [41]. Detailed hyper-parameters for training are shown in the supplementary material. GPU hours were calculated on GTX1080Ti, and 1 to 2 GPUs were utilized, depending on the architecture.

4.1   Empirical Analysis for Loss Landscape-Based Evaluation

This subsection proves that the proposed method effectively searches the sub-network of smoother loss landscape. Unfortunately, direct comparison with existing supernet-based search algorithms is unreasonable. Instead, we assigned a

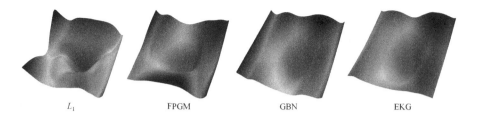

Fig. 4. Visualization of loss landscapes of sub-networks searched by various filter importance scoring algorithms

Table 1. Comparison of validation loss, condition number, and test accuracy of various scoring methods on CIFAR10 and CIFAR100

	Score	L_1 [13]	FPGM [16]	GBN [43]	EKG
CIFAR10	Validation loss	0.1569	0.1066	**0.0710**	0.0713
	Condition number	0.4078	0.1461	0.0414	**0.0211**
	Test accuracy	93.57	93.57	93.70	**93.85**
CIFAR100	Validation loss	1.2050	0.7635	**0.6816**	0.7733
	Condition number	0.2535	0.0838	0.0747	**0.0649**
	Test accuracy	71.43	71.60	71.60	**71.82**

score to each filter using a specific filter scoring technique, found a sub-network with the minimum validation loss, and employed the result for comparison.

First, we visualized the loss landscapes of the sub-networks for each method. As a network for visualization, ResNet-56 trained on CIFAR10 was used. Two directions for constructing the loss landscape were set to a gradient direction and its orthogonal random direction. Figure 4 is the visualization result. We can observe that the sub-network of EKG has the least fluctuation. In particular, it is noteworthy that L_1-norm-based pruning, which has been adopted in many studies, is in fact not so effective. GBN [43] has the lowest fluctuation among conventional techniques. GBN has a similar structure to the proposed method because it is also based on Taylor expansion. However, since a reward of GBN relies on performance only and does not consider the information of the architecture sufficiently, its loss landscape fluctuation is larger than that of EKG.

Next, let's compare the performance when sub-networks are trained with the same native fine-tuning (see Table 1). We could find that the correlation between the CN and the test accuracy is high. Even though EKG has a higher validation loss than GBN, its potential performance is expected to be higher because the CN or the loss landscape fluctuation of EKG is smaller. Actually, EKG's test accuracy on CIFAR10 was the highest as 93.85%. Therefore, we can find that the proposed search process based on ensemble KD effectively finds the sub-network of smoother loss landscape. Also, the loss landscape fluctuation is a very accurate indicator for evaluating potential performance.

Table 2. Ablation study to verify each part of the proposed method. Here, None, Single, and Ensemble refer to three methods of using the teacher network. GPU hours were measured in search and fine-tuning phases, respectively

Teacher		ResNet-56			MobileNet-v2		
Search	Fine-tune	CIFAR10	CIFAR100	GPU hours	CIFAR10	CIFAR100	GPU hours
Baseline		93.84	72.62	0.44	94.21	76.07	1.83
None	None	93.78 ($\pm$0.07)	71.63 ($\pm$0.21)	0.19/0.50	93.69 ($\pm$0.06)	74.27 ($\pm$0.18)	0.31/1.35
Single	None	93.54 ($\pm$0.09)	71.66 ($\pm$0.17)	0.21/0.50	93.73 ($\pm$0.04)	74.10 ($\pm$0.12)	0.35/1.35
Ensemble	None	93.85 ($\pm$0.10)	71.82 ($\pm$0.20)	0.21/0.50	93.89 ($\pm$0.04)	74.53 ($\pm$0.16)	0.35/1.35
Ensemble	Single	94.02 ($\pm$0.08)	72.62 ($\pm$0.15)	0.22/0.88	94.44 ($\pm$0.09)	76.11 ($\pm$0.15)	0.38/2.36
Ensemble	Ensemble	**94.09** ($\pm$0.07)	**72.93** ($\pm$0.16)	0.22/0.68	**94.52** ($\pm$0.05)	**76.29** ($\pm$0.17)	0.38/1.64

4.2 Ablation Study

This section analyzes the effects of EKG on the search and fine-tuning phases. In this experiment, three scenarios of the teacher network are compared: None, Single (when only the pre-trained network is used), and Ensemble (when the proposed ensemble teacher is used). ResNet-56 and MobileNet-v2 trained with CIFAR10 and CIFAR100 were used as datasets, and the FLOPs reduction rate was set to 50%. The experimental results are shown in Table 2.

First, let's look at the sub-network search phase. Although the search process of 'None' has almost no difference from that of GBN except for the score calculation position, 'None' achieved some performance improvement over GBN. In the case of 'Single,' over-constraint occurs due to the far performance gap between the pre-trained and interim sub-networks. Since interim sub-networks did not undergo fine-tuning to recover the information loss, this phenomenon is further

Table 3. Comparison with various techniques for ResNet-56 trained on CIFAR10. Here, Acc is the accuracy, and FLOPs $\downarrow$ and Param $\downarrow$ are the reduction rates of FLOPs and the number of parameters, respectively

Method	Acc	FLOPs $\downarrow$	Param $\downarrow$
TAS [6]	93.69	52.7	–
ABC [22]	93.23	54.13	52.20
GAL [23]	91.58	60.2	65.9
GBN-60 [43]	93.43	60.1	53.5
FPGM [16]	93.49	52.6	–
Hrank [21]	93.17	50	57.6
LFPC [15]	93.34	52.9	–
DSA [28]	92.93	52.6	–
ManiDP [39]	93.64	62.4	–
NPPM [8]	93.40	50.0	–
SRR-GR [40]	93.75	53.8	–
ResRep [5]	93.71	52.91	–
GDP [12]	93.97	53.35	–
EKG	93.92	55.22	33.69
	93.69	65.11	46.52

highlighted. Thus, we can observe no performance improvement or rather a decrease in most configurations. For example, in the case of ResNet-56 on CIFAR10, the performance of 'None' and 'Single' was 93.78% and 93.54%, respectively. That is, using knowledge rather causes degradation of 0.24%. On the other hand, if bridging the pre-trained and pruned networks through interim sub-networks, i.e., 'Ensemble,' a sub-network of high potential performance can be selected because of more effective knowledge-based guidance. For example, in the case of ResNet-56 on CIFAR10, the performance of 'Ensemble' is 0.31% higher than 'Single' and outperforms 'None.' Here, it is noteworthy that the performance of 'None' is already comparable than conventional algorithms (see Fig. 5). Therefore, performance improvement of 'Ensemble' is sufficiently acceptable considering that it requires almost no cost.

Next, because the fine-tuning phase re-learns the sub-network, the over-constraints caused by the far performance gap are somewhat mitigated. Accordingly, even if only a single teacher is used, the performance of the pruned network is sufficiently improved, reaching 94.02% in the case of ResNet-56 on CIFAR10. However, as the time for forwarding the teacher network is added, the training time increases by 71.09% compared to 'None.' On the other hand, when training with the proposed memory bank, which ensembles the knowledge of interim sub-networks, that is, in the case of 'Ensemble' knowledge, we can achieve similar performance improvement with only 29.03% increase in training time. This is because the memory bank provides qualified guidance by encapsulating information from many networks. Therefore, each part of the proposed method contributes to effectively providing a lightweight network with high performance at a low cost.

4.3   Performance Evaluation

Experiments using ResNet family [14] prove that the proposed method reaches SOTA performance. The datasets used in the experiments are CIFAR10 and ImageNet.

First, Table 3 shows the experimental results for ResNet-56 trained on CIFAR10. For example, when 65.11% of FLOPs are removed by the proposed method, the performance reaches 93.69%. This proves that EKG out-

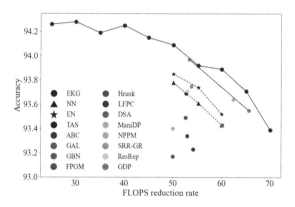

Fig. 5. FLOPS reduction rate-accuracy of various pruning techniques for ResNet-56 on CIFAR10

performs the other methods with similar FLOPs reduction rates. To analyze the performance of EKG in detail, Fig. 5 plots the performances at various FLOPs reduction rates. We can observe that EKG provides higher performance than other methods at all FLOPs reduction rates. Note that NN (i.e., 'None'), where any knowledge is not utilized in both phases, already shows a better or comparable performance than most conventional methods. Here, EN utilizing 'Ensemble' knowledge only in the search phase gives sufficient performance improvement. This is because EKG accurately searches for a sub-network with high potential performance even at high reduction rates. In particular, the higher performance than the pre-trained network even at low FLOPs reduction rates shows that the proposed fine-tuning has a similar regularization effect even though the actual

Table 4. Comparison with various techniques for ResNet family trained on ImageNet

ResNet	Method	Top-1 (diff.)	Top-5 (diff.)	FLOPs ↓
18	TAS [6]	69.15 (−1.50)	89.19 (−0.68)	33.3
	ABC [22]	67.80 (−1.86)	88.00 (−1.08)	46.9
	FPGM [16]	68.41 (−1.87)	88.48 (−1.15)	41.8
	DSA [28]	68.62 (−1.11)	88.25 (−0.82)	40.0
	DMCP [11]	69.20	N/A	43.0
	ManiDP [39]	68.88 (−0.88)	88.76 (−0.32)	51.0
	EKG	69.39 (−0.99)	88.65 (−0.87)	50.1
34	FPGM [22]	72.63 (−1.28)	91.08 (−0.54)	41.1
	SFP [16]	71.84 (−2.09)	89.70 (−1.92)	41.1
	NPPM [8]	73.01 (−0.29)	91.30 (−0.12)	44.0
	ManiDP [39]	73.30 (−0.01)	91.42 (−0.00)	46.8
	EKG	73.51 (−0.34)	91.27 (−0.19)	45.1
50	DSA [29]	75.1 (−0.92)	92.45 (−0.41)	40.5
	FPGM [22]	75.59 (−0.56)	92.63 (−0.24)	42.7
	BNP [25]	75.51 (−1.01)	92.43 (−0.66)	45.6
	GBN [43]	76.19 (+0.31)	92.83 (−0.16)	41.0
	TAS [6]	76.20 (−1.26)	92.06 (−0.81)	44.1
	SRR-GR [40]	75.76 (−0.37)	92.67 (−0.19)	45.3
	NPPM [8]	75.96 (−0.19)	92.75 (−0.12)	56.2
	ResRep [5]	76.15 (−0.00)	92.89 (+0.02)	54.9
	Autoslim [44]	75.6	N/A	51.6
	DMCP [11]	76.20	N/A	46.7
	CafeNet [35]	76.90	93.3	52.0
	BCNet [36]	76.90	93.3	52.0
	EKG	76.43 (−0.02)	93.13 (−0.02)	45.0
		75.93 (−0.52)	92.82 (−0.33)	55.0
	EKG-BYOL	76.60 (−0.40)	93.23 (−0.31)	55.0

teacher network is not used. Also, this phenomenon indicates that better fine-tuning strategy is more crucial than better sub-network search. Next, Table 4 shows the experimental results for ResNet family trained on ImageNet. EKG achieved SOTA performance in all architectures. For instance, top-1 accuracy of EKG reached 73.51% at 45.1% FLOPs reduction rate for ResNet-34. Since EKG can be plugged-in easily, we adopted a primitive search algorithm, i.e., the greedy search of Autoslim [44] to verify EKG. As a result, EKG improved Autoslim by 0.34% in top-1 accuracy and by 3.5% in FLOPs reduction rate for ResNet-50. On the other hand, in the case of ResNet-50, EKG showed lower performance than the latest supernet-based algorithms, i.e., CafeNet and BCNet.

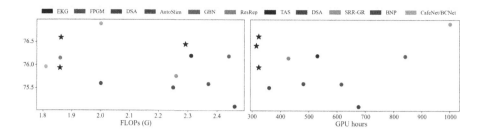

Fig. 6. Performance analysis for ResNet-50 trained on ImageNet. The left plot is the FLOPs reduction rate-Top-1 accuracy, and the right plot is the GPU hours-Top-1 accuracy

However, since EKG can employ well-trained networks which are not allowed for supernet-based algorithms, it can overcome even CafeNet and BCNet. For example, when using BYOL [10], which is the best among available Tensorflow-based ResNet-50, the top-1 accuracy of EKG was improved up to 76.6% without cost increase. Therefore, if a better pre-trained network is released, EKG will be able to overwhelm the supernet in performance.

For further analysis, the GPU hours-accuracy plot is given in Fig. 6, together with the results for ResNet-50. The cost of pruning and fine-tuning ResNet-50 with EKG is about 315 GPU hours, which is much lower than the other methods with similar performance. In particular, EKG is computationally efficient because we used just mid-range GPUs mounted on general-purpose workstations. In other words, EKG democratizes DNNs because it allows more users to learn high-performance lightweight DNNs.

4.4 Limitation

EKG improves the performance of conventional NAS-based pruning by effectively searching and fine-tuning sub-networks. However, some limitations still remain. First, as mentioned above, CN looks impractical because it cannot directly replace performance due to computational complexity. In this paper, we indirectly induced the sub-network of the smoother loss landscape to be selected, but additional research is needed on how the CN can be used as a direct metric. Second, the fact that EKG has not been evaluated in sufficiently many configurations obscures its generalization a little bit. In order to analyze the performance of EKG more clearly, it is necessary to verify light-weight architectures on a large dataset or to attach to other search algorithms, e.g., evolutionary search.

5 Conclusion

This paper proposes EKG as an efficient and effective solution for search reward and fine-tuning strategy, which have been rarely considered in existing NAS-based pruning approaches. In particular, sub-network evaluation based on loss landscape fluctuations can reflect generality more accurately than validation performance which may cause a high risk of overfitting. Furthermore, EKG is pretty valuable because it can be easily plugged-in to the existing NAS-based pruning. Despite some unresolved limitations, we expect this paper to provide meaningful insight to filter pruning researchers. In particular, we pointed out and corrected the fact that inaccurate validation performance has been adopted without any doubt.

Acknowledgments. This work was supported by IITP grants (No.2021-0-02068, AI Innovation Hub and RS-2022-00155915, Artificial Intelligence Convergence Research Center(Inha University)), and the NRF grants (No. 2022R1A2C2010095 and No. 2022R1A4A1033549) funded by the Korea government (MSIT).

References

1. Abadi, M., et al.: TensorFlow: a system for large-scale machine learning. In: 12th {USENIX} Symposium on Operating Systems Design and Implementation ({OSDI} 16), pp. 265–283 (2016)
2. Cai, H., Gan, C., Wang, T., Zhang, Z., Han, S.: Once-for-all: Train one network and specialize it for efficient deployment. In: International Conference on Learning Representations (2019)
3. Chen, T., Zhang, Z., Liu, S., Chang, S., Wang, Z.: Robust overfitting may be mitigated by properly learned smoothening. In: International Conference on Learning Representations (2021). https://openreview.net/forum?id=qZzy5urZw9
4. Chen, T., Kornblith, S., Norouzi, M., Hinton, G.: A simple framework for contrastive learning of visual representations (2020)
5. Ding, X., et al.: ResRep: lossless CNN pruning via decoupling remembering and forgetting. In: Proceedings of the IEEE/CVF International Conference on Computer Vision, pp. 4510–4520 (2021)
6. Dong, X., Yang, Y.: Network pruning via transformable architecture search. In: Advances in Neural Information Processing Systems, pp. 760–771 (2019)
7. Frankle, J., Carbin, M.: The lottery ticket hypothesis: finding sparse, trainable neural networks. In: International Conference on Learning Representations (2018)
8. Gao, S., Huang, F., Cai, W., Huang, H.: Network pruning via performance maximization. In: Proceedings of the IEEE/CVF Conference on Computer Vision and Pattern Recognition, pp. 9270–9280 (2021)
9. Gotmare, A., Keskar, N.S., Xiong, C., Socher, R.: A closer look at deep learning heuristics: Learning rate restarts, warmup and distillation. arXiv preprint arXiv:1810.13243 (2018)
10. Grill, J.B., et al.: Bootstrap your own latent: a new approach to self-supervised learning. arXiv preprint arXiv:2006.07733 (2020)
11. Guo, S., Wang, Y., Li, Q., Yan, J.: DMCP: differentiable Markov channel pruning for neural networks. In: Proceedings of the IEEE/CVF Conference on Computer Vision and Pattern Recognition, pp. 1539–1547 (2020)
12. Guo, Y., Yuan, H., Tan, J., Wang, Z., Yang, S., Liu, J.: GDP: stabilized neural network pruning via gates with differentiable polarization. In: Proceedings of the IEEE/CVF International Conference on Computer Vision, pp. 5239–5250 (2021)

13. Han, S., Mao, H., Dally, W.J.: Deep compression: compressing deep neural networks with pruning, trained quantization and Huffman coding. arXiv preprint arXiv:1510.00149 (2015)
14. He, K., Zhang, X., Ren, S., Sun, J.: Deep residual learning for image recognition. In: Proceedings of the IEEE Conference on Computer Vision and Pattern Recognition, pp. 770–778 (2016)
15. He, Y., Ding, Y., Liu, P., Zhu, L., Zhang, H., Yang, Y.: Learning filter pruning criteria for deep convolutional neural networks acceleration. In: Proceedings of the IEEE/CVF Conference on Computer Vision and Pattern Recognition (CVPR) (2020)
16. He, Y., Liu, P., Wang, Z., Hu, Z., Yang, Y.: Filter pruning via geometric median for deep convolutional neural networks acceleration. In: Proceedings of the IEEE Conference on Computer Vision and Pattern Recognition, pp. 4340–4349 (2019)
17. Hinton, G., Vinyals, O., Dean, J.: Distilling the knowledge in a neural network. arXiv preprint arXiv:1503.02531 (2015)
18. Krizhevsky, A., Hinton, G.: Learning multiple layers of features from tiny images. Technical Report Citeseer (2009)
19. Lee, S., Heo, B., Ha, J.W., Song, B.C.: Filter pruning and re-initialization via latent space clustering. IEEE Access 8, 189587–189597 (2020). https://doi.org/10.1109/ACCESS.2020.3031031
20. Li, H., Xu, Z., Taylor, G., Studer, C., Goldstein, T.: Visualizing the loss landscape of neural nets. In: Proceedings of the 32nd International Conference on Neural Information Processing Systems, pp. 6391–6401 (2018)
21. Lin, M., et al.: HRank: filter pruning using high-rank feature map. In: Proceedings of the IEEE/CVF Conference on Computer Vision and Pattern Recognition (CVPR) (2020)
22. Lin, M., Ji, R., Zhang, Y., Zhang, B., Wu, Y., Tian, Y.: Channel pruning via automatic structure search. arXiv preprint arXiv:2001.08565 (2020)
23. Lin, S., et al.: Towards optimal structured CNN pruning via generative adversarial learning. In: Proceedings of the IEEE Conference on Computer Vision and Pattern Recognition, pp. 2790–2799 (2019)
24. Liu, Z., et al.: Metapruning: meta learning for automatic neural network channel pruning. In: Proceedings of the IEEE/CVF International Conference on Computer Vision, pp. 3296–3305 (2019)
25. Lu, X., Huang, H., Dong, W., Li, X., Shi, G.: Beyond network pruning: a joint search-and-training approach. In: IJCAI, pp. 2583–2590 (2020)
26. Mirzadeh, S.I., Farajtabar, M., Li, A., Levine, N., Matsukawa, A., Ghasemzadeh, H.: Improved knowledge distillation via teacher assistant. In: Proceedings of the AAAI Conference on Artificial Intelligence, vol. 34, pp. 5191–5198 (2020)
27. Molchanov, P., Tyree, S., Karras, T., Aila, T., Kautz, J.: Pruning convolutional neural networks for resource efficient inference. In: International Conference on Learning Representations (2017)
28. Ning, X., Zhao, T., Li, W., Lei, P., Wang, Yu., Yang, H.: DSA: more efficient budgeted pruning via differentiable sparsity allocation. In: Vedaldi, A., Bischof, H., Brox, T., Frahm, J.-M. (eds.) ECCV 2020. LNCS, vol. 12348, pp. 592–607. Springer, Cham (2020). https://doi.org/10.1007/978-3-030-58580-8_35
29. Ning, X., Zhao, T., Li, W., Lei, P., Wang, Y., Yang, H.: DSA: more efficient budgeted pruning via differentiable sparsity allocation. arXiv preprint arXiv:2004.02164 (2020)
30. Nocedal, J., Wright, S.: Numerical Optimization. Springer, Heidelberg (2006)

31. Russakovsky, O., et al.: ImageNet large scale visual recognition challenge. Int. J. Comput. Vis. **115**(3), 211–252 (2015). https://doi.org/10.1007/s11263-015-0816-y

32. Sandler, M., Howard, A., Zhu, M., Zhmoginov, A., Chen, L.C.: MobileNetV 2: inverted residuals and linear bottlenecks. In: Proceedings of the IEEE Conference on Computer Vision and Pattern Recognition, pp. 4510–4520 (2018)

33. Santurkar, S., Tsipras, D., Ilyas, A., Madry, A.: How does batch normalization help optimization? In: Proceedings of the 32nd International Conference on Neural Information Processing Systems, pp. 2488–2498 (2018)

34. Soper, H., Young, A., Cave, B., Lee, A., Pearson, K.: On the distribution of the correlation coefficient in small samples. appendix ii to the papers of "student" and RA fisher. Biometrika **11**(4), 328–413 (1917)

35. Su, X., et al.: Locally free weight sharing for network width search. In: International Conference on Learning Representations (2021). https://openreview.net/forum?id=S0UdquAnr9k

36. Su, X., You, S., Wang, F., Qian, C., Zhang, C., Xu, C.: BCNet: searching for network width with bilaterally coupled network. In: Proceedings of the IEEE/CVF Conference on Computer Vision and Pattern Recognition (CVPR), pp. 2175–2184 (2021)

37. Tan, X., Ren, Y., He, D., Qin, T., Liu, T.Y.: Multilingual neural machine translation with knowledge distillation. In: International Conference on Learning Representations (2019). https://openreview.net/forum?id=S1gUsoR9YX

38. Tang, J., et al.: Understanding and improving knowledge distillation. arXiv preprint arXiv:2002.03532 (2020)

39. Tang, Y., et al.: Manifold regularized dynamic network pruning. In: Proceedings of the IEEE/CVF Conference on Computer Vision and Pattern Recognition, pp. 5018–5028 (2021)

40. Wang, Z., Li, C., Wang, X.: Convolutional neural network pruning with structural redundancy reduction. In: Proceedings of the IEEE/CVF Conference on Computer Vision and Pattern Recognition, pp. 14913–14922 (2021)

41. Wu, Y., et al.: Tensorpack. https://github.com/tensorpack/ (2016)

42. Yalniz, I.Z., Jégou, H., Chen, K., Paluri, M., Mahajan, D.: Billion-scale semi-supervised learning for image classification (2019)

43. You, Z., Yan, K., Ye, J., Ma, M., Wang, P.: Gate decorator: global filter pruning method for accelerating deep convolutional neural networks. In: Advances in Neural Information Processing Systems, pp. 2133–2144 (2019)

44. Yu, J., Huang, T.: AutoSlim: towards one-shot architecture search for channel numbers. arXiv preprint arXiv:1903.11728 (2019)

45. Yu, J., Huang, T.S.: Universally slimmable networks and improved training techniques. In: Proceedings of the IEEE/CVF International Conference on Computer Vision, pp. 1803–1811 (2019)

46. Yu, J., Yang, L., Xu, N., Yang, J., Huang, T.: Slimmable neural networks (2018)

Network Binarization via Contrastive Learning

Yuzhang Shang[1], Dan Xu[2], Ziliang Zong[3], Liqiang Nie[4], and Yan Yan[1(✉)]

[1] Illinois Institute of Technology, Chicago, USA
yshang4@hawk.iit.edu, yyan34@iit.edu
[2] Hong Kong University of Science and Technology, Clear Water Bay, Hong Kong
danxu@cse.ust.hk
[3] Texas State University, San Marcos, USA
ziliang@txstate.edu
[4] Harbin Institute of Technology, Shenzhen, China

Abstract. Neural network binarization accelerates deep models by quantizing their weights and activations into 1-bit. However, there is still a huge performance gap between Binary Neural Networks (BNNs) and their full-precision (FP) counterparts. As the quantization error caused by weights binarization has been reduced in earlier works, the activations binarization becomes the major obstacle for further improvement of the accuracy. BNN characterises a unique and interesting structure, where the binary and latent FP activations exist in the same forward pass (*i.e.* Binarize($\mathbf{a}_F$) = $\mathbf{a}_B$). To mitigate the information degradation caused by the binarization operation from FP to binary activations, we establish a contrastive learning framework while training BNNs through the lens of Mutual Information (MI) maximization. MI is introduced as the metric to measure the information shared between binary and the FP activations, which assists binarization with contrastive learning. Specifically, the representation ability of the BNNs is greatly strengthened via pulling the positive pairs with binary and FP activations from the same input samples, as well as pushing negative pairs from different samples (the number of negative pairs can be exponentially large). This benefits the downstream tasks, not only classification but also segmentation and depth estimation, *etc.* The experimental results show that our method can be implemented as a pile-up module on existing state-of-the-art binarization methods and can remarkably improve the performance over them on CIFAR-10/100 and ImageNet, in addition to the great generalization ability on NYUD-v2. The code is available at https://github.com/42Shawn/CMIM.

Keywords: Neural network compression · Network binarization · Contrastive learning · Mutual information maximization

Supplementary Information The online version contains supplementary material available at https://doi.org/10.1007/978-3-031-20083-0_35.

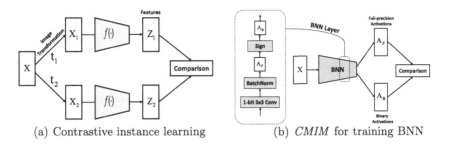

(a) Contrastive instance learning   (b) *CMIM* for training BNN

Fig. 1. (a): In contrastive instance learning, the features produced by different transformations of the same sample are contrasted to each other. **(b)**: However BNN can yield the binary activations A_B and full-precision activations A_F (*i.e.* two transformations of an image both from the same BNN) in the same forward pass, thus the BNN can act as two image transformations in the literature of contrastive learning.

1   Introduction

Although deep learning [27] has achieved remarkable success in various computer vision tasks such as image classification [25] and semantic image segmentation [5], its over-parametrization problem makes its computationally expensive and storage excessive. To advance the development of deep learning in resource-constrained scenarios, several neural network compression paradigms have been proposed, such as network pruning [15,28], knowledge distillation [19,39] and network quantization [22]. Among the network quantization methods, the network binarization method stands out for quantizing weights and activations (*i.e.* intermediate feature maps) to ±1, compressing the full-precision counterpart $32\times$, and replacing time-consuming inner-product in full-precision networks with efficient xnor-bitcount operation in the BNNs [22].

However, severe accuracy drop-off always exists between full-precision models and their binary counterparts. To tackle this problem, previous works mainly focus on reducing the quantization error induced by weights binarization [29,38], and elaborately approximating binarization function to alleviate the gradient mismatch issue in the backward propagation [31,37]. Indeed, they achieve the SoTA performance. Yet narrowing down the quantization error and enhancing the gradient transmission reach their bottlenecks [4,23], since the 1W32A (only quantizing the weights into 1-bit, remaining the activations 32-bit) models are capable of performing as well as the full-precision models [18,29], implying that the activations binarization becomes the main issue for further performance improvement.

To address the accuracy degradation caused by the activations binarization, a few studies are proposed to regulate the distributions of the binary activations, *e.g.* researchers in [10] design a distribution loss to explicitly regularize the activation flow; researchers in [23] propose to shift the thresholds of binary activation functions to unbalance the distribution of the binary activations. They heuristically design low-level patterns to analyze the distributions of binary activations, such as minimum of the activations and the balanced property of distributions. Nevertheless, they neglect the high-level indicators of the distribution and the unique characteristics of BNN, where the binary activations and latent full-precision activations co-exist in the same forward pass. Thus, we argue that the high-level properties of distributions, such as correlations and dependencies between binary and full-precision activations should be captured and utilized.

In this work, we explore introducing mutual information for BNNs, in which the mutual information acts as a metric to quantify the information amount shared by the binary and latent real-valued activations in BNNs. In contrast to the works mentioned above focusing on learning the distribution of binary activations, mutual information naturally captures statistical dependencies between variables, quantifying the degree of the dependence [11]. Based on this metric, we propose a novel method, termed as Network Binarization via Contrastive Learning for Mutual Information Maximization (*CMIM*). Specifically, we design a highly effective optimization strategy using contrastive estimation for mutual information maximization. As illustrated in Fig. 1, we replace the data transformation module in contrastive learning with the exclusive structure in BNNs, where full-precision and binary activations are in the same forward pass. In this way, contrastive learning contributes to inter-class decorrelation of binary activations, and avoids collapse solutions. In other words, our method is built upon a contrastive learning framework to learn representative binary activations, in which we pull the binary activation closer to the full-precision activation and push the binary activation further away from other binary activations in the contrastive space. Moreover, by utilizing an additional MLP module to extract representations of activations, our method can explicitly capture higher-order dependencies in the contrastive space. To the best of our knowledge, it is the first work aiming at maximizing the mutual information of the activations in BNNs within a contrastive learning framework.

Overall, the contributions of this paper are three-fold:

- Considering the distributions of activations, we propose a novel contrastive framework to optimize BNNs, via maximizing the mutual information between the binary activation and its latent real-valued counterpart;
- We develop an effective contrastive learning strategy to achieve the goal of mutual information maximization for BNNs, and benefited from it, the representation ability of BNNs is strengthened for not only the classification task but also downstream CV tasks;
- Experimental results show that our method can significantly improve the existing SoTA methods over the classification task on CIFAR-10/100 and ImageNet, *e.g.* 6.4% on CIFAR-100 and 3.0% on ImageNet. Besides, we also demonstrate the great generalization ability of the proposed *CMIM* on other challenging CV tasks such as depth estimation and semantic segmentation.

2   Related Work

In [22], the researchers introduce the sign function to binarize weights and activations to 1-bit, initiating the studies of BNNs. In this work, the straight-through estimator (STE) [2] is utilized to approximate the derivative of the sign function. Following the seminal art, copious studies contribute to improving the performance of BNNs. For example, Rastegari *et al.* [38] disclose that the quantization error between the full-precision weights and the corresponding binarized weights is one of the major obstacles degrading the representation capabilities of BNNs. Reducing the quantization error thus becomes a fundamental research direction to improve the performance of BNNs. Researchers propose XNOR-Net [38] to introduce a scaling factor calculated by L1 norm for both weights and activation functions to minimize the quantization error. Inspired by XNOR-Net, XNOR++ [3] further learns both spatial and channel-wise scaling factors to improves the performances. Bi-Real [31] proposes double residual connections with full-precision downsampling layers to mitigate the excessive gradient vanishing issue caused by binarization. ProxyBNN [18] designs a proxy matrix as a basis of the latent parameter space to guide the alignment of the weights with different bits by recovering the smoothness of BNNs. ReActNet [32] implements binarization with MobileNet [21] instead of ResNet, and achieves SoTA performance.

Nevertheless, we argue that those methods focusing on narrowing down the quantization error and enhancing the gradient transmission reach their bottleneck (*e.g.* 1W32A ResNet-18 trained by ProxyBNN achieves 67.7% Top-1 accuracy on ImageNet, while full-precision version is only 68.5%). Because they neglect the activations in BNNs, especially the relationship between the binary and latent full-precision activations. We treat them as discrete variables and investigate them under the metric of mutual information. By maximizing the mutual information via contrastive learning, the performance of BNNs is further improved. The experimental results show that *CMIM* can consistently improve the aforementioned methods by directly adding our *CMIM* module on them.

3   Training BNNs via Contrastive Learning for Mutual Information Maximization

3.1   Preliminaries

We define a K-layer Multi-Layer Perceptron (MLP). For simplification, we discard the bias term of this MLP. Then the network $f(\mathbf{x})$ can be denoted as:

$$f(\mathbf{W}^1, \cdots, \mathbf{W}^K; \mathbf{x}) = (\mathbf{W}^K \cdot \sigma \cdot \mathbf{W}^{K-1} \cdot \cdots \cdot \sigma \cdot \mathbf{W}^1)(\mathbf{x}), \qquad (1)$$

where $\mathbf{x}$ is the input sample and $\mathbf{W}^k : \mathbb{R}^{d_{k-1}} \longmapsto \mathbb{R}^{d_k} (k = 1, ..., K)$ stands for the weight matrix connecting the $(k-1)$-th and the k-th layer, with d_{k-1} and d_k representing the sizes of the input and output of the k-th network layer, respectively. The $\sigma(\cdot)$ function performs element-wise activation operation on the input feature maps.

Based on those predefined notions, the sectional MLP $f^k(\mathbf{x})$ with the front k layers of the $f(\mathbf{x})$ can be represented as:

$$f^k(\mathbf{W}^1, \cdots, \mathbf{W}^k; \mathbf{x}) = (\mathbf{W}^k \cdot \sigma \cdots \sigma \cdot \mathbf{W}^1)(\mathbf{x}). \qquad (2)$$

And the MLP f can be seen as a special case in the function sequence $\{f^k\}(k \in \{1, \cdots, K\})$, *i.e.* $f = f^K$.

Binary Neural Networks. Here, we review the general binarization method in [8,22], which maintains latent full-precision weights $\{\mathbf{W}_F^k\}(k \in \{1, \cdots, K\})$ for gradient updates, and the k-th weight matrix $\mathbf{W}_F^k$ is binarized into ± 1, obtaining the binary weight matrix $\mathbf{W}_B^k$ by a binarize function (normally $sgn(\cdot)$), *i.e.* $\mathbf{W}_B^k = sgn(\mathbf{W}_F^k)$. Then the intermediate activation map (full-precision) of the k-th layer is produced by $\mathbf{A}_F^k = \mathbf{W}_B^k \mathbf{A}_B^{k-1}$. Finally, the same sign function is used to binarize the full-precision activations into binary activations as $\mathbf{A}_B^k = sgn(\mathbf{A}_F^k)$ (see Fig. 1b), and the whole forward pass of a BNN is performed by iterating this process for L times.

Mutual Information and Contrastive Learning. For two discrete variables $\mathbf{X}$ and $\mathbf{Y}$, their mutual information (MI) can be defined as [26]:

$$I(\mathbf{X}, \mathbf{Y}) = \sum_{x,y} P_{\mathbf{XY}}(x, y) \log \frac{P_{\mathbf{XY}}(x, y)}{P_{\mathbf{X}}(x) P_{\mathbf{Y}}(y)}, \tag{3}$$

where $P_{\mathbf{XY}}(x, y)$ is the joint distribution, $P_{\mathbf{X}}(x) = \sum_y P_{\mathbf{XY}}(x, y)$ and $P_{\mathbf{Y}}(y) = \sum_x P_{\mathbf{XY}}(x, y)$ are the marginals of $\mathbf{X}$ and $\mathbf{Y}$, respectively.

Mutual information quantifies the amount of information obtained about one random variable by observing the other random variable. It is a dimensionless quantity with (generally) units of bits, and can be considered as the reduction in uncertainty about one random variable given knowledge of another. High mutual information indicates a large reduction in uncertainty and *vice versa* [26]. In the content of binarization, considering the binary and full-precision activations as random variables, we would like them share as much information as possible, since the binary activations are proceeded from their corresponding full-precision activations. Theoretically, the mutual information between those two variables should be maximized.

Our motivation can also be testified from the perspective of RBNN [29]. In RBNN, Lin *et al.* devise a rotation mechanism leading to around 50% weight flips which maximizes the information gain, $H(\mathbf{a}_B^{k,i})$. As MI can be written in another form as $I(\mathbf{X}, \mathbf{Y}) = H(\mathbf{X}) - I(\mathbf{X} \mid \mathbf{Y})$, the MI between binary and FP activations can be formulated as:

$$I(\mathbf{a}_B^{k,i}, \mathbf{a}_F^{k,j}) = H(\mathbf{a}_B^{k,i}) - I(\mathbf{a}_B^{k,i} \mid \mathbf{a}_F^{k,j}), \tag{4}$$

in which maximizing the first term on the right can partially lead to maximizing the whole MI. In this work, we aim to universally maximize the targeted MI.

Recently, contrastive learning is proven to be an effective approach to MI maximization, and many methods based on contrastive loss for self-supervised learning are proposed, such as Deep InfoMax [20], Contrastive Predictive Coding [34], MemoryBank [42], Augmented Multiscale DIM [1], MoCo [16] and SimSaim [7]. These methods are generally rooted in NCE [13] and InfoNCE [20] which can serve as optimizing the lower bound of mutual information [36]. Intuitively, the key idea of contrastive learning is to pull representations in positive

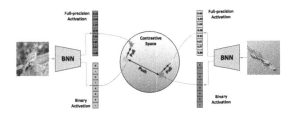

Fig. 2. Feeding two images into a BNN, and obtaining the three pairs of binary and full-precision activations. Our goal is to embed the activations into a contrastive space, then learn from the pair correlation with the contrastive learning task in Eq. 13.

pairs close and push representations in negative pairs apart in a contrastive space, and thus the major obstacle for resorting to the contrastive loss is to define the negative and positive pairs.

3.2 Contrastive Learning for Mutual Information Maximization

In this section, we formalize the idea of constructing a contrastive loss based on Noise-Contrastive Estimation (NCE) to maximize the mutual information between the binary and the full-precision activations. Particularly, we derive a novel *CMIM* loss for training BNNs, where NCE is introduced to avoid the direct mutual information computation by estimating it with its lower bound in Eq. 9. Straightforwardly, the binary and full-precision activations from samples can be pull close, and activations from different samples can be pushed away, which corresponds to the core idea of contrastive learning.

For binary network f_B and its latent full-precision counterpart f_F in the same training iteration, the series of their activations $\{\mathbf{a}_B^k\}$ and $\{\mathbf{a}_F^k\}(k \in \{1, \cdots, K\})$, where $\mathbf{A}_B^k = (\mathbf{a}_B^{k,1}, \cdots, \mathbf{a}_B^{k,N})$ and $\mathbf{A}_F^k = (\mathbf{a}_F^{k,1}, \cdots, \mathbf{a}_F^{k,N})$ can be considered as a series of variables. The corresponding variables $(\mathbf{a}_B^k, \mathbf{a}_F^k)$ should share more information, *i.e.* the mutual information of the same layer's output activations $I(\mathbf{a}_B^k, \mathbf{a}_F^k)(k \in \{1, \cdots, K\})$ should be maximized to enforce them mutually dependent.

To this end, we introduce the contrastive learning framework into our targeted binarization task. The basic idea of contrastive learning is to compare different views of the data (usually under different data augmentations) to calculate similarity scores [1,7,16,20,34]. This framework is suitable for our case, since the binary and full-precision activations can be seen as two different views. For a training batch with N samples, the samples can be denoted as: $\{\mathbf{x}_i\}(i \in \{1, \cdots, N\})$. We feed a batch of samples to the BNN and obtain KN^2 pairs of activations $(\mathbf{a}_B^{k,i}, \mathbf{a}_F^{k,j})$, which augments the data for the auxiliary task. We define a pair containing two activations from the same sample as positive pair, *i.e.* if $i = j$, $(\mathbf{a}_B^{k,i}, \mathbf{a}_F^{k,j})_+$ and *vice versa*. The core idea of contrastive learning is to discriminate whether a given pair of activation $(\mathbf{a}_B^{k,i}, \mathbf{a}_F^{k,j})$ is positive or negative, *i.e.*, inferring the distribution $P(D \mid \mathbf{a}_B^{k,i}, \mathbf{a}_F^{k,j})$, in which D

is the variable decides whether $i = j$ or $i \neq j$. However, we can not directly compute the distribution $P(D \mid \mathbf{a}_B^{k,i}, \mathbf{a}_F^{k,j})$ [13], and we introduce its variational approximation

$$q(D \mid \mathbf{a}_B^{k,i}, \mathbf{a}_F^{k,j}), \tag{5}$$

which can be calculated by our models. Intuitively, $q(D \mid \mathbf{a}_B^{k,i}, \mathbf{a}_F^{k,j})$ can be treated as a binary classifier, which can classify a given pair $(\mathbf{a}_B^{k,i}, \mathbf{a}_F^{k,j})$ into positive or negative.

With the Bayes' theorem, the posterior probability of two activations from the positive pair can be formalized as:

$$q(D = 1 \mid \mathbf{a}_B^{k,i}, \mathbf{a}_F^{k,j}) = \frac{q(\mathbf{a}_B^{k,i}, \mathbf{a}_F^{k,j} \mid D = 1)\frac{1}{N}}{q(\mathbf{a}_B^{k,i}, \mathbf{a}_F^{k,j} \mid D = 1)\frac{1}{N} + q(\mathbf{a}_B^{k,i}, \mathbf{a}_F^{k,j} \mid D = 1)\frac{N-1}{N}}. \tag{6}$$

The probability of activations from negative pair is $q(D = 0 \mid \mathbf{a}_B^{k,i}, \mathbf{a}_F^{k,j}) = 1 - q(D = 1 \mid \mathbf{a}_B^{k,i}, \mathbf{a}_F^{k,j})$. To simplify the NCE derivative, several works [13, 41,42] build assumption about the dependence of the variables, we also use the assumption that the activations from positive pairs are dependent and the ones from negative pairs are independent, $i.e.$ $q(\mathbf{a}_B^{k,i}, \mathbf{a}_F^{k,j} \mid D = 1) = P(\mathbf{a}_B^{k,i}, \mathbf{a}_F^{k,j})$ and $q(\mathbf{a}_B^{k,i}, \mathbf{a}_F^{k,j} \mid D = 0) = P(\mathbf{a}_B^{k,i})P(\mathbf{a}_F^{k,j})$. Hence, the above equation can be simplified as:

$$q(D = 1 \mid \mathbf{a}_B^{k,i}, \mathbf{a}_F^{k,j}) = \frac{P(\mathbf{a}_B^{k,i}, \mathbf{a}_F^{k,j})}{P(\mathbf{a}_B^{k,i}, \mathbf{a}_F^{k,j}) + P(\mathbf{a}_B^{k,i})P(\mathbf{a}_F^{k,j})(N-1)}. \tag{7}$$

Performing logarithm to Eq. 7 and arranging the terms, we can achieve

$$\log q(D = 1 \mid \mathbf{a}_B^{k,i}, \mathbf{a}_F^{k,j}) \leq \log \frac{P(\mathbf{a}_B^{k,i}, \mathbf{a}_F^{k,j})}{P(\mathbf{a}_B^{k,i})P(\mathbf{a}_F^{k,j})} - \log(N-1). \tag{8}$$

Taking expectation on both sides with respect to $P(\mathbf{a}_B^{k,i}, \mathbf{a}_F^{k,j})$, and combining the definition of mutual information in Eq. 3, we can derive the form of mutual information as:

$$\overbrace{I(\mathbf{a}_B^k, \mathbf{a}_F^k)}^{\text{targeted MI}} \geq \overbrace{\mathbb{E}_{P(\mathbf{a}_B^{k,i}, \mathbf{a}_F^{k,j})}\left[\log q(D = 1 \mid \mathbf{a}_B^{k,i}, \mathbf{a}_F^{k,j})\right] + \log(N-1)}^{\text{optimized lower bound}}, \tag{9}$$

where $I(\mathbf{a}_B^k, \mathbf{a}_F^k)$ is the mutual information between the binary and full-precision distributions of our targeted object. Instead of directly maximizing the mutual information, maximizing the lower bound in the Eq. 9 is a practical solution.

However, $q(D = 1 \mid \mathbf{a}_B^{k,i}, \mathbf{a}_F^{k,j})$ is still hard to estimate. Thus, we introduce critic function h with parameter ϕ ($i.e.$ $h(\mathbf{a}_B^{k,i}, \mathbf{a}_F^{k,j}; \phi)$) as previous contrastive learning works [1,6,20,34,41]. Basically, the critic function h needs to map $\mathbf{a}_B^k, \mathbf{a}_F^k$ to $[0, 1]$ ($i.e.$ discriminate whether a given pair is positive or negative). In practice, we design our critic function for our BNN case based on the critic function in [41]:

$$h(\mathbf{a}_B^{k,i}, \mathbf{a}_F^{k,j}) = \exp(\frac{<\mathbf{a}_B^{k,i}, \mathbf{a}_F^{k,j}>}{\tau})/C, \tag{10}$$

in which $C = \exp(\frac{<\mathbf{a}_B^{k,i}, \mathbf{a}_F^{k,j}>}{\tau}) + N/M$, M is the number of all possible pairs, as well as τ is a temperature parameter that controls the concentration level of the distribution [19].

The activations of BNN have their properties can be used here, *i.e.*

$$sgn(\mathbf{a}_F^{k,i}) = \mathbf{a}_B^{k,i} \quad \text{and} \quad <\mathbf{a}_B^{k,i}, \mathbf{a}_F^{k,i}> = \|\mathbf{a}_F^{k,i}\|_1 \tag{11}$$

Thus, the critic function in Eq. 10 can be further simplified as follows:

$$h(\mathbf{a}_B^{k,i}, \mathbf{a}_F^{k,j}) = \exp(\frac{<sgn(\mathbf{a}_F^{k,i}), \mathbf{a}_F^{k,j}>}{\tau}) = \begin{cases} \exp(\frac{\|\mathbf{a}_F^{k,i}\|_1}{\tau}) & i = j, \\ \exp(\frac{<sgn(\mathbf{a}_F^{k,i}), \mathbf{a}_F^{k,j}>}{\tau}) & i \neq j \end{cases} \tag{12}$$

Critic in the View of Activation Flip. Equation 12 reveals the working mechanism of *CMIM* from a perspective of activation flip. Specifically, by turning the $+$ activation into $-$, binary activation in the critic can pull the activations in positive pair close and push the ones in the negative pair away via inner product. For example, suppose $\mathbf{a}_F^{k,1} = (0.3, -0.4, -0.6)$ and $\mathbf{a}_F^{k,2} = (0.6, -0.9, 0.7)$, and then $\mathbf{a}_B^{k,1} = (+1, -1, -1)$ is the anchor. Thus, for the positive pair, $<sgn(\mathbf{a}_B^{k,1}), \mathbf{a}_F^{k,1}> = 0.3 \times (+1) + (-0.4) \times (-1) + (-0.6) \times (-1) = \|\mathbf{a}_F^{k,1}\|_1$ maximizing their similarity score; and for the negative pair, $<sgn(\mathbf{a}_B^{k,1}), \mathbf{a}_F^{k,2}> = $

$$0.6 \times (+1) + (-0.9) \times (-1) + \overbrace{(0.6) \times (-1)}^{flipped}$$ gradually minimizing the score, where the flipped term serve as a penalty for the negative pair. In this way, the binary anchor pull the positive full-precision activation close, and push the negative full-precision ones away by flipping numbers in the full-precision activations. Note that the process is iteratively operated during training, and thus all the binary activations can play the role as anchor, which eventually leads to better representation capacity in the contrastive space.

Loss Function. We define the contrastive loss function $\mathcal{L}_{NCE}^k$ between the k-th layer's activations $\mathbf{A}_B^k$ and $\mathbf{A}_F^k$ as: $\mathcal{L}_{NCE}^k =$

$$\mathbb{E}_{q(\mathbf{a}_B^{k,i}, \mathbf{a}_F^{k,j}|D=1)} \left[\log h(\mathbf{a}_B^{k,i}, \mathbf{a}_F^{k,j}) \right] + N\mathbb{E}_{q(\mathbf{a}_B^{k,i}, \mathbf{a}_F^{k,j}|D=0)} \left[\log(1 - h(\mathbf{a}_B^{k,i}, \mathbf{a}_F^{k,j})) \right]. \tag{13}$$

We would comment on the above loss function from the perspective of contrastive learning. The first term of positive pairs is optimized for capturing more intra-class correlations and the second term of negative pairs is for inter-class decorrelation. Because the pair construction is instance-wise, the number of negative samples theoretically can be the size of the entire training set, *e.g.* 1.2 million for ImageNet. With those additional hand-craft designed contrastive pairs for the proxy optimization problem in Eq. 13, the representation capacity of

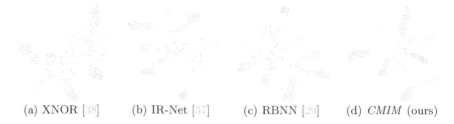

(a) XNOR [38]   (b) IR-Net [37]   (c) RBNN [29]   (d) *CMIM* (ours)

Fig. 3. t-SNE [33] visualization of the activations representing for random 10 classes in CIFAR-100. Every color represents a different class. We can clearly witness the improvement of our method for learning better binary representations.

BNNs can be further improved, as many contrastive learning methods demonstrated [1,7,20,34].

Combining the series of NCE loss from different layers $\{\mathcal{L}_{NCE}^k\}, (k = 1, \cdots, K)$, the overall loss $\mathcal{L}$ can be defined as:

$$\mathcal{L} = \lambda \sum_{k=1}^{K} \frac{\mathcal{L}_{NCE}^k}{\beta^{K-1-k}} + \mathcal{L}_{cls}, \tag{14}$$

where $\mathcal{L}_{cls}$ is the classification loss respect to the ground truth, λ is used to control the degree of NCE loss, β is a coefficient greater than 1, and we denote the *CMIM* loss as $\mathcal{L}_{CMIM} = \sum_{k=1}^{K} \frac{\mathcal{L}_{NCE}^k}{\beta^{K-1-k}}$. Hence, the β^{K-1-k} decreases with k increasing and consequently the $\frac{\mathcal{L}_{NCE}^k}{\beta^{K-1-k}}$ increases. In this way, the activations of latter layer can be substantially retained, which leads to better performance in practice. The complete training process of *CMIM* is presented in Algorithm 1 in the Supplemental Materials.

3.3 Discussion on CMIM

Besides the theoretical formulation from the perspective of mutual information maximization, we also provide an intuitive explanation about *CMIM*. As illustrated in Fig. 2, we strengthen the representation ability of binary activations (see Fig. 3) via designing a proxy task under the contrastive learning framework. By embedding the activations to the contrastive space and pull-and-push the paired embeddings, the BNNs can learn better representations from this difficult yet effective auxiliary contrastive learning task. Note that even though we only pick up two images to formulate Fig. 2, the actual number of negative samples can be huge in practice (*e.g.* 16,384 for training ResNet-18 on ImageNet), benefit from the MemoryBank [42] technique.

With this property, we speculate that the contrastive pairing works as the data augmentation, which contributes to our method. This additional pairing provides more information for training the BNNs, thus *CMIM* can be treated as an overfitting-mitigated module. We also conduct experiments in the Sect. 4.2 and 4.3 to validate our speculation.

Comparison with Other Contrastive Learning Methods. The key idea of contrastive learning is to pull representations close in positive pairs and push representations apart in negative pairs in a contrastive space. Several self-supervised learning methods are rooted in well-established idea of the mutual information maximization, such as Deep InfoMax [20], Contrastive Predictive Coding [34], MemoryBank [42], Augmented Multiscale DIM [1], MoCo [16] and SimSaim [7]. These are based on NCE [13] and InfoNCE [20] which can be seen as a lower bound on mutual information [36]. In the meantime, Tian *et al.* [41] and Chen *et al.* [6] generalize the contrastive idea into the content of knowledge distillation (KD) to pull-and-push the representations of teacher and student.

Our formulation of *CMIM*-BNN absorbs the core idea (*i.e.* construct the appropriate positive and negative pairs for contrastive loss) of the existing contrastive learning methods, especially the contrastive knowledge distillation methods, CRD [41] and WCoRD [6]. However, our approach has several differences from those methods. Firstly, our work can not be treated as a simply application with the teacher-and-student framework. In KD, the teacher is basically fixed to offer additional supervision signals and is not optimizable. But in our formulation, we leverage the exclusive structure of BNN, where FP and binary activations exist in the same forward pass, *i.e.* only one BNN is involved, without using another network as a teacher. Therefore, the accuracy improvement of the BNN trained by our method is purely benefited from the activation alignment in a contrastive way, rather than a more accurate teacher network. Secondly, due to the particular structure of BNNs (Eq. 11), our critic function is largely different from the normal critic in contrastive learning (see Eq. 11 and Eq. 12). Importantly, the critic functions of CRD and WCoRD must utilize a fully-connected layer over the representations to transform them into the same dimension and further normalize them by L_2 norm before the inner product, but ours does not. In the literature of binarization, our designed critic function act as an activation flip as we discussed below Eq. 12. Thirdly, instead of only using the activation of the final layer, we align the activations layer-by-layer with a hyperparameter to adjust the weight of each layer as shown in Eq. 14, which is a more suitable design for BNN. In conclusion, using contrastive objective as a tool to realize mutual information maximization for our network binarization is new.

4   Experiments

In this section, we first conduct experiments to compare with existing state-of-the-art methods in image classification. Following popular settings in most studies, we use CIFAR-10/100 [24] and ImageNet ILSVRC-2012 [9] to validate the effectiveness of our proposed binarization method. Besides comparing our method with the SoTA methods, we design experiments in semantic segmentation and depth estimation tasks on the NYUD-v2 [40] dataset to testify the generalization ability of our method. Meanwhile, we conduct a series of ablation studies to verify the effectiveness of our proposed technique, and we empirically explain the efficacy of *CMIM* from the perspective of mitigating overfitting. All

Table 1. Top-1 accuracy (%) on CIFAR-10 (C-10) and CIFAR-100 (C-100) test set. The higher the better. W/A denotes the bit number of weights/activations.

Topology	Method	Bit-width (W/A)	Acc. (%) (C-10)	Acc. (%) (C-100)
ResNet-20	Full-precision	32/32	92.1	70.7
	DoReFa [44]	1/1	79.3	–
	QSQ [12]	1/1	84.1	–
	SLB [43]	1/1	85.5	–
	LNS [14]	1/1	85.8	–
	IR-Net [37]	1/1	86.5	65.6
	RBNN [29]	1/1	87.0	66.0
	IR-Net + *CMIM*	1/1	**87.3**	**68.1**
	RBNN + *CMIM*	1/1	**87.6**	**68.2**
ResNet-18	Full-precision	32/32	93.0	72.5
	RAD [10]	1/1	90.5	–
	Proxy-BNN [18]	1/1	91.8	67.2
	IR-Net [37]	1/1	91.6	64.5
	RBNN [29]	1/1	92.2	65.3
	IR-Net + *CMIM*	1/1	**92.2**	**71.2**
	RBNN + *CMIM*	1/1	**92.8**	**71.4**
VGG-small	Full-precision	32/32	94.1	73.0
	XNOR [38]	1/1	90.5	–
	DoReFa [44]	1/1	90.2	–
	RAD [10]	1/1	90.5	–
	QSQ [12]	1/1	90.0	–
	SLB [43]	1/1	92.0	–
	Proxy-BNN [18]	1/1	91.8	67.2
	IR-Net [37]	1/1	90.4	67.0
	RBNN [29]	1/1	91.3	67.4
	IR-Net + *CMIM*	1/1	**92.0**	**70.0**
	RBNN + *CMIM*	1/1	**92.2**	**71.0**

Table 2. Top-1 and Top-5 accuracy on ImageNet. † represents the architecture which varies from the standard ResNet architecture but in the same FLOPs level.

Topology	Method	BW (W/A)	Top-1 (%)	Top-5 (%)
ResNet-18	Full-precision	32/32	69.6	89.2
	ABC-Net [30]	1/1	42.7	67.6
	XNOR-Net [38]	1/1	51.2	73.2
	BNN+ [22]	1/1	53.0	72.6
	DoReFa [44]	1/2	53.4	–
	XNOR++ [3]	1/1	57.1	79.9
	BiReal [31]	1/1	56.4	79.5
	IR-Net [37]	1/1	58.1	80.0
	RBNN [29]	1/1	59.9	81.0
	BiReal + CMIM	1/1	**60.1**	**81.3**
	IR-Net + CMIM	1/1	**61.2**	**83.0**
	RBNN + CMIM	1/1	**62.5**	**84.2**
	ReActNet [32]†	1/1	69.4	85.5
	ReActNet + CMIM†	1/1	**71.0**	**86.3**
ResNet-34	Full-precision	32/32	73.3	91.3
	ABC-Net [30]	1/1	52.4	76.5
	XNOR-Net [38]	1/1	53.1	76.2
	BiReal [31]	1/1	62.2	83.9
	XNOR++ [3]	1/1	57.1	79.9
	IR-Net [37]	1/1	62.9	84.1
	LNS [14]	1/1	59.4	81.7
	RBNN [29]	1/1	63.1	84.4
	IR-Net + CMIM	1/1	**64.9**	**85.8**
	RBNN + CMIM	1/1	**65.0**	**85.7**

experiments are implemented using PyTorch [35] with one NVIDIA RTX 6000 while training on CIFAR-10/100 and NYUD-v2, and four GPUs on ImageNet.

Experimental Setup. On CIFAR-10/100, the BNNs are trained by *CMIM* for 400 epochs with batch size of 256, initial learning rate of 0.1 and cosine learning rate scheduler. We adopt SGD optimizer with momentum of 0.9 and weight decay of 1e−4. On ImageNet, binary models are trained for 100 epochs with batch size of 256. SGD optimizer is applied with momentum of 0.9, weight decay of 1e−4, initial learning rate of 0.1 with cosine learning rate scheduler (for fair comparison, we also use ADAM optimizer in some ResNet-variant settings).

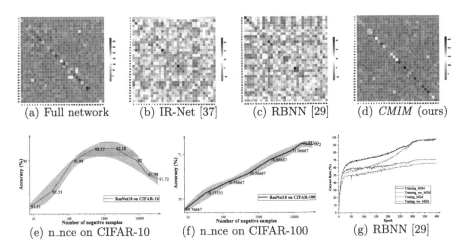

Fig. 4. In-depth analysis on different aspects of the proposed approach including correlation maps (a–d) the effect of number of negative samples in contrastive mutual information maximization (e, f), and training and testing curves (g). (Color figure online)

4.1   Experimental Results

CIFAR-10/100 are widely-used image classification datasets, where each consists of 50K training images and 10K testing images of size 32×32 divided into 10/100 classes. 10K training images are randomly sampled for cross-validation and the rest images are utilized for training. Data augmentation strategy includes random crop and random flipping as in [17] during training.

For ResNet-20, we compare with DoReFa [44], QSQ [12], SLB [43], LNS [14], IR-Net [37] and RBNN [29]. For ResNet-18, RAD [10], Proxy-BNN [18], IR-Net and RBNN are chosen to be the benchmarks. For VGG-small, our method is compared with IR-Net and RBNN, *etc.*

As presented in Table 1, *CMIM* constantly outperforms other SOTA methods. On CIFAR-100, our method achieves 2.5%, 6.1% and 4.0% performance improvement with ResNet-20, ResNet-18 and VGG-small architectures, respectively. To show the pile-up property, we add *CMIM* on different baseline methods, and we can obviously observe the accuracy gain with *CMIM*.

ImageNet is a dataset with 1.2 million training images and 50k validation images equally divided into 1K classes. ImageNet has greater diversity, and its image size is 469×387 (average). We report the single-crop evaluation result using 224×224 center crop from images.

For ResNet-18, we compare our method with XNOR-Net [38], ABC-Net [30], DoReFa [44], BiReal [31], XNOR++ [3], IR-Net [37], RBNN [29]. For ResNet-34, we compare our method with BiReal, IR-Net and RBNN, *etc.* All experimental results are either taken from their published papers or reproduced by ourselves using their code. As demonstrated in Table 2, our proposed method exceeds all the methods in both top-1 and top-5 accuracy. Particularly, *CMIM* achieves around 1.3% Top-1 accuracy gain with ResNet-18 architecture, as well as 1.9% Top-1 accuracy improvement with ResNet-34 architecture, compared with the SoTA RBNN method.

4.2 Number of Negative Samples in *CMIM*

The number of negative samples n_nce is an important hyper-parameter in our method, which ensures the estimation accuracy level of the optimized distribution in Eq. 9. We perform experiments with ResNet18 on CIFAR-100 for parameter analysis of n_nce, with range from 2^0 to 2^{15}. As the results in Fig. 4(f) and 4(e) presented, the accuracy arises with increasing n_nce, which also validates our speculation in the Sect. 3.2 that the contrastive pairing module, serving as a data augmentation module in training, contributes to the performance improvement of *CMIM*.

4.3 Mitigate Overfitting

A good training objective should consistently improve the model performance in testing set [42]. We investigate the relation between the training and the testing performance *w.r.t.* training iterations. Figure 4(g) shows that (1) the binary ResNet-18 can reach 100% on training set of CIFAR-100, which means its representative ability is enough for this dataset; (2) the testing performance of the BNN trained with *CMIM* loss is much better on the final stage, while the training performance is relatively lower. This is a clear sign of mitigating overfitting. In addition, as the results shown in the Table 3, we can observe the phenomenon that the accuracy gain on CIFAR-100 is more noticeable than the gain on ImageNet. This phenomenon can also be explained from the perspective of mitigating overfitting. Since the contrastive pairing (data augmentation for the proxy contrastive learning task) plays a significant role in improving the performance of BNNs, and the data for training is sufficient on ImageNet than on CIFAR. The overfitting issue is not that severe on ImageNet. Hence, our binarization method could be more suitable for relatively data-deficient tasks.

4.4 Ablation Study

We conduct a series of ablative studies of our proposed method in CIFAR-10/100 and ImageNet datasets with the ResNet18 architecture. By adjusting the coefficient λ in the loss function $\mathcal{L}_{CMIM}$ (Eq. 14), where $\lambda = 0$ equals to no *CMIM* loss are added as our baseline. In the ablative studies, we introduce IR-Net [37] as our baseline on all the datasets. The results are shown in Table 3. With λ increasing, the improving performance validates the efficacy of *CMIM* loss.

Table 3. Ablation study of *CMIM*. The results are presented in the form of accuracy rate (%). $\lambda = 0$ denotes no *CMIM* loss added, serving as our baseline.

Dataset	λ							
	0 (baseline)	0.2	0.4	0.8	1.6	3.2	6.4	12.8
CIFAR-10	87.59	90.92	91.63	92.06	**92.18**	91.89	91.32	91.01
CIFAR-100	64.53	68.21	69.31	70.67	70.86	71.09	**71.19**	71.17
ImageNet-1K	58.03	59.29	59.99	**61.22**	61.17	61.02	60.64	59.7

4.5 Generalization Ability

To study the dependence of the binary activations from the same layer, we visualize the correlation matrix of those activations by using the shade of the color to represent the cosine similarity of two activations. Red stands for two activations are similar and blue *vice versa*. As shown in Figs. 4(a)–4(d), *CMIM* captures more intra-class correlations (diagonal boxes are redder) and alleviates more inter-class correlations (non-diagonal boxes are bluer). Those intensified representative activations are constructive for fine-tuning down-stream tasks.

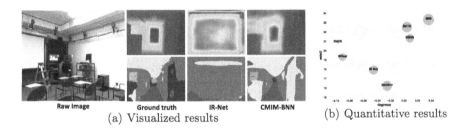

(a) Visualized results   (b) Quantitative results

Fig. 5. Results of depth estimation and segmentation on NYUD-v2

To further evaluate the generalization capacity of the learned binary features, we transfer the learned binary backbone to the image segmentation and depth estimation on NYUD-v2 dataset. We follow the standard pipeline for fine-tuning. A prevalent practice is to pre-train the backbone network on ImageNet and fine-tune it for the downstream tasks. Thus, we conduct experiments with DeepLab heads with the binary ResNet18 backbone. While fine-tuning, the learning rate is initialized to 0.001 and scaled down by 10 times after every 10K iterations and we fix the binary backbone weights, only updating the task-specific heads layers. The results are presented in Fig. 5(b), X-axis is the depth estimation accuracy (-logrmse, higher is better), Y-axis is segmentation performance (mIoU, higher is better) and the size of dot denotes the performance of classification (bigger is better). The visualization results are presented in Fig. 5(a). We can observe that the models with backbone pre-trained by *CMIM* outperform other methods on both segmentation and depth estimation tasks.

5   Conclusion

In this paper, we investigate the activations of BNNs by introducing mutual information to measure the distributional similarity between the binary and full-precision activations. We establish a proxy task via contrastive learning to maximize the targeted mutual information between those binary and real-valued activations. We name our method *CMIM*-BNN. Because of the push-and-pull scheme in the contrastive learning, the BNNs optimized by our method have better representation ability, benefiting downstream tasks, such as classification and segmentation, *etc.* We conduct experiments on CIFAR, ImageNet (for classification) and NYUD-v2 (fine-tuning for depth estimation and segmentation). The results show that CMIM outperforms several state-of-the-art binarization methods on those tasks.

Acknowledgements. This research was partially supported by NSF CNS-1908658 (ZZ, YY), NeTS-2109982 (YY), Early Career Scheme of the Research Grants Council (RGC) of the Hong Kong SAR under grant No. 26202321 (DX), HKUST Startup Fund No. R9253 (DX) and the gift donation from Cisco (YY). This article solely reflects the opinions and conclusions of its authors and not the funding agents.

References

1. Bachman, P., Hjelm, R.D., Buchwalter, W.: Learning representations by maximizing mutual information across views. In: NeurIPS (2019)
2. Bengio, Y., Léonard, N., Courville, A.: Estimating or propagating gradients through stochastic neurons for conditional computation. arXiv:1308.3432 (2013)
3. Bulat, A., Tzimiropoulos, G.: XNOR-Net++: improved binary neural networks. In: BMVC (2019)
4. Cai, Z., He, X., Sun, J., Vasconcelos, N.: Deep learning with low precision by half-wave gaussian quantization. In: CVPR (2017)
5. Chen, L.C., Papandreou, G., Kokkinos, I., Murphy, K., Yuille, A.L.: DeepLab: semantic image segmentation with deep convolutional nets, atrous convolution, and fully connected CRFs. TPAMI **40**(4), 834–848 (2017)
6. Chen, L., Wang, D., Gan, Z., Liu, J., Henao, R., Carin, L.: Wasserstein contrastive representation distillation. In: CVPR (2021)
7. Chen, X., He, K.: Exploring simple siamese representation learning. In: CVPR (2021)
8. Courbariaux, M., Bengio, Y., David, J.P.: BinaryConnect: training deep neural networks with binary weights during propagations. In: NeurIPS (2016)
9. Deng, J., Dong, W., Socher, R., Li, L.J., Li, K., Fei-Fei, L.: ImageNet: a large-scale hierarchical image database. In: CVPR (2009)
10. Ding, R., Chin, T.W., Liu, Z., Marculescu, D.: Regularizing activation distribution for training binarized deep networks. In: CVPR (2019)
11. Gao, S., Ver Steeg, G., Galstyan, A.: Efficient estimation of mutual information for strongly dependent variables. In: AISTATS (2015)
12. Gong, R., et al.: Differentiable soft quantization: bridging full-precision and low-bit neural networks. In: ICCV (2019)

13. Gutmann, M., Hyvärinen, A.: Noise-contrastive estimation: a new estimation principle for unnormalized statistical models. In: AISTATS (2010)
14. Han, K., Wang, Y., Xu, Y., Xu, C., Wu, E., Xu, C.: Training binary neural networks through learning with noisy supervision. In: ICML (2020)
15. Han, S., Mao, H., Dally, W.J.: Deep compression: compressing deep neural networks with pruning, trained quantization and Huffman coding. In: ICLR (2016)
16. He, K., Fan, H., Wu, Y., Xie, S., Girshick, R.: Momentum contrast for unsupervised visual representation learning. In: CVPR (2020)
17. He, K., Zhang, X., Ren, S., Sun, J.: Deep residual learning for image recognition. In: CVPR (2016)
18. He, X., et al.: ProxyBNN: learning binarized neural networks via proxy matrices. In: CVPR (2020)
19. Hinton, G., Vinyals, O., Dean, J.: Distilling the knowledge in a neural network. In: NeurIPS (2014)
20. Hjelm, R.D., et al.: Learning deep representations by mutual information estimation and maximization. arXiv preprint arXiv:1808.06670 (2018)
21. Howard, A.G., et al.: MobileNets: efficient convolutional neural networks for mobile vision applications. arXiv preprint arXiv:1704.04861 (2017)
22. Hubara, I., Courbariaux, M., Soudry, D., El-Yaniv, R., Bengio, Y.: Binarized neural networks. In: NeurIPS (2016)
23. Kim, H., Park, J., Lee, C., Kim, J.J.: Improving accuracy of binary neural networks using unbalanced activation distribution. In: CVPR (2021)
24. Krizhevsky, A., Hinton, G., et al.: Learning multiple layers of features from tiny images (2009)
25. Krizhevsky, A., Sutskever, I., Hinton, G.E.: ImageNet classification with deep convolutional neural networks. In: NeurIPS (2012)
26. Kullback, S.: Information Theory and Statistics. Courier Corporation (1997)
27. LeCun, Y., Bengio, Y., Hinton, G.: Deep learning. Nature **521**(7553), 436–444 (2015)
28. LeCun, Y., Denker, J., Solla, S.: Optimal brain damage. In: NeurIPS (1989)
29. Lin, M., et al.: Rotated binary neural network. In: NeurIPS (2020)
30. Lin, X., Zhao, C., Pan, W.: Towards accurate binary convolutional neural network. In: NeurIPS (2017)
31. Liu, Z., Luo, W., Wu, B., Yang, X., Liu, W., Cheng, K.T.: Bi-real net: binarizing deep network towards real-network performance. IJCV **128**, 202–219 (2020). https://doi.org/10.1007/s11263-019-01227-8
32. Liu, Z., Shen, Z., Savvides, M., Cheng, K.-T.: ReActNet: towards precise binary neural network with generalized activation functions. In: Vedaldi, A., Bischof, H., Brox, T., Frahm, J.-M. (eds.) ECCV 2020. LNCS, vol. 12359, pp. 143–159. Springer, Cham (2020). https://doi.org/10.1007/978-3-030-58568-6_9
33. Van der Maaten, L., Hinton, G.: Visualizing data using t-SNE. JMLR **9**, 2579–2605 (2008)
34. van den Oord, A., Li, Y., Vinyals, O.: Representation learning with contrastive predictive coding. arXiv preprint arXiv:1807.03748 (2018)
35. Paszke, A., et al.: PyTorch: an imperative style, high-performance deep learning library. In: NeurIPS (2019)
36. Poole, B., Ozair, S., Van Den Oord, A., Alemi, A., Tucker, G.: On variational bounds of mutual information. In: ICML (2019)
37. Qin, H., et al.: Forward and backward information retention for accurate binary neural networks. In: CVPR (2020)

38. Rastegari, M., Ordonez, V., Redmon, J., Farhadi, A.: XNOR-Net: ImageNet classification using binary convolutional neural networks. In: Leibe, B., Matas, J., Sebe, N., Welling, M. (eds.) ECCV 2016. LNCS, vol. 9908, pp. 525–542. Springer, Cham (2016). https://doi.org/10.1007/978-3-319-46493-0_32

39. Shang, Y., Duan, B., Zong, Z., Nie, L., Yan, Y.: Lipschitz continuity guided knowledge distillation. In: ICCV (2021)

40. Silberman, N., Hoiem, D., Kohli, P., Fergus, R.: Indoor segmentation and support inference from RGBD images. In: Fitzgibbon, A., Lazebnik, S., Perona, P., Sato, Y., Schmid, C. (eds.) ECCV 2012. LNCS, vol. 7576, pp. 746–760. Springer, Heidelberg (2012). https://doi.org/10.1007/978-3-642-33715-4_54

41. Tian, Y., Krishnan, D., Isola, P.: Contrastive representation distillation. In: ICLR (2021)

42. Wu, Z., Xiong, Y., Yu, S.X., Lin, D.: Unsupervised feature learning via non-parametric instance discrimination. In: CVPR (2018)

43. Yang, Z., et al.: Searching for low-bit weights in quantized neural networks. In: NeurIPS (2020)

44. Zhou, S., Wu, Y., Ni, Z., Zhou, X., Wen, H., Zou, Y.: DoReFa-Net: training low bitwidth convolutional neural networks with low bitwidth gradients. arXiv preprint arXiv:1606.06160 (2016)

Lipschitz Continuity Retained Binary Neural Network

Yuzhang Shang[1], Dan Xu[2], Bin Duan[1], Ziliang Zong[3], Liqiang Nie[4], and Yan Yan[1(✉)]

[1] Illinois Institute of Technology, Chicago, USA
{yshang4,bduan2}@hawk.iit.edu, yyan34@iit.edu
[2] Hong Kong University of Science and Technology, Hong Kong, Hong Kong
danxu@cse.ust.hk
[3] Texas State University, San Marcos, USA
ziliang@txstate.edu
[4] Harbin Institute of Technology, Shenzhen, China

Abstract. Relying on the premise that the performance of a binary neural network can be largely restored with eliminated quantization error between full-precision weight vectors and their corresponding binary vectors, existing works of network binarization frequently adopt the idea of model robustness to reach the aforementioned objective. However, robustness remains to be an ill-defined concept without solid theoretical support. In this work, we introduce the Lipschitz continuity, a well-defined functional property, as the rigorous criteria to define the model robustness for BNN. We then propose to retain the Lipschitz continuity as a regularization term to improve the model robustness. Particularly, while the popular Lipschitz-involved regularization methods often collapse in BNN due to its extreme sparsity, we design the Retention Matrices to approximate spectral norms of the targeted weight matrices, which can be deployed as the approximation for the Lipschitz constant of BNNs without the exact Lipschitz constant computation (NP-hard). Our experiments prove that our BNN-specific regularization method can effectively enhance the robustness of BNN (testified on ImageNet-C), achieving SoTA on CIFAR10 and ImageNet. Our code is available at https://github.com/42Shawn/LCR_BNN.

Keywords: Neural network compression · Network binarization · Lipschitz continuity

1 Introduction

Recently, Deep Neural Networks achieve significant accomplishment in computer vision tasks such as image classification [26] and object detection [27,42]. However, their inference-cumbersome problem hinders their broader implementations. To develop deep models in resource-constrained edge devices, researchers

Supplementary Information The online version contains supplementary material available at https://doi.org/10.1007/978-3-031-20083-0_36.

propose several neural network compression paradigms, *e.g.*, knowledge distillation [20,21], network pruning [16,28] and network quantization [24,40]. Among the network quantization methods, the network binarization [24] stands out, as it extremely quantizes weights and activations (*i.e.* intermediate feature maps) to ±1. Under this framework, the full-precision (FP) network is compressed 32× more, and the time-consuming inner-product operations are replaced with the efficient Xnor-bitcount operations.

However, BNNs can hardly achieve comparable performance to the original models due to the loss of FP weights and activations. A major reason for the performance drop is that the inferior robustness comes from the error amplification effect, where the binarization operation degrades the distance induced by amplified noise [30]. The destructive manner of $sgn(\cdot)$ severely corrupts the robustness of the BNN, and thus undermines their representation capacity [6,18,34].

As some theoretical works validated, robustness is a significant property for functions (neural networks in our context), which further influences their generalization ability [3,35]. In the above-mentioned binarization works, researchers investigate the effectiveness of their methods via the ill-defined concepts of function robustness without solid theoretical support, such as observing the visualized distributions of weights and activations [18,30,31,34]. However, they rarely introduced the well-defined mathematical property, Lipschitz continuity, for measuring the robustness of functions into BNN. Lipschitz continuity has been proven to be a powerful and strict tool for systematically analyzing deep learning models. For instance, Miyato *et al.* propose the well-known Spectral Normalization [36,49] utilizing the Lipschitz constant to regularize network training, which is initially designed for GAN and then extended to other network architectures, achieving great success [37]; Lin *et al.* [30] design a Lipschitz-based regularization method for network (low-bit) quantization, and testify that Lipschitz continuity is significantly related to the robustness of the low-bit network. But simply bridging those existing Lipschitz-based regularization methods with the binary neural networks (1-bit) is sub-optimal, as the exclusive property of BNN, *e.g.*, the extreme sparsity of binary weight matrix [24] impedes calculating the singular values, which is the core module in those Lipschitz-involved methods.

To tackle this problem, we analyze the association between the structures and the Lipschitz constant of BNN. Motivated by this analysis, we design a new approach to effectively retain the Lipschitz constant of BNNs and make it close to the Lipschitz constant of its latent FP counterpart. Particularly, we develop a Lipschitz Continuity Retention Matrix (**RM**) for each block and calculate the spectral norm of **RM** via the iterative power method to avoid the high complexity of calculating exact Lipschitz constants. It is worth to note that the designed loss function for retaining the Lipschitz continuity of BNNs is differentiable *w.r.t.* the binary weights.

Overall, the contributions of this paper are three-fold:

- We propose a novel network binarization framework, named as **L**ipschitz **C**ontinuity **R**atined Binary Neural Network (***LCR***-BNN), to enhance the robustness of binary network optimization process. To the best of our knowledge, we are the first on exploring the Lipschitz continuity to enhance the representation capacity of BNNs;

– We devise a Lipschitz Continuity Retention Matrix to approximate the Lipschitz constant with activations (instead of directly using weights as SN [36] and DQ [30] devised) of networks in the BNN forward pass;

– By adding our designed regularization term on the existing state-of-the-art methods, we observe the enhanced robustness are validated on ImageNet-C and promising accuracy improvement on CIAFR and ImageNet datasets.

2   Related Work

2.1   Network Binarization

In the pioneer art of BNNs, Hubara *et al.* [24] quantize weights and activations to ±1 via sign function. Due to the non-differentiability of the sign function, the straight-through estimator (STE) [4] is introduced for approximating the derivative of the sign function. Inspired by this archetype, numerous researchers dig into the field of BNNs and propose their modules to improve the performance of BNNs. For instance, Rastegari *et al.* [41] reveal that the quantization error between the FP weights and corresponding binarized weights is one of the obstacles degrading the representation capabilities of BNNs. Then they propose to introduce a scaling factor calculated by the L1-norm for both weights and activation functions to minimize the quantization error. XNOR++ [6] absorbs the idea of scaling factor and proposes learning both spatial and channel-wise scaling factors to improve performance. Furthermore, Bi-Real [33] proposes double residual connections with full-precision downsampling layers to lessen the information loss. ProxyBNN [18] designs a proxy matrix as a basis of the latent parameter space to guide the alignment of the weights with different bits by recovering the smoothness of BNNs. Those methods try to lessen the quantization error and investigate the effectiveness from the perspective of model smoothness (normally via visualizing the distribution of weights). A more detailed presentation and history of BNNs can be found in the Survey [39].

However, none of them take the functional property, Lipschitz continuity, into consideration, which is a well-developed mathematical tool to study the robustness of functions. Bridging Lipschitz continuity with BNNs, we propose to retain the Lipschitz continuity of BNNs, which can serve as a regularization term and further improve the performance of BNNs by strengthening their robustness.

2.2   Lipschitz Continuity in Neural Networks

The Lipschitz constant is an upper bound of the ratio between input perturbation and output variation within a given distance. It is a well-defined metric to quantify the robustness of neural networks to small perturbations [45]. Also, the Lipschitz constant $\|f\|_{Lip}$ can be regarded as a functional norm to measure the Lipschitz continuity of given functions. Due to its property, the

Lipschitz constant is the primary concept to measure the robustness of functions [3,35,37]. In the deep learning era, previous theoretical arts [37,47] disclose the regularity of deep networks via Lipschitz continuity. Lipschitz continuity is widely introduced into many deep learning topics for achieving the SoTA performance [36,46,49,50]. For example, in image synthesis, Miyato *et al.* [36,49] devise spectral normalization to constrain the Lipschitz constant of the discriminator for optimizing a generative adversarial network, acting as a regularization term to smooth the discriminator function; in knowledge distillation, Shang *et al.* [46] propose to utilize the Lipschitz constant as a form of knowledge to supervise the training process of student network; in neural network architecture design, Zhang *et al.* [50] propose a novel L_∞-dist network using naturally 1-Lipschitz functions as neurons.

The works above highlight the significance of Lipschitz constant in expressiveness and robustness of deep models. Particularly, retaining Lipschitz continuity at an appropriate level is proven to be an effective technique for enhancing the model robustness. Therefore, the functional information of neural networks, Lipschitz constant, should be introduced into network binarization to fill the robustness gap between BNN and its real-valued counterpart.

Relation to Spectral Normalization (SN) [36]. We empirically implement the SN in BNN but fail. By analyzing the failure of the implementation, we conclude that the SN is not suitable for BNNs. The reasons are: (i) One of the key modules in SN is spectral norm computation based on singular value calculatiuon, which is directly implemented on the weight matrix (*e.g.*, the matrices of convolutional and linear layers). But the binarization enforcing the FP weight into 1 or -1 makes the weight matrix extremely sparse. Thus, applying the existing algorithm to binary matrices collapses. (ii) In contrast to normal networks, the forward and backward passes of BNN are more complex, *e.g.*, FP weights (after backpropagation) and binary weights (after binarization) exist in the same training iteration. This complexity problem impedes broader implementations of SN on BNNs as the number of structures in a BNN exceeds the number in a normal network. To tackle those problems, we propose a novel Lipschitz regularization technique targeted to train BNNs. We elaborate more technical comparisons between our method and SN in the following Sect. 3.3.

3   Lipschitz Continuity Retention for BNNs

3.1   Preliminaries

We first define a general neural network with L fully-connected layers (without bias term for simplification). This network $f(\mathbf{x})$ can be denoted as:

$$f(\mathbf{W}^1, \cdots, \mathbf{W}^L; \mathbf{x}) = (\mathbf{W}^L \cdot \sigma \cdot \mathbf{W}^{L-1} \cdots \cdot \sigma \cdot \mathbf{W}^1)(\mathbf{x}), \qquad (1)$$

where $\mathbf{x}$ is the input sample and $\mathbf{W}^k \in \mathbb{R}^{d_{k-1} \times d_k} (k = 1, ..., L-1)$ stands for the weight matrix connecting the $(k-1)$-th and the k-th layer, with d_{k-1} and d_k representing the sizes of the input and output of the k-th network layer, respectively. The $\sigma(\cdot)$ function performs element-wise activation for the activations.

Binary Neural Networks. Here, we revisit the general gradient-based method in [7], which maintains full-precision latent variables $\mathbf{W}_F$ for gradient updates, and the k-th weight matrix $\mathbf{W}_F^k$ is binarized into ± 1 binary weight matrix $\mathbf{W}_B^k$ by a binarize function (normally $sgn(\cdot)$) as $\mathbf{W}_B^k = sgn(\mathbf{W}_F^k)$. Then the activation map of the k-th layer is produced by $\mathbf{A}^k = \mathbf{W}_B^k \mathbf{A}^{k-1}$, and a whole forward pass of binarization is performed by iterating this process for L times.

Lipschitz Constant (Definition 1). A function $g : \mathbb{R}^n \longmapsto \mathbb{R}^m$ is called Lipschitz continuous if there exists a constant L such that:

$$\forall \mathbf{x}, \mathbf{y} \in \mathbb{R}^n, \|g(\mathbf{x}) - g(\mathbf{y})\|_2 \leq L\|\mathbf{x} - \mathbf{y}\|_2, \tag{2}$$

where $\mathbf{x}, \mathbf{y}$ represent two random inputs of the function g. The smallest L holding the inequality is the Lipschitz constant of function g, denoted as $\|g\|_{Lip}$. By Definition 1, $\|\cdot\|_{Lip}$ can upper bound of the ratio between input perturbation and output variation within a given distance (generally L2 norm), and thus it is naturally considered as a metric to evaluate the robustness of neural networks [43,45,46].

In the following section, we propose our Lipschitz Continuity Retention Procedure (Sect. 3.2), where the a BNN is enforced to close to its FP counterpart in term of Lipschitz constant. In addition, we introduce the proposed loss function and gradient approximation for optimizing the binary network (Sect. 3.3). Finally, we discuss the relation between LCR and Lipschitz continuity, and compare our method to the well-known Spectral Normalization [36] (Sect. 3.3).

3.2 Lipschitz Continuity Retention Procedure

We aim to retain the Lipschitz constants in an appropriate level. In practice, we need to pull $\|f_B\|_{Lip}$ and $\|f_F\|_{Lip}$ closely to stabilize the Lipschitz constant of the BNNs. However, it is NP-hard to compute the exact Lipschitz constant of neural networks [47], especially involving the binarization process. To solve this problem, we propose to bypass the exact Lipschitz constant computation by introducing a sequence of Retention Matrices produced by the adjacent activations, and then compute their spectral norms via power iteration method to form a LCR loss for retaining the Lipschitz continuity of the BNN as demonstrated in Fig. 1.

Lipschitz Constant of Neural Networks. We fragment an affine function for the k-th layer with weight matrix $\mathbf{W}^k$, $f^k(\cdot)$ mapping $\mathbf{a}^{k-1} \longmapsto \mathbf{a}^k$, in which $\mathbf{a}^{k-1} \in \mathbb{R}^{d_{k-1}}$ and $\mathbf{a}^k \in \mathbb{R}^{d_k}$ are the activations produced from the $(k-1)$-th and the k-th layer, respectively. Based on Lemma 1 in the Supplemental Materials, $\|f^k\|_{Lip} = \sup_{\mathbf{a}} \|\nabla \mathbf{W}^k(\mathbf{a})\|_{SN}$, where $\|\cdot\|_{SN}$ is the matrix spectral norm formally defined as:

$$\|\mathbf{W}^k\|_{SN} \triangleq \max_{\mathbf{x}:\mathbf{x}\neq 0} \frac{\|\mathbf{W}^k\mathbf{x}\|_2}{\|\mathbf{x}\|_2} = \max_{\|\mathbf{x}\|_2 \leq 1} \|\mathbf{W}^k\mathbf{x}\|_2, \tag{3}$$

where the spectral norm of the matrix $\mathbf{W}$ is equivalent to its largest singular value. Thus, for the f^k, based on Lemma 2 in the Supplemental Materials, its

Lipschitz constant can be derived as:

$$\|\mathbf{W}^k\|_{Lip} = \sup_{\mathbf{a}}\|\nabla\mathbf{W}^k(\mathbf{a})\|_{SN} = \|\mathbf{W}^k\|_{SN}. \tag{4}$$

Moreover, as for the most functional structures in neural network such as ReLU, Tanh, Sigmoid, Sign, batch normalization and other pooling layers, they all have simple and explicit Lipschitz constants [14,36,46]. Note that for the sign function in BNN, though it is not theoretically differentiable, it still has an explicit Lipschitz constant as its derivative is numerically approximated by Hard-Tanh function [4]. This fixed Lipschitz constant property renders our derivation to be applicable to most network architectures, such as binary ResNet [17,24] and variant binary ResNet [5,34].

By the inequality of norm, *i.e.* $\|\mathbf{W}^k \cdot \mathbf{W}^{k+1}\|_{Lip} \leq \|\mathbf{W}^k\|_{Lip} \cdot \|\mathbf{W}^{k+1}\|_{Lip}$, we obtain the following upper bound of the Lipschitz constant of network f, *i.e.*,

$$\|f\|_{Lip} \leq \|\mathbf{W}^L\|_{Lip} \cdot \|\sigma\|_{Lip} \cdots \|\mathbf{W}^1\|_{Lip} = \prod_{k=1}^{L} \|\mathbf{W}^k\|_{SN}. \tag{5}$$

In this way, we can retain the Lipschitz constant through maintaining a sequence of spectral norms of intermediate layers in the network.

Construction of Lipschitz Continuity Retention Matrix. We now aim to design a novel optimization loss to retain Lipschitz continuity by narrowing the distance between the spectral norms of corresponding weights of full-precision and binary networks. Moreover, we need to compute the spectral norm of binarized weight matrices. Nevertheless, it is inaccessible to calculate the spectral norm of the binary weight matrix $\mathbf{W}_B^k$ in BNNs by popular SVD-based methods [1]. Therefore, we design the Lipschitz Continuity Retention Matrix (**RM**) to bypass the complex calculation of the spectral norm of $\mathbf{W}_B^k$. Approaching the final goal through the bridge of the Retention Matrix allows feasible computation to retain the Lipschitz constant and facilitates its further use as a loss function.

For training data with a batch size of N, we have a batch of corresponding activations after a forward process for the $(k$-1)-th layer as

$$\mathbf{A}^{k-1} = (\mathbf{a}_1^{k-1}, \cdots, \mathbf{a}_n^{k-1}) \in \mathbb{R}^{d_{k-1} \times N}, \tag{6}$$

where $\mathbf{W}^k\mathbf{A}^{k-1} = \mathbf{A}^k$ for each $k \in \{1, \ldots, L-1\}$.

Studies about similarity of activations illustrate that for well-trained networks, their batch of activations in the same layer (*i.e.* $\{\mathbf{a}_i^{k-1}\}, i \in \{1, \ldots, n\}$) have strong mutual linear independence. We formalize the independence of the activations as follows:

$$\begin{aligned}(\mathbf{a}_i^{k-1})^\mathsf{T}\mathbf{a}_j^{k-1} &\approx 0, \quad \forall i \neq j \in \{1, \cdots, N\}, \\ (\mathbf{a}_i^{k-1})^\mathsf{T}\mathbf{a}_i^{k-1} &\neq 0, \quad \forall i \in \{1, \cdots, N\}.\end{aligned} \tag{7}$$

We also empirically and theoretically discuss the validation of this assumption in the Sect. 4.4.

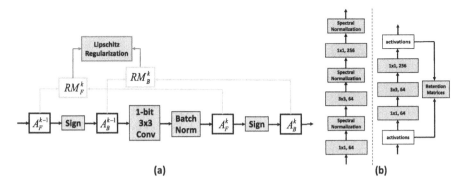

(a)   **(b)**

Fig. 1. (a) An overview of our Lipschitz regularization for a binary convolutional layer: regularizing the BNN via aligning the Lipschitz constants of binary network and its latent full-precision counterpart is the goal of our work. To reach this goal, the input and output activations of the k-th layer compose the Retention Matrix ($\mathbf{RM}^k$) for approximating the Lipschitz constant of this layer. $\mathbf{RM}_F^k$ and $\mathbf{RM}_B^k$ are then used to calculate the Lipschitz constant of this layer (the validation of this approximation is elaborated in 3.2). Finally, the Lipschitz continuity of the BNN is retained under a regularization module. (b) Difference between Spectral Normalization (Left) and LCR (Right). More details are discussed in 3.3.

With the above assumption, we formalize the devised Retention Matrix $\mathbf{RM}^k$ for estimating the spectral norm of matrix $\mathbf{W}^k$ as:

$$
\begin{aligned}
\mathbf{RM}^k &\triangleq \left[(\mathbf{A}^{k-1})^\mathsf{T}\mathbf{A}^k\right]^\mathsf{T}\left[(\mathbf{A}^{k-1})^\mathsf{T}\mathbf{A}^k\right] \\
&= (\mathbf{A}^{k-1})^\mathsf{T}(\mathbf{W}^k)^\mathsf{T}(\mathbf{A}^{k-1})(\mathbf{A}^{k-1})^\mathsf{T}\mathbf{W}^k\mathbf{A}^{k-1}.
\end{aligned}
\tag{8}
$$

Incorporating independence assumption in Eq. 7 (*i.e.*, $(\mathbf{A}^{k-1})(\mathbf{A}^{k-1}) = \mathbf{I}$)) with Eq. 8, we can transfer the $\mathbf{RM}^k$ as follows:

$$
\mathbf{RM}^k = (\mathbf{A}^{k-1})^\mathsf{T}(\mathbf{W}^{k^\mathsf{T}}\mathbf{W}^k)\mathbf{A}^{k-1}.
\tag{9}
$$

Based on Theorem 1 in supplemental material and Eq. 9, $\sigma_1(\mathbf{RM}^k) = \sigma_1(\mathbf{W}^{k^\mathsf{T}}\mathbf{W}^k)$ where $\sigma_1(\cdot)$ is the function for computing the largest eigenvalue, *i.e.*, Retention Matrix $\mathbf{RM}^k$ has the same largest eigenvalue with $\mathbf{W}^{k^\mathsf{T}}\mathbf{W}^k$. Thus, with the definition of spectral norm $\|\mathbf{W}^k\|_{SN} = \sigma_1(\mathbf{W}^{k^\mathsf{T}}\mathbf{W}^k)$, the spectral norm of the matrix $\mathbf{W}^k$ can be yielded through calculating the largest eigenvalue of $\mathbf{RM}^k$, *i.e.* $\sigma_1(\mathbf{RM}^k)$, which is solvable [46].

For networks with more complex layers, such as the residual block and block in MobileNet [17,22], we can also design such a Retention Matrix to bypass the Lipschitz constant computation layer-wisely. By considering the block as an affine mapping from front to back activations, the proposed Retention Matrix can also be produced block-wisely, making our spectral norm calculation more efficient. Specifically, we define the Retention Matrix $\mathbf{RM}$ for the residual blocks as follows:

$$\mathbf{RM}_m \triangleq \left[(\mathbf{A}^f)^\mathsf{T} \mathbf{A}^l \right]^\mathsf{T} \left[(\mathbf{A}^f)^\mathsf{T} \mathbf{A}^l \right], \tag{10}$$

where $\mathbf{A}^f$ and $\mathbf{A}^l$ denote the front-layer activation maps and the back-layer activation maps of the residual block, respectively.

Calculation of Spectral Norms. Here, to calculate the spectral norms of two matrices, an intuitive way is to use SVD to compute the spectral norm, which results in overloaded computation. Rather than SVD, we utilize **Power Iteration** method [12,36] to approximate the spectral norm of the targeted matrix with a small trade-off of accuracy. By Power Iteration Algorithm (see Supplemental Material), we can obtain the spectral norms of the binary and corresponding FP Retention Matrices, respectively (*i.e.* $\|\mathbf{RM}_F^k\|_{SN}$ and $\|\mathbf{RM}_B^k\|_{SN}$ for each $k \in \{1, \ldots, L-1\}$). And then, we can calculate the distance between these two spectral norms to construct the loss function.

3.3   Binary Neural Network Optimization

Optimization Losses. We define the Lipschitz continuity retention loss function $\mathcal{L}_{Lip}$ as

$$\mathcal{L}_{Lip} = \sum_{k=1}^{L-1} \left[\left(\frac{\|\mathbf{RM}_B^k\|_{SN}}{\|\mathbf{RM}_F^k\|_{SN}} - 1 \right) \beta^{k-L} \right]^2, \tag{11}$$

where β is a coefficient greater than 1. Hence, with k increasing, the $\left[\left(\frac{\|\mathbf{RM}_B^k\|_{SN}}{\|\mathbf{RM}_F^k\|_{SN}} - 1 \right) \beta^{k-L} \right]^2$ increases. In this way, the spectral norm of latter layer can be more retained.

Combined with the cross entropy loss $\mathcal{L}_{CE}$, we propose a novel loss function for the overall optimization objective as

$$\mathcal{L} = \frac{\lambda}{2} \cdot \mathcal{L}_{Lip} + \mathcal{L}_{CE}, \tag{12}$$

where λ is used to control the degree of retaining the Lipschitz constant. We analyze the effect of the coefficient λ in the supplementary material. After we define the overall loss function, our method is finally formulated. The forward and backward propagation processes of LCR are elaborated in Algorithm 1.

Gradient Approximation. Several works [30,36,44] investigate the robustness of neural networks by introducing the concept of Lipschitzness. In this section, we differentiate the loss function of our proposed method, and reveal the mechanism of how Lipschitzness effect the robustness of BNNs.

The derivative of the loss function $\mathcal{L}$ w.r.t $\mathbf{W}_B^k$ is:

$$\begin{aligned}
\frac{\partial \mathcal{L}}{\partial \mathbf{W}_B} &= \frac{\partial(\mathcal{L}_{CE})}{\partial \mathbf{W}_B} + \frac{\partial(\mathcal{L}_{Lip})}{\partial \mathbf{W}_B^k} \\
&\approx \mathbf{M} - \lambda \sum_{k=1}^{L-1} \beta^{k-L} \left(\frac{\|\mathbf{RM}_F^k\|_{SN}}{\|\mathbf{RM}_B^k\|_{SN}} \right) \mathbf{u}_1^k (\mathbf{v}_1^k)^\mathsf{T},
\end{aligned} \tag{13}$$

Algorithm 1. Forward and Backward Propagation of LCR-BNN

Require: A minibatch of data samples $(\mathbf{X}, \mathbf{Y})$, current binary weight $\mathbf{W}_B^k$, latent full-precision weights $\mathbf{W}_F^k$, and learning rate η.
Ensure: Update weights $\mathbf{W}_F^{k\,\prime}$.
 1: **Forward Propagation**:
 2: **for** $k = 1$ to $L - 1$ **do**
 3:     Binarize latent weights: $\mathbf{W}_B^k \leftarrow \mathrm{sgn}(\mathbf{W}_F^k)$;
 4:     Perform binary operation with the activations of last layer: $\mathbf{A}_F^k \leftarrow \mathbf{W}_B^k \cdot \mathbf{A}_B^{k-1}$;
 5:     Binarize activations: $\mathbf{A}_B^k \leftarrow \mathrm{sgn}(\mathbf{A}_F^k)$;
 6:     Produce the Retention Matrices $\mathbf{RM}_F^k$ and $\mathbf{RM}_B^k$ by Eq. 9;
 7: **end for**
 8: Approximate the spectral norm of a series of **RM**s by Algorithm ?? in the Supplemental Material, and obtain $\|\mathbf{RM}_F^k\|_{SN}$ and $\|\mathbf{RM}_B^k\|_{SN}$ for each $k \in \{1, \ldots, L-1\}$;
 9: Compute the Lipschitz continuity retention loss $\mathcal{L}_{Lip}$ by Eq. 11;
10: Combine the cross entropy loss $\mathcal{L}_{CE}$ and the quantization error loss $\mathcal{L}_{QE}$ for the overall loss $\mathcal{L}$ by Eq. 12;
11: **Backward Propagation**: compute the gradient of the overall loss function, *i.e.* $\frac{\partial \mathcal{L}}{\partial \mathbf{W_B}}$, using the straight through estimator (STE) [4] to tackle the sign function;
12: **Parameter Update**: update the full-precision weights: $\mathbf{W}_F^{k\,\prime} \leftarrow \mathbf{W}_F^k - \eta \frac{\partial \mathcal{L}}{\partial \mathbf{W}_B^k}$.

---

where $\mathbf{M} \triangleq \frac{\partial(\mathcal{L}_{CE})}{\partial \mathbf{W}_B}$, $\mathbf{u}_1^k$ and $\mathbf{v}_1^k$ are respectively the first left and right singular vectors of $\mathbf{W}_B^k$. In the content of SVD, $\mathbf{W}_B^k$ can be re-constructed by a series of singular vector, *i.e.*

$$\mathbf{W}_B^k = \sum_{j=1}^{d_k} \sigma_j(\mathbf{W}_B^k)\mathbf{u}_j^k\mathbf{v}_j^k, \tag{14}$$

where d_k is the rank of $\mathbf{W}_B^k$, $\sigma_j(\mathbf{W}_B^k)$ is the j-th biggest singular value, $\mathbf{u}_j^k$ and $\mathbf{v}_j^k$ are left and singular vectors, respectively [46]. In Eq. 13, the first term $\mathbf{M}$ is the same as the derivative of the loss function of general binarization method with reducing quantization error. As for the second term, based on Eq. 14, it can be seen as the regularization term penalizing the general binarization loss with an adaptive regularization coefficient $\gamma \triangleq \lambda \beta^{k-L} \left(\frac{\|\mathbf{RM}_F^k\|_{SN}}{\|\mathbf{RM}_B^k\|_{SN}}\right)$ (More detailed derivation can be found in the supplemental materials). Note that even we analyze the regularization property under the concept of SVD, we do not actually use SVD in our algorithm. And Eqs. 13 and 14 only demonstrate that LCR regularization is related to the biggest singular value and its corresponding singular vectors. The LCR Algorithm 1 only uses the Power Iteration (see Algorithm in the Supplemental Materials) within less iteration steps (5 in practice) to approximate the biggest singular value.

Discussion on Retention Matrix. Here, we would like to give a straightforward explanation of why optimizing LCR Loss in Eq. 11 is equivalent to retaining Lipschitz continuity of BNN. Since the Lipschitz constant of a network $\|f\|_{Lip}$ can be upper-bounded by a set of spectral norms of weight matrices, *i.e.* $\{\|\mathbf{W}_F^k\|_{SN}\}$ (see Eqs. 3–5), we aim at retaining the spectral norms of binary weight matrices, instead of targeting on the network itself. And because Eqs. 7 to 9 derive $\|\mathbf{RM}_F^k\|_{SN} = \|\mathbf{W}_F^k\|_{SN}$ and $\|\mathbf{RM}_B^k\|_{SN} = \|\mathbf{W}_B^k\|_{SN}$, we only need to calculate the spectral norm of our designed Retention Matrix $\|\mathbf{RM}_B^k\|_{SN}$. Finally, minimizing Eq. 11 equals to enforcing $\|\mathbf{RM}_B^k\|_{SN} \longrightarrow \|\mathbf{RM}_F^k\|_{SN}$, which retains the spectral norm (Lipschitz continuity) of BNN. Therefore, the BNNs trained by our method have better performance, because the Lipschitz continuity is retained, which can smooth the BNNs.

Differences with Spectral Normalization (SN) and Defensive Quantization (DQ). There are two major differences: (i) In contrast to SN and DQ directly calculating the spectral norm with weight matrix, our method compute the spectral norm of specifically designed Retention Matrix to approximate the targeted spectral norms by leveraging the activations in BNNs. In this way, we can approximate the targeted yet inaccessible Lipschitz constant of binary networks as shown in Fig. 1(a), in which the weight matrix is extremely sparse. Particularly, instead of layer-wisely calculating the spectral norm of weight matrix proposed in SN, our method does *not rely on weight matrix* since the calculation can be done using only the in/out activations (Eq. 8). (ii) To tackle the training architecture complexity, our designed Retention Matrix gives flexibility to regularize BNNs via utilizing Lipschitz constant in a module manner (*e.g.*, residual blocks in ResNet [17]), instead of calculating the spectral norm and normalizing the weight matrix to 1 for each layer as shown in Fig. 1(b). Benefit from module-wise simplification, total computation cost of our method is much lower compared with SN and DQ.

4   Experiments

In this section, we conduct experiments on the image classification. Following popular setting in most studies [31,40], we use the CIFAR-10 [26] and the ImageNet ILSVRC-2012 [26] to validate the effectiveness of our proposed binarization method. In addition to comparing our method with the state-of-the-art methods, we design a series of ablative studies to verify the effectiveness of our proposed regularization technique. All experiments are implemented using PyTorch [38]. We use one NVIDIA GeForce 3090 GPU when training on the CIFAR-10 dataset, and four GPUs on the ImageNet dataset.

Experimental Setup. On CIFAR-10, the BNNs are trained for 400 epochs, batch size is 128 and initial learning rate is 0.1. We use SGD optimizer with the momentum of 0.9, and set weight decay is 1e−4. On ImageNet, the binary models are trained the for 120 epochs with a batch size of 256. We use cosine learning rate scheduler, and the learning rate is initially set to 0.1. All the training and testing settings follow the codebases of IR-Net [40] and RBNN [31].

4.1   CIFAR

CIFAR-10 [25] is the most widely-used image classification dataset, which consists of 50 K training images and 10 K testing images of size 32×32 divided into

Table 1. Top-1 and Top-5 accuracy on ImageNet. † represents the architecture which varies from the standard ResNet architecture but in the same FLOPs level.

Topology	Method	BW (W/A)	Top-1 (%)	Top-5 (%)
ResNet-18	Baseline	32/32	69.6	89.2
	ABC-Net [32]	1/1	42.7	67.6
	XNOR-Net [41]	1/1	51.2	73.2
	BNN+ [8]	1/1	53.0	72.6
	DoReFa [51]	1/2	53.4	–
	BiReal [33]	1/1	56.4	79.5
	XNOR++ [6]	1/1	57.1	79.9
	IR-Net [40]	1/1	58.1	80.0
	ProxyBNN [18]	1/1	58.7	81.2
	Ours	1/1	**59.6**	**81.6**
	Baseline	32/32	69.6	89.2
	SQ-BWN [11]	1/32	58.4	81.6
	BWN [41]	1/32	60.8	83.0
	HWGQ [29]	1/32	61.3	83.2
	SQ-TWN [11]	2/32	63.8	85.7
	BWHN [23]	1/32	64.3	85.9
	IR-Net [40]	1/32	66.5	85.9
	Ours	1/32	**66.9**	**86.4**
ResNet-34	Baseline	32/32	73.3	91.3
	ABC-Net [32]	1/1	52.4	76.5
	Bi-Real [33]	1/1	62.2	83.9
	IR-Net [40]	1/1	62.9	84.1
	ProxyBNN [18]	1/1	62.7	84.5
	Ours	1/1	**63.5**	**84.6**
Variant ResNet	ReActNet† [34]	1/1	69.4	85.5
	Ours†	1/1	**69.8**	**85.7**

Table 2. Top-1 accuracy (%) on CIFAR-10 (C-10) test set. The higher the better. W/A denotes the bit number of weights/activations.

Topology	Method	Bit-width (W/A)	Acc. (%)
ResNet-18	Baseline	32/32	93.0
	RAD [10]	1/1	90.5
	IR-Net [40]	1/1	91.5
	Ours	1/1	**91.8**
ResNet-20	Baseline	32/32	91.7
	DoReFa [51]	1/1	79.3
	DSQ [13]	1/1	84.1
	IR-Net [40]	1/1	85.5
	IR-bireal [40]	1/1	86.5
	LNS [15]	1/1	85.7
	SLB [48]	1/1	85.5
	Ours	1/1	**86.0**
	Ours-bireal	1/1	**87.2**
	Baseline	32/32	91.7
	DoReFa [51]	1/32	90.0
	DSQ [13]	1/32	90.1
	IR-Net [40]	1/32	90.2
	LNS [15]	1/32	90.8
	SLB [48]	1/32	90.6
	Ours	1/32	**91.2**

10 classes. For training, 10,000 training images are randomly sampled for validation and the rest images are for training. Data augmentation strategy includes random crop and random flipping as in [17] during training. For testing, we evaluate the single view of the original image for fair comparison.

For ResNet-18, we compare with RAD [10] and IR-Net [40]. For ResNet-34, we compare with LNS [15] and SLB [48], *etc.* As the Table 1 presented, our method constantly outperforms other methods. *LCR*-BNN achieves 0.3%, 0.7% and 0.6% performance improvement over ResNet-18, ResNet-20 and ResNet-20 (without binarizing activations), respectively. In addition, our method also validate the effectiveness of bi-real structure [33]. When turning on the bi-real module, IR-Net achieves 1.0% accuracy improvements yet our method improves 1.2%.

4.2 ImageNet

ImageNet [9] is a larger dataset with 1.2 million training images and 50k validation images divided into 1,000 classes. ImageNet has greater diversity, and its image size is 469×387 (average). The commonly used data augmentation strategy including random crop and flipping in PyTorch examples [38] is adopted for training. We report the single-crop evaluation result using 224×224 center crop from images.

For ResNet-18, we compare our method with XNOR-Net [41], ABC-Net [32], DoReFa [51], BiReal [33], XNOR++ [6], IR-Net [40], ProxyBNN [18]. For ResNet-34, we compare our method with ABC-Net [32], BiReal [33], IR-Net [40], ProxyBNN [18]. As demonstrated in Table 2, our proposed method also outperforms other methods in both top-1 and top-5 accuracy on the ImageNet. Particularly, *LCR*-BNN achieves 0.9% Top-1 accuracy improvement with ResNet-18 architecture, compared with STOA method ProxyBNN [18], as well as 0.6% Top-1 accuracy improvement with ResNet-34 architecture, compared with state-of-the-art method ProxyBNN [40]. Apart from those methods implemented on standard ResNet architectures, by adding our Lipschitz regularization module on ResNet-variant architecture, ReActNet [34], we also observe the accuracy improvement. Note that the training setting of adding our *LCR* module on ReActNet is also different based on the codebase of ReActNet.

4.3 Ablation Study

In this section, the ablation study is conducted on CIFAR-10 with ResNet-20 architecture and on ImageNet with ResNet-18. The results are presented in Table 4. By piling up our regularization term on IR-Net [40] and ReActNet [34], our method achieves 1.2% and 0.4% improvement on ImageNet, respectively. Note that ReActNet is a strong baseline with a variant ResNet architecture. We also study the effect of hyper-parameter λ in loss function on CIFAR. As shown in Fig 3, we can observe that the performance improves with λ increasing. Both experiments validate the effectiveness of our method. Apart from that, to investigate the regularization property of our method, we visualize several training and testing curves with various settings. Due to the space limitation, we put those demonstrations in the supplemental materials.

4.4 Further Analysis

Computational Cost Analysis. In Table 5, we separate the number of binary operations and floating point operations, including all types of operations such as skip structure, max pooling, *etc.* It shows that our method leaves the number of BOPs and number of FLOPs constant in the model inference stage, even though our method is more computational expensive in the training stage. Thus, our Lipschitz regularization term does not undermine the main benefit of the network binarization, which is to speed up the inference of neural networks.

Weight Distribution Visualization. To validate the effectiveness of our proposed method from the perspective of weight distribution, we choose our *LCR*-BNN and IR-Net to visualize the distribution of weights from different layers. For fair comparison, we randomly pick up 10,000 parameters in each layer to

Table 3. Effect of hyper-parameter λ in loss function. Higher is better.

Topology	$\log_2 \lambda$					
	$\lambda = 0$	-1	0	1	2	3
ResNet-18	85.9	86.2	87.9	90.1	91.2	**91.8**
ResNet-20	83.9	83.7	84.5	85.9	**87.2**	86.5

Table 4. Ablation Study of LCR-BNN.

Dataset	Method	Acc(%)
CIFAR	Full precision	91.7
	IR-Net [40] (w/o BiReal)	85.5
	IR-Net + LCR (w/o BiReal)	86.0
	IR-Net [40] (w/ BiReal)	86.5
	IR-Net + LCR (w/o BiReal)	87.2
ImageNet	Full precision	69.6
	IR-Net [40] (w/o BiReal)	56.9
	IR-Net + LCR (w/o BiReal)	58.4
	IR-Net [40] (w/ BiReal)	58.1
	IR-Net + LCR	59.6
	ReActNet	69.4
	ReActNet + LCR	69.8

Table 5. FLOPS and BOPS for ResNet-18

Method	BOPS	FLOPS
BNN [24]	1.695×10^9	1.314×10^8
XNOR-Net [41]	1.695×10^9	1.333×10^8
ProxyBNN [18]	1.695×10^9	1.564×10^8
IR-Net [40]	1.676×10^9	1.544×10^8
Ours	1.676×10^9	1.544×10^8
Full Precision	0	1.826×10^9

Table 6. mCE on ImageNet-C. Lower is better.

Method	mCE (%)
IR-Net [40]	89.2
IR-Net + LCR (ours)	84.9 ↓
RBNN [31]	87.5
RBNN + LCR (ours)	84.8 ↓
ReActNet [34]	87.0
IR-Net + LCR (ours)	84.9 ↓

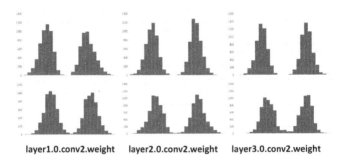

layer1.0.conv2.weight   layer2.0.conv2.weight   layer3.0.conv2.weight

Fig. 2. Histograms of weights (before binarization) of the IR-Net [40] and *LCR*-BNN with ResNet-18 architecture. The first row shows the results of the IR-Net, and the second row shows the results of ours. The BNN trained by our method has smoother weight distribution.

formulate the Fig. 2. Compared with IR-Net, the BNN trained by our method possesses smoother weight distribution, which correspondingly helps our method achieve 1.6% accuracy improvement on ImageNet as listed in Table 2. More precisely, the standard deviation of the distribution of the IR-Net is 1.42, 28% higher than ours 1.11, in the layer3.0.conv2 layer.

Robustness Study on ImageNet-C. ImageNet-C [19] becomes the standard dataset for investigation of model robustness, which consists of 19 different types of corruptions with five levels of severity from the noise, blur, weather and digital

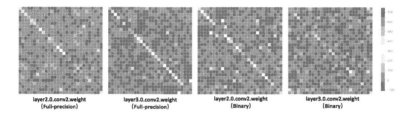

Fig. 3. Correlation maps for reflecting independence assumption in Eq. 7.

categories applied to the validation images of ImageNet (see Samples in Supplemental Materials). We consider all the 19 corruptions at the highest severity level (severity = 5) and report the mean top-1 accuracy. We use Mean Corruption Error (mCE) to measure the robustness of models on this dataset. We freeze the backbone for learning the representations of data *w.r.t.* classification task, and only fine-tune the task-specific heads over the backbone (*i.e.* linear protocol). The results in Table 6 prove that add *LCR* on the existing methods can improve the robustness of binary models.

Independence Assumption Reflection. The assumption used in Eq. 7 is the core of our method derivation, as it theoretically supports the approximation of the spectral norms of weight matrix with the designed retention matrix. Thus, we investigate this assumption by visualizing the correlation matrix of feature maps in the same batch. Specifically, we visualise the correlation matrices of full-precision and binary activations, where red stands for two activations are similar and blue *vice versa*. As shown in Fig 3, we can clearly observe that an activation is only correlated with itself, which largely testify this assumption. Besides, we also design another mechanism to use this assumption properly. We set a coefficient β greater than 1 to give more weight on latter layer's features such that they contribute more to $\mathcal{L}_{Lip}$ (Eq. 11). As in neural network, the feature maps of latter layers have stronger mutual linear independence [2].

5   Conclusion

In this paper, we introduce Lipschitz continuity to measure the robustness of BNN. Motivated by this, we propose *LCR*-BNN to retain the Lipschitz constant serving as a regularization term to improve the robustness of binary models. Specifically, to bypass the NP-hard Lipschitz constant computation in BNN, we devise the Retention Matrices to approximate the Lipschitz constant, and then constrain the Lipschitz constants of those Retention Matrices. Experimental results demonstrate the efficacy of our method.

Ethical Issues. All datasets used in our paper are open-source datasets and do not contain any personally identifiable or sensitive personally identifiable information. **Limitations.** Although our method achieve SoTA, adding it on existing method costs more time (around 20% more) to train BNN, which is the obvious limitation of our method.

Acknowledgements. This research was partially supported by NSF CNS-1908658 (ZZ,YY), NeTS-2109982 (YY), Early Career Scheme of the Research Grants Council (RGC) of the Hong Kong SAR under grant No. 26202321 (DX), HKUST Startup Fund No. R9253 (DX) and the gift donation from Cisco (YY). This article solely reflects the opinions and conclusions of its authors and not the funding agents.

References

1. Aharon, M., Elad, M., Bruckstein, A.: K-svd: an algorithm for designing over-complete dictionaries for sparse representation. IEEE Trans. Sig. Process. **54**(11), 4311–4322 (2006)
2. Alain, G., Bengio, Y.: Understanding intermediate layers using linear classifier probes. arXiv preprint arXiv:1610.01644 (2016)
3. Bartlett, P.L., Foster, D.J., Telgarsky, M.J.: Spectrally-normalized margin bounds for neural networks. In: NeurIPS (2017)
4. Bengio, Y., Léonard, N., Courville, A.: Estimating or propagating gradients through stochastic neurons for conditional computation. arXiv:1308.3432 (2013)
5. Bulat, A., Martinez, B., Tzimiropoulos, G.: BATS: binary architecture search. In: Vedaldi, A., Bischof, H., Brox, T., Frahm, J.-M. (eds.) ECCV 2020. LNCS, vol. 12368, pp. 309–325. Springer, Cham (2020). https://doi.org/10.1007/978-3-030-58592-1_19
6. Bulat, A., Tzimiropoulos, G.: Xnor-net++: improved binary neural networks. In: BMVC (2019)
7. Courbariaux, M., Bengio, Y., David, J.P.: Binaryconnect: training deep neural networks with binary weights during propagations. In: NeurIPS (2016)
8. Darabi, S., Belbahri, M., Courbariaux, M., Nia, V.P.: Bnn+: improved binary network training. CoRR (2018)
9. Deng, J., Dong, W., Socher, R., Li, L.J., Li, K., Fei-Fei, L.: Imagenet: a large-scale hierarchical image database. In: CVPR (2009)
10. Ding, R., Chin, T.W., Liu, Z., Marculescu, D.: Regularizing activation distribution for training binarized deep networks. In: CVPR (2019)
11. Dong, Y., Ni, R., Li, J., Chen, Y., Zhu, J., Su, H.: Learning accurate low-bit deep neural networks with stochastic quantization. In: BMVC (2017)
12. Golub, G.H., Van der Vorst, H.A.: Eigenvalue computation in the 20th century. In: JCAM (2000)
13. Gong, R., et al.: Differentiable soft quantization: bridging full-precision and low-bit neural networks. In: ICCV (2019)
14. Goodfellow, I., Bengio, Y., Courville, A., Bengio, Y.: Deep learning (2016)
15. Han, K., Wang, Y., Xu, Y., Xu, C., Wu, E., Xu, C.: Training binary neural networks through learning with noisy supervision. In: ICML (2020)
16. Han, S., Mao, H., Dally, W.J.: Deep compression: Compressing deep neural networks with pruning, trained quantization and huffman coding. In: ICLR (2016)
17. He, K., Zhang, X., Ren, S., Sun, J.: Deep residual learning for image recognition. In: CVPR (2016)

18. He, X., et al.: ProxyBNN: learning binarized neural networks via proxy matrices. In: Vedaldi, A., Bischof, H., Brox, T., Frahm, J.-M. (eds.) ECCV 2020. LNCS, vol. 12348, pp. 223–241. Springer, Cham (2020). https://doi.org/10.1007/978-3-030-58580-8_14

19. Hendrycks, D., Dietterich, T.: Benchmarking neural network robustness to common corruptions and perturbations. In: ICLR (2019)

20. Heo, B., Kim, J., Yun, S., Park, H., Kwak, N., Choi, J.Y.: A comprehensive overhaul of feature distillation. In: ICCV (2019)

21. Hinton, G., Vinyals, O., Dean, J.: Distilling the knowledge in a neural network. In: NeurIPS (2014)

22. Howard, A.G., et al.: Mobilenets: efficient convolutional neural networks for mobile vision applications. In: NeurIPS (2017)

23. Hu, Q., Wang, P., Cheng, J.: From hashing to cnns: training binary weight networks via hashing. In: AAAI (2018)

24. Hubara, I., Courbariaux, M., Soudry, D., El-Yaniv, R., Bengio, Y.: Binarized neural networks. In: NeurIPS (2016)

25. Krizhevsky, A., Hinton, G., et al.: Learning multiple layers of features from tiny images (2009)

26. Krizhevsky, A., Sutskever, I., Hinton, G.E.: Imagenet classification with deep convolutional neural networks. In: NeurIPS (2012)

27. LeCun, Y., Bengio, Y., Hinton, G.: Deep learning. In: Nature (2015)

28. LeCun, Y., Denker, J., Solla, S.: Optimal brain damage. In: NeurIPS (1989)

29. Li, Z., Ni, B., Zhang, W., Yang, X., Gao, W.: Performance guaranteed network acceleration via high-order residual quantization. In: ICCV (2017)

30. Lin, J., Gan, C., Han, S.: Defensive quantization: when efficiency meets robustness. arXiv preprint arXiv:1904.08444 (2019)

31. Lin, M., et al.: Rotated binary neural network. In: NeurIPS (2020)

32. Lin, X., Zhao, C., Pan, W.: Towards accurate binary convolutional neural network. In: NeurIPS (2017)

33. Liu, Z., Luo, W., Wu, B., Yang, X., Liu, W., Cheng, K.-T.: Bi-Real Net: binarizing deep network towards real-network performance. Int. J. Comput. Vis. **128**(1), 202–219 (2019). https://doi.org/10.1007/s11263-019-01227-8

34. Liu, Z., Shen, Z., Savvides, M., Cheng, K.-T.: ReActNet: towards precise binary neural network with generalized activation functions. In: Vedaldi, A., Bischof, H., Brox, T., Frahm, J.-M. (eds.) ECCV 2020. LNCS, vol. 12359, pp. 143–159. Springer, Cham (2020). https://doi.org/10.1007/978-3-030-58568-6_9

35. Luxburg, U.v., Bousquet, O.: Distance-based classification with lipschitz functions. In: JMLR (2004)

36. Miyato, T., Kataoka, T., Koyama, M., Yoshida, Y.: Spectral normalization for generative adversarial networks. In: ICLR (2018)

37. Neyshabur, B., Bhojanapalli, S., McAllester, D., Srebro, N.: Exploring generalization in deep learning. In: NeurIPS (2017)

38. Paszke, A., et al.: Pytorch: an imperative style, high-performance deep learning library. In: NeurIPS (2019)

39. Qin, H., Gong, R., Liu, X., Bai, X., Song, J., Sebe, N.: Binary neural networks: a survey. In: PR (2020)

40. Qin, H., et al.: Forward and backward information retention for accurate binary neural networks. In: CVPR (2020)

41. Rastegari, M., Ordonez, V., Redmon, J., Farhadi, A.: XNOR-Net: imagenet classification using binary convolutional neural networks. In: Leibe, B., Matas, J., Sebe,

N., Welling, M. (eds.) ECCV 2016. LNCS, vol. 9908, pp. 525–542. Springer, Cham (2016). https://doi.org/10.1007/978-3-319-46493-0_32

42. Ren, S., He, K., Girshick, R., Sun, J.: Faster r-cnn: towards real-time object detection with region proposal networks. In: NeurIPS (2015)

43. Rosca, M., Weber, T., Gretton, A., Mohamed, S.: A case for new neural network smoothness constraints. In: NeurIPS Workshop (2020)

44. Santurkar, S., Tsipras, D., Ilyas, A., Madry, A.: How does batch normalization help optimization? In: NeurIPS (2018)

45. Scaman, K., Virmaux, A.: Lipschitz regularity of deep neural networks: analysis and efficient estimation. In: NeurIPS (2018)

46. Shang, Y., Duan, B., Zong, Z., Nie, L., Yan, Y.: Lipschitz continuity guided knowledge distillation. In: ICCV (2021)

47. Virmaux, A., Scaman, K.: Lipschitz regularity of deep neural networks: analysis and efficient estimation. In: NeurIPS (2018)

48. Yang, Z., et al.: Searching for low-bit weights in quantized neural networks. In: NeurIPS (2020)

49. Yoshida, Y., Miyato, T.: Spectral norm regularization for improving the generalizability of deep learning. arXiv:1705.10941 (2017)

50. Zhang, B., Cai, T., Lu, Z., He, D., Wang, L.: Towards certifying robustness using neural networks with l-dist neurons. In: ICML (2021)

51. Zhou, S., Wu, Y., Ni, Z., Zhou, X., Wen, H., Zou, Y.: Dorefa-net: training low bitwidth convolutional neural networks with low bitwidth gradients. arXiv:1606.06160 (2016)

SPViT: Enabling Faster Vision Transformers via Latency-Aware Soft Token Pruning

Zhenglun Kong[1], Peiyan Dong[1], Xiaolong Ma[2], Xin Meng[3], Wei Niu[4], Mengshu Sun[1], Xuan Shen[1], Geng Yuan[1], Bin Ren[4], Hao Tang[5], Minghai Qin[1], and Yanzhi Wang[1]([envelope])

[1] Northeastern University, Boston, MA 02115, USA
{kong.zhe,dong.pe,yanz.wang}@northeastern.edu
[2] Clemson University, Clemson, SC 29634, USA
[3] Peking university, Beijing 100871, China
[4] College of William and Mary, Williamsburg, VA 23185, USA
[5] CVL, ETH Zürich, 8092 Zürich, Switzerland

Abstract. Recently, Vision Transformer (ViT) has continuously established new milestones in the computer vision field, while the high computation and memory cost makes its propagation in industrial production difficult. Considering the computation complexity, the internal data pattern of ViTs, and the edge device deployment, we propose a latency-aware soft token pruning framework, **SPViT**, which can be set up on vanilla Transformers of both flatten and hierarchical structures, such as DeiTs and Swin-Transformers (Swin). More concretely, we design a dynamic attention-based multi-head token selector, which is a lightweight module for adaptive instance-wise token selection. We further introduce a soft pruning technique, which integrates the less informative tokens chosen by the selector module into a package token rather than discarding them completely. SPViT is bound to the trade-off between accuracy and latency requirements of specific edge devices through our proposed latency-aware training strategy. Experiment results show that SPViT significantly reduces the computation cost of ViTs with comparable performance on image classification. Moreover, SPViT can guarantee the identified model meets the latency specifications of mobile devices and FPGA, and even achieve the real-time execution of DeiT-T on mobile devices. For example, SPViT reduces the latency of DeiT-T to 26 ms ($26\%-41\%$ superior to existing works) on the mobile device with $0.25\%-4\%$ higher top-1 accuracy on ImageNet. Our code is released at https://github.com/PeiyanFlying/SPViT.

Keywords: Vision transformer · Model compression · Hardware acceleration · Mobile devices · FPGA

Z. Kong and P. Dong—Both authors contributed equally.

Supplementary Information The online version contains supplementary material available at https://doi.org/10.1007/978-3-031-20083-0_37.

S. Avidan et al. (Eds.): ECCV 2022, LNCS 13671, pp. 620–640, 2022.
https://doi.org/10.1007/978-3-031-20083-0_37

1   Introduction

Recently, a new trend of leveraging Transformer architecture [80] into the computer vision domain has emerged [13,21,33,38,41,75,90,109]. The Vision Transformer (ViT), which solely exploits the self-attention mechanism that inherits from the Transformer architecture, has set up many state-of-the-art (SOTA) records in image classifications [7,22,79], object detection [1,3,19,60], tracking [15,59,91], semantic segmentation [16,110], depth estimation [45,94], image retrieval [23], and image enhancement [8,50,93]. However, despite the impressive general results, ViTs have sacrificed lightweight model capacity, portability, and trainability in return for high accuracy. The mass amount of computations brought by operations (e.g. Conv, MatMul, Add) in existing models remains a setback for edge device deployment.

Pruning has been proved as the one of the most effective methods to reduce network dimensions in convolution-based neural networks [4,11,36,47,52,54,55, 57,62,70,72,101,107,108]. However, when huge amount of AI-powered applications are benefiting from the network pruning advantages [17,25–27,30,44, 51,53,56,63,77,98–100], the applications of self-attention-based neural network pruning remain scarce [32,43,61,74,81]. There still exists a gap between the actual device deployment and acceleration in the ViT pruning frameworks. For instance, attention head pruning [12] performs weight pruning on the transformation matrix (W_Q, W_K, W_V) before the multi-head self-attention (MSA) operation. It is an inefficient way for computation reduction because only part of the ViT computations (i.e., MSA) can be alleviated (see Sect. 3 for justification). In a lightweight model, head pruning cannot guarantee an ideal pruning rate without significant accuracy deterioration. Static token pruning [69] reduces

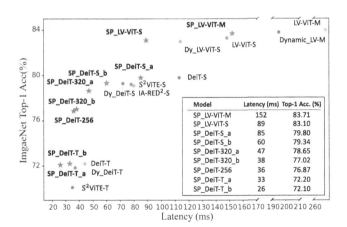

Fig. 1. Comparison of different pruning methods with various accuracy-latency trade-offs. We can increase the accuracy of light weight models at similar latency, and expedite larger models with negligible decrease of accuracy. Models are tested on Samsung Galaxy S20.

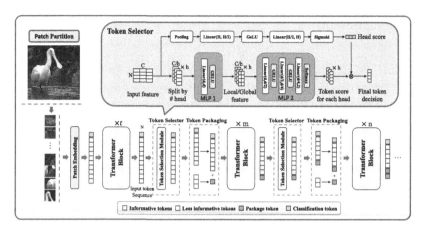

Fig. 2. Overall workflow. Bottom figure: Token selector is inserted multiple times throughout the model, along with the token packaging technique to generate a package token from the less informative tokens. The package token is concatenated with the informative tokens to be fed in the following transformer blocks. Upper figure: Our attention-based multi-head token selector to obtain token scores for keep/prune decisions.

the number of input tokens by a fixed ratio for different images, which restricts the image pruning rate, ignoring the fact that the high-level information of each image varies both in the region size and location. Furthermore, it is difficult for the deployment on edge devices since newly introduced operations (e.g., Argsort) are currently not well supported by many frameworks [66]. In contrast, dynamic token pruning [64] deletes redundant tokens based on the inherent image characteristics to achieve a per-image adaptive pruning rate. However, this method implies a potentially huge search space, which will easily cause a limited overall pruning rate or undermined accuracy if the token selection mechanism is not carefully designed. In addition, the pruning mechanism in [64] unreservedly discards less informative tokens, which results in the loss of the informative part of the removed tokens.

In this paper, we manage to overcome the above limitations. Specifically, as shown in Fig. 2, we propose a latency-aware Soft Pruning framework (SPViT), which simultaneously optimizes ViT accuracy and maximizes per-image dynamic pruning rate while maintaining actual computation constraints on edge devices. In ViT, each head encodes the visual receptive field independently [35,58,64], which implies that each token has a different influence in different heads [22, 29,96,106]. We thus propose a token selector to evaluate the importance score of each token based on its characteristic statistics in all heads. Then, through an attention-based branch [37] in the selector, we calculate the weighted sum of each score to obtain the final score of a token, which determines whether the token should be pruned. With the token selector, all tokens generated from the input images can be precisely ranked and pruned based on their importance scores and thus achieving a high overall pruning rate.

The token representations [5, 10, 84, 87] in early and middle layers are insufficiently encoded, which makes token pruning quite difficult. To mitigate the challenge, we introduce a package token technique, which compresses the less-informative tokens, picked out by the token selector, into a package token. Then, we concatenate the package token to the remaining tokens for subsequent blocks. On the one hand, although informative tokens may be discarded due to the poor encoding ability in earlier blocks of ViT [89], this error will be partly corrected by the residual information stored in the package token. On the other hand, background features can help emphasize foreground features [92]. Completely removing less informative (negative) tokens will weaken the ability of self-attention to capture key information. Therefore, the package token can serve as a way to help preserve background features. By adding minimal computation cost, the token pruning rate can be increased significantly.

In addition, we elaborate a latency-aware training strategy, which consists of two parts: latency-aware loss function and layer-to-phase progressive training. The former bridges the token pruning rates with latency specifications of diverse edge devices. The latter indicates that we progressively insert one selector in each block and train the new selector under the latency budget of the target device. Next, we group adjacent blocks with similar pruning rates into a phase, keep the first selector in this phase and remove others. While maintaining high accuracy, it can search for the appropriate pruning rate for each block and the desired insertion position of the selector. Figure 1 shows the on device performance of our model compared with other pruned or scaled models.

Our contributions are summarized as follows:

- We provide a detailed analysis on the computational complexity of ViT and different compression strategies. Based on our analysis, token pruning holds a greater computation reduction compared to the compression of other dimensions.
- Considering the vision pattern inside ViT, we propose SPViT, a novel method which includes the attention-based multi-head token selector and the token packaging technique to achieve per-image adaptive pruning. We design a latency-aware training strategy, which efficiently explores the SPViT design space given the hardware latency budget, and maximizes the per-image pruning rate without any accuracy degradation.
- SPViT enables a higher pruning rate than other state-of-the-art with comparable accuracy. For lightweight models, SPViT allows the DeiT-S and DeiT-T to reduce inference latency by 40%–60% within 0.5% accuracy loss. It can further generate more efficient PiTs and Swins with negligible performance drops. In particular, SPViT is superior in the compression of lightweight models.
- We demonstrate a real-time realization of DeiT-T on mobile phones (e.g., 26 ms on a Samsung Galaxy S20) and DeiT-S on a Xilinx FPGA (13.2 ms on a Xilinx ZCU102). To the best of our knowledge, it is the first time that the ViT models perform inference on the edge devices beyond real-time[1].

---

[1] Real-time inference usually means 30 frames per second, which is approximately 33 ms/image.

2   Related Work

Vision Transformers. ViT [22] is a pioneering work that uses only a Transformer to solve various vision tasks. Compared to traditional CNN structures, ViT allows all the positions in an image to interact through transformer blocks, whereas CNNs operate on a fixed-sized window with restricted spatial interactions, which can have trouble capturing relations at the pixel level in both spatial and time domains [68]. Since then, many variants have been proposed [2,9,24,31,34,48,49,76,82,83,86,102]. For example, DeiT [79], T2T-ViT [103] and Mixer [14] tackle the data-inefficiency problem in ViT by training only with ImageNet. PiT [35] replaces the uniform structure of Transformer with depth-wise convolution pooling layer to reduce spacial dimension and increase channel dimension. LV-ViT [40] introduces a token labeling method to improve training. PS-ViT [105] applied progressive sampled tokens.

Efficient ViT. The huge memory usage and computation cost of the self-attention mechanism serve as the roadblock to the efficient deployment of ViT models on edge devices. Many works aim at accelerating the inference speed of ViT [6]. For instance, S^2ViTE [12] prunes token and attention head in a structured way via sparse training. VTP [112] reduces the input feature dimension by learning their associated importance scores with L1 regularization. IA-RED^2 [64] drops redundant tokens with a multi-head interpreter. PS-ViT (T2T) [78] discards useless patches in a top-down paradigm. DynamicViT [69] removes redundant tokens by estimating their importance score with a MLP [80] based prediction module. Evo-ViT [89] develops a slow-fast token evolution method to preserve more image information during pruning. TokenLearner [73] and PATCH-MERGER [71] uses spatial attention to generate a small set of token vectors adaptive to the input. However, to the best of our knowledge, our idea of considering actual edge device deployment and acceleration has not been investigated by any existing ViT pruning methods.

3   Computational Complexity Analysis

Given an input sequence $N \times D$, where N is the input sequence length or the token number and D is the embedding dimension [79] of each token, some works [64,112] address the computational complexity of ViT as $(12ND^2 + 2N^2D)$. However, D represents different dimensions and should be written as $(4ND_{ch}D_{attn} + 2N^2D_{attn} + 8ND_{ch}D_{fc})$. Neglecting the difference may cause misleading conclusions, especially when analyzing the validity of pruning methods such as token pruning and dimension pruning.

Table 1 shows an analysis of each operation in a Transformer block. There are three main branches of ViT pruning. (i) Token channel pruning: The sequence tokens are pruned along D_{ch} dimension. D_{ch} is non-transmissible, which means reducing input dimension only affects the computation of the current matrix multiplication. To reduce computation for all layers, a mask layer is added to multiply with the input before going through the linear layer [112]. (ii) Token

Table 1. The computational complexity of each operation in a ViT block. The input $N \times D_{ch}$ goes through three linear transformation layers with $D_{ch} \times D_{attn}$ to generate Query (Q), Key (K), and Value (V) matrices of size $N \times D_{attn}$. N is transitive, while D_{ch} is not.

#	Module	Input size	Operation	Layer size	Output size	Computation
①	MSA	$N \times D_{ch}$	Linear transformation	$D_{ch} \times D_{attn}$	$N \times D_{attn}$	$N D_{ch} D_{attn} \times 3$
②		$N \times D_{attn}$	Q Multiplying K^T	-	$N \times N$	$N^2 D_{attn}$
③		$N \times N$	Multiplying V	-	$N \times D_{attn}$	$N^2 D_{attn}$
④		$N \times D_{attn}$	Projection	$D_{attn} \times D_{ch}$	$N \times D_{ch}$	$N D_{attn} D_{ch}$
⑤	FNN	$N \times D_{ch}$	FC layer	$D_{ch} \times 4D_{fc}$	$N \times 4D_{fc}$	$4N D_{ch} D_{fc}$
⑥		$N \times 4D_{fc}$	FC layer	$4D_{fc} \times D_{ch}$	$N \times D_{ch}$	$4N D_{fc} D_{ch}$
Total computational complexity						$4N D_{ch} D_{attn} +$ $2N^2 D_{attn} + 8N D_{ch} D_{fc}$

pruning: N is transitive, so directly pruning tokens will contribute to the linearly or even quadratically (N^2 in ② and ③) reduction of all operations. (iii) Attention head pruning (or attention channel pruning): The pruning operations are performed on weight tensors of each attention head in the MSA module. However, only the D_{attn} in the MSA module can be counted towards computation reduction, which usually contributes less than 40% of the total computation in most ViT architectures. Therefore, with the same pruning rate, pruning tokens (reducing N) can reduce more overall computation than pruning channels (reducing D_{ch} or D_{attn}).

4   Latency-Aware Soft Pruning

In this section, we first introduce our soft token pruning framework. Then, we show an elaborate design of each module. Finally, we give a detailed discussion of our latency-aware training strategy.

4.1   Framework Overview

Our soft pruning framework includes a token selector and a token packaging technique. We propose a hierarchical pruning scheme, where these two modules are inserted between multiple blocks throughout the model. As shown in Fig. 2, the input token sequence first goes through a token selector, where each token is scored and defined as either informative or less informative. After that, less informative tokens are separated from the sequence and integrated into a package token. This package token then concatenates to the informative tokens to involve in subsequent calculations in the blocks. In the next phase, a newly generated package token will connect with the existing package token.

For ViT training with our framework, we devise a latency-aware sparsity loss for the hardware's maximum computation bandwidth. We perform a layer-to-phase progressive training schedule to compress the search space, where model

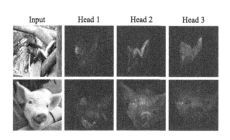

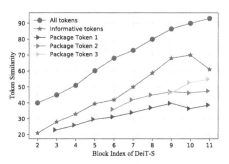

Fig. 3. Heatmaps showing the informative region detected by each head in DeiT-T. Each attention head focuses on encoding different image features and visual receptive fields.

Fig. 4. The CKA between the final CLS token and other tokens.

accuracy optimization and hardware computation reduction can be simultaneously achieved. The overall framework is hardware friendly with no unsupported operations and miniature computation cost.

Multi-head Token Selector. We propose a fine-grained approach to evaluate token scores. As shown in Fig. 3, in ViT's multi-head vision pattern, each head focus on encoding different features and respective fields of an image. This implies that the importance of each token towards each head is different. Our multi-head selector generates a list of token scores for each head. Let one head dimension be $d = C/H$, where C is the input dimension and H is the number of head. We split the input $X \in \mathbb{R}^{N \times C}$ by the number of attention head into $\{x_i\}_{i=1}^{H} \in \mathbb{R}^{N \times d}$, and obtain local f_i^{local} and global f_i^{global} features separately through an MLP layer with a pipeline of $LayerNorm \rightarrow Linear(d, d/2) \rightarrow GELU$:

$$f_i^{local} = \text{MLP}(x_i) \in \mathbb{R}^{N \times d/2}, \tag{1}$$

$$f_i^{global} = \text{AvgPool}(\text{MLP}(x_i), D) \in \mathbb{R}^{1 \times d/2}, \tag{2}$$

where D is the keep/prune decision of the current tokens evaluated by Eq. (7). We then pass the combined feature $f_i = [f_i^{local}, f_i^{global}] \in \mathbb{R}^{N \times d}$ through a MLP pipeline of $Linear(d, d/2) \rightarrow GELU \rightarrow Linear(d/2, d/4) \rightarrow GELU \rightarrow Linear(d/4, 2)$ to produce a series of token score maps $\{t_i\}_{i=1}^{H} \in \mathbb{R}^{N \times 2}$, with t_i indicating the token score from each attention head:

$$t_i = \text{Softmax}(\text{MLP}(f_i)) \in \mathbb{R}^{N \times 2}, \tag{3}$$

where $N \times 2$ represents the keep and prune probabilities of N number of tokens.

Head Attention Branch. We merge the individual score maps by the weights of each attention head to get the overall token score. As shown in Fig. 2, we add an attention-based branch along the selector backbone to synthesis the importance of each head:

$$\bar{X} = \text{AvgPool}(X) = Concat\{\frac{1}{C}\sum_{i=1}^{C} x_i\}_{j=1}^{H} \in \mathbb{R}^{N \times H}, \tag{4}$$

$$A = \text{Sigmoid}(\text{Linear}(\text{GeLU}(\text{Linear}(\bar{X})))) \in \mathbb{R}^{N \times H}, \tag{5}$$

where $\bar{X}$ is a head-wise statistic generated by shrinking X through its channel dimension C with global average pooling. In Eq. (5), the attention head score vector A is obtained by feeding $\bar{X}$ into the $Linear(H, H/2) \rightarrow GeLU \rightarrow Linear(H/2, H) \rightarrow Sigmoid$ pipeline to fully capture head-wise dependencies. The overall token score is calculated by adding the token scores from each individual attention head, multiplying by their individual head score $\{a_i\}_{i=1}^{H} \in \mathbb{R}^{N \times 1}$:

$$\tilde{T} = \frac{\sum_{i=1}^{H} t_i * a_i}{\sum_{i=1}^{H} a_i} \in \mathbb{R}^{N \times 2}, \tag{6}$$

where $\tilde{T}$ is the final token probability score. To make the token removing differentiable, we apply the Gumbel-Softmax technique to generate the token keep/prune decision during training:

$$D = \text{GumbelSoftmax}(\tilde{T}) \in \{0, 1\}^{N}. \tag{7}$$

Next, D passes on to the following layers until reaching the next token selector, where it will be updated by applying Hadamard product with the new token keep decision $D \odot D'$ during our hierarchical pruning scheme.

Self-attention matrices-based methods [46,89] usually require sorting and evaluating the importance of tokens by a Top-k operation, which is currently not supported in many frameworks for edge devices [66]. On the contrary, our selector generates binary matrices with the help of gumble softmax and FC layers to perform pruning instead of Top-k ordering. For hardware efficiency, our token selector mainly leverages the FC layers to reuse the GEMM hardware engine already built for the backbone ViT.

4.2 Token Packaging Technique

As discussed before, ViT is less accurate for evaluating token values in earlier blocks. Poor scoring may cause important tokens to be removed. Moreover, completely removing background (negative) tokens will weaken self-attention's ability to capture key information [92]. Instead of completely discarding tokens that are considered less informative, we apply a token packaging technique that integrates them into a package token. Assume there are Q less informative tokens $\hat{X} = \{n_i\}_{i=1}^{Q}$, $n_i \in \mathbb{R}^C$, along with their token scores $\hat{T} = \{m_i\}_{i=1}^{Q}$, $m_i \in \mathbb{R}^2$ These tokens are combined into one token by:

$$P = \frac{\sum_{i=1}^{Q} n_i \cdot m_i[0]}{\sum_{i=1}^{Q} m_i[0]} \in \mathbb{R}^C, \tag{8}$$

Table 2. Latency of one DeiT block on the Xilinx ZCU102 FPGA board.

Pruning rate	0.0	0.1	0.2	0.3	0.4	0.5
DeiT-T latency (ms)	0.689	0.630	0.587	0.509	0.468	0.424
DeiT-S latency (ms)	2.107	1.891	1.710	1.503	1.315	1.121

where P is the package token; $m_i[0]$ is the probability of keeping the token. Token P will participate in the subsequent calculations along with the informative tokens, enabling the model to correct scoring mistakes. Our overall framework is efficient, with miniature computation cost (less than 1% of the total model GFLOPs). All the operations (MLP, Softmax, Pooling, Sigmoid, etc.) are well supported on edge platforms.

4.3 Latency-Aware Training Strategy

Our latency-aware training strategy includes two parts: (1) the training objective where we introduce the latency-aware sparsity loss to obtain the pruning rate of token constrained by the latency specifications of the target devices; (2) the layer-to-phase progressive training schedule by which we can determine the location of inserted selectors and their suitable pruning rates.

Latency-Sparsity Table. In order to bridge the inference of ViT model produced by SPViT to the actual latency bound of hardware operation, we measure the latency-sparsity table of the target device, shown in Table 2. Note that the computation amount of one selector is less than 1% of one ViT block and the specific latency can be disregarded.

Latency-Aware Sparsity Loss. Based on the relationship between the pruning rate and latency in Table 2, we introduce a latency-aware sparsity loss $\mathcal{L}_{ratio}$:

$$\text{Block\_lat}(\rho_i) = latency\_sparsity\_table(\rho_i), \tag{9}$$

$$\sum_{i=1}^{L} \text{Block\_lat}(\rho_i) \leq \text{LatencyLimit}, \tag{10}$$

$$\mathcal{L}_{ratio} = \sum_{i=1}^{L} (1 - \rho_i - \frac{1}{B} \sum_{b=1}^{B} \sum_{j=1}^{N} D_j^{i,b})^2, \tag{11}$$

where Eq. (9) is a look-up-table which aims to find the latency of one block Block_lat under the corresponding ratio ρ_i with Table 2. Equation (10) guarantees that the inference latency of the model should be under the limit of target edge devices after token pruning. $LatencyLimit$ is the latency constraints of the target device. With i being the block index, ρ_i is the corresponding pruning rate. Through Eqs. (9) and (10), we derive appropriate ρ_i and feed it to the final sparsity loss (11), where B is the training batch size, and $D_*^{i,*}$ (Eq. (7)) is token keep decision. In order to achieve per-image adaptive pruning, we set the average pruning rate of all images in one batch as the convergence target of the Eq. (11).

Training Objective. It includes the standard cross-entropy loss, soft distillation loss, and latency-aware sparsity loss. The former two are the same as the loss strategy used in DeiT [79].

$$\mathcal{L} = \mathcal{L}_{cls} + \lambda_{KL}\mathcal{L}_{KL} + \lambda_{distill}\mathcal{L}_{distill} + \lambda_{ratio}\mathcal{L}_{ratio}, \tag{12}$$

where we set λ_{KL}=0.5, $\lambda_{distill}$=0.5, λ_{ratio}=2 in all our experiments.

Layer-to-Phase Progressive Training Schedule. Based on [111], we assume that the final CLS token is strongly correlated with classification. And we use centered kernel alignment (CKA) similarity [42] to calculate the similarity of the token features in each block and the final CLS token. As shown in Fig. 4, the final CLS token feature is quite different from token features in earlier blocks. It shows that the representations in earlier blocks are encoded inadequately, which proves the difficulty of pruning tokens in the earlier blocks. Combined with this encoding pattern, we design a latency-aware progressive training strategy to find the optimal accuracy-pruning rate trade-offs and proper locations for token selectors. In a ViT, tokens can be more effectively encoded in later blocks. Hence, we adopt progressive training on the token selector from later blocks to earlier ones. Specifically, each time we insert a token selector, we train the current selector and finetune the other parts (backbone and other selectors) by increasing the pruning rate of the current block until accuracy decreases noticeably (>0.5%). We repeat the insertion until there is one selector for each block. Then if the adjacent selectors have a similar pruning rate (difference <8.5%), we combine them as one selection phase and solely keep the first selector of the phase. Finally, if the final computations are lower than the target latency of specific edge devices, we reduce the pruning rate of the first selector. This is because we observe that earlier blocks are more sensitive to pruning.

5   Experiments

Datasets and Implementation Details. Our experiments are conducted on ImageNet-1K [20] with different backbones including DeiT-T, DeiT-S [79]; LV-ViT-S, LV-ViT-M [40]; PiT-T, PiT-XS, PiT-S [35]; Swin-T, Swin-S [49]. The image resolution is 224 × 224. We follow most of the training settings as in DeiT and train all backbone models for 60 epochs. Through our layer-to-phase training, we observe that inserting three token pruning selectors is best for the computation-accuracy tradeoff. For DeiT-T/S, we insert the token selector after the 3rd, 6th, and 9th layers. For LV-ViT-S, we insert the token selector after the 4th, 8th, and 12th layers. For LV-ViT-M, we insert the token selector after the 5th, 10th, and 15th layers. For PiT-T/XS/S, we insert the token selector after the 1st, 5th, and 10th layers. For Swin-T/S, we insert the token selector after each patch merging layer at the 2nd, 3rd, and 4th stage. Our batch size is

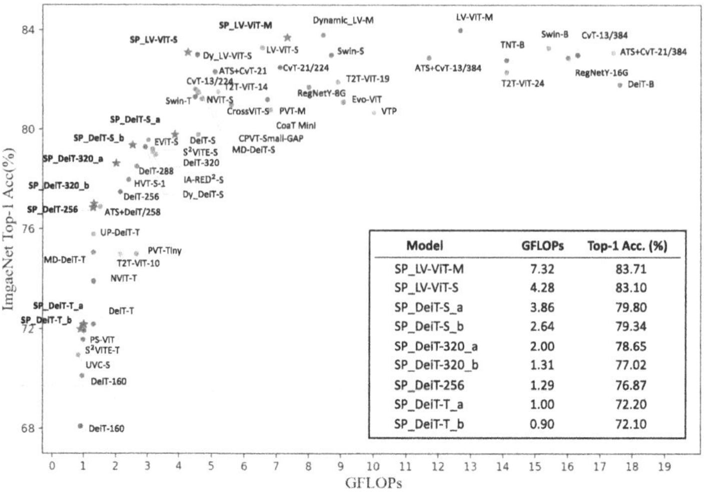

Fig. 5. Computation (GFLOPs) and top-1 accuracy trade-offs on ImageNet. Our models can achieve better trade-offs compared to other pruned or scaled models.

256 for DeiT-T, DeiT-S, and LV-ViT-S; and 128 for LV-ViT-M, PiT-T, PiT-XS, and PiT-S. We set an initial learning rate to be 5e-4 for the soft pruning module and 5e-6 for the backbone. The final model has three token selectors. All models are trained on 8 NVIDIA A100-SXM4-40GB GPUs. The latency is measured on a Samsung Galaxy S20 cell phone that has Snapdragon 865 processor, which consists of an Octa-core Kryo 585 CPU.

5.1 Experimental Results

Main Results. We compare our method with several representative methods including DynamicViT [69], IA-RED^2 [64], RegNetY [67], CrossViT [7], VTP [112], ATS [28], CvT [85], PVT [83], T2T-ViT [104], UP-DeiT [95], PS-ViT [78], Evo-ViT [89], TNT [34], HVT [65], Swin [49], CoaT [88], CPVT [18], EViT [46], UVC [97], MD-DeiT [39],and S^{2}ViTE [12]. Figure 5 demonstrates that our models achieve better accuracy-computation trade-offs compared to other pruned or scaled models. Our SPViT reduces the computation cost by 31%–43% for various backbones with negligible 0.1%–0.5% accuracy degradation, which outperforms existing methods on both accuracy and efficiency. On lightweight ViT, DeiT-T, the proposed SPViT still reduces GFLOPs by 31% with a negligible 0.1% decrease of accuracy (72.10% vs. 72.20%). To explore model scaling on ViT, we train more DeiT models with the embedding dimension of 160/256/288/320 as our baselines. On DeiT-T and DeiT-S under the same or similar GFLOPs, the accuracy improvement of SPViT over DeiT-160 is 4% (72.1% vs. 68.1% with $\sim$ 0.9 GFLOPs), 4.67% (76.87% vs. 72.20% with $\sim$1.3 GFLOPs) of SPViT-256 over DeiT-T-192, 4.82% (77.02% vs. 72.20% with $\sim$1.3 GFLOPs) of SPViT-320 over DeiT-T-192, and 0.81% (79.34% vs. 78.53% with

Table 3. Evaluation results on Hierarchical Architectures with SPViT.

Model	GFLOPs	Top1 Acc (%)
Swin-S	8.70	83.20
SPViT (Ours)	6.35 (26.4% ↓)	**82.71** (0.49% ↓)
Swin-T	4.50	81.20
SPViT (Ours)	3.47 (23.0% ↓)	**80.70** (0.50% ↓)
PiT-S	2.90	80.90
SPViT (Ours)	2.22 (23.3% ↓)	**80.32** (0.58% ↓)
PiT-XS	1.40	78.10
SPViT (Ours)	1.13 (18.7% ↓)	**77.86** (0.24% ↓)

Table 4. Evaluation results on Samsung Galaxy S20 with Snapdragon 865 processor and Xilinx ZCU102 FPGA board.

Model	Method	Top-1 Acc. (%)	Latency (ms)
Samsung Galaxy S20			
DeiT-T	Baseline	72.20	44
	SPViT (Ours)	**72.10**	**26**
DeiT-S	Baseline	79.80	113
	SPViT (Ours)	**79.34**	**60**
Xilinx ZCU102 FPGA			
DeiT-T	Baseline	72.20	8.81
	SPViT (Ours)	**72.10**	**5.60**
DeiT-S	Baseline	79.80	22.31
	SPViT (Ours)	**79.34**	**13.23**

~2.65 GFLOPs) of SPViT over DeiT-S-288. Additionally, our method can prune up to 23.1% on DeiT-T and 16.1% on DeiT-S without any accuracy degradation.

Results on Hierarchical Architectures. We also perform SPViT on lightweight hierarchical ViTs: Swin-Transformer and PiT, and present the results in Table 3. Our SPViT reduces the computation cost by 23%–27% for Swin with a slight accuracy degradation of 0.4%–0.5%, and by 18%–24% for PiT with a degradation of 0.2%–0.6%. Even though Swin has scaled down the computation complexity to $O(N)$ through window-based self-attention, and PiT is already a lightweight ViT model, we still can achieve a fair amount of compression while keeping the accuracy intact.

5.2 Deployment on Edge Devices

To evaluate the hardware performance, we implement a framework that runs the ViT model on edge devices. The evaluation is conducted on a Samsung Galaxy S20 cell phone that has a Snapdragon 865 processor, which consists of an Octa-core Kryo 585 CPU carrying high performance with good power efficiency. We use all eight cores on mobile CPUs. We report the average latency of over 100 inferences. As shown in Fig. 1, our method outperforms existing pruning methods on both latency and accuracy. The deficiencies of other methods mainly lie in three categories: limited pruning capability (low pruning rate) [64], non-optimal pruning dimension (number of heads) [12], and less efficient operators (e.g., Argsort.) [69]. As shown in Table 4, on the one hand, our models can outperform lightweight models such as DeiT-T by up to 4.8% under similar latency. On the other hand, we are able to reduce the latency of larger models such as DeiT-S by up to 47% (60 ms vs. 113 ms) with only 0.46% decrease of accuracy. Especially, for DeiT-T, we achieve 26 ms per inference on mobile CPUs, which meets the real-time requirement. As far as we know, this is the first demonstration of ViT inference over 30 fps on edge devices.

Additionally, SPViT is evaluated on an embedded FPGA platform, Xilinx ZCU102. To maintain the model accuracy on hardware, 16-bit fixed-point preci-

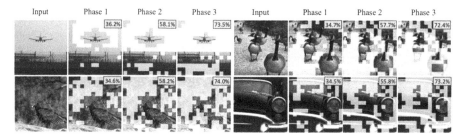

Fig. 6. Visualization of each pruning phase. In the 1st phase, the selector removes part of the background. In the 2nd phase, it targets the object of interest closely. In the 3rd phase, it localizes the informative features of the objects. The top right corner of each image shows the pruning rate after each phase.

Table 5. Token selector number/ location evaluation on DeiT-S.

Location	Params (M)	GFLOPs	Top-1 Acc. (%)
3-6-9	22.13	**2.65**	**79.34**
1-6-9	22.13	2.70	76.10
3-6-11	22.13	2.72	78.76
6-9	22.10	2.71	78.53
3-5-7-9	22.16	2.66	79.34

Table 6. Comparison of different pruning methods.

Model	Method	GFLOPs	Top-1 Acc. (%)
DeiT-T	Random	0.90	69.87
	Structure	0.90	70.32
	Token selector	**0.90**	**72.10**
DeiT-S	Random	2.64	77.25
	Structure	2.64	77.86
	Token selector	**2.64**	**79.34**

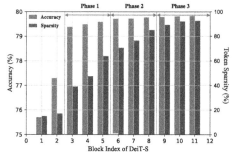

Fig. 7. The accuracy and the token sparsity distribution after the Layer-to-Phase Progressive Training. We do the insertion behind the $Block_{index}$. Our final phase plan is demonstrated above.

sion is adopted to represent all the model parameters and activation data. The comparison results with baseline models are shown in Table 4. In addition to the total latency, the average latency of the multi-head attention and MLP modules in each model is listed. Compared with the baseline, DeiT-T and DeiT-S, SPViT could achieve 1.57× and 1.69× acceleration in the total latency, respectively.

5.3   Token Pruning Visualization

We further visualize the hierarchical token reduction process of SPViT within Fig. 6. We show the input images along with their sparsification results after each phase. The masked regions represent the tokens that have been soft pruned. Our SPViT can gradually drop less informative tokens and preserve the tokens that contain representative regions with an adaptive pruning rate for each image.

5.4   Ablation Analysis

Token Selector Number and Location. After progressive training (each selector is fine-tuned by 25 epochs), we can get the pruning rate of each block as shown in Fig. 7. Based on the trend of the figure, we can divide the evolution of the pruning rate into 2 phases, 3 phases, and 4 phases. We keep the appropriate selectors accordingly and re-finetuning the whole model. In Table 5, the 3-6-9 division style has the highest accuracy and the lowest computation cost, just like 3-5-7-9. According to the test on Samsung Galaxy S20, each selector and corresponding package token will introduce a delay of 1.67 ms, so we choose 3-6-9 as the best. For another 3-phase style, 1-6-9, the accuracy and computation cost are both not ideal. This shows that due to insufficient encoding, it is difficult to perform token pruning in the earlier blocks of ViTs. Meanwhile, for the 3-6-11 style, both the accuracy and computation cost are slightly inferior to the 3–6-9 style. The possible reason is the pruning rate of the second phase should be smaller than the third phase and the coverage of the second phase is too wide. As a result, there is still a lot of redundancy in the tokens of the third phase, restricting the accuracy and computation efficiency of the model at the same time. Furthermore, because of a similar reason, the 2-phase style, 3-6, cannot achieve a better trade-off between accuracy and the computation cost.

Comparison of Different Pruning Methods. To further prove the effectiveness of our score-based dynamic token pruning method, we compare with some general pruning methods: random pruning and structure pruning. For random pruning, we randomly remove the input token, neglecting the token importance. For structure pruning, we prune the input feature map by dimension, which will impair every token. Results are shown in Table 6. Under the same computational complexities (0.9 GFLOPs for DeiT-T and 2.64 GFLOPs for DeiT-S), our proposed method achieves the best accuracy.

5.5   Limitations

For the algorithm design, it might be more effective to combine our framework with the weight pruning strategy for larger ViTs. For the hardware deployment, large amounts of data movement bring much pressure to the memory due to multiple blocks and many intermediate results, which will be optimized in our further work.

6   Conclusion

In this paper, we propose a dynamic, latency-aware soft token pruning framework called SPViT. Our attention-based multi-head token selector and token packaging technique, along with the latency-aware training strategy can well balance the tradeoff between accuracy and specific hardware constraints. We deploy our model on mobile and FPGA, which both meet the real-time requirement.

Acknowledgments. The research reported here was funded in whole or in part by the Army Research Office/Army Research Laboratory via grant W911-NF-20-1-0167 to Northeastern University. Any errors and opinions are not those of the Army Research Office or Department of Defense and are attributable solely to the author(s). This research is also partially supported by National Science Foundation CCF-1919117 and CMMI-2125326.

References

1. Amini, A., Periyasamy, A.S., Behnke, S.: T6d-direct: transformers for multi-object 6d pose direct regression. arXiv preprint arXiv:2109.10948 (2021)
2. Bao, H., Dong, L., Piao, S., Wei, F.: BEit: BERT pre-training of image transformers. In: International Conference on Learning Representations (2022). https://openreview.net/forum?id=p-BhZSz59o4
3. Carion, N., Massa, F., Synnaeve, G., Usunier, N., Kirillov, A., Zagoruyko, S.: End-to-end object detection with transformers. In: Vedaldi, A., Bischof, H., Brox, T., Frahm, J.-M. (eds.) ECCV 2020. LNCS, vol. 12346, pp. 213–229. Springer, Cham (2020). https://doi.org/10.1007/978-3-030-58452-8_13
4. Chang, S.E., et al.: Mix and match: a novel fpga-centric deep neural network quantization framework. In: 2021 IEEE International Symposium on High-Performance Computer Architecture (HPCA), pp. 208–220. IEEE (2021)
5. Chefer, H., Gur, S., Wolf, L.: Transformer interpretability beyond attention visualization. In: Proceedings of the IEEE/CVF Conference on Computer Vision and Pattern Recognition, pp. 782–791 (2021)
6. Chen, B., et al.: Psvit: better vision transformer via token pooling and attention sharing. arXiv preprint arXiv:2108.03428 (2021)
7. Chen, C.F.R., Fan, Q., Panda, R.: Crossvit: cross-attention multi-scale vision transformer for image classification. In: Proceedings of the IEEE/CVF International Conference on Computer Vision, pp. 357–366 (2021)
8. Chen, H., et al.: Pre-trained image processing transformer. In: Proceedings of the IEEE/CVF Conference on Computer Vision and Pattern Recognition, pp. 12299–12310 (2021)
9. Chen, M., Peng, H., Fu, J., Ling, H.: Autoformer: searching transformers for visual recognition. In: Proceedings of the IEEE/CVF International Conference on Computer Vision, pp. 12270–12280 (2021)
10. Chen, P., Chen, Y., Liu, S., Yang, M., Jia, J.: Exploring and improving mobile level vision transformers. arXiv preprint arXiv:2108.13015 (2021)
11. Chen, T., Chen, X., Ma, X., Wang, Y., Wang, Z.: Coarsening the granularity: towards structurally sparse lottery tickets. In: Proceedings of the International Conference on Machine Learning (ICML) (2022)
12. Chen, T., Cheng, Y., Gan, Z., Yuan, L., Zhang, L., Wang, Z.: Chasing sparsity in vision transformers: an end-to-end exploration. In: Advances in Neural Information Processing Systems (2021)
13. Chen, T., Saxena, S., Li, L., Fleet, D.J., Hinton, G.: Pix2seq: a language modeling framework for object detection. arXiv preprint arXiv:2109.10852 (2021)
14. Chen, X., Hsieh, C.J., Gong, B.: When vision transformers outperform resnets without pre-training or strong data augmentations. In: International Conference on Learning Representations (2022). https://openreview.net/forum?id=LtKcMgGOeLt

15. Chen, X., Yan, B., Zhu, J., Wang, D., Yang, X., Lu, H.: Transformer tracking. In: Proceedings of the IEEE/CVF Conference on Computer Vision and Pattern Recognition, pp. 8126–8135 (2021)

16. Cheng, B., Schwing, A., Kirillov, A.: Per-pixel classification is not all you need for semantic segmentation. In: Beygelzimer, A., Dauphin, Y., Liang, P., Vaughan, J.W. (eds.) Advances in Neural Information Processing Systems (2021). https://openreview.net/forum?id=0lz69oI5iZP

17. Chu, C., et al.: Pim-prune: fine-grain dcnn pruning for crossbar-based process-in-memory architecture. In: 2020 57th ACM/IEEE Design Automation Conference (DAC), pp. 1–6. IEEE (2020)

18. Chu, X., et al.: Conditional positional encodings for vision transformers. arXiv preprint arXiv:2102.10882 (2021)

19. Dai, Z., Cai, B., Lin, Y., Chen, J.: Up-detr: unsupervised pre-training for object detection with transformers. In: Proceedings of the IEEE/CVF Conference on Computer Vision and Pattern Recognition, pp. 1601–1610 (2021)

20. Deng, J., Dong, W., Socher, R., Li, L.J., Li, K., Fei-Fei, L.: Imagenet: a large-scale hierarchical image database. In: 2009 IEEE Conference on Computer Vision and Pattern Recognition, pp. 248–255 (2009). https://doi.org/10.1109/CVPR.2009.5206848

21. Deng, J., Yang, Z., Chen, T., Zhou, W., Li, H.: Transvg: end-to-end visual grounding with transformers. In: Proceedings of the IEEE/CVF International Conference on Computer Vision (ICCV), pp. 1769–1779 (2021)

22. Dosovitskiy, A., et al.: An image is worth 16×16 words: transformers for image recognition at scale. In: International Conference on Learning Representations (2021). https://openreview.net/forum?id=YicbFdNTTy

23. El-Nouby, A., Neverova, N., Laptev, I., Jégou, H.: Training vision transformers for image retrieval. arXiv preprint arXiv:2102.05644 (2021)

24. El-Nouby, A., et al.: XCit: Cross-covariance image transformers. In: Beygelzimer, A., Dauphin, Y., Liang, P., Vaughan, J.W. (eds.) Advances in Neural Information Processing Systems (2021). https://openreview.net/forum?id=kzPtpIpF8o

25. Fang, H., Mei, Z., Shrestha, A., Zhao, Z., Li, Y., Qiu, Q.: Encoding, model, and architecture: systematic optimization for spiking neural network in fpgas. In: 2020 IEEE/ACM International Conference On Computer Aided Design (ICCAD), pp. 1–9. IEEE (2020)

26. Fang, H., Shrestha, A., Zhao, Z., Qiu, Q.: Exploiting neuron and synapse filter dynamics in spatial temporal learning of deep spiking neural network. In: Proceedings of the Twenty-Ninth International Joint Conference on Artificial Intelligence. IJCAI 2020 (2021)

27. Fang, H., Taylor, B., Li, Z., Mei, Z., Li, H.H., Qiu, Q.: Neuromorphic algorithm-hardware codesign for temporal pattern learning. In: 2021 58th ACM/IEEE Design Automation Conference (DAC), pp. 361–366. IEEE (2021)

28. Fayyaz, M., et al.: Ats: adaptive token sampling for efficient vision transformers. arXiv preprint arXiv:2111.15667 (2021)

29. Gao, P., Lu, J., Li, H., Mottaghi, R., Kembhavi, A.: Container: context aggregation network. arXiv preprint arXiv:2106.01401 (2021)

30. Gong, Y., et al.: A privacy-preserving-oriented dnn pruning and mobile acceleration framework. In: Proceedings of the 2020 on Great Lakes Symposium on VLSI, pp. 119–124 (2020)

31. Graham, B., et al.: Levit: a vision transformer in convnet's clothing for faster inference. In: Proceedings of the IEEE/CVF International Conference on Computer Vision (ICCV), pp. 12259–12269 (2021)

32. Guo, C., et al.: Accelerating sparse dnn models without hardware-support via tile-wise sparsity. In: SC20: International Conference for High Performance Computing, Networking, Storage and Analysis, pp. 1–15. IEEE (2020)
33. Guo, M.H., Cai, J.X., Liu, Z.N., Mu, T.J., Martin, R.R., Hu, S.M.: Pct: point cloud transformer. Comput. Visual Media **7**(2), 187–199 (2021)
34. Han, K., Xiao, A., Wu, E., Guo, J., Xu, C., Wang, Y.: Transformer in transformer. In: Advances in Neural Information Processing Systems (2021)
35. Heo, B., Yun, S., Han, D., Chun, S., Choe, J., Oh, S.J.: Rethinking spatial dimensions of vision transformers. In: International Conference on Computer Vision (ICCV) (2021)
36. Hou, Z., et al.: Chex: channel exploration for cnn model compression. In: Proceedings of the IEEE/CVF Conference on Computer Vision and Pattern Recognition (CVPR), pp. 12287–12298 (2022)
37. Hu, J., Shen, L., Sun, G.: Squeeze-and-excitation networks. In: Proceedings of the IEEE Conference on Computer Vision and Pattern Recognition, pp. 7132–7141 (2018)
38. Hudson, D.A., Zitnick, C.L.: Generative adversarial transformers. In: Proceedings of the 38th International Conference on Machine Learning, ICML 2021 (2021)
39. Jia, D., et al.: Efficient vision transformers via fine-grained manifold distillation. arXiv preprint arXiv:2107.01378 (2021)
40. Jiang, Z., et al.: All tokens matter: token labeling for training better vision transformers. arXiv preprint arXiv:2104.10858 (2021)
41. Kim, B., Lee, J., Kang, J., Kim, E.S., Kim, H.J.: Hotr: end-to-end human-object interaction detection with transformers. In: Proceedings of the IEEE/CVF Conference on Computer Vision and Pattern Recognition, pp. 74–83 (2021)
42. Kornblith, S., Norouzi, M., Lee, H., Hinton, G.: Similarity of neural network representations revisited. In: International Conference on Machine Learning, pp. 3519–3529. PMLR (2019)
43. Li, B., et al.: Efficient transformer-based large scale language representations using hardware-friendly block structured pruning. In: Findings of the Association for Computational Linguistics: EMNLP 2020, pp. 3187–3199 (2020)
44. Li, Y., Fang, H., Li, M., Ma, Y., Qiu, Q.: Neural network pruning and fast training for drl-based uav trajectory planning. In: 2022 27th Asia and South Pacific Design Automation Conference (ASP-DAC), pp. 574–579. IEEE (2022)
45. Li, Z., et al.: Revisiting stereo depth estimation from a sequence-to-sequence perspective with transformers. In: Proceedings of the IEEE/CVF International Conference on Computer Vision, pp. 6197–6206 (2021)
46. Liang, Y., GE, C., Tong, Z., Song, Y., Wang, J., Xie, P.: EVit: expediting vision transformers via token reorganizations. In: International Conference on Learning Representations (2022). https://openreview.net/forum?id=BjyvwnXXVn_
47. Liu, N., et al.: Lottery ticket preserves weight correlation: is it desirable or not? In: International Conference on Machine Learning (ICML), pp. 7011–7020. PMLR (2021)
48. Liu, Y., Sangineto, E., Bi, W., Sebe, N., Lepri, B., De Nadai, M.: Efficient training of visual transformers with small-size datasets. arXiv preprint arXiv:2106.03746 (2021)
49. Liu, Z., et al.: Swin transformer: hierarchical vision transformer using shifted windows. In: International Conference on Computer Vision (ICCV) (2021)
50. Lu, Z., Liu, H., Li, J., Zhang, L.: Efficient transformer for single image super-resolution. arXiv preprint arXiv:2108.11084 (2021)

51. Ma, X., et al.: PCONV: the missing but desirable sparsity in DNN weight pruning for real-time execution on mobile devices. In: Proceedings of the AAAI Conference on Artificial Intelligence (AAAI), vol. 34, pp. 5117–5124 (2020)

52. Ma, X., et al.: Non-structured dnn weight pruning-is it beneficial in any platform? In: IEEE Transactions on Neural Networks and Learning Systems (TNNLS) (2021)

53. Ma, X., et al.: An image enhancing pattern-based sparsity for real-time inference on mobile devices. In: Proceedings of the European conference on computer vision (ECCV). pp. 629–645. Springer (2020). https://doi.org/10.1007/978-3-030-58601-0_37

54. Ma, X., et al.: Effective model sparsification by scheduled grow-and-prune methods. In: Proceedings of the International Conference on Learning Representations (ICLR) (2021)

55. Ma, X., et al.: Blcr: Towards real-time dnn execution with block-based reweighted pruning. In: International Symposium on Quality Electronic Design (ISQED), pp. 1–8. IEEE (2022)

56. Ma, X., et al.: Tiny but accurate: a pruned, quantized and optimized memristor crossbar framework for ultra efficient dnn implementation. In: 2020 25th Asia and South Pacific design automation conference (ASP-DAC), pp. 301–306. IEEE (2020)

57. Ma, X., et al.: Sanity checks for lottery tickets: Does your winning ticket really win the jackpot? In: Advances in Neural Information Processing Systems (NeurIPS) 34 (2021)

58. Mao, M., et al.: Dual-stream network for visual recognition. In: Advances in Neural Information Processing Systems (2021)

59. Meinhardt, T., Kirillov, A., Leal-Taixe, L., Feichtenhofer, C.: Trackformer: multi-object tracking with transformers. arXiv preprint arXiv:2101.02702 (2021)

60. Misra, I., Girdhar, R., Joulin, A.: An end-to-end transformer model for 3d object detection. In: ICCV (2021)

61. Niu, W., et al.: A compression-compilation framework for on-mobile real-time bert applications. arXiv preprint arXiv:2106.00526 (2021)

62. Niu, W., et al.: Grim: A general, real-time deep learning inference framework for mobile devices based on fine-grained structured weight sparsity. In: IEEE Transactions on Pattern Analysis and Machine Intelligence (TPAMI) (2021)

63. Niu, W., et al.: Patdnn: achieving real-time dnn execution on mobile devices with pattern-based weight pruning. In: Proceedings of the Twenty-Fifth International Conference on Architectural Support for Programming Languages and Operating Systems (ASPLOS), pp. 907–922 (2020)

64. Pan, B., Jiang, Y., Panda, R., Wang, Z., Feris, R., Oliva, A.: Ia-red[2]: Interpretability-aware redundancy reduction for vision transformers. In: Advances in Neural Information Processing Systems (2021)

65. Pan, Z., Zhuang, B., Liu, J., He, H., Cai, J.: Scalable vision transformers with hierarchical pooling. In: Proceedings of the IEEE/CVF International Conference on Computer Vision (ICCV), pp. 377–386 (2021)

66. Prillo, S., Eisenschlos, J.: Softsort: a continuous relaxation for the argsort operator. In: International Conference on Machine Learning, pp. 7793–7802. PMLR (2020)

67. Radosavovic, I., Kosaraju, R.P., Girshick, R., He, K., Dollár, P.: Designing network design spaces. In: Proceedings of the IEEE/CVF Conference on Computer Vision and Pattern Recognition (CVPR), pp. 10428–10436 (2020)

68. Raghu, M., Unterthiner, T., Kornblith, S., Zhang, C., Dosovitskiy, A.: Do vision transformers see like convolutional neural networks? arXiv preprint arXiv:2108.08810 (2021)
69. Rao, Y., Zhao, W., Liu, B., Lu, J., Zhou, J., Hsieh, C.J.: Dynamicvit: efficient vision transformers with dynamic token sparsification. In: Advances in Neural Information Processing Systems (2021)
70. Ren, A., et al.: Admm-nn: an algorithm-hardware co-design framework of dnns using alternating direction methods of multipliers. In: Proceedings of the Twenty-Fourth International Conference on Architectural Support for Programming Languages and Operating Systems, pp. 925–938 (2019)
71. Renggli, C., Pinto, A.S., Houlsby, N., Mustafa, B., Puigcerver, J., Riquelme, C.: Learning to merge tokens in vision transformers. arXiv preprint arXiv:2202.12015 (2022)
72. Rumi, M.A., Ma, X., Wang, Y., Jiang, P.: Accelerating sparse cnn inference on gpus with performance-aware weight pruning. In: Proceedings of the ACM International Conference on Parallel Architectures and Compilation Techniques (PACT), pp. 267–278 (2020)
73. Ryoo, M.S., Piergiovanni, A., Arnab, A., Dehghani, M., Angelova, A.: Token-learner: what can 8 learned tokens do for images and videos? In: Advances in Neural Information Processing Systems (2021)
74. Sanh, V., Wolf, T., Rush, A.M.: Movement pruning: adaptive sparsity by fine-tuning. arXiv preprint arXiv:2005.07683 (2020)
75. Srinivas, A., Lin, T.Y., Parmar, N., Shlens, J., Abbeel, P., Vaswani, A.: Bottleneck transformers for visual recognition. In: Proceedings of the IEEE/CVF Conference on Computer Vision and Pattern Recognition, pp. 16519–16529 (2021)
76. Steiner, A., Kolesnikov, A., Zhai, X., Wightman, R., Uszkoreit, J., Beyer, L.: How to train your vit? data, augmentation, and regularization in vision transformers. arXiv preprint arXiv:2106.10270 (2021)
77. Tan, Z., et al.: Pcnn: pattern-based fine-grained regular pruning towards optimizing cnn accelerators. In: 2020 57th ACM/IEEE Design Automation Conference (DAC), pp. 1–6. IEEE (2020)
78. Tang, Y., et al.: Patch slimming for efficient vision transformers (2021)
79. Touvron, H., Cord, M., Douze, M., Massa, F., Sablayrolles, A., J'egou, H.: Training data-efficient image transformers & distillation through attention. In: ICML (2021)
80. Vaswani, A., et al.: Attention is all you need. In: Advances in Neural Information Processing Systems, pp. 5998–6008 (2017)
81. Wang, H., Zhang, Z., Han, S.: Spatten: efficient sparse attention architecture with cascade token and head pruning. In: 2021 IEEE International Symposium on High-Performance Computer Architecture (HPCA), pp. 97–110. IEEE (2021)
82. Wang, P., et al.: Kvt: k-nn attention for boosting vision transformers. arXiv preprint arXiv:2106.00515 (2021)
83. Wang, W., et al.: Pyramid vision transformer: a versatile backbone for dense prediction without convolutions. In: IEEE ICCV (2021)
84. Wu, B., et al.: Visual transformers: token-based image representation and processing for computer vision. arXiv preprint arXiv:2006.03677 (2020)
85. Wu, H., et al.: Cvt: introducing convolutions to vision transformers. In: Proceedings of the IEEE/CVF International Conference on Computer Vision (ICCV), pp. 22–31 (2021)

86. Wu, K., Peng, H., Chen, M., Fu, J., Chao, H.: Rethinking and improving relative position encoding for vision transformer. In: Proceedings of the IEEE/CVF International Conference on Computer Vision, pp. 10033–10041 (2021)
87. Xu, C., et al.: You only group once: efficient point-cloud processing with token representation and relation inference module. arXiv preprint arXiv:2103.09975 (2021)
88. Xu, W., Xu, Y., Chang, T., Tu, Z.: Co-scale conv-attentional image transformers. arXiv preprint arXiv:2104.06399 (2021)
89. Xu, Y., et al.: Evo-vit: slow-fast token evolution for dynamic vision transformer. In: Proceedings of the AAAI Conference on Artificial Intelligence (2022)
90. Xue, F., Wang, Q., Guo, G.: Transfer: learning relation-aware facial expression representations with transformers. In: Proceedings of the IEEE/CVF International Conference on Computer Vision, pp. 3601–3610 (2021)
91. Yan, B., Peng, H., Fu, J., Wang, D., Lu, H.: Learning spatio-temporal transformer for visual tracking. arXiv preprint arXiv:2103.17154 (2021)
92. Yang, C., Wu, Z., Zhou, B., Lin, S.: Instance localization for self-supervised detection pretraining. In: Proceedings of the IEEE/CVF Conference on Computer Vision and Pattern Recognition, pp. 3987–3996 (2021)
93. Yang, F., Yang, H., Fu, J., Lu, H., Guo, B.: Learning texture transformer network for image super-resolution. In: Proceedings of the IEEE/CVF Conference on Computer Vision and Pattern Recognition, pp. 5791–5800 (2020)
94. Yang, G., Tang, H., Ding, M., Sebe, N., Ricci, E.: Transformer-based attention networks for continuous pixel-wise prediction. In: ICCV (2021)
95. Yu, H., Wu, J.: A unified pruning framework for vision transformers. arXiv preprint arXiv:2111.15127 (2021)
96. Yu, Q., Xia, Y., Bai, Y., Lu, Y., Yuille, A., Shen, W.: Glance-and-gaze vision transformer. In: Advances in Neural Information Processing Systems (2021)
97. Yu, S., et al.: Unified visual transformer compression. In: International Conference on Learning Representations (2022). https://openreview.net/forum?id=9jsZiUgkCZP
98. Yuan, G., et al.: Tinyadc: Peripheral circuit-aware weight pruning framework for mixed-signal dnn accelerators. In: 2021 Design, Automation & Test in Europe Conference & Exhibition (DATE), pp. 926–931. IEEE (2021)
99. Yuan, G., et al.: Improving dnn fault tolerance using weight pruning and differential crossbar mapping for reram-based edge ai. In: 2021 22nd International Symposium on Quality Electronic Design (ISQED), pp. 135–141. IEEE (2021)
100. Yuan, G., et al.: An ultra-efficient memristor-based dnn framework with structured weight pruning and quantization using admm. In: 2019 IEEE/ACM International Symposium on Low Power Electronics and Design (ISLPED), pp. 1–6. IEEE (2019)
101. Yuan, G., et al.: Mest: accurate and fast memory-economic sparse training framework on the edge. In: Advances in Neural Information Processing Systems (NeurIPS) 34 (2021)
102. Yuan, K., Guo, S., Liu, Z., Zhou, A., Yu, F., Wu, W.: Incorporating convolution designs into visual transformers. In: Proceedings of the IEEE/CVF International Conference on Computer Vision (ICCV), pp. 579–588 (2021)
103. Yuan, L., et al.: Tokens-to-token vit: training vision transformers from scratch on imagenet. In: Proceedings of the IEEE/CVF International Conference on Computer Vision (ICCV), pp. 558–567 (2021)
104. Yuan, L., et al.: Tokens-to-token vit: training vision transformers from scratch on imagenet. arXiv preprint arXiv:2101.11986 (2021)

105. Yue, X., Sun, S., Kuang, Z., Wei, M., Torr, P.H., Zhang, W., Lin, D.: Vision transformer with progressive sampling. In: Proceedings of the IEEE/CVF International Conference on Computer Vision (ICCV), pp. 387–396 (2021)
106. Zhai, X., Kolesnikov, A., Houlsby, N., Beyer, L.: Scaling vision transformers. arXiv preprint arXiv:2106.04560 (2021)
107. Zhang, T., et al.: A unified dnn weight pruning framework using reweighted optimization methods. In: 2021 58th ACM/IEEE Design Automation Conference (DAC), pp. 493–498. IEEE (2021)
108. Zhang, T., et al.: Structadmm: achieving ultrahigh efficiency in structured pruning for dnns. In: IEEE Transactions on Neural Networks and Learning Systems (TNNLS) (2021)
109. Zhao, H., Jiang, L., Jia, J., Torr, P.H., Koltun, V.: Point transformer. In: Proceedings of the IEEE/CVF International Conference on Computer Vision, pp. 16259–16268 (2021)
110. Zheng, S., et al.: Rethinking semantic segmentation from a sequence-to-sequence perspective with transformers. In: Proceedings of the IEEE/CVF Conference on Computer Vision and Pattern Recognition, pp. 6881–6890 (2021)
111. Zhou, D., et al.: Refiner: refining self-attention for vision transformers (2021)
112. Zhu, M., Han, K., Tang, Y., Wang, Y.: Visual transformer pruning. In: KDD 2021 Workshop on Model Mining (2021)

Soft Masking for Cost-Constrained Channel Pruning

Ryan Humble[1]([✉]) [iD], Maying Shen[2], Jorge Albericio Latorre[2], Eric Darve[1] [iD],
and Jose Alvarez[2] [iD]

[1] Stanford University, Stanford, CA 94305, USA
{ryhumble,darve}@stanford.edu
[2] NVIDIA, Santa Clara, CA 95051, USA
{mshen,jalbericiola,josea}@nvidia.com

Abstract. Structured channel pruning has been shown to significantly accelerate inference time for convolution neural networks (CNNs) on modern hardware, with a relatively minor loss of network accuracy. Recent works permanently zero these channels during training, which we observe to significantly hamper final accuracy, particularly as the fraction of the network being pruned increases. We propose Soft Masking for cost-constrained Channel Pruning (SMCP) to allow pruned channels to adaptively return to the network while simultaneously pruning towards a target cost constraint. By adding a soft mask re-parameterization of the weights and channel pruning from the perspective of removing input channels, we allow gradient updates to previously pruned channels and the opportunity for the channels to later return to the network. We then formulate input channel pruning as a global resource allocation problem. Our method outperforms prior works on both the ImageNet classification and PASCAL VOC detection datasets.

Keywords: Neural network pruning · Model compression

1 Introduction

Deep neural networks have rapidly developed over the last decade and come to dominate many traditional algorithms in a wide range of tasks. In particular, convolutional neural networks (CNNs) have shown state-of-the-art results on a range of computer vision tasks, including classification, detection, and segmentation. However, modern CNNs have grown in size, computation, energy requirement, and prediction latency, as researchers push for accuracy improvements. Unfortunately, these models can now easily exceed the capabilities of many edge computing devices and requirements of real-time inference tasks, such as those found in autonomous vehicle applications.

R. Humble—Work performed during a NVIDIA internship.

Supplementary Information The online version contains supplementary material available at https://doi.org/10.1007/978-3-031-20083-0_38.

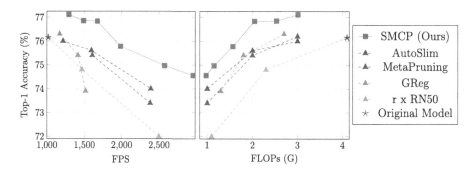

Fig. 1. Top-1 accuracy tradeoff curve for pruning ResNet50 on the ImageNet classification dataset using a latency cost constraint. Baseline is from PyTorch [34] model hub. Accuracy against FPS speed (left) and FLOPs (right) show the benefit of our method, particularly at high pruning ratios. For FPS, top-right is better. For FLOPs, top-left is better. FPS measured on an NVIDIA TITAN V GPU.

Since neural networks have been shown to be heavily over-parameterized [52], one popular method for reducing the computation and prediction latency is to prune (or remove) portions of the neural network, ultimately yielding a model with fewer parameters. Due to the strict requirements for many deployment applications, a large fraction of the parameters often must be removed; we focus on this regime, which we refer to as the high pruning ratio regime. Towards this aim, many pruning methods have been proposed to identify and remove those parameters that are least important for inference [1,15,20,27,29,47,51]. Since each layer of the network involves a different computation and associated computational burden, each parameter does not contribute equally to the final network inference cost, typically measured as FLOPs or latency, so more recent works have focused on pruning the network subject to explicit cost constraints. To maximize inference speedup on modern hardware (e.g., GPUs), these works largely focus on channel pruning [21,27,29,37,42,50].

However, in general, existing pruning works permanently remove the network parameters along these channels, zeroing the network weights and preventing the channel from being used during the rest of training. Particularly at high pruning ratios, where a significant fraction of the total channels in the network must be removed, the decisions on which channels to remove early during pruning are potentially myopic. Moreover, as a large number of channels are removed, the gradients to the remaining channels in each layer are significantly disrupted and can grow quite substantially due to the batch normalization layers ubiquitous in modern CNNs. This interferes with both network training and the identification of which further channels to remove.

In this work, we introduce a novel channel pruning approach for neural networks that is particularly suitable for large pruning ratios. The core of our approach relies on regularly rewiring the network sparsity, through soft masking of the network weights, to minimize the accuracy drop for large pruning ratios.

The introduction of soft masking allows previously pruned channels to later be restored to the network, instead of being permanently pruned. Additionally, to mitigate the effect of large gradient magnitudes caused by removing many channels, we incorporate a new batch normalization scaling approach. Lastly, we formulate channel pruning under a cost constraint as a resource allocation problem and show it can be efficiently solved. All together, we refer to this method as Soft Masking for cost-constrained Channel Pruning (SMCP).

Our main contributions are:

1. We demonstrate that a network's channel sparsity can be adaptively rewired, using a soft mask re-parameterization of the network weights, and that this requires channel pruning to be performed along input, instead of output, channels, see Sect. 3.1.
2. We propose a new scaling technique for the batch normalization weights to mitigate a gradient instability at high channel pruning ratios, see Sect. 3.2.
3. We perform channel pruning subject to a cost constraint by encoding it as a resource allocation problem, which automatically allocates cost across the network instead of relying on manual or heuristic-based layer-wise pruning ratios. We show this allocation problem is a variant of the classic 0–1 knapsack problem, called the multiple-choice knapsack problem [38], which can be efficiently solved for our experiments, see Sect. 3.3.
4. We analyze our method's accuracy and cost improvements for the ImageNet and PASCAL VOC datasets for ResNet, MobileNet, and SSD architectures. We outperform prior pruning approaches, as shown in Fig. 1 and more extensively in Sect. 4. In particular, at high pruning ratios for ResNet50/ResNet101 on ImageNet, SMCP can achieve up to an additional 20% speedup at the same Top-1 accuracy level or up to a 0.6% Top-1 accuracy improvement at the same FPS (frames per second). SMCP can also prune an SSD512 with a ResNet50 backbone to achieve a speedup of 2.12×, exceeding the FPS (frames per second) of the smaller SSD300-ResNet50 model by 12%, while simultaneously improving on the mAP of the baseline model.

2   Related Work

2.1   Soft Pruning

Most pruning methods start with a dense pretrained network and prune iteratively over a schedule to obtain a final network with the desired cost, where at each pruning step parameters are permanently zeroed (or masked). This effectively limits the model capacity as pruning occurs. Stosic and Stosic [39] argue that preserving the larger model capacity is critical to sparse model training by forming new paths for optimization that are not available for permanently pruned networks; they suggest it is important to allow gradient flow to previously pruned parameters and to rewire the sparsity occasionally.

Along these lines, several works have proposed soft pruning methods where parameters can be pruned and later unpruned if desirable. He et al. [14] zero weights during pruning but allows gradients to update them in an effort to maintain model capacity. Dettmers and Zettlemoyer [6], Evci et al. [8], Mostafa and Wang [31], and Wortsman et al. [43] allow previously pruned weights to be regrown. Kusupati et al. [19] used a soft thresholding operator to achieve state-of-the-art results for unstructured and low-rank structured pruning. Kang and Han [18] introduces soft channel pruning by adding a differentiable mask in the batch normalization layers; however, their approach is limited to an implicit cost constraint on the total number of neurons. Our approach though is most similar to Guo et al. [10], Lin et al. [23], De Jorge et al. [17], and Zhou et al. [53], which explicitly or implicitly use the Straight-through Estimator (STE) [2] to adaptively prune parameters during training. The first three target unstructured sparsity, and the last targets N:M structured sparsity. In our work, we extend the use of the STE to channel pruning, show this requires pruning to be formulated along input channels, and embed this soft masking into a general-purpose, explicit cost-constrained formulation.

2.2   Cost-Constrained and Structured Pruning

The goal of most pruning methods is to maximize network accuracy subject to low memory, computation, and/or latency requirements. Although unstructured sparsity approaches have proven to be very successfully in removing upwards of 95% of weights without affecting network accuracy [12], modern hardware has poor support for unstructured sparsity and therefore this rarely translates to actual speedup. Therefore, it is common to choose a pruning sparsity structure that can actually be accelerated in hardware, typically channel pruning for CNNs. There is now some hardware support for other sparsity structures, such as the N:M structured sparsity of [28], but we limit our focus to channel pruning in this work. Both Li et al. [21] and Yang et al. [45,46] select the best constraint-abiding network from a large number of candidate networks, which can be prohibitively expensive. Yu and Huang et al. [50], Tan et al. [41], and Wu et al. [44] pose cost-constrained optimization problems but use a greedy selection or cost-aware importance score to approximately select the best channels to prune. Chen et al. [3] presents a Bayesian optimization approach to determine compression hyperparameters that satisfy a cost constraint while maximizing network accuracy. Liu et al. [26] linked network pruning to Neural Architecture Search (NAS), arguing that the resulting pruned architectures are the novel contribution instead of the trained weights themselves. However, most NAS methods, such as those in [5,7,40], remain more computationally expensive than network pruning approaches. Our approach is most similar to the concurrent work of Shen et al. [37], called HALP, which also poses a cost-constrained resource allocation problem. There are however several major differences. First, we reduce our allocation problem to the multiple choice knapsack problem [38] and solve it with a meet-in-the-middle algorithm, which provides both optimality guarantees and efficient (<1 s) solutions for general cost-constraints. HALP solves their allocation problem with a custom augmented knapsack solver, which gives no optimality guarantees and requires significant extra computation (1+ minute for each pruning step on ResNet50 [13], even after a large GPU-specific neuron grouping step). Second, our method uses soft input channel masking as opposed to the permanent output channel pruning of HALP; we show this change alone yields performance gains in Sect. 4.3. Lastly, we use a new batch normalization scaling technique to stabilize training at high pruning ratios.

2.3   Pruning Impact on Batch Normalization Layers

Channel pruning can have a significant impact on the batch normalization statistics, which therefore strongly affects the network gradients to the remaining channels. This effect is particularly pronounced at high pruning ratios, since a large number of channels are being removed from most layers. Several pruning methods note this phenomenon and describe mitigation strategies. Li et al. [21] demonstrated the need to update the batch normalization statistics after pruning, as they can be significantly impacted, before evaluating possible pruned candidate networks. This approach does not however alleviate the issue of large gradients. Instead of immediately removing pruned weights and incurring the disruption, Wang et al. [42] slowly regularized them away, noticing significant performance gains particularly at high pruning ratios. They do not connect this to a sudden change in batch normalization statistics and gradients caused by pruning. They also use a non-gradient based importance so the impact on the importance of the remaining parameters is somewhat subdued. Since we are adaptively adjusting the sparsity and want to preserve the ability for pruned weights to become later unpruned, we do not want to regularize away pruned weights. We instead adopt a scaling technique on the batch normalization weights to stabilize training at high pruning ratios.

2.4   Parameter Importance Scoring

In order to decide which parameters of the network can be pruned while least harming network accuracy, most pruning methods define an importance for each parameter (or set of parameters) that approximates the effect of removal on the network's loss. Many importance scores have been proposed, largely falling into three groups: (i) based on weight magnitude [1,11,22,25,47,51]; (ii) based on a reconstruction-based objective [15,27]; and (iii) based on network gradients [20, 29,30]. We adopt the Taylor first-order importance [29] due to its computational simplicity and its strong correlation with the true impact on the network's loss.

3   Soft Masking for Cost-Constrained Channel Pruning

We propose a novel input channel pruning approach targeted towards high pruning ratios. Our method is initialized with a pretrained CNN model, and the desired network cost function and target cost constraint. We first re-parameterize the network weights with input channel masking variables, as shown in Sect. 3.1, to enable adaptive channel pruning. Then, after a warmup period, we iteratively

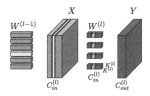

Fig. 2. Input channel pruning of a convolutional layer. Removing an input channel from weight $W^{(l)} \in \mathbb{R}^{C_{out}^{(l)} \times C_{in}^{(l)} \times K^{(l)} \times K^{(l)}}$ in layer l removes the corresponding channel in the input feature map X and the corresponding output channel in the previous weight $W^{(l-1)}$. The shape of the output feature map Y is unaffected.

prune every r minibatches by solving a resource allocation optimization problem, discussed in Sect. 3.3, to update the channel masks. After each mask update, we apply the batch normalization scaling described in Sect. 3.2, which stabilizes training at high pruning ratios. Finally, we fix the masks for a cooldown and fine-tuning period. We present the full algorithm and pseudocode in Sect. 3.4.

3.1   Soft Input Channel Pruning

We specifically consider input channel pruning, as previously done in [15] and shown in Fig. 2, where we mask and later remove input channels to sparsify the CNN. As we will shortly show, channel pruning with a soft mask re-parameterization requires it to be done along input channels, as this approach does not work when performing output channel pruning. This is a departure from the many output channel pruning approaches. From a global view of network sparsity, pruning one layer's input channel is equivalent to pruning the previous layer's output channel; however, the approaches are distinct when considering the effect on each individual layer.

For soft input channel masking, we consider a neural network with weights $W = \{W^{(l)}\}$, where $W^{(l)} \in \mathbb{R}^{C_{out}^{(l)} \times C_{in}^{(l)} \times K^{(l)} \times K^{(l)}}$ is the weight for layer l of the network and has $C_{in}^{(l)}$ input channels and $C_{out}^{(l)}$ output channels. To allow input channels to be pruned and later unpruned, we introduce an input channel mask $m^{(l)} \in \{0,1\}^{C_{in}^{(l)}}$ for each layer l. Using these masks, we re-parameterize the weights so that the network's sparse weights are

$$\widetilde{W}^{(l)} = W^{(l)} \odot m^{(l)}. \tag{1}$$

where $m^{(l)}$ is broadcasted to match the shape of $W^{(l)}$. Instead of permanently zeroing a channel when pruning, the underlying network weights can be preserved and merely the masks set to zero. This has two distinct advantages. First, it helps preserve the full capacity of the original model while training towards a sparse model. Second, by allowing channels to be restored to their original values at a later time, poor early decisions on where to allocate the sparsity across the layers can be undone. This is particularly important for high pruning ratios where a large portion of the network's channels must be removed.

As written though, our masking definition would define the gradient with respect to $W^{(l)}$ as $g_{W^{(l)}} = g_{\widetilde{W}^{(l)}} \odot m^{(l)}$. This masks the gradients as they flow back to the completely dense weights W, rendering masked weights unused in the forward pass and left untouched by the backward pass. Following the argument by Stosic and Stosic [39] that updating parameters not currently participating in the forward pass offers additional optimization paths that improve training of sparse networks, we adopt the Straight-through Estimator (STE) [2]. The STE has been successfully used in model quantization [35] and Ampere 2:4 structured pruning [53] for sparse parameter updates. The STE defines the gradient as

$$g_{W^{(l)}} = g_{\widetilde{W}^{(l)}}, \tag{2}$$

where gradients on the sparse weights pass straight through to the underlying, dense weights. Note that we still use the masks when computing the gradient with respect to the input feature map of the layer.

However, for this STE to have a useful impact in a modern CNN with the ubiquitous Conv-BN-ReLU pattern, it requires that channel pruning must be posed as input-oriented. Since $g_{\widetilde{W}^{(l)}}$ is defined by a matrix multiplication using the input feature map and the gradient of the output feature map, a masked input channel still receives non-zero gradients, except under a few edge cases. If we had instead masked output channels, the elements of $g_{W^{(l)}}$ would be either 0 or ∞, depending on the value of the batch normalization bias. Alternatively, if we instead tried to directly mask the batch normalization weight $\gamma^{(l)}$ and bias $\beta^{(l)}$ to emulate pruning the channel, we would get $g_{\gamma^{(l)}} = g_{\beta^{(l)}} = 0$ due to the ReLU. In either of these cases, the gradient $g_{W^{(l)}}$ is not useful.

Finally then, for input channel pruning with soft masking, we define the importance of each input channel, a proxy for the effect of removing this channel on the network's loss, according to the group first-order Taylor importance of [29]:

$$\mathcal{I}_i^{(l)} = \left| \sum_{o,r,s} W_{o,i,r,s}^{(l)} g_{W_{o,i,r,s}^{(l)}} \right| \tag{3}$$

where $\mathcal{I}_i^{(l)}$ is the importance of the ith input channel to layer l. Under certain conditions, this is in fact equivalent to the first-order batch normalization-based Taylor importance of [29], as shown in the supplementary materials.

3.2 Batch Normalization Scaling

When channel pruning at high ratios, there are many layers where a significant number of channels must be pruned. As a result of pruning these channels, either by zeroing them out or by applying masking, the subsequent gradient magnitudes to the remaining unpruned channels can be excessively large, which we show in the supplementary materials. We propose a batch normalization scaling technique that adjusts the batch normalization weight $\gamma^{(l)}$ of layer l to mitigate large gradients and stabilize the network sparsity and training. Specifically, we scale $\gamma^{(l)}$ according to the fraction of channels left unpruned by the current input channel mask $m^{(l)} \in \{0,1\}^{C_{in}^{(l)}}$

$$\gamma^{(l)} \leftarrow \gamma_{orig}^{(l)} \frac{\sum_i m_i^{(l)}}{C_{in}^{(l)}}. \tag{4}$$

In practice, we always treat $\gamma_{orig}^{(l)}$ as the parameter under optimization and vary a scaling variable $s^{(l)}$ to adjust the weight used by the network.

Moderating gradient magnitudes is particularly consequential since we employ the gradient-based importance score shown in Eq. (3). Even without soft masking and the STE, the large gradients cause importance accumulation in the remaining channels as pruning iteratively proceeds, artificially inhibiting additional channels in the layer from being pruned. When employing soft masking without this scaling technique, the large gradients cause large network sparsity thrashing. For example, if at one pruning iteration a large number of the channels are pruned, the importance to every channel, not only those left unpruned, is boosted by the resulting large gradient magnitudes. At the very next pruning iteration, those channels appear quite important and are restored to the network, causing other portions of the network to be pruned to still meet the cost constraint. This can oscillate, inhibiting network convergence and the final network accuracy. Moreover, for architectures in which pruning entire layers is possible, such as ResNet due to the skip connections, the infinite gradient magnitudes cause numerical overflow in updating the weights or even calculating the importance of channels. As shown in our experiments in Sect. 4, the proposed batch normalization scaling is crucial to overcome these training issues.

3.3   Cost-Constrained Channel Pruning

At each pruning iteration, we seek to both minimize the impact on the network's loss as a result of pruning and sparsify the network towards the final cost constraint (e.g., latency constraints). We therefore formulate pruning as a cost-constrained importance maximization problem

$$\max_{m^{(2)},\ldots,m^{(L)}} \sum_{l=1}^{L} \sum_{i=1}^{C_{in}^{(l)}} \mathcal{I}_i^{(l)} m_i^{(l)} \tag{5}$$

$$\text{s.t.} \quad \sum_{l=1}^{L} \mathcal{T}^{(l)} \left(\left\| m^{(l)} \right\|_1, \left\| m^{(l+1)} \right\|_1 \right) \leq \tau$$

$$\left\| m^{(l)} \right\|_1 \in \mathcal{P}^{(l)},$$

where L is the number of layers in the network, layer l has $C_{in}^{(l)}$ input channels, $\mathcal{I}_i^{(l)}$ is the importance of input channel i of layer l, $m^{(l)} \in \{0,1\}^{C_{in}^{(l)}}$ is the input channel mask for layer l, $\mathcal{T}^{(l)}$ is the cost function for layer l, τ is the cost constraint, and $\mathcal{P}^{(l)}$ is the set of permitted values for the number of channels kept by mask $m^{(l)}$. By definition, $m_i^{(1)} = 1$ and $m_i^{(L+1)} = 1$ since those are the unprunable inputs and outputs of the network. A complete derivation of Eq. (5) can be found in the supplementary materials, as well as a discussion on how to handle skip connections in architectures like ResNet [13].

The final constraint, on the set of permitted values $\mathcal{P}^{(l)}$, is optional but useful in several situations. First, it can be used to disallow pruning the entire layer: by omitting 0 from $\mathcal{P}^{(l)}$ we prevent $m^{(l)} = 0$. As explained in the supplementary materials, layer pruning violates a key assumption of the derivation of Eq. (5). Second, it can be used to ensure the number of remaining channels is hardware-friendly, such as 8× multiples for GPU tensorcores [32] with $\mathcal{P}^{(l)} = \{0, 8, 16, \ldots, \lfloor C_{in}^{(l)}/8 \rfloor\}$.

We can further reduce this to an optimization over only the number of channels $p^{(l)}$, as the most important channels will always be kept in each layer:

$$\max_{p^{(2)},\ldots,p^{(L)}} \sum_{l=1}^{L} \sum_{i=1}^{p^{(l)}} \mathcal{I}_{(i)}^{(l)} \tag{6}$$

$$\text{s.t.} \quad \sum_{l=1}^{L} \mathcal{T}^{(l)}\left(p^{(l)}, \overline{p^{(l+1)}}\right) \leq \tau$$

$$p^{(l)} \in \mathcal{P}^{(l)}$$

where $p^{(l)} = \left\|m^{(l)}\right\|_1$ and $\mathcal{I}_{(i)}^{(l)}$ is the ith largest value in $\mathcal{I}^{(l)}$. We also approximated the constraint using the current channel counts $\overline{p^{(l)}}$ to decouple the cost impact of masks in consecutive layers, which is required to pose this as an example of the following class of optimization problems.

Multiple-Choice Knapsack Problem. The optimization problem in Eq. (6) is an example of a generalization of the classic 0–1 knapsack problem called the multiple-choice knapsack (MCK) problem [38]. We show this connection explicitly in the supplementary materials. The MCK problem takes the form

$$\max_{x} \sum_{l=1}^{L} \sum_{i=1}^{n_l} v_{l,i} x_{l,i} \tag{7}$$

$$\text{s.t.} \quad \sum_{l=1}^{L} \sum_{i=1}^{n_l} c_{l,i} x_{l,i} \leq C$$

$$x_{l,i} \in \{0,1\}, \quad \sum_{i=1}^{n_l} x_{l,i} = 1$$

where L is the number of groups, group l has size n_l, and the items have value $v_{l,i}$ and cost $c_{l,i} \geq 0$. The additional constraint relative to the classic 0–1 knapsack problem enforces that we select exactly one item from each group.

Algorithm 1. Soft masking for cost-constrained channel pruning.

Inputs: Pretrained network weights W, training set $\mathcal{D}$, total number of epochs E, pruning schedule (r, K_w, K_t, K_c), target cost τ

1: Initialize masks $m^{(l)} = 1$
2: Re-parameterize the weights (Eq. (1))
3: Train the network as usual for K_w epochs
4: Calculate the pruning schedule $\{\tau_e\}$
5: **for** epoch $e \in [K_w, E - K_c)$ **do**
6:   **for** step s in epoch e **do**
7:     Perform the forward pass and backward pass, using Eqs. (1) and (2)
8:     Calculate and accumulate $\mathcal{I}_i^{(l)}$ (Eq. (3))
9:     **if** $s\%r = 0$ **then**
10:       Solve the optimization problem (Eq. (6)) using target cost τ_e
11:       Update the masks $m^{(l)}$ accordingly
12:       Scale the BN weights $\gamma^{(l)}$ (Eq. (4))
13:       Reset the accumulated importance
14:     **end if**
15:   **end for**
16: **end for**
17: Train the network as usual for K_c epochs
18: Apply the masks to the weights permanently
19: **return** Sparse network weights W

We solve Eq. (7) with a GPU-implemented meet-in-the-middle algorithm, presented in full in the supplementary materials. Our approach generalizes the standard meet-in-the-middle algorithm for the classic 0–1 knapsack problem, does not require integer costs, and very efficiently solves the MCK problem for our use cases. For example, for a ResNet50 [13], our approach solves the MCK problem in under 1 s. We present more complete timing details in the supplementary materials.

3.4 Overall Method

We present our full method in Algorithm 1. We start with a pretrained network, layer-wise cost functions $\mathcal{T}^{(l)}$, and a global cost constraint τ. We define our pruning schedule by: (i) K_w: the number of warmup epochs before starting pruning; (ii) K_t: the number of epochs after the warmup to reach the target cost τ; (iii) r: the number of steps between recomputing the channel masks; and (iv) K_c: the number of cooldown epochs where the masks are kept fixed. During those K_t epochs to reach the target cost, we define intermediate cost constraints $\{\tau_e\}$ using the exponential scheduler of [17]. Additionally, to stabilize the importance scores, which can be noisy due to the stochastic minibatches, we calculate and accumulate the importance score in Eq. (3) every minibatch between pruning iterations according to the exponential momentum approach of [29].

4   Results

We evaluate our method on both the ImageNet and PASCAL VOC benchmark datasets[1]. Full details on training settings and architectures can be found in the supplementary materials. We use a latency cost constraint, defined by a layer-wise lookup table (LUT) as previously described in [37,45,48]. We target and measure latency speed on a NVIDIA TITAN V GPU with cudNN V7.6.5 [4].

4.1   ImageNet Results

We compare SMCP with several prior works on the ImageNet ILSVRC2012 [36] classification dataset. In Table 1, we compare the results of pruning ResNet50, ResNet101 [13], and MobileNet-V1 [16] at a number of pruning thresholds. We refer to SMCP-$X\%$ as retaining $X\%$ of the full model's original latency and calculate the frames per second (FPS) and speedup of the final network. For ResNet50, we show results for two different baseline models for a better comparison with prior works. The first baseline is from the PyTorch [34] model hub, with a Top-1 accuracy of 76.15%; the second baseline is the one used as a baseline for EagleEye [21] and has a Top-1 accuracy of 77.2%. We prune and fine-tune following the training setup of [33].

Our method performs comparably to prior works at low pruning ratios and outperforms them for large pruning ratios. For the PyTorch ResNet50 baseline model, we achieve a 0.3% higher Top-1 accuracy with a higher FPS at 2G and 1G FLOPs with an additional 0.04× and 0.19× speedup respectively. For the EagleEye [21] baseline, our method produces models near 1G FLOPs that have a 0.6% higher Top-1 accuracy for nearly the same FPS or a similar Top-1 accuracy while being 19% (or 0.5×) faster. The results are similar for ResNet101, which is based on the PyTorch model hub baseline model. At 2G FLOPs, we get a 0.3% higher Top-1 accuracy and an additional 0.03× speedup. On the already compact MobileNet-V1 model, where the desired pruning ratios are smaller, our method performs comparably to prior works; at the highest pruning ratio, we show a minor FPS improvement of 0.07× despite a higher FLOPs count, demonstrating the ability of the optimization problem in Sect. 3.3 to choose cost-constraint aware masks.

The benefits of our method, particularly at high pruning ratios, are possibly more easily seen when plotting the tradeoff curve for Top-1 accuracy versus FPS, as shown in Fig. 1 for the PyTorch baseline and Fig. 3 for the EagleEye baseline. For example in Fig. 3, at the 75% latency reduction level (or 3102 FPS), our method outperforms the nearest HALP [37] model with a 0.2% higher Top-1 accuracy and a 15% higher FPS; compared to EagleEye [21], we show a 0.23% higher Top-1 accuracy and a 26% higher FPS.

Moreover, our method can aggressively prune large, over-parameterized models to outperform smaller unpruned models. As shown in Table 1 and Fig. 3, a 50% pruned ResNet101 achieves a 1.6% Top-1 improvement over a baseline

---

[1] Our code can be accessed at https://github.com/NVlabs/SMCP.

Table 1. Pruning results on the ImageNet classification dataset considering two different ResNet50 baseline models as well as ResNet101 and MobileNetV1. We group results by those with similar FLOP counts, and refer to SMCP-X% as retaining X% of the full model's original latency. Results for prior works are as shown in [37].

Method	FLOPs (G)	Top1 (%)	Top5 (%)	FPS (im/s)	Speedup
ResNet50					
No pruning	4.1	76.2	92.87	1019	1×
ThiNet-70 [27]	2.9	75.8	90.67	–	–
AutoSlim [50]	3.0	76.0	–	1215	1.14×
MetaPruning [26]	3.0	76.2	–	–	–
GReg-1 [42]	2.7	76.3	~	1171	1.15×
HALP-80% [37]	3.1	**77.2**	**93.47**	1256	1.23×
SMCP-80% (Ours)	3.0	77.1	93.43	**1292**	**1.27×**
0.75× ResNet50 [13]	2.3	74.8	–	1467	1.44×
ThiNet-50 [27]	2.1	74.7	90.02	–	–
AutoSlim [50]	2.0	75.6	–	1592	1.56×
MetaPruning [26]	2.0	75.4	–	1604	1.58×
GBN [49]	2.4	76.2	92.83	–	–
GReg-2 [42]	1.8	75.4	–	1414	1.39×
HALP-55% [37]	2.0	76.5	93.05	1630	1.60×
SMCP-55% (Ours)	2.0	**76.8**	**93.22**	**1673**	**1.64×**
0.50× ResNet50 [13]	1.1	72.0	–	2498	2.45×
ThiNet-30 [27]	1.2	72.1	88.30	–	–
AutoSlim [50]	1.0	74.0	–	2390	2.45×
MetaPruning [26]	1.0	73.4	–	2381	2.34×
GReg-2 [42]	1.3	73.9	–	1514	1.49×
HALP-30% [37]	1.0	74.3	91.81	2755	2.70×
SMCP-30% (Ours)	1.0	**74.6**	**92.00**	**2947**	**2.89×**
ResNet50 - EagleEye [21] baseline					
No pruning	4.1	77.2	93.70	1019	1×
EagleEye-3G [21]	3.0	77.1	93.37	1165	1.14×
HALP-80% [37]	3.0	77.5	93.60	1203	1.18×
SMCP-80% (Ours)	3.1	**77.6**	**93.61**	**1263**	**1.23×**
EagleEye-2G [21]	2.1	76.4	92.89	1471	1.44×
HALP-55% [37]	2.1	**76.6**	93.16	1672	**1.64×**
SMCP-50% (Ours)	1.9	**76.6**	**93.17**	**1706**	**1.67×**
EagleEye-1G [21]	1.0	74.2	91.77	2429	2.38×
HALP-30% [37]	1.2	74.5	91.87	2597	2.55×
SMCP-30% (Ours)	1.1	**75.1**	**92.29**	2589	2.51×
SMCP-25% (Ours)	0.9	74.4	91.98	**3102**	**3.01×**

Method	FLOPs (G)	Top1 (%)	FPS (im/s)	Speedup
ResNet101				
No pruning	7.8	77.4	620	1×
Taylor-75% [29]	4.7	77.4	750	1.21×
HALP-60% [37]	4.3	**78.3**	847	1.37×
SMCP-60% (Ours)	4.0	78.1	951	1.53×
HALP-50% [37]	**3.6**	77.8	994	1.60×
SMCP-50% (Ours)	**3.6**	77.8	**1016**	**1.64×**
Taylor-55% [29]	2.9	76.0	908	1.47×
HALP-40% [37]	2.7	77.2	1180	1.90×
SMCP-30% (Ours)	2.6	**77.3**	1273	2.05×
HALP-30% [37]	**2.0**	76.5	1521	2.45×
SMCP-25% (Ours)	**2.0**	76.8	**1535**	**2.48×**

Method	FLOPs (M)	Top1 (%)	FPS (im/s)	Speedup
MobileNet-V1				
No pruning	569	72.6	3415	1×
0.75× MobileNetV1	325	68.4	4678	1.37×
NetAdapt [45]	284	69.1	–	–
MetaPruning [26]	316	70.9	4838	1.42×
EagleEye [21]	284	70.9	5020	1.47×
HALP-60% [37]	297	**71.3**	5754	1.68×
SMCP-60% (Ours)	356	71.0	**5870**	**1.72×**
MetaPruning [26]	142	66.1	7050	2.06×
AutoSlim [50]	150	67.9	7743	2.27×
HALP-42% [37]	171	**68.3**	7940	2.32×
SMCP-40% (Ours)	208	**68.3**	**8163**	**2.39×**

ResNet50, with no performance loss, and a 80% pruned ResNet50 achieves a similar Top-1 to an unpruned MobileNet-V1 while achieving a 10% FPS speedup.

Lastly, the accuracy and performance gains are in part due to the final network architecture chosen by our method. In particular, since we solve a global resource allocation problem during training, our method automatically deter-

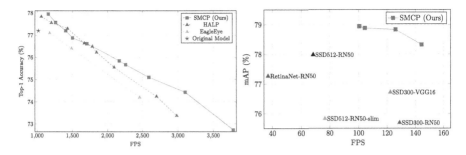

Fig. 3. (Left) Top-1 accuracy tradeoff curve for pruning ResNet50 on the ImageNet classification dataset using a latency cost constraint. Baseline model is from Eagle-Eye [21]. (Right) mAP accuracy tradeoff curve for pruning SSD512-RN50 on the PAS-CAL VOC object detection dataset using a latency cost constraint. Top-right is better.

mines the layer-wise pruning ratios for the given cost function and constraint. For example, on ResNet50, we find that SMCP is aggressive in pruning the early convolution layers and leaves the later layers better preserved; we provide additional analysis and figures in the supplementary materials.

4.2 PASCAL VOC Results

To analyze our method beyond image classification, we also analyze SMCP on the PASCAL VOC object detection dataset [9]. Specifically, we consider whether a large model, such as SSD512 [24] with a ResNet50 backbone, can be pruned at a high ratio to match the FPS of smaller models while retaining a superior mAP (mean average precision). We use the "07+12" train and test setup of [24] and prune both the backbone and feature layers.

As shown in Fig. 3, our method can prune an SSD512-RN50 to have a higher mAP than the pretrained model and a faster FPS than the much smaller SSD300-RN50 model, again showing the ability of our method to aggressively prune large over-parameterized models to outperform smaller models. In particular, our fastest pruned model has a 2.63 point higher mAP score while achieving 12% higher FPS. Critically, the latency reduction to achieve this is 75%, demonstrating the strength of our approach in the high pruning ratio regime. We also compare to and outperform a number of other common detector models.

4.3 Ablation Study

We also study the effect of our contributions on the accuracy results shown above, specifically at high pruning ratios. We run our method again on the ImageNet classification dataset, starting from the ResNet50 EagleEye [21] baseline. We first remove the batch normalization scaling technique from Sect. 3.2 while keeping the soft input channel masking re-parameterization of Sect. 3.1. We

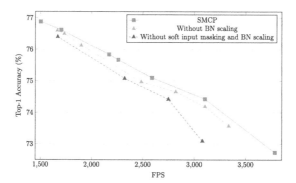

Fig. 4. Ablation study for SMCP at high pruning ratios on ResNet50 using the Eagle-Eye [21] baseline. We remove consecutively two major components of our method, soft input masking and batch normalization scaling, and observed worse Top-1 accuracy and FPS than the full SMCP method.

then additionally remove the soft input channel masking, reverting to permanent pruning. We keep the solver and latency constraint in Sect. 3.3 unchanged. The ablation results are shown in Fig. 4. Removing the batch normalization scaling generally leads to marginally worse results, due to the training instability described in Sect. 3.2. Additionally removing the soft input masking, thereby using permanent channel pruning, degrades accuracy and performance further.

4.4 Choice of Latency Cost Constraint

Although our cost-constrained formulation is general to any number of cost functions, the benefits of our approach are most pronounced under challenging, nonlinear latency cost landscapes (i.e., latency cliffs for GPUs). Linear constraints (i.e., parameter/FLOP constraints) lessen the need for soft masking and an efficient and global resource allocation: removed channels are more likely to stay pruned once removed and the number of remaining channels in each layer tends to change slowly. Despite training against a latency constraint, Table 1 shows that SMCP is comparable to or even outperforms previous methods under low FLOP constraints.

5 Conclusion

By applying channel pruning, modern CNNs can be significantly accelerated, with a smaller memory footprint, computational cost, and inference time. In this work, we presented a novel structured input channel pruning approach, called SMCP, that combines soft masking of input channels, a batch normalization scaling technique, and the solution to a resource allocation problem to outperform prior works. We motivate the use of each component of our method and demonstrate their effectiveness on both the ImageNet and PASCAL VOC datasets. Although we only consider channel pruning in this work, our approach can be extended to jointly consider both channel and N:M structured pruning [28] to satisfy an explicit cost-constraint. This can be viewed as an extension of both this work and that of [53] and is left for a future work.

References

1. Alvarez, J.M., Salzmann, M.: Learning the number of neurons in deep networks. In: NeurIPS, pp. 2262–2270 (2016)
2. Bengio, Y., Léonard, N., Courville, A.C.: Estimating or propagating gradients through stochastic neurons for conditional computation. CoRR abs/1308.3432 (2013)
3. Chen, C., Tung, F., Vedula, N., Mori, G.: Constraint-aware deep neural network compression. In: Ferrari, V., Hebert, M., Sminchisescu, C., Weiss, Y. (eds.) ECCV 2018. LNCS, vol. 11212, pp. 409–424. Springer, Cham (2018). https://doi.org/10.1007/978-3-030-01237-3_25
4. Chetlur, S., et al.: CUDNN: efficient primitives for deep learning. CoRR abs/1410.0759 (2014)
5. Dai, X., et al.: ChamNet: towards efficient network design through platform-aware model adaptation. In: CVPR, pp. 11398–11407 (2019)
6. Dettmers, T., Zettlemoyer, L.: Sparse networks from scratch: faster training without losing performance. CoRR abs/1907.04840 (2019)
7. Dong, J.-D., Cheng, A.-C., Juan, D.-C., Wei, W., Sun, M.: DPP-Net: device-aware progressive search for pareto-optimal neural architectures. In: Ferrari, V., Hebert, M., Sminchisescu, C., Weiss, Y. (eds.) ECCV 2018. LNCS, vol. 11215, pp. 540–555. Springer, Cham (2018). https://doi.org/10.1007/978-3-030-01252-6_32
8. Evci, U., Gale, T., Menick, J., Castro, P.S., Elsen, E.: Rigging the lottery: making all tickets winners. In: ICML, pp. 2943–2952 (2020)
9. Everingham, M., Gool, L.V., Williams, C.K.I., Winn, J.M., Zisserman, A.: The pascal visual object classes (VOC) challenge. Int. J. Comput. Vis. **88**(2), 303–338 (2010)
10. Guo, Y., Yao, A., Chen, Y.: Dynamic network surgery for efficient DNNs. In: Lee, D.D., Sugiyama, M., von Luxburg, U., Guyon, I., Garnett, R. (eds.) NeurIPS, pp. 1379–1387 (2016)
11. Han, S., Mao, H., Dally, W.J.: Deep compression: compressing deep neural network with pruning, trained quantization and huffman coding. In: ICLR (2016)
12. Han, S., Pool, J., Tran, J., Dally, W.J.: Learning both weights and connections for efficient neural network. In: NeurIPS, pp. 1135–1143 (2015)
13. He, K., Zhang, X., Ren, S., Sun, J.: Deep residual learning for image recognition. In: CVPR, pp. 770–778 (2016)
14. He, Y., Kang, G., Dong, X., Fu, Y., Yang, Y.: Soft filter pruning for accelerating deep convolutional neural networks. In: IJCAI, pp. 2234–2240 (2018)
15. He, Y., Zhang, X., Sun, J.: Channel pruning for accelerating very deep neural networks. In: ICCV, pp. 1398–1406 (2017)
16. Howard, A.G., et al.: MobileNets: efficient convolutional neural networks for mobile vision applications. CoRR abs/1704.04861 (2017)
17. de Jorge, P., Sanyal, A., Behl, H.S., Torr, P.H.S., Rogez, G., Dokania, P.K.: Progressive skeletonization: trimming more fat from a network at initialization. In: ICLR (2021)

18. Kang, M., Han, B.: Operation-aware soft channel pruning using differentiable masks. In: ICML, pp. 5122–5131 (2020)
19. Kusupati, A., et al.: Soft threshold weight reparameterization for learnable sparsity. In: ICML, pp. 5544–5555 (2020)
20. LeCun, Y., Denker, J.S., Solla, S.A.: Optimal brain damage. In: NeurIPS, pp. 598–605 (1989)
21. Li, B., Wu, B., Su, J., Wang, G.: EagleEye: fast sub-net evaluation for efficient neural network pruning. In: Vedaldi, A., Bischof, H., Brox, T., Frahm, J.-M. (eds.) ECCV 2020. LNCS, vol. 12347, pp. 639–654. Springer, Cham (2020). https://doi.org/10.1007/978-3-030-58536-5_38
22. Li, H., Kadav, A., Durdanovic, I., Samet, H., Graf, H.P.: Pruning filters for efficient convnets. In: ICLR (2017)
23. Lin, T., Stich, S.U., Barba, L., Dmitriev, D., Jaggi, M.: Dynamic model pruning with feedback. In: ICLR (2020)
24. Lui, W., et al.: SSD: single shot MultiBox detector. In: Leibe, B., Matas, J., Sebe, N., Welling, M. (eds.) ECCV 2016. LNCS, vol. 9905, pp. 21–37. Springer, Cham (2016). https://doi.org/10.1007/978-3-319-46448-0_2
25. Liu, Z., Li, J., Shen, Z., Huang, G., Yan, S., Zhang, C.: Learning efficient convolutional networks through network slimming. In: ICCV, pp. 2755–2763 (2017)
26. Liu, Z., Sun, M., Zhou, T., Huang, G., Darrell, T.: Rethinking the value of network pruning. In: ICLR (2019)
27. Luo, J., Wu, J., Lin, W.: ThiNet: a filter level pruning method for deep neural network compression. In: ICCV, pp. 5068–5076 (2017)
28. Mishra, A.K., et al.: Accelerating sparse deep neural networks. CoRR abs/2104.08378 (2021)
29. Molchanov, P., Mallya, A., Tyree, S., Frosio, I., Kautz, J.: Importance estimation for neural network pruning. In: CVPR, pp. 11264–11272 (2019)
30. Molchanov, P., Tyree, S., Karras, T., Aila, T., Kautz, J.: Pruning convolutional neural networks for resource efficient inference. In: ICLR (2017)
31. Mostafa, H., Wang, X.: Parameter efficient training of deep convolutional neural networks by dynamic sparse reparameterization. In: ICML, pp. 4646–4655 (2019)
32. NVIDIA deep learning performance guide: convolutional layers user guide. https://docs.nvidia.com/deeplearning/performance/dl-performance-convolutional/index.html. Accessed 15 Nov 2021
33. NVIDIA Deep Learning Examples: ResNet50 v1.5 For Pytorch. https://github.com/NVIDIA/DeepLearningExamples/blob/master/PyTorch/Classification/ConvNets/resnet50v1.5/README.md. Accessed 15 Nov 2021
34. Paszke, A., et al.: PyTorch: an imperative style, high-performance deep learning library. In: 2019 Wallach, H.M., Larochelle, H., Beygelzimer, A., d'Alché-Buc, F., Fox, E.B., Garnett, R. (eds.) NeurIPS, pp. 8024–8035 (2019)
35. Rastegari, M., Ordonez, V., Redmon, J., Farhadi, A.: XNOR-Net: imagenet classification using binary convolutional neural networks. In: Leibe, B., Matas, J., Sebe, N., Welling, M. (eds.) ECCV 2016. LNCS, vol. 9908, pp. 525–542. Springer, Cham (2016). https://doi.org/10.1007/978-3-319-46493-0_32
36. Russakovsky, O., et al.: Imagenet large scale visual recognition challenge. Int. J. Comput. Vis. **115**(3), 211–252 (2015)
37. Shen, M., Yin, H., Molchanov, P., Mao, L., Liu, J., Alvarez, J.M.: HALP: hardware-aware latency pruning. CoRR abs/2110.10811 (2021)
38. Sinha, P., Zoltners, A.A.: The multiple-choice knapsack problem. Oper. Res. **27**(3), 503–515 (1979)

39. Stosic, D., Stosic, D.: Search spaces for neural model training. CoRR abs/2105.12920 (2021)
40. Su, X., You, S., Wang, F., Qian, C., Zhang, C., Xu, C.: BCNet: searching for network width with bilaterally coupled network. In: CVPR, pp. 2175–2184 (2021)
41. Tan, M., et al.: MnasNet: platform-aware neural architecture search for mobile. In: CVPR, pp. 2820–2828 (2019)
42. Wang, H., Qin, C., Zhang, Y., Fu, Y.: Neural pruning via growing regularization. In: ICLR (2021)
43. Wortsman, M., Farhadi, A., Rastegari, M.: Discovering neural wirings. In: NeurIPS, pp. 2680–2690 (2019)
44. Wu, Y., Liu, C., Chen, B., Chien, S.: Constraint-aware importance estimation for global filter pruning under multiple resource constraints. In: CVPR, pp. 2935–2943 (2020)
45. Yang, T.-J., et al.: NetAdapt: platform-aware neural network adaptation for mobile applications. In: Ferrari, V., Hebert, M., Sminchisescu, C., Weiss, Y. (eds.) ECCV 2018. LNCS, vol. 11214, pp. 289–304. Springer, Cham (2018). https://doi.org/10.1007/978-3-030-01249-6_18
46. Yang, T., Liao, Y., Sze, V.: Netadaptv2: efficient neural architecture search with fast super-network training and architecture optimization. In: CVPR, pp. 2402–2411. Computer Vision Foundation/IEEE (2021)
47. Ye, J., Lu, X., Lin, Z., Wang, J.Z.: Rethinking the smaller-norm-less-informative assumption in channel pruning of convolution layers. In: ICLR (2018)
48. Yin, H., et al.: Dreaming to distill: data-free knowledge transfer via DeepInversion. In: CVPR, pp. 8712–8721 (2020)
49. You, Z., Yan, K., Ye, J., Ma, M., Wang, P.: Gate decorator: global filter pruning method for accelerating deep convolutional neural networks. In: NeurIPS, pp. 2130–2141 (2019)
50. Yu, J., Huang, T.S.: Network slimming by slimmable networks: towards one-shot architecture search for channel numbers. CoRR abs/1903.11728 (2019)
51. Yu, R., et al.: NISP: pruning networks using neuron importance score propagation. In: CVPR, pp. 9194–9203 (2018)
52. Zhang, C., Bengio, S., Hardt, M., Recht, B., Vinyals, O.: Understanding deep learning requires rethinking generalization. In: ICLR (2017)
53. Zhou, A., et al.: Learning N: M fine-grained structured sparse neural networks from scratch. In: ICLR (2021)

Non-uniform Step Size Quantization for Accurate Post-training Quantization

Sangyun Oh[1], Hyeonuk Sim[2], Jounghyun Kim[3], and Jongeun Lee[1,3](✉)

[1] Department of Electrical Engineering, UNIST, Ulsan, Korea
{syoh,jlee}@unist.ac.kr
[2] Department of Computer Science and Engineering, UNIST, Ulsan, Korea
detective@unist.ac.kr
[3] Artificial Intelligence Graduate School, UNIST, Ulsan, Korea
maxedset@unist.ac.kr

Abstract. Quantization is a very effective optimization technique to reduce hardware cost and memory footprint of deep neural network (DNN) accelerators. In particular, post-training quantization (PTQ) is often preferred as it does not require a full dataset or costly retraining. However, performance of PTQ lags significantly behind that of quantization-aware training especially for low-precision networks ($\leq$4-bit). In this paper we propose a novel PTQ scheme (Code will be publicly available at https://github.com/sogh5/SubsetQ) to bridge the gap, with minimal impact on hardware cost. The main idea of our scheme is to increase arithmetic precision while retaining the same representational precision. The excess arithmetic precision enables us to better match the input data distribution while also presenting a new optimization problem, to which we propose a novel search-based solution. Our scheme is based on logarithmic-scale quantization, which can help reduce hardware cost through the use of shifters instead of multipliers. Our evaluation results using various DNN models on challenging computer vision tasks (image classification, object detection, semantic segmentation) show superior accuracy compared with the state-of-the-art PTQ methods at various low-bit precisions.

Keywords: Deep neural networks · Logarithmic-scale quantization · Post-training quantization · Subset quantization

1 Introduction

As deep learning becomes the highest performing method for many machine learning tasks, there is a growing interest in hardware DNN (Deep Neural Network) accelerators. DNNs of computer vision tasks such as image enhancement and super-resolution applications [2,20] often have very high compute and memory requirement. To reduce the hardware cost and memory footprint of such

---

Supplementary Information The online version contains supplementary material available at https://doi.org/10.1007/978-3-031-20083-0_39.

S. Avidan et al. (Eds.): ECCV 2022, LNCS 13671, pp. 658–673, 2022.
https://doi.org/10.1007/978-3-031-20083-0_39

DNN accelerators, quantization can be a very effective approach [6,13,14]. In particular, post-training quantization (PTQ) is preferred for DNN deployment as it requires no costly retraining or large dataset. However, the performance of PTQ lags behind that of quantization-aware training especially for low-precision networks (e.g., <4-bit).

In this paper we propose a novel PTQ scheme to bridge the gap, with minimal impact on the cost of a DNN hardware accelerator. Our scheme is based on logarithmic-scale quantization [16–18], which can help minimize hardware cost through the use of shifters instead of multipliers. The main idea of our scheme is to use an arithmetic precision that is higher than the representational precision. In linear (i.e., uniform step size) quantization, the two are the same. For instance, 3-bit quantized neural network would use 3-bit multipliers. However, for logarithmic-scale quantization [16], they are not always the same. For instance, 3-bit log-quantized data can have the same arithmetic precision as 4-bit linear-quantized data. Moreover, recent advanced logarithmic-scale quantization methods [8,17,18,25] often represent input data as a sum of two log-quantized words, which further disconnects arithmetic precision from representational precision.

Thus there is a new optimization problem: given a representational precision (P_r) and an arithmetic precision (P_a) where $P_a > P_r$, how to design a quantizer function such that it can maximize quantization performance for a given data distribution. Quantization performance can be defined in terms of quantization error or inference accuracy of a quantized DNN. We study this problem in the context of two-word log-scale quantization [8,17,18,25], for which we find the set of all possible *quantization points*, with each subset of the quantization points defining a new quantizer function. In other words, a quantizer may be specified as a *subset* of quantization points instead of an arithmetic function, and this extended view leads to a new category of quantizers that can make a more efficient use of the limited arithmetic precision than is possible otherwise. We call our quantization scheme *subset quantization*, which is more formally defined in Defining Quantizer for Subset Quantization Section.

In this paper we make the following contributions.

- We propose subset quantization (SQ), a novel quantization scheme for PTQ.
- We present a method to find the best subset and scale factor for subset quantization, for a given data distribution.
- We evaluate our method on image classification using ResNet models showing that our method outperforms state-of-the-art PTQ methods in most cases.
- We evaluate our method on another vision tasks such as object detection and semantic segmentation, demonstrating our method can achieve close to FP32 accuracy at ultra-low-bit weight precision.

2   Related Work

2.1   Uniform vs. Non-uniform Quantization

A quantizer is defined as a function from the set of real numbers $\mathcal{R}$ to the set of integers. But in the context of DNN quantization, we are often interested in a *simulated quantizer*, which is a function that simulates the effect of quantization by applying a quantizer and a de-quantizer in a row [15]:

$$Q : \mathcal{R} \to \mathcal{R} = dequantizer \circ quantizer. \tag{1}$$

where $\circ$ represents function composition, and the input and output values have the same domain and the same scale. In this case, the essential role of a quantizer is to approximate a real value to one of a finite set of values, which we call *quantization points*.[1]

Linear quantization [6,9,13,14] uses quantization points that are spaced linearly. An example linear quantizer is:

$$q = Q_u(x) = \text{clip}\left(\left\lfloor \frac{x}{s} \right\rfloor, L, U\right). \tag{2}$$

where s is step size, L, U are the lower and upper limits of the quantized values, and $\text{clip}(x, a, b) = \min(\max(x, a), b)$. In the above quantizer, the step size, or the distance from one quantization point to the next, is constant for all quantization points; hence, it is *uniform step size* quantization. By using clip we can focus on the more interesting range of input, i.e., $[sL, sU]$, which can result in a more efficient allocation of the limited number of quantization points as in ACIQ-Mix [1].

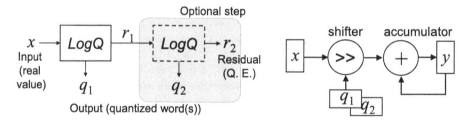

Fig. 1. Non-uniform step-size quantization example: selective two-word log-scale quantization [8,17,25]. Its processing element hardware (right) uses shifters instead of multipliers as in log-scale quantization.

There are a few quantization schemes that have non-uniform step size. Logarithmic-scale (*log-scale* for short) quantization [16] is an example, in which quantization points are spaced geometrically. Usually the base of exponent is set to 2, which enables very efficient arithmetic hardware using shifters instead of costly multipliers. Log-scale quantization is also motivated by the bell-shaped distribution of weight/activation values [16,18], which can, in certain cases, lead

---

[1] Quantization points are similar to quantization levels but there are some differences. Whereas quantization levels are often integers and may have a different scale than quantization thresholds, quantization points have the same scale as quantization thresholds and can be used as a substitute for them.

to lower expected quantization error than in linear quantization, since the former allocates more quantization points for common, low-magnitude values.

On the other hand, log-scale quantization may allocate too many quantization points near zero, which is especially true at high precision. Selective two-word log-scale quantization (STLQ) schemes [8,17,25] address this problem by employing another round of quantization step for those values that have high quantization error in the first round. In other words, each input value may be quantized with one or two quantized words, depending on the residual in the first round (see Fig. 1). APoT [18] is another solution, which is based on a similar idea but always uses two shifters. The APoT quantizer maps an input value to the sum of two terms, each of which is a power-of-two. Our subset quantization, which may be seen as an evolution of log-based quantization schemes, further extends the set of quantization points while not affecting representational precision.

2.2   Determining Quantization Parameters

The main problem of quantization is that of determining quantization parameters such as step size s and lower and upper bounds L, U in the case of the linear quantizer in (2). Quantization parameters may be determined through training or other means (e.g., statistics). State-of-the-art performances in DNN quantization are often obtained by determining quantization parameters through training. This kind of training usually involves training of both weight and quantization parameters, hence called *joint-training*. However, even if quantization parameters are not trained, quantization-aware training, which trains weight parameters only (in this case, quantization parameters may be fixed or determined by other means, e.g., by statistics) can still outperform post-training quantization [15]. Post-training quantization (PTQ) is any method that determines quantization parameters for a given set of (fixed) weight parameters. Despite its lower performance, PTQ is often preferred for deployment as it does not require retraining or a full dataset.

2.3   Post-training Quantization Methods

Here we briefly review recent PTQ methods. For image classification tasks, recent state-of-the-art PTQ methods have shown accuracy close to that of FP32 (32-bit floating-point precision) at low bits such as 4-bit for both weight and activation [10,12,19,23,28].

BitSplit [28] splits the n-bit binary sequence (e.g., power-of-two value of weight including sign bit) into $n - 1$ ternary bits, then minimizes quantization error via calibration set and alpha scaling factor. It shows highly competitive results at various precision settings including 3-bit. AdaRound [23] subdivides the round operation of quantization process into floor and ceiling, and finds the favorable selection in terms of quantization error. It proposes an approximated per-layer quadratic task loss for the relaxation of selections, and shows result close to FP32 at 4-bit precision. AdaQuant [12] proposes a joint optimization method for weight and activation including layer-wise integer programming that searches bit combinations, followed by parameter tuning for batch normalization and bias. It shows performance close to FP32 at 4-bit for image classification.

PWLQ [10] divides bell-shaped distribution into two symmetric regions by using a breakpoint, and equally divides quantization points into each region, so that step size can be determined according to data density. It can search multiple breakpoints (up to 3) and shows close to FP32 performance at 4-bit precision. BRECQ [19] minimizes quantization error in various granularities such as model, stage, block, and layer. It uses a sigmoid-like trainable parameter for the round operation, and searches bit combinations using the genetic algorithm.

3   Our Proposed Method

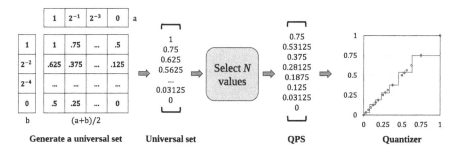

Fig. 2. Overview of our proposed quantization scheme ($N = 8$, 4-bit quantization including the sign bit). In the quantizer graph, black circles and red plus signs represent a universal set and a quantization point set (QPS), respectively. (Color figure online)

3.1   Overview

Figure 2 illustrates our proposed quantization scheme. First we design a universal set S_U such that the set is rich enough to represent any given input distribution but each element in the set can be efficiently generated by hardware. We present one such set in Sect. 3.4. Due to the encoding restriction (i.e., limited weight precision) we cannot use all the elements in S_U at least simultaneously. Instead we select N values, where N is determined by the quantization precision (i.e., representational precision).

For n-bit quantization (including the sign bit), we select $N = 2^{n-1}$ values on the non-negative side, which may or may not include zero. (If zero is included, it means that we are wasting one code word, since the negation of zero is also zero, but we see that this rarely happens.) For instance, if $n = 4$, then $N = 8$, and if $n = 3$, $N = 4$. There are many ways to choose N elements out of S_U, and this flexibility affords us a degree of freedom by which we can better approximate any input data distribution.

Once we select the set of N points, which we call *quantization point set* (QPS), the negative values are defined simply as the negation of the non-negative values. The symmetricity in the quantizer reduces QPS exploration time (see Optimization for Subset Quantization Section) as well as hardware complexity. Finally, from the chosen QPS directly follows a quantizer/dequantizer definition, in which we include a scale factor α, which plays a similar role as in previous work [18,28]. Note that scale factors can be implemented as part of the succeeding layer, where it is combined with input quantization computation, thus having negligible cost [13,30]. We next present a generalized quantizer framework based on QPS, followed by our quantization parameter determination algorithm and universal set design.

3.2   Defining Quantizer for Subset Quantization

Quantization Point Set: Quantization point set (QPS) of a quantizer is the *range* of its simulated quantizer function. Since simulated quantizer [15] is a function that applies a quantizer and a de-quantizer in a row (see (1)), its input domain has the same scale as the output domain. Therefore, QPS is also the set of points in the input domain that can be quantized with no quantization error.

Then one can view all quantization schemes as approximating an input to the nearest[2] element in a QPS, with the only difference among different quantization schemes being the definition of QPS. Accordingly, we can write a fairly general quantizer definition that takes as parameter a QPS (S_Q) in addition to an input value (x). Its simulated quantizer is as follows:

$$Q(x, S_Q) = \arg\min_{p \in S_Q} |x - p|. \tag{3}$$

Examples: The QPS for linear quantization such as (2) can be defined as follows, where α is a step size and $2N$ is the number of quantization points; that is, for k-bit quantization, $N = 2^{k-1}$. For brevity we assume that input is symmetric around zero.

$$S_Q^{lin}(\alpha) = \{\alpha \cdot i \,|\, i = -N, -N+1, \cdots, N-1\}. \tag{4}$$

Note that the above is the QPS of (2) if we set $\alpha = s$, $L = -sN$, $U = s(N-1)$. Similarly, the QPS for log-scale quantization can be defined as follows, where α is a scale factor.

$$S_Q^{log}(\alpha) = \{-\alpha\, 2^{-i} \,|\, i = 0, 1, \cdots, N-1\} \cup \{0\}$$
$$\cup \{\alpha\, 2^{-i} \,|\, i = 0, 1, \cdots, N-2\}. \tag{5}$$

---

[2] One may design a quantizer to output a non-nearest element, which is suboptimal but may be motivated by computational efficiency. An example is log-scale quantization, which was defined [16] as doing a round operation in the logarithmic domain, which is not necessarily the nearest one in the linear domain.

One way to enhance the accuracy of log-scale quantization is to use two words [17,18], the QPS for which can be defined as follows.

$$S_Q^{2log}(\alpha) = \{q_1 + q_2 \mid q_1 \in S_Q^{log}(\alpha), q_2 \in S_Q^{log}(\alpha)\}. \tag{6}$$

Subset Quantization: The idea of subset quantization (SQ) is to define S_Q not as a fixed set of numbers or variables but as an arbitrary subset (with a size limit of $2N$) of a larger set. The larger set is called *universal set*, denoted by S_U.

$$S_Q^{sq}(\alpha) = \{\text{any } q_i \in S_U(\alpha) \mid i = 1, \cdots, 2N\}. \tag{7}$$

which says S_Q^{sq} is any subset of S_U with $2N$ or fewer elements. Note that quantization precision restricts only the size of a QPS, but not that of a universal set. While (7) and (3) define the simulated quantizer for SQ, the definition is undeterministic. Exactly which subset to choose is left for optimization (see next section), much like determining the value of α. A quantization scheme may choose a different subset for each layer (per-layer quantization) or for each channel (per-channel quantization).

While any set may be used as the universal set, in this paper we use one that is similar to the QPS of the two-word log-scale quantization scheme as our universal set (i.e., S_Q^{2log}) due to its low hardware complexity and rich expressiveness (see Sect. 3.4).

3.3   Optimization for Subset Quantization

Minimizing Quantization Error. Optimization for SQ needs to determine (i) the value of α and (ii) which subset of S_U to choose as QPS. For PTQ we are given all weight values ($\{w_i\}$), and the objective is to minimize L2 quantization error.

Finding Optimal Scale Factor. Given a QPS, the optimal scale factor can be found efficiently. Let $\{q_j\}$ be the given QPS (before applying a scale factor). After applying scale factor α, the final QPS can be written as $S_Q = \{\alpha q_j\}$. To calculate quantization loss between $\{w_i\}$ and S_Q, let αq_i be the nearest quantization point in S_Q for w_i. Then,

$$\forall i, \quad \alpha q_i = \arg\min_{p \in S_Q} |p - w_i|. \tag{8}$$

$$\mathcal{L} = \sum_i (w_i - \alpha q_i)^2. \tag{9}$$

To find α that minimizes $\mathcal{L}$, we set the derivative of $\mathcal{L}$ w.r.t. α to zero, which gives

$$\alpha^* = \frac{\sum_i w_i \cdot q_i}{\sum_i q_i^2}. \tag{10}$$

Since finding αq_i (the nearest quantization point for w_i) in (8) depends on the current value of α, the above procedure (8)–(10) is repeated until α converges. We initialize α to 1. We find empirically that this iterative method converges usually fast, at the average within 17 iterations (when tolerance for α is 1e−5).

Proof of Convergence. To find the optimal scale factor α for a given quantization point set $S(\alpha) = \{\alpha q_j\}$ that minimizes L2 quantization error, we use this iterative procedure, as explained in previous section.

– First, initialize α to 1.
– Second, update α using the (10) until α does not change any more.

Note that q_i on the right hand side of the (10) is a short-hand notation of a function $q(w_i, \alpha)$ defined below:

$$q_i := q(w_i, \alpha) = \frac{1}{\alpha} \cdot \arg\min_{p \in S(\alpha)} |p - w_i|. \tag{11}$$

We argue that updating α using (10) converges. Recall (9) which is the L2 quantization error. When α is updated according to (10), $q_i = q(w_i, \alpha)$ may or may not be updated. If no q_i is updated (i.e., $\forall i$, $q(w_i, \alpha) = q(w_i, \alpha^*)$), then (9) is a simple 2nd order function on α, and the optimal value of α can be found in one step by (10), after which α remains unchanged.

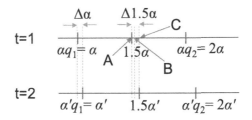

Fig. 3. The nearest quantization point may change when α is increased by $\Delta\alpha = \alpha' - \alpha$. In this example, $q_1 = 1$, $q_2 = 2$.

If some q_i are updated (i.e., $\exists i, q(w_i, \alpha) \neq q(w_i, \alpha^*)$), it leads to either (a) a decreased or the same value of $\mathcal{L}$ or (b) temporarily increased value of $\mathcal{L}$ but afterwards $\mathcal{L}$ is reduced. To see this, let us consider an example illustrated in Fig. 3. In this example, we only consider one weight value w, which is quantized in the first iteration ($t = 1$) using $S(\alpha) = \{\alpha q_1, \alpha q_2\}$ as the quantization point set. The first iteration updates α to α' (assume $\alpha' > \alpha$). Then in the second iteration ($t = 2$), w is quantized using $S(\alpha') = \{\alpha' q_1, \alpha' q_2\}$. To further simplify, let us assume that the two quantization points q_1, q_2 are given as follows: $q_1 = 1, q_2 = 2$. In the figure, $A = 1.5\alpha$, $C = 1.5\alpha'$, and $B = A + 0.5\Delta\alpha$, where $\Delta\alpha = \alpha' - \alpha$. There are three cases as follow.

Case 1: All the points between α' and 1.5α $(= A)$ will be quantized to αq_1 (at $t = 1$) or $\alpha' q_1$ (at $t = 2$). In this case, $q(w, \alpha)$ is not updated.

Case 2: Similarly, all the points between $1.5\alpha'$ $(= C)$ and 2α will be quantized to αq_2 (at $t = 1$) or $\alpha' q_2$ (at $t = 2$). In this case also, $q(w, \alpha)$ is not updated.

Case 3: If w is between points A and C, $q(w, \alpha)$ is updated from q_2 to q_1. There are two sub-cases. If w is greater than B, then the quantization error at $t = 2$ is *reduced* compared with that of $t = 1$. However, if w is less than B, the quantization error is *increased* compared with that of $t = 1$. Because of this last case, we cannot say that (10) monotonically decreases $\mathcal{L}$. Now, in order for our procedure to oscillate and not converge, the update rule (10) should decrease α, so that we can go back to the situation of $t = 1$, which however cannot happen. This is because when w is less than B, the quantization error is $w - \alpha'$, which is minimized when α *increases*. Therefore, there cannot be an indefinite oscillation between two α values, and $\mathcal{L}$ is only temporarily increased, but eventually converges to a minimum value.

The above is for the case of $\alpha' > \alpha$, but the convergence of the other case, $\alpha' < \alpha$, can be shown in a similar way.

Algorithm 1: FINDBESTQPS

Input: w: pretrained weight, S_U: universal set, N: 2^{k-1}, loss(w, S): loss function
Result: S_Q: the QPS that minimizes loss
1   $l_{\min} \leftarrow \infty$
2   **for** S **in** each N-element subset of S_U **do**
3      $\alpha \leftarrow$ FINDSCALEFACTOR(S, w)
4      $l_{\text{curr}} \leftarrow$ loss$(w, \alpha S)$
5   **if** $l_{\text{curr}} < l_{\min}$ **then**
6         $S_Q \leftarrow \alpha S$
7         $l_{\min} \leftarrow l_{\text{curr}}$
8   **end**
9   **end**
10 **return** S_Q

Finding the Best Subset. There are only $\binom{|S_U|}{|S_Q|}$ number of subsets to consider. Thus we can search exhaustively all cases for the one that gives the minimum quantization error (see Algorithm 1). We do this for each layer or each output channel, depending on the quantization granularity.

3.4 Designing a Universal Set

Our initial candidate for S_U is S_Q^{2log}, which is rich and simple enough to implement in hardware, requiring just two shifters and one adder for the multiplication. But in order to further optimize the multiplication hardware design, we explore the following design options. They all have the form of $a + b$, where a and b are either zero or 2^k as listed in Table 1. Note that thanks to a scale factor the sets are equivalent to their scaled versions, thus we assume that they are scaled such that the maximum value is one.

Table 1. Exploring universal set design, $S_U = \{a + b \,|\, a \in A, b \in B\}$.

Option	A	B
1	$\{2^{-1}, 2^{-2}, 2^{-3}, 0\}$	$\{2^{-1}, 2^{-2}, 2^{-4}, 0\}$
2	$\{1, 2^{-2}, 2^{-4}, 0\}$	$\{2^{-1}, 2^{-3}, 2^{-5}, 0\}$
3	$\{1, 2^{-1}, 2^{-2}, 2^{-3}, 2^{-4}, 0\}$	$\{1, 2^{-1}, 2^{-2}, 2^{-3}, 2^{-4}, 0\}$
4	$\{1, 2^{-1}, 2^{-2}, 2^{-3}, 2^{-4}, 2^{-5}, 0\}$	$\{1, 2^{-1}, 2^{-2}, 2^{-3}, 2^{-4}, 2^{-5}, 0\}$
5 (chosen)	$\{1, 2^{-1}, 2^{-3}, 0\}$	$\{1, 2^{-2}, 2^{-4}, 0\}$

The universal set designs differ in terms of hardware complexity as well as expressiveness. For instance, Option 4 generates a universal set with 23 elements whereas Option 3 and Option 5 have only 17 and 16 elements, respectively, but Option 4 also has the highest complexity. After a careful comparison (see Supplementary Material), we have chosen Option 5, which is rich enough yet has the lowest hardware complexity among the options.

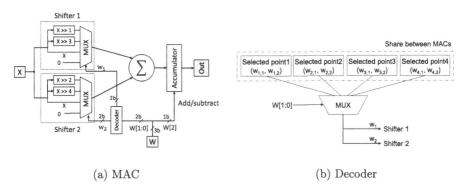

(a) MAC   (b) Decoder

Fig. 4. MAC design for the proposed subset quantization (3-bit case including sign-bit).

3.5   Hardware Design

Figure 4 illustrates the MAC (multiply-and-accumulate) hardware design for our quantizer. We assume that activation is linear-quantized and weight is quantized with our subset quantization. From the definition of our universal set (S_U) (i.e., sum of two power-of-two's), one straightforward implementation of multiplication is to use two shifters and one adder, in which shifters are variable-amount shifters (called *barrel shifters*). This naïve design, which would work for any universal set definition, can be optimized as shown in Fig. 4 exploiting our universal set definition. In our optimized design shown in Fig. 4a, each barrel shifter is replaced with a MUX and two constant-amount shifters. Note that a MUX is much cheaper than a barrel shifter and constant-amount shifters are just wires,

Table 2. Image classification results for *weight-only* PTQ methods. * indicates results with the same model but different baseline. For quantized cases, we report *performance degradation* with absolute performance numbers in parentheses (same in later tables).

Network	W-bits/A-bits	32/32 (FP)	4/32	3/32	2/32
ResNet-18	OMSE+opt [7]	69.64	2.52 (67.12)	–	–
	AdaRound [23]	69.68/71.08*	0.97 (68.71)	3.01 (68.07*)	15.12 (55.96*)
	AdaQuant [12]	71.08	2.26 (68.82)	12.96 (58.12)	70.78 (0.30)
	BitSplit [28]	69.76	0.65 (69.11)	3.01 (66.75)	–
	BRECQ [19]	71.08	0.38 (70.70)	1.27 (69.81)	4.78 (66.30)
	SQ (Ours)	69.76	**0.30 (69.46)**	**1.00 (68.76)**	**4.14 (65.62)**
ResNet-50	OMSE+opt [7]	76.01	1.34 (74.67)	–	–
	AdaRound [23]	76.07/77.00*	0.84 (75.23)	3.58 (73.42*)	29.05 (47.95*)
	AdaQuant [12]	77.20/77.00*	3.50 (73.70)	12.39 (67.61*)	76.51 (0.49*)
	BitSplit [28]	76.13	0.55 (75.58)	2.89 (73.24)	–
	BRECQ [19]	77.00	0.71 (76.29)	1.39 (75.61)	4.60 (72.40)
	SQ (Ours)	76.13	**0.38 (75.75)**	**0.99 (75.14)**	**3.86 (72.27)**
InceptionV3	OMSE+opt [7]	77.40	3.74 (73.66)	–	–
	AdaRound [23]	77.40	1.64 (75.76)	–	–
	OCS [29]	77.40	72.60 (4.80)	–	–
	SQ (Ours)	77.24	**0.58 (76.66)**	3.61 (73.63)	11.36 (65.87)

taking no logic gates. This is possible because of the way our universal set is constructed—each term has only four cases including zero.

Each MUX can select inputs independently, resulting in 16 different combinations, out of which only four cases are actually used due to the limited width of quantized weight value (which is 2 except the sign-bit in the figure). Thus the role of QPS is to select the four cases that will be actually used, and is implemented as the decoder. The decoder consists of a 16-bit register ($= 4 \times 4$-bit) and a 4-bit 4-to-1 MUX. The 16-bit register stores the QPS chosen by Algorithm 1 (actually their 2-bit logarithm values), and can be shared among all the MACs in the same layer or channel, depending on the quantization granularity, thus having negligible area. Then a decoder is practically reduced to a simple MUX, but even this MUX can often be shared among a number of MACs within a MAC array, depending on the dataflow of the MAC array [5]. In summary, the hardware cost of our optimized MAC is very small: two 4-to-1 MUXes, one adder, and an accumulator, plus a 4-to-1 MUX, which could be shared among a number of MACs depending on the hardware dataflow.

4   Experiments

4.1   Experimental Setup

For evaluation we use three applications: image classification using ImageNet dataset [11,26], object detection [21], and semantic segmentation [4]. We perform PTQ using pretrained weights; we have not modified a network or performed

Table 3. Image classification results for *fully-quantized* PTQ methods. * indicates results with the same model but different baseline.

Network	W-bits/A-bits	32/32 (FP)	4/8	4/4	3/3	2/4
ResNet-18	AdaQuant [12]	71.97/71.08*	–	4.57 (67.40)	–	71.08 (0.21*)
	Seq. AdaQuant [12]	71.97	–	2.57 (69.40)	–	–
	AdaRound [23]	69.68	1.13 (68.55)	–	–	–
	ACIQ-Mix [1]	69.70	–	2.70 (67.00)	–	–
	LAPQ [24]	69.76/71.08*	–	9.46 (60.30)	–	70.90 (0.18*)
	BitSplit [28]	69.76	0.66 (69.10)	2.20 (67.56)	8.46 (61.30)	–
	BRECQ [19]	71.08	0.50 (70.58)	1.48 (69.60)	5.05 (66.03)	6.28 (64.80)
	SQ (Ours)	69.76	**0.37 (69.39)**	**1.21 (68.55)**	**4.62 (65.14)**	**5.06 (64.70)**
ResNet-50	AdaQuant [12]	77.20/77.00*	–	3.50 (73.70)	–	76.88 (0.12*)
	Seq. AdaQuant [12]	77.20	–	2.10 (75.10)	–	–
	AdaRound [23]	76.07	1.06 (75.01)	–	–	–
	OMSE+opt [7]	76.01	1.03 (74.98)	3.41 (72.60)	–	–
	ACIQ-Mix [1]	76.10	0.80 (75.30)	2.30 (73.80)	–	–
	LAPQ [24]	76.10/77.00*	–	6.10 (70.00)	–	76.86 (0.14*)
	BitSplit [28]	76.13	–	2.42 (73.71)	9.91 (66.22)	–
	PWLQ [10]	76.13	0.51 (75.62)	**1.28 (74.85)**	–	–
	BRECQ [19]	77.00	–	1.95 (75.05)	8.04 (68.96)	6.71 (70.29)
	SQ (Ours)	76.13	**0.46 (75.67)**	1.48 (74.65)	**6.80 (69.33)**	**5.43 (70.70)**
InceptionV3	AdaRound [23]	77.40	1.68 (75.72)	–	–	–
	ACIQ-Mix [1]	77.20	9.00 (68.20)	–	–	–
	OMSE+opt [7]	76.23	1.44 (74.79)	–	–	–
	PWLQ [10]	77.49	1.04 (76.45)	–	–	–
	SQ (Ours)	77.24	**0.66 (76.58)**	–	–	–

retraining before PTQ in any way. We apply SQ to weight only; activation is quantized using BRECQ [19] unless noted otherwise. We have used the framework, models, and pretrained weights mainly from the official PyTorch 1.6.0 and used Nvidia RTX Titan GPUs for experiments (CUDA 9.2, cuDNN 7.6.5).

4.2   Comprehensive Results

For all the results, we have repeated the PTQ experiments five times for each case by changing the random seed value, and the first and last layers are set to 8-bit linear quantization. In all the cases, the granularity of SQ is per-channel but we only use one QPS per layer. In other words, we use a scale factor for each channel in our PTQ process, but we map all weight values with one QPS for each layer. The per-layer QPS not only shows superior performance, but also allows for simpler hardware implementation.

Image Classification. For image classification, we use the ImageNet dataset and present the results for ResNet series [11] and InceptionV3 [27]. We compare with various state-of-the-art PTQ methods [10,12,19,23,28], and our SQ, as applied to weight quantization. We have mainly referred to BRECQ[3] for our code implementation.

---

[3] https://github.com/yhhli/BRECQ.

Table 2 and 3 show the results. When only weight is quantized and activation is floating-point, our SQ shows consistently and significantly better performance than other PTQ methods and we see a similar trend when activation is quantized as well.[4] In particular, our SQ shows superior results in the ultra-low-bit condition for both 3-bit and 2-bit.

Semantic Segmentation. For semantic segmentation task, we have applied our SQ method to the official DeepLabV3+ [4] source code[5] and compared it with other PTQ methods. We use MobileNetV2 [26] as the encoder backbone, and use ASPP [3] as the decoder. SQ is applied to all layers with weight parameters, and activation is quantized to 8-bit using linear quantization. Table 4 shows the results. We have conducted experiments on low-bit cases less than or equal 4-bit in consideration of performance and search time efficiency of SQ. The 4-bit result clearly shows the superior performance of our method compared with the previous PTQ method. The previous PTQ methods do not report the result of 3-bit or lower, but our 3-bit SQ result shows relatively small performance degradation, and is even better than that of 4-bit uniform quantization.

Table 4. PTQ methods on semantic segmentation with the MobileNetV2 [26] backbone.

Network	W-bits/A-bits	32/32 (FP)	4/8	3/8
DeepLabV3+(mIoU%)	Uniform	70.81	20.76 (50.05)	–
	PWLQ [10]	70.81	3.15 (67.66)	–
	SQ (ours)	70.81	**1.85 (68.96)**	**8.87 (61.94)**

Object Detection. We have evaluated our SQ method using object detection task, SSD-Lite [21]. We have modified the author's official PyTorch source code[6], and used MobileNetV2 [26] as the backbone network. Again we have applied SQ to all layers with weight parameters and 8-bit linear quantization is applied to activation. Table 5 shows the results. In the case of 4-bit, SQ shows a similar level of performance as that of the current state-of-the-art. In the 3-bit case, the performance trend is similar to that of the semantic segmentation results, but the performance degradation is smaller and again our 3-bit SQ result shows lower performance degradation than 4-bit uniform quantization.

---

[4] For InceptionV3 4-bit in Table 3, we only present the result with 8-bit linear quantization because our implementation for low-bit activations [19] did not work properly in this case.

[5] https://github.com/jfzhang95/pytorch-deeplab-xception.

[6] https://github.com/qfgaohao/pytorch-ssd.

Table 5. PTQ methods on object detection with the MobileNetV2 [26] backbone.

Network	W-bits/A-bits	32/32 (FP)	4/8	3/8
SSD-Lite (mAP%)	Uniform	68.70	3.91 (64.79)	–
	DFQ [22]	68.47	0.56 (67.91)	–
	PWLQ [10]	68.70	**0.38 (68.32)**	–
	SQ (ours)	68.59	**0.38 (68.21)**	**3.87 (64.72)**

5   Conclusion

We presented a novel non-uniform quantization method called subset quantization (SQ), which is a high-performing PTQ methods at extreme low-precision while remaining hardware-friendly. Subset quantization adds a new dimension to quantizer definition by allowing quantization points to be defined as a subset of a larger pool called universal set. This view allows a quantization point set to be defined in a more flexible fashion so that it can adapt to diverse input statistics, which is very useful for deep neural networks. Our experimental results with challenging vision tasks demonstrate that our SQ results for ultra low-bit weight quantization outperform state-of-the-art quantizers of the same precision.

Acknowledgements. This work was supported by the Samsung Advanced Institute of Technology, Samsung Electronics Co., Ltd., by IITP grants (No. 2020-0-01336, Artificial Intelligence Graduate School Program (UNIST), and No. 1711080972, Neuromorphic Computing Software Platform for Artificial Intelligence Systems) and NRF grant (No. 2020R1A2C2015066) funded by MSIT of Korea, and by Free Innovative Research Fund of UNIST (1.170067.01).

References

1. Banner, R., Nahshan, Y., Soudry, D.: Post training 4-bit quantization of convolutional networks for rapid-deployment. In: Advances in Neural Information Processing Systems, vol. 32. Curran Associates, Inc. (2019). https://proceedings.neurips.cc/paper/2019/file/c0a62e133894cdce435bcb4a5df1db2d-Paper.pdf

2. Chen, C., Chen, Q., Xu, J., Koltun, V.: Learning to see in the dark. In: Proceedings of the IEEE Conference on Computer Vision and Pattern Recognition, pp. 3291–3300 (2018)

3. Chen, L., Papandreou, G., Kokkinos, I., Murphy, K., Yuille, A.L.: DeepLab: semantic image segmentation with deep convolutional nets, Atrous convolution, and fully connected CRFs. CoRR abs/1606.00915 (2016). arxiv.org/abs/1606.00915

4. Chen, L.C., Zhu, Y., Papandreou, G., Schroff, F., Adam, H.: Encoder-decoder with atrous separable convolution for semantic image segmentation. In: European Conference on Computer Vision (ECCV), pp. 801–818 (2018)

5. Chen, Y., Krishna, T., Emer, J.S., Sze, V.: Eyeriss: an energy-efficient reconfigurable accelerator for deep convolutional neural networks. IEEE J. Solid-State Circuits **52**(1), 127–138 (2017). https://doi.org/10.1109/JSSC.2016.2616357

6. Choi, J., Wang, Z., Venkataramani, S., Chuang, P.I.J., Srinivasan, V., Gopalakr-ishnan, K.: Pact: parameterized clipping activation for quantized neural networks. arXiv preprint arXiv:1805.06085 (2018)

7. Choukroun, Y., Kravchik, E., Yang, F., Kisilev, P.: Low-bit quantization of neural networks for efficient inference. In: 2019 IEEE/CVF International Conference on Computer Vision Workshop (ICCVW), pp. 3009–3018 (2019). https://doi.org/10.1109/ICCVW.2019.00363

8. Ding, R., Liu, Z., Chin, T.W., Marculescu, D., Blanton, R.D.S.: FlightNNs: lightweight quantized deep neural networks for fast and accurate inference. In: Proceedings of the 56th Annual Design Automation Conference 2019, DAC 2019, Association for Computing Machinery, New York (2019). https://doi.org/10.1145/3316781.3317828

9. Esser, S.K., McKinstry, J.L., Bablani, D., Appuswamy, R., Modha, D.S.: Learned step size quantization. In: International Conference on Learning Representations (2019)

10. Fang, J., Shafiee, A., Abdel-Aziz, H., Thorsley, D., Georgiadis, G., Hassoun, J.H.: Post-training piecewise linear quantization for deep neural networks. In: Vedaldi, A., Bischof, H., Brox, T., Frahm, J.-M. (eds.) ECCV 2020. LNCS, vol. 12347, pp. 69–86. Springer, Cham (2020). https://doi.org/10.1007/978-3-030-58536-5_5

11. He, K., Zhang, X., Ren, S., Sun, J.: Deep residual learning for image recognition. In: IEEE Conference on Computer Vision and Pattern Recognition (2016)

12. Hubara, I., Nahshan, Y., Hanani, Y., Banner, R., Soudry, D.: Improving post training neural quantization: layer-wise calibration and integer programming. CoRR abs/2006.10518 (2020). arxiv.org/abs/2006.10518

13. Jacob, B., et al.: Quantization and training of neural networks for efficient integer-arithmetic-only inference. In: Proceedings of the IEEE Conference on Computer Vision and Pattern Recognition, pp. 2704–2713 (2018)

14. Jung, S., et al.: Learning to quantize deep networks by optimizing quantization intervals with task loss. In: Proceedings of the IEEE Conference on Computer Vision and Pattern Recognition, pp. 4350–4359 (2019)

15. Krishnamoorthi, R.: Quantizing deep convolutional networks for efficient inference: a whitepaper. arXiv preprint arXiv:1806.08342 (2018)

16. Lee, E.H., Miyashita, D., Chai, E., Murmann, B., Wong, S.S.: LogNet: energy-efficient neural networks using logarithmic computation. In: 2017 IEEE International Conference on Acoustics, Speech and Signal Processing (ICASSP), pp. 5900–5904 (2017). https://doi.org/10.1109/ICASSP.2017.7953288

17. Lee, S., Sim, H., Choi, J., Lee, J.: Successive log quantization for cost-efficient neural networks using stochastic computing. In: 2019 Proceedings of the 56th Annual Design Automation Conference. DAC '19, Association for Computing Machinery, New York, NY, USA (2019). https://doi.org/10.1145/3316781.3317916

18. Li, Y., Dong, X., Wang, W.: Additive powers-of-two quantization: an efficient non-uniform discretization for neural networks. In: International Conference on Learning Representations (2020)

19. Li, Y., et al.: BRECQ: pushing the limit of post-training quantization by block reconstruction. In: International Conference on Learning Representations (2021)

20. Lim, B., Son, S., Kim, H., Nah, S., Mu Lee, K.: Enhanced deep residual networks for single image super-resolution. In: Proceedings of the IEEE Conference on Computer Vision and Pattern Recognition Workshops, pp. 136–144 (2017)

21. Liu, W., et al.: SSD: single shot MultiBox detector. In: Leibe, B., Matas, J., Sebe, N., Welling, M. (eds.) ECCV 2016. LNCS, vol. 9905, pp. 21–37. Springer, Cham (2016). https://doi.org/10.1007/978-3-319-46448-0_2

22. Nagel, M., van Baalen, M., Blankevoort, T., Welling, M.: Data-free quantization through weight equalization and bias correction. In: IEEE/CVF International Conference on Computer Vision (2019)
23. Nagel, M., Amjad, R.A., van Baalen, M., Louizos, C., Blankevoort, T.: Up or down? Adaptive rounding for post-training quantization. CoRR abs/2004.10568 (2020). arxiv.org/abs/2004.10568
24. Nahshan, Y., et al.: Loss aware post-training quantization. CoRR abs/1911.07190 (2019). arxiv.org/abs/1911.07190
25. Oh, S., Sim, H., Lee, S., Lee, J.: Automated log-scale quantization for low-cost deep neural networks. In: Proceedings of the IEEE Conference on Computer Vision Pattern Recognition (CVPR), pp. 742–751 (2021)
26. Sandler, M., Howard, A.G., Zhu, M., Zhmoginov, A., Chen, L.: Inverted residuals and linear bottlenecks: mobile networks for classification, detection and segmentation. In: IEEE Conference on Computer Vision and Pattern Recognition (2018)
27. Szegedy, C., Vanhoucke, V., Ioffe, S., Shlens, J., Wojna, Z.: Rethinking the inception architecture for computer vision. CoRR abs/1512.00567 (2015). arxiv.org/abs/1512.00567
28. Wang, P., Chen, Q., He, X., Cheng, J.: Towards accurate post-training network quantization via bit-split and stitching. In: International Conference on Machine Learning (2020)
29. Zhao, R., Hu, Y., Dotzel, J., Sa, C.D., Zhang, Z.: Improving neural network quantization without retraining using outlier channel splitting. In: International Conference on Machine Learning (2019)
30. Zhao, X., Wang, Y., Cai, X., Liu, C., Zhang, L.: Linear symmetric quantization of neural networks for low-precision integer hardware. In: International Conference on Learning Representations (2020). https://openreview.net/forum?id=H1lBj2VFPS

SuperTickets: Drawing Task-Agnostic Lottery Tickets from Supernets via Jointly Architecture Searching and Parameter Pruning

Haoran You[1], Baopu Li[2,3](✉), Zhanyi Sun[1], Xu Ouyang[1], and Yingyan Lin[1](✉)

[1] Rice University, Houston, USA
{haoran.you,zs19,xo2,yingyan.lin}@rice.edu
[2] Baidu, Sunnyvale, USA
[3] Oracle Health and AI, Redwood City, USA
baopu.li@oracle.com

Abstract. Neural architecture search (NAS) has demonstrated amazing success in searching for efficient deep neural networks (DNNs) from a given supernet. In parallel, lottery ticket hypothesis has shown that DNNs contain small subnetworks that can be trained from scratch to achieve a comparable or even higher accuracy than the original DNNs. As such, it is currently a common practice to develop efficient DNNs via a pipeline of first search and then prune. Nevertheless, doing so often requires a tedious and costly process of search-train-prune-retrain and thus prohibitive computational cost. In this paper, we discover for the first time that both efficient DNNs and their lottery subnetworks (i.e., lottery tickets) can be directly identified from a supernet, which we term as **SuperTickets**, via a two-in-one training scheme with jointly architecture searching and parameter pruning. Moreover, we develop a progressive and unified SuperTickets identificationcesstab strategy that allows the connectivity of subnetworks to change during supernet training, achieving better accuracy and efficiency trade-offs than conventional sparse training. Finally, we evaluate whether such identified SuperTickets drawn from one task can transfer well to other tasks, validating their potential of simultaneously handling multiple tasks. Extensive experiments and ablation studies on three tasks and four benchmark datasets validate that our proposed SuperTickets achieve boosted accuracy and efficiency trade-offs than both typical NAS and pruning pipelines, regardless of having retraining or not. Codes and pretrained models are available at https://github.com/RICE-EIC/SuperTickets.

Keywords: Lottery ticket hypothesis · Efficient training/inference · Neural architecture search · Task-agnostic DNNs

---

H. You—Work done while interning at Baidu USA.

Supplementary Information The online version contains supplementary material available at https://doi.org/10.1007/978-3-031-20083-0_40.

S. Avidan et al. (Eds.): ECCV 2022, LNCS 13671, pp. 674–690, 2022.
https://doi.org/10.1007/978-3-031-20083-0_40

1   Introduction

While deep neural networks (DNNs) have achieved unprecedented performance in various tasks and applications like classification, segmentation, and detection [8], their prohibitive training and inference costs limit their deployment on resource-constrained devices for more pervasive intelligence. For example, one forward pass of the ResNet50 [15] requires 4 GFLOPs (FLOPs: floating point operations) and its training requires 10^{18} FLOPs [47]. To close the aforementioned gap, extensive attempts have been made to compress DNNs from either macro-architecture (e.g., NAS [8,35,42]) or fine-grained parameter (e.g., network pruning [11,14]) levels. A commonly adopted DNN compression pipeline following a coarse-to-fine principle is to first automatically search efficient and powerful DNN architectures from a larger supernet and then prune the searched DNNs via costly train-prune-retrain process [9,10,20] to derive smaller and sparser subnetworks with a comparable or degraded accuracy but largely reduced inference costs. However, such pipeline requires a tedious search-train-prune-retrain process and thus still prohibitive training costs.

To address the above limitation for simplifying the pipeline and further improve the accuracy-efficiency trade-offs of the identified networks, we advocate a **two-in-one training** framework for simultaneously identifying both efficient DNNs and their lottery subnetworks via jointly architecture searching and parameter pruning. We term the identified small subnetworks as **SuperTickets** if they achieve comparable or even superior accuracy-efficiency trade-offs than previously adopted search-then-prune baselines, because they are drawn from supernets and represent both coarse-grained DNN architectures and fine-grained DNN subnetworks. We make non-trivial efforts to explore and validate the potential of SuperTickets by answering three key questions: *(1) whether such SuperTickets can be directly found from a supernet via two-in-one training? If yes, then (2) how to effectively identify such SuperTickets? and (3) can SuperTickets found from one task/dataset transfer to another, i.e., have the potential to handle different tasks/datasets?* To the best of our knowledge, this is the first attempt taken towards identifying both DNN architectures and their corresponding lottery ticket subnetworks through a unified two-in-one training scheme. Our contributions can be summarized as follows:

- We **for the first time** discover that efficient DNN architectures and their lottery subnetworks, i.e., SuperTickets, can be simultaneously identified from a supernet leading to superior accuracy-efficiency trade-offs.
- We develop an unified progressive identification strategy to effectively find the SuperTickets via a two-in-one training scheme which allows the subnetworks to iteratively reactivate the pruned connections during training, offering better performance than conventional sparse training. Notably, our identified SuperTickets *without retraining* already outperform previously adopted first-search-then-prune baselines, and thus can be directly deployed.
- We validate the transferability of identified SuperTickets across different tasks/datasets, and conduct extensive experiments to compare the proposed SuperTickets with those from existing search-then-prune baselines, typical

NAS techniques, and pruning works. Results on three tasks and four datasets demonstrate the consistently superior accuracy-efficiency trade-offs and the promising transferability for handling different tasks offered by SuperTickets.

2   Related Works

Neural Architecture Search (NAS). NAS has achieved an amazing success in automating the design of efficient DNN architectures and boosting accuracy-efficiency trade-offs [16,36,54]. To search for task-specific DNNs, early works [16,35,36] adopt reinforcement learning based methods that require a prohibitive search time and computing resources, while recent works [23,38,42,45] update both the weights and architectures during supernet training via differentiable search that can greatly improve the search efficiency as compared to prior NAS works. More recently, some works adopt one-shot NAS [3,13,39,50] to decouple the architecture search from supernet training. Such methods are generally applicable to search for efficient CNNs [2,13] or Transformers [4,34,40] for solving both vision and language tasks. To search for multi-task DNNs, recently emerging works like HR-NAS [8] and FBNetv5 [43] advocate supernet designs with multi-resolution branches so as to accommodate both image classification and other dense prediction tasks that require high-resolution representations. In this work, we propose to directly search for not only efficient DNNs but also their lottery subnetworks from supernets to achieve better accuracy-efficiency trade-offs while being able to handle different tasks.

Lottery Ticket Hypothesis (LTH). Frankle et al. [11,12] showed that winning tickets (i.e., small subnetworks) exist in randomly initialized dense networks, which can be retrained to restore a comparable or even better accuracy than their dense network counterparts. This finding has inspired lots of research directions as it implies the potential of sparse subnetworks. For efficient training, You et al. [47] consistently find winning tickets at early training stages, largely reducing DNNs' training costs. Such finding has been extended to language models (e.g., BERT) [5], generative models (e.g., GAN) [30], and graph neural networks [48]; Zhang et al. [52] recognize winning tickets more efficiently by training with only a specially selected subset of data; and Ramanujan et al. [31] further identify winning tickets directly from random initialization that perform well even without retraining. In contrast, our goal is to simultaneously find both efficient DNNs and their lottery subnetworks from supernets, beyond the scope of sparse training or drawing winning tickets from dense DNN models.

Task-Agnostic DNNs Design. To facilitate designing DNNs for different tasks, recent works [16,24,41] propose to design general architecture backbones for various computer vision tasks. For example, HR-Net [41] maintains high-resolution representations through the whole network for supporting dense prediction tasks, instead of connecting high-to-low resolution convolutions in series like ResNet or VGGNet; Swin-Transformer [24] adopts hierarchical vision transformers to serve as a general-purpose backbone that is compatible with a broad range of vision tasks; ViLBERT [26,27] proposes a multi-modal two-stream model to learn task-agnostic joint representations of both image and language; Data2vec [1] designs a general framework for self-supervised learning in speech, vision and language. Moreover, recent works [8,43,44] also leverage NAS to automatically search for task-agnostic and efficient DNNs from hand-crafted supernets. In this work, we aim to identify task-agnostic SuperTickets that achieve better accuracy-efficiency trade-offs.

3 The Proposed SuperTickets Method

In this section, we address the three key questions of SuperTickets. First, we develop a two-in-one training scheme to validate our hypothesis that SuperTickets exist and can be found directly from a supernet. Second, we further explore more effective SuperTickets identification strategies via iterative neuron reactivation and progressive pruning, largely boosting the accuracy-efficiency trade-offs. Third, we evaluate the transferability of the identified SuperTickets across different datasets or tasks, validating their potential of being task-agnostic.

3.1 Do SuperTickets Exist in Supernets?

SuperTickets Hypothesis. We hypothesize that both efficient DNN architectures and their lottery subnetworks can be directly identified from a supernet, and term these subnetworks as **SuperTickets** if they achieve on par or even better accuracy-efficiency trade-offs than those from first-search-then-prune counterparts. Considering a supernet $f(x; \theta_S)$, various DNN architectures a are sampled from it whose weights are represented by $\theta_S(a)$, then we can define SuperTickets as $f(x; m \odot \theta_S(a))$, where $m \in \{0,1\}$ is a mask to indicate the pruned and unpruned connections in searched DNNs. The SuperTickets Hypothesis implies that jointly optimizing DNN architectures a and corresponding sparse masks m works better, i.e., resulting in superior accuracy-efficiency trade-offs, than sequentially optimizing them.

Experiment Settings. To perform experiments for exploring whether SuperTickets generally exist, we need (1) a suitable supernet taking both classical efficient building blocks and task-agnostic DNN design principles into consideration and (2) corresponding tasks, datasets, and metrics. We elaborate our settings below. NAS and Supernets: We consider a multi-branch search space containing both efficient convolution and attention building blocks following one state-of-the-art (SOTA) work of HR-NAS [8], whose unique hierarchical multi-resolution search space for handling multiple vision tasks stands out compared to others. In general, it contains two paths: MixConv [37] and lightweight Transformer for extracting both local and global context information. Both the number of convolutional channels with various kernel sizes and the number of tokens in the Transformer are searchable parameters. Tasks, Datasets, and Metrics: We consider semantic segmentation on Cityscapes [6] and human pose estimation on COCO keypoint [21] as two representative tasks for illustrative purposes. For Cityscapes, the mean Intersection over Union (mIoU), mean Accuracy (mAcc),

---

Algorithm 1: Two-in-One Framework for Identifying SuperTickets.

Input: The supernet weights θ_S, drop threshold ϵ, and pruning ratio p;
Output: Efficient DNNs and their lottery subnetworks $f(x; m \odot \theta_S(a))$.

1 **while** t *(epoch)* $< t_{max}$ **do**
2     $t = t + 1$;
3   Update weights θ_S and importance factor r using SGD training;
4   **if** t mod $t_s = 0$ **then**   ▷ `Search for DNNs`
5   Remove search units whose importance factors $r < \epsilon$;
6   Recalibrate the running statistics of BN layers to obtain subnet a;
          // If enabling the iterative reactivation technique
7
8   **else if** t mod $t_p = 0$ **then**   ▷ `Prune for subnetworks`
          // If enabling the progressive pruning technique
9
10   Perform magnitude-based pruning towards the target ratio;
11   Keep the sparse mask m_t and disable pruned weights' gradients;
12   **end**
13 **end**
14 **return** $f(x; m_t \odot \theta_S(a))$;   ▷ `SuperTickets`

---

and overall Accuracy (aAcc) are evaluation metrics. For COCO keypoint, we train the model using input size 256×192, an initial learning rate of 1e−3, a batch size of 384 for 210 epochs. The average precision (AP), recall scores (AR), AP^M and AP^L for medium or large objects are evaluation metrics. All experiments are run on Tesla V100*8 GPUs.

Two-in-One Training. To validate the SuperTickets hypothesis, we propose a two-in-one training algorithm that simultaneously searches and prunes during supernet training of NAS. As shown in Alg. 1 and Fig. 1, for searching for efficient DNNs, we adopt a progressive shrinking NAS by gradually removing unimportant search units that can be either convolutional channels or Transformer

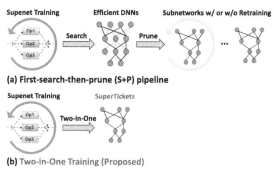

(a) **First-search-then-prune (S+P) pipeline**

(b) **Two-in-One Training (Proposed)**

Fig. 1. Illustrating first-search-then-prune (S+P) vs. our two-in-One training.

tokens. After every p_s training epochs, we will detect and remove the unimportant search units once their corresponding importance factors r (i.e., the scales in Batch Normalization (BN) layers) are less than a predefined drop threshold ϵ. Note that r can be jointly learned with supernet weights, such removing will not affect the remaining search units since channels in depth-wise convolutions are

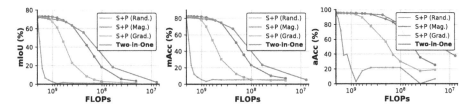

Fig. 2. Comparing the mIoU, mAcc, aAcc and inference FLOPs of the resulting networks from the proposed two-in-one training and first-search-then-prune (S+P) baselines on semantic segmentation task and Cityscapes dataset, where Rand., Mag., and Grad. represent random, magnitude, and gradient-based pruning, respectively. Note that each method has a series of points for representing different pruning ratios ranging from 10% to 98%. All accuracies are averaged over three runs.

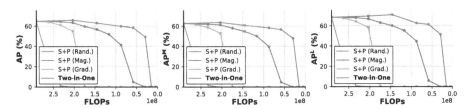

Fig. 3. Comparing the AP, AP^M, AP^L and inference FLOPs of the resulting networks from the proposed two-in-one training and baselines on human pose estimation task and COCO keypoint dataset. Each method has a series of points for representing different pruning ratios ranging from 10% to 98%. All accuracies are averaged over three runs.

independent among each other, as also validated by [8,29]. In addition, we follow network slimming [25] to add a l_1 penalty as a regularization term for polarizing the importance factors to ease the detection of unimportant units. After removing them, the running statistics in BN layers are recalibrated in order to match the searched DNN architecture a for avoiding covariate shift [18,46]. For pruning of searched DNNs, we perform magnitude-based pruning towards the given pruning ratio per t_p epochs, the generated spare mask m_t will be kept so as to disable the gradients flow of the pruned weights during the following training. Note that we do not incorporate the iterative reactivation and progressive pruning techniques (highlighted with colors/shadows in Algorithm 1, which will be elaborated later) as for now. Such vanilla two-in-one training algorithm can be regarded as the first step towards answering the puzzle whether SuperTickets generally exist.

Existence of SuperTickets. We compare the proposed two-in-one training with first-search-then-prune (S+P) baselines and report the results on Cityscapes and COCO keypoint at Fig. 2 and Fig. 3, respectively. We see that the proposed two-in-one training consistently generates comparable or even better accuracy-efficiency trade-offs as compared to S+P with various pruning criteria (random, magnitude, and gradient) since our methods demonstrate much better performance of segmentation or human pose estimation under different FLOPs reductions as shown

Table 1. Breakdown analysis of the proposed SuperTickets identification strategy. We report the performance of found subnetworks under 90%/80% sparsity on two datasets.

Methods	2-in-1	PP	IR-P	IR-S	Retrain	Cityscapes			COCO Keypoint			
						mIoU	mAcc	aAcc	AP	AP^M	AP^L	AR
S+P (Mag.)						42.12	50.49	87.45	5.04	4.69	5.89	10.67
S+P (Mag.)				✓		51.03	59.61	90.88	48.63	46.82	51.74	53.38
Ours	✓					55.84	67.38	92.97	58.38	56.68	61.26	62.23
Ours	✓	✓				63.89	73.56	94.17	60.14	57.93	63.70	63.79
Ours	✓	✓	✓			45.73	55.52	89.36	5.48	7.43	4.36	10.85
Ours	✓	✓		✓		**66.61**	**76.30**	**94.63**	**61.02**	**58.80**	**64.64**	**64.78**
Ours	✓	✓		✓	✓	**67.17**	**77.03**	**94.73**	**61.48**	**59.30**	**65.19**	**65.20**

in the above two figures, indicating that SuperTickets generally exist in a supernet and have great potential to outperform the commonly adopted approaches, i.e., sequentially optimizing DNN architectures and sparse masks.

3.2 How to More Effectively Identify SuperTickets?

We have validated the existence of SuperTickets, the natural next question is how to more effectively identify them. To this end, we propose two techniques that can be seamlessly incorporated into the two-in-one training framework to more effectively identify SuperTickets and further boost their achievable performance.

Progressive Pruning (PP). Although simultaneously searching and pruning during supernet training enables the opportunity of cooperation between coarse-grained search units removal and fine-grained weights pruning, i.e., NAS helps to refine the pruned networks as a compensation by removing over-pruned units for avoiding bottlenecked layers, we find that over-pruning at the early training stages inevitably hurts the networks' generalizability, and further propose a progressive pruning (PP) techniques to overcome this shortcoming. As highlighted in the cyan part of Algorithm 1, the pruning ratio is defined as $\min\{p, 10\% \times \lfloor t/t_p \rfloor\}$, which means that the network sparsity will gradually increase from 10% to the target ratio p, by 10% per t_p epochs. The PP technique helps to effectively avoid over-pruning at early training stages and thus largely boosts the final performance. As demonstrated in Table 1, two-in-one training with PP achieves 8.05%/6.18%/1.2% mIoU/mAcc/aAcc and 1.76%/1.25%/2.44%/1.56% $AP/AP^M/AP^L/AR$ improvements on Cityscapes and COCO keypoint datasets, respectively, as compared to the vanilla two-in-one training under 90% sparsity.

Iterative Reactivation (IR). Another problem in the two-in-one framework is that the pruned weights will never get gradients updates throughout the remaining training. To further boost the performance, we design an iterative reactivation (IR) strategy to facilitate the effective SuperTickets identification by allowing the connectivity of subnetworks to change during supernet training.

Specifically, we reactivate the gradients of pruned weights as highlighted in the orange part of Algorithm. 1. Note that we reactivate during searching instead of right after pruning, based on a hypothesis that sparse training is also essential to the two-in-one training framework. In practice, the pruning interval p_t is different from the searching interval p_s in order to allow a period of sparse training. To validate the hypothesis, we design two variants: IR-S and IR-P that reactivate pruned weights' gradients during searching and pruning, respectively, and show the comparisons in Table 1. We observe that: (1) IR-P leads to even worse accuracy than vanilla two-in-one training, validating that sparse training is essential; (2) IR-S further leads to 2.72%/2.74%/0.46% mIoU/mAcc/aAcc and 0.88%/0.87%/0.94%/0.99% AP/AP^M/ AP^L/AR improvements on Cityscapes and COCO keypoint, respectively, on top of two-in-one training with PP.

SuperTickets w/ or w/o Retraining. Since the supernet training, architecture search, and weight pruning are conducted in an unified end-to-end manner, the resulting SuperTickets can be deployed directly without retraining, achieving better accuracy-efficiency trade-offs than S+P baselines (even with retraining) as indicated by Table 1. To investigate whether retraining can further boost the performance, we retrain the found SuperTickets for another 50 epochs and report the results at Table 1. We see that retraining further leads to 0.56%/0.73%/0.10% mIoU/mAcc/aAcc and 0.46%/0.50%/0.55%/0.42% AP/AP^M/ AP^L/AR improvements on Cityscapes and COCO keypoint datasets, respectively.

3.3   Can the Identified SuperTickets Transfer?

To validate the potential of identified SuperTickets for handling different tasks and datasets, we provide empirical experiments and analysis as follows. Note that we adjust the final classifier to match target datasets during transfer learning.

SuperTickets Transferring Among Datasets. We first test the transferability of the identified SuperTickets among different datasets within the same task, i.e., Cityscapes and ADE20K as two representatives in the semantic segmentation task. Table 2 shows that SuperTickets identified from one dataset can transfer to another dataset while leading to comparable or even better performance than S+P baselines with (denoted as "w/ RT") or without retraining (by

Table 2. Supertickets transfer validation tests under 90% sparsity.

Methods	Params	FLOPs	Cityscapes		
			mIoU	mAcc	aAcc
S+P (Grad.)	0.13M	203M	8.41	12.39	56.77
S+P (Mag.)	0.13M	203M	42.12	50.49	87.45
S+P (Mag.) w/ RT	0.13M	203M	60.76	70.40	93.38
ADE20K Tickets	0.20M	247M	**62.91**	**73.32**	**93.82**
ImageNet Tickets	0.18M	294M	61.64	71.78	93.75

Methods	Params	FLOPs	ADE20K		
			mIoU	mAcc	aAcc
S+P (Grad.)	0.11M	154M	0.79	1.50	25.58
S+P (Mag.)	0.11M	154M	3.37	4.70	39.47
Cityscapes Tickets	0.13M	119M	20.83	29.95	69.00
ImageNet Tickets	0.21M	189M	**22.42**	**31.87**	**70.21**

default). For example, when tested on Cityscapes, SuperTickets identified from ADE20K after fine-tuning lead to 2.2% and 20.8% higher mIoU than S+P (Mag.) w/ and w/o RT baselines which are directly trained on target Cityscapes dataset. Likewise, the SuperTickets transferred from Cityscapes to ADE20K also outperform baselines on target dataset.

SuperTickets Transferring Among Tasks. To further investigate whether the identified SuperTickets can transfer among different tasks. We consider to transfer SuperTickets's feature extraction modules identified from ImageNet on classification task to Cityscapes and ADE20K on segmentation tasks, where the dense prediction heads and final classifier are still inherited from the target datasets. The results are presented in the last row of the two sub-tables in Table 2. We observe that such transferred networks still perform well on downstream tasks. Sometimes, it even achieves better performance than transferring within one task, e.g., ImageNet → ADE20K works better (1.6% higher mIoU) than Cityscapes → ADE20K. We supply more experiments on various pruning ratios in Sect. 4.3.

4 Experiment Results

4.1 Experiment Setting

Tasks, Datasets, and Supernets. Tasks and Datasets. We consider four benchmark datasets and three representative vision tasks to demonstrate the effectiveness of SuperTickets, including image classification on ImageNet [7] dataset with 1.2 million training images and 50K validation images; semantic segmentation on Cityscapes [6] and ADE20K [53] datasets with 2975/500/1525 and 20K/2K/3K images for training, validation, and testing, respectively; human pose estimation on COCO keypoint [21] dataset with 57K images and 150K person instances for training, and 5K images for validation. These selected datasets require different receptive fields and global/local contexts, manifesting themselves as proper test-beds for SuperTickets on multiple tasks. Supernets. For all experiments, we adopt the same supernet as HR-NAS [8] thanks to the task-agnostic multi-resolution supernet design. It begins with two 3×3 convolutions with stride 2, which is followed by five parallel modules to gradually divide it into four branches of decreasing resolutions, the learned features from all branches are then merged together for classification or dense prediction.

Search and Training Settings. For training supernets on ImageNet, we adopt a RMSProp optimizer with 0.9 momentum and 1e−5 weight decay, exponential moving average (EMA) with 0.9999 decay, and exponential learning rate decay with an initial learning rate of 0.016 and 256 batch size for 350 epochs. For Cityscapes and ADE20K, we use an AdamW optimizer, an initial learning rate of 0.04 with batch size 32 due to larger input image sizes, and train for 430 and 200 epochs, respectively, following [8]. For COCO keypoint, we follow [41] to use an Adam optimizer for 210 epochs, the initial learning rate is set to 1e−3, and is divided by 10 at the 170th and 200th epochs, respectively. In addition, we perform architecture search during supernet training. For all search units, we use the scales from their attached BN layers as importance factors r; search units with $r < 0.001$ are regarded as unimportant and removed every 10 epochs (i.e., $t_s = 10$); Correspondingly, magnitude-based pruning will be performed per 25 epochs for ImageNet and Cityscapes, or per 15 epochs for ADE20K and COCO keypoint (i.e., $t_p = 25/15$), resulting intervals for sparse training as in Sect. 3.2.

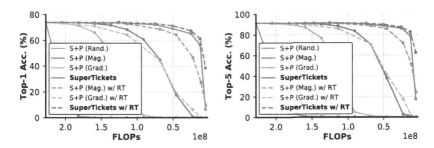

Fig. 4. Comparing the top-1/5 accuracy and FLOPs of the proposed SuperTickets and S+P baselines on ImageNet. Each method has a series of points to represent different pruning ratios ranging from 10% to 98%. All accuracies are averaged over three runs. We also benchmark all methods with retraining (denoted as w/ RT).

Baselines and Evaluation Metrics. Baselines. For all experiments, we consider the S+P pipeline as one of our baselines, where the search method follows [8]; the pruning methods can be chosen from random pruning, magnitude pruning [11,14], and gradient pruning [19]. In addition, we also benchmark with hand-crafted DNNs, e.g., ShuffleNet [28,51] and MobiletNetV2 [32], and prior typical NAS resulting task-specific DNNs, e.g., MobileNetV3 [16] and AutoDeepLab [22]. We do not compare with NAS/tickets works with SOTA accuracy due to different goals and experimental settings. All baselines are benchmarked under similar FLOPs or accuracy for fair comparisons. Evaluation Metrics. We evaluate the SuperTickets and all baselines in terms of accuracy-efficiency trade-offs. Specifically, the accuracy metrics refer to top-1/5 accuracy for classification tasks; mIoU, mAcc, and aAcc for segmentation tasks; AP, AR, AP^M, and AP^L for human pose estimation tasks. For efficiency metrics, we evaluate and compare both the number of parameters and inference FLOPs.

4.2   Evaluating SuperTickets over Typical Baselines

4.2.1 SuperTickets on the Classification Task.

We show the overall comparisons between SuperTickets and some typical baselines in terms of accuracy-efficiency trade-offs in Fig. 4 and Table. 3, from which we have **two observations**. First, SuperTickets consistently outperform all baselines by reducing the inference FLOPs while achieving a comparable or even better accuracy. Specifically, SuperTickets reduce 61.4% ∼ 81.5% FLOPs while offering a comparable or

Table 3. SuperTickets vs. some typical methods on ImageNet. FLOPs is measured with the input size of 224×224.

Model	Params	FLOPs	Top-1 Acc.
CondenseNet [17]	2.9M	274M	71.0%
ShuffleNetV1 [51]	3.4M	292M	71.5%
ShuffleNetV2 [28]	3.5M	299M	72.6%
MobileNetV2 [32]	3.4M	300M	72.0%
FBNet [42]	4.5M	295M	74.1%
S+P (Grad.)	2.7M	114M	64.3%
S+P (Mag.)	2.7M	114M	72.8%
SuperTickets	2.7M	125M	**74.2%**

better accuracy (+0.1%∼+4.6%) as compared to both S+P and some task-specific DNNs; Likewise, when comparing under comparable number of param-

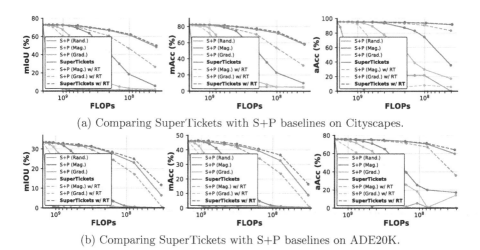

(a) Comparing SuperTickets with S+P baselines on Cityscapes.

(b) Comparing SuperTickets with S+P baselines on ADE20K.

Fig. 5. Comparing the mIoU, mAcc, aAcc and inference FLOPs of the proposed SuperTickets and S+P baselines on Cityscapes and ADE20K datasets. Each method has a series of points to represent different pruning ratios ranging from 10% to 98%.

eters or FLOPs, SuperTickets lead to on average 26.5% (up to 64.5%) and on average 41.3% (up to 71.9%) top-1 accuracy improvements as compared to S+P (Mag.) and S+P (Grad.) across various pruning ratios, e.g., under 50% pruning ratios, SuperTickets achieve 74.2% top-1 accuracy, +1.4% and +9.9% over S+P (Mag.) and S+P (Grad.), respectively. Second, SuperTickets w/o retraining even surpass S+P baselines with retraining as demonstrated in Fig. 4, leading to on average 6.7% (up to 29.2%) higher top-1 accuracy under comparable FLOPs across various pruning ratios (10% ~ 98%). Furthermore, SuperTickets w/ retraining achieve 0.1%~31.9% (on average 5.3%) higher accuracy than the counterparts w/o retraining, pushing forward the frontier of accuracy-efficiency trade-offs.

4.2.2 SuperTickets on the Segmentation Task.

Experiments on Cityscapes. We compare SuperTickets with typical baselines on Cityscapes as shown in Fig. 5(a) and Table 4. We see that SuperTickets consistently outperform all baselines in terms of mIoU/mAcc/aAcc and FLOPs. Specifically, SuperTickets reduce 60%~80.86% FLOPs while offering a comparable or better mIoU (0.28 %~43.26%) as compared to both S+P and task-specific DNNs; Likewise, when comparing under comparable number of parameters or FLOPs, SuperTickets lead to on average 17.70% (up to 42.86%)

Table 4. SuperTickets vs. some typical methods on Cityscapes. FLOPs is measured with the input size of 512×1024.

Model	Params	FLOPs	mIoU
BiSeNet [49]	5.8M	6.6G	69.00%
MobileNetV3 [16]	1.5M	2.5G	72.36%
ShuffleNetV2 [28]	3.0M	6.9G	71.30%
Auto-DeepLab [22]	3.2M	27.3G	71.21%
SqueezeNAS [33]	0.73M	8.4G	72.40%
S+P (Grad.) w/RT	0.63M	1.0G	60.66%
S+P (Mag.) w/RT	0.63M	1.0G	72.31%
SuperTickets	0.63M	1.0G	**72.68%**

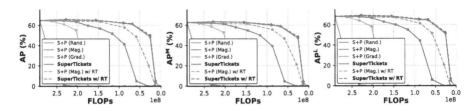

Fig. 6. Comparing the AP, AP^M, AP^L and inference FLOPs of the proposed SuperTickets and baselines on human pose estimation task and COCO keypoint dataset. Each method has a series of points for representing different pruning ratios ranging from 10% to 98%. All accuracies are averaged over three runs.

and 33.36% (up to 58.05%) mIoU improvements as compared to S+P (Mag.) and S+P (Grad.) across various pruning ratios, e.g., under 50% pruning ratios, SuperTickets achieve 72.68% mIoU, +0.37% and +12% over S+P (Mag.) and S+P (Grad.), respectively. We also report the comparison among methods after retraining at Fig. 5, as denoted by "w/ RT". We find that S+P (Grad.) w/ RT suffers from overfitting and even leads to worse performance; In contrast, SuperTickets w/ retraining further achieve 0.51%~1.64% higher accuracy than the counterparts w/o retraining, pushing forward the frontier of accuracy-efficiency trade-offs.

Experiments on ADE20K. Similarly, we test the superiority of SuperTickets on ADE20K as shown in Fig. 5(b) and Table 5. The proposed SuperTickets consistently outperform all baselines in terms of accuracy-efficiency trade-offs, reducing 38.46%~48.53% FLOPs when comparing under similar mIoU. When compared under com-

Table 5. SuperTickets vs. typical methods on ADE20K. FLOPs is measured with the input size of 512×512.

Model	Params	FLOPs	mIoU
MobileNetV2 [32]	2.2M	2.8G	32.04%
MobileNetV3 [16]	1.6M	1.3G	32.31%
S+P (Grad.)	1.0M	0.8G	24.14%
S+P (Mag.)	1.0M	0.8G	31.59%
SuperTickets	1.0M	0.8G	**32.54%**

parable number of parameters or FLOPs, SuperTickets lead to an average of 9.43% (up to 22.6%) and 14.17% (up to 27.61%) mIoU improvements as compared to S+P (Mag.) and S+P (Grad.), respectively, across various pruning ratios. In addition, SuperTickets w/ retraining further achieve 0.01% ~ 5.3% higher accuracy than the counterparts w/o retraining on ADE20K.

4.2.3 SuperTickets on the Human Pose Estimation Task.

We compare SuperTickets with a few typical baselines on COCO keypoint as shown in Fig. 6 and Table 6. We see that SuperTickets consistently outperform

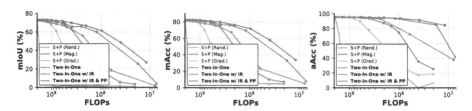

Fig. 7. Ablation studies of the SuperTickets identified from two-in-one framework w/ or w/o the proposed iterative activation (IR) and progressive pruning (PP) techniques.

all the related baselines in terms of AP/AP^M/AP^L/AR and FLOPs. Specifically, SuperTickets reduce 30.3%~78.1% FLOPs while offering a comparable or better AP (+0.8%~11.79%) as compared to both S+P and task-specific DNNs; Likewise, when comparing under comparable number of

Table 6. SuperTickets vs. typical algorithms on COCO. FLOPs is measured with the input size of 256×192.

Model	Params	FLOPs	AP	AP^M	AP^L	AR
ShuffleNetV1 [51]	1.0M	0.16G	58.5	55.2	64.6	65.1
ShuffleNetV2 [28]	1.3M	0.17G	59.8	56.5	66.2	66.4
MobileNetV2 [32]	2.3M	0.33G	64.6	61.0	71.1	70.7
S+P (Mag.)	0.6M	0.23G	63.4	61.2	66.8	67.3
SuperTickets	0.6M	0.23G	**65.4**	**63.4**	69.0	68.9

parameters or FLOPs, SuperTickets lead to on average 17.4% (up to 55.9%) AP improvements. In addition, SuperTickets w/ retraining further achieve on average 1.1% higher accuracy than the counterparts w/o retraining on COCO keypoint.

4.3 Ablation Studies of the Proposed SuperTickets

4.3.1 Ablation Studies of SuperTickets' Identification.

We provide comprehensive ablation studies to show the benefit breakdown of the proposed two-in-one training framework and more effective identification techniques, i.e., progressive pruning (PP) and iterative reactivation (IR). As shown in Fig. 7, we report the complete mIoU-FLOPs trade-offs with various pruning ratios ranging from 10% to 99% when testing on Cityscapes dataset, where x axis is represented by log-scale for emphasizing the improvements when pruning ratio reaches high. As compared to S+P (Mag.), SuperTickets identified from vanilla two-in-one framework achieve up to 40.17% FLOPs reductions when comparing under similar mIoU, or up to 13.72% accuracy improvements when comparing under similar FLOPs; Adopting IR during two-in-one training further leads to up to 68.32% FLOPs reductions or up to 39.12% mIoU improvements; On top of the above, adopting both IR and PP during two-in-one training offers up to 80.86% FLOPs reductions or up to 43.26% mIoU improvements. This set of experiments validate the effectiveness of the general two-in-one framework and each of the proposed techniques.

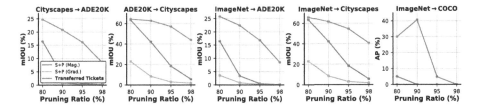

Fig. 8. Ablation studies of transferring identified SuperTickets from one dataset/task to another dataset/task under various pruning ratios ranging from 80% to 98%.

4.3.2 Ablation Studies of SuperTickets' Transferability.

We previously use one set of experiments under 90% sparsity in Sect. 3.3 to validate that the identified SuperTickets can transfer well. In this section, we supply more comprehensive ablation experiments under various pruning ratios and among several datasets/tasks. As shown in Fig. 8, the left two subplots indicate the transfer between different datasets (Cityscapes ↔ ADE20K) generally works across four pruning ratios. In particular, transferred SuperTickets lead to 76.14%~81.35% FLOPs reductions as compared to the most competitive S+P baseline, while offering comparable mIoU (0.27%~1.85%). Furthermore, the right three subplots validate that the identified SuperTickets from classification task can transfer well to other tasks (i.e., segmentation and human pose estimation). Specifically, it leads to 68.67%~69.43% FLOPs reductions as compared to the S+P (Mag.) baseline, when achieving comparable mIoU or AP.

5   Conclusion

In this paper, we advocate a two-in-one framework where both efficient DNN architectures and their lottery subnetworks (i.e., SuperTickets) can be identified from a supernet simultaneously, resulting in better performance than first-search-then-prune baselines. Also, we develop two techniques during supernet training to more effectively identify such SuperTickets, pushing forward the frontier of accuracy-efficiency trade-offs. Moreover, we test the transferability of SuperTickets to reveal their potential for being task-agnostic. Results on three tasks and four datasets consistently demonstrate the superiority of proposed two-in-one framework and the resulting SuperTickets, opening up a new perspective in searching and pruning for more accurate and efficient networks.

Acknowledgement. We would like to acknowledge the funding support from the NSF NeTS funding (Award number: 1801865) and NSF SCH funding (Award number: 1838873) for this project.

References

1. Baevski, A., Hsu, W.N., Xu, Q., Babu, A., Gu, J., Auli, M.: Data2vec: a general framework for self-supervised learning in speech, vision and language. arXiv preprint arXiv:2202.03555 (2022)

2. Bender, G., Kindermans, P.J., Zoph, B., Vasudevan, V., Le, Q.: Understanding and simplifying one-shot architecture search. In: International Conference on Machine Learning, pp. 550–559. PMLR (2018)

3. Cai, H., Gan, C., Wang, T., Zhang, Z., Han, S.: Once-for-all: train one network and specialize it for efficient deployment. arXiv preprint arXiv:1908.09791 (2019)

4. Chen, M., Peng, H., Fu, J., Ling, H.: AutoFormer: searching transformers for visual recognition. In: Proceedings of the IEEE/CVF International Conference on Computer Vision (ICCV) (2021)

5. Chen, X., Cheng, Y., Wang, S., Gan, Z., Wang, Z., Liu, J.: EarlyBERT: efficient BERT training via early-bird lottery tickets. arXiv preprint arXiv:2101.00063 (2020)

6. Cordts, M., et al.: The cityscapes dataset for semantic urban scene understanding. In: Proceedings of the IEEE Conference on Computer Vision and Pattern Recognition, pp. 3213–3223 (2016)

7. Deng, J., Dong, W., Socher, R., Li, L.J., Li, K., Fei-Fei, L.: ImageNet: a large-scale hierarchical image database. In: 2009 IEEE Conference on Computer Vision and Pattern Recognition, pp. 248–255. IEEE (2009)

8. Ding, M., et al.: HR-NAS: searching efficient high-resolution neural architectures with lightweight transformers. In: Proceedings of the IEEE/CVF Conference on Computer Vision and Pattern Recognition, pp. 2982–2992 (2021)

9. Ding, Y., et al.: NAP: neural architecture search with pruning. Neurocomputing **477**, 85–95 (2022)

10. Feng, Q., Xu, K., Li, Y., Sun, Y., Wang, D.: Edge-wise one-level global pruning on NAS generated networks. In: Ma, H., et al. (eds.) PRCV 2021. LNCS, vol. 13022, pp. 3–15. Springer, Cham (2021). https://doi.org/10.1007/978-3-030-88013-2_1

11. Frankle, J., Carbin, M.: The lottery ticket hypothesis: finding sparse, trainable neural networks. In: International Conference on Learning Representations (2019). https://openreview.net/forum?id=rJl-b3RcF7

12. Frankle, J., Dziugaite, G.K., Roy, D., Carbin, M.: Linear mode connectivity and the lottery ticket hypothesis. In: International Conference on Machine Learning, pp. 3259–3269. PMLR (2020)

13. Guo, Z., et al.: Single path one-shot neural architecture search with uniform sampling. In: Vedaldi, A., Bischof, H., Brox, T., Frahm, J.-M. (eds.) ECCV 2020. LNCS, vol. 12361, pp. 544–560. Springer, Cham (2020). https://doi.org/10.1007/978-3-030-58517-4_32

14. Han, S., Mao, H., Dally, W.J.: Deep compression: compressing deep neural networks with pruning, trained quantization and Huffman coding. arXiv preprint arXiv:1510.00149 (2015)

15. He, K., Zhang, X., Ren, S., Sun, J.: Deep residual learning for image recognition. In: Proceedings of the IEEE Conference on Computer Vision and Pattern Recognition, pp. 770–778 (2016). https://github.com/facebookarchive/fb.resnet.torch

16. Howard, A., et al.: Searching for mobilenetv3. In: Proceedings of the IEEE/CVF International Conference on Computer Vision, pp. 1314–1324 (2019)

17. Huang, G., Liu, S., Van der Maaten, L., Weinberger, K.Q.: CondenseNet: an efficient DenseNet using learned group convolutions. In: Proceedings of the IEEE Conference on Computer Vision and Pattern Recognition, pp. 2752–2761 (2018)

18. Ioffe, S., Szegedy, C.: Batch normalization: accelerating deep network training by reducing internal covariate shift. In: International Conference on Machine Learning, pp. 448–456. PMLR (2015)

19. Lee, N., Ajanthan, T., Torr, P.H.: Snip: single-shot network pruning based on connection sensitivity. arXiv preprint arXiv:1810.02340 (2018)

20. Li, Z., et al.: NPAS: a compiler-aware framework of unified network pruning and architecture search for beyond real-time mobile acceleration. In: Proceedings of the IEEE/CVF Conference on Computer Vision and Pattern Recognition, pp. 14255–14266 (2021)

21. Lin, T.-Y., et al.: Microsoft COCO: common objects in context. In: Fleet, D., Pajdla, T., Schiele, B., Tuytelaars, T. (eds.) ECCV 2014. LNCS, vol. 8693, pp. 740–755. Springer, Cham (2014). https://doi.org/10.1007/978-3-319-10602-1_48

22. Liu, C., et al.: Auto-Deeplab: hierarchical neural architecture search for semantic image segmentation. In: Proceedings of the IEEE/CVF conference on computer vision and pattern recognition, pp. 82–92 (2019)

23. Liu, H., Simonyan, K., Yang, Y.: Darts: differentiable architecture search. arXiv preprint arXiv:1806.09055 (2018)

24. Liu, Z., et al.: Swin transformer: hierarchical vision transformer using shifted windows. In: Proceedings of the IEEE/CVF International Conference on Computer Vision, pp. 10012–10022 (2021)

25. Liu, Z., Li, J., Shen, Z., Huang, G., Yan, S., Zhang, C.: Learning efficient convolutional networks through network slimming. In: Proceedings of the IEEE International Conference on Computer Vision, pp. 2736–2744 (2017)

26. Lu, J., Batra, D., Parikh, D., Lee, S.: ViLBERT: pretraining task-agnostic visiolinguistic representations for vision-and-language tasks. In: Advances in Neural Information Processing Systems, vol. 32 (2019)

27. Lu, J., Goswami, V., Rohrbach, M., Parikh, D., Lee, S.: 12-in-1: Multi-task vision and language representation learning. In: Proceedings of the IEEE/CVF Conference on Computer Vision and Pattern Recognition. pp. 10437–10446 (2020)

28. Ma, N., Zhang, X., Zheng, H.T., Sun, J.: ShuffleNet v2: practical guidelines for efficient CNN architecture design. In: Proceedings of the European Conference on Computer Vision (ECCV), pp. 116–131 (2018)

29. Mei, J., et al.: ATOMNAS: fine-grained end-to-end neural architecture search. arXiv preprint arXiv:1912.09640 (2019)

30. Kalibhat, N.M., Balaji, Y., Feizi, S.: Winning lottery tickets in deep generative models. arXiv e-prints pp. arXiv-2010 (2020)

31. Ramanujan, V., Wortsman, M., Kembhavi, A., Farhadi, A., Rastegari, M.: What's hidden in a randomly weighted neural network? In: Proceedings of the IEEE/CVF Conference on Computer Vision and Pattern Recognition, pp. 11893–11902 (2020)

32. Sandler, M., Howard, A., Zhu, M., Zhmoginov, A., Chen, L.C.: MobileNetv 2: inverted residuals and linear bottlenecks. In: Proceedings of the IEEE Conference on Computer Vision and Pattern Recognition, pp. 4510–4520 (2018)

33. Shaw, A., Hunter, D., Landola, F., Sidhu, S.: SqueezeNAS: fast neural architecture search for faster semantic segmentation. In: Proceedings of the IEEE/CVF International Conference on Computer Vision Workshops (2019)

34. Su, X., et al.: Vision transformer architecture search. arXiv preprint arXiv:2106.13700 (2021)

35. Tan, M., et al.: MnasNet: platform-aware neural architecture search for mobile. In: Proceedings of the IEEE/CVF Conference on Computer Vision and Pattern Recognition, pp. 2820–2828 (2019)

36. Tan, M., Le, Q.: EfficientNet: rethinking model scaling for convolutional neural networks. In: International Conference on Machine Learning, pp. 6105–6114. PMLR (2019)

37. Tan, M., Le, Q.V.: MixConv: mixed depthwise convolutional kernels. arXiv preprint arXiv:1907.09595 (2019)

38. Wan, A., et al.: FBNetV2: differentiable neural architecture search for spatial and channel dimensions. In: Proceedings of the IEEE/CVF Conference on Computer Vision and Pattern Recognition, pp. 12965–12974 (2020)

39. Wang, D., Gong, C., Li, M., Liu, Q., Chandra, V.: AlphaNet: improved training of superNet with alpha-divergence. arXiv preprint arXiv:2102.07954 (2021)

40. Wang, H., et al.: Hat: hardware-aware transformers for efficient natural language processing. arXiv preprint arXiv:2005.14187 (2020)

41. Wang, J., et al.: Deep high-resolution representation learning for visual recognition. IEEE Trans. Pattern Anal. Mach. Intell. **43**(10), 3349–3364 (2020)

42. Wu, B., et al.: FBNet: hardware-aware efficient convnet design via differentiable neural architecture search. In: Proceedings of the IEEE/CVF Conference on Computer Vision and Pattern Recognition, pp. 10734–10742 (2019)

43. Wu, B., Li, C., Zhang, H., Dai, X., Zhang, P., Yu, M., Wang, J., Lin, Y., Vajda, P.: Fbnetv5: Neural architecture search for multiple tasks in one run. arXiv preprint arXiv:2111.10007 (2021)

44. Xu, J., et al.: NAS-BERT: task-agnostic and adaptive-size BERT compression with neural architecture search. In: Proceedings of the 27th ACM SIGKDD Conference on Knowledge Discovery & Data Mining, pp. 1933–1943 (2021)

45. Yang, Y., You, S., Li, H., Wang, F., Qian, C., Lin, Z.: Towards improving the consistency, efficiency, and flexibility of differentiable neural architecture search. In: Proceedings of the IEEE/CVF Conference on Computer Vision and Pattern Recognition (CVPR), pp. 6667–6676 (2021)

46. You, F., Li, J., Zhao, Z.: Test-time batch statistics calibration for covariate shift. arXiv preprint arXiv:2110.04065 (2021)

47. You, H., et al.: Drawing early-bird tickets: toward more efficient training of deep networks. In: International Conference on Learning Representations (2020). https://openreview.net/forum?id=BJxsrgStvr

48. You, H., Lu, Z., Zhou, Z., Fu, Y., Lin, Y.: Early-bird GCNs: graph-network co-optimization towards more efficient GCN training and inference via drawing early-bird lottery tickets. In: Association for the Advancement of Artificial Intelligence (2022)

49. Yu, C., Wang, J., Peng, C., Gao, C., Yu, G., Sang, N.: BiSeNet: bilateral segmentation network for real-time semantic segmentation. In: Proceedings of the European Conference on Computer Vision (ECCV), pp. 325–341 (2018)

50. Yu, J., et al.: BigNAS: scaling up neural architecture search with big single-stage models. In: Vedaldi, A., Bischof, H., Brox, T., Frahm, J.-M. (eds.) ECCV 2020. LNCS, vol. 12352, pp. 702–717. Springer, Cham (2020). https://doi.org/10.1007/978-3-030-58571-6_41

51. Zhang, X., Zhou, X., Lin, M., Sun, J.: ShuffleNet: an extremely efficient convolutional neural network for mobile devices. In: Proceedings of the IEEE Conference on Computer Vision and Pattern Recognition, pp. 6848–6856 (2018)

52. Zhang, Z., Chen, X., Chen, T., Wang, Z.: Efficient lottery ticket finding: less data is more. In: International Conference on Machine Learning, pp. 12380–12390. PMLR (2021)

53. Zhou, B., Zhao, H., Puig, X., Fidler, S., Barriuso, A., Torralba, A.: Scene parsing through ade20k dataset. In: Proceedings of the IEEE Conference on Computer Vision and Pattern Recognition, pp. 633–641 (2017)

54. Zoph, B., Vasudevan, V., Shlens, J., Le, Q.V.: Learning transferable architectures for scalable image recognition. In: Proceedings of the IEEE Conference on Computer Vision and Pattern Recognition, pp. 8697–8710 (2018)

Meta-GF: Training Dynamic-Depth Neural Networks Harmoniously

Yi Sun, Jian Li, and Xin Xu[✉]

The College of Intelligence Science and Technology,
National University of Defense Technology, Changsha 410000, China
{sunyi13,lijian,xinxu}@nudt.edu.cn

Abstract. Most state-of-the-art deep neural networks use static inference graphs, which makes it impossible for such networks to dynamically adjust the depth or width of the network according to the complexity of the input data. Different from these static models, depth-adaptive neural networks, e.g. the multi-exit networks, aim at improving the computation efficiency by conducting adaptive inference conditioned on the input. To achieve adaptive inference, multiple output exits are attached at different depths of the multi-exit networks. Unfortunately, these exits usually interfere with each other in the training stage. The interference would reduce performance of the models and cause negative influences on the convergence speed. To address this problem, we investigate the gradient conflict of these multi-exit networks, and propose a novel meta-learning based training paradigm namely **Meta-GF** (meta gradient fusion) to harmoniously train these exits. Different from existing approaches, Meta-GF takes account of the importances of the shared parameters to each exit, and fuses the gradients of each exit by the meta-learned weights. Experimental results on CIFAR and ImageNet verify the effectiveness of the proposed method. Furthermore, the proposed Meta-GF requires no modification on the network structures and can be directly combined with previous training techniques. The code is available at https://github.com/SYVAE/MetaGF.

Keywords: Dynamic networks · Gradient conflict · Meta learning

1 Introduction

Deep neural networks have achieved tremendous progress in many applications such as recognition [8,18] and detection [3,13]. Due to the implicit regularization caused by over-parameterization [9,16], the very deep networks with large number of parameters empirically have stronger representation capacity and more

---

Y. Sun and J. Li—Co-first authors with equal contributions.

Supplementary Information The online version contains supplementary material available at https://doi.org/10.1007/978-3-031-20083-0_41.

robust generalization ability compared with the small ones. Yet they are always much more computationally expensive, especially when deployed on resource-constrained platforms. Furthermore, due to the varying complexity of input data, hence not every input requires the same network depth. To realize better accuracy with efficient inference speed, network pruning approaches [37] and various lightweight networks [41, 45] are proposed for obtaining lightweight models. However, the capacity of the resulting small networks is limited, and they cannot dynamically extend their depth or width for dealing with more complicated input. Both the large networks and the small ones perform static inference graphs, which limits the efficiency of the large models and the representation ability of the small models respectively. To sum up, the large networks cannot dynamically reduce their computation complexity to efficiently deal with easy inputs, and the capacity of the small ones are limited which cannot be dynamically extended to handle challenging inputs.

Recently, numerous researches [20, 25, 26, 39, 47–49] propose solutions to achieving robust and efficient prediction via implementing depth-adaptive inference, such as the MSDnet [20] and SDN [26]. The inference depth of the above mentioned adaptive networks are conditioned on the input. As the most popular adaptive architecture, multi-exit networks [20, 25, 39] realize depth-adaptive inference by early exiting, i.e. allowing the result of "easy" inputs to be output from the shallow exits without executing deeper layers [17]. Toward this end, multiple intermediate output exits are attached to the networks at different depths. In the inference stage, the multi-exit models dynamically decide to stop inference at which output exit based on the predefined exit-policy such as the confidence [20] or learned policy [4]. Depth-adaptive inference is a valuable mechanism for networks to save unnecessary computation costs while keeping prediction performance.

Existing works mainly focus on designing more excellent adaptive structures [23, 26, 39, 54] or better inference algorithms [4], but the interference between different intermediate exits have attracted less attention. To be specific, large numbers of model parameters are shared by these exits, and in the training stage, these shared parameters always receive conflicted gradients from different exits. As defined in [55], "gradients conflict" in this work represents that two gradients have a negative cosine similarity value i.e. the conflicted update directions. The interference between different exits degrades the overall performance and the convergence speed of the multi-exit networks.

To alleviate such a interference, specially designed networks such as the MSDnet [20] were proposed. In [20], they added dense connections between the deeper exits and the early modules of the networks to reduces the negative impact of the early exits on the deeper exits. Knowledge-distillation based approaches [40, 49] are also proposed to align the learning targets of shallow exits with the deeper exits. Another kind of solution is performed by gradient adjustment. The Gradient Equilibrium(GE) proposed in [29] reduces the gradient variance of the deeper exits, which is useful for reducing the negative impact of the deep exits on shallow exits. In the works of the [55] and [48], when two gradients conflict with each other, they project each gradient onto the normal plane of the other for suppressing the interfering components of the gradients. This kind of gradient surgery method is termed Pcgrad, and it was verified in [48,55] that Pcgrad can reduce the conflict between different tasks.

Despite the effectiveness of existing approaches, there's room for improvement. Firstly, approaches based on knowledge-distillation and network architecture designs conduct less necessary analysis about the interference at gradient level, and it's considered that gradient directly participates in the model optimization. Secondly, GE controls the gradient variance but doesn't take account of adjusting gradient direction. However, gradient direction decides the update trend of the networks, and the gradient direction conflict is one of the essence of the interference. Thirdly, we find that not all shared parameters are equally important to each exit. The over-parameterization of networks have adequate capacity to allow different exits to own their preferred parameters. Unconstrained gradient re-projection policy might hinder the convergence of the networks. To be specific, when gradients of two exits conflict with each other, the re-projection policy is supposed to take account of the importances of the shared parameters for different exits, instead of simply suppressing the interfering components in the two conflicted gradients.

To tackle the above mentioned issues, we propose a novel gradient fusion method named Meta-GF for training the multi-exit networks. In contrast to the previous approaches, the proposed Meta-GF takes a meta-learned weighted fusion policy to combine the gradients of each exit, while taking account of the different importances of the shared parameters for different exits. Due to the over-parameterization, there are adequate model capacity, which makes the Meta-GF could implicitly disentangle the shared parameters of different tasks as more as possible, e.g. finding exit-specific parameters for each exit by the learned fusion weights. By fusing the gradient of different exits with the meta-weight fusion policy, the proposed approach achieves more harmonious training of the multi-exit networks. Extensive experiments have been conducted on CIFAR and ImageNet datasets. The experimental results demonstrate the effectiveness of the proposed approach. We investigate the gradient conflict problem and introduce Meta-GF in Sect. 3. The experimental results are introduced in Sect. 4.

2   Related Works

2.1   Dynamic Deep Neural Networks

Dynamic deep neural network is a promising research field, where the networks take conditional computation by using adaptive parameters [5,31,36,53,56], networks width or networks depth [23,25,26,39]. Specially, the depth-adaptive deep neural networks aim at achieving trade-off between the robustness and efficiency by dynamically adjusting the network inference depth. To be specific, the networks conditionally adjust their inference depth according to the complexity of inputs. There are mainly two kinds of depth-adaptive neural network structures: the multi-exit networks and the skip-style networks.

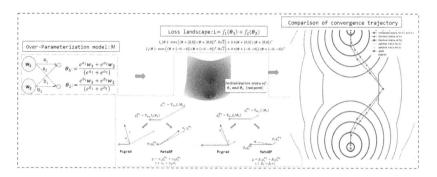

Fig. 1. Meta-Gradient Fusion: given an over-parameterization networks, of which two task exit:$\{\boldsymbol{\theta_1}, \boldsymbol{\theta_2}\}$ are linearly combined by six learnable parameters: $\{\mathbf{w}_1, \mathbf{w}_2, (a_1, a_2), (b_1, b_2)\}$. The total loss surface of both two tasks is $f_1(\boldsymbol{\theta_1}) + f_2(\boldsymbol{\theta_2})$, which is shown in the middle subfigure. When jointly training the two tasks, the gradients of two tasks conflict with each other, i.e. $g_1^{\mathbf{w}_1} \cdot g_2^{\mathbf{w}_1} < 0$ and $g_1^{\mathbf{w}_2} \cdot g_2^{\mathbf{w}_2} < 0$. The proposed Meta-GF takes a meta-weight fusion policy to combine the expected gradients of two tasks, which takes account of the different importances of the shared parameters for different exits. Due to the over-parameterization, there are adequate model capacity, which makes the Meta-GF could disentangle the shared parameters of different tasks as more as possible, and achieve harmonious joint training process. In the right subfigure, the model convergence trajectory when using the Meta-GF are more close to the trajectories when training the two tasks independently. It indicates the Meta-GF achieves more harmoniously training, and verifies the effectiveness of Meta-GF.

Multi-exit Networks. Multi-exit structure is commonly adopted to construct inference depth-adaptive networks. The most intuitive approach to designing multi-exit networks are attaching different output exits at different depth of the model. [23,25,26,39]. By designing early-exiting policy such as the confidence-based criterion [20,26,42,50,52] or the learned policy networks [19,19,25] for evaluating the complexity of inputs, the multi-exit networks can adaptively select the output exits and thus adjust the inference depth according to the input complexity. In [57], they force the networks to exit when the predicted confidence score doesn't change for a predefined depth. Compared with the confidence-based criterion, the learned policy networks trained by reinforcement learning [2,7,19,25] or variational Bayes optimization [4] have better expandability that can be transferred to other tasks such as the object tracking [19]. Besides, Liu et al. [35] propose an approach to estimate the complexity of the input according to the reconstruction loss of the data. To make the shallow module of the multi-exit networks can obtain multi-scale receptive fields, the MSDnet proposed in [20] adopts parallel multi-resolution calculation. Recently, the transformer-style networks have shown powerful and robust representation ability, and it's also can be modified into multi-exit style [1,11] to improve the efficiency of the transformer-style networks.

However, different exits might interfere with each other. Specifically speaking, the model parameters shared by these intermediate exits always receive conflicted update gradients from different exits in the training stage. To address this interference, some recent works [32, 40, 49] proposed training algorithms based on knowledge distillation to align the learning objective of each exit. Differently, the gradient-surgery approaches such as the Pcgrad [48, 55] reduce the conflict between different exits by performing gradient re-projection policy. Gradient equilibrium [29] or weighted-loss [10] is applied to adjust the gradient norm of each exit for optimizing the training progress.

Skip-Style Networks. In addition to the multi-exit networks, the skip-style networks dynamically adjust the inference depth of the networks by adaptively skip the non-linear neural modules. Early works like the stochastic depth networks [21] randomly skip the residual modules in ResNet [18] in the training stage. In [12], Angela et al. applied a learned drop-out rate to randomly skip transformer layers in the testing stage. To achieve more controllable depth-adaptive inference in the testing stage, the networks proposed in [24, 46, 47, 51] set series of gating modules in the position of skip-connections. The Skipnet [47] and Block-drop [51] adopt reinforcement learning to make the gating modules learn discrete decision strategies, i.e. "skip" or "not skip". Instead of skipping the whole layer, the fractional skipping policies proposed in [28, 30, 44] dynamically select part of the layer channels to execute, which can be also regarded as dynamic channel pruning approaches [14]. In this work, we mainly focus on the multi-exit depth adaptive networks.

2.2 Gradient De-Conflict in Multi-task Learning

The gradient conflict problems also exist in the regime of multi-task learning, because different task heads shared the models. If the gradients of different task objectives are not well aligned [33], the average gradients would not provide a well convergence direction for the multi-task networks. To tackle this issue, different gradient adjustment approaches were proposed [6, 15, 22, 33, 34, 43]. Magnitude-rescaling algorithm such as the GradNorm [6] or IMTL [34], is one kind of the gradient adjustment strategies, which aims at balancing the gradient magnitudes of different tasks. In [15], they propose an adaptive loss-weighting policy to prioritize more difficult tasks. The MGDA-UB [43] and CAGrad [33] optimize the overall objective of the multi-task models to find a Pareto optimal solution. In order to reduce the level of gradient conflict, the PCgrad [55] projects each conflict gradient onto the normal plane of the other for suppressing the interfering components. These methods mentioned above all concentrate on solving the task conflicts on the shared parts of the multi-task models. In contrast to these gradient-adjustment approaches, the proposed Meta-GF takes a meta-learned weighted fusion policy to combine the gradients of each exit, while taking account of the different importances of the shared parameters for different exits.

3   Method

In this section, we firstly investigate the gradient conflict by using a toy experiment as shown in Fig. 1. Then we introduce details of the proposed Meta-GF.

3.1   Gradient Conflict

Without loss of generality, we make analysis of the gradient conflict problems on the two-exit over-parameterization networks: $M = \{\boldsymbol{\theta}_1, \boldsymbol{\theta}_2\}$. The $\{\boldsymbol{\theta}_1, \boldsymbol{\theta}_2\}$ are the parameters of the two exits respectively. Defining the shared parameters are $\{\mathbf{w}_1, \mathbf{w}_2\}$, and the objective function of the two exits are $\{f_1(\boldsymbol{\theta}), f_2(\boldsymbol{\theta})\}$ as shown in Fig. 1. We initialize the model M with $\{\mathbf{w}_1, \mathbf{w}_2, a_1, a_2, b_1, b_2\} = \{(5, 1.5), (5, 0.5), 1, 1, 1, 1\}$. **We take the gradient conflict on $\mathbf{w}_1$ as example** when $\nabla_{\mathbf{w}_1} f_1 \cdot \nabla_{\mathbf{w}_1} f_2 < 0$. The loss degradation of f_1 can be approximately by the First-order Taylor expansion as shown in Eq. (1) when the learning rate ϵ is small:

$$\begin{aligned} \Delta f_1^{g_1+g_2} &\approx -\epsilon \left(g_1^2 + g_1 g_2\right) + o(\epsilon^2), \\ &\leq -\epsilon \left(g_1^2\right) + o(\epsilon^2) \approx \Delta f_1^{g_1} \end{aligned}, \qquad g_1 \cdot g_2 < 0, \tag{1}$$

where (g_1, g_2) denote $(\nabla_{\mathbf{w}_1} f_1, \nabla_{\mathbf{w}_1} f_2)$, and ϵ is the low learning rate ($\epsilon = 0.001$). Obviously, the convergence of f_1 is negatively influenced by the gradient of exit-2: g_2. The total loss degradation when updating $\mathbf{w}_1$ is calculated as follows:

$$\Delta L = \Delta f_1^{g_1+g_2} + \Delta f_2^{g_1+g_2} \approx -\epsilon \left(g_1^2 + g_2^2 + 2 g_1 g_2\right) + o(\epsilon^2). \tag{2}$$

The gradient de-conflict approach Pcgrad proposed in [55] projects each conflict gradient onto the normal plane of the other for suppressing the interfering components, which improves the performance of multi-task networks. The gradient re-projection can be formulated as:

$$\hat{g}_1 = (g_1 - \frac{g_1 g_2}{\|g_2\|^2} g_2), \hat{g}_2 = (g_2 - \frac{g_1 g_2}{\|g_1\|^2} g_1). \tag{3}$$

The loss degradation ΔL by re-projecting both two gradients is:

$$\Delta L_{g_2 \rightleftharpoons g_1} = -\epsilon \left(g_1^2 + g_2^2 - \frac{(g_1 g_2)^2}{\|g_1\|^2} - \frac{(g_1 g_2)^2}{\|g_2\|^2} + 2 g_1 g_2 ((cos\alpha)^2 - 1) \right) + o(\epsilon^2). \tag{4}$$

The $(cos\alpha)$ is the cosine similarity between g_1 and $g_2((cos\alpha) < 0)$.

If we only re-project the gradient g_1, then the total loss degradation $\Delta L_{g_1 \rightarrow g_2}$ is:

$$\Delta L_{g_1 \rightarrow g_2} = -\epsilon \left(g_1^2 + g_2^2 - \frac{(g_1 g_2)^2}{\|g_2\|^2} \right) + o(\epsilon^2). \tag{5}$$

Similarly, when we only re-project the g_2:

$$\Delta L_{g_2 \rightarrow g_1} = -\epsilon \left(g_1^2 + g_2^2 - \frac{(g_1 g_2)^2}{\|g_1\|^2} \right) + o(\epsilon^2). \tag{6}$$

It's obvious that $\Delta L_{g_2 \leftrightharpoons g_1}$ isn't always larger than $\Delta L_{g_1 \to g_2}$ or $\Delta L_{g_2 \to g_1}$. Specifically, assuming $\|g_1\| > \|g_2\|$, the following inequation only holds when the norms of two gradients have a limited difference (please refer to ??):

$$\Delta L_{g_2 \leftrightharpoons g_1} > \Delta L_{g_2 \to g_1}, \qquad \text{when } \frac{\|g_2\|}{\|g_1\|} > \frac{-0.5(cos\alpha)}{(sin\alpha)^2}. \qquad (7)$$

Otherwise, re-projecting the small one gradient g_2 is better than adjusting both gradients. This analysis indicates that the shared parameters are sometimes not actually "shared" equally, i.e. the importances of the shared parameters for different exits are different. In conclusion, it would be better to take account of the importances of the shared parameters for different exits when jointly training multi-exit networks. The toy experiment in Fig. 1 verifies this assumption.

3.2   Meta Weighted Gradient Fusion

Assuming $\|g_1\| > \|g_2\|$ and $g_1 \cdot g_2 < 0$, our previous analysis indicates that the optimal gradient fusion policy for $\mathbf{w}_1$ is:

$$g_f = \begin{cases} (1 - \frac{\|g_1\|}{\|g_2\|}cos\alpha)g_1 + (1 - \frac{\|g_2\|}{\|g_1\|}cos\alpha)g_2, & \text{if } \frac{\|g_2\|}{\|g_1\|} > \frac{-0.5(cos\alpha)}{(sin\alpha)^2} \\ (1 - \frac{\|g_1\|}{\|g_2\|}cos\alpha)g_1 + g_2, & \text{otherwise.} \end{cases} \qquad (8)$$

It can be seen in Eq. (8) that the larger gradient is always enhanced, which illustrates the preference of the networks to the dominant gradient in fusing gradients of different exits. The above mentioned analysis suggests that it's an potential solution to gradient conflict by weighting the gradients from different exits according to their importances. Hence, we further describe the fusion policy in a weighted-fusion policy:

$$g_f = \eta_1 g_1 + \eta_2 g_2, \qquad \eta_1 > \eta_2 > 0. \qquad (9)$$

The Meta-GF Algorithm. Considering there are always millions of parameters in nowadays deep neural networks, it's computationally expensive to calculate the inner production between gradients and their norms. Inspired by the meta-learning approaches, we instead learn the fusion weights of gradients in a data-driven manner. The number of exits is referred to as n, and the parameter set of the networks is referred to as $\mathbf{W}$. For a given shared parameters $\mathbf{w} \in \mathbf{W}$, the fusion gradient is:

$$g_f^{\mathbf{w}} = \frac{\sum_{i=1}^n e^{\eta_i^{\mathbf{w}}} g_i^{\mathbf{w}}}{\sum_{i=1}^n e^{\eta_i^{\mathbf{w}}}}. \qquad (10)$$

The $\eta = \{\eta_1^{\mathbf{w}}, ..., \eta_n^{\mathbf{w}} | \mathbf{w} \in \mathbf{W}\}$ are the learnable fusion weights, which are dynamically optimized to minimize the cost of the objective functions F:

$$\eta = \underset{\eta}{\arg\min} \, F(\mathbf{W} - \epsilon \frac{\sum_{i=1}^n e^{\eta_i} g_i}{\sum_{i=1}^n e^{\eta_i}}). \qquad (11)$$

The default values of the fusion weights are all set to 1 in the initialization stage. The objective function F is the cross-entropy loss function because we verify the proposed approach on the multi-exit classification network: MSDnet [20] and SDN [26].

It's also important to reduce the noises in the gradients of each exit. In the current mini-batch training settings, the gradient variance in each mini-batch should be considered. The uncertainty of each gradient will cause negative influence on the gradient de-conflict performance. There exists an easy method to estimate the expected gradient of each exit. Inspired by the Reptile algorithm [38], the expected gradient Eg of each task is obtained by training the task independently for one epoch, where the independent training process share the same initial model $\mathbf{W}_0$. This method can be formulated as:

$$\mathbf{W}_{task} = \arg\min_{\mathbf{W}} F_{task}(\mathbf{W}; \mathbf{W}_0, D), \tag{12}$$

$$Eg_{task} = \mathbf{W}_{task} - \mathbf{W}_0, \tag{13}$$

where D is the training set and F_{task} is the objective function. The details of the proposed Meta-GF approaches are illustrated in Algorithm 1.

Algorithm 1. Meta Weighted Gradient Fusion:

Input: Initial parameters:$\mathbf{W}_0$, training dataset:D, learning rate:ϵ, fusion weight:η. The number of exits is n. $F = \{F_1, ..., F_n\}$ is the objective function of the exits.

Output: W

1: **while** $i < $ MaxIter **do**
2:     ▼ *1. Calculating expected gradients:*
3:     **for** j=1,...,n **do**
4:         $\hat{\mathbf{W}} = \mathbf{W}_0$
5:         $\mathbf{W}_j = \arg\min_{\mathbf{W}} F_j(\mathbf{W}; \hat{\mathbf{W}}, D)$
6:         $g_j = \mathbf{W}_j - \hat{\mathbf{W}}$,
7:     **end for**
8:     ▼ *2. Meta Weighted Gradient Fusion:*
9:     $\eta = \arg\min_{\eta} F(\mathbf{W}_0 - \epsilon \frac{\sum_{i=1}^n e^{\eta_i} g_i}{\sum_{i=1}^n e^{\eta_i}}; D)$.
10:     $\mathbf{W} = \mathbf{W}_0 - \epsilon \frac{\sum_{i=1}^n e^{\eta_i} g_i}{\sum_{i=1}^n e^{\eta_i}}$
11:     $\mathbf{W}_0 = \mathbf{W}$
12: **end while**
13: **return** Y

---

4   Experiments

To verify the effectiveness of the proposed Meta-GF, we conduct extensive experiments on the representative image classification dataset CIFAR [27] and ILSVRC 2012(ImageNet). Besides, we make detailed analysis of the proposed meta-fusion method. For fair comparison, all of the methods for comparison in this work use the same multi-exit networks: MSDnet [20] and SDN [26].

Datasets. CIFAR100 and CIFAR10 both contain 60000 RGB images of size 32×32. 50000 of them are applied for training and 10000 for test in the two datasets. The images in CIFAR10 and CIFAR100 are corresponding to 10 classes and 100 classes respectively. We adopt the same data augmentation policy as introduced in [29], which includes random crop, random flip and data normalization. We select 5000 of the training sets on CIFAR100 and CIFAR10 respectively for validation. The ImageNet dataset contains 1000 classes, where the input size of images is set to 224×224. The training set have 1.2 million images and we select 50000 of them for validation, where the public validation set of the ImageNet is referred to as the test set in this work because the true test set has not been made public.

Implementation Details. We optimize all models by using stochastic gradient descent with batch size of 64 on CIFAR and 512 on ImageNet. The momentum weight and weight decay are set to 0.9 and 10^{-4} respectively. We train the MSDnet for $maxiter = 300$ epochs on both CIFAR datasets and for $maxiter = 90$ epochs on ImageNet. For the SDN, the maximum epoch is set to 100 on CIFAR datasets. The adjustment of the learning rate is achieved by multi-step policy, where we divide the learning rate by a factor of 10 after $0.5 \times maxiter$ and $0.75 \times maxiter$ epochs. The initial learning rate is 0.1.

The fusion weights are all initialized as 1. Those fusion weights are optimized by Adam optimizer, of which initial learning rate is set to 10^{-1}, and we take the same multi-step policy as above to adjust the learning rate. During an training epoch, we first train each exit independently for estimating the expected updating directions. However, we find that independently training each exit without optimizing other exits might cause negative influence on the Batch Normalization layers of the deeper exits. Therefore, except for the current selected exit, we actually train other exits with a very small learning rate. Then we train the fusion weights with the training sets for one iteration. Finally, we merge the expected gradients of each task with the proposed Meta-GF.

Compared Methods. We compare the proposed Meta-GF with four representative approaches. The Pcgrad [55] and Cagrad [33] are proposed for the multi-task learning problem, and we apply them to the multi-exit neural networks.

- MSDnet [20]/SDN [26]. In this work, the proposed method and other compared training methods all adopt the MSDNet/SDN as the network structure. It serves as a baseline in the experiments. It takes the SGD as the optimizer for training.
- Gradient Equilibrium (GE) [29]. It rescales the magnitude of gradients along its backward propagation path. It helps to reduce gradient variance and stabilize the training procedure.
- Pcgrad [55]. When two gradient conflict with each other, it projects each gradient onto the normal plane of the other for suppressing the interfering components of the gradients.
- Cagrad [33]. To regularize the algorithm trajectory, it looks for an gradient fusion policy that maximizes the worst local improvement of any objective

in the neighborhood of the average gradient. Different from Pcgrad, it aims at forcing the models to converge to a Pareto set of the joint objectives. As same as Pcgrad, we implement the Cagrad on the MSDnet and the SDN to adjust the direction of gradients.

On the CIFAR datasets, we set the exit number of the MSDnet to 7. The depth of 7 exits are $\{4, 6, 8, 10, 12, 14, 16\}$ respectively. The input size of the image is 32×32. We train the MSDnet with three-scale features, i.e., 32×32, 16×16 and 8×8. On the ImageNet, there are 5 exits in the MSDnet, which are respectively inserted at the depth of $\{4, 8, 12, 16, 20\}$. The input size on ImageNet datasets is 224×224, and we use four-scale MSDnet, i.e. the multi-scale feature maps are $\{56 \times 56, 28 \times 28, 14 \times 147 \times 7\}$. All of the compared methods use the same MSDnet architectures on the CIFAR and ImageNet. For the SDN-style networks, we take the same training settings as described in [26], and conduct experiments on the Resnet-SDN and Vgg-SDN.

4.1   Prediction Accuracy of Each Exit

In the anytime prediction setting [20], the model maintains a progressively updated distribution over classes, and it can be forced to output its most up-to-date prediction at an arbitrary time [19]. Therefore, the prediction accuracy of different exits is much significant for the performance of anytime prediction.

As illustrated in Table 1, we compare the prediction top-1 accuracy of the MSDnet when using different training approaches on CIFAR. On CIFAR100, the previous gradient adjustment approaches: GE, Cagrad and Pcgrad perform better than the baseline model, which indicates that the adjustment of gradient can effectively alleviate the conflicts between different exits. Especially, Cagrad performs better than other methods at shallow exits which demonstrates that Cagrad successfully maximizes the worst local improvement of any objective in the neighborhood of the average gradient. But it hurts the performance of the deeper exits.

Table 1. Classification accuracy of individual classifiers in multi-exit MSDnet on CIFAR-100 and CIFAR10.

	Params(M)	Flops(M)	CIFAR100					CIFAR10				
			MSDnet	GE	Cagrad	Pcgrad	Ours	MSDnet	GE	Cagrad	Pcgrad	Ours
Exit-1	0.90	56.43	66.41	67.74	**68.78**	67.06	67.97	91.13	92.02	92.19	91.66	**92.38**
Exit-2	1.84	101.00	70.48	71.87	**72.55**	71.37	72.27	92.91	93.53	93.49	93.59	**94.22**
Exit-3	2.80	155.31	73.25	73.81	74.23	74.86	**75.06**	93.98	94.14	94.47	94.32	**94.49**
Exit-4	3.76	198.10	74.02	75.13	74.97	**75.78**	75.77	94.46	94.49	94.45	94.60	**94.96**
Exit-5	4.92	249.53	74.87	75.86	75.35	76.25	**76.38**	94.68	94.73	94.48	94.81	**94.82**
Exit-6	6.10	298.05	75.33	76.23	75.82	76.95	**77.11**	94.78	94.89	94.53	94.83	**94.97**
Exit-7	7.36	340.64	75.42	75.98	76.08	76.71	**77.47**	94.64	94.96	94.48	94.82	**94.97**
Average	-	-	72.83	73.80	73.96	74.14	**74.57**	93.80	94.11	94.01	94.09	**94.54**

Different from the Pcgrad and Cagrad, which treat all the shared parameters as a whole and manipulate the gradients, the Meta-GF aims at softly weighting the gradients of each shared parameter by considering its importance for each exit. As shown in Table 1, despite that our approach doesn't always achieve the highest score at each exit, the proposed Meta-GF enables the multi-exit MSDnet obtain the best overall accuracy on both the CIFAR10 and CIFAR100 datasets.

To further verify the effectiveness of the proposed Meta-GF, we compare the existing approaches and our method on the ImageNet. The MSDnet trained on the ImageNet have 5 exits. As shown in Table 2, the overall performance of the proposed Meta-GF outperforms GE, Pcgrad and the baseline. The shallow exits of the baseline achieve the best prediction accuracy, but interfere the performance of the deeper exits.

Table 2. Classification accuracy of individual classifiers on ImageNet.

	Params(M)	Flops(M)	ImageNet			
			MSDnet	GE	Pcgrad	Ours
Exit-1	4.24	339.90	**58.48**	57.75	57.62	57.43
Exit-2	8.77	685.46	**65.96**	65.54	64.87	64.82
Exit-3	13.07	1008.16	68.66	**69.24**	68.93	69.08
Exit-4	16.75	1254.47	69.48	70.27	71.05	**71.67**
Exit-5	23.96	1360.53	71.03	71.89	72.45	**73.27**
Average	-	-	66.72	66.94	66.98	**67.25**

We also conduct experiments on two shallow depth networks proposed in [26], i.e. the Resnet-SDN and the Vgg-SDN. We take the same training settings as described in [26]. The results shown in Table 3 and Table 4 demonstrate that the proposed Meta-GF not only works in the well-designed multi-exit networks–MSDnet, but can also improve the performance of other multi-exit networks. It's worthy noting that we regarded all the filters of each layer as one parameter in the MSDNet and Resnet-SDN, however in the Vgg-SDN, we regard the filters belong to each output channel as a parameter. Because the structure of Vgg is flat-style without skip-connection, if the middle-layers become very task-specific, the performance of the deeper exits will inevitably influenced.

Table 3. Classification accuracy of individual classifiers in multi-exit Vgg-SDN [26] on CIFAR-100 and CIFAR10.

	Params(M)	Flops(M)	CIFAR100					CIFAR10				
			SDN-vgg	GE	Cagrad	Pcgrad	Ours	SDN-vgg	GE	Cagrad	Pcgrad	Ours
Exit-1	0.05	39.76	44.42	44.46	**53.08**	43.59	49.91	69.03	68.97	**76.27**	67.41	74.92
Exit-2	0.29	96.52	61.08	61.0	61.39	**63.02**	61.09	84.72	84.52	86.3	85.28	**88.69**
Exit-3	1.22	153.25	69.8	69.54	70.9	70.04	**71.38**	92.15	92.02	92.4	91.8	**92.75**
Exit-4	1.85	191.08	72.23	72.11	71.55	73.14	**75.77**	92.5	92.62	92.79	92.74	**93.07**
Exit-5	5.47	247.81	72.48	72.32	72.41	72.59	**74.12**	92.46	92.78	92.99	92.75	**93.13**
Exit-6	7.86	285.68	72.63	72.38	72.45	72.54	**74.23**	**93.59**	92.83	93.07	92.7	93.12
Exit-7	15.47	314.45	71.76	71.58	71.43	71.39	**73.1**	92.61	92.85	93.0	**93.69**	93.07
Average	-	-	66.34	66.19	67.60	66.61	**68.51**	88.15	88.08	89.54	88.05	**89.82**

Table 4. Classification accuracy of individual classifiers in multi-exit Resnet-SDN [26] on CIFAR-100 and CIFAR10.

	Params(M)	Flops(M)	CIFAR100					CIFAR10				
			SDN-Resnet	GE	Cagrad	Pcgrad	Ours	SDN-Resnet	GE	Cagrad	Pcgrad	Ours
Exit-1	0.02	19.50	40.20	42.10	**48.73**	40.10	44.41	71.64	71.37	**80.94**	69.74	76.04
Exit-2	0.04	38.54	45.45	46.91	47.05	45.67	**47.17**	78.10	77.11	**80.24**	77.24	78.11
Exit-3	0.10	56.47	59.08	**59.85**	57.77	60.04	59.70	87.32	87.21	86.31	**87.75**	86.43
Exit-4	0.18	75.43	62.40	**63.81**	62.62	63.47	63.25	**89.85**	89.63	88.62	89.79	89.09
Exit-5	0.36	93.32	67.88	**68.52**	67.16	67.78	68.38	91.45	91.51	90.73	**91.53**	91.48
Exit-6	0.67	112.25	70.06	69.88	69.26	69.70	**70.25**	92.26	92.33	91.31	92.17	**92.33**
Exit-7	0.89	126.44	70.02	69.63	68.40	70.07	**70.08**	92.33	92.21	91.19	92.09	**92.87**
Average	-	-	59.29	60.10	60.14	59.54	**60.32**	86.13	85.91	**87.04**	85.76	86.62

Table 5. Classification accuracy of individual classifiers on CIFAR100(150).

	Params(M)	Flops(M)	CIFAR100			
			MSDnet	Meta-GF	KD	Meta-GF(KD)
Exit-1	0.90	56.43	50.82	52.44	56.66	57.43
Exit-2	1.84	101.00	54.38	55.37	58.35	59.05
Exit-3	2.80	155.31	56.29	57.51	59.39	60.10
Exit-4	3.76	198.10	57.54	58.83	60.05	60.23
Exit-5	4.92	249.53	58.42	60.28	60.35	60.78
Exit-6	6.10	298.05	58.28	60.55	60.2	61.02
Exit-7	7.36	340.64	58.96	60.55	59.66	60.54
Average	-	-	56.38	57.93	59.23	59.87

We take further comparisons with the distillation-based works proposed in [3] as shown in Table 5. The knowledge distillation-based methods are mainly applied to provide soft target distributions, or in other words, distill the knowledge by deeper networks for improving the generalization ability of shallow networks. The distillation-based works is complementary to the gradient-adjust approaches [29,33,55], and thus is also complementary to the proposed Meta-GF. For simplicity, we takes 150 samples per class on CIFAR100 datasets for training. As shown in Table 5, by integrating the knowledge distillation with the proposed Meta-GF, the performance of the multi-exit networks can be further improved.

4.2   Adaptive Inference

In budgeted batch prediction mode [20], the computational budget is given in advance and the model is supposed to allocate different resources according to the complexity of inputs. For example, "easy" inputs are usually predicted by the shallow exits for saving the computation resources. When the multi-exit network

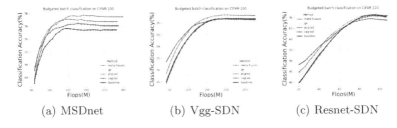

Fig. 2. Performance comparison: classification accuracy of budgeted batch classification as a function of average computational budget per image on the CIFAR-100.

Fig. 3. The accuracy of the multi-exit networks in the training stage on CIFAR100.

conducts budgeted batch prediction, it forwards the input through the intermediate exits from the shallow ones to the deep ones. If the prediction confidence at certain exit, which is the highest softmax probability in this experiment, is higher than a threshold, then the inference stops at this exit and the network outputs the prediction of this exit as the result. Otherwise, the subsequent exits is evaluated, until a sufficient high confidence has been obtained, or the last exit is evaluated [29]. The threshold is calculated on the validation set as described in [20]. We refer the readers to the work proposed in [20] for more details. As shown in Fig. 2, the proposed Meta-GF achieves competitive performance on three kinds of multi-exit networks when compared with other approaches. These results are consistent with the anytime prediction experiments, and demonstrate that the Meta-GF effectively balance the learning behavior of each exit.

4.3 Analysis About Meta-GF

In this section, we first make analysis about the convergence speed of the multi-exit networks when using the proposed Meta-GF. Then we investigate whether the learned fusion weights have the ability to reflect the importances of the shared parameters for the corresponding exits.

The Convergence Speed When Using the Proposed Approach. As shown in Fig. 3, the solid line represents the accuracy of the last exit on the training set. Note that the convergence speed in this work is evaluated by the

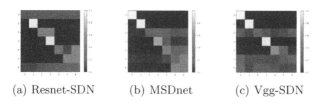

(a) Resnet-SDN   (b) MSDnet   (c) Vgg-SDN

Fig. 4. The accuracy degradations when pruning the important shared parameters of different exits.

training accuracy at the specific training iterations. Compared with the previous methods, the proposed Meta-GF obviously improve the convergence speed of the model, which benefits from the meta-learned expected gradient fusion. Yet for the Vgg-SDN, the convergence speed of the model when using the Meta-GF falls behind using the Cagrad at the first 50 epochs, but exceed it at the last 50 epochs. The final performance by using the Meta-GF stills surpass the Cagrad as shown in Table 3.

Analysis About the Learned Fusion Weights. We further make analysis of the learned fusion weights on CIFAR100. As mentioned above, the gradient fusion weights η_i of each exit are supposed to reflect the importances of the shared-parameters for the exit. We preliminarily define the parameter w with large η_i^w as the important parameter for the ith exit, where $\frac{e^{\eta_i^w}}{\sum_{j=1}^{n} e^{\eta_j^w}} > 0.5$. As shown in Fig. 4, we iteratively prune the important parameters of each exit from the 1st exit to the 7th exit, the relative accuracy degradation is shown along the horizontal axis. It can be seen that when we prune the important parameters of one exit, it mainly reduces the accuracy of this exit. This result in Fig. 4 indicates that the learned fusion weights effectively capture the importances of the share-parameters for each exit. Therefore, it's feasible to use the Meta-GF to alleviate the interference between different exits. It's worthy noting that not all exits in these multi-networks can own sufficient task-specific shared-parameters. Hence in Fig. 4, we can still see that pruning the important parameters of some exits doesn't cause the largest accuracy degradation to the associated exits.

5   Conclusion and Discussion

In this work, we propose a meta gradient fusion approach to tackle the gradient conflict problems when training multi-exit depth adaptive networks. The proposed Meta-GF takes a meta-learned weighted fusion policy to combine the expected gradients from each exit. We conduct extensive experiments on CIFAR and ImageNet, and the experimental results demonstrate the effectiveness of our approach.

However there is still room for improvement, we will further develop the Meta-GF through three aspects. Firstly, the meta fusion progress, in the current settings, doesn't take account of the parameters of the BatchNorm layer, and we will extend the Meta-GF to effectively fuse the BatchNorm layer either. Secondly, we plan to explicitly design objective functions for alleviating the gradient conflicts between different exits, which is not used in this work. Finally, we believe that the idea of the Meta-GF can be relevant to the network pruning researches to some extent. Though the learned fusion weights by the Meta-GF cannot be directly applied to prune the multi-exit networks for each exit, the results in Fig. 4 demonstrate a promising direction for our future researches, which means that we can combine the Meta-GF with the network pruning approaches for better disentangling the shared parts of the multi-exit networks, and also for alleviating the interference between different exits.

Acknowledgement. This work was supported by the National Natural Science Foundation of China (NSFC) under Grants 61825305 and 61973311.

References

1. Bakhtiarnia, A., Zhang, Q., Iosifidis, A.: Multi-exit vision transformer for dynamic inference (2021)
2. Bolukbasi, T., Wang, J., Dekel, O., Saligrama, V.: Adaptive neural networks for efficient inference. In: International Conference on Machine Learning, pp. 527–536. PMLR (2017)
3. Carion, N., Massa, F., Synnaeve, G., Usunier, N., Kirillov, A., Zagoruyko, S.: End-to-end object detection with transformers. In: Vedaldi, A., Bischof, H., Brox, T., Frahm, J.-M. (eds.) ECCV 2020. LNCS, vol. 12346, pp. 213–229. Springer, Cham (2020). https://doi.org/10.1007/978-3-030-58452-8_13
4. Chen, X., Dai, H., Li, Y., Gao, X., Song, L.: Learning to stop while learning to predict. In: International Conference on Machine Learning, pp. 1520–1530. PMLR (2020)
5. Chen, Y., Dai, X., Liu, M., Chen, D., Yuan, L., Liu, Z.: Dynamic convolution: attention over convolution kernels. In: Proceedings of the IEEE/CVF Conference on Computer Vision and Pattern Recognition, pp. 11030–11039 (2020)
6. Chen, Z., Badrinarayanan, V., Lee, C.Y., Rabinovich, A.: Gradnorm: gradient normalization for adaptive loss balancing in deep multitask networks. In: International Conference on Machine Learning, pp. 794–803. PMLR (2018)
7. Dai, X., Kong, X., Guo, T.: Epnet: learning to exit with flexible multi-branch network. In: Proceedings of the 29th ACM International Conference on Information & Knowledge Management, pp. 235–244 (2020)
8. Dosovitskiy, A., et al.: An image is worth 16x16 words: transformers for image recognition at scale. In: International Conference on Learning Representations (2020)
9. Du, S., Lee, J.: On the power of over-parametrization in neural networks with quadratic activation. In: International Conference on Machine Learning, pp. 1329–1338. PMLR (2018)
10. Duggal, R., Freitas, S., Dhamnani, S., Chau, D.H., Sun, J.: Elf: an early-exiting framework for long-tailed classification. arXiv preprint arXiv:2006.11979 (2020)
11. Elbayad, M., Gu, J., Grave, E., Auli, M.: Depth-adaptive transformer. In: ICLR 2020-Eighth International Conference on Learning Representations, pp. 1–14 (2020)

12. Fan, A., Grave, E., Joulin, A.: Reducing transformer depth on demand with structured dropout. In: International Conference on Learning Representations (2019)
13. Farhadi, A., Redmon, J.: Yolov3: an incremental improvement. In: Computer Vision and Pattern Recognition, pp. 1804–2767. Springer, Heidelberg (2018)
14. Gao, X., Zhao, Y., Dudziak, L., Mullins, R., Xu, C.z.: Dynamic channel pruning: feature boosting and suppression. In: International Conference on Learning Representations (2018)
15. Guo, M., Haque, A., Huang, D.A., Yeung, S., Fei-Fei, L.: Dynamic task prioritization for multitask learning. In: Proceedings of the European Conference on Computer Vision (ECCV), pp. 270–287 (2018)
16. Guo, S., Alvarez, J.M., Salzmann, M.: Expandnets: linear over-parameterization to train compact convolutional networks. In: Advances in Neural Information Processing Systems, vol. 33 (2020)
17. Han, Y., Huang, G., Song, S., Yang, L., Wang, H., Wang, Y.: Dynamic neural networks: a survey. IEEE Trans. Pattern Anal. Mach. Intell. (2021). https://doi.org/10.1109/TPAMI.2021.3117837
18. He, K., Zhang, X., Ren, S., Sun, J.: Deep residual learning for image recognition. In: Proceedings of the IEEE Conference on Computer Vision and Pattern Recognition, pp. 770–778 (2016)
19. Huang, C., Lucey, S., Ramanan, D.: Learning policies for adaptive tracking with deep feature cascades. In: Proceedings of the IEEE International Conference on Computer Vision, pp. 105–114 (2017)
20. Huang, G., Chen, D., Li, T., Wu, F., van der Maaten, L., Weinberger, K.Q.: Multi-scale dense networks for resource efficient image classification. arXiv preprint arXiv:1703.09844 (2017)
21. Huang, G., Sun, Yu., Liu, Z., Sedra, D., Weinberger, K.Q.: Deep networks with stochastic depth. In: Leibe, B., Matas, J., Sebe, N., Welling, M. (eds.) ECCV 2016. LNCS, vol. 9908, pp. 646–661. Springer, Cham (2016). https://doi.org/10.1007/978-3-319-46493-0_39
22. Javaloy, A., Valera, I.: Rotograd: gradient homogenization in multi-task learning (2021)
23. Jeon, G.-W., Choi, J.-H., Kim, J.-H., Lee, J.-S.: LarvaNet: hierarchical super-resolution via multi-exit architecture. In: Bartoli, A., Fusiello, A. (eds.) ECCV 2020. LNCS, vol. 12537, pp. 73–86. Springer, Cham (2020). https://doi.org/10.1007/978-3-030-67070-2_4
24. Jiang, Y.G., Cheng, C., Lin, H., Fu, Y.: Learning layer-skippable inference network. IEEE Trans. Image Process. **29**, 8747–8759 (2020)
25. Jie, Z., Sun, P., Li, X., Feng, J., Liu, W.: Anytime recognition with routing convolutional networks. IEEE Trans. Pattern Anal. Mach. Intell. **43**(6), 1875–1886 (2019)
26. Kaya, Y., Hong, S., Dumitras, T.: Shallow-deep networks: understanding and mitigating network overthinking. In: International Conference on Machine Learning, pp. 3301–3310. PMLR (2019)
27. Krizhevsky, A., Hinton, G., et al.: Learning multiple layers of features from tiny images (2009)
28. Li, F., Li, G., He, X., Cheng, J.: Dynamic dual gating neural networks. In: Proceedings of the IEEE/CVF International Conference on Computer Vision, pp. 5330–5339 (2021)
29. Li, H., Zhang, H., Qi, X., Yang, R., Huang, G.: Improved techniques for training adaptive deep networks. In: Proceedings of the IEEE/CVF International Conference on Computer Vision, pp. 1891–1900 (2019)

30. Li, Y., Song, L., Chen, Y., Li, Z., Zhang, X., Wang, X., Sun, J.: Learning dynamic routing for semantic segmentation. In: Proceedings of the IEEE/CVF Conference on Computer Vision and Pattern Recognition, pp. 8553–8562 (2020)
31. Li, Y., Chen, Y.: Revisiting dynamic convolution via matrix decomposition. In: International Conference on Learning Representations (2021)
32. Liu, B., Rao, Y., Lu, J., Zhou, J., Hsieh, C.-J.: MetaDistiller: network self-boosting via meta-learned top-down distillation. In: Vedaldi, A., Bischof, H., Brox, T., Frahm, J.-M. (eds.) ECCV 2020. LNCS, vol. 12359, pp. 694–709. Springer, Cham (2020). https://doi.org/10.1007/978-3-030-58568-6_41
33. Liu, B., Liu, X., Jin, X., Stone, P., Liu, Q.: Conflict-averse gradient descent for multi-task learning. In: Advances in Neural Information Processing Systems, vol. 34 (2021)
34. Liu, L., et al.: Towards impartial multi-task learning. In: ICLR (2021)
35. Liu, Y., Meng, F., Zhou, J., Chen, Y., Xu, J.: Faster depth-adaptive transformers. In: Proceedings of the AAAI Conference on Artificial Intelligence, vol. 35, pp. 13424–13432 (2021)
36. Ma, N., Zhang, X., Huang, J., Sun, J.: WeightNet: revisiting the design space of weight networks. In: Vedaldi, A., Bischof, H., Brox, T., Frahm, J.-M. (eds.) ECCV 2020. LNCS, vol. 12360, pp. 776–792. Springer, Cham (2020). https://doi.org/10.1007/978-3-030-58555-6_46
37. Molchanov, P., Tyree, S., Karras, T., Aila, T., Kautz, J.: Pruning convolutional neural networks for resource efficient inference. In: 5th International Conference on Learning Representations, ICLR 2017-Conference Track Proceedings (2019)
38. Nichol, A., Achiam, J., Schulman, J.: On first-order meta-learning algorithms. arXiv preprint arXiv:1803.02999 (2018)
39. Passalis, N., Raitoharju, J., Tefas, A., Gabbouj, M.: Efficient adaptive inference for deep convolutional neural networks using hierarchical early exits. Pattern Recogn. 105, 107346 (2020)
40. Phuong, M., Lampert, C.H.: Distillation-based training for multi-exit architectures. In: Proceedings of the IEEE/CVF International Conference on Computer Vision, pp. 1355–1364 (2019)
41. Sandler, M., Howard, A., Zhu, M., Zhmoginov, A., Chen, L.C.: MobileNetV2: inverted residuals and linear bottlenecks. In: Proceedings of the IEEE Conference on Computer Vision and Pattern Recognition, pp. 4510–4520 (2018)
42. Schwartz, R., Stanovsky, G., Swayamdipta, S., Dodge, J., Smith, N.A.: The right tool for the job: matching model and instance complexities. In: Proceedings of the 58th Annual Meeting of the Association for Computational Linguistics, pp. 6640–6651 (2020)
43. Sener, O., Koltun, V.: Multi-task learning as multi-objective optimization. In: Advances in Neural Information Processing Systems, vol. 31 (2018)
44. Shen, J., Wang, Y., Xu, P., Fu, Y., Wang, Z., Lin, Y.: Fractional skipping: towards finer-grained dynamic CNN inference. In: Proceedings of the AAAI Conference on Artificial Intelligence, vol. 34, pp. 5700–5708 (2020)
45. Tan, M., Le, Q.: Efficientnet: rethinking model scaling for convolutional neural networks. In: International Conference on Machine Learning, pp. 6105–6114. PMLR (2019)
46. Veit, A., Belongie, S.: Convolutional networks with adaptive inference graphs. In: Proceedings of the European Conference on Computer Vision (ECCV), pp. 3–18 (2018)

47. Wang, X., Yu, F., Dou, Z.Y., Darrell, T., Gonzalez, J.E.: Skipnet: learning dynamic routing in convolutional networks. In: Proceedings of the European Conference on Computer Vision (ECCV), pp. 409–424 (2018)

48. Wang, X., Li, Y.: Gradient deconfliction-based training for multi-exit architectures. In: 2020 IEEE International Conference on Image Processing (ICIP), pp. 1866–1870. IEEE (2020)

49. Wang, X., Li, Y.: Harmonized dense knowledge distillation training for multi-exit architectures. In: Proceedings of the AAAI Conference on Artificial Intelligence, vol. 35, pp. 10218–10226 (2021)

50. Wołczyk, M., et al.: Zero time waste: recycling predictions in early exit neural networks. Adv. Neural. Inf. Process. Syst. **34**, 2516–2528 (2021)

51. Wu, Z., et al.: Blockdrop: dynamic inference paths in residual networks. In: Proceedings of the IEEE Conference on Computer Vision and Pattern Recognition, pp. 8817–8826 (2018)

52. Xin, J., Tang, R., Lee, J., Yu, Y., Lin, J.: Deebert: dynamic early exiting for accelerating bert inference. In: Proceedings of the 58th Annual Meeting of the Association for Computational Linguistics, pp. 2246–2251 (2020)

53. Yang, B., Bender, G., Le, Q.V., Ngiam, J.: Condconv: conditionally parameterized convolutions for efficient inference. In: Advances in Neural Information Processing Systems, pp. 1307–1318 (2019)

54. Yang, L., Han, Y., Chen, X., Song, S., Dai, J., Huang, G.: Resolution adaptive networks for efficient inference. In: Proceedings of the IEEE/CVF Conference on Computer Vision and Pattern Recognition, pp. 2369–2378 (2020)

55. Yu, T., Kumar, S., Gupta, A., Levine, S., Hausman, K., Finn, C.: Gradient surgery for multi-task learning. In: Advances in Neural Information Processing Systems, vol. 33 (2020)

56. Zhou, J., Jampani, V., Pi, Z., Liu, Q., Yang, M.H.: Decoupled dynamic filter networks. In: Proceedings of the IEEE/CVF Conference on Computer Vision and Pattern Recognition, pp. 6647–6656 (2021)

57. Zhou, W., Xu, C., Ge, T., McAuley, J., Xu, K., Wei, F.: Bert loses patience: fast and robust inference with early exit. Adv. Neural. Inf. Process. Syst. **33**, 18330–18341 (2020)

Towards Ultra Low Latency Spiking Neural Networks for Vision and Sequential Tasks Using Temporal Pruning

Sayeed Shafayet Chowdhury$^{(\boxtimes)}$, Nitin Rathi , and Kaushik Roy

Purdue University, West Lafayette, IN 47907, USA
{chowdh23,rathi2,kaushik}@purdue.edu

Abstract. Spiking Neural Networks (SNNs) can be energy efficient alternatives to commonly used deep neural networks (DNNs). However, computation over multiple timesteps increases latency and energy and incurs memory access overhead of membrane potentials. Hence, latency reduction is pivotal to obtain SNNs with high energy efficiency. But, reducing latency can have an adverse effect on accuracy. To optimize the accuracy-energy-latency trade-off, we propose a temporal pruning method which starts with an SNN of T timesteps, and reduces T every iteration of training, with threshold and leak as trainable parameters. This results in a continuum of SNNs from T timesteps, all the way up to unit timestep. Training SNNs directly with 1 timestep results in convergence failure due to layerwise spike vanishing and difficulty in finding optimum thresholds. The proposed temporal pruning overcomes this by enabling the learning of suitable layerwise thresholds with backpropagation by maintaining sufficient spiking activity. Using the proposed algorithm, we achieve top-1 accuracy of 93.05%, 70.15% and 69.00% on CIFAR-10, CIFAR-100 and ImageNet, respectively with VGG16, in just 1 timestep. Note, SNNs with leaky-integrate-and-fire (LIF) neurons behave as Recurrent Neural Networks (RNNs), with the membrane potential retaining information of previous inputs. The proposed SNNs also enable performing sequential tasks such as reinforcement learning on Cartpole and Atari pong environments using only 1 to 5 timesteps.

Keywords: Spiking neural networks · Unit timestep · Energy efficiency · Temporal pruning · Reinforcement learning

1 Introduction

Deep neural networks (DNNs) have revolutionized the fields of object detection, classification and natural language processing [5,14,21]. However, such performance boost comes at the cost of extremely energy intensive DNN architectures [23]. Therefore, edge deployment of such DNNs remains a challenge. One approach to counter this is to use bio-inspired Spiking Neural Networks (SNNs)

Supplementary Information The online version contains supplementary material available at https://doi.org/10.1007/978-3-031-20083-0_42.

[25,33], which perform computations using spikes instead of analog activations used in standard networks. In this paper, standard networks are referred to as Artificial Neural Networks (ANNs) in contrast to SNNs having spike activation. As such, the sparse event-driven nature of SNNs makes them an attractive alternative to ANNs [9].

The applicability of SNNs was initially limited due to the unavailability of suitable training algorithms. At first, ANN-SNN conversion methods [7,36] were adopted, but incurred high latency. Recently, direct training using surrogate gradients [27,44] has resulted in low latency. Most of the commonly used SNNs use Poisson rate-coding [7,36] where a large number of timesteps[1] is usually required to obtain high performance [11,36]. However, multi-timestep processing causes twofold challenges in SNNs - (i) too long a latency might be unsuitable for real-time applications, (ii) the need for accumulation of membrane potential (V_{mem}) over numerous timesteps results in higher number of operations, thereby reducing efficiency. Moreover, in contrast to ANNs, additional memory is required to store the intermediate V_{mem} and memory access cost is incurred for fetching V_{mem} at each timestep. Note, the memory access cost can be significantly higher compared to floating point add operations [12]. So, reducing inference latency is critical for widespread deployment of SNNs.

To leverage the full potential of SNNs, we propose a temporal pruning technique to obtain a continuum of SNNs optimizing the associated accuracy-energy-latency trade-offs. SNNs, unlike ANNs have a temporal dimension. As a result, low latency SNNs might be obtained if suitable compression can be performed along the temporal axis, which we refer to as 'temporal pruning' here. Such temporal compression is required since training SNNs directly with very few timesteps results in convergence failure due to significant decay of spiking activity in the deeper layers. To infer with very low latency, sufficient spikes must be propagated till the final layer in only a few forward passes. To achieve that, the layerwise neuron thresholds and leaks must be adjusted properly. The optimum approach to set the thresholds and leaks is learning them using backpropagation (BP). However, without having enough spikes at the output, the optimization gets stuck and no learning can occur for extremely low latency. To circumvent this issue, we adopt a temporal pruning method which starts with an SNN of T (T > 1) timesteps, and gradually reduces T at every iteration of training using threshold and leak as trainable parameters alongside the weights. At each stage of timestep reduction, the network trained at previous stage with higher timestep is used as initialization for subsequent training with lower timestep. Using such a pruning process, we obtain a continuum of SNNs, starting from T timesteps, eventually leading up to unit timestep. Note, under unit timestep, if a neuron receives an input, it updates V_{mem} and in the event of crossing a given threshold, it outputs a spike in a single shot, similar to binary activated ANNs [1,35].

As mentioned, the proposed temporal pruning starts from an initial SNN trained for T timesteps. We obtain this initial SNN using a hybrid training method

---

[1] 1 timestep is defined as the time taken to perform 1 forward pass through the network.

[32]. First an ANN is trained, followed by ANN-SNN conversion, and SNN training with surrogate gradient based backpropagation [27]. The initial design using the above approach leads to an SNN with T = 5 [31] and our proposed approach reduces the timestep to 1. We use direct input encoding [34], with the first convolutional layer of the network acting as spike generator. We evaluate the performance of the proposed method on image classification and achieve top-1 accuracy of 93.05%, 70.15% and 69.00% on CIFAR-10, CIFAR-100 and ImageNet, respectively with VGG16, in just 1 timestep. These results demonstrate that for static vision tasks, SNNs can perform comparable to ANNs using single shot inference. However, a key distinction between ANNs and SNNs is the time axis which can potentially enhance the performance of SNNs for sequential tasks if the temporal information can be leveraged suitably using the membrane potential of neurons. To validate this hypothesis, we apply our technique to SNN-based reinforcement learning (RL) agents on Cartpole and Atari pong environments. Though the proposed method can provide RL agents with unit timestep, performance improves significantly for SNNs with larger timesteps. For cartpole, we obtain mean reward of 38.7 and 52.2 for 1 and 3 timesteps, respectively. Similarly, the mean reward increased from 17.4 to 19.4 as the timesteps are increased from 1 to 5 for Atari pong. It is noteworthy that even unit timestep SNNs can handle dynamic inputs. However, increasing timesteps leads to enhancement in performance for sequential tasks. Overall, the proposed continuum of SNNs is able to provide optimum solutions for both static and dynamic tasks. While SNNs with unit timestep obtain satisfactory performance with highest efficiency for static inputs, choosing SNNs having only a few timesteps from the continuum of SNNs can provide ANN-like performance for dynamic RL tasks with much lower energy cost. To summarize, the main contributions of this work are-

- We present a temporal pruning technique to obtain a continuum of SNNs with varying latency, starting from T timesteps up to 1. To the best of our knowledge, this is the first SNN work to achieve competitive classification performance (top-1 accuracy of 69%) on ImageNet using unit timestep.
- Our approach does not incur the memory access cost of accumulated membrane potentials, unlike previously proposed SNNs.
- One timestep SNNs infer with up to 5X lower latency compared to state-of-the-art SNNs, while achieving comparable accuracy. This also leads to efficient computation, resulting in SNNs which are up to 33X more energy efficient compared to ANNs with equivalent architecture.
- The proposed method enables deep-Q reinforcement learning on Cartpole and Atari Pong with SNNs having few (1–5) timesteps, thereby showing the efficacy of multi-timestep SNNs (as RNNs) for sequential tasks. Compared to unit timestep, 3–5 timesteps enhance the performance of SNNs considerably for such tasks by leveraging the inherent recurrence of spiking neurons.

2   Related Works

ANN-SNN Conversion. A widely used approach for training deep SNNs involves training an ANN and converting to SNN for fine-tuning [2,7,36]. Proper layerwise threshold adjustment is critical to convert ANNs to SNNs successfully. One approach is choosing the thresholds as the maximum pre-activation of the neurons [36]. While it provides high accuracy, the associated drawback is high inference latency (about 1000 timesteps). Alternatively, [34] suggests choosing a certain percentile of the pre-activation distribution as the threshold to reduce latency. However, these methods [11,34] still require few hundred timesteps.

Backpropagation from Scratch and Hybrid Training. An alternate route of training SNNs with reduced latency is learning from scratch using backpropagation (BP). To circumvent the non-differentiability of the spike function, surrogate gradient based optimization has been proposed [27] to implement BP in SNNs [16,22]. A related approach is using surrogate gradient based BP on membrane potential [38,47]. Overall, these approaches obtain SNNs with high accuracy, but the latency is still significant ($\sim$100–125 timesteps). Recently, surrogate gradient-based deep residual learning has been employed to directly train SNNs with just 4 timesteps [8]. A hybrid approach is proposed in [32] where a pre-trained ANN is used as initialization for subsequent SNN learning. Such a hybrid approach improves upon conversion by reducing latency and speeds up convergence of direct BP from scratch method.

Temporal Encoding. Temporal coding schemes such as phase [18] or burst [28] coding attempt to capture temporal information into learning; a related method is time-to-first-spike (TTFS) coding [29], where each neuron is allowed to spike just once. While these techniques enhance efficiency by reducing the spike count, issues regarding high latency and memory access overhead persist.

Direct Encoding. The analog pixels are directly applied to the 1st layer of the network in direct encoding [31,34,49]. Using direct coding and utilizing the first layer as spike generator, authors in [31] achieve competitive performance on ImageNet with 5 timesteps. Threshold-dependent batch normalization is employed with direct encoding by [49] to obtain high performing SNNs on ImageNet with 6 timesteps. Inspired by such performance, we adopt the direct encoding method. The difference between our work and [31] is the temporal pruning aspect, which enables to reduce the timestep to lowest possible limit while maintaining performance on complex datasets. This was infeasible for state-of-the-art SNNs [8,31,49], even with direct encoding. Moreover, our approach is able to enhance performance by incorporating batch-normalization unlike [31,32].

Binary Neural Networks. Unit timestep SNN is closely related to binary neural networks (BNN) [30,40], as both infer in a single shot with binary activation. However, there are quite a few distinctions. While BNNs binarize the outputs as ± 1, SNNs give spike (i.e., $\{0, 1\}$) output, which leads to higher sparsity. Additionally, our model uses LIF neurons with trainable thresholds, whereas, BNNs employ a Heaviside activation where the firing threshold is zero. The activation

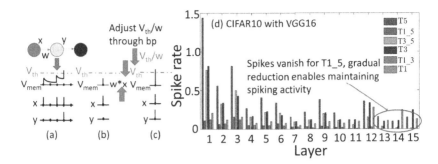

Fig. 1. (a) Schematic of an SNN with dynamics shown for yellow neuron with x as input and y as output, (b) No output spike for direct transition from 5 to 1 timestep, (c) Output spike in just 1 timestep through V_{th}/w lowering, (d) Layerwise spike rates; Tx represents an SNN trained with 'x' timesteps; Tx_y represents an SNN trained with 'x' timesteps but initialized with an SNN that was trained for 'y' timesteps. (Color figure online)

of the proposed SNN reduces to Heaviside function with tunable threshold for $T \rightarrow 1$. However, the use of LIF neurons enables us to use the same model for sequential processing using V_{mem}, which is non-trivial in BNNs.

3 Background

Spiking Neuron Model. The LIF neuron model [17] is described as-

$$\tau_m \frac{dU}{dt} = -(U - U_{rest}) + RI, \quad U \leq V_{th} \tag{1}$$

where U, I, τ_m, R, V_{th} and U_{rest} denote membrane potential, input, time constant for membrane potential decay, leakage resistance, firing threshold and resting potential, respectively. We employ a discretized version of Eq. 1-

$$u_i^t = \lambda_i u_i^{t-1} + \sum_j w_{ij} o_j^t - v_i o_i^{t-1}, \tag{2}$$

$$z_i^{t-1} = \frac{u_i^{t-1}}{v_i} \quad \text{and} \quad o_i^{t-1} = \begin{cases} 1, & \text{if } u_i^{t-1} > v_i \\ 0, & \text{otherwise} \end{cases} \tag{3}$$

where u is the membrane potential, subscripts i and j represent the post and pre-neuron, respectively, t denotes timestep, λ is the leak constant$= e^{\frac{-1}{\tau_m}}$, w_{ij} represents the weight between the i-th and j-th neurons, o is the output spike, and v is the threshold. The detailed methodology of training SNN with a certain timestep is provided in supplementary section 1. In particular, Algorithm S1 of supplementary depicts the training scheme for one iteration.

4   Proposed Latency Reduction Method

We start by training an ANN with batch-norm (BN) and subsequently the BN parameters are fused with the layerwise weights as done in [34]. With such pretrained ANN, the weights are copied to an iso-architecture SNN and we select the 90.0 percentile of the pre-activation distribution at each layer as its threshold. Then the SNN is trained for T timesteps using BP which serves as our baseline; training steps are detailed in Algorithm S1 of supplementary. Our starting point is a $T = 5$ timesteps trained network, since 4–6 is the minimum latency range that state-of-the-art (SOTA) SNNs have reported for ImageNet training [8,31,49] with high performance. However, our method is generic and can be applied to SNNs with higher T as starting point too, but that increases training overhead which motivates our choice of starting point as $T = 5$.

Direct Inference with Unit Timestep. As mentioned in Sect. 1, our goal is to reduce latency of SNNs as much as possible. To that effect, next we explore the feasibility of directly reducing latency from $T = 5$ to 1. Figure 1(a) schematically depicts an SNN with 3 neurons (one per layer), we focus on the yellow neuron. Suppose it receives x as input and y is its output. With the weight (w) and threshold (V_{th}) trained for 5 timesteps, there is enough accumulation of membrane potential (V_{mem}) to cross V_{th} and propagate spikes to next layer within that 5 timestep window. However, when we try to infer with 1 timestep (Fig. 1(b)), there is no output spike as the V_{mem} is unable to reach V_{th} instantly. Therefore, w and V_{th} need to be adjusted such that information can propagate even within 1 step. Balancing the thresholds properly is critical for SNNs to perform well [49]. Hence, our goal is adjusting V_{th}/w through learning using BP, so that only neurons salient for information propagation are able to spike in 1 timestep (Fig. 1(c)), while other neurons remain dormant.

Direct Transition to Training with Unit Timestep. Next, we begin training with 1 timestep initialized with the 5 timestep trained SNN. However, the network fails to train. To investigate this, we plot the layerwise spike rates for a VGG16 on CIFAR10 in Fig. 1(d). Here, with Tx denotes an SNN trained with spike based back propagation for x timesteps, and Tx_y denotes an SNN trained with x timesteps, starting from an initial SNN that was trained for y timesteps ($y > x$). While T5 has sufficient spiking activity till final layer, for T1_5, spikes die out in the earlier layers. Due to such spike vanishing, all outputs at the deeper layers (layers 11–16) are 0, thus BP fails to start training. This happens since the initial V_{th} and w are trained for 5 timesteps, and retraining directly for 1 timestep hinders spike propagation to the later layers.

Gradual Temporal Pruning. To mitigate the above issue, we propose a gradual latency reduction approach. We observe that despite significant layerwise spike activity decay, some spikes still reach the final layer for T3_5. Hence, we

---

Algorithm 1. Pseudo-code of training.

Input: Trained SNN with N timesteps (TN), timesteps reduction step size (b), number of epochs to train (e)
Initialize: new SNN initialized with trained parameters of TN, reduced latency $T_r = N - b$
while $T_r > 0$ **do**
    // **Training Phase**
    for *epoch* $\leftarrow$ 1 **to** e **do**
        //Train network with T_r timesteps using algorithm S1 of supplementary
    end for
    // Initialize another iso-architecture SNN with parameters of above trained network
    // Temporal pruning
    $T_r = T_r - b$
end while

---

start training with 3 timesteps using T5 as initialization and training is able to converge through BP. The spiking activity is recovered when learning converges as shown in Fig. 1(d), case T3. Subsequently, we train a network with just 1 timestep by initializing it with T3, and successfully attain convergence. The results for this case is shown as T1 in Fig. 1(d). Motivated by these observations, we propose an iterative temporal pruning method which enables training a continuum of SNNs starting from T timesteps (T > 1) up to 1. Since in our case, we compress the temporal dimension of SNNs, it is termed as 'temporal pruning'. A pseudo-code of training is given in Algorithm 1. Beginning with T5, we gradually reduce the latency by 1 at each step and train till convergence.

How does Temporal Pruning Help in Learning? In this section, we investigate the effect of the temporal pruning on SNN learning. Without temporal pruning, spikes may not propagate to the final layers in few timesteps. As a result, no gradient can be propagated back to train the SNN. In terms of the optimization landscape, the neural network parameters remain stuck at their initialization point since the gradient updates remain zero. However, using gradual temporal pruning, we facilitate the propagation of spikes (and suitable gradient flow is enabled) by properly learning the threshold and leak parameters, leading to training convergence with very few timesteps. This can be visualized as a form of curriculum training, where the SNN is first trained with a comparatively easier learning case (higher T) and then more stringent constraints (training with lower T) are incorporated gradually. Similar curriculum training has been applied for ANNs [20] where directly imposing a tougher learning scenario leads to optimization failure; however, gradual training alleviates the problem. The effect of temporal pruning can also be analyzed by comparing it to spatial pruning in ANNs [12,13]. In case of spatial pruning, a more complex network is first trained and then spatial compression is performed while maintaining performance. Our proposed approach is similar with the exception

Table 1. Top-1 classification accuracy (%), Tx denotes SNN trained with 'x' timestep

Architecture	Dataset	ANN	T5	T4	T3	T2	T1
VGG6	CIFAR10	91.59	90.61	90.52	90.40	90.05	89.10
VGG16	CIFAR10	94.10	93.90	93.87	93.85	93.72	93.05
ResNet20	CIFAR10	93.34	92.62	92.58	92.56	92.11	91.10
VGG16	CIFAR100	72.46	71.58	71.51	71.46	71.43	70.15
ResNet20	CIFAR100	65.90	65.57	65.37	65.16	64.86	63.30
VGG16	ImageNet	70.08	69.05	69.03	69.01	69.00	69.00

that we leverage the time axis of SNNs to perform pruning. SNNs with multiple timesteps are trained with BP through time (BPTT) [27], like RNNs. If we unroll SNNs in time, it becomes obvious that each timestep adds a new hidden state. In essence, temporal pruning helps latency reduction in SNNs through compression. From a related perspective, we can also perceive gradual temporal pruning as training by generations. Similar sequential compression (training by generations) has been implemented in ANNs [10, 46] to obtain better performing compressed networks.

5   Experiments and Results

Datasets and Models. We perform experiments on CIFAR10, CIFAR100 and ImageNet using VGG16 and ResNet20, with some studies involving VGG6. The proposed method is also evaluated on reinforcement learning (RL) using Cartpole and Atari-pong. Supplementary section 2 includes architectural details and hyperparameters. The code is submitted as part of the supplementary material.

Results on CIFAR and ImageNet. The experimental results using the proposed scheme are shown in Table 1. We achieve top-1 accuracy of 93.05% and 70.15% on CIFAR-10 and CIFAR-100, respectively using VGG16, with just 1 timestep (T1); results with ResNet20 are also shown in this Table. With reduction of latency from 5 to 1, there is a slight accuracy degradation, however that is due to the inherent accuracy versus latency trade-off in SNNs. Next, to investigate the scalability of the proposed algorithm, we experiment with ImageNet where we obtain 69.00% top-1 accuracy with T1. Notably, the proposed technique allows us to reduce the SNN latency to lowest possible limit.

Performance Comparison. Next, we compare our performance with different state-of-the-art SNNs in Table 2. T1 SNN performs better than or comparably to all these methods, while achieving significantly lower inference latency. In particular, previously it was challenging to obtain satisfactory performance with low latency on ImageNet, with lowest reported timesteps of 4 [8], 5 [31] and 6 [49]. In contrast, we report 69.00% top-1 accuracy on ImageNet using T1. Overall,

Table 2. Comparison of T1 with other SNN results. SGB, hybrid and TTFS denote surrogate-gradient based backprop, pretrained ANN followed by SNN fine-tuning, and time-to-first-spike, respectively and (qC, dL) denotes q conv layers and d linear layers.

Method	Dataset	Architecture	Accuracy(%)	Timesteps (T)
ANN-SNN conversion [17]	CIFAR10	2C, 2L	82.95	6000
ANN-SNN conversion [2]	CIFAR10	3C, 2L	77.43	400
ANN-SNN conversion [36]	CIFAR10	VGG16	91.55	2500
SGB [22]	CIFAR10	VGG9	90.45	100
ANN-SNN conversion [34]	CIFAR10	4C, 2L	90.85	400
Hybrid [32]	CIFAR10	VGG9	90.5	100
TTFS [29]	CIFAR10	VGG16	91.4	680
Burst-coding [28]	CIFAR10	VGG16	91.4	1125
Phase-coding [18]	CIFAR10	VGG16	91.2	1500
SGB [43]	CIFAR10	2C, 2L	50.7	30
SGB [44]	CIFAR10	5C, 2L	90.53	12
Tandem Learning [42]	CIFAR10	5C, 2L	90.98	8
SGB [48]	CIFAR10	5C, 2L	91.41	5
Direct Encoding [31]	CIFAR10	VGG16	92.70	5
STBP-tdBN [49]	CIFAR10	ResNet-19	93.16	6
Temporal pruning (ours)	**CIFAR10**	**VGG16**	**93.05**	1
ANN-SNN conversion [24]	CIFAR100	VGG15	63.2	62
Hybrid [32]	CIFAR100	VGG11	67.9	125
TTFS [29]	CIFAR100	VGG16	68.8	680
Burst-coding [28]	CIFAR100	VGG16	68.77	3100
Phase-coding [18]	CIFAR100	VGG16	68.6	8950
Direct Encoding [31]	CIFAR100	VGG16	69.67	5
Temporal pruning (ours)	**CIFAR100**	**VGG16**	**70.15**	1
ANN-SNN conversion [36]	ImageNet	VGG16	69.96	2500
ANN-SNN conversion [34]	ImageNet	VGG16	49.61	400
Hybrid [32]	ImageNet	VGG16	65.19	250
Tandem Learning [42]	ImageNet	AlexNet	50.22	10
ANN-SNN conversion [24]	ImageNet	VGG15	66.56	64
Direct Encoding [31]	ImageNet	VGG16	69.00	5
SGB [8]	ImageNet	ResNet-34	67.04	4
STBP-tdBN [49]	ImageNet	ResNet-34	67.05	6
Temporal pruning (ours)	**ImageNet**	**VGG16**	**69.00**	1

T1 SNN demonstrates 4-2500X improvement in inference latency compared to other works while maintaining iso or better classification performance. Table 2 also demonstrates the gradual progression of SNN training from ANN-SNN conversion to T1 SNN. Initial ANN-SNN conversion methods required latency on the order of thousands [17,36]. Surrogate-gradient based BP [22,44] reduced it to few tens to hundred. Next, hybrid training [32] combined these two methods to bring the latency down to few hundreds on ImageNet. Subsequently, direct input encoding enabled convergence on ImageNet with latency of ~ 5 [31,49]. The proposed method leverages all these previously proposed techniques and improves upon them by incorporating the temporal pruning approach. We also achieve better performance compared to different SNN encoding schemes such as TTFS [29], phase [18], burst [28]; detailed comparison with these methods is provided in supplementary section 3.

Inference Efficiency. Next, we compare the energy efficiency of T1 SNNs with ANNs and multi-timestep SNNs. In SNNs, the floating-point (FP) additions replace the FP MAC operations. This results in higher compute efficiency as the cost of a MAC (4.6 pJ) is 5.1× to an addition (0.9 pJ) [15] in 45 nm CMOS technology (as shown in Fig. 2(e)). Supplementary section 4 contains the equations of computational cost in the form of operations per layer in an ANN, $\#ANN_{ops}$. For an SNN, the number of operations is given as $\#SNN_{ops, q}$ = spike rate$_q$ × $\#ANN_{ops, q}$; where spike rate$_q$ denotes the average number of spikes per neuron per inference over all timesteps in layer q. The layerwise spike rates across T5 to T1 are shown in Fig. 2(a–c). Note, the spike rates decrease significantly with latency reduction from 5 to 1, leading to considerable reduction in operation count. The overall average spike rates using T1 for CIFAR10, CIFAR100 and ImageNet on VGG16 are 0.13, 0.15 and 0.18, respectively; all significantly below 5.1 (relative cost of MAC to addition). The first layer with direct input encoded SNNs receive analog inputs, hence the operations are same as an ANN at this layer. Considering it, we calculate the compute energy benefits of T1 SNN over ANN, α as,

$$\alpha = \frac{E_{ANN}}{E_{SNN}} = \frac{\sum_{q=1}^{L} \#ANN_{ops,q} * 4.6}{\# SNN_{ops,1} * 4.6 + \sum_{q=2}^{L} \# SNN_{ops,q} * 0.9}. \qquad (4)$$

The values of α for different datasets and architectures are given in Fig. 2(d). For VGG16, we obtain α of 33.0, 29.24 and 24.61 on CIFAR-10, CIFAR-100 and ImageNet, respectively. Besides compute energy, another significant overhead in SNNs occurs due to memory access costs which can be significantly higher [12] compared to FP adds as shown in Fig. 2(e). However, most previous works [29,31,32,49] did not consider this cost while comparing the energy benefits of SNN to ANN. To obtain a fairer comparison, we analyze the costs taking the memory access issue into consideration. For multi-timestep SNNs, in addition to the weights, V_{mem} needs to be stored and fetched at each timestep. However, the memory requirements for T1 SNNs are same as ANNs. The actual improvements in energy depends on the hardware architecture and system configurations. Hence, we compare the reduction in terms of number of memory access. For a VGG16, the proposed T1 reduces the number of memory

Table 3. Comparison of T1 SNN with BNN

Method	Dataset	Accuracy(%)
Binary activation [35]	CIFAR10	89.6
STE-BNN [1]	CIFAR10	85.2
BN-less BNN [3]	CIFAR10	92.08
Trained binarization [45]	CIFAR10	92.3
BBG-Net [37]	CIFAR10	92.46
CI-BCNN [41]	CIFAR10	92.47
Binary activation [6]	CIFAR10	91.88
This work (T1)	CIFAR10	93.05
Binary activation [6]	CIFAR100	70.43
BN-less BNN [3]	CIFAR100	68.34
BBG-Net [37]	CIFAR100	69.38
This work (T1)	CIFAR100	70.15

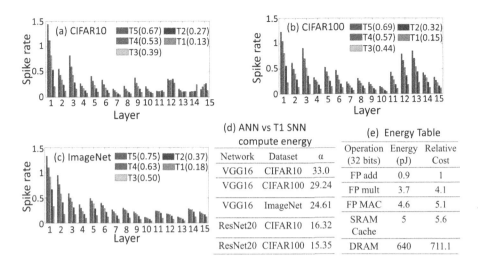

Fig. 2. Layerwise spike rates for a VGG16 (average spike rate in parenthesis) on (a) CIFAR10, (b) CIFAR100, (c) ImageNet, (d) relative cost of compute between ANN and T1, (e) operation-wise energy consumption in 45 nm CMOS [15].

accesses by 5.03× compared to 5 timestep SNN of [31]. More generally, our scheme with T1 reduces the number of memory accesses by approximately T× compared to an SNN trained with T timesteps.

Comparison with Binary Activated ANNs. A key distinction between ANNs and SNNs is the notion of time. However, T1 SNN and binary neural networks (BNNs) both infer in single shot using binary activations. But, there are some differences which have been discussed in Sect. 2. Here, we compare the performance of T1 SNN and BNNs. We observe that T1 SNN performs on par or better than other BNN approaches on CIFAR10 and CIFAR100 datasets, as depicted in Table 3. Furthermore, training large scale datasets such as ImageNet using BNN methods [1,3,4,35,37,45] has been challenging, while the proposed method scales well to ImageNet. Notably, Xnor-net [30] and Dorefa-net [50] achieve 44.2% and 43.6% top-1 accuracy on ImageNet, respectively, with 1-bit activations, while T1 achieves 69%. However, SNNs like some of the BNNs in Table 3 [6,35] use real-valued weights, while other BNNs use binary weights [4,30,50]. Although SNNs use full precision weights, they provide computational efficiency like BNNs by replacing MACs with adds, while maintaining accuracy. Note, we performed energy comparison of T1 with ANNs following [31,32]. However, BNNs infer using XNOR and popcount ops instead of adds (used in SNNs), which might further improve efficiency. Therefore, the proposed SNNs can provide suitable trade-off between accuracy and efficiency between ANN and BNNs.

Table 4. Accuracy(%) on CIFAR10

T	VGG6g	VGG6d
5	90.61	90.15
4	90.52	90.08
3	90.40	89.91
2	90.05	89.35
1	89.10	88.64

Table 5. Accuracy(%) with VGG16, D_w and D_wo denote dataset D with and without batch-norm respectively

T	CIFAR10_w	CIFAR10_wo	CIFAR100_w	CIFAR100_wo
5	93.90	92.15	71.58	69.86
4	93.87	91.95	71.51	69.84
3	93.85	91.90	71.46	69.76
2	93.72	91.88	71.43	69.30
1	93.05	91.03	70.15	67.76

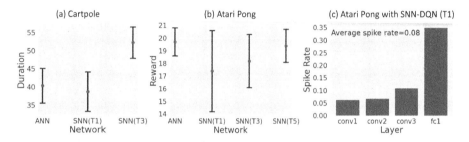

Fig. 3. Average reward (errorbars depict mean ± std), using DQN with ANN and SNN on (a) Cartpole and (b) Atari Pong, (c) layerwise spike rate with T1 on Atari Pong.

Efficacy of Temporal Pruning with Shallow Networks. Temporal pruning is required due to spike vanishing at later layers if direct transition from T5 to T1 is attempted. However, for shallow networks, training with T1 following direct conversion from ANN might be possible. To investigate this, we experiment with a VGG6 on CIFAR10 and the results are shown in Table 4, where VGG6g denotestrained with temporal pruning and VGG6d denotes directly converted from ANN and trained using that particular timestep. The proposed scheme provides slightly higher accuracy compared to direct training in case of shallower networks. This is consistent with [10,13], where the authors achieve better performing networks using sequential model compression.

Proposed Method With and Without Batch-Norm. Recent works have achieved low latency by adopting batch-normalization (BN) suitably in SNNs [19,49]. To disentangle the effect of BN from the proposed scheme and ensure that achieving convergence with 1 timestep is orthogonal to using BN, we perform ablation studies as shown in Table 5. For both CIFAR10 and CIFAR100, we are able to perform training with T5 to T1 irrespective of using BN during ANN training. Using BN enhances accuracy, but sequential temporal pruning can be performed independently from it. Also, [49] reports that using threshold-dependent BN allows reducing latency up to a minimum of 6 timesteps, but we can go up to 1. Note, BN is used only during ANN training and the BN parameters are fused with the weights during ANN-SNN conversion as proposed in [34]; BN is not used in the SNN domain.

Skipping Intermediate Timestep Reduction Steps. Since the proposed scheme increases training overhead due to sequential temporal pruning, it is worth investigating if this cost can be reduced by skipping in between timestep reduction steps. To that effect, we experiment with 2 cases- (i) training T5 followed by T3_5, followed by T1_3, (ii) including all intermediate timesteps with the sequence- T5, T4_5, T3_4, T2_3 and T1_2. Interestingly, both these cases perform comparably; for CIFAR10, we obtain 93.01% and 93.05% accuracy, respectively with 1 timestep for cases (i) and (ii). These values for CIFAR100 are 69.92% and 70.15%, respectively, and for ImageNet, the values are 68.98% and 69%, respectively. This indicates that if the end goal is obtaining T1 SNN, training overhead can be reduced by skipping intermediate steps.

T1 with Spatial Pruning. With T1, we obtain maximum compression in the temporal domain of SNNs. Additionally, we hypothesize that T1 might be amenable to spatial pruning as there is redundancy in the weights of DNNs [12,13]. To investigate this, we perform experiments on T1 with magnitude based weight pruning [12]. Our result indicate that we can remove up to 90% of the total weights of a VGG16 without large drop in performance. Using VGG16 (T1), we obtain 91.15% accuracy on CIFAR10 and 68.20% accuracy on CIFAR100, while retaining just 10% of the original spatial connections. This provides evidence that spatial pruning techniques can be combined with T1 SNNs.

Reinforcement Learning (RL) with Proposed SNNs. Due to their inherent recurrence, SNNs with multiple timesteps might be more useful in sequential decision-making (such as RL tasks) than static image classification. However, application of SNNs in RL may be limited if the latency is too high, since the agent has to make decisions in real-time. The authors in [39] obtain high performing RL agents on Atari games with SNNs, but with 500 timesteps. In this section, we investigate if our technique can enable training SNNs for RL tasks with low latency. The training pipeline is similar to that used for image classification tasks (hybrid SNN training with temporal pruning). Experiments are performed using deep Q-networks (DQN) [26] with SNNs (SNN-DQN) on cartpole and Atari pong environments. As shown in Fig. 3(a) and (b), for both cases, we can train SNNs up to 1 timestep. The rewards are obtained by averaging over 20 trials and plotted with error-bars showing mean ± std. In (a), the reward is the duration of the cartpole remaining balanced. DQN with ANN, SNN(T1) and SNN(T3) achieve 40.3 ± 4.8, 38.7 ± 5.4, 52.2 ± 4.3 reward, respectively. While T1 SNN performs slightly worse compared to ANN-based DQN, T3 outperforms the ANN. Similar performance improvement over ANN using SNN for some RL tasks has been reported in [39]. Next, we experiment with a more complex task, Atari pong game. In this case, T1 achieves reward of 17.4 ± 3.2 compared to

ANN based DQN's 19.7 ± 1.1. However, with T5, we obtain comparable reward (19.4 ± 1.3) to ANN. These results demonstrate that even T1 can handle dynamic inputs, albeit with lower accuracy than T3 or T5 (Fig. 3). Number of timestep is defined as the number of forward passes through the model. As such, T1 does not prevent usage with temporal inputs, rather each frame is processed just once, like an RNN. However, for sequential tasks, SNNs with multiple timesteps indeed provide enhanced performance using the inherent memory of membrane potential in neurons. Notably, we can obtain SNNs for RL task (pong) with significantly lower latency compared to prior art. On pong, [39] reports reward of 19.8 ± 1.3 using 500 timesteps, whereas, we obtain 19.4 ± 1.3 reward using just 5 timesteps. Our training method enables SNNs for RL tasks with comparable performance to ANNs with ~ 5 timesteps, and up to 1 timestep with slightly lower performance. Moreover, such latency reduction translates to considerable energy savings which is critical for agents operating in the real world. The layerwise spike rates for SNN-DQN (T1) are shown in Fig. 3(c), the average spike rate is 0.08, which results in 7.55X higher computational energy efficiency compared to ANN-DQN. Furthermore, if we compare iso-performing networks to ANN-DQN, SNN-DQN (T5) infers with average spike rate of 0.42, thus providing 5.22X higher computational energy efficiency compared to ANN-DQN. Details of training and additional results are given in supplementary section 6.

Is Direct Training Feasible? Though the overhead due to iterative training does not affect our primary goal (inference efficiency), we explore the feasibility of directly obtaining T1 SNNs here. We reproduce the results of [1,35] where the networks are trained as BNN; but as shown in Table 3, the iterative process provides better results. Furthermore, it is challenging to scale these networks with purely binary activation to ImageNet. More importantly, the continuum of SNNs cannot be obtained using this process, which is required to have optimal SNN solutions for both static and dynamic tasks. Additionally, we investigate if direct conversion from ANN to T1 is feasible and find that this case leads to convergence failure (details provided in supplementary section 5).

Limitations. A limitation of the proposed scheme is the training overhead due to sequential training. However, this occurs only in training (can be performed offline) and our end goal is inference which is not impacted by this iterative process. Moreover, this cost can be reduced by skipping latency reduction steps. Furthermore, if the end goal is T1, training can be performed with equivalent overhead to T5 by early stopping training of parent SNNs. In this case, we do not train the SNNs with higher timesteps till full convergence. Rather, we keep the number of overall training epochs fixed and divide the epochs equally among the temporal pruning stages. Let, the number of training epochs used for training only T5 is v. Then, for the case of training T5 followed by T3_5, followed by T1_3, we use (v/3) number of training epochs at each stage. Interestingly, this causes negligible performance drop ($<0.3\%$) for CIFAR10 with VGG16.

6   Conclusion

Bio-plausible SNNs hold promise as energy efficient alternatives to ANNs. However, mitigating high inference latency is critical for their edge deployment. To that end, we propose a temporal pruning approach which results in a continuum of SNNs from T timesteps up to unity (T1). The proposed low-latency SNNs use direct input coding with threshold and leak as trainable parameters along with weights. This leads to a spectrum of optimum SNN solutions for both static and sequential tasks. For static vision applications, T1 SNNs provide the best efficiency and latency. In particular, the T1 SNNs enhance computational efficiency by 25X on ImageNet compared to ANNs using VGG16 and are able to reduce the memory access overhead compared to SNNs with higher T. On the other hand, the presented approach also provides SNN based solutions for sequential tasks using few (1–5) timesteps. Notably, these SNN based deep Q-networks (DQNs) perform comparably to ANNs while providing significantly higher efficiency. Thus, the proposed technique enables training highly efficient SNNs on static tasks (up to unit timestep) as well as on sequential tasks (with few timesteps) while maintaining satisfactory performance.

Acknowledgement. The work was supported in part by, Center for Brain-inspired Computing (C- BRIC), a DARPA sponsored JUMP center, Semiconductor Research Corporation, National Science Foundation, Intel Corporation, the DoD Vannevar Bush Fellowship and U.S. Army Research Laboratory.

References

1. Bengio, Y., Léonard, N., Courville, A.: Estimating or propagating gradients through stochastic neurons for conditional computation. arXiv preprint arXiv:1308.3432 (2013)
2. Cao, Y., Chen, Y., Khosla, D.: Spiking deep convolutional neural networks for energy-efficient object recognition. Int. J. Comput. Vision **113**(1), 54–66 (2015)
3. Chen, T., Zhang, Z., Ouyang, X., Liu, Z., Shen, Z., Wang, Z.: "bnn-bn=?": training binary neural networks without batch normalization. In: Proceedings of the IEEE/CVF Conference on Computer Vision and Pattern Recognition, pp. 4619–4629 (2021)
4. Courbariaux, M., Hubara, I., Soudry, D., El-Yaniv, R., Bengio, Y.: Binarized neural networks: training deep neural networks with weights and activations constrained to+1 or −1. arXiv preprint arXiv:1602.02830 (2016)
5. Deng, L., Liu, Y.: Deep Learning in Natural Language Processing. Springer, Heidelberg (2018). https://doi.org/10.1007/978-981-10-5209-5
6. Deng, S., Gu, S.: Optimal conversion of conventional artificial neural networks to spiking neural networks. In: International Conference on Learning Representations (2020)
7. Diehl, P.U., Neil, D., Binas, J., Cook, M., Liu, S.C., Pfeiffer, M.: Fast-classifying, high-accuracy spiking deep networks through weight and threshold balancing. In: 2015 International Joint Conference on Neural Networks (IJCNN), pp. 1–8. IEEE (2015)

8. Fang, W., Yu, Z., Chen, Y., Huang, T., Masquelier, T., Tian, Y.: Deep residual learning in spiking neural networks. arXiv preprint arXiv:2102.04159 (2021)
9. Frenkel, C.: Sparsity provides a competitive advantage. Nat. Mach. Intell. **3**(9), 742–743 (2021)
10. Furlanello, T., Lipton, Z., Tschannen, M., Itti, L., Anandkumar, A.: Born again neural networks. In: International Conference on Machine Learning, pp. 1607–1616. PMLR (2018)
11. Han, B., Srinivasan, G., Roy, K.: Rmp-snn: residual membrane potential neuron for enabling deeper high-accuracy and low-latency spiking neural network. In: Proceedings of the IEEE/CVF Conference on Computer Vision and Pattern Recognition, pp. 13558–13567 (2020)
12. Han, S., Mao, H., Dally, W.J.: Deep compression: compressing deep neural networks with pruning, trained quantization and huffman coding. arXiv preprint arXiv:1510.00149 (2015)
13. Han, S., Pool, J., Tran, J., Dally, W.J.: Learning both weights and connections for efficient neural networks. arXiv preprint arXiv:1506.02626 (2015)
14. Hinton, G., et al.: Deep neural networks for acoustic modeling in speech recognition: the shared views of four research groups. IEEE Signal Process. Mag. **29**(6), 82–97 (2012)
15. Horowitz, M.: 1.1 computing's energy problem (and what we can do about it). In: 2014 IEEE International Solid-State Circuits Conference Digest of Technical Papers (ISSCC), pp. 10–14. IEEE (2014)
16. Huh, D., Sejnowski, T.J.: Gradient descent for spiking neural networks. Adv. Neural Inf. Process. Syst., 1433–1443 (2018)
17. Hunsberger, E., Eliasmith, C.: Spiking deep networks with lif neurons. arXiv preprint arXiv:1510.08829 (2015)
18. Kim, J., Kim, H., Huh, S., Lee, J., Choi, K.: Deep neural networks with weighted spikes. Neurocomputing **311**, 373–386 (2018)
19. Kim, Y., Panda, P.: Revisiting batch normalization for training low-latency deep spiking neural networks from scratch. arXiv preprint arXiv:2010.01729 (2020)
20. Krishnapriyan, A., Gholami, A., Zhe, S., Kirby, R., Mahoney, M.W.: Characterizing possible failure modes in physics-informed neural networks. Adv. Neural Inf. Process. Syst. **34** (2021)
21. Krizhevsky, A., Sutskever, I., Hinton, G.E.: Imagenet classification with deep convolutional neural networks. Adv. Neural Inf. Process. Syst., 1097–1105 (2012)
22. Lee, C., Sarwar, S.S., Panda, P., Srinivasan, G., Roy, K.: Enabling spike-based backpropagation for training deep neural network architectures. Front. Neurosci. **14** (2020)
23. Li, D., Chen, X., Becchi, M., Zong, Z.: Evaluating the energy efficiency of deep convolutional neural networks on cpus and gpus. In: 2016 IEEE International Conferences on Big Data and Cloud Computing (BDCloud), Social Computing and Networking (SocialCom), Sustainable Computing and Communications (SustainCom) (BDCloud-SocialCom-SustainCom), pp. 477–484. IEEE (2016)
24. Lu, S., Sengupta, A.: Exploring the connection between binary and spiking neural networks. arXiv preprint arXiv:2002.10064 (2020)
25. Maass, W.: Networks of spiking neurons: the third generation of neural network models. Neural Netw. **10**(9), 1659–1671 (1997)
26. Mnih, V., et al.: Human-level control through deep reinforcement learning. Nature **518**(7540), 529–533 (2015)
27. Neftci, E.O., Mostafa, H., Zenke, F.: Surrogate gradient learning in spiking neural networks. IEEE Signal Process. Mag. **36**, 61–63 (2019)

28. Park, S., Kim, S., Choe, H., Yoon, S.: Fast and efficient information transmission with burst spikes in deep spiking neural networks. In: 2019 56th ACM/IEEE Design Automation Conference (DAC), pp. 1–6. IEEE (2019)

29. Park, S., Kim, S., Na, B., Yoon, S.: T2fsnn: deep spiking neural networks with time-to-first-spike coding. arXiv preprint arXiv:2003.11741 (2020)

30. Rastegari, M., Ordonez, V., Redmon, J., Farhadi, A.: XNOR-Net: ImageNet classification using binary convolutional neural networks. In: Leibe, B., Matas, J., Sebe, N., Welling, M. (eds.) ECCV 2016. LNCS, vol. 9908, pp. 525–542. Springer, Cham (2016). https://doi.org/10.1007/978-3-319-46493-0_32

31. Rathi, N., Roy, K.: Diet-snn: direct input encoding with leakage and threshold optimization in deep spiking neural networks. arXiv preprint arXiv:2008.03658 (2020)

32. Rathi, N., Srinivasan, G., Panda, P., Roy, K.: Enabling deep spiking neural networks with hybrid conversion and spike timing dependent backpropagation. In: International Conference on Learning Representations (2020). https://openreview.net/forum?id=B1xSperKvH

33. Roy, K., Jaiswal, A., Panda, P.: Towards spike-based machine intelligence with neuromorphic computing. Nature **575**(7784), 607–617 (2019)

34. Rueckauer, B., Lungu, I.A., Hu, Y., Pfeiffer, M., Liu, S.C.: Conversion of continuous-valued deep networks to efficient event-driven networks for image classification. Front. Neurosci. **11**, 682 (2017)

35. Sakr, C., Choi, J., Wang, Z., Gopalakrishnan, K., Shanbhag, N.: True gradient-based training of deep binary activated neural networks via continuous binarization. In: 2018 IEEE International Conference on Acoustics, Speech and Signal Processing (ICASSP), pp. 2346–2350. IEEE (2018)

36. Sengupta, A., Ye, Y., Wang, R., Liu, C., Roy, K.: Going deeper in spiking neural networks: Vgg and residual architectures. Front. Neurosci. **13**, 95 (2019)

37. Shen, M., Liu, X., Gong, R., Han, K.: Balanced binary neural networks with gated residual. In: ICASSP 2020–2020 IEEE International Conference on Acoustics, Speech and Signal Processing (ICASSP), pp. 4197–4201. IEEE (2020)

38. Shrestha, S.B., Orchard, G.: Slayer: spike layer error reassignment in time. Adv. Neural Inf. Process. Syst., 1412–1421 (2018)

39. Tan, W., Patel, D., Kozma, R.: Strategy and benchmark for converting deep q-networks to event-driven spiking neural networks. arXiv preprint arXiv:2009.14456 (2020)

40. Wang, P., He, X., Li, G., Zhao, T., Cheng, J.: Sparsity-inducing binarized neural networks. In: Proceedings of the AAAI Conference on Artificial Intelligence, vol. 34, pp. 12192–12199 (2020)

41. Wang, Z., Lu, J., Tao, C., Zhou, J., Tian, Q.: Learning channel-wise interactions for binary convolutional neural networks. In: Proceedings of the IEEE/CVF Conference on Computer Vision and Pattern Recognition, pp. 568–577 (2019)

42. Wu, J., Chua, Y., Zhang, M., Li, G., Li, H., Tan, K.C.: A tandem learning rule for efficient and rapid inference on deep spiking neural networks. arXiv pp. arXiv-1907 (2019)

43. Wu, Y., Deng, L., Li, G., Zhu, J., Shi, L.: Spatio-temporal backpropagation for training high-performance spiking neural networks. Front. Neurosci. **12**, 331 (2018)

44. Wu, Y., Deng, L., Li, G., Zhu, J., Xie, Y., Shi, L.: Direct training for spiking neural networks: faster, larger, better. In: Proceedings of the AAAI Conference on Artificial Intelligence, vol. 33, pp. 1311–1318 (2019)

45. Xu, Z., Cheung, R.C.: Accurate and compact convolutional neural networks with trained binarization. In: 30th British Machine Vision Conference (BMVC 2019) (2019)
46. Yang, C., Xie, L., Qiao, S., Yuille, A.L.: Training deep neural networks in generations: a more tolerant teacher educates better students. In: Proceedings of the AAAI Conference on Artificial Intelligence, vol. 33, pp. 5628–5635 (2019)
47. Zenke, F., Ganguli, S.: Superspike: supervised learning in multilayer spiking neural networks. Neural Comput. **30**(6), 1514–1541 (2018)
48. Zhang, W., Li, P.: Temporal spike sequence learning via backpropagation for deep spiking neural networks. Adv. Neural Inf. Process. Syst. **33** (2020)
49. Zheng, H., Wu, Y., Deng, L., Hu, Y., Li, G.: Going deeper with directly-trained larger spiking neural networks. In: Proceedings of the AAAI Conference on Artificial Intelligence, vol. 35, pp. 11062–11070 (2021)
50. Zhou, S., Wu, Y., Ni, Z., Zhou, X., Wen, H., Zou, Y.: Dorefa-net: training low bitwidth convolutional neural networks with low bitwidth gradients. arXiv preprint arXiv:1606.06160 (2016)

Towards Accurate Network Quantization with Equivalent Smooth Regularizer

Kirill Solodskikh[1(✉)], Vladimir Chikin[1], Ruslan Aydarkhanov[1], Dehua Song[1], Irina Zhelavskaya[2], and Jiansheng Wei[1]

[1] Huawei Noah's Ark Lab, Hong Kong, China
{solodskikh.kirill1,vladimir.chikin,ruslan.aydarkhanov,
dehua.song,weijiansheng}@huawei.com
[2] Skolkovo Institute of Science and Technology (Skoltech), Moscow, Russia
irina.zhelavskaya@skolkovotech.ru

Abstract. Neural network quantization techniques have been a prevailing way to reduce the inference time and storage cost of full-precision models for mobile devices. However, they still suffer from accuracy degradation due to inappropriate gradients in the optimization phase, especially for low-bit precision network and low-level vision tasks. To alleviate this issue, this paper defines a family of equivalent smooth regularizers for neural network quantization, named as SQR, which represents the equivalent of actual quantization error. Based on the definition, we propose a novel QSin regularizer as an instance to evaluate the performance of SQR, and also build up an algorithm to train the network for integer weight and activation. Extensive experimental results on classification and SR tasks reveal that the proposed method achieves higher accuracy than other prominent quantization approaches. Especially for SR task, our method alleviates the plaid artifacts effectively for quantized networks in terms of visual quality.

Keywords: Network quantization · Smooth regularizer · Equivalence · Gradient · Low-level vision task

1 Introduction

Deep Neural Network (DNN) has dramatically boosted the performance of various practical tasks due to its strong representation capacity, for example, image classification [19], image translation [7] and speech recognition [14]. Along with the requirements of deploying DNN into mobile devices increasing, it has been necessary to develop low-latency, efficient and compact networks. Recently, large amounts of approaches have been proposed to solve this problem, including network pruning [11,25], quantization [9,18] and adder neural network [2].

K. Solodskikh and V. Chikin—These authors contributed equally to this work.

Supplementary Information The online version contains supplementary material available at https://doi.org/10.1007/978-3-031-20083-0_43.

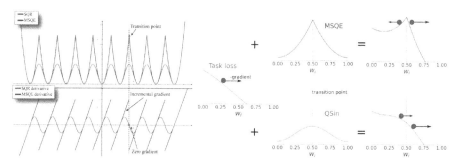

(a) Function and gradient curve.   (b) Effect of gradients around transition point.

Fig. 1. The comparison of SQR and MSQE [4] regularizer. SQR is smooth everywhere instead of the unsmoothness of MSQE in each transition point, and represents the equivalent of actual quantization error, which allows to obtain better gradient behavior in neighborhood of transition points.

Network quantization is one of the most appealing way to reduce the inference latency, energy consumption and memory cost of neural networks. Since low-bit integer tensors (weight/activation) and integer arithmetics are employed in quantized network, the model size and computation consumption could be decreased significantly. The advantages of quantization network on low precision hardware has been demonstrated with multiple systems [10,18], but it still suffers from accuracy degradation due to inappropriate gradients in the optimization phase, especially for low-bit precision network and low-level vision tasks.

Minimization of objective function for quantized neural networks in general case is a hard optimization problem since the gradient is either zero or undefined. The prominent Quantization Aware Training (QAT) algorithms [5,16,31] usually adopted the Straight-Through Estimator (STE) [1,3,18] strategy to solve this gradient issue, which approximates the gradient of the rounding operator as 1. Although several further approaches [22,28] have been proposed to refine the gradient approximation, such kind of algorithms still suffer from the gradient error, especially for lower-bit quantization. Another alternative way is to train the network with regularizer [4,8] of quantization error to generate the quantized model, where gradients from accuracy loss could be propagated effectively.

Unfortunately, the gradient of the most conventional regularizer for quantization, mean square quantization error (MSQE) [4], is undefined in each transition point which is illustrated in Fig. 1. It hinders the quantization error from being propagated to the weights of each layer. What's worse, steep gradients around transition points would dominate the direction of update step for the joint objective, which is prone to reach the closest grid point instead of the optimal point of the accuracy loss (see Fig. 1(b)). SinReQ [8] explored a smooth regularizer for quantization to alleviate the gradient issue. However, its variation trend outside of quantization segment is quite different from the actual quantization error, which results in high clamping error and significant accuracy degradation. To reduce the quantization error, the prime regularizer should not only be smooth everywhere but also represent the equivalent of actual quantization error. Hence, this paper defined a family of equivalent smooth regularizers for neural network quantization, called SQR. Based on this definition, we proposed a novel QSin regularizer as an instance to evaluate the performance of SQR, and also built up an algorithm to train the network for integer weight and activation. The quantization error could be reflected effectively and propagated to weights smoothly. To evaluate the performance and generality of our approach, extensive experiments on classification and SR tasks were conducted. The results reveal that the proposed method achieves higher accuracy than other prominent approaches. Especially for SR task, our method alleviates the plaid artifacts effectively for quantized networks in terms of visual quality, since the pixel value regression is more easily affected by the quantization error.

The main contributions of this paper are threefold:

1. We defined a novel family of equivalent smooth regularizer for quantization and analyzed its properties theoretically.
2. We proposed a novel QSin regularizer belonging to SQR and built up a general algorithm to train the network with weight and activation quantization for any bit-width. It is important to note that our regularizer allows to train quantized network without weights rounding comparing with the most quantization algorithms.
3. Our method works stable and achieves state-of-the-art results on wide spectra of computer vision tasks, including image classification and super-resolution tasks. What's more, our method could alleviate the plaid artifacts effectively for quantized networks.

2   Related Works

Quantization is one of the most important technique for model compression, which attracts many researchers to investigate it. In the last decade, many quantization approaches were proposed to improve the performance of quantization network. According to the criterion of whether training the quantized network or not, the quantization methods could be roughly divided into two categories: Post Training Quantization (PTQ) [23] and Quantization Aware Training (QAT) [16].

Post Training Quantization. Post training quantization algorithms aim at quantizing the trained full precision network into low-precision one with compact unlabeled calibration set or even without any data. Nowadays such algorithms have achieved significant progresses in quantization of classification networks. Nagel *et al.* [23] employed the minimum and maximum values of weights to define weight quantization parameters and moving average of minimum and maximum values to define activations quantization parameters, respectively. Based on this method, Hubara *et al.* [17] further explored tuning batch normalization layer and boosted the performance of quantized network significantly. Such kind of methods are attractive because of quick implementation, setup and application but usually lead to accuracy drop comparing with full precision networks.

Quantization Aware Training. To reduce the accuracy drop of quantized network, numerous quantization aware training algorithms [3,5,31] have been proposed, which utilize stochastic gradient descent technique with quantized weights and activations on forward pass stage but full precision weights on backward pass procedure. Since the gradient of round function is zero or undefined everywhere, Straight Through Estimator [1] has been proposed to propagate the derivative. LSQ [9] method was proposed to further improve the accuracy via learnable step size. More quantization parameters were suggested to learn with end-to-end optimization manner [30]. To alleviate the gradient error problem, various approaches (DSQ [12], PACT [3], QuantNoise [28], *etc.*) were introduced with progressive way to train the quantization network. Unfortunately, these methods still cannot solve this issue thoroughly.

Quantization Through Regularization. Another alternative way of generating quantized network is to train the network with regularizer [4,8,24] of quantization error, where gradients from accuracy loss could be propagated smoothly. Choi *et al.* [4] firstly proposed regularizer term of mean squared quantization error (MSQE) for weight and activation quantization. However, the gradients of MSQE regularizer in transition points are undefined, which prevents the quantization error from propagating. SinReQ [8] explored periodic functions as regularizer for weight quantization. Unfortunately, its variation trend outside of quantization segment is quite different from the actual quantization error, which results in high clamping error and significant accuracy degradation.

3   Preliminaries and Motivation

Here we firstly briefly introduce the basic principles of neural network quantization, and then discuss the difficulty of quantization network training.

We consider neural network $\mathcal{F}(\mathbf{W}, \mathbf{X})$ as an ordered graph with n layers, and each layer corresponds to the function $\mathcal{F}_i(\mathbf{W}_i, \mathbf{A}_i)$, where $\mathbf{W}_i$ and $\mathbf{A}_i$ denote the parameter tensor and input features of the i-th layer, respectively. For convenience, we denote the set $\{\mathbf{W}_i\}_{i=1}^N$ as $\mathbf{W}$ and an input data tensor with $\mathbf{X}$. The $\mathbf{X}$ could be simply modeled by continuous distribution ξ.

Quantization. Network quantization aims at reducing the precision of both parameters and activations with minimal impact on the representation ability of full-precision models. Firstly, we need to define a function which can quantize real value set into a finite set. The conventional uniform quantization function $\mathcal{Q}_U$ is defined as follows:

$$\mathcal{Q}_U(x) = \begin{cases} \lfloor x \rceil, & \text{if } r_b \leq x \leq r_t, \\ r_b, & \text{if } x < r_b, \\ r_t, & \text{if } x > r_t, \end{cases} \tag{1}$$

where x is the input of function. r_b and r_t denote the minimum and maximum of clipping range, respectively. $\lfloor \cdot \rceil$ is the round-to-nearest operator. To quantize a real value into an integer, we usually need three quantization parameters: scale factor s, zero-point and bit-width. For convenience, here we employ the symmetric uniform quantization to analyze problems. Then, Eq. 2 could be utilized to quantize the weights and activations.

$$q = \mathcal{Q}_U\left(\frac{x}{s}\right), \tag{2}$$

where x is the real value, and q is the quantized integer. When both weights and activations are quantized into integer, computation could be executed with an integer-arithmetic way, which results in significant acceleration on hardware. The full-precision layer $\mathcal{F}_i(\mathbf{W}_i, \mathbf{A}_i)$ is replaced by the quantized layer:

$$\mathcal{F}_i^q = s_{w_i} s_{a_i} \mathcal{F}_i\left(\mathcal{Q}_U\left(\frac{\mathbf{W}_i}{s_{w_i}}\right), \mathcal{Q}_U\left(\frac{\mathbf{A}_i}{s_{a_i}}\right)\right), \tag{3}$$

where s_{w_i} and s_{a_i} denote the scale factor of weight and activation of the i-th layer, respectively.

Mean Squared Quantization Error. Considering network quantization problem as an optimization problem with special constraints, Choi *et al.* [4] proposed the mean-squared-quantization-error (MSQE) as regularization term for weight and activation quantization, which is defined as follows:

$$MSQE(\mathbf{V}; s) = \frac{1}{K} \sum_{x_j \in \mathbf{V}} |x_j - s \cdot \mathcal{Q}_U\left(\frac{x_j}{s}\right)|^2, \tag{4}$$

where V denotes the input tensor with K components. It reflects the error between original full-precision value and its quantized value. To constraint the weights and activations of the whole network, the regularizer term should contains the quantization error of each layer. The complete MSQE regularizer terms for weights ($MSQE_w$) and activations ($MSQE_a$) are defined as Eq. 5.

$$MSQE_w = \frac{1}{N} \sum_{i=1}^{N} MSQE(\mathbf{W}_i, s_{w_i}), \quad MSQE_a = \frac{1}{N} \sum_{i=1}^{N} MSQE(\mathbf{A}_i, s_{a_i}). \tag{5}$$

Let $\mathcal{L}$ is the original objective function of full-precision neural network $\mathcal{F}(\mathbf{W}, \mathbf{X})$. Then, we consider the quantization network training issue as a *neural network optimization problem with quantization constraints*:

$$\begin{cases} E\left[\mathcal{L}\left(\mathcal{F}(\mathbf{W}, \mathbf{X})\right)\right] \to \min, \\ MSQE_w < C_w, \\ E[MSQE_a] < C_a. \end{cases} \tag{6}$$

where C_w and C_a denote the thresholds restricting the quantization error for weights and activations. $E[\cdot]$ is the expectation among the whole database. Theoretically, we can not effectively reach the minimization of Lagrange function 6 since MSQE is not smooth. From Fig. 1, we can see that the gradient of MSQE is undefined in each transition point, which limits the performance of quantization network. To address this issue, we define a class of Smooth Quantization Regularizers (SQR) which represents the equivalent of actual quantization error.

4   SQR: Equivalent Smooth Quantization Regularizer

Smooth property of regularizer is friendly to network optimization, which is helpful to solve the gradient problem. Here, we propose a family of smooth quantization regularizer to replace the MSQE regularizer, which represents the equivalent of actual quantization error and allows to obtain better gradient behavior in neighborhood of transition points.

4.1   Definition of SQR

From Eq. 4 and Fig. 1, we can observe that the MSQE regularizer is not a smooth function due to the quantization operator. To acquire the smooth regularizer for quantization, we should deal with the transition points carefully. Besides the smooth property, we also hope that the smooth regularizer could effectively reflect the trend of quantization error and preserve the same number of minimum with MSQE. Then, we can define the ideal smooth quantization regularizers (SQR) as follows. Here we abbreviate the quantization regularizer $MSQE(x; s)$ as $MSQE(x)$ for simplicity.

Definition 1. *With the same constant scale factor s with* $MSQE(x)$, *function* $\phi(x)$ *is a Smooth Quantization Regularizer (SQR) for the uniform grid of integers with the segment* $[r_b, r_t]$ *when it satisfies the following three properties:*

1) **Order preserving.** *Function* $\phi(x)$ *preserves the order of* $MSQE(x)$, *i.e.:*

$$MSQE(x_1) \leq MSQE(x_2) \Leftrightarrow \phi(x_1) \leq \phi(x_2),$$

$\forall\ x_1, x_2 \in [r_b, r_t]\ or\ x_1, x_2 \in \mathbb{R} \setminus [r_b, r_t].$
2) **Equivalence.** *There exists* $a, b \in \mathbb{R}$, *and* $0 < a < b$, *such that*

$$a\,MSQE(x) \leq \phi(x) \leq b\,MSQE(x),\ \forall x \in \mathbb{R}. \tag{7}$$

3) **Smoothness.** $\phi(x) \in \mathcal{C}^2(\mathbb{R})$.

where $\mathcal{C}^2(\mathbb{R})$ *denotes the twice differentiable function family for the domain of all real numbers.*

According to the definition, we could further infer that SQRs are periodic within the domain of quantization segment $[r_b, r_t]$. In addition, SQRs could not only preserve the same minima points with MSQE, but also acquire the close asymptotic around the quantization grid points and at infinity. In other words, for arbitrary SQR ϕ and $s > 0$, there exists $B > 0$ such that the following relation holds for $MSQE(x; s) \to 0$:

$$s^2 \phi\left(\frac{x}{s}\right) = B \cdot MSQE(x; s) + o(MSQE(x)). \tag{8}$$

These admirable characteristics guarantee that we could employ SQRs to replace the conventional MSQE with negligible relaxation.

Following the notation of MSQE, we extend the SQR for tensor $\mathbf{X}$ with the average of all the components' $\phi(x)$ values. Then, Lagrange function minimization of quantization network in the definition domain of parameters $(\mathbf{W}, \mathbf{s}_w, \mathbf{s}_a)$ could be rewritten as follows:

$$\mathcal{L}_Q = \mathrm{E}\big[\mathcal{L}\big(\mathcal{F}(\mathbf{W}, \mathbf{X})\big)\big] + \lambda_w \mathcal{L}_w + \lambda_a \mathcal{L}_a \tag{9}$$

$$\mathcal{L}_w(\mathbf{W}; \mathbf{s}_w) = \frac{1}{N}\sum_{i=1}^{N} s_{w_i}^2 \phi(\mathbf{W}_i, s_{w_i}), \ \ \mathcal{L}_a(\mathbf{A}; \mathbf{s}_a) = \frac{1}{N}\sum_{i=1}^{N} \mathrm{E}[s_{a_i}^2 \phi(\mathbf{A}_i, s_{a_i})]. \tag{10}$$

This objective function becomes smooth and amenable to optimize. According to the property defined in Eq. 7, it also effectively constrains the solution of network in the compact domain which belongs to the solution domain with MSQE regularizer. Indeed, if SQR $\phi(x)$ is less than $c \in \mathbb{R}$ in some domain $x \in \Omega$, therefore MSQE is less than ac for some $a \in \mathbb{R}$. This means that while we minimize SQR to zero then MSQE also converges to zero. More details and proofs could be seen in Appendix A.

4.2   QSin Regularizer

According to the definition of SQR, we proposed a novel smooth regularizer, QSin, to improve the performance of quantization network. The definition of QSin is showed in Eq. 11. QSin is a sinusoidal periodic function within the domain of quantization segment $[r_b, r_t]$, while it is a quadratic function beyond the quantization segment domain.

$$QSin(\mathbf{V}; s) = \frac{s^2}{K}\sum_{x_j \in \mathbf{V}} QSin_{on}\left(\frac{x_j}{s}\right), \tag{11}$$

where the $QSin_{on}$ is defined as follows:

$$QSin_{on}(x) = \begin{cases} \sin^2(\pi x), & \text{if } r_b \leq x \leq r_t, \\ \pi^2(x - r_b)^2, & \text{if } x < r_b, \\ \pi^2(x - r_t)^2, & \text{if } x > r_t. \end{cases} \tag{12}$$

QSin is a twice differentiable function for the whole domain of definition. Its function curve and gradient curve with $s = 1$ for scalar input are illustrated in

Fig. 1. QSin is smooth everywhere since there is no quantization operator, which is quite different from MSQE. As for quantization network optimization, the scale factor s usually needs to be optimized. Hence, we also compared the best solution of s for QSin and MSQE regularizer. As for the uniform quantization of random variable ξ, we randomly sampled M (e.g.128) values from standard normal distribution, and then computed the quantization error $MSQE(\xi; s)$ and $QSin(\xi; s)$ for each scale value. The best solution of s from QSin is quite close to that from MSQE for various distributions of ξ. More details could be seen in Appendix B. Therefore, we can employ smooth QSin to replace the MSQE regularizer while preserving sufficient constraint for quantization error.

4.3   Quantization Network Optimization

Considering network quantization problem as an optimization problem with special constraints, we substitute Eq. 11 into Eq. 9 and acquire the final objective function. This method employs the full precision network $\mathcal{F}(\mathbf{W}, \mathbf{X})$ without quantization operations on the forward pass during training stage, called Round Free (RF). Then, SGD technique could be utilized to optimize this objective function $\mathcal{L}_Q$ of network directly. Inspired by LSQ [9] method, we set the scale factor of weights and activations as a learnable parameter to improve the performance of quantization network. s_a and w_a will be updated during each step of gradient descent to minimize the SQR. More details could be seen in Appendix B. During the validation or test stage, the quantized network will be computed according to Eq. 3 with quantization operations and integer arithmetics.

Although our novel regularizer could constraint the activation effectively, there still exists differences between constrained activations and quantized activations to some degree, especially for low-bit quantization. Hence, as for the extreme low-bit quantization, an optional way is to add the quantization operation Eq. 2 to the activations to alleviate the quantization error accumulation problem. The activations before getting through the quantization operation are utilized to calculate the QSin regularizer. In this scheme, STE [1] should be employed to propagate the gradients through round function Q_U.

As for the coefficient of regularizer λ_w and λ_a, we set them as a power of 10 to normalize the regularizer loss $\mathcal{L}_a$ and $\mathcal{L}_w$ for acquiring the same order with the main loss. Weight quantization is achieved with a progressive way. The weight regularizer coefficient λ_w is adjusted with multiplying by 10 gradually during the training stage. The value of λ_a usually does not change for stable training. The whole quantized network training procedure is summarized in the Algorithm 1.

4.4   Discussion

In addition to MSQE, SinReQ [8] explored sinusoidal functions as regularizer for weight quantization. Unfortunately, its variation trend outside of quantization segment is quite different from the actual quantization error (see Fig. 2 and

---

Algorithm 1. Quantization with smooth regularizers

Require: W, $\mathbf{s}_w$, $\mathbf{s}_a$ – learnable parameters (weights and scale factors).
   λ_w, λ_a – regularization coefficients.
   lr, N_{train}, N_{init} – learning rate, epoch size, initialization batches number.
1: Initialize $\mathbf{s}_a$ by sample statistics evaluating $\mathcal{F}$ on N_{init} batches from **X**.
2: **for** N_{train} times **do**
3:     Sample random batch from train dataset.
4:     Evaluate quantization Lagrange function $\mathcal{L}_Q$.
5:     Calculate the gradients of learnable parameters: $\mathbf{G}_w = \frac{\partial \mathcal{L}}{\partial \mathbf{W}} + \lambda_w \frac{\partial \mathcal{L}_w}{\partial \mathbf{W}}$, $\mathbf{G}_{s_w} = \lambda_w \frac{\partial \mathcal{L}_w}{\partial \mathbf{s}_w}$,
6:     **If RF**: $\mathbf{G}_{s_a} = \lambda_a \frac{\partial \mathcal{L}_a}{\partial \mathbf{s}_a}$,
7:     **If STE**: $\mathbf{G}_{s_a} = \frac{\partial(\mathcal{L} + \lambda_a \mathcal{L}_a)}{\partial \mathbf{s}_a}$,
8:     Update learnable parameters using calculated gradients and learning rate lr.
9: **end for**
10: Validate quantized model $\mathcal{F}^q$.
Return: W, $\mathbf{s}_w$, $\mathbf{s}_a$

Fig. 1(a)), which results in high clamping error and significant accuracy degradation. Our QSin regularizer represents the equivalent of the actual quantization error effectively. QSin introduces penalty for clamping values which allows to train the quantization scale and improve the accuracy of quantization network. Besides, QSin also adds multiplier π^2 and preserves the twice differentiable property successfully. At last, the original SinReQ is only utilized on weight quantization for model compression. Our QSin method is employed on both weights and activations to acquire a fully quantized network. Hence, the proposed QSin method is quite different from the other quantization regularizers.

5   Experiment

Extensive experiments are conducted to demonstrate the effectiveness of the proposed QSin method, including classification task and Super-Resolution task.

5.1   Implementation Details

Database. As for classification task, we employed two popular datasets: CIFAR-10 and ImageNet (ILSVRC12) [6]. The CIFAR-10 database contains 50K training images and 10K test images with the size of 32×32, which belong to 10 classes. ImageNet database consists of about 1.2 million training images and 50K test images belonging to 1000 classes. For Super-Resolution task, we employ DIV2K [29] database to train the standard SR network. DIV2K database consists of 800 training images and 100 validation images. The low-resolution images are generated with bicubic degradation operator. During the test stage, Set5, Set14 and Urban100 [15] database are utilized to evaluate the performance of quantized SR network.

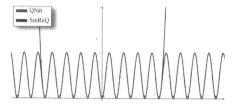

Fig. 2. Comparison of QSin and SinReQ [8] regularizer.

Settings. In following experiments, the weights and activations of all layers except for the first and last layers are quantized into low-precision integer. We employed Round Free mode for 8-bit quantization and STE mode on activations for 4-bit quantization during the training stage. The quantized network is trained by SGD optimizer with learning rate $lr = 0.001$ and momentum $m = 0.9$. The coefficient of activation regularizer λ_a is setted as a constant value 1. The coefficient of weight regularizer λ_w is initialized as 1 and adjusted each 30 epochs by multiplication on 10. Conventional data augmentation strategies [19,21] are also utilized in these experiments. More details about training configurations could be seen in Appendix D.

5.2   Comparison with State-of-the-arts

To compare with other SOTA quantization algorithm, we select four mainstream approaches: MSQE regularization [4], SinReQ [8] regularization, QAT in Tensorflow [18], LSQ [9] and DSQ [12]. By comparing with MSQE and SinReQ methods, we show that QSin achieves better performance since it is smooth and reflects the actual quantization error effectively. The benefits of QSin could be demonstrated by comparing with all these SOTA quantization approaches.

Image Classification. To facilitate comparison, we select conventional classification models including ResNet18 [13] and MibleNet-V2 [26]. We trained the quantized networks with the QSin regularizer on ImageNet database for 4-bit and 8-bit. The experimental results summarized in Table 1 show that QSin method achieves higher top-1 accuracy than other prominent quantization approaches for 4-bit and 8-bit with the architectures considered here. For 8-bit, the quantization networks even achieves slightly better performance than its full-precision model in some cases. Compact neural networks are usually hard to quantize while preserving the accuracy of full-precision model. It is interesting to note that the quantized 8-bit MobileNet-V2 network could achieve close results with full-precision model through QSin approaches. Moreover, QSin results were obtained without weights round during training what lead to higher accuracy on MobileNet-V2 comparing with STE approaches like LSQ.

Image Super-Resolution. To evaluate the performance of QSin method on low-level vision tasks, we consider the single image super-resolution (SISR) as a

Table 1. Quantitative results in comparison with state-of-the-art quantization methods on the ImageNet database. The best results are highlighted in bold.

Network	Method	Top-1 accuracy (%)	
		4-bit	8-bit
ResNet-18 [13]	Full-precision	69.8 (fp32)	
	QAT TF [18]	68.9	69.7
	PACT [30]	69.2	69.8
	DSQ [12]	69.4	69.8
	LSQ [9]	**69.8**	69.8
	SinReQ [8]	64.63	69.7
	MSQE [4]	67.3	68.1
	QSin	**69.7**	**70.0**
MobileNet-V2 [26]	Full-precision	71.8 (fp32)	
	PACT [30]	61.4	71.5
	DSQ [12]	64.8	71.6
	LSQ [9]	68.1	71.6
	SinReQ [8]	61.1	71.2
	MSQE [4]	67.4	71.2
	QSin	**68.7**	**71.9**

conventional task. EDSR [21] and ESPCN [27] are selected typical SR networks to facilitate the comparison. We trained these quantization networks with L_1 loss and QSin regularizer on DIV2K database and summarized the 8-bit quantization results in Table 2. It shows that the proposed QSin method achieves better performance than other quantization approaches, especially for MSQE. Besides, we also illustrate the visual quality of output images from quantized networks in Fig. 3. It is obvious that our QSin method could preserve the texture details of the full-precision model effectively since the smooth property of regularizer. In addition, we also provided comparison with PAMS quantization method [20] which is specially designed for SR network. It could be seen that, without any specific setups and initialization for SISR task, QSin could achieve the same or even slightly better performance as PAMS. More image examples could be seen in Appendix E.

5.3 Ablation Study

Effect of the Coefficient of Regularizer. The coefficients of regularizer λ_w and λ_a take an important role in the performance of quantized networks. Here we conducted ablation experiments on the CIFAR-10 database to explore the affect of coefficients of regularizer. A series of combinations of λ_w and λ_a were utilized

Table 2. Quantitative results of SR task in comparison with state-of-the-art quantization methods on benchmark databases. The best 8-bit quantization results are highlighted in bold.

Network	Method	PSNR (dB)		
		Set5	Set14	Urban100
4x EDSR [21]	Full-precision	32.2	28.5	26
	QAT TF [18]	31.9	28.4	25.7
	PACT [30]	31.5	28.2	25.25
	LSQ [20]	32.1	28.5	25.9
	PAMS [20]	32.1	**28.6**	26
	SinReQ [8]	32.1	28.3	25.3
	MSQE [4]	32.1	28.5	25.9
	QSin	**32.2**	28.5	**26**
3x ESPCN [27]	Full-precision	32.5	29	26.1
	QAT TF [18]	32.35	28.8	25.9
	LSQ [9]	32.4	28.9	26
	SinReQ [8]	32.2	28.9	26
	MSQE [4]	32.45	28.95	26
	QSin	**32.5**	**29**	**26.1**

to train the ResNet-20 quantization network. The experimental results summarized in Table 3 reveal that the coefficients has a significant influence on the final performance. Influence of λ_a is not so significant as λ_w but still obvious. The best results from our empirical evaluation shows that multistep scheduling of λ_w allows significantly improve quality. The intuition behind are follows: less values of λ_w helps to task loss makes more contribution, during training we increase λ_w to slightly decrease quantization error.

Table 3. Experimental results with various coefficients of regularizer on the CIFAR-10 database. Last line shows multistep scheduling on λ_w.

λ_w for weights	λ_a for activations	Accuracy (%)	
		4-bit	8-bit
0	0	88.9	91.6
0	1	89.2	91.6
0	10	89.2	91.6
1	0	90.0	91.6
1	1	90.2	91.7
10	1	90.6	91.7
100	1	91.1	91.8
(1, 10, 100)	1	91.7	91.9

Fig. 3. Visual Comparison of 3x ESPCN with state-of-the-art methods in terms visual quality. δ is the residual map of the corresponding image of quantized network and full-precision network.

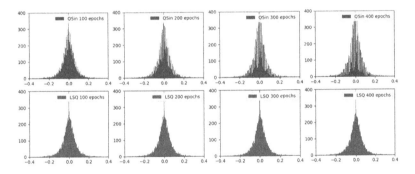

Fig. 4. Histograms of the weights distribution from the third convolution layer of ESPCNN model for SR task. The dynamic evolution of weights distribution from QSin and LSQ approaches are compared here.

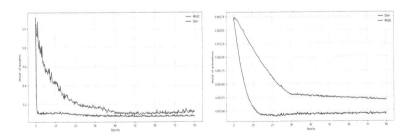

Fig. 5. Quantization error comparison of QSin and MSQE for ResNet-20 on CIFAR-10 in 4-bit quantization.

Quantization Error and Convergence Analysis. To analyze the effect of Qsin in terms of reducing the quantization error, we explored the variation of the mean square quantization error of activations and weights along with the increasing of training iterations. Figure 5 illustrates the results of 4-bit quantization on ResNet-20 with QSin and MSQE regularizers. It is obvious that QSin leads to much lower quantization error compared with MSQE regularizer since its smooth property. It reveals that QSin not only achieved higher accuracy than MSQE but also acquired more stable convergence. This phenomenon should thanks to the smooth property of QSin regularizer which is more helpful to optimization.

Weight Analysis, To analyze the weight quantization, we provide the histograms of weights distribution from the model which was trained with QSin regularizer in Fig. WeightDistribution. Multiple histograms from various epochs are illustrated together to explore the dynamic evolution of weights distribution. We have compared weights distributions of networks trained through QSin, MSQE and LSQ method [9]. LSQ employs trainable scale factor with straight through estimator to propagate through the round function. From Fig. 4, we can see that the weights distribution of QSin method becomes more and more similar to the categorical distribution along with rising of training epochs. In addition, the histograms of weights distribution from the network trained by QSin are closer to categorical distribution than weights histogram obtained from the LSQ method. During the evaluation and test stage, the quantized network would generate less quantization error on weights for QSin method. This is one reason why QSin method could achieve better performance than LSQ in Table 1 and Table 3. More distributions could be find in Appendix C.

6   Conclusion

This paper defined a family of equivalent smooth quantization regularizer to alleviate the accuracy degradation problem in network quantization. Then, we proposed a novel QSin regularizer belonging to SQR which represents the equivalent of actual quantization error and allows to obtain better gradient behavior in the neighborhood of transition points. In addition, we built up the corresponding algorithm to train the quantization network. The extensive experimental results show that the proposed SQR method could achieve much better performance than other prominent quantization approaches on image classification and super-resolution task. What's more, in terms of visual quality, SQR approach would not generate the grid artifact compared with other quantization methods due to its smooth property. The ablation study reveals that SQR could reduce the quantization error significantly and acquire stable convergence. Furthermore, distributions of the learned weights from SQR regularizer are more close to categorical distribution, which is helpful to booting the performance of quantized network.

References

1. Bengio, Y., Léonard, N., Courville, A.: Estimating or propagating gradients through stochastic neurons for conditional computation. arXiv preprint arXiv:1308.3432 (2013)
2. Chen, H., et al.: Addernet: do we really need multiplications in deep learning? In: Proceedings of the IEEE/CVF Conference on Computer Vision and Pattern Recognition, pp. 1468–1477 (2020)
3. Choi, J., Wang, Z., Venkataramani, S., Chuang, P.I.J., Srinivasan, V., Gopalakrishnan, K.: Pact: parameterized clipping activation for quantized neural networks. arXiv preprint arXiv:1805.06085 (2018)
4. Choi, Y., El-Khamy, M., Lee, J.: Learning low precision deep neural networks through regularization. arXiv preprint arXiv:1809.00095 2 (2018)
5. Courbariaux, M., Hubara, I., Soudry, D., El-Yaniv, R., Bengio, Y.: Binarized neural networks: training deep neural networks with weights and activations constrained to +1 or −1 (2016)
6. Deng, J., Dong, W., Socher, R., Li, L.J., Li, K., Fei-Fei, L.: Imagenet: a large-scale hierarchical image database. In: 2009 IEEE Conference on Computer Vision and Pattern Recognition, pp. 248–255. IEEE (2009)
7. Dong, C., Loy, C.C., He, K., Tang, X.: Image super-resolution using deep convolutional networks. IEEE Trans. Pattern Anal. Mach. Intell. **38**(2), 295–307 (2015)
8. Elthakeb, A.T., Pilligundla, P., Esmaeilzadeh, H.: Sinreq: generalized sinusoidal regularization for low-bitwidth deep quantized training. arXiv preprint arXiv:1905.01416 (2019)
9. Esser, S.K., McKinstry, J.L., Bablani, D., Appuswamy, R., Modha, D.S.: Learned step size quantization. In: International Conference on Learning Representations (2019)
10. Esser, S.K.: Convolutional networks for fast, energy-efficient neuromorphic computing. Proc. Natl. Acad. Sci. **113**(41), 11441–11446 (2016)
11. Frankle, J., Carbin, M.: The lottery ticket hypothesis: finding sparse, trainable neural networks. In: International Conference on Learning Representations (2018)
12. Gong, R., et al.: Differentiable soft quantization: bridging full-precision and low-bit neural networks. In: Proceedings of the IEEE International Conference on Computer Vision, pp. 4852–4861 (2019)
13. He, K., Zhang, X., Ren, S., Sun, J.: Deep residual learning for image recognition. In: Proceedings of the IEEE Conference on Computer Vision and Pattern Recognition, pp. 770–778 (2016)
14. Hinton, G., et al.: Deep neural networks for acoustic modeling in speech recognition: the shared views of four research groups. IEEE Signal Process. Mag. **29**(6), 82–97 (2012)
15. Huang, J.B., Singh, A., Ahuja, N.: Single image super-resolution from transformed self-exemplars. In: Proceedings of the IEEE Conference on Computer Vision and Pattern Recognition, pp. 5197–5206 (2015)
16. Hubara, I., Courbariaux, M., Soudry, D., El-Yaniv, R., Bengio, Y.: Quantized neural networks: training neural networks with low precision weights and activations. J. Mach. Learn. Res. **18**(1), 6869–6898 (2017)
17. Hubara, I., Nahshan, Y., Hanani, Y., Banner, R., Soudry, D.: Improving post training neural quantization: Layer-wise calibration and integer programming. arXiv preprint arXiv:2006.10518 (2020)

18. Jacob, B., et al.: Quantization and training of neural networks for efficient integer-arithmetic-only inference. In: Proceedings of the IEEE Conference on Computer Vision and Pattern Recognition, pp. 2704–2713 (2018)
19. Krizhevsky, A., Sutskever, I., Hinton, G.E.: Imagenet classification with deep convolutional neural networks. Adv. Neural Inf. Process. Syst. **25**, 1097–1105 (2012)
20. Li, H., Yan, C., Lin, S., Zheng, X., Zhang, B., Yang, F., Ji, R.: PAMS: quantized super-resolution via parameterized max scale. In: Vedaldi, A., Bischof, H., Brox, T., Frahm, J.-M. (eds.) ECCV 2020. LNCS, vol. 12370, pp. 564–580. Springer, Cham (2020). https://doi.org/10.1007/978-3-030-58595-2_34
21. Lim, B., Son, S., Kim, H., Nah, S., Mu Lee, K.: Enhanced deep residual networks for single image super-resolution. In: Proceedings of the IEEE Conference on Computer Vision and Pattern Recognition Workshops, pp. 136–144 (2017)
22. McKinstry, J.L., et al.: Discovering low-precision networks close to full-precision networks for efficient inference. In: 2019 Fifth Workshop on Energy Efficient Machine Learning and Cognitive Computing-NeurIPS Edition (EMC2-NIPS), pp. 6–9. IEEE (2019)
23. Nagel, M., Fournarakis, M., Amjad, R.A., Bondarenko, Y., van Baalen, M., Blankevoort, T.: A white paper on neural network quantization. arXiv preprint arXiv:2106.08295 (2021)
24. Naumov, M., Diril, U., Park, J., Ray, B., Jablonski, J., Tulloch, A.: On periodic functions as regularizers for quantization of neural networks. arXiv preprint arXiv:1811.09862 (2018)
25. Peng, H., Wu, J., Chen, S., Huang, J.: Collaborative channel pruning for deep networks. In: International Conference on Machine Learning, pp. 5113–5122. PMLR (2019)
26. Sandler, M., Howard, A., Zhu, M., Zhmoginov, A., Chen, L.C.: Mobilenetv 2: inverted residuals and linear bottlenecks. In: Proceedings of the IEEE Conference on Computer Vision and Pattern Recognition, pp. 4510–4520 (2018)
27. Shi, W., et al.: Real-time single image and video super-resolution using an efficient sub-pixel convolutional neural network. In: Proceedings of the IEEE Conference on Computer Vision and Pattern Recognition, pp. 1874–1883 (2016)
28. Stock, P., et al.: Training with quantization noise for extreme model compression. In: International Conference on Learning Representations (2021)
29. Timofte, R., Agustsson, E., Van Gool, L., Yang, M.H., Zhang, L.: Ntire 2017 challenge on single image super-resolution: methods and results. In: Proceedings of the IEEE Conference on Computer Vision and Pattern Recognition Workshops, pp. 114–125 (2017)
30. Zhang, D., Yang, J., Ye, D., Hua, G.: Lq-nets: learned quantization for highly accurate and compact deep neural networks. In: Proceedings of the European Conference on Computer Vision (ECCV), pp. 365–382 (2018)
31. Zhou, S., Wu, Y., Ni, Z., Zhou, X., Wen, H., Zou, Y.: Dorefa-net: training low bitwidth convolutional neural networks with low bitwidth gradients (2016)

Author Index